SMOOTHED POINT INTERPOLATION METHODS
G Space Theory and Weakened Weak Forms

SMOOTHED POINT INTERPOLATION METHODS

G Space Theory and Weakened Weak Forms

G. R. Liu

University of Cincinnati, USA

G. Y. Zhang

The University of Western Australia, Australia

 World Scientific

NEW JERSEY · LONDON · SINGAPORE · BEIJING · SHANGHAI · HONG KONG · TAIPEI · CHENNAI

Published by

World Scientific Publishing Co. Pte. Ltd.

5 Toh Tuck Link, Singapore 596224

USA office: 27 Warren Street, Suite 401-402, Hackensack, NJ 07601

UK office: 57 Shelton Street, Covent Garden, London WC2H 9HE

Library of Congress Cataloging-in-Publication Data
Liu, G. R. (Gui-Rong)
    Smoothed point interpolation methods : G space theory and weakened weak forms / G. R. Liu
(University of Cincinnati, USA) & G. Y. Zhang (The University of Western Australia, Australia).
    pages cm
    Includes bibliographical references and index.
    ISBN 978-9814452847 (hardcover : alk. paper)
    1. G-spaces. 2. Interpolation. I. Zhang, G. Y. (Guiyong) II. Title.
    QA689.L58 2013
    516'.1--dc23
                                             2013011473

British Library Cataloguing-in-Publication Data
A catalogue record for this book is available from the British Library.

Printed in Singapore by World Scientific Printers.

Dedication

To Zuona

Yun, Kun, Run,

and my extended family

for the time, support and love they gave to me

G. R. Liu

To my parents

and my elder sister

for their love and support

G. Y. Zhang

Preface

The finite element method (FEM) has been an essential and most important tool for modeling and simulation for practical engineering problems for solids and structures with complex geometry. The first author is one of the FEM users in the past some 30 years, and he wrote his first FEM code to solve a nonlinear mechanics problem for frame structures in 1979, as his university final year research project. In the past two decades he has participated in and directed many engineering projects of very large scale with millions of degrees of freedom. In those practices, we have frequently encountered problems with mesh related issues in using FEM. For accuracy reasons, we want to use quality quadrilateral (for 2D) or hexahedron (for 3D) elements, but such a mesh is quite difficult to generate and require a lot of manual operations to cut the domain into proper subdomains. The time spent on such a manual operations has been very significant. When using triangular (for 2D) or tetrahedron (for 3D) elements, the mesh generation becomes much easier and it can often be done automatically without manual operations. However, the accuracy of the results from the FEM can often be quite poor.

In searching for alternatives, the authors' group has started to learn and to develop meshfree methods, and quite good progress has been made in the past 15 years. We can now quite safely say that using proper meshfree techniques we can do pretty much what we want using only a background mesh of triangles/tetrahedrons. However, the operations in meshfree methods are generally more complicated, and can be quite costly in terms of computational efforts and resources. We have also found that proper combinations of the meshfree and FEM techniques can be of advantages, and the so-called smoothed finite element method or S-FEM is a typical such a combined approach. It needs only a triangular/tetrahedral mesh, but can deliver excellent solutions in efficient manner. The S-FEM with triangular mesh, however, uses mostly linear interpolations, and the assumed displacement function is still continuous (in a H^1 space). When higher order interpolations are used, the assumed displacement function is no longer continuous, and requires a fundamental change in theory, leading to the so-called G space theory. This class of methods is termed as smoothed point interpolation method or S-PIM. It is a typical weakened weak (W^2) formulation. It offers additional flexibility in formulation, and can deliver much better solutions. This book is dedicated to the S-PIM.

The large part of the research work related to this book was carried out at the Centre for Advanced Computations in Engineering Science (ACES), Department of Mechanical Engineering, National University of Singapore (NUS). In preparing this book, a number of my colleagues and students at NUS have supported and contributed to the writing. We express sincere thanks to all of them. Special thanks to L. Chen, Y. Jian, Nourbakhshnia N, Q. Tang, S. C. Wu, A. L. Yang, Z. Q. Zhang and many others. Many of these individuals have contributed examples to this book in addition to their hard work in carrying out a number of projects related to the S-PIM covered in this book.

G. R. Liu and G. Y. Zhang

Contents

The Authors

G. R. Liu received his PhD from Tohoku University, Japan in 1991. He was a Postdoctoral Fellow at Northwestern University, U. S. A. He was a Professor at the Department of Mechanical Engineering, National University of Singapore. He founded the Centre for Advanced Computations in Engineering Science (ACES), National University of Singapore, and served as the Director of ACES from 1998-2010. He founded the Association for Computational Mechanics (Singapore) or SACM and served as the President of the SACM from 2002-2010. He is an executive council member of the International Associate for Computational Mechanics, and the President of the Asia-Pacific Association for Computational Mechanics (APACM). He is currently a Professor, School Faculty Chair and Ohio Eminent Scholar at the School of Aerospace Systems, University of Cincinnati, USA. He has provided consultation services to many national and international organizations. He authored more than 500 technical publications including more than 380 international journal papers and eight authored books, including two bestsellers: "Meshfree Method: moving beyond the finite element method" and "Smoothed Particle Hydrodynamics: a Meshfree Particle Methods". He is the Editor-in-Chief of the *International Journal of Computational Methods*, Associate Editor of *Inverse Problems in Science and*

Engineering, and served as an editorial member of five other journals including the IJNME. He is the recipient of the Outstanding University Researchers Awards, the Defence Technology Prize (National award), Silver Award at CrayQuest competition, the Excellent Teachers Awards, the Engineering Educator Awards, the APACM Award for Computational Mechanics, and the JSME Award for Computational Mechanics. His research interests include Aerospace Systems, Bio-mechanical Systems, Computational Mechanics, Mesh Free Methods, Nano-scale Computation, Micro bio-system computation, Vibration and Wave Propagation in Composites, Mechanics of Composites and Smart Materials, Inverse Problems and Numerical Analysis.

G. Y. Zhang (Guiyong ZHANG) received his B.E. degree from Dalian University of Technology (DUT), China in 2001, and his Ph.D. degree from National University of Singapore (NUS), Singapore in 2007. He worked as postdoctoral Research Fellow at Singapore-MIT Alliance (SMA) under NUS from 2007 to 2010. He once served as the Manager of the Centre for Advanced Computations in Engineering Science (ACES) at NUS in 2010 and the Trustee of the Association for Computational Mechanics (Singapore) or SACM from 2009 to 2010. Currently he is a Research Assistant Professor at the School of Mechanical and Chemical Engineering, the University of Western Australia (UWA), Australia. He has conducted a number of research projects and has authored more than 40 peer-reviewed journal papers. In recognition of his outstanding accomplishments through research publications, he was awarded the prestigious *APACM Award for Young Investigators in Computational Mechanics* by Asia-Pacific Association for Computational Mechanics in 2010. His research interests include Computational Mechanics, Finite Element Analysis and Modeling, Meshfree (Meshless) Methods, Static and Dynamic Analysis of Structures, Numerical Simulation of Biological Soft Tissues, Error Estimation and Adaptive Analysis, and Acoustic Analysis.

Chapter 1

Preliminaries

In designing an advanced engineering system, modeling and simulation have become one of the essential tools to ensure the safety and reliability of the designed system with desired and optimal performance. This book presents a family of powerful *meshfree methods* called *smoothed point interpolation methods* (S-PIMs) for modeling and simulating mechanics problems of solids and structures in engineering systems.

This chapter addresses first the basic equations for solid mechanics problems, general and preliminary matters on the basic ideas, properties, settings, and the overall procedure and steps of the S-PIMs. For easy comprehension, our discussions are in comparison with the well-known and widely used finite element method (FEM) [1-6] that are familiar to many readers. We start with a briefing on the basic equations of solid mechanics problems, through which we hope also to clarify a number of basic but very important *concepts* and *terminologies* that are often overlooked.

1.1 Basic equations for solid mechanics

In this book *solid mechanics* problems are considered as the *default* problem, because it is one of the most fundamental problems in all engineering systems made of solid materials, and is dealt with in most of the chapters in this book. The basic settings of solid mechanics problems are briefed as follows.

FIGURE 1.1 shows a three-dimensional (3D) solid that occupies a domain Ω bounded by $\Gamma = \Gamma_u \bigcup \Gamma_t$. The movement of the solid is constrained on a part of the boundary (Γ_u), and the solid is subjected to given surface forces on another parts of the boundary (Γ_t) and *body forces* distributed over portions of the volume. These forces are often called *external forces*, and hence are given before the analysis. Under such a condition, the solid undergoes a *deformation*, and any point within the solid can be displaced and stressed. To mathematically formulate the problem, we need to "quantify" precisely the status of the solid under given conditions. The status of the deformation is measured with the so-called *displacement field* that can have three independent displacement components in, respectively, x, y, and z directions. The stressed status is measured by variables called *stress field* that has (in general) six stress components. By "field" of a variable we mean that the variable is distributed over domain Ω, implying the variable is a function of the x, y, and z coordinates. Thus displacement field is also often called *displacement function*. The solutions of the displacement and stress fields are unknown before the analysis, and it is the task of the analysis to find out these fields.

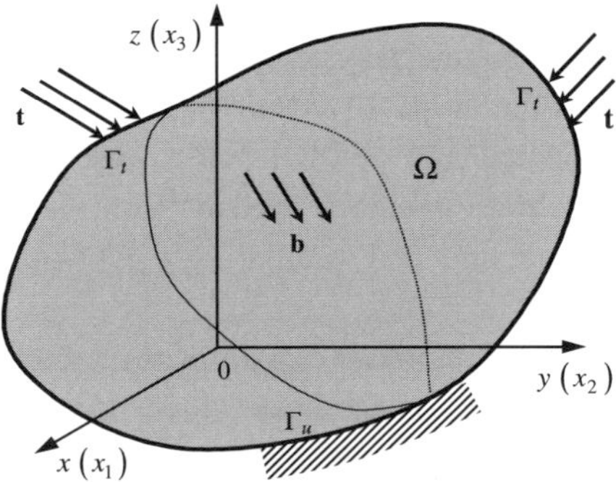

FIGURE 1.1 A 3D solid occupying a domain Ω bounded by $\Gamma = \Gamma_u \bigcup \Gamma_t$. It is constrained at the boundary Γ_u and subjected to external forces on Γ_t and over a volume of the solid.

There are often cases, such that 3D problems can be *simplified* as two-dimensional (2D) problems. FIGURE 1.2 shows a simplified case of a 2D solid that is very thin in the z-direction, and stress components in the z-direction are assumed zero, which is known as the *plane stress* problem. In this case the displacement has only two independent components in, respectively, x and y directions; and the stress has only three in-plane components: two normal stresses in x and y directions asingd one shear stress in x-y plane.

FIGURE 1.3 shows another simple case of 2D solid that is very thick in the z-direction and the external forces and constraints are independent of z, resulting in zero strain components in the z-direction. This problem is known as the *plane strain* problem. In this case the displacement field has two independent displacement components in, respectively, x and y directions, which is the same as the plane stress case. In the plane strain case, however, the strain field has three in-plane components: two normal strains in x and y directions and one shear strain in x-y plane.

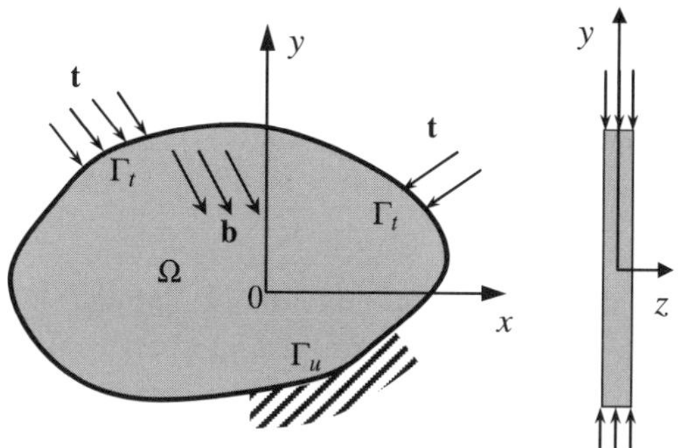

FIGURE 1.2 Plane stress problem: A 2D solid that is very thin in the z-direction subject to loadings within the x-y plane.

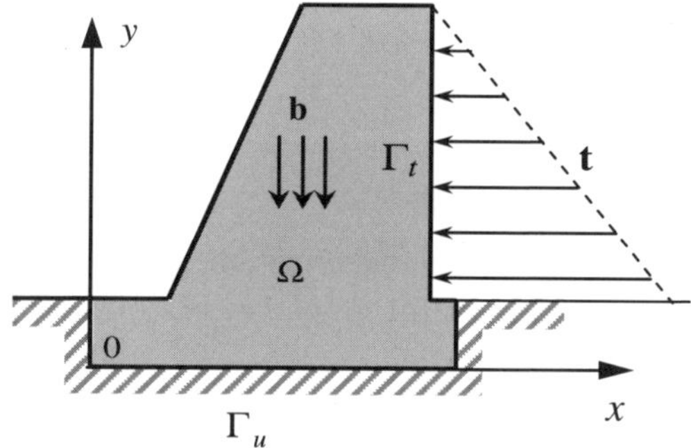

FIGURE 1.3 Plane strain problem: A 2D solid with all the strain components in the z-direction being zero (cross-section of a very long body).

It is understood that a practically useful numerical method found (proven and tested) workable for 2D problems, needs to be extendable in theory also to 3D problems. Of course the setting of the problem, the discretization of the problem domain, the coding, and the analysis of the computed data are far more complicated for 3D problems. Therefore, in the development of a numerical method, 2D problems are often used for intensive tests for a numerical method, before it is extended for 3D problems.

1.1.1 Equilibrium equations in terms of stresses

Consider, in general, a *d*-dimensional solid mechanics problem with a physical domain of $\Omega \in \mathbb{R}^d$ bounded by Γ. The static *equilibrium equation* governing the solid can be written in terms of the stress field:

$$\frac{\partial \sigma_{ij}}{\partial x_j} + b_i = 0, \quad i,j = 1,\cdots,d \quad \text{in } \Omega \tag{1.1}$$

where b_i is the component of the externally applied body force in the x_i-direction, and σ_{ij} is the component of the (internal) stress in which the first subscript letter indicating the direction of the normal to the plane under consideration and the second indicating the direction of the stress component. Equation (1.1) states simply that the total force (externally applied body force and the gradient of all the internal stresses) at any point in the solids must vanish: in equilibrium. In Equation (1.1) the stress is the primary field variable. A method that solves Equation (1.1) primarily for the stress fields is called stress method or equilibrium method. Equation (1.1) is a typical set of partial differential equations (PDEs).

1.1.2 Constitutive equations

For ideal situations of linear elastic materials, the stresses relate to the strains ε_{ij} via the *constitutive equation* (or the generalized Hook's law):

$$\sigma_{ij} = C_{ijkl}\varepsilon_{kl} \tag{1.2}$$

where C_{ijkl} is the elasticity tensor that is the material constants of the solid. The elasticity tensor is symmetrical meaning that

$$C_{ijkl} = C_{jikl} = C_{ijlk} = C_{klij} \tag{1.3}$$

For isotropic Saint-Venant Kirchhoff elastic materials, we shall have

$$C_{ijkl} = \lambda \delta_{ji}\delta_{kl} + \mu(\delta_{ik}\delta_{jl} + \delta_{il}\delta_{jk}) \tag{1.4}$$

where λ and μ are the Lame's elastic constants. In the formulations of this book, we consider mostly isotropic materials.

1.1.3 Compatibility equations

For "small" displacement fields, the strain tensor ε_{ij} relates to the displacements in the form of the *compatibility equations* (also known as the *kinematic equations*).

$$\varepsilon_{ij} = \frac{1}{2}\left(\frac{\partial u_i}{\partial x_j} + \frac{\partial u_j}{\partial x_i}\right) \tag{1.5}$$

where u_i $(i=1,\cdots,d)$ is the displacement component in the x_i-direction at a point in Ω. When a strain field satisfies Equation (1.5) in a domain, it is called a *compatible strain* field and is termed as $\tilde{\varepsilon}_{ij}$ in later chapters.

1.1.4 Equilibrium equations in terms of displacements

Substituting Equations (1.2) and (1.5) into (1.1), we arrived at

$$\frac{\partial}{\partial x_j}\left(C_{ijkl}\frac{\partial u_k}{\partial x_l}\right) + b_i = 0 , \quad i,j,k,l=1,\cdots,d \quad \text{in } \Omega \tag{1.6}$$

where the displacement becomes the primary field variable. A method that solves Equation (1.6) primarily for the displacements is called *displacement method*. A natural and typical displacement method is to first assume *properly* a displacement function with a set of "tunable" unknown coefficients. Using a *proper* set of conditions, we can then determine these coefficients and hence the displacement solution. The art of a displacement method is essential to ensure the "properness" in the process.

In this book, we often use equations in matrix form for concise expressions. The equilibrium equation (1.1) becomes much simpler as

$$\mathbf{L}_d^{\mathrm{T}}\boldsymbol{\sigma} + \mathbf{b} = \mathbf{0} \tag{1.7}$$

where the externally applied body force vector is written in the form of

$$\mathbf{b} = \begin{Bmatrix} b_1 \\ b_2 \end{Bmatrix} \text{ for 2D}, \quad \mathbf{b} = \begin{Bmatrix} b_1 \\ b_2 \\ b_3 \end{Bmatrix} \text{ for 3D} \tag{1.8}$$

The matrix of differential operators $\mathbf{L}_d$ with the first order derivatives is given by

$$\mathbf{L}_d = \begin{bmatrix} \partial/\partial x_1 & 0 \\ 0 & \partial/\partial x_2 \\ \partial/\partial x_2 & \partial/\partial x_1 \end{bmatrix}_{3\times 2} \qquad \text{for 2D}$$

$$\mathbf{L}_d = \begin{bmatrix} \partial/\partial x_1 & 0 & 0 \\ 0 & \partial/\partial x_2 & 0 \\ 0 & 0 & \partial/\partial x_3 \\ 0 & \partial/\partial x_3 & \partial/\partial x_2 \\ \partial/\partial x_3 & 0 & \partial/\partial x_1 \\ \partial/\partial x_2 & \partial/\partial x_1 & 0 \end{bmatrix}_{6\times 3} \qquad \text{for 3D} \tag{1.9}$$

The constitutive equation (1.2) becomes

$$\boldsymbol{\sigma} = \mathbf{c}\boldsymbol{\varepsilon} \tag{1.10}$$

where $\boldsymbol{\sigma}$ is a vector that collects these stress components in the form of

$$\boldsymbol{\sigma} = \begin{Bmatrix} \sigma_{11} \\ \sigma_{22} \\ \sigma_{12} \end{Bmatrix} \text{ for 2D,} \qquad \boldsymbol{\sigma} = \begin{Bmatrix} \sigma_{11} \\ \sigma_{22} \\ \sigma_{33} \\ \sigma_{23} \\ \sigma_{13} \\ \sigma_{12} \end{Bmatrix} \text{ for 3D} \tag{1.11}$$

and $\boldsymbol{\varepsilon}$ is a vector that collects these strain components:

$$\boldsymbol{\varepsilon} = \begin{Bmatrix} \varepsilon_{11} \\ \varepsilon_{22} \\ 2\varepsilon_{12} \end{Bmatrix} = \begin{Bmatrix} \varepsilon_{11} \\ \varepsilon_{22} \\ \gamma_{12} \end{Bmatrix} \text{ for 2D,} \qquad \boldsymbol{\varepsilon} = \begin{Bmatrix} \varepsilon_{11} \\ \varepsilon_{22} \\ \varepsilon_{33} \\ 2\varepsilon_{23} \\ 2\varepsilon_{13} \\ 2\varepsilon_{12} \end{Bmatrix} = \begin{Bmatrix} \varepsilon_{11} \\ \varepsilon_{22} \\ \varepsilon_{33} \\ \gamma_{23} \\ \gamma_{13} \\ \gamma_{12} \end{Bmatrix} \text{ for 3D} \tag{1.12}$$

In Equation (1.10), the matrix of material stiffness constants $\mathbf{c}$ can be written explicitly in more familiar engineering notations as

$$\mathbf{c} = \begin{bmatrix} c_{11} & c_{12} & c_{13} & c_{14} & c_{15} & c_{16} \\ & c_{22} & c_{23} & c_{24} & c_{25} & c_{26} \\ & & c_{33} & c_{34} & c_{35} & c_{36} \\ & & & c_{44} & c_{45} & c_{46} \\ & sy. & & & c_{55} & c_{56} \\ & & & & & c_{66} \end{bmatrix} \qquad (1.13)$$

where "*sy.*" stands for symmetry: $c_{ij} = c_{ji}$. Thus, there are in general 21 independent material constants c_{ij}. For isotropic materials, $\mathbf{c}$ can be greatly reduced to

$$\mathbf{c} = \begin{bmatrix} c_{11} & c_{12} & c_{12} & 0 & 0 & 0 \\ & c_{11} & c_{12} & 0 & 0 & 0 \\ & & c_{11} & 0 & 0 & 0 \\ & & & \dfrac{c_{11}-c_{12}}{2} & 0 & 0 \\ & sy. & & & \dfrac{c_{11}-c_{12}}{2} & 0 \\ & & & & & \dfrac{c_{11}-c_{12}}{2} \end{bmatrix} \qquad (1.14)$$

These constants in the forgoing equation can be written as

$$c_{11} = \frac{E(1-v)}{(1-2v)(1+v)} \; ; \quad c_{12} = \frac{Ev}{(1-2v)(1+v)} \; ; \quad \frac{c_{11}-c_{12}}{2} = \mu \qquad (1.15)$$

where E, v, and μ are, respectively, Young's modulus, Poisson's ratio, and shear modulus of the material, known as material constants. Note that these three constants have the following relationship:

$$\mu = \frac{E}{2(1+v)} \qquad (1.16)$$

Hence, there are only two independent constants for isotropic materials. Giving any two of these three constants, the other can then be calculated using Equation (1.16). For 2D *plane stress* problems we further have

$$\mathbf{c} = \frac{E}{1-v^2} \begin{bmatrix} 1 & v & 0 \\ v & 1 & 0 \\ 0 & 0 & (1-v)/2 \end{bmatrix} \quad \text{(Plane stress)} \tag{1.17}$$

For 2D *plane strain* problems, the matrix of material stiffness constants $\mathbf{c}$ can be obtained by simply replacing E and v, respectively, with $E/(1-v^2)$ and $v/(1-v)$, which leads to

$$\mathbf{c} = \frac{E(1-v)}{(1+v)(1-2v)} \begin{bmatrix} 1 & \dfrac{v}{1-v} & 0 \\ \dfrac{v}{1-v} & 1 & 0 \\ 0 & 0 & \dfrac{1-2v}{2(1-v)} \end{bmatrix} \quad \text{(Plane strain)} \tag{1.18}$$

The compatibility equation (1.5) can also be written in the matrix form of

$$\boldsymbol{\varepsilon} = \mathbf{L}_d \mathbf{u} \tag{1.19}$$

where

$$\mathbf{u} = \begin{Bmatrix} u_1 \\ u_2 \end{Bmatrix} \text{ for 2D,} \qquad \mathbf{u} = \begin{Bmatrix} u_1 \\ u_2 \\ u_3 \end{Bmatrix} \text{ for 3D} \tag{1.20}$$

is the displacement vector. Substituting Equation (1.19) into (1.10) and (1.7) gives

$$\mathbf{L}_d^{\mathrm{T}} \mathbf{c} \mathbf{L}_d \mathbf{u} + \mathbf{b} = \mathbf{0} \tag{1.21}$$

which is the equilibrium equation written in matrix form with the displacements as the primary field variables.

1.1.5 Boundary conditions

The boundary conditions (BCs) can have two types: *Dirichlet* (*essential, displacement*) boundary condition and *Neumann* (*natural, stress*) boundary condition. Let Γ_u denotes a part of Γ, on which Dirichlet boundary condition is specified, we then have

$$\mathbf{u} = \mathbf{u}_\Gamma, \quad \text{on } \Gamma_u \in \Gamma \tag{1.22}$$

where $\mathbf{u}_\Gamma$ is the vector of specified displacement components on Γ_u. In this book, for simplicity we consider by default *homogeneous essential boundary conditions* ($\mathbf{u}_\Gamma = \mathbf{0}$). This type of problems is also called *force-driving problem* (in contrary to the displacement-driving problem), meaning that the displacement field (and hence the stress/strain field) is caused only by the externally applied forces. For non-homogeneous essential boundary conditions, the displacement field is caused also by the "moving" boundary. In such cases, simple treatments used in the standard FEM shall apply [1-6].

Let Γ_t denotes a part of Γ, on which Neumann boundary conditions are specified,

$$\mathbf{L}_n^{\mathrm{T}} \boldsymbol{\sigma} = \mathbf{t}_\Gamma, \quad \text{on } \Gamma_t \in \Gamma \tag{1.23}$$

where $\mathbf{t}_\Gamma$ is the vector of specified boundary stress components and $\mathbf{L}_n$ is the matrix of the components of the unit *outward normal* in the form of

$$\mathbf{L}_n = \begin{bmatrix} n_1 & 0 \\ 0 & n_2 \\ n_2 & n_1 \end{bmatrix}_{3 \times 2} \text{ for 2D,} \qquad \mathbf{L}_n = \begin{bmatrix} n_1 & 0 & 0 \\ 0 & n_2 & 0 \\ 0 & 0 & n_3 \\ 0 & n_3 & n_2 \\ n_3 & 0 & n_1 \\ n_2 & n_1 & 0 \end{bmatrix}_{6 \times 3} \text{ for 3D} \tag{1.24}$$

1.1.6 Strain energy in solids

For stressed elastic solids, there is so-called *strain energy* stored in the solids. Such strain energy is also called potential energy and can be quantified using

$$U_{PE} = \int_\Omega \frac{1}{2} \boldsymbol{\varepsilon}^{\mathrm{T}}(\mathbf{x}) \boldsymbol{\sigma}(\mathbf{x}) d\Omega = \int_\Omega \frac{1}{2} \boldsymbol{\varepsilon}^{\mathrm{T}}(\mathbf{x}) \mathbf{c} \boldsymbol{\varepsilon}(\mathbf{x}) d\Omega \tag{1.25}$$

It is clear that the strain energy is a global measure of the level of the stressed solid. The integrand in the forgoing equation is often termed as the *strain energy density* that is a local measure of the level of the stressed solid.

1.1.7 Some notations and conventions

This book uses both the matrix and the indicial notations from time to time for more concise presentation. Therefore, we allow the vectors and matrices to have the following forms for easy conversion between these two notations whenever it is needed, as shown in TABLE 1.1.

TABLE 1.1 Matrix and indicial expressions

Matrix notation	Indicial notation
$\mathbf{x} = \begin{Bmatrix} x_1 \\ x_2 \\ x_3 \end{Bmatrix} = \begin{Bmatrix} x \\ y \\ z \end{Bmatrix}$	$x_i, i = 1,2,3$
$\mathbf{u} = \begin{Bmatrix} u_1 \\ u_2 \\ u_3 \end{Bmatrix} = \begin{Bmatrix} u_x \\ u_y \\ u_z \end{Bmatrix} = \begin{Bmatrix} u \\ v \\ w \end{Bmatrix}$	$u_i, i = 1,2,3$
$\mathbf{b} = \begin{Bmatrix} b_1 \\ b_2 \\ b_3 \end{Bmatrix} = \begin{Bmatrix} b_x \\ b_y \\ b_z \end{Bmatrix}$	$b_i, i = 1,2,3$
$\mathbf{n} = \begin{Bmatrix} n_1 \\ n_2 \\ n_3 \end{Bmatrix} = \begin{Bmatrix} n_x \\ n_y \\ n_z \end{Bmatrix}$	$n_i, i = 1,2,3$
$\boldsymbol{\sigma}^{\mathrm{T}} = \{ \sigma_{11} \quad \sigma_{22} \quad \sigma_{33} \quad \sigma_{23} \quad \sigma_{13} \quad \sigma_{12} \}$ $= \{ \sigma_{xx} \quad \sigma_{yy} \quad \sigma_{zz} \quad \sigma_{yz} \quad \sigma_{xz} \quad \sigma_{xy} \}$	$\sigma_{ij}, i,j = 1,2,3$
$\boldsymbol{\varepsilon}^{\mathrm{T}} = \{ \varepsilon_{11} \quad \varepsilon_{22} \quad \varepsilon_{33} \quad 2\varepsilon_{23} \quad 2\varepsilon_{13} \quad 2\varepsilon_{12} \}$ $= \{ \varepsilon_{xx} \quad \varepsilon_{yy} \quad \varepsilon_{zz} \quad \gamma_{yz} \quad \gamma_{xz} \quad \gamma_{xy} \}$	$\varepsilon_{ij}, i,j = 1,2,3$

Some often used mathematic notations are also listed in TABLE 1.2.

TABLE 1.2 Some often used mathematic notations

Notations	In words	Notations	In words
$\subset$	a subset (or space) of	$\forall$	for all
$\supset$	a superset of	$\exists$	there exists
$\cup$	union of	$\rightarrow$	map to
$\cap$	intersection of	$\Rightarrow$	if ... then
$\backslash$	set minus	$\Leftrightarrow$	if and only if
$\mid$	s.t. (subjected to) or restricted within	$\propto$	proportional to

1.1.8 Some basic concepts

We now note the following remarks for future references.

Remark 1.1 Stable materials

Physically stable materials require that *any* finite amount of strain in the solid material will result in a finite amount of stress and hence a finite amount positive strain energy. Physically, it requires the material being able to response to any finite strain with a finite stress, or vice versa. Mathematically, it requires that these material constants are *positive definite* or the matrix of the material constants $\mathbf{c}$ is *symmetric positive definite* (SPD). This can be echoed from Equation (1.25): when $\mathbf{c}$ is SPD, U_{PE} will always be strictly positive for any nonzero $\boldsymbol{\varepsilon}$.

In this book, unless specified otherwise, we consider only solids and structures made of materials that are physically *stable*. For stable materials, the stress-strain relation can also be written in the following reverse form:

$$\boldsymbol{\varepsilon} = \mathbf{s}\boldsymbol{\sigma} \tag{1.26}$$

where $\mathbf{s}$ is the matrix of material *flexibility* (or compliance) *constants* that can be obtained from the standard experiments for a given stable material. Because $\mathbf{c}$ is SPD, $\mathbf{s}$ must also be SPD, and hence they are both invertible. We then have the following simple relations.

$$\mathbf{s} = \mathbf{c}^{-1} \quad \text{or} \quad \mathbf{c} = \mathbf{s}^{-1} \tag{1.27}$$

Remark 1.2 Volumetric locking

For a material with Poisson's ratio (ν) approaches to 0.5, the denominator $(1-2\nu)$ in Equations (1.15) and (1.18) becomes zero, suggesting a possible singularity problem. This type of material is called "incompressible", such as some rubbers, metals at some stages of plastic deformations, and some composites. This can have numerical implications for displacement methods, known as *volumetric locking*: the material is not capable to response to the finite pressure with a finite amount of volume deformation. In such cases, the numerical solution can be erroneous. Special techniques have been developed in FEM, smoothed finite element method (S-FEM) and meshfree methods to overcome this numerical problem for accurate solutions [6-9].

Remark 1.3 Volumetric locking free

The volumetric locking will not show as a numerical problem when an *equilibrium model* is used, because the stress is used as the primary field variable. In these methods, the strain is obtained using Equation (1.26) instead of Equations (1.15) or (1.18). Therefore, the term $(1-2\nu)$ will not be in the denominator and hence will not cause numerical problem. This type of formulation is known as a volumetric locking free formulation. The node-based smoothed point interpolation method (NS-PIM) discussed in Chapter 6 has the similar features of an equilibrium model, and is found being volumetric locking free.

Remark 1.4 Constrained solid

If the displacement BCs imposed on the solid are sufficient to at least constrain all the rigid body movements of the solid, the solid is said a "constrained solid". Any finite amount of externally applied force will results in stresses in a constrained solid of stable materials.

Remark 1.5 Well-posed problem

The setting of the solid mechanics problem (PDEs with boundary and initial conditions) must be well-posed in the Hadamard sense, by which we mean that there exists a unique solution that depends continuously on the data. For example, for static linear elasticity problems, we essentially require 1) the solid is constrained, and 2) the material is stable.

This book deals with only well-posed problems. In other words, we deal with problems that are set or posed physically stable. Certain types of ill-posed problems are treatable [7], but it requires a special consideration and is beyond the scope of this book.

Remark 1.6 Force-driving problem

If the displacement BCs are homogeneous, the problem is called force-driving problem. In such cases, the resultant displacement field (and hence the stress/strain field) in the constrained solid of stable material is caused only by the externally applied forces.

Remark 1.7 Strong form

Equation (1.21) is called the *strong-form* system of equations that governs the mechanics behavior of solids with the displacement as the primary dependent field variables. The displacement functions are required to have at least the same order of consistency in the entire problem domain as the order of the differentiations, which is 2nd for the PDE given in (1.21). Such a requirement on consistency for the displacement functions is said strong, in contrast to the weak requirements to be briefed in a moment.

1.2 Numerical techniques: FEM vs. S-PIM

1.2.1 An overview

We now give a brief discussion on the *discrete* numerical methods for practical solutions to the solid mechanics problems defined in Section 1.1. In using a discrete numerical method for practical engineering problems, the problem domain is usually discretized using a set of nodes and the field function is approximated using its nodal values. Most practical methods use *local approximation*, meaning that the approximation of the field function at a point is performed using only the local surrounding nodes. This is largely because the methods using local approximation produce "banded" system matrix, hence the system equations can be solved efficiently with small storage. In addition, it is much more efficient for nonlinear and non-smooth problems, such as contact problems, fracture problem, plastic problems, etc. Therefore, this book discusses only discrete numerical methods using local approximations.

There are largely two types of numerical methods: direct and indirect approaches. The direct approach is known as *strong-form* methods, such as the finite difference method or FDM, collocation method (e.g., [10]), and smoothed particle hydrodynamics or SPH (e.g., [11]). The strong form methods discretize and solve the PDEs directly, and the procedure is quite simple. However, care must be taken in using strong-form methods with local approximations using local irregular nodes, because of stability and convergence issues [8, 10].

Indirect approach such as *weak-form* methods establishes first an alternative weak-form equation that governs the same physics and then solves it. Such weak formulation is widely used in FEM and meshfree methods [1-6, 12-22]. The weak-form equations are usually in an integral form, implying that they need to be satisfied only in an integral sense. The essential idea of a weak-form method is to assess a global behavior (such as the strain energy) of the entire solid and then find a solution that is the best in simulating the global behavior.

For example, we examine the energy potentials in the solid, and try to find a displacement solution that makes the total energy potential minimum. At such a minimum or stationary status, we know that the solid will be stable, and hence in an equilibrium state. Such a formulation is called a weak formulation, because the evaluation of strain energy potential requires only strains and stresses that can all be obtained by the 1st differentiation of the displacement function, and hence no need for the 2nd order derivatives (see, Remark 1.7): a *weak requirement*. The FEM is a typical weak-form method, and is known for good stability and convergence that can be properly proven mathematically.

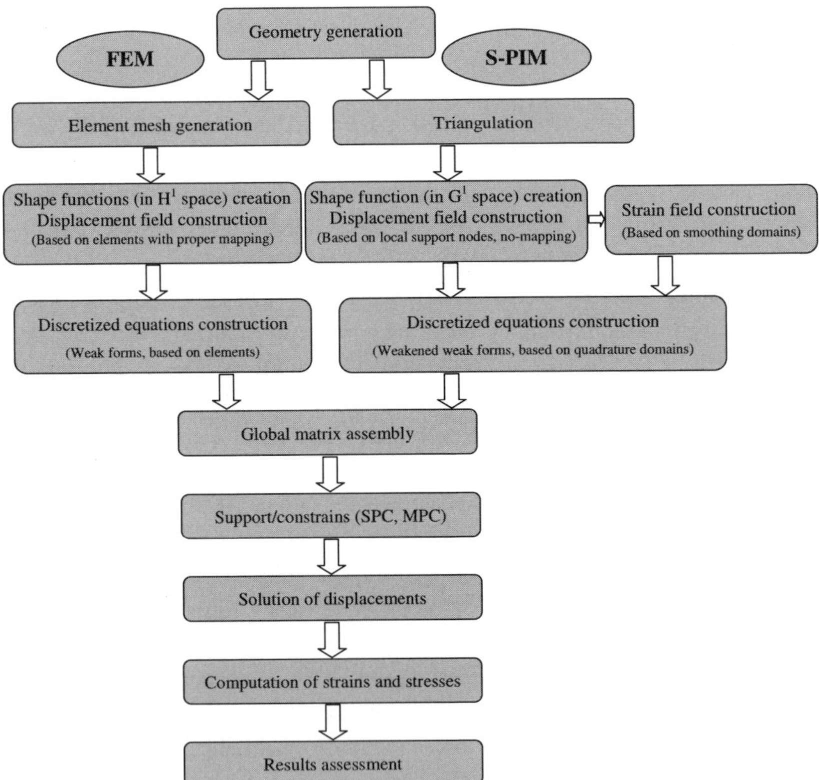

FIGURE 1.4 Flowchart for the FEM and S-PIM methods procedures.

The weak form can be further weakened leading to the so-called weakened weak (W^2) formulation [23, 26, 27]. This is achieved by approximating the strain field using only the displacements (not the derivatives). Therefore, the strain energy can be evaluated using only the displacement, meaning even the 1st derivatives of the displacement are not required: a *weakened weak requirement*. The stability and convergence of a W^2 formulation is ensured by the G space theory [28]. The S-PIM is a typical W^2 form method with excellent stability and convergence that can be proven based on the G space theory, and will be the focal point of discussion in this book.

The procedures in the FEM and S-PIM can be outlined in FIGURE 1.4. It is seen that we need a number of key techniques: 1) shape function creation, 2) numerical integration, and 3) establishment of weak forms. The differences between the FEM and S-PIM methods are elaborated in the following sections, in terms of these key techniques. Note in FIGURE 1.4, SPC and MPC stand for single point constrain and multi-point constrain, respectively.

Remark 1.8 On stability

We just mentioned for times *stability*, which requires some clarification. The word "stable" is used frequently in the literature on numerical methods, and it is to certain extend quite "abused". Rigorously, a numerical solution said "stable" implies that for any admissible set of input data (loads, excitations, BCs, etc.), the solution (in a proper norm measure) can always be bounded by the input data (in a norm) for a well-posed physical problem (Remark 1.5). A method that always produces stable solutions is said a stable method (accuracy aside). This definition essentially concerns with 1) the property of the numerical method, and 2) the property of the input data. When we say a method is stable, we shall specify with respect to what types of input data. For some types of input data, we may formulate many stable numerical methods; but for some types of input data, one may never find a stable numerical method in a specified norm measure. In this book, we consider by default only "real life" inputs that are at least square integrable (in $\mathbb{L}^2$ space, see Chapter 2). We do not usually consider mathematically idealized forces like "point load" of Dirac type that is not square integrable and never exists in reality. Under this consideration on the input data, our definition of "stability" in this book concerns only the numerical methods. Essentially, it relates directly to the positivity of the bilinear forms for weak-form methods, which will be discussed in detail in Chapter 5.

Remark 1.9 Spatial stability and temporal stability

Further on stability, we can have two types: spatial stability and temporal stability. The former refers to a numerical method for solving static problems, and the latter for solving dynamic problems. When a method produces stable solutions for static problems, it is said spatially stable. However, a spatially stable method is not necessarily stable when solving dynamic problems (even if a stable time-integration scheme is used). The temporal instability is often observed as spurious modes in free vibration analysis or unphysical oscillations (in transient vibration) in the solution to a transient dynamic problem. A temporally stable method should produce 1) only "legal" zero-energy modes that correspond to rigid body movements of the solid, and 2) all the nonzero energy modes must correspond to physical deformation of the solid. It should be emphasized that the temporal stability concept is different from the concept of "time-integration stability". The latter relates to time-integration schemes and the use of the time step. The former relates to the property of the numerical model (spatial discretization, local approximation, weak-form evolution, etc.). In other words, for a temporally instable model, one cannot find a time-integration scheme to produce numerically stable dynamic solution of frequency beyond that of the spurious modes. This concept is a little more involved and particularly important for W^2 models [23, 26, 27].

Remark 1.10 On convergence

We mentioned also for times "convergence" that is a quite involved terminology in numerical methods. The original meaning of convergence refers to the numerical solution approaches to a certain value, when the discretization is properly refined to the limit. However, the "certain" value may not be the exact solution of the original physical problem. If a convergent solution is not to the exact solution, the meaningfulness of the solution can be in question, and hence special care may be needed. In this book by "convergence", we mean by default that the numerical solution converges to the exact solution. Note also that for a method to produce convergent solutions, it must be stable.

1.2.2 On computational efficiency

Once a stable and convergent numerical method is developed, we have to examine its computational efficiency, because it is the ultimate measure for a numerical method to "survive" and its applicability to engineering problems. When different methods are used to analyze a problem using the same set of nodes, the CPU time used to solve the problem and the quality (accuracy) of the solutions will all be generally different. Therefore, a fair comparison must be on the computational efficiency, in examining which we measure the CPU time for solving the same problem for a solution of the same accuracy. The less the CPU time taken for the solution of same accuracy, the more efficient the method is. Unfortunately, the comparison of computational efficiency is quite involved and can be problem dependent. Therefore, a *particular* comparison using one example may not tell the whole story. On the other hand, a rigorous *general* analysis is very difficult. In the following we suggest an intuitive approach for a rough estimation of the computational efficiency of numerical methods. This approach will be used in this book.

TABLE 1.3 Estimation of the leading term of the complexity of linear algebraic equation solvers for symmetric system matrices

Solver	Estimated complexity	Applications
Full-matrix	N_{DOF}^3	Up to thousands DOF
		Fully populated matrices
Bandwidth solver	$N_{DOF}b_w^2$	2D Problems
1D-storage solver	$N_{DOF}n_{ze}^2$	2D and 3D Problems
Iterative solver	$n_{iter}N_{DOF}n_{ze}$	2D and 3D Problems with large
		scale linear algebraic equations

Notes: DOF: degree of freedom; N_{DOF}: the total number of DOF; b_w: bandwidth of system matrix **K**; n_{iter}: the number of iterations needed to get a converged solution in an iterative solver; n_{ze}: the (average) nonzero entries in rows of **K**.

First, we have to decide the type of solver used to solve the linear algebraic system equations with symmetric system matrix. Largely, we have three types of solvers widely used: full-matrix solver, bandwidth solver, and iterative solver, as shown in TABLE 1.3. The *full-matrix solver* stores the system (stiffness) matrix **K** in full, and solves the equations as **K** is full. It is simple but very expensive for large matrix systems. The *complexity* (or number of computational operations) of a full-matrix solver is approximately N_{DOF}^3, where N_{DOF} is the total number of degrees of freedom (DOFs). Therefore, when the model is large with more than thousands of DOFs, it is not practical at all to use a full-matrix solver. Because of the simplicity, it is often used in the studies for small models, where the CPU time is not much a concern.

We know that for numerical methods using local approximations such as the FEM and the major meshfree methods, the system matrix is very sparse and often banded if the nodes are properly numbered. In this case, we often use the *bandwidth solver* that stores only a small diagonal "band" of the system matrix **K** and solve the equations by taking the advantage of the banded **K**. The complexity of a bandwidth solver is approximately $N_{DOF}b_w^2$, where b_w is the bandwidth of system matrix and is usually much smaller than N_{DOF}. Therefore, the bandwidth solver is clearly much more efficient compared to the full-matrix solver. For 2D problems, it works quite well and is widely used in practice. For 3D problem, $\mathbf{K}$ is extremely sparse, there can be a lot of zeroes even within the "band", and hence 1D-storage solver that stores and then operates only on the none-zero entries should be used. In such a case the complexity is related to the square of n_{ze} that is the (average) number of the nonzero entries in the system matrix. Usually, n_{ze} is smaller than b_w, and hence 1D-storage solver is superior to the bandwidth solver. Note that it is very difficult to provide an accurate value of n_{ze} for different methods. However, n_{ze} is proportional to the number of local support nodes for the element or the integration cell used in computing $\mathbf{K}$, which can be used later to roughly estimate the relative n_{ze} values of different methods.

For most 3D problems, however, the DOFs and the bandwidth or the number of the nonzero entries are usually very large. In these cases, the *iterative solver* is used and the calculation involves only nonzero entries, is the most preferred and is used in the commercial FEM packages. In this case, the complexity of the solver is approximately $n_{iter}N_{DOF}n_{ze}$, where n_{iter} is the number of iterations needed to get a converged solution in the iterative solver. Typically n_{iter} can be much smaller (depends on the conditioning) than b_w. Therefore, the iterative solver is much more efficient than the bandwidth and even the 1D-storage solver. In practice, we often prefer using an iterative solver for 2D and 3D problems with large scale linear algebraic equations. The discussion on computational efficiency for methods examined in this book is hence based on the iterative solver.

Second, when the iterative solver is used, n_{iter} depends on the condition number of the system matrix $\mathbf{K}$, and we have

$$n_{iter} \propto \sqrt{cond(\mathbf{K})} = \sqrt{\lambda_{\max}/\lambda_{\min}} \qquad (1.28)$$

where $\lambda_{\max}$ and $\lambda_{\min}$ are the maximum and minimum eigenvalues of $\mathbf{K}$. The condition number of a matrix measures how sensitive the eigenvalues and eigenspaces of the matrix are to a small change in it [29], and hence indicates the numerical stability and the conditioning of the matrix. Because the condition number of $\mathbf{K}$ depends on the numerical method that creates $\mathbf{K}$, therefore, when we examine the efficiency of a numerical method, it is also important to examine the condition number of the system matrix that creates. Such examinations will be frequently performed for various methods in later chapters.

Note that we do need *overhead* CPU time to create the system matrix $\mathbf{K}$, and it depends also on the method used and a precise analysis is very difficult. Fortunately, the overhead time is only relatively significant for small models. In usual situations for large models, the solver CPU time dominates, and the overhead time is a very small fraction of the solver time. Even if a method is twice less efficient in creating $\mathbf{K}$, it will not affect the total CPU time very much. Therefore, in this book we do not consider the overhead time in our efficiency analysis for numerical methods.

Finally we can give a quantitative study on the efficiency which is measured by the reciprocal of the product of solution error and computational cost.

1.2.3 On shape function creation

In the FEM, the shape function creation is based on the elements that are the basic building-blocks of an FEM model. Proper coordinate mapping is needed for complicated domains to ensure the *compatibility* of the shape functions between the adjacent elements [6]. These shape functions are usually predetermined for types of elements before the finite element analysis starts.

In S-PIM, however, the shape functions are created essentially in a meshfree nature, using a small number of local support nodes selected beyond the cells/elements. The shape function generally changes with the location of the point of interest, and is usually created during the analysis, not before the analysis. The procedure to create the shape functions uses the so-called point interpolation method (PIM) [8, 10, 14, 15]. It is very flexible, based on nodes, a direct interpolation without coordinate mapping, and hence the Jacobian matrix is not needed. The nodal PIM shape function is in general not continuous, and hence the displacement field is not compatible, but is admissible in S-PIM models. Chapter 3 provides a very detailed discussion on this issue.

1.2.4 On integration over the problem domain

The integration in the FEM is also based on the element in the natural coordinate system. The integral over the domain becomes a summation of all these integrals over the elements that are evaluated using a numerical integration technique such as the well-known Gauss quadrature rule.

The integration in S-PIM is, in general, based on quadrature/integration cells that are usually different from the elements. These integration cells can be created in the various means, such as:

1) Using directly triangular background cells created for the problem domain, which is quite similar as in the FEM using triangular elements, and is used in the cell-based smoothed point interpolation method (CS-PIM) [23-25].

2) Using node-based smoothing domains created based upon the triangular background cells, as in the NS-PIM [30-35].

3) Using face-based smoothing domains created based upon the tetrahedral background cells, as in the face-based smoothed point interpolation method (FS-PIM) [36].

4) Using edge-based smoothing domains created based upon the triangular background cells, as in the edge-based smoothed point interpolation method (ES-PIM) [37].

5) Using triangular sub-cells that are created by further dividing the background cells, which is used in the strain-constructed point interpolation method (SC-PIM) [38-43].

6) Combination of above [44, 45].

The integral over the problem domain in S-PIM models is a simple summation of all the integrals over each of the *integration cells*. The integration over the cells is usually very simple and can be carried out exactly. The means of integration in S-PIM are also much diversified for desired performance and properties.

1.2.5 On the use of weak forms

The FEM uses the variational or energy principles such as the minimum total potential energy principle, the principle of virtual work, Hamilton's principle, and so on. These principles can lead to weak form known as the standard Galerkin weak form.

The S-PIM is based on the weakened weak (W^2) formulation. In general, it uses the strain-constructed Galerkin or SC-Galerkin weak forms [26, 27, 39, 40]. In particular, it often uses the generalized smoothed Galerkin or GS-Galerkin weak form [46].

Remark 1.11 Elements vs. triangular background cells

It is clear that all the FEM operations are based on various types of elements, and numerical operations (interpolation, integration, etc.) are within an element.

It is known that FEM requires quality quadrilateral elements for accuracy reasons. Although FEM uses also triangular elements, there are a number of problems, such as overly stiff, poor accuracy, locking, etc., due to the element-confined operations [6].

All the numerical operations in the S-PIM are based on triangular/tetrahedral background cells. They are called "background cells", because the numerical operations are usually performed beyond a cell. The standard element concept is virtually lost. By using different types of smoothing domains, and different ways to construct the displacement and strain fields, S-PIM models work well with triangular cells and can have special properties including, ultra-accuracy, superconvergence, volumetric locking free, and upper bound properties. Because the triangular types of cells can be generated automatically, S-PIM works well for problems with arbitrarily complicated geometries, and adaptive analysis can be performed with ease.

1.2.6 Sampling principle

The *sampling principle* states that in general for a given discrete model with finite number of degrees of freedom and certain order of consistency in field function approximation, the *stability* of the discretized system equations increases with the increase of the number of independent *samplings* used to create the discretized equations.

For collocation methods (strong form), the number of samplings is the number of collocation points. However, whether or not these samplings are independent depends on how the collocation is performed. For FEM (weak form), the number of samplings is the number of total Gauss integration points used in the model. The independence of these samplings is ensued by the element-based operations and the rules of the Gauss integration. For S-PIM (weakened weak form), the number of samplings is the number of the integration cells used in the model. The independence of these samplings is ensued by a proper creation of the integration cells, which will be detailed in Chapter 4.

1.2.7 Summarized remarks

We summarized the similarities and differences between the standard FEM and the present S-PIMs and listed them in the following TABLE 1.4.

TABLE 1.4 Comparison between the FEM and the S-PIMs

No.	Items	FEM	S-PIMs
1.	Element mesh	Yes	Triangular/tetrahedral mesh/cell
2.	Mesh creation and automation	Difficult due to the need for quality elements	Relatively easy and less demand on mesh quality
3.	Mesh automation, Adaptive analysis	Difficult for 3D cases	Much easier to perform
4.	Shape function creation	Element based	Node based
5.	Shape function property	Kronecker Delta; valid for all elements of the same type; compatible	Kronecker Delta; may vary from location to location; may not compatible
6.	Strain field construction	Not performed (use compatible strain directly)	Smoothed or constructed strains based on smoothing domains
7.	Integration	Element based	Integration cell based
8.	Discretized system stiffness matrix	Sparse, banded, symmetrical	Sparse, banded and symmetrical
9.	Imposition of essential BCs	Easy and standard	As same as the FEM: easy and standard
10.	Computation speed (for same number of nodes)	Fast	1.0 to 15 times slower compared to the linear FEM depending on the method and measure used
11.	Accuracy (for same number of nodes)	Accurate compared with FDM	Can be much more accurate compared with the linear FEM
12.	Computation efficiency	Efficient	0.3 to 100 times more efficient compared to the linear FEM depending on the method and measure used
13.	Retrieval of results	Special technique required	Standard routine
14.	Stage of development	Very well developed	Fast developing, promising and with challenging problems
15.	Commercial software package	Widely available	Only codes released in this book.

1.3 Basic ideas of S-PIM

The most important basic ideas of the S-PIM are summarized as:

1) **Background cells**: We use triangular/tetrahedral background cells that can be generated easily and even automatically for complicated practical problems in engineering. All the sequential numerical operations can be

coded and performed automatically, as long as a set of background cells are provided. Based on this setting adaptive analysis can be performed with ease.

2) **Theory**: The G space theory is used to ensure stability and convergence for the discrete model and allowing simple and effective shape function creation and strain field construction without worrying about compatibility issues.

3) **Formulation**: We use W^2 form that is *symmetric* to establish the discrete equation system, which provides effective "knobs" to tune the discrete models for desired important and unique properties, such as upper bound property, ultra-accuracy, SPD, etc.

4) **Interpolation**: Lower order interpolations are used that are often linear and bilinear (trilinear), so that we require only minimum regularity on the solution and hence the S-PIM models are applicable to the most wide range of practical problems.

5) **Local support nodes**: Minimum local support nodes are used for interpolation for maximum sparseness in the global system matrices and hence the optimal computational efficiency (storage and CPU time).

It is clear that the S-PIM aims for simplicity, robustness, efficiency, automation, applicability to practical problems, adaptation to complicated geometry, automation in modeling and simulation, ultra-accuracy, outstanding performance, certified solutions, and other desired unique properties.

1.4   Basic properties of S-PIM

In terms of properties and performance, we further note:

1) The operations in an S-PIM model are usually more sophisticated than the standard linear FEM model, but all the S-PIM operations are performed automatically, no additional DOFs are needed.

2) Essentially, the S-PIM method uses some carefully designed manipulations based on the simplest setting of triangular background cells to establish the best possible lower order models, or models with desired properties.

3) To the programmer, a slightly more delicate coding is required.

4) To the users, an S-PIM code looks exactly like the simplest linear FEM model using the simplest triangular/tetrahedral elements. It works for problems of arbitrary complexity as long as a set of triangular/tetrahedral cells can be built.

5) To the computer, a slightly more "overhead" and CPU time are needed in S-PIM, but the solution accuracy can be drastically improved compared to the linear FEM using the same mesh.

6) In terms of computational efficiency which is measured by the reciprocal of the product of solution error and computational cost, the S-PIM can stand out clearly compared to the linear FEM.

7) Some of the S-PIM models have unique properties that the standard displacement-based FEM will not possible to have, such as upper bound and free from volumetric locking, etc.

8) The overall performance of the S-PIM will be clearly superior to the linear FEM model in many ways.

1.5 Basic steps in S-PIM

The general procedure and basic steps for the S-PIM are briefly introduced in this section, using the solid mechanics problem as an example.

Step 1   Domain discretization

The geometry of the solid can be first created in a CAE tool or a preprocessor. It is then properly triangulated to a set of triangular (2D) and tetrahedral (3D) cells with a set of nodes scattered in the problem domain and on its boundary, as described in Section 1.6.1. The density and the distribution of the nodes depend on the requirements on the accuracy of the geometry presentation, the solution accuracy, and the limits of the computer resources. Because adaptive algorithms can be used in S-PIM, the nodal distribution can be eventually controlled automatically and adaptively in the S-PIM code [47-52]. Therefore, we do not need to worry too much about the distribution quality of the initial nodes used in usual situations. In addition, the S-PIM does not demand too much for the quality of nodal distribution. It works within reason for arbitrarily distributed nodes. If needed, the nodes can be practically randomly distributed. Because the nodes carry the values of the field variables in an S-PIM formulation, they are often called *field nodes*.

Step 2   Displacement field construction

In the S-PIM, the field variable, say a component of the displacement vector, u at any point of interest $\mathbf{x}=(x, y, z)$ (that is usually a quadrature point) within the problem domain is approximated via interpolation using the displacements:

$$u^h(\mathbf{x}) = \sum_{i \in S_n} \phi_i(\mathbf{x})u_i = \mathbf{\Phi}_s(\mathbf{x})\mathbf{d}_s \qquad (1.29)$$

where S_n is the set of *support nodes* included in a small local *support domain* of $\mathbf{x}$, u_i is the nodal displacement at the ith node, $\mathbf{d}_s$ is the vector collecting all the nodal displacements at these local support nodes, ϕ_i is the nodal *shape function* for the ith node, and $\mathbf{\Phi}_s$ is a row-matrix with entries of these nodal shape functions for these local support nodes.

A support domain determines the number of support nodes to be used to create the nodal shape functions. It can have different shapes and dimension that can change with $\mathbf{x}$ [8]. Alternatively, we can use *influence domain*, which is defined as a domain that a node exerts an influence upon [8]. The support/ influence domain is used in many meshfree methods.

Since triangular/tetrahedral background cells are used in our S-PIM, the so-called T-schemes are much more often used for node selection, which is practical, efficient, robust, and better-controlled, as detailed in Section 1.6.3. The T-schemes are used by default in all the S-PIM models.

Note also that the interpolation, defined in Equation (1.29), is generally performed for all the displacement components. Taking a 3D solid mechanics problem as an example, the displacement should have three components: displacements in the x, y, and z directions. The same set of shape functions is usually used for all three displacement components.

Step 3   Strain field construction

In an S-PIM model, we need an additional "unconventional" step of constructing the strain field comparing to the standard FEM, as shown in FIGURE 1.4. This is because S-PIM uses the so-called smoothed strain (or constructed strain) field instead of the compatible strain field [46]. The use of smoothed strains can provide crucial softening effects and hence various important properties to the S-PIM models. The construction of the strain field is performed based on a set of smoothing domains properly constructed on top of the triangular/tetrahedral background cells. The construction of the smoothing domains can be done with ease and automatically.

Step 4   Weakened weak form: formulation of system equations

The discrete equations of the S-PIM can be established using the constructed displacement and strain fields using a weakened weak form. These equations are usually written in matrix form based on the integration cells and assembled into the global system matrices for the entire problem domain. The SPD property of the stiffness matrix is ensured in the S-PIM model, and the condition number is usually reduced. Hence these equations can be solved with ease for a unique solution using standard routines.

Step 5   Solving system equations

The above-mentioned equations can be solved using the standard routines depending on the type of analysis.

1) For static analyses, the algebraic equations are solved using a standard solver for the displacements at all the nodes in the entire problem domain. The standard algebraic equation solver includes the Gauss elimination, LU decomposition, and often iterative methods.

2) For free vibration analyses, eigenvalues and the corresponding eigenvectors can be obtained using the standard eigenvalue solvers including the Householder's method, the bisection method, QR method, sub-space iteration, Lanczos's method, and so on.

3) For transient dynamics problems, the time history of displacement, velocity, and acceleration can be obtained following standard routines:

   - The modal superposition method may be used for vibration types of problems and problems of far field response to low speed impact with a large number of load cases.

   - For problems with a few load cases, the direct integration method can be used, which uses the standard finite difference method for time stepping with implicit and explicit schemes.

For nonlinear problems one needs an additional iteration loop to obtain the results. The strain and stress can then be retrieved easily after the displacements are obtained, as detailed in Chapter 5.

1.6   Basic settings in S-PIM

1.6.1   Triangulation

All the numerical operations (local node selection and interpolation, smoothing domain construction, integration, error estimation, refinement, etc.) in S-PIM models are performed in automatic fashion based on a set of triangular/tetrahedral background cells. All the theoretical proof, analysis and examination on the stability, convergence and error estimation are also based on the background cells. Therefore, it is essential to properly define the background cells via a generic term: *triangulation*.

Problem domain

Consider a d-dimensional problem domain $\Omega \in \mathbb{R}^d$ bounded by Γ. By default in this book, we speak of "Lipschitzian" where no "singular" points such as reentrants and cracks. For domain with singular points, we shall deal with it separately using special techniques. We also speak of "open" domain by default that does not include the boundary of the domain. When we refer to a "closed" domain we will specifically use a box: $\boxed{\Omega} = \Omega \cup \Gamma$.

Triangulation such as the widely used Delaunay triangulation algorithm is the most flexible way to create triangular background cells for S-PIM models. The process can be almost fully automated for 2D and even 3D domains of complicated geometry. Therefore, it is used in most commercial preprocessors. For 1D problems, a cell is defined in $\mathbb{R}^1$ and is simply a line segment, for 2D problems it is defined in $\mathbb{R}^2$ and becomes a triangle, and for 3D problems it is a tetrahedron defined in $\mathbb{R}^3$. Thus, "triangulation" is really a generic term for all dimensions.

Background cells

The triangulation leading to a set of N_c background cells Ω_i^c, $i = 1, 2, \cdots, N_c$, that must be nonoverlapping and seamless, such that $\boxed{\Omega} = \bigcup_{i=1}^{N_c} \boxed{\Omega}_i^c$ and $\Omega_i^c \cap \Omega_j^c = \varnothing$, $\forall i \neq j$. The problem domain is then represented by this set of background cells. In such an triangulation process, N_{eg} straight line segments L_i^c called "edges of the cells" will be produced. These cell edges are the interfaces of the triangular cells. We assume that a given problem domain can always be divided into N_c cells with N_n nodes (vertices of the triangles) and N_{eg} edges. This means that we assume there is no error in this geometry representation via triangulation.

In theory, any inner angle θ of the triangular cells should be strictly larger than zero and strictly less than 180 degree, and in practice we often require $15<\theta<120$. We also do not allow having any unconnected *free* nodes and any *duplicated* nodes. Under such conditions, we shall have $A_i^c > 0, i = 1, 2, \cdots, N_c$ and $h_i > 0, i = 1, 2, \cdots, N_c$, where A_i^c is the area and h_i is the edge length of the ith cell. The geometrical modeling for our S-PIM is then properly set.

Examples of background cells

The triangulation can be easily performed using standard algorithms such as the Delaunay algorithm or the advanced front algorithm with proper "cosmetic" treatments to improve mesh quality. These algorithms have been very well

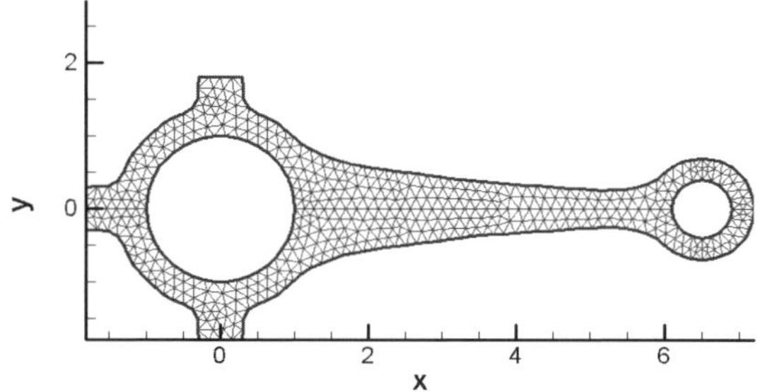

FIGURE 1.5 A triangular mesh of elements or background cells for a 2D domain.

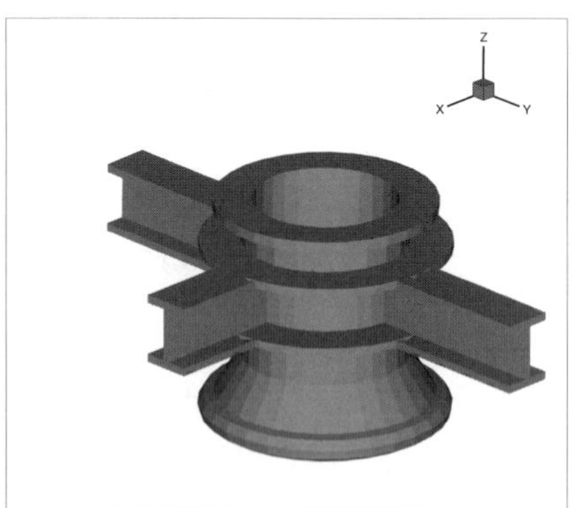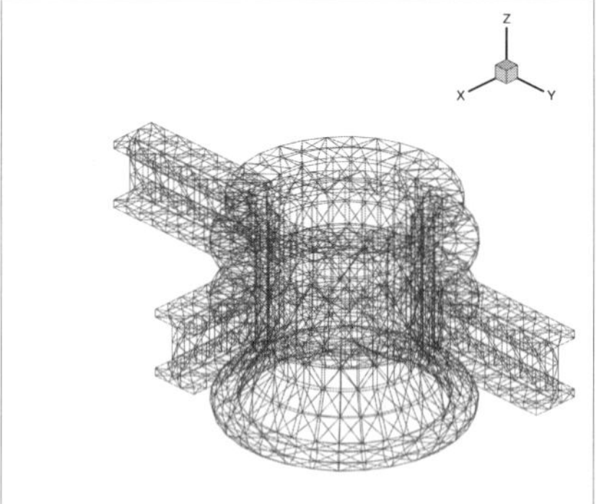

FIGURE 1.6 A tetrahedral mesh of elements or background cells for a 3D domain.

studied and developed. Most commercial pre-processes offer such triangulation functions. An example of triangulation of 2D domains to triangular elements/cells is given in FIGURE 1.5. For 3D domains, the triangulation leads to tetrahedral elements/cells, as shown in FIGURE 1.6.

Note that this book does not exclude any other types of mesh, as long as it does not give any problem in creating such a mesh for the problem to be solved. If other type of mesh is used, we assume that similar geometric controls on mesh as the triangulation defined above must be in place.

1.6.2 Characteristic length

A characteristic length can now be defined to measure the "dimension" of the cells. For uniform discretization of isolateral triangles, the edge length of any cell h is defined as the characteristic length of cells. In the case of non-uniform discretization of arbitrary triangles, there are a number of ways to define h. First, we can let

$$h_{\max} = \max_{i=1,\cdots,N_{eg}} (h_i), \quad h_{\min} = \min_{i=1,\cdots,N_{eg}} (h_i) \tag{1.30}$$

be the maximum and minimum edge lengths among all the N_{eg} edges, and the ratio between these two extreme lengths should be bounded as follows

$$h_{\max}/h_{\min} = c_{rh} < \infty \tag{1.31}$$

In this case, the characteristic length of the cells can be defined as

$$h = h_{\max} \quad \text{or} \quad h = h_{\min} \tag{1.32}$$

The characteristic length of cells can also be defined as some kind of averaged cell length, such as

$$h = \sqrt{4\bar{A}^c / \sqrt{3}} \tag{1.33}$$

where $\bar{A}^c$ is the "average" area of the triangular cells defined as $\bar{A}^c = A/N_c$ in which A is the area of the entire problem domain. In Equations (1.33), we assume the cells are all isolateral triangles. Alternatively, we may simply use

$$h = \sqrt{2\bar{A}^c} \tag{1.34}$$

where we assume the cells are all right-isosceles triangles.

In this book, the definition in Equation (1.33) is used to compute the characteristic length of triangular cells. Similarly, the characteristic length of tetrahedral cells can be obtained using

$$h = \sqrt[3]{12\overline{V}^c / \sqrt{2}} \qquad (1.35)$$

where $\overline{V}^c$ is the average volume of the tetrahedral cells defined as $\overline{V}^c = V/N_c$ in which V is the volume of the problem domain.

Note that Equations (1.32), (1.33) and (1.34) are in fact "equivalent" in the sense of that "controlling h defined in any of these three ways can put the entire mesh under control". When we say h approaches to zero, the dimensions of all the cells in the entire problem domain approach to zero, ensured by Equation (1.31). Therefore, in the convergence study of an S-PIM model (solution approaches to exact solution when h approaches to zero), any of these equations can be used, and they all should deliver the same convergence rate for the same set of meshes used to examine the same problem, although the actual value of the solution error may be different.

1.6.3 T-schemes for node selection

Since the background cells are needed for the integration for meshfree methods using weak or weakened weak form, background cells are often already made available. Therefore, it is natural to make use of them also for the selection of support nodes for shape function creation. Background cells of triangular type generated by triangulation detailed in Section 1.6.1 have been found most practical, robust, reliable and efficient for local support nodes selection. It works particularly well for the family of point interpolation methods (PIMs). Triangular/tetrahedral-mesh-based node selection schemes are termed as T-schemes that are detailed in this section.

First, we note that for any point on the boundary of the problem domain, we use only *linear interpolation* using the two adjacent nodes of the point on the boundary for 2D problems. Similarly for 3D problems, linear interpolation using three nodes will be used for the point located on the boundary surfaces. This rule is applied for all the T-schemes and for all the S-PIM models. This rule is necessary to ensure the S-PIM model passing the standard patch tests (see, Remarks 6.3 and 6.6).

Second, for an interior point within the problem domain, a proper T-scheme should be selected for different S-PIM models of desired properties. TABLE 1.5 lists the T-schemes used for support nodes selection based on triangular background cells for PIM shape function creation of 2D problems and TABLE 1.6 lists the T-schemes selecting support nodes based on tetrahedral background cells for 3D problems.

TABLE 1.5 T-schemes for support nodes selection based on triangular background cells for PIM shape function creation of 2D problems

Name	Node selection for interpolation at any point within the problem domain	Applications
Cell-based T3-scheme	3 nodes (vertices) of the home cell	Polynomial PIM
Cell-based T6/3-scheme	For an interior home cell, 3 nodes of the home cell and 3 remote nodes of the three neighboring cells. For a boundary home cell, only 3 nodes of the home cell.	Polynomial PIM Radial PIM
Cell-based T6-scheme	For an interior home cell, 3 nodes of the home cell and 3 remote nodes of the three neighboring cells. For a boundary home cell, 3 nodes of the home cell, 2 (or 1) remote nodes of the neighboring cells plus 1 (or 2) field node which is nearest to the centroid of the home cell.	Radial PIM
Cell-based T2L-scheme	Nodes of the home cell plus one layer of nodes which are the nodes connected directly to the vertices of the home cell; thus two layers of nodes, corresponding to the home cell, are selected.	Radial PIM
Edge-based T2-scheme	2 end nodes of the home edge	Polynomial PIM
Edge-based T4-scheme	For an interior edge, 2 end nodes of the home edge plus 2 remote nodes of the neighboring cells of the edge.	Polynomial PIM with CT Radial PIM
Edge-based T2L-scheme	For an interior edge, 4 nodes selected using the edge-based T4-scheme plus another layer of nodes connecting to these 4 nodes directly.	Radial PIM

TABLE 1.6 T-schemes for support nodes selection based on tetrahedral background cells for PIM shape function creation of 3D problems

Name	Node selection for interpolation at any point within the problem domain	Applications
Cell-based T4-scheme	4 nodes of the home tetrahedral cell	Polynomial PIM
Cell-based T2L-scheme	Nodes of the home cell plus one layer of nodes connected directly to the nodes of the home cell. Thus two layers of nodes surrounding the home cell are selected.	Radial PIM
Face-based T3-scheme	3 nodes of the home triangular face	Polynomial PIM
Face-based T5-scheme	For an interior face, 3 nodes of the face plus 2 remote nodes of the neighboring cells of the face.	Polynomial PIM with CT Radial PIM
Face-based T2L-scheme	For an interior face, 5 nodes selected using the face-based T5-scheme plus another layer of nodes connecting directly to these 5 nodes.	Radial PIM

Notes: a *home cell* refers to the cell which hosts the point of interest (usually the quadrature sampling point). An *interior home cell* is a home cell that has no edge on the boundary of the problem domain and a *boundary home cell* is a home cell which has at least one edge on the boundary. An *interior edge* is an edge located inside the problem domain and a *boundary edge* refers one edge located on the domain boundaries. An *interior face* is a triangular surface of a tetrahedral element and located inside the problem domain; a *boundary face* refers to a triangular face located on the boundary surfaces of the domain. A *neighboring cell of a home cell* refers to the cell which shares one edge/face with the home cell. A *neighboring cell of an edge* refers to the cell containing this edge. A *neighboring cell of a face* is a cell containing this face. "CT" stands for Coordinate Transformation.

According to the location of the point of interest, i.e. the quadrature point $(\mathbf{x}_Q)$, the T-schemes are generally classified into the three types:

1) cell-based T-schemes for the point of interest located inside background cells (2D or 3D);

2) edge-based T-schemes for the point of interest located on the edges of triangular cells (2D); and

3) face-based T-schemes for the point of interest located on the faces of tetrahedral cells (3D).

The following gives detailed descriptions of these T-schemes.

Cell-based T3-scheme

In the cell-based T3-scheme, we simply select three nodes of the home cell of the point of interest [30]. As illustrated in FIGURE 1.7a, no matter the point of interest ($\mathbf{x}_Q$) located in an interior home cell (cell i) or a boundary home cell (cell j), only the three nodes of the home cell (i_1-i_3 or j_1-j_3) are selected. The cell-based T3-scheme is used only for creating linear PIM shape functions by using polynomial basis functions. Note that the linear PIM shape functions so created are exactly the same as those in FEM using linear triangular elements. The shape functions can always be created (the moment matrix will never be singular), as long as the background cells of triangular type are generated by triangulation detailed in Section 1.6.1.

Cell-based T6/3-scheme

The cell-based T6/3-scheme selects six nodes to interpolate a point of interest located in an interior cell and three nodes for those located in boundary home cells [30]. As illustrated in FIGURE 1.7b, when the point of interest ($\mathbf{x}_Q$) is located in an interior home cell (cell i), we select six nodes: three nodes of the home cell (i_1-i_3) and another three nodes located at the remote vertices of the three neighboring cells (i_4-i_6). When the point of interest is located in a boundary home cell (cell j), we select only three nodes of the home cell, i.e., j_1-j_3.

The cell-based T6/3-scheme was purposely devised for creating high-order PIM shape functions, where quadratic interpolations are performed for the interior home cells and linear interpolations for boundary home cells. This scheme was first used in the NS-PIM [30]. It not only successfully overcomes the singular problem but also improves the efficiency of the method. Using this scheme the shape functions can always be created, as long as such six nodes can be found for all the interior home cells. Note the cell-based T6/3-scheme can also be used for creating radial PIM (RPIM) shape functions without the singularity problem.

Cell-based T6-scheme

Same as the cell-based T6/3-scheme, the cell-based T6-scheme, as shown in FIGURE 1.7c, also selects six nodes for an interior home cell: three nodes of the home cell and three vertices at the remote vertices of the three neighboring cells

(i_1-i_6 for cell i). However, for a boundary cell (cell j), the cell-based T6-scheme still selects six nodes: three nodes of the home cell (j_1-j_3), two remote nodes of the neighboring cells (j_4 and j_5) and one field node (j_6) which is nearest to the centroid of the home cell excepting the five nodes that have been selected [37].

The cell-based T6-scheme was purposely devised for creating RPIM shape functions on considering both accuracy and efficiency. Different form the T6/3-scheme, this scheme selects six nodes for all home cells containing the point of interest. The shape functions can always be created because the radial moment matrix is always invertible for arbitrary scattered nodes as long as to avoid using some specific shape parameters (see, Chapter 3).

Cell-based T2L-scheme

The cell-based T2L-scheme selects two layers of nodes to perform interpolation based on triangular (tetrahedral) meshes [27, 31, 37]. As shown in FIGURE 1.7d where the 2D domain is presented using triangular background cells, the first layer of nodes refers to the three nodes (vertices) of the home cell, and the second layer contains those nodes which are directly connected to the three nodes of the first layer. The cell-based T2L-scheme for 3D problems is the analogy of the cell-based T2L-scheme for 2D problems, but used for nodes selection for 3D domains presented using Te4 background cells.

The cell-based T2L-schme was devised for creating RPIM shape functions. This scheme usually selects much more nodes than the cell-based T6-scheme and leads to more time consumption. The RPIM shape functions can always be created because the radial moment matrix is always invertible for arbitrarily scattered nodes. We can use this scheme to create RPIM shape functions with high order of consistence and for extremely irregularly distributed nodes. Such RPIM shape functions can also be used for strong form meshfree method methods where higher order of consistence is required [53, 54].

Edge-based T2-scheme

When the point of interest is located on edges of triangular background cells, we may use various edge-based T-schemes to select support nodes [23]. The edge-based T2-scheme selects the two nodes at the ends of the edge to perform the (linear) interpolation at the point of interest located on this edge, as shown in FIGURE 1.7e. This scheme can only be used to create linear PIM shape functions by using linear polynomial basis functions. The T2-scheme is also used in all the 2D S-PIMs for points of interest at all boundary edges.

Edge-based T4-scheme

The edge-based T4-scheme selects four nodes to perform the interpolation for a point of interest located on an interior edge [23-25]. Note for points located on boundary edges we still use two nodes (edge-based T2-scheme), as illustrated in FIGURE 1.7f.

The edge-based T4-scheme can be used for creating both PIM and RPIM shape functions. For PIM shape functions creation, bilinear interpolations are performed for interior edges and linear interpolations for boundary edges. Noted that to overcome the singularity problem occurring in the process of PIM shape functions creation, additional measures should be conducted, such as the efficient coordinates transformation (CT) operation detailed in Section 3.4. For RPIM shape functions creation, this scheme can be directly used to select support nodes without additional measures.

Edge-based T2L-scheme

The edge-based T2L-scheme selects two layers of nodes to perform the interpolation at points located on an interior edge of triangular cells [23, 24], as shown in FIGURE 1.7g. The first layer of nodes is exactly the four nodes selected using the edge-based T4-scheme and the second layer of nodes includes those nodes directly connected to the first layer of nodes. When the point of interest is located on a boundary edge, we still use the edge-based T2-scheme.

This scheme is used to select support nodes for creating RPIM shape functions, which can always be obtained because the radial moment matrix is always invertible for arbitrarily scattered nodes. Similar as the cell-based T2L-scheme, this scheme will select much more support nodes and can be used to create RPIM shape functions with high order of consistence and for extremely irregularly distributed nodes.

Cell-based T4-scheme

The cell-based T4-scheme is the analogy of the cell-based T3-scheme, but used for nodes selection for 3D domains presented using four-node tetrahedral (Te4) background cells [31]. In this scheme, we select four nodes of the home tetrahedral cell of the point of interest located inside the cell, as shown in FIGURE 1.7h.

Face-based T3-scheme

For 3D problem presented using four-node tetrahedral cells, we use the face-based T-scheme to select support nodes for those points of interest located on the triangular faces of the background cell [25], as shown in FIGURE 1.7i. The face-based T3-scheme selects three nodes (vertices) of a triangular face as the support nodes. This scheme can only be used to create linear PIM shape functions. The face-based T3-scheme is also used in all the 3D S-PIMs for points of interest at all boundary faces.

Face-based T5-scheme

The face-based T5-scheme selects five nodes for a point of interest located on an interior face and three nodes for those located on boundary faces [25], as shown in FIGURE 1.7j. The five nodes include three nodes (vertices) of the triangular face and two remote nodes of the neighboring cells sharing this face. The face-based T5-scheme is used for creating RPIM shape functions for 3D problems. Note for points located on boundary faces we still use the face-based T3-scheme.

Face-based T2L-scheme

The face-based T2L-scheme selects two layers of nodes for a point of interest located on a triangular face of a tetrahedral cell [25]. For an interior face, nodes of the first layer are exactly the five nodes selected using the face-based T5-scheme and the second layer of nodes includes those connected directly to these five nodes. For points on the boundary face, only three nodes (vertices) of the triangular face are selected as support nodes. Note again that for points located on boundary faces we still use the face-based T3-scheme.

This scheme is used to select support nodes for creating RPIM shape functions. It selects much more support nodes and can be used to create RPIM shape functions with high order of consistence. Because many nodes used, this scheme work well for extremely irregular nodes.

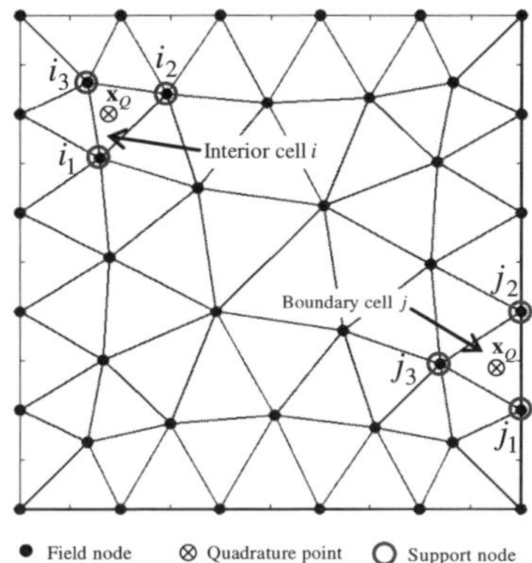

(a) Cell-based T3-scheme: i_1, i_2 and i_3 are support nodes for $\mathbf{x}_Q$ located in the interior cell i; j_1, j_2 and j_3 are support nodes for $\mathbf{x}_Q$ in the boundary cell j.

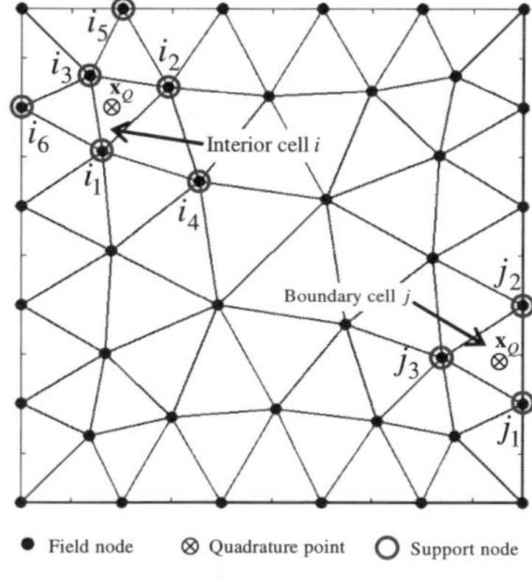

(b) Cell-based T6/3-scheme: i_1~i_6 are support nodes for $\mathbf{x}_Q$ located in the interior cell i; j_1, j_2 and j_3 are support nodes for $\mathbf{x}_Q$ in the boundary cell j.

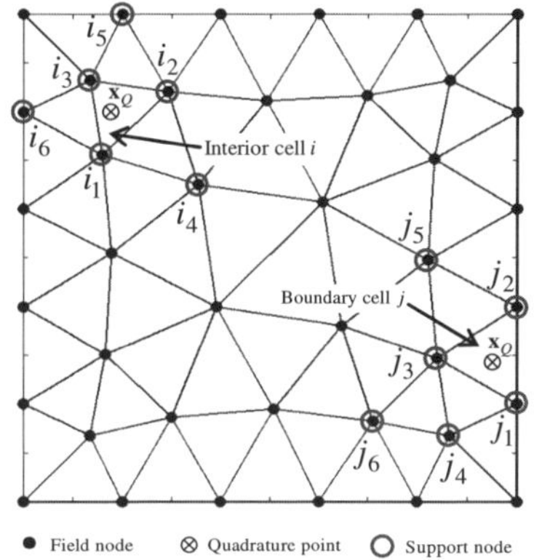

(c) Cell-based T6-scheme: i_1~i_6 are support nodes for $\mathbf{x}_Q$ located in the interior cell i; j_1~j_6 are support nodes for $\mathbf{x}_Q$ in the boundary cell j.

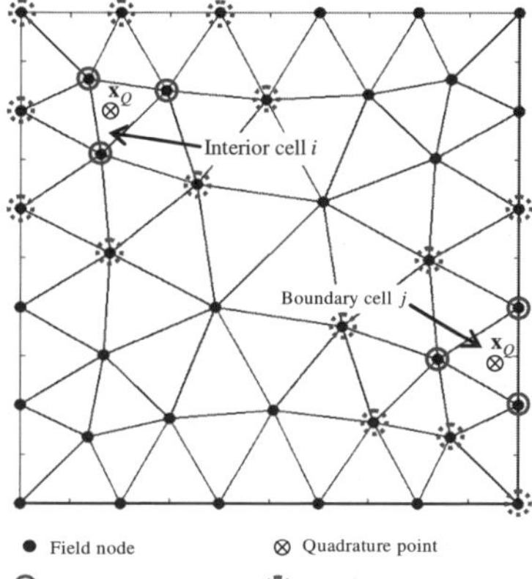

(d) Cell-based T2L-scheme for an interior home cell i and boundary cell j: two layers of nodes of the cells are selected as support nodes.

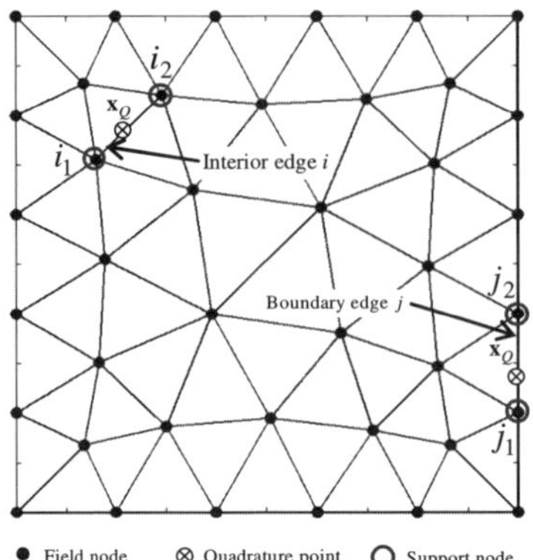

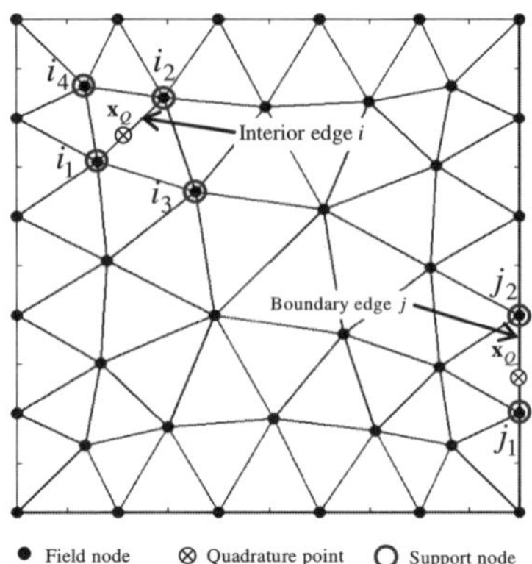

(e) Edge-based T2-scheme: i_1 and i_2 are support nodes for $\mathbf{x}_Q$ located on the interior edge i; j_1 and j_2 are support nodes for $\mathbf{x}_Q$ on the boundary edge j.

(f) Edge-based T4-scheme: $i_1 \sim i_4$ are support nodes for $\mathbf{x}_Q$ located on the interior edge i; j_1 and j_2 are support nodes for $\mathbf{x}_Q$ on the boundary edge j.

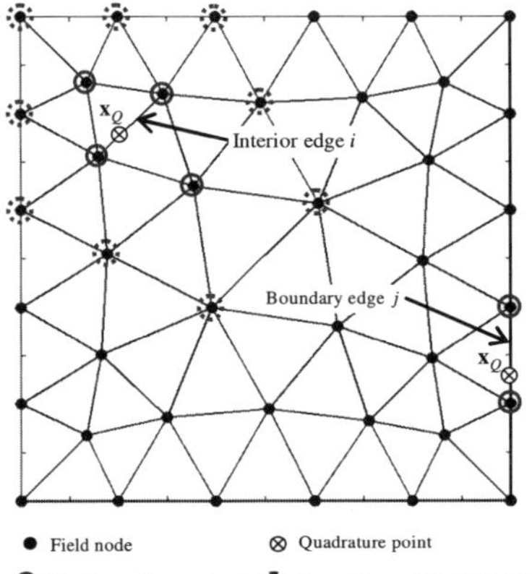

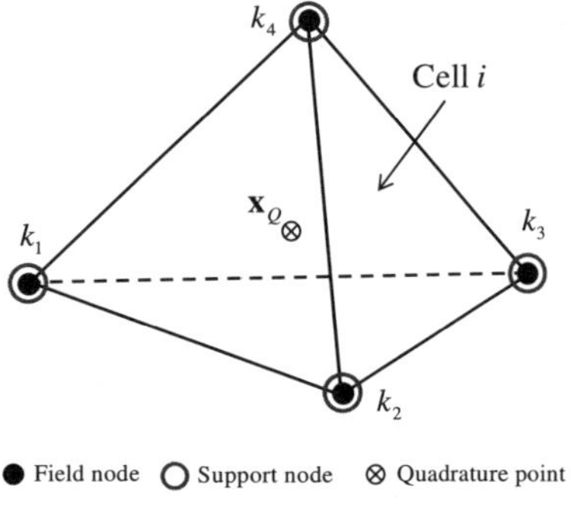

(g) Edge-based T2L-scheme: two layers of nodes are support nodes for $\mathbf{x}_Q$ located on the interior edge i; j_1 and j_2 are the support nodes for $\mathbf{x}_Q$ located on the boundary edge j.

(h) Cell-based T4-scheme: four nodes of cell i including k_1, k_2, k_3 and k_4 are the support nodes for $\mathbf{x}_Q$ located inside the cell.

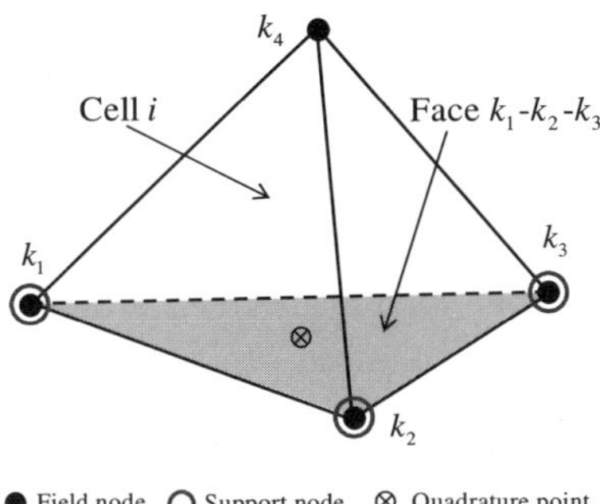
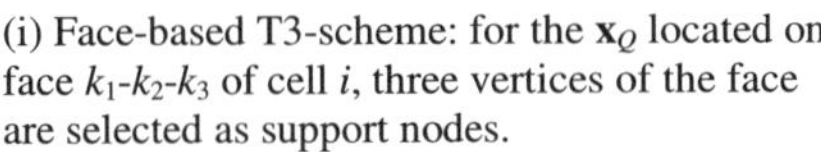

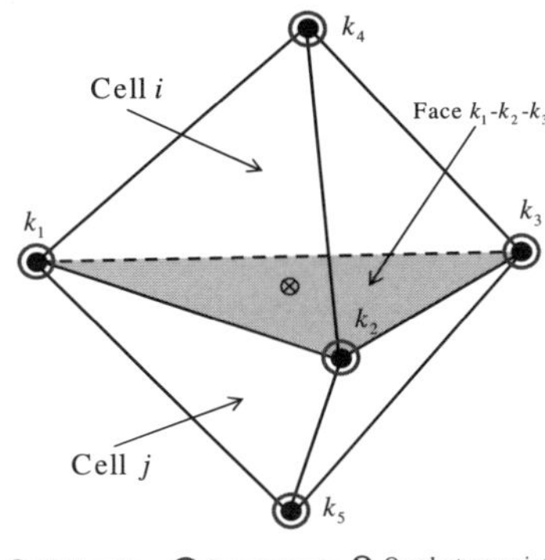

(i) Face-based T3-scheme: for the $\mathbf{x}_Q$ located on face k_1-k_2-k_3 of cell i, three vertices of the face are selected as support nodes.

(j) Face-based T5-scheme: for the $\mathbf{x}_Q$ located on the common face of cells i and j, three vertices of the face plus two remote nodes are selected as support nodes.

FIGURE 1.7 Illustration of selecting local support nodes using T-schemes based on three-node triangular (Tr3) cells for 2D problems and four-node tetrahedral (Te4) cells for 3D problems.

1.7 Outline of the book

This book provides a detailed introduction to meshfree smoothed point interpolation methods (S-PIMs), and their applications to the following types of mechanics problems:

- Mechanics for solids (1D, 2D and 3D)
- Error estimation and adaptive analysis in 2D and 3D domains
- Fracture mechanics in 2D solids
- Heat transfer and thermoelastic problems (1D, 2D and 3D)

The bulk of the material in the book is the result of the intensive research work by the authors' research team at the Centre for Advanced Computations in Engineering and Science (ACES) in the past eight years. The significance of this book is as follows:

1) This is the first book published that intensively introduces the family of S-PIMs, which is the only group of proven stable methods based on the G space theory.

2) The book presents, in a systematic manner, the newly developed weakened weak (W^2) formulation and G space theory. It covers all the principles, techniques, and procedures in solving mechanics problems using the S-PIMs. This important piece of research has opened an important window for the development of novel and powerful numerical methods.

3) Readers will benefit from the research outcome of G. R. Liu's research team and their long term research projects on meshfree methods founded by the Singapore government and other organizations. Many materials in this book are the results of the ongoing projects, and have not been previously published.

4) A large number of examples with illustrations are provided for validating, benchmarking, and demonstrating meshfree methods. These examples can be useful reference materials for other researchers.

The book is written for senior university students, graduate students, researchers, and professionals in engineering and science. Mechanical engineers and practitioners and structural engineers and practitioners will also find the book useful. Knowledge of FEM is not required but would help a great deal in understanding many concepts and procedures of smoothed point interpolation methods. The closest method to the S-PIM is the smoothed finite element method (or S-FEM) [9]. In fact, the S-FEM is a watered-down version of the S-PIM. Knowledge on any of the S-FEM models would be helpful in understanding S-PIM. Also, basic knowledge of mechanics is certainly helpful in reading this book more smoothly.

The S-PIM is a relatively new numerical method and is in a rapidly developing and growing stage. There still exist some problems, in both theoretical and practical levels, that offer ample opportunities for researchers to improve the present method and develop the next generation of numerical methods. This book addresses some of the current important issues, both positive and negative, which should prove beneficial to researchers, engineers, and students who are interested in venturing into this area of research. The chapter-by-chapter description of this book is as follows:

Chapter 1: Provides the basic equations, basic idea, background, overall procedures, and common preliminary techniques for S-PIM.

Chapter 2: Introduces function spaces useful for general numerical methods and particularly the S-PIM. The standard spaces that are widely used in FEM and meshfree are first briefed, followed by the relatively new G spaces that are the foundation of the weakened weak formulations, are particularly useful for both meshfree and FEM settings.

Chapter 3: Provides a detailed description on creating shape functions using PIMs in proper G spaces, which is one of the most important issues for numerical methods. Detailed discussions on the properties of PIM shape functions and the issues related to use of these shape functions are provided.

Chapter 4: Introduces some techniques used for strain (gradient) field constructions, which is particularly important for S-PIM, including strain gradient scaling, strain smoothing, generalized smoothing, point interpolation, and least square projection techniques.

Chapter 5: Introduces the principles and weak forms that will be used for creating discretized system equations, including the standard Galerkin weak forms, and various weakened weak forms such as the SC-Galerkin, GS-Galerkin and Galerkin-like weak forms.

Chapter 6: Introduces the node-based smoothed point interpolation methods (NS-PIMs), which are formulated based on the GS-Galerkin weak form using both polynomial and radial PIM shape functions. Properties of the NS-PIMs, such as the softening effects, upper bound and free of volumetric locking, are discussed in detail. A comparison study is conducted against other numerical methods. The computer programs coded in FORTRAN 90 for various NS-PIM models are provided (for 2D cases) at the end of the Chapter.

Chapter 7: Introduces the edge-based smoothed point interpolation methods (ES-PIMs) formulated based on the GS-Galerkin weak form using both polynomial and radial PIM shape functions. Properties of the ES-PIMs are detailed and a comparison study is carried against other methods. The computer programs for various ES-PIM models are provided (for 2D cases) at the end of the Chapter.

Chapter 8: Introduces the cell-based smoothed point interpolation methods (CS-PIMs), which are formulated based on the GS-Galerkin weak form using both polynomial and radial PIM shape functions. Properties of the

CS-PIMs are presented and compared with other numerical methods. The computer programs for various CS-PIM models are provided (for 2D cases) at the end of the Chapter.

Chapter 9: Introduces the cell-based smoothed alpha radial point interpolation method (CS-αRPIM) formulated based on the GS-Galerkin weak form using αRPIM shape functions. A general procedure has been proposed for determining the particular value of α leading to "nearly exact" solution in strain energy.

Chapter 10: Introduces the strain-constructed point interpolation methods (SC-PIMs) formulated based on the strain-constructed Galerkin weak form using polynomial PIM shape functions. A number of schemes for further constructing linear strain fields using the smoothed or compatible strains are presented.

Chapter 11: Introduces S-PIMs developed for heat transfer and thermoelastic problems in 1D, 2D and 3D domains. Formulations of S-PIMs for thermal analysis are further derived. Properties, such as upper bound solutions, superconvergence and ultra-accuracy are discussed.

Chapter 12: Formulates the singular CS-RPIM (sCS-RPIM) for linear fracture analysis. A five-node singular element has been devised for simulating the singular stress field near the crack tip. The interaction integral method is used to compute the stress intensity factors. A number of attractive properties, such as upper and lower band solutions, and accuracy are discussed.

Chapter 13: Presents the adaptive analysis procedures using NS-PIM/ES-PIM models. An efficient error indicator has been presented, together with a simple h-type refinement strategy. The adaptive procedure is developed for both 2D and 3D problems using triangular and tetrahedral meshes.

Appendix: Provides a code library of common subroutines used in the programs for various S-PIM models. The explanation of the subroutine functions, the definition of the major parameters and arrays, and the input file for the demonstration case are also provided.

1.8 References

1. Zienkiewicz, O. C. and Taylor R. L., *The Finite Element Method*, 5th ed., Butterworth Heimemann, Oxford, 2000.

2. Hughes, T. J. R., *The Finite Element Method: Linear Static and Dynamic Finite Element Analysis*, Prentice-Hall, 1987.

3. Reddy, J. N., *Finite Element Method*, John Wiley & Sons Inc., New York, 1993.

4. Bathe, K. J., *Finite Element Procedures*, Prentice-Hall, 1996.

5. Belytschko, T., Liu, W. K. and Moran, B., *Nonlinear Finite Elements for Continua and Structures*, John Wiley & Sons, Ltd, 2000.

6. Liu, G. R. and Quek, S. S., *The finite element method: a practical course.* Butterworth Heinemann: Oxford, 2002.

7. Liu, G R and X Han, *Computational Inverse techniques in nondestructive evaluation*, CRC press, 2003.

8. Liu, G. R., *Meshfree Methods: Moving beyond the Finite Element Method*, 2nd ed., CRC press, Boca Taton, USA, 2009.

9. Liu, G. R. and Nguyen T. T, *Smoothed Finite Element Method*, CRC press, Boca Taton, USA, 2010.

10. Liu, G. R. and Gu, Y. T., *An introduction to meshfree method methods and their programming*, Springer, 2005.

11. Liu, G. R. and Liu, M. B., *Smoothed Particle Hydrodynamics—A Meshfree Practical Method.* World Scientific: Singapore, 2003.

12. Belytschko, T., Lu, Y. Y. and Gu, L., Element-free Galerkin methods, *International Journal for Numerical Methods in Engineering*, 37: 229-256, 1994.

13. Liu, W. K., Jun, S. and Zhang, Y. F., Reproducing kernel particle methods. *International Journal for Numerical Methods in Engineering*, 20: 1081-1106, 1995.

14. Liu, G. R. and Gu, Y. T., A point interpolation method for two-dimensional solids, *International Journal for Numerical Methods in Engineering*, 50: 937-951, 2001.

15. Wang, J. G. and Liu, G. R., A point interpolation meshless method based on radial basis functions, *International Journal for Numerical Methods in Engineering*, 54: 1623-1648, 2002.

16. Atluri, S. N. and Zhu, T., A new meshless local Petrov-Galerkin (MLPG) approach in computational mechanics. *Computational Mechanics*, 22: 117-127, 1998.

17. Onate, E., Idelsohn, S., Zienkiewicz, O. C. and Taylor, R. L., A finite point method in computational mechanics. Applications to convective transport and fluid flow. *International Journal for Numerical Methods in Engineering*, 39(22): 3839-3866, 1996.

18. Gu, Y. T. and Liu, G. R., A meshless local Petrov-Galerkin (MLPG) method for free and forced vibration analyses for solids, *Computational Mechanics,* 27(3): 188-198, 2001.

19. Armando, D. C. and Oden, J. T., *Hp clouds-a meshless method to solve boundary value problems*. TICAM Report 95-05, University of Texas at Austin, 1995.

20. Melenk, J. M. and Babuska, I., The Partition of Unity Finite Element Method: Basic Theory and Applications. *Computer Methods in Applied Mechanics and Engineering*, 139: 289-314, 1996.

21. Yagawa, G. and Yamada, T., Free mesh method, a kind of meshless finite element method. *Computational Mechanics*, 18: 383-386, 1996.

22. De, S. and Bathe, K. J., The method of finite spheres. *Computational Mechanics*, 25: 329-345, 2000.

23. Liu, G. R. and Zhang, G. Y., A normed G space and weakened weak (W^2) formulation of cell-based smoothed point interpolation method. *International Journal of Computational Methods*, 6(1): 147-179, 2009.

24. Zhang, G. Y. and Liu, G. R., Meshfree cell-based smoothed point interpolation method using isoparametric PIM shape functions and condensed RPIM shape functions. *International Journal of Computational Methods*, 8(4): 705-730, 2011.

25. Liu, G. R., Zhang, G. Y., Zong, Z. and Li, M., Meshfree cell-based smoothed alpha radial point interpolation method (CS-αRPIM) for solid mechanics problems. *International Journal of Computational Methods*, 10(4), 31 pages, DOI: 10.1142/S0219876213500205, 2013.

26. Liu, G. R., A G space theory and a weakened weak (W^2) form for a unified formulation of compatible and incompatible methods: Part I theory. *International Journal for Numerical Methods in Engineering*, 81: 1093-1126, 2010.

27. Liu, G. R., A G space theory and a weakened weak (W^2) form for a unified formulation of compatible and incompatible methods: Part II applications to solid mechanics problems. *International Journal for Numerical Methods in Engineering*, 81: 1127-1156, 2010.

28. Liu, G. R., On a G space theory. *International Journal of Computational Methods*, 6(2): 257–289, 2009.

29. Bai, Z, Demmel, J., Dongarra, J., Ruhe, A. and Vorst, H. van der, editors. *Templates for the Solution of Algebraic Eigenvalue Problems: A Practical Guide.* SIAM, Philadelphia, 2000.

30. Liu, G. R., Zhang, G. Y., Dai, K. Y., Wang, Y. Y., Zhong, Z. H., Li, G. Y. and Han, X., A linearly conforming point interpolation method (LC-PIM) for 2D solid mechanics problems, *International Journal of Computational Methods*, 2(4): 645-665, 2005.

31. Zhang, G. Y., Liu, G. R., Wang, Y. Y., Huang, H. T., Zhong, Z. H., Li, G. Y. and Han, X., A linearly conforming point interpolation method (LC-PIM) for three-dimensional elasticity problems, *International Journal for Numerical Methods in Engineering*, 72: 1524-1543, 2007.

32. Liu, G. R. and Zhang, G. Y., Upper bound solution to elasticity problems: A unique property of the linearly conforming point interpolation method (LC-PIM). *International Journal for Numerical Methods in Engineering*, 74: 1128-1161, 2008.

33. Liu, G. R, Li, Y., Dai, K. Y., Luan, M. T. and Xue, W., A linearly conforming radial point interpolation method for solid mechanics problems. *International Journal of Computational Methods,* 3(4): 401-428, 2006.

34. Li, Y., Liu, G. R., Luan, M. T., Dai, K. Y., Zhong, Z. H., Li, G. Y. and Han, X., Contact analysis for solids based on linearly conforming radial point interpolation method, *Computational Mechanics*, 39, 537-554, 2007.

35. Zhang, G. Y., Liu, G. R., Nguyen-Thoi, T., Song, C. X., Han, X., Zhong, Z. H. and Li, G. Y., The upper bound property for solid mechanics of the linearly conforming radial point interpolation method (LC-RPIM). *International Journal of Computational Methods*, 4(3): 521-541, 2007.

36. Nguyen-Thoi, T., Liu, G. R., Lam, K. Y. and Zhang, G. Y., A Face-based Smoothed Finite Element Method (F-SFEM) for 3D linear and nonlinear solid mechanics problems using 4-node tetrahedral elements. *International Journal for Numerical Methods in Engineering*, 78: 324-353, 2009.

37. Liu, G. R. and Zhang, G. Y., Edge-based smoothed point interpolation methods. *International Journal of Computational Methods*, 5(4): 621-646, 2008.

38. Liu, G. R. and Zhang, G. Y., A novel scheme of strain-constructed point interpolation method for static and dynamic mechanics problems. *International Journal of Applied Mechanics,* 1(1): 233-258, 2009.

39. Zhang, G. Y., Liu, G. R. and Xu, X., A strain-constructed point interpolation method (SC-PIM) and strain field construction schemes for solid mechanics problems using triangular mesh. *Applied Mathematics and Computation,* 219: 2067-2086, 2012.

40. Liu, G. R., Xu, X., Zhang, G. Y. and Gu, Y. T., An extended Galerkin weak form and a point interpolation method with continuous strain field and superconvergence (PIM-CS) using triangular mesh, *Computational Mechanics*, 43: 651-673, 2009.

41. Liu, G. R., Xu, X., Zhang, G. Y. and Nguyen-Thoi, T., A superconvergence point interpolation method (SC-PIM) with piecewise linear strain field using triangular mesh, *International Journal for Numerical Methods in Engineering*, 77: 1439-1467, 2009.

42. Xu, X., Liu, G. R. and Zhang, G.Y., A point interpolation method with least square strain field (PIM-LSS) for solution bounds and ultra-accurate solutions using triangular mesh. *Computer Methods in Applied Mechanics and Engineering*, 198: 1486-1499, 2009.

43. Xu, X., Liu, G. R., Gu, Y. T., Zhang, G. Y., Luo J. W. and Peng J. X, A point interpolation method with locally smoothed strain field (PIM-LS2) for mechanics

problems using triangular mesh. *Finite Elements in Analysis and Design*, 46(10): 862-874, 2010.

44. Liu, G. R., Nguyen-Thoi, T. and Lam, K. Y., A novel Alpha Finite Element Method (αFEM) for exact solution to mechanics problems using triangular and tetrahedral elements. *Computer Methods in Applied Mechanics and Engineering*, 197: 3883-3897, 2008.

45. Nguyen-Thoi, T., Liu, G. R., Dai, K. Y. and Lam, K. Y., Selective smoothed finite element method. *Tsinghua Science and Technology*, 12(5): 497-508, 2007.

46. Liu, G. R., A generalized Gradient smoothing technique and the smoothed bilinear form for Galerkin formulation of a wide class of computational methods. *International Journal of Computational Methods*, 5(2): 199-236, 2008.

47. Liu, G. R., Tu, Z. H. and Wu, Y. G., MFree2D[©]: an adaptive stress analysis package based on mesh free technology, presented at *3rd South Africa Conference on Applied Mechanics*, 11-13 January, Durban, South Africa, 2000.

48. Liu, G. R. and Tu, Z. H., An adaptive procedure based on background cells for meshless methods, *Computer Method in Applied Mechanics and Engineering*, 191, 1923-1943, 2002.

49. Zhang, G. Y., Liu, G. R. and Li, Y., An efficient adaptive analysis procedure for certified solutions with exact bounds of strain energy for elasticity problems. *Finite Elements in Analysis and Design*, 44(14): 831-841, 2008.

50. Tang, Q., Liu, G. R., Zhong, Z. H. and Zhang, G. Y., A three-dimensional adaptive analysis using the meshfree node-based smoothed point interpolation method (NS-PIM). *Engineering Analysis with Boundary Elements*, 35(10): 1123-1135, 2011.

51. Tang, Q., Zhong, Z. H., Zhang, G. Y. and Xu, X., An efficient adaptive analysis procedure for node-based smoothed point interpolation method (NS-PIM). *Applied Mathematics and Computation*, 217(21): 8387-8402, 2011.

52. Tang, Q., Zhang, G. Y., Liu, G. R., Zhong, Z. H. and He, Z. C., An efficient adaptive analysis procedure using the edge-based smoothed point interpolation method (ES-PIM) for 2D and 3D problems. *Engineering Analysis with Boundary Elements,* 36: 1424-1443, 2012.

53. Liu, G. R. and Kee, B. B. T., A stabilized least-squares radial point collocation method (LS-RPCM) for adaptive analysis, *Computer Method in Applied Mechanics and Engineering*, 195: 4843-4861, 2006.

54. Kee, B. B. T, Liu, G. R. and Lu. C., A regularized least-squares radial point collocation method (RLS-RPCM) for adaptive analysis. *Computational Mechanics*, 40: 837-853, 2007.

Chapter 2

G Spaces

To properly formulate a numerical method that ensures *stable* and *convergent* solutions, and to assess the *quality* and the *property* of the solutions, theories on *function spaces* and *functional analysis* are necessary. In this chapter, we thus discuss some essential topics related to *space theory* and *functional analysis*, with the focus on the relatively new G space theory. We start with a briefing on the standard *function spaces* that are widely used in the usual weak formulations such as in the finite element method (FEM) [1-3], and some meshfree methods [4]. We then provide a detailed discussion on the G spaces that are the foundation of a weakened weak (W^2) formulation, and are particularly useful for numerical methods with both meshfree and FEM settings, including the S-PIM.

Topics related to spaces can be quite mathematical and hence difficult for an engineer to comprehend, but for reasonably accurate and concise presentations of a weak or weakened weak (W^2) formulation procedure, it is quite difficult to avoid. We thus choose to present these related topics in "engineering" languages for easier comprehension. We focus only on the minimum necessary ingredients that are sufficient for the use in later chapters. Readers who are only interested in the application of the S-PIM may simply skip this chapter, but encouraged to skim thorough. It may be discovered that the theory of spaces presented in this chapter is not that difficult to understand after all.

2.1 General issues on function spaces

We first briefly introduce some of the important basic definitions and terminologies related to functional spaces. For more intensive and systematic descriptions, readers may refer to [5].

2.1.1 Linear spaces

A space $\mathbb{S}$ of functions is said linear, if the summation of any two functions from the space still belongs to the same space, and any function from the space still belongs to the same space after a scaling. In mathematical expression, a *linear space* satisfies

$$
\begin{aligned}
\forall w, v \in \mathbb{S}, && (w+v) \in \mathbb{S} && \textit{Addition} \\
\forall \alpha \in \mathbb{R} \text{ and } \forall v \in \mathbb{S}, && \alpha v \in \mathbb{S} && \textit{Scaling}
\end{aligned}
\tag{2.1}
$$

where $\mathbb{R}$ is the real (algebraic) field of all the real numbers.

A group of the assumed displacement functions used for weak or W^2 formulations for linear elasticity problems shall form a linear space, so that new functions in the same space can be constructed simply by linear combinations of some of these functions in the space.

2.1.2 Functional

A *functional J* takes a function from a space as input (or argument) and returns a scalar as an output. It is expressed as

$$
J: \underbrace{\mathbb{S}}_{\text{space offers the input}} \rightarrow \underbrace{\mathbb{R}}_{\text{space hosts the output}}
\tag{2.2}
$$

which means that J takes a function from $\mathbb{S}$ as input, and then produces a real number as output. A functional is also visually said "a function of functions" that produces a real number.

The strain energy in a linear elastic solid is a typical functional that takes in a displacement function and returns a real number of energy.

2.1.3 Norm

A functional $\lVert \cdot \rVert_{\mathbb{S}}$ from a linear space $\mathbb{S}$ to $\mathbb{R}$ is called a "norm" *if and only if* it has the following properties:

(a) *Positivity*

$$
\forall v \in \mathbb{S}: \lVert v \rVert_{\mathbb{S}} \geq 0, \ \lVert v \rVert_{\mathbb{S}} = 0 \ \Leftrightarrow \ v = 0
\tag{2.3}
$$

(b) *Scalar multiplication*

$$\|\alpha v\|_{\mathbb{S}} = |\alpha| \|v\|_{\mathbb{S}}, \qquad \forall \alpha \in \mathbb{R} \tag{2.4}$$

(c) *Triangular inequality*

$$\|w + v\|_{\mathbb{S}} \leq \|w\|_{\mathbb{S}} + \|v\|_{\mathbb{S}}, \qquad \forall w, v \in \mathbb{S} \tag{2.5}$$

where "$\Leftrightarrow$" stands for "if and only if". A norm is used in functional analysis to measure how "big" a function is.

Strain energy in a *constrained solid* of stable linear elastic material can be used as a norm of the displacements, known as the *energy norm*. Physically, it measures globally how much the solid is stressed.

2.1.4 Semi-norm

If a functional satisfies *conditions* (2.4) and (2.5), and the *semi*-positivity

$$\|v\|_{\mathbb{S}} \geq 0, \qquad \forall v \in \mathbb{S} \tag{2.6}$$

instead of the (full) positivity Equation (2.3), it is called a *semi-norm* and denoted as $|\cdot|_{\mathbb{S}}$. A semi-norm is often used to measure the strength of the variation such as the gradient/derivatives of a function. For solid mechanics problems, it is used to measure the strength of the strain field resulted from a displacement field. The *energy norm* for a solid (constrained or unconstrained) is a typical semi-norm. It is a *global* measure on how much is the entire solid being stressed under the displacement field. Because when a solid is unconstrained, any nonzero displacement field of a rigid motion does not create any strain or stress in the solid, the strain energy is zero for such a nonzero input. The strain energy should be semi-positive for such a solid.

2.1.5 Linear forms

A linear form $L(v)$ is a functional such that

$$L: \underbrace{\mathbb{S}}_{\text{space offers the input}} \rightarrow \underbrace{\mathbb{R}}_{\text{space hosts the output}} \tag{2.7}$$

with the following property.

$$L(\alpha w + v) = \alpha L(w) + L(v), \quad \forall \alpha \in \mathbb{R}, \forall w, v \in \mathbb{S} \qquad (2.8)$$

Work done by externally applied forces over a given displacement field in a linear elastic solid is a typical linear form.

2.1.6 Bilinear forms

A bilinear form often denoted as $a(w, v)$ is a functional that takes two inputs of functions, respectively, from two spaces and returns a scalar as an output:

$$a: \quad \underbrace{\mathbb{S}_1}_{\text{space offers } w} \times \underbrace{\mathbb{S}_2}_{\text{space offers } v} \to \underbrace{\mathbb{R}}_{\text{space hosts the output}} \qquad (2.9)$$

with the following bilinear property.

$$a(w, \cancel{v}) \text{ is a linear form in } w \text{ for a fixed } v$$
$$a(\cancel{w}, v) \text{ is a linear form in } v \text{ for a fixed } w \qquad (2.10)$$

Work done by the internal stresses resulted from a displacement function v in a solid undergoes an admissible virtual displacement function w is a typical bilinear form.

2.1.7 Inner product

An inner product $(w, v)_{\mathbb{S}}$ on $\mathbb{S}$ is a scalar valued functional on $\mathbb{S} \times \mathbb{S}$ that satisfies the following three conditions.

(a) *Symmetry*

$$(w, v)_{\mathbb{S}} = (v, w)_{\mathbb{S}}, \quad \forall w, v \in \mathbb{S} \qquad (2.11)$$

(b) *Positive definite*

$$(w, w)_{\mathbb{S}} \geq 0, \forall w \in \mathbb{S} \text{ and } (w, w)_{\mathbb{S}} = 0 \iff w = 0 \qquad (2.12)$$

(c) *Bilinear*

$$(v + w, u)_{\mathbb{S}} = (v, u)_{\mathbb{S}} + (w, u)_{\mathbb{S}}, \quad \forall v, w \in \mathbb{S} \quad \text{and}$$
$$(\alpha v, w)_{\mathbb{S}} = \alpha(v, w)_{\mathbb{S}}, \quad \forall v, w \in \mathbb{S}, \forall \alpha \in \mathbb{R} \qquad (2.13)$$

A linear space $\mathbb{S}$ on which an inner product can be defined is called an *inner product space*. The norm of such a space can be associated with the inner product as

$$\|v\|_{\mathbb{S}} = \sqrt{(v,v)_{\mathbb{S}}} \tag{2.14}$$

A space of admissible displacement functions can have an inner product defined and hence is an inner product space.

2.1.8 Cauchy-Schwarz inequality

For a given inner product space $\mathbb{S}$, we then have the well-known Cauchy-Schwarz inequality:

$$(w,v)_{\mathbb{S}} \le \|w\|_{\mathbb{S}} \cdot \|v\|_{\mathbb{S}} \tag{2.15}$$

Let $a(w,v)$ be a semi-definite bilinear form. The Cauchy-Schwarz inequality can then be expressed in a more general form of

$$|a(w,v)| \le \sqrt{a(w,w)}\sqrt{a(v,v)} \tag{2.16}$$

The Cauchy-Schwarz inequality is often used to derive the bound properties of a bilinear form.

2.1.9 General notation of derivatives

For concise expressions, we first define the following general notation of differentiations:

$$D^{\alpha} = \frac{\partial^{|\alpha|}}{\partial x_1^{\alpha_1} \cdots \partial x_d^{\alpha_d}} \tag{2.17}$$

where d is the dimension of the problem domain, α is an d-tuple of nonnegative integers $\alpha = (\alpha_1, \cdots, \alpha_d)$, and $|\alpha| = \sum_{i=1}^{d} \alpha_i$. For example, for a 3D problem governed by a 2nd order PDE, we shall have $d=3$, and $D^2 v$ represents one of the six possible 2nd order differentiations all with $|\alpha| = 2$ is given below.

TABLE 2.1 The six possible 2nd order differentiations of $D^{\alpha}v$ with $|\alpha|=2$

$D^{\alpha}v$	$\alpha=(\alpha_1,\alpha_2,\alpha_3)$	$D^{\alpha}v$	$\alpha=(\alpha_1,\alpha_2,\alpha_3)$
$D^2v=\dfrac{\partial^2 v}{\partial x_1^2}$	$(2,0,0)$	$D^2v=\dfrac{\partial^2 v}{\partial x_1\partial x_2}$	$(1,1,0)$
$D^2v=\dfrac{\partial^2 v}{\partial x_2^2}$	$(0,2,0)$	$D^2v=\dfrac{\partial^2 v}{\partial x_1\partial x_3}$	$(1,0,1)$
$D^2v=\dfrac{\partial^2 v}{\partial x_3^2}$	$(0,0,2)$	$D^2v=\dfrac{\partial^2 v}{\partial x_2\partial x_3}$	$(0,1,1)$

2.2 Useful spaces in weak formulation

In the formulation of FEM based on weak forms for stable and convergent solutions, we have to be very "choosy": we cannot use anyhow constructed displacement functions. Therefore, we have to classify the "qualified" functions in a proper space. This section lists some of the often used spaces in the standard weak formulations, because these spaces are also used or referenced in our S-PIM formulations.

2.2.1 Lebesgue spaces

The *Lebesgue* space $\mathbb{L}^p(\Omega)$ is defined as

$$\mathbb{L}^p(\Omega)=\left\{v\left|\ \int_{\Omega}|v|^p\,\mathrm{d}\Omega<\infty\right.\right\},\ \ 1\leq p<\infty \tag{2.18}$$

meaning that all functions with a bounded integral $\int_{\Omega}|v|^p\,\mathrm{d}\Omega<\infty$ belong to the $\mathbb{L}^p(\Omega)$ space. A Lebesgue space is a *normed* space associated with the norm of

$$\|v\|_{\mathbb{L}^p(\Omega)}^p=\int_{\Omega}|v|^p\,\mathrm{d}\Omega,\ \ 1\leq p<\infty \tag{2.19}$$

This book uses only the $\mathbb{L}^2$ space that is a Lebesgue space with $p=2$:

$$\mathbb{L}^2(\Omega)=\left\{v\left|\ \int_{\Omega}|v|^2\,\mathrm{d}\Omega<\infty\right.\right\} \tag{2.20}$$

which means that any function v in $\mathbb{L}^2(\Omega)$ must be *square integrable* (integral is bounded) over Ω. The function can be discontinuous at finite number of locations, but the function values must be finite over the problem domain Ω. The norm for functions in an $\mathbb{L}^2$ space is induced from the inner product of

$$(w,v)_{\mathbb{L}^2(\Omega)} = \int_\Omega wv\,\mathrm{d}\Omega \tag{2.21}$$

The induced $\mathbb{L}^2$ norm of a function v is then defined as

$$\|v\|^2_{\mathbb{L}^2(\Omega)} = \underbrace{\int_\Omega v^2\,\mathrm{d}\Omega}_{(v,v)_{\mathbb{L}^2(\Omega)}} \tag{2.22}$$

Example 2.2.1 A piecewise linear and continuous function

Consider the following function v defined by

$$v(x) = \begin{cases} c(1+x) & -1 < x \le 0 \\ c(1-x) & 0 < x < 1 \end{cases} \tag{2.23}$$

where $c \in \mathbb{R}$ is arbitrary but finite. It is clear that the function is piecewise linear and continuous on the finite 1D domain $\Omega = (-1,1)$, as plotted in FIGURE 2.1a. It lives in $\mathbb{L}^2(\Omega)$ space.

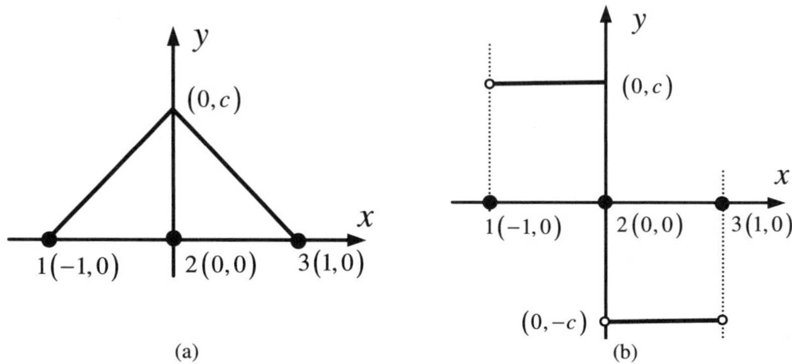

FIGURE 2.1 Example functions living in $\mathbb{L}^2$ space: (a) a piecewise continuous and linear function; (b) a Heaviside type function (finite and discontinuous at one point).

To show this, we examine the integration of v^2. Because this function has two pieces linear function seamlessly connected at $x=0$, the integration (in Lebesgue sense) is broken into two pieces:

$$\int_{-1}^{1} v^2 dx = \int_{-1}^{0} c^2 \left(1+x\right)^2 dx + \lim_{\varepsilon \to 0+} \int_{\varepsilon}^{1} c^2 \left(1-x\right)^2 dx$$

$$= \frac{1}{3}c^2 + \lim_{\varepsilon \to 0+} \left(\frac{1}{3} - \varepsilon + \varepsilon^2 - \frac{\varepsilon^3}{3} \right) c^2 \tag{2.24}$$

$$= \frac{2}{3}c^2 < \infty$$

which is finite (and thus bounded), meaning that the function v in Equation (2.23) belongs to $\mathbb{L}^2\left(\Omega\right)$.

The $\mathbb{L}^2$ norm of the function v in Equation (2.23) is

$$\left\| v \right\|_{\mathbb{L}^2(\Omega)} = \left(\int_{-1}^{1} v^2 dx \right)^{1/2} = \sqrt{\frac{2}{3}} \left| c \right| \tag{2.25}$$

The displacement functions constructed using the liner PIM for 1D (see, Example 3.2.1), 2D and 3D problems are all piecewise linear functions with a finite number of seamlessly connected points (lines and surfaces). These functions are all in $\mathbb{L}^2\left(\Omega\right)$ space. This example shows that piecewise continuous functions that are finite over Ω are in $\mathbb{L}^2\left(\Omega\right)$ space.

As the field variables of solid mechanics problems are vector fields of d-dimensions, we denote more precisely the spaces for the field functions as

$$\left(\mathbb{L}^2\left(\Omega\right) \right)^d = \left\{ \mathbf{v} = \left(v_i, \cdots, v_d\right); \ v_i \in \mathbb{L}^2\left(\Omega\right), i = 1, \cdots, d \right\} \tag{2.26}$$

Equation (2.26) means that each of the displacement components has to be square integrable over Ω, and the dimension of the space is expanded by d times. This is because of the independence of the displacement components. The corresponding norm becomes

$$\left\| \mathbf{v} \right\|_{\left(\mathbb{L}^2(\Omega)\right)^d} \equiv \left\| \mathbf{v} \right\|_{\mathbb{L}^2(\Omega)} = \left(\sum_{i=1}^{d} \left\| v_i \right\|_{\mathbb{L}^2(\Omega)}^2 \right)^{1/2} \tag{2.27}$$

Example 2.2.2 Heaviside type function

A Heaviside type function defined as

$$v(x) = \begin{cases} c & -1 < x \leq 0 \\ -c & 0 < x < 1 \end{cases} \tag{2.28}$$

belongs to $\mathbb{L}^2(\Omega)$ space, where c is an arbitrary but finite real number.

To show this, we observe that the function has two continuous pieces and a finite discontinuous point at $x = 0$, it is all defined and finite over the entire domain $\Omega = (-1,1)$, as plotted in FIGURE 2.1b. The integration of v^2 can be performed in two pieces $\Omega_1 = (-1,0]$ and $\Omega_2 = (\varepsilon,1)$ (again in Lebesgue sense), where ε is a positive infinitely small real number. We then have

$$\begin{aligned} \int_{-1}^{1} v^2 dx &= \int_{-1}^{0} c^2 dx + \lim_{\varepsilon \to 0+} \int_{\varepsilon}^{1} (-c)^2 \, dx \\ &= c^2 + \underbrace{\lim_{\varepsilon \to 0+} (1-\varepsilon)c^2}_{\to 1} \\ &= c^2 + c^2 \\ &= 2c^2 < \infty \end{aligned} \tag{2.29}$$

which shows clearly that the integration is bounded. Therefore, the Heaviside type function in Equation (2.28) belongs to $\mathbb{L}^2(\Omega)$. We observed that the discontinuity of the function value at $x=0$ is allowed in the integration.

The $\mathbb{L}^2$ norm of the function defined in Equation (2.28)

$$\|v\|_{\mathbb{L}^2(\Omega)} = \left(\int_{-1}^{1} v^2 dx \right)^{1/2} = \sqrt{2}\,|c| = \sqrt{2}c \tag{2.30}$$

The function of a strain component field in a linear FEM model for 1D, 2D and 3D problems are all essentially a Heaviside types of functions with multiple (but finite number of) discontinuous points (lines and surfaces). The strain component field functions in an FEM model are in a $\mathbb{L}^2(\Omega)$ space.

Example 2.2.3 A general discontinuous function

A more general discontinues function is defined as

$$v(x) = \begin{cases} 1+x & -1 < x < 0 \\ c & x = 0 \\ -(x+1) & 0 < x < 1 \end{cases} \tag{2.31}$$

where c is a given finite real number, lives in $\mathbb{L}^2(\Omega)$ space.

To show this, we observe that the function defined in Equation (2.31) has two continuous linear pieces in domain $\Omega = (-1,1)$ and one "jumping" point at $x=0$, as plotted in FIGURE 2.2. The integration of v^2 can be performed in three pieces as follows

$$\begin{aligned}
\int_{-1}^{1} v^2 dx &= \lim_{\varepsilon \to 0-} \int_{-1}^{\varepsilon} (1+x)^2 dx + \lim_{\substack{\varepsilon \to 0- \\ \beta \to 0+}} \int_{\varepsilon}^{\beta} c^2 dx + \lim_{\beta \to 0+} \int_{\beta}^{1} (-1-x)^2 \, dx \\
&= \lim_{\varepsilon \to 0-} \underbrace{\left(\frac{1}{3} + \varepsilon + \varepsilon^2 + \frac{\varepsilon^3}{3} \right)}_{\to 1/3} + \underbrace{\lim_{\substack{\varepsilon \to 0- \\ \beta \to 0+}} \underbrace{c^2}_{\text{finite}} (\beta - \varepsilon)}_{=0} \\
&\quad + \lim_{\beta \to 0+} \underbrace{\left(\frac{7}{3} - \beta - \beta^2 - \frac{\beta^3}{3} \right)}_{\to 1/3} \\
&= \frac{1}{3} + \frac{7}{3} = \frac{8}{3} < \infty
\end{aligned} \tag{2.32}$$

which it is finite. Therefore, the general discontinuous function in Equation (2.31) belongs to $\mathbb{L}^2(\Omega)$. We observed again that the function value at $x=0$ can be omitted from the integral, regardless of the given value of c, as long as it is finite.

The $\mathbb{L}^2$ norm of the function defined in Equation (2.31) is

$$\|v\|_{\mathbb{L}^2(\Omega)} = \left(\int_{-1}^{1} v^2 dx \right)^{1/2} = 2\sqrt{\frac{2}{3}} \tag{2.33}$$

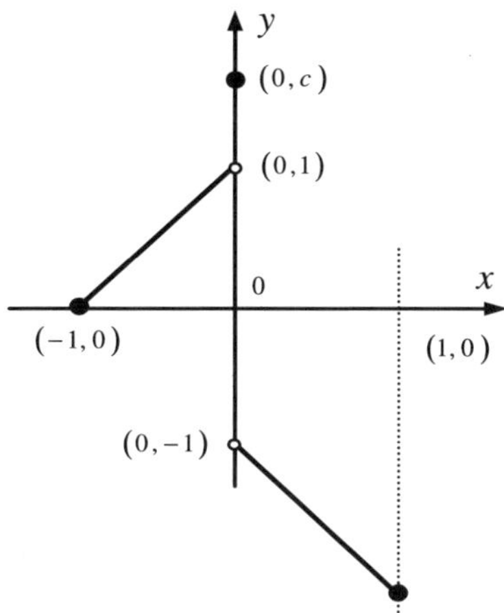

FIGURE 2.2 A general discontinuous function.

Example 2.2.2 and Example 2.2.3 show that functions that are continuous or discontinuous are both in $\mathbb{L}^2(\Omega)$ space, as long as the functions are finite over Ω.

Remark 2.1 Lebesgue integration

The Lebesgue integration value will not change if we make "finite" changes to the finite values of the integrand at a finite set of discrete points. The Lebesgue integration is thus very forgiving of "occasionally omissions". This is of fundamental importance to weak-form methods (see, Chapter 5) such as FEM, because it essentially allows us to use piecewise continuous displacement functions constructed using an mesh, even though such functions result in discontinuities along these element interfaces in the strain field that are squarely integrated in the weak form. We can practically omit the discontinuity of the strain fields, as long as our integration is performed based on the element. The integration over the entire problem can be changed to a summation of integrations over all the elements, via omissions on the elemental boundaries.

In our S-PIM models, however, the integration of the weakened weak form (see, Chapter 5) is performed over the integration/quadrature domains, and we use the so-called constructed strain field (see, Chapter 4). Hence we need to make sure that the constructed strain field must be continuous over the quadrature domains. The discontinuities of the strain field on the interfaces of the quadrature domains are admissible.

2.2.2 Hilbert spaces

We now define the Hilbert space. For a non-negative integer m,

$$\mathbb{H}^m(\Omega) = \left\{ v \,\middle|\, D^\alpha v \in \mathbb{L}^2(\Omega), \ \ \forall |\alpha| \le m \right\} \tag{2.34}$$

which means that $\mathbb{H}^m(\Omega)$ space hosts all functions whose derivatives up to mth order are all square integrable. The associated inner product is given by

$$(w, v) = \sum_{|\alpha| \le m} \int_\Omega (D^\alpha w) \cdot (D^\alpha v) \mathrm{d}\Omega \tag{2.35}$$

The induced (full) norm becomes

$$\|v\|_{\mathbb{H}^m(\Omega)} = \left(\sum_{|\alpha| \le m} \int_\Omega \left| D^\alpha v \right|^2 \mathrm{d}\Omega \right)^{1/2} \tag{2.36}$$

It measures all the first m (including zero) derivatives of v are in the $\mathbb{L}^2$ norm. The semi-norm includes only the mth derivative:

$$|v|_{\mathbb{H}^m(\Omega)} = \left(\int_\Omega \left| D^\alpha v \right|^2 \mathrm{d}\Omega \right)^{1/2}, \ \ |\alpha| = m \tag{2.37}$$

By the definition, it is clear that $\mathbb{H}^0(\Omega) = \mathbb{L}^2(\Omega)$. This book uses mostly the $\mathbb{H}^1$ space. It is defined for d-dimensional problem domains as

$$\mathbb{H}^1(\Omega) = \left\{ v \,\middle|\, v \in \mathbb{L}^2(\Omega), \ \partial v / \partial x_i \in \mathbb{L}^2(\Omega), \ i = 1, \cdots, d \right\} \tag{2.38}$$

which means that any function in the space and its first derivatives of the function are all square integrable. Clearly, $\mathbb{H}^1$ space is much smaller than the $\mathbb{L}^2$ space, because the additional conditions imposed on the derivatives. The associated inner product is

$$(w, v)_{\mathbb{H}^1(\Omega)} = \underbrace{\int_\Omega wv \mathrm{d}\Omega}_{(w,v)_{\mathbb{L}^2(\Omega)}} + \underbrace{\int_\Omega (\nabla w) \cdot (\nabla v) \mathrm{d}\Omega}_{(\nabla w, \nabla v)_{\mathbb{L}^2(\Omega)}} \tag{2.39}$$

where

$$\nabla v = \left(\frac{\partial v}{\partial x_1} \quad \frac{\partial v}{\partial x_2} \quad \cdots \quad \frac{\partial v}{\partial x_d} \right) \tag{2.40}$$

The full norm of a function v in $\mathbb{H}^1$ is given by

$$\|v\|_{\mathbb{H}^1(\Omega)}^2 = \underbrace{\int_\Omega v^2 d\Omega}_{\|v\|_{\mathbb{L}^2(\Omega)}^2} + \underbrace{\int_\Omega (\nabla v) \cdot (\nabla v) d\Omega}_{|v|_{\mathbb{H}^1(\Omega)}^2} \tag{2.41}$$

The $\mathbb{H}^1(\Omega)$ semi-norm is

$$|v|_{\mathbb{H}^1(\Omega)}^2 = \int_\Omega (\nabla v) \cdot (\nabla v) d\Omega \tag{2.42}$$

It is seen that we can only expect semi-positivity for the semi-norm, because a nonzero constant v in $\mathbb{H}^1$ space produces a zero semi-norm. Functions in $\mathbb{H}^1$ that satisfy the essential (displacement) boundary conditions form a sub-space:

$$\mathbb{H}_0^1(\Omega) = \left\{ v \in \mathbb{H}^1(\Omega) \middle| v = 0 \text{ on } \Gamma_u \right\} \tag{2.43}$$

where Γ_u stands for the essential boundary on which the displacement are constrained. A function in $\mathbb{H}_0^1(\Omega)$ is not "floating", as long as all possible rigid motions are constrained. Once the functions cannot float, the $\mathbb{H}^1$ semi-norm becomes (full) positive, and we have that the $\mathbb{H}^1$ semi-norm is *equivalent* to the $\mathbb{H}^1$ full norm: there exists a positive constant c_{PF} such that

$$c_{PF} \|w\|_{H^1(\Omega)}^2 \leq |w|_{H^1(\Omega)}^2, \quad \forall w \in \mathbb{H}_0^1 \tag{2.44}$$

which is known as the Poincare-Friedrichs inequality and c_{PF} is known as the Poincare-Friedrichs constant. Equation (2.44) is one of the most important inequalities in the weak formulation, because it ensures fundamentally the stability of a weak-form model for a well-posed problem.

A sub-space in $\mathbb{H}^1$ space created using a set of FEM shape functions can then be defined as

$$\mathbb{H}_h^1(\Omega) = \left\{ v \in \mathbb{H}^1(\Omega) \middle| \ v(\mathbf{x}) = \mathbf{\Phi}^H(\mathbf{x})\mathbf{d}, \ \ \mathbf{d} \in \mathbb{R}^{N_n} \right\} \tag{2.45}$$

where $\mathbf{\Phi}^H(\mathbf{x})$ is the matrix of all the (compatible) nodal shape functions created using an FEM model written as

$$\mathbf{\Phi}^H(\mathbf{x}) = \left[\phi_1^H(\mathbf{x}) \quad \phi_2^H(\mathbf{x}) \quad \cdots \quad \phi_{N_n}^H(\mathbf{x}) \right] \tag{2.46}$$

Because $\mathbb{H}_h^1(\Omega)$ is a linear space, each of the nodal shape functions $\phi_i^H(\mathbf{x})$ must also be in $\mathbb{H}_h^1(\Omega)$. The linear independence of shape functions $\phi_i^H(\mathbf{x})$ $(i = 1, 2, \cdots, N_n)$ are ensured by the element-based (with proper mapping) procedure. In Equation (2.45), $\mathbf{d}$ is the vector of all the nodal functions values:

$$\mathbf{d} = \left\{ u_1 \quad u_2 \quad \cdots \quad u_{N_n} \right\}^{\mathrm{T}} \tag{2.47}$$

Since the value at each node can change independently, they forms an N_n dimensional space, and we have $\mathbf{d} \in \mathbb{R}^{N_n}$ where $\mathbb{R}^{N_n}$ stands for a real field of N_n dimensions. The norm of $\mathbf{d}$ is defined as

$$\|\mathbf{d}\|_2^2 \equiv \sum_{i=1}^{N_n} (u_i)^2 \tag{2.48}$$

Because $\mathbb{H}_h^1$ is constructed in a discrete form with finite dimensions, it is marked with a subscript "h". Functions in $\mathbb{H}_h^1$ that satisfy the essential boundary conditions form a sub-space

$$\mathbb{H}_{h,0}^1(\Omega) = \left\{ v \in \mathbb{H}_h^1(\Omega) \middle| v = 0 \ \ \text{on} \ \ \Gamma_u \right\} \tag{2.49}$$

A very special sub-space constructed using only linear interpolation with three-node triangular element mesh (or polynomial PIM with cell-based T3-scheme) can be defined as

$$\mathbb{H}_b^1(\Omega) = \left\{ v \in \mathbb{H}^1(\Omega) \middle| \ v(\mathbf{x}) = \mathbf{\Phi}^b(\mathbf{x})\mathbf{d}, \ \mathbf{d} \in \mathbb{R}^{N_n} \right\} \tag{2.50}$$

where $\mathbf{\Phi}^b(\mathbf{x})$ is the matrix of all the nodal shape functions written as

$$\mathbf{\Phi}^b(\mathbf{x}) = \begin{bmatrix} \phi_1^b(\mathbf{x}) & \phi_2^b(\mathbf{x}) & \cdots & \phi_{N_n}^b(\mathbf{x}) \end{bmatrix} \qquad (2.51)$$

In this book, we often use $\mathbb{H}_b^1(\Omega)$ to create a *base* model, called linear FEM or FEM-Tr3 model. In such a case, the triangular cells defined in Section 1.6.1 become the usual triangular elements. Similarly, functions in $\mathbb{H}_b^1$ that satisfy the essential (displacement) boundary conditions form a space:

$$\mathbb{H}_{b,0}^1(\Omega) = \left\{ v \in \mathbb{H}_b^1(\Omega) \middle| v = 0 \quad \text{on} \ \Gamma_u \right\} \qquad (2.52)$$

An $\mathbb{H}_h^1$ space is indeed very exclusive. The methods that can create functions in an $\mathbb{H}_h^1$ space are very much limited: typically FEM technique and the moving least square (MLS) approximation [4].

Because field variables in solid mechanics problems are vector fields, we denote more precisely the spaces as

$$\left(\mathbb{H}^1(\Omega)\right)^d = \left\{ \mathbf{v} = (v_1, \cdots, v_d); \ \ v_i \in \mathbb{H}^1(\Omega), \ i = 1, \cdots, d \right\} \qquad (2.53)$$

The corresponding full norm becomes

$$\|\mathbf{v}\|_{\left(\mathbb{H}^1(\Omega)\right)^d} \equiv \|\mathbf{v}\|_{\mathbb{H}^1(\Omega)} = \left(\sum_{i=1}^{d} \|v_i\|_{\mathbb{H}^1(\Omega)}^2 \right)^{\!1/2} \qquad (2.54)$$

and the semi-norm is

$$|\mathbf{v}|_{\left(\mathbb{H}^1(\Omega)\right)^d} \equiv |\mathbf{v}|_{\mathbb{H}^1(\Omega)} = \left(\sum_{i=1}^{d} |v_i|_{\mathbb{H}^1(\Omega)}^2 \right)^{\!1/2} \qquad (2.55)$$

Likewise, the space $\left(\mathbb{H}_0^1(\Omega)\right)^d = \left\{ \mathbf{v} \in \left(\mathbb{H}^1(\Omega)\right)^d \middle| \mathbf{v} = \mathbf{0} \ \text{on} \ \Gamma_u \right\}$ is the subset of $\left(\mathbb{H}^1(\Omega)\right)^d$ with vanishing values on Γ_u, and $\left(\mathbb{H}_0^1(\Omega)\right)^d$ is equipped with the same inner product and norms as $\left(\mathbb{H}^1(\Omega)\right)^d$.

Example 2.2.4 A piecewise linear and continuous function

The function v defined in Equation (2.23) over the 1D domain $\Omega = (-1,1)$ belongs to $\mathbb{H}^1(\Omega)$, because 1) the function v belongs to $\mathbb{L}^2(\Omega)$ as already shown in Example 2.2.1, and 2) the derivative of the function

$$\frac{\partial v}{\partial x} = \begin{cases} c & -1 < x < 0 \\ -c & 0 < x < 1 \end{cases} \tag{2.56}$$

is a function of Heaviside type, and also belongs to $\mathbb{L}^2(\Omega)$, as shown in Example 2.2.2.

The $\mathbb{H}^1(\Omega)$ (full) norm of the function v in Example 2.2.1 is

$$\begin{aligned}
\|v\|_{\mathbb{H}^1(\Omega)} &= \left[\int_{-1}^{1} \left(v^2 + |\nabla v|^2 \right) dx \right]^{1/2} \\
&= \left[\int_{-1}^{0} \left((1+x)^2 + 1^2 \right) c^2 dx + \lim_{\varepsilon \to 0+} \int_{\varepsilon}^{1} \left((1-x)^2 + (-1)^2 \right) c^2 dx \right]^{1/2} \\
&= |c| \sqrt{\frac{4}{3} + \lim_{\varepsilon \to 0+} \left(\frac{4}{3} - 2\varepsilon + \varepsilon^2 - \frac{\varepsilon^3}{3} \right)} \\
&= 2|c| \sqrt{\frac{2}{3}}
\end{aligned} \tag{2.57}$$

The semi-norm of the function becomes

$$\begin{aligned}
|v|_{\mathbb{H}^1(\Omega)} &= \left[\int_{-1}^{1} |\nabla v|^2 \, dx \right]^{1/2} = \left[\int_{-1}^{0} c^2 dx + \lim_{\varepsilon \to 0+} \int_{\varepsilon}^{1} |-1c|^2 \, dx \right]^{1/2} \\
&= |c| \sqrt{2 - \lim_{\varepsilon \to 0+} \varepsilon} \\
&= |c| \sqrt{2}
\end{aligned} \tag{2.58}$$

The analysis given in Example 2.2.1 and Example 2.2.4 shows that the displacement functions in a linear FEM model for 1D, 2D and 3D problems all live in an $\mathbb{H}^1(\Omega) \subset \mathbb{L}^2(\Omega)$ space. This conclusion can be easily extended to higher order elements, because the change of these two linear pieces to higher order ones does not affect the proofs given in these examples. These two pieces, however, have to be connected together: continuous. The linear PIM shape functions (see, Example 3.2.1) are also in an $\mathbb{H}^1(\Omega) \subset \mathbb{L}^2(\Omega)$ space, because they are continuous.

Example 2.2.5 A general discontinuous function

The function v defined in Equation (2.31) does not belong to $\mathbb{H}^1(\Omega)$.

To examine this, we first note that function defined in Equation (2.31) belongs to $\mathbb{L}^2(\Omega)$ as shown in Example 2.2.3. Let's now examine whether or not $\partial v/\partial x$ belongs to $\mathbb{L}^2(\Omega)$, using the following simple *intuitive* (may not be very rigorous) procedure. We know that $\partial v/\partial x$ does not even exist at $x=0$, and hence it is difficult to examine it directly. We then construct a "milder" function (that is at least piecewise differentiable) in the following form:

$$v_m(x) = \begin{cases} 1+x & -1 < x < -\varepsilon \\ \dfrac{-x}{\varepsilon} & -\varepsilon \le x \le \varepsilon \\ -(x+1) & \varepsilon < x < 1 \end{cases} \tag{2.59}$$

where ε is a positive infinitely small real number. This milder function is continuous for any given ε, and should be smoother than the original function v for any *arbitrary* small positive ε. We now examine the property of v_m defined in Equation (2.59). The derivative of v_m can be easily found as

$$\frac{\partial v_m}{\partial x} = \begin{cases} 1 & -1 < x < -\varepsilon \\ \dfrac{-1}{\varepsilon} & -\varepsilon < x < \varepsilon \\ -1 & \varepsilon < x < 1 \end{cases} \tag{2.60}$$

The integration of $(\partial v_m/\partial x)^2$ is performed in three pieces $\Omega_1 = (-1, -\varepsilon)$, $\Omega_2 = (-\varepsilon, \varepsilon)$, and $\Omega_3 = (\varepsilon, 1)$:

$$\int_{-1}^{1} \left(\frac{\partial v_m}{\partial x}\right)^2 dx = \int_{-1}^{-\varepsilon} 1^2 dx + \int_{-\varepsilon}^{\varepsilon} \left(\frac{-1}{\varepsilon}\right)^2 dx + \int_{\varepsilon}^{1} (-1)^2 dx$$

$$= (-\varepsilon + 1) + 2\frac{1}{\varepsilon^2}\varepsilon + (1 - \varepsilon) \tag{2.61}$$

$$= 2 - 2\varepsilon + 2\frac{1}{\varepsilon}$$

It is easy to see that $\int_{-1}^{1} (\partial v_m/\partial x)^2 dx$ cannot be bounded at the limit of $\varepsilon \to 0$, due to the presence of the last term in Equation (2.61) and hence $\partial v_m/\partial x$ does not belongs to $\mathbb{L}^2(\Omega)$, at the limit of $\varepsilon \to 0$. Therefore, the function v_m does not belong to $\mathbb{H}^1(\Omega)$ at the limit of $\varepsilon \to 0$. Hence, we can conclude that the function v defined in Equation (2.31) does not belong to $\mathbb{H}^1(\Omega)$, because it is even less smoother than v_m with any arbitrary small ε.

Example 2.2.5 is a typical example that discontinuous functions in the form of Equation (2.31) do not belong to the $\mathbb{H}^1(\Omega)$ space. The root of the un-boundedness is at the discontinuous point. These two continuous pieces of the function on both sides of this discontinuous point is immaterial. Therefore, for this type of functions, the boundedness is all determined by whether or not the function is "locally integrable" in the vicinity of the discontinuous points. Essentially, if the function is discontinuous at a point inside the domain, the derivative does not exist at that point. Therefore, there are plenty of means to construct "milder" functions that are smoother than the original one. All we need is to find "a" milder function that is not locally integrable as a counter example, and hence the un-boundedness can be shown.

Discontinuous functions in the form of Equation (2.31) cannot be used directly in a weak formulation based on the Hilbert space theory (as in the FEM). To make use of *some* (not all) types of the discontinuous functions in constructing stable and convergent numerical models, we need to use the $\mathbb{G}$ space theory to be discussed in the Section 2.3.

2.2.3　Sobolev spaces

The Sobolev spaces $\mathbb{W}^{m,p}(\Omega)$ for integer $m \geq 0$ and $p \geq 1$ is defined as

$$\mathbb{W}^{m,p}(\Omega) = \left\{ v \,\middle|\, D^\alpha v \in \mathbb{L}^p(\Omega), \ \forall |\alpha| \leq m \right\} \tag{2.62}$$

with norm of

$$\left\| v \right\|_{\mathbf{W}^{m,p}(\Omega)} = \left(\sum_{|\alpha| \leq m} \int_\Omega \left| D^\alpha v \right|^p \, \mathrm{d}\Omega \right)^{1/p} \tag{2.63}$$

The $\mathbb{W}^{m,p}$ norm measures the first m derivatives of functions in the $\mathbb{L}^p$ norm. We note that $\mathbb{W}^{m,2}(\Omega) = \mathbb{H}^m(\Omega)$. When $p \neq 2$, however, Sobolev spaces are not Hilbert spaces. We also note $\mathbb{W}^{0,p}(\Omega) = \mathbb{L}^p(\Omega)$.

2.2.4 Spaces of continuous functions

The spaces of continuous functions, $\mathbb{C}^m(\Omega)$ for any integer m, is defined as

$$\mathbb{C}^m(\Omega) = \left\{ v \,\middle|\, D^\alpha v \text{ continuous and bounded over } \Omega, \ \forall |\alpha| \le m \right\} \qquad (2.64)$$

For example,

$\mathbb{C}^0(\Omega)$: the space of functions that are continuous over Ω;

$\mathbb{C}^{-1}(\Omega)$: the space of functions whose anti-derivatives are continuous over Ω;

$\mathbb{C}^1(\Omega)$: the space of functions whose first derivatives are continuous over Ω;

$\mathbb{C}^\infty(\Omega)$: the space of functions of which all derivatives exist and are continuous over Ω.

2.3 G spaces: definition

G spaces are defined for discrete models, following largely the work reported in [6-10]. Consider a problem domain Ω triangulated following the procedure given in Section 1.6.1, such that $\boxed{\Omega} = \bigcup_{i=1}^{N_c} \boxed{\Omega}_i^c$ and $\Omega_i^c \cap \Omega_j^c = \varnothing$, $\forall i \ne j$ where N_c is the total number of triangular cells. Each of the cells is bounded by Γ_i^c.

2.3.1 Smoothing domain creation

We next divide, in a basically independent way, the domain Ω into N_s nonoverlapping and no-gap sub-domains called smoothing domains: $\boxed{\Omega} = \bigcup_{k=1}^{N_s} \boxed{\Omega}_k^s$ with line interfaces Γ_i^s $(i=1,\cdots,N_s)$ between the smoothing domains, as shown in FIGURE 2.3. The division of Ω into $\boxed{\Omega}_k^s$ is performed following the "no-sharing rule": the interfaces Γ_i^s do not share any *finite* portion of the lines on which the function is not square integrable. The interfaces of $\boxed{\Omega}_k^s$ can go across these lines. FIGURE 2.4 shows an example of such a division for a 1D domain, and FIGURE 6.2 shows that for 2D domains, where node-based smoothing domains are created upon a set of triangular background cells.

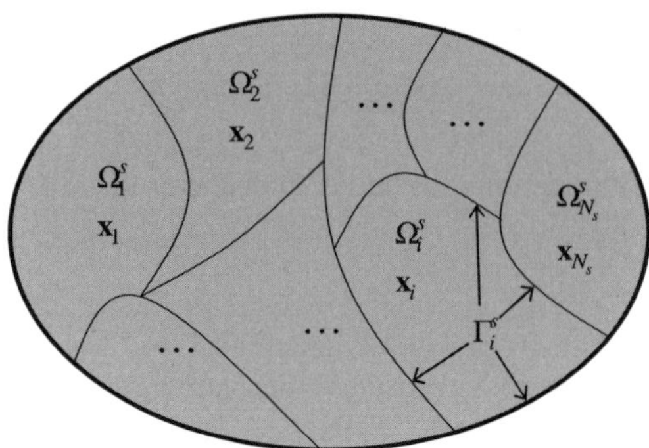

FIGURE 2.3 Division of problem domain into nonoverlapping and no-gap stationary smoothing domains Ω_i^s for $\mathbf{x}_i$ bounded by Γ_i^s. The smoothing domain is also used as basis for integration.

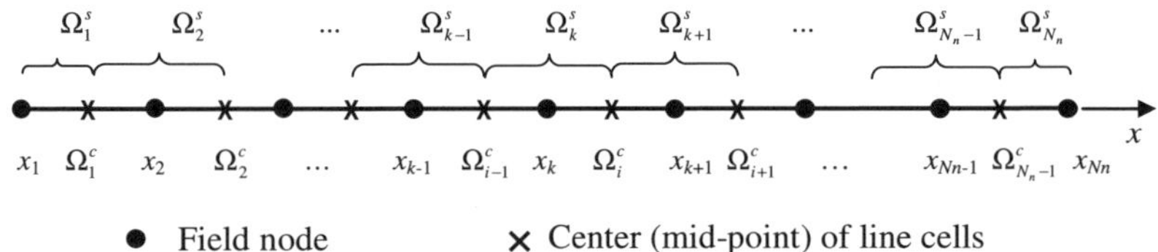

FIGURE 2.4 Background line cells and node-based smoothing domains created by sequentially connecting the centers of the two neighboring cells.

2.3.2 Linearly independent smoothing domains

Linearly independent smoothing domains are defined as a set of smoothing domains that ensure the full rank of the (global) smoothed strain-displacement matrix. The detailed definition and analysis can be found in [9]. A proven practical means to ensure the linear independence is to have the smoothing domains associated with the nodes, edges, or cells of the background mesh. Typical divisions that give sufficient number of independent smoothing domains are cell/element based, node-based, edge-based, as well as face-based, which will be discussed in later chapters.

2.3.3 Integral representation of function derivatives

The integral representation of a function is a useful way to approximate the function and its derivatives, using a convolution performed for the function with a predefined local smoothing function [4, 11]. The integral representation for the derivatives of a function can be written as:

$$\widehat{\frac{\partial w_l}{\partial x_i}}(\mathbf{x}) = \int_{\Omega_{\mathbf{x}}^s} \frac{\partial w_l(\xi)}{\partial x_i} \hat{W}(\mathbf{x}-\xi)\,d\xi \tag{2.65}$$

where $\widehat{\dfrac{\partial w_l}{\partial x_i}}(\mathbf{x})$ denotes the smoothed derivatives of $w_l(\mathbf{x})$ and $\hat{W}(\mathbf{x}-\xi)$ is smoothing function. When both $w_l(\mathbf{x})$ and $\hat{W}(\mathbf{x}-\xi)$ are *continuous* and differentiable in the *closure* of $\Omega_{\mathbf{x}}^s$, the integration by parts can be applied, and Equation (2.65) becomes

$$\widehat{\frac{\partial w_l}{\partial x_i}}(\mathbf{x}) = \int_{\Gamma_{\mathbf{x}}^s} w_l(\xi)n_i\hat{W}(\mathbf{x}-\xi)\,d\Gamma - \int_{\Omega_{\mathbf{x}}^s} w_l(\xi)\frac{\partial \hat{W}(\mathbf{x}-\xi)}{\partial x_i}\,d\xi \tag{2.66}$$

where n_i is the directional cosine of the outwards normal on $\Gamma_{\mathbf{x}}^s$.

2.3.4 Derivatives approximation

When $w_l(\mathbf{x})$ is discontinuous in $\Omega_{\mathbf{x}}^s$, the integration by parts is no longer applicable, and hence we *approximate* the gradient of $w_l(\mathbf{x})$ in the form of

$$\widehat{\frac{\partial w_l}{\partial x_i}}(\mathbf{x}) \approx \int_{\Gamma_{\mathbf{x}}^s} w_l(\xi)n_i\hat{W}(\mathbf{x}-\xi)\,d\Gamma - \int_{\Omega_{\mathbf{x}}^s} w_l(\xi)\frac{\partial \hat{W}(\mathbf{x}-\xi)}{\partial x_i}\,d\xi \tag{2.67}$$

This generalization given in Equation (2.67) is not rigorous in theory. It is fortunately possible to implement, because no differentiation on w_l is required on the right-hand-side of Equation (2.67). It was first performed in [6] and provides the foundation for all the S-PIM methods using incompatible nodal shape functions. A discussion on the physical implications can be found in [7]. The generalization given in Equation (2.67) is very important for the establishment of the G space theory. We thus note

Remark 2.2 "Smoothed" derivatives

The "smoothed derivatives" defined in Equation (2.67) is a generalized concept. It is NOT "the derivative obtained by smoothing the derivatives of the function", because such a gradient does not in general exist, as the function may not be continuous! Rigorously speaking, the "smoothed derivative" is the outward flux of the function across the smoothing domain boundary Γ_x^s. The smoothed derivative of a function can be approximated using only the function values, and no differentiation is needed. Hence the consistency requirement on the function is reduced, if only the approximate derivative is required.

2.3.5   On the treatment of the discontinuity

The understanding of the approximation of Equation (2.67), may not be so straightforward. Below is a mathematical derivation revealing some of the details. Let's consider a simple (but without the loss of generality) case of a local smoothing domain Ω_x^s for point $\mathbf{x}$, in which the function w_l is discontinuous along a line Γ_{xd} as shown in FIGURE 2.5. We now divide the smoothing domain into three enclosed sub-domains: the sub-domain on the left Ω_{xL}^s bounded by $\Gamma_{xLeft} = \Gamma_{xL}^s \bigcup \Gamma_{xT}^- \bigcup \Gamma_{xd}^- \bigcup \Gamma_{xB}^-$, the sub-domain on the right Ω_{xR}^s bounded by $\Gamma_{xRight} = \Gamma_{xR}^s \bigcup \Gamma_{xT}^+ \bigcup \Gamma_{xd}^+ \bigcup \Gamma_{xB}^+$, and the sub-domain Ω_{xd}^s enveloping (counter-clock-wise) the discontinuous line Γ_{xd} bounded by $\Gamma_{xd}^+ \bigcup \Gamma_{xd}^-$ (Because Γ_{xd}^+ and Γ_{xd}^- are of infinite lose, the area of Ω_{xd}^s is zero). In such a sub-domain division, w_l and $\widehat{W}(\mathbf{x}-\xi)$ are now *continuous* and differentiable in the *closure* of Ω_{xL}^s and Ω_{xR}^s, respectively. Therefore, Equation (2.66) can now be used for these two sub-domains, and we shall have

$$
\begin{aligned}
\widehat{\frac{\partial w_l}{\partial x_i}}(\mathbf{x}) &= \int_{\Omega_x^s} \frac{\partial w_l(\xi)}{\partial x_i} \widehat{W}(\mathbf{x}-\xi)\,d\xi \\[2mm]
&= \int_{\Omega_{xL}^s} \frac{\partial w_l(\xi)}{\partial x_i} \widehat{W}(\mathbf{x}-\xi)\,d\xi + \int_{\Omega_{xR}^s} \frac{\partial w_l(\xi)}{\partial x_i} \widehat{W}(\mathbf{x}-\xi)\,d\xi \\[2mm]
&\quad + \underbrace{\int_{\Omega_{xd}^s} \frac{\partial w_l(\xi)}{\partial x_i} \widehat{W}(\mathbf{x}-\xi)\,d\xi}_{I_3} \\[2mm]
&= \underbrace{\int_{\Gamma_{xLeft}} w_l(\xi)\,n_i\,\widehat{W}(\mathbf{x}-\xi)\,d\Gamma + \int_{\Gamma_{xRight}} w_l(\xi)\,n_i\,\widehat{W}(\mathbf{x}-\xi)\,d\Gamma}_{I_1} \\[2mm]
&\quad - \underbrace{\left(\int_{\Omega_{xL}^s} w_l(\xi)\frac{\partial \widehat{W}(\mathbf{x}-\xi)}{\partial x_i}\,d\xi + \int_{\Omega_{xR}^s} w_l(\xi)\frac{\partial \widehat{W}(\mathbf{x}-\xi)}{\partial x_i}\,d\xi \right)}_{I_2} + I_3
\end{aligned}
\tag{2.68}
$$

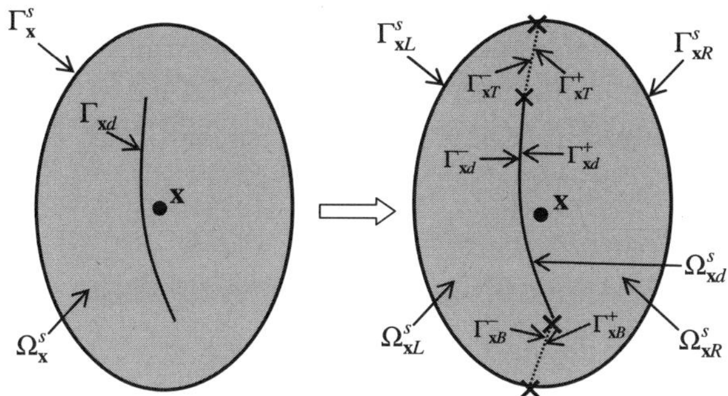

FIGURE 2.5 Division of a local smoothing domain with a discontinuous line into three sub-domains.

Since function w_l is continuous along Γ_{xT}^- and Γ_{xB}^-, the summation of these two line-integration on Γ_{xLeft} and Γ_{xRight} in Equation (2.68) leads to the vanish of the line-integrations on Γ_{xT} and Γ_{xB}, and we shall have

$$
\begin{aligned}
I_1 &= \int_{\Gamma_{xL}^s \cup \Gamma_{xd}^-} w_l(\xi)\, n_i \widehat{W}(\mathbf{x}-\xi)\, d\Gamma + \int_{\Gamma_{xR}^s \cup \Gamma_{xd}^+} w_l(\xi)\, n_i \widehat{W}(\mathbf{x}-\xi)\, d\Gamma \\
&= \int_{\Gamma_x^s} w_l(\xi)\, n_i \widehat{W}(\mathbf{x}-\xi)\, d\Gamma + \int_{\Gamma_{xd}^-} w_l(\xi)\, n_i \widehat{W}(\mathbf{x}-\xi)\, d\Gamma \\
&\quad + \int_{\Gamma_{xd}^+} w_l(\xi)\, n_i \widehat{W}(\mathbf{x}-\xi)\, d\Gamma
\end{aligned}
\tag{2.69}
$$

Integration I_2 in Equation (2.68) becomes (in *Lebesgue* sense)

$$
\begin{aligned}
I_2 &= \int_{\Omega_{xL}^s} w_l(\xi)\, \frac{\partial \widehat{W}(\mathbf{x}-\xi)}{\partial x_i}\, d\xi + \int_{\Omega_{xR}^s} w_l(\xi)\, \frac{\partial \widehat{W}(\mathbf{x}-\xi)}{\partial x_i}\, d\xi \\
&= \int_{\Omega_x^s} w_l(\xi)\, \frac{\partial \widehat{W}(\mathbf{x}-\xi)}{\partial x_i}\, d\xi
\end{aligned}
\tag{2.70}
$$

In Equation (2.68), however, I_3 needs to be approximated, because the derivative of w_l is not defined in Ω_{xd}^s. We then simply assume

$$
\begin{aligned}
I_3 &= \int_{\Omega_{xd}^s} \frac{\partial w_l(\xi)}{\partial x_i}\, \widehat{W}(\mathbf{x}-\xi)\, d\xi \\
&\approx -\int_{\Gamma_{xd}^+} w_l(\xi)\, n_i \widehat{W}(\mathbf{x}-\xi)\, d\Gamma - \int_{\Gamma_{xd}^-} w_l(\xi)\, n_i \widehat{W}(\mathbf{x}-\xi)\, d\Gamma
\end{aligned}
\tag{2.71}
$$

where the minus signs come from the fact that $\Gamma_{\mathbf{x}d}^{+}$ for sub-domain $\Omega_{\mathbf{x}d}^{s}$ is the reverse in direction with respect to that for sub-domain $\Omega_{\mathbf{x}R}^{s}$, and the same is also true for $\Gamma_{\mathbf{x}d}^{-}$. Equation (2.71) may be viewed as *gap smoothing* to be analyzed further later. Finally, submitting Equations (2.69), (2.70) and (2.71) into Equation (2.68), we immediately have Equation (2.67). $\qquad\square$

2.3.6 On physical meaning of the gap smoothing

Let's now examine the physical meaning of the "gap smoothing" approximation given in Equation (2.71). Because the smoothing function is continuous on $\Gamma_{\mathbf{x}d}$, Equation (2.71) becomes

$$
\begin{aligned}
I_3 &\approx -\int_{\Gamma_{\mathbf{x}d}^{+}} w_l(\boldsymbol{\xi})\, n_i \widehat{W}(\mathbf{x}-\boldsymbol{\xi})\,\mathrm{d}\Gamma - \int_{\Gamma_{\mathbf{x}d}^{-}} w_l(\boldsymbol{\xi})\, n_i \widehat{W}(\mathbf{x}-\boldsymbol{\xi})\,\mathrm{d}\Gamma \\
&= \left(\int_{\Gamma_{\mathbf{x}d}^{+}} w_l(\boldsymbol{\xi})\, n_{i\mathbf{x}d}^{-}\,\mathrm{d}\Gamma - \int_{\Gamma_{\mathbf{x}d}^{-}} w_l(\boldsymbol{\xi})\, n_{i\mathbf{x}d}^{-}\,\mathrm{d}\Gamma \right) \widehat{W}(\mathbf{x}-\boldsymbol{\xi}_{\mathbf{x}d})
\end{aligned}
\tag{2.72}
$$

For 1D case along coordinate x as shown in FIGURE 2.4, we shall have $n_{i\mathbf{x}d}^{-}=1$. Further if $\widehat{W}(\mathbf{x}-\boldsymbol{\xi}_{\mathbf{x}d})=\dfrac{1}{l_{\mathbf{x}}^{s}}$ (see later for more precise definition for 2D cases) where $l_{\mathbf{x}}^{s}$ is the length of the smoothing domain $\Omega_{\mathbf{x}}^{s}$, Equation (2.72) is simplified to

$$
I_3 \approx \frac{w_l\left(\boldsymbol{\xi}_{\mathbf{x}d}^{+}\right) - w_l\left(\boldsymbol{\xi}_{\mathbf{x}d}^{-}\right)}{l_{\mathbf{x}}^{s}}
\tag{2.73}
$$

which means that we are approximating the "gradient" of the discontinuity of the function over the smoothed domain by simply dividing the discontinuous "gap" with the length of the smoothing domain. In other words, the gap over an infinitely thin discontinuous line has been "smeared" over the finite dimension of the smoothing domain.

This analysis reveals intuitively that although we allow discontinuity in functions, the gap cannot be arbitrary. It cannot be finite and should approach zero when the dimension of the smoothing domains approaching zero. Since the smoothing domains are somehow tied with the field nodes and hence tied with the cell dimension h, the dimension of the smoothing domain relates also to h. Such a "tie" cannot be too "loose", and thus we require the "minimum number of linearly independent smoothing domains" for a given set of cells/nodes as discussed in detail in [6, 9]. Therefore, with such a tie, we now only require the gap approaching zero when h approaching zero. This can be always ensured when the nodal shape functions are properly created using the point interpolation method (PIM) or RPIM methods [4]. In summary, as long as a set of linearly independent nodal shape functions are created properly using the PIM for a given set of cells/nodes, the compatibility of these nodal shape functions is no longer a concern in our W^2 formulation based on the G space theory as demonstrated in [9, 10]. The stability and convergence are always ensured, even with the presence of the discontinuity in the assumed displacement functions.

2.3.7 Heaviside smoothing function

For simplicity, we use in this book the following special smoothing function that is a local constant (Heaviside type):

$$\widehat{W}\left(\mathbf{x}-\xi\right)=\overline{W}\left(\mathbf{x}-\xi\right)=\begin{cases}1/A_{\mathbf{x}}^{s} & \xi\in\boxed{\Omega}_{\mathbf{x}}^{s}\\[2mm]0 & \xi\notin\boxed{\Omega}_{\mathbf{x}}^{s}\end{cases}\tag{2.74}$$

where $A_{\mathbf{x}}^{s}$ is the area of smoothing domain associated with the point at $\mathbf{x}$. It is clear that the smoothing function given above satisfies the conditions of unity and positivity. Using the smoothing domains defined in Section 2.3.1, we have for the nth smoothing domain associated with the point at $\mathbf{x}_n$

$$\overline{\frac{\partial w_l}{\partial x_i}}\left(\mathbf{x}_n\right)=\begin{cases}\dfrac{1}{A_n^{s}}\displaystyle\int_{\Omega_n^{s}}\frac{\partial w_l}{\partial x_i}\,\mathrm{d}\Omega=\dfrac{1}{A_n^{s}}\displaystyle\int_{\Gamma_n^{s}}w_l\left(s\right)n_i\mathrm{d}s\\[4mm]\qquad\text{when }w_l\left(\mathbf{x}\right)\in\mathbb{C}^{0}(\Omega_n^{s})\text{ (continous)}\\[4mm]\dfrac{1}{A_n^{s}}\displaystyle\int_{\Gamma_n^{s}}w_l\left(s\right)n_i\mathrm{d}s\\[4mm]\qquad\text{when }w_l\left(\mathbf{x}\right)\in\mathbb{C}^{-1}(\Omega_n^{s})\text{ (discontinous)}\end{cases}\tag{2.75}$$

In carrying out the line integrations on Γ_n^{s}, we simply use the standard Gauss integration that are widely used in standard FEM [1-3].

Let's now examine the physical meaning of Equation (2.75), using a 1D case with coordinate x. In such a case, $n_i=1$, and Equation (2.75) can be rewritten as

$$\overline{\frac{\partial w_l}{\partial x_i}}(\mathbf{x}_k) = \frac{1}{A_k^s}\int_{\Gamma_k^s} w_l(s)n_i \mathrm{d}s = \frac{w_l\left(x_{k+\frac{1}{2}}\right) - w_l\left(x_{k-\frac{1}{2}}\right)}{l_k^s} \tag{2.76}$$

which is the "averaged" gradient of the function evaluated using the central difference method.

We further assume the derivative is constant in Ω_n^s and equals to that at point $\mathbf{x}_n$:

$$\overline{\frac{\partial w_l}{\partial x_i}}(\mathbf{x}) = \overline{\frac{\partial w_l}{\partial x_i}}(\mathbf{x}_n) = \underbrace{\frac{1}{A_n^s}\int_{\Gamma_n^s} w_l(s)n_i \mathrm{d}s}_{\bar{g}_{l,i}:\text{ constant in } \Omega_n^s}, \quad \forall \mathbf{x} \in \Omega_n^s \tag{2.77}$$

The 2nd derivatives of w_l are defined (for 2D cases) as

$$\overline{\frac{\partial^2 w_l}{\partial x_i \partial x_j}} = \underbrace{\frac{1}{2A_n^s}\int_{\Gamma_n^s}\left(\frac{\partial w_l}{\partial x_i}n_j + \frac{\partial w_l}{\partial x_j}n_i\right)\mathrm{d}s}_{\bar{g}_{l,ij}:\text{ constant in } \Omega_n^s}, \quad l=1,2;\ i,j=1,2 \tag{2.78}$$

Here we assumed the first derivatives for the function exist.

The 3rd derivatives of w_l can always be expressed as the 2nd derivative of a 1st derivative of w_l, and hence it can be obtained easily using Equation (2.78) by treating the 2nd derivative of w_l as a *new* function. For example

$$\begin{aligned}
\overline{\frac{\partial^3 w_l}{\partial x_1^3}} &= \overline{\frac{\partial^2}{\partial x_1 \partial x_1}\left(\frac{\partial w_l}{\partial x_1}\right)} \\
&= \frac{1}{2A_n^s}\int_{\Gamma_n^s}\left(\frac{\partial}{\partial x_1}\left(\frac{\partial w_l}{\partial x_1}\right)n_1 + \frac{\partial}{\partial x_1}\left(\frac{\partial w_l}{\partial x_1}\right)n_1\right)\mathrm{d}s \\
&= \underbrace{\frac{1}{A_n^s}\int_{\Gamma_n^s}\left(\frac{\partial^2 w_l}{\partial x_1^2}n_1\right)\mathrm{d}s}_{\bar{g}_{l,iii}:\text{ constant in } \Omega_n^s}
\end{aligned} \tag{2.79}$$

In this case we assumed that all the 2nd derivatives for the function exist. The same token can be used for defining any αth smoothed derivatives of w_l, and we should have in general a concise form of

$$\overline{D^\alpha w_l} = \overline{\frac{\partial^{|\alpha|} w_l}{\partial x_1^{\alpha_1} \cdots \partial x_d^{\alpha_d}}}, \quad |\alpha| = \sum_{i=1}^{d} \alpha_i \tag{2.80}$$

as long as all $D^{\alpha-1} w_l$ exist.

2.3.8 General definition of G space

The general expression of G spaces is defined as follows. For a non-negative m

$$\mathbb{G}_h^m(\Omega) = \begin{cases} v \middle| \; v(\mathbf{x}) = \sum_{n=1}^{N_n} \phi_n(\mathbf{x})u_n = \mathbf{\Phi}(\mathbf{x})\mathbf{d}, \quad \mathbf{d} \in \mathbb{R}^{N_n} \\ \quad D^\alpha v \in \mathbb{L}^2(\Omega), \quad \forall \alpha : |\alpha| \le (m-1), \\ \quad \sum_{k=1}^{N_s} \left(\int_{\Gamma_k^s} (D^\alpha v) n_i \mathrm{d}s \right)^2 > 0 \Leftrightarrow v \ne c \in \mathbb{R}; \; i=1,\cdots,d, \forall \alpha : |\alpha| \le (m-1) \end{cases} \tag{2.81}$$

where $\mathbf{d} = \{ u_1 \quad u_2 \quad \cdots \quad u_{N_n} \}^{\mathrm{T}}$ is the vector of all the nodal function values, and $\mathbf{\Phi}(\mathbf{x})$ is the row-matrix of all nodal shape functions created using a general point interpolation method discussed in Chapter 3, and can be written as

$$\mathbf{\Phi}(\mathbf{x}) = \begin{bmatrix} \phi_1(\mathbf{x}) & \phi_2(\mathbf{x}) & \cdots & \phi_{N_n}(\mathbf{x}) \end{bmatrix} \tag{2.82}$$

It is observed that the G space is spanned by a set of linear independent shape functions. The derivatives of the functions up to the $(m-1)$th orders are square integrable in Ω. Because of the discretized nature of a G space, it is marked with a subscript "h". The major difference between a $\mathbb{G}_h$ space and the corresponding *discrete* Hilbert space or $\mathbb{H}_h$ space is that the latter require $D^\alpha v \in \mathbf{L}^2(\Omega)$ for $|\alpha| = m$, but in the $\mathbb{G}_h$ space we require only $D^\alpha v \in \mathbf{L}^2(\Omega)$ for $|\alpha| = (m-1)$. It is therefore clear now that a $\mathbb{H}_h^m(\Omega)$ is also a $\mathbb{G}_h^m(\Omega)$ space: any function in $\mathbb{H}_h^m(\Omega)$ is surely qualified as a member in $\mathbb{G}_h^m(\Omega)$.

The inner product associated with a $\mathbb{G}_h^m(\Omega)$ space is then defined as

$$(w,v)_{\mathbb{G}^m(\Omega)} = \sum_{|\alpha| \le m-1} \left(\int_\Omega D^\alpha w \cdot D^\alpha v \mathrm{d}\Omega \right) + \sum_{|\alpha|=m} \sum_{k=1}^{N_s} A_k^s \overline{D^\alpha w} \cdot \overline{D^\alpha v} \tag{2.83}$$

The inner product induced (full) norm is next defined as

$$\|w\|_{\mathbb{G}^m(\Omega)} = \left[\sum_{|\alpha|\leq m-1}\left(\int_\Omega\left|D^\alpha w\right|^2 d\Omega\right) + \sum_{|\alpha|=m}\left(\sum_{k=1}^{N_s} A_k^s\left|\overline{D^\alpha w}\right|^2\right)\right]^{1/2} \tag{2.84}$$

The $\mathbb{G}_h^m(\Omega)$ semi-norm is finally defined using only the smoothed αth derivatives:

$$|w|_{\mathbb{G}^m(\Omega)} = \left[\sum_{|\alpha|=m}\left(\sum_{k=1}^{N_s} A_k^s\left|\overline{D^\alpha w}\right|^2\right)\right]^{1/2} \tag{2.85}$$

In this book we use only the $\mathbb{G}_h^1$ spaces, and therefore more details are given in the following sections.

2.3.9 G^1 space and norms

The $\mathbb{G}_h^1$ space used in this book can be then defined as follows.

$$\mathbb{G}_h^1(\Omega) = \left\{ \begin{array}{l} v\,\Big|\; v(\mathbf{x}) = \displaystyle\sum_{n=1}^{N_n}\phi_n(\mathbf{x})u_n = \mathbf{\Phi}(\mathbf{x})\mathbf{d}, \quad \mathbf{d}\in\mathbb{R}^{N_n} \\[2mm] v\in\mathbb{L}^2(\Omega), \\[2mm] \displaystyle\sum_{k=1}^{N_s}\left(\int_{\Gamma_k^s} v(s)n_i \, ds\right)^2 > 0 \Leftrightarrow v\neq c\in\mathbb{R};\; i=1,\cdots,d \end{array} \right\} \tag{2.86}$$

In creating functions in $\mathbb{G}_h^1$ spaces, we must use the nodal shape functions. We do not restrict on how these shape functions are created, as long as they satisfy the following conditions:

1) *Linearly independent condition*: all these nodal shape functions used for creating functions in $\mathbb{G}_h^1$ spaces must be linearly independent over Ω and hence are capable to form a basis. Anyhow created Heaviside type functions may not be included.

2) *Boundedness condition*: all the functions constructed using these shape function must be square integrable over the problem domain. Clearly, Dirac type Delta functions are excluded. This is to ensure the convergence of a numerical model to be created.

3) *Positivity condition*: there exists a fixed division of Ω_k^s for the discrete model such that $\sum_{k=1}^{N_s}\left(\int_{\Gamma_k^s} v(s)\,n_i\,\mathrm{d}s\right)^2 > 0 \iff v \neq c \in \mathbb{R}$, for $i = 1,\cdots,d$, where c is a real constant. This (together with the linearly independent condition) is to ensure the stability of a numerical model to be created.

When PIM or RPIM shape functions created in Chapter 3 are used, the functions constructed will in general not be continuous over the entire problem domain and hence are not compatible. Such an interpolant is not in an $\mathbb{H}_h^1$ space, but in a $\mathbb{G}_h^1$ space, because all these above-listed three conditions can be satisfied, as long as a proper set of smoothing domains can be constructed. Note that since $\phi_i(\mathbf{x}), (i = 1, 2, \cdots, N_n)$ are created using nodes selected using a T-scheme given in Section 1.6.3 and at least the three nodes of any home cell are always used. Hence, all PIM and RPIM shape functions have necessary order of consistency for passing the standard patch test.

Remark 2.3 $\mathbf{G}^1$ space: reduced consistency requirement

The major difference between a $\mathbb{G}_h^1$ space and $\mathbb{H}_h^1$ space is that the $\mathbb{H}_h^1$ space requires the first gradient of the function square integrable, but in the $\mathbb{G}_h^1$ space only the function itself is required square integrable. Therefore, the consistency requirement on function is now further weakened compared to the requirement for functions in an $\mathbb{H}_h^1$ space.

Remark 2.4 $\mathbf{G}^1$ space: naturally bounded

In an $\mathbb{H}_h^1$ space, the boundedness condition is achieved by the imposing the first derivatives of the function being square integrable, which can be difficult to satisfy and often requires special measures to ensure the function is at least piecewise continuous. In the $\mathbb{G}_h^1$ space, it is controlled by imposing the function being square integrable, and can be naturally satisfied, because any finite function that is continuous or discontinuous at finite number of points are all in $\mathbb{L}^2(\Omega)$ space as shown in Example 2.2.2 and Example 2.2.3.

Remark 2.5 $\mathbf{G}^1$ space: care needed for stability

On the other hand, the stability is automatically ensured for functions in an $\mathbb{H}_h^1$ space as long as the smoothness is satisfied, due to the Poincare-Friedrichs inequality Equation (2.44). The stability in the $\mathbb{G}_h^1$ space, however, is ensured by imposing the positivity condition, which requires a proper construction of smoothing domains.

Remark 2.6 G^1 space: a space between H^1 and L^2

Because a member in a $\mathbb{G}_h^1$ space is also a member of the $\mathbb{L}^2$ space, therefore a $\mathbb{G}_h^1$ space is a sub-space of $\mathbb{L}^2$ space: $\mathbb{G}_h^1(\Omega) \subset \mathbb{L}^2(\Omega)$. With a set of proper smoothing domains, a $\mathbb{H}_h^1(\Omega)$ space is also a $\mathbb{G}_h^1(\Omega)$ space: any function in $\mathbb{H}_h^1(\Omega)$ is surely qualified as a member in $\mathbb{G}_h^1(\Omega)$. A $\mathbb{G}_h^1$ space is very accommodating and inclusive compared to $\mathbb{H}_h^1(\Omega)$: all the methods discussed in Chapter 3, used in FEM and meshfree methods can be used to create functions for $\mathbb{G}_h^1$ space. This is very important in the formulation of various numerical methods. In conclusion, we have $\mathbb{H}_h^1(\Omega) \subset \mathbb{G}_h^1(\Omega) \subset \mathbb{L}^2(\Omega)$.

Remark 2.7 Proper space and stable and convergent numerical methods

Remark 2.6 is very important. We knew for long that using functions in $\mathbb{L}^2$ spaces to construct a discrete numerical method using local approximations is quite problematic: sometime it is stable, sometime it is not [12] and often special techniques are needed to stabilize the solution [13, 14]. Even it is stable, the solution may not converge to the exact solution of the original problem defined in Section 1.1. On the other hand, using functions in $\mathbb{H}_h^1$ spaces can always construct a stable and convergent discrete numerical method (such as FEM). In between $\mathbb{H}_h^1$ and $\mathbb{L}^2$ spaces, we now found $\mathbb{G}_h^1$ spaces that can be used to contract a stable and convergent discrete numerical method (such as S-PIM).

Remark 2.8 Normed or unnormed G space

G space can either be normed or un-normed. Un-normed G spaces are used for the strain-constructed Galerkin (or SC-Galerkin) models (Chapter 10), where admissible conditions for the constructed strains are defined properly in a separated manner. The normed G space is used to the generalized smoothed Galerkin (or GS-Galerkin) models that are special cases of W^2 formulations. Normed G spaces require a proper construction of smoothing domains following the rules detailed in Section 2.3.1.

Note that un-normed G spaces can be used for establishing strong form meshfree methods by, for example, simple collocation. In such cases, the stability is left "uncontrolled" when the assumed functions are used to create a discrete model, and we need additional procedure such as the regularization techniques [e.g. 13, 14] or proper using of different types of smoothing domains [15, 16] to restore the stability.

2.3.10 Minimum number of smoothing domains

To ensure the positivity condition in Equation (2.86) for a normed $\mathbb{G}_h^1$ space, we have to use at least the minimum number of smoothing domains N_s, based on the study in [6]. The consideration is that the minimum number of smoothing domains times the number of equations that can be established by each smoothing domain should be at least the same as the number of the un-prescribed nodal unknowns N_u, and therefore it depends also on the type of physical problems.

For 1D solid mechanics problem models with one node being fixed, we have immediately $N_s^{\min} = N_u = (N_n - 1)$.

TABLE 2.2 Minimum number of smoothing domains $N_s^{\min}$ for problems with n_t total nodal unknowns

Dimension of the problem	Minimum number of smoothing domains Vector field (e.g., solid mechanics problems)
1D	$N_s^{\min} = n_t$
2D	$N_s^{\min} = 2n_t / 3$
3D	$N_s^{\min} = 3n_t / 6 = n_t / 2$

For 2D solid mechanics problem models with n_t (unconstrained) nodes used for displacement field construction, the total number of unknowns in the model should be $N_u = 2n_t$, because one node carries two unknowns (displacement components in x and y directions). On the other hand, the total number of equations that can be sampled from all the smoothing domains should be $3N_s$, because one smoothing domain gives three equations for measuring the strain energy norm (each of three strain components produces strain energy independently). Therefore, we must have $N_s^{\min} = 2n_t/3$.

The same analysis can be done exactly for 3D vector fields of mechanics problems. We now summarize the discussions to TABLE 2.2.

2.3.11 $\mathbf{G}^1$ norms for 1D scalar fields

For normed $\mathbb{G}_h^1$ spaces, the norms are *induced* from the inner products defined as follows for various cases. The associated inner product is given by

$$
\begin{aligned}
\left(w,v\right)_{G^1(\Omega)} &= \int_\Omega wv\,d\Omega + \sum_{k=1}^{N_s} A_k^s\,\overline{w'}\cdot\overline{v'} \\
&= \underbrace{\int_\Omega wv\,d\Omega}_{(w,v)_{L^2(\Omega)}} + \underbrace{\sum_{k=1}^{N_s} A_k^s\,\overline{g}(w)\overline{g}(v)}_{\left(\overline{w'},\overline{v'}\right)_{L^2(\Omega)}}
\end{aligned}
\tag{2.87}
$$

Note the summation is possible because the division of Ω into Ω_i^s is performed in such a way that the interfaces Γ_i^s of Ω_i^s do not share any *finite* portion of the lines on which the function is not square integrable: no energy loss in the interface of the smoothing domains. In Equation (2.87) the (approximated) smoothed derivative is denoted as

$$
\overline{w'} = \overline{\frac{\partial w}{\partial x}} = \frac{1}{A_k^s}\int_{\Gamma_k^s} w(s)\,n_x\,ds = \underbrace{\frac{1}{l_k^s}\left(w_{k+\frac{1}{2}} - w_{k-\frac{1}{2}}\right)}_{=\,\overline{g},\ \text{constant in }\Omega_k^s} = \overline{g}(w)
\tag{2.88}
$$

where $\overline{g}(w)$ denotes the smoothed derivatives of w with respect to x, and the smoothing domains Ω_i^s "centered" at $\mathbf{x}_i$ ($=x_k$, in this 1D case) bounded by $x_{k-\frac{1}{2}}$ and $x_{k+\frac{1}{2}}$, as shown in FIGURE 2.4. The G^1 semi-norm is next defined as

$$
\left|w\right|^2_{G^1(\Omega)} = \sum_{k=1}^{N_s} A_k^s\left|\overline{w'}\right|^2 = \underbrace{\sum_{k=1}^{N_s} l_k^s\,\overline{g}^2(w)}_{\left(\overline{w'},\overline{w'}\right)}
\tag{2.89}
$$

and the G^1 full norm becomes

$$
\left\|w\right\|^2_{G^1(\Omega)} = \underbrace{\int_\Omega w^2\,d\Omega}_{(w,w)=\|w\|^2_{L^2}} + \underbrace{\left|w\right|^2_{G^1(\Omega)}}_{\left(\overline{w'},\overline{w'}\right)=|w|^2_{G^1(\Omega)}} = \left\|w\right\|^2_{L^2} + \left|w\right|^2_{G^1(\Omega)}
\tag{2.90}
$$

which is induced from the inner product defined in Equation (2.87).

2.3.12 G^1 norms for 2D scalar fields

For 2D problems, division of problem domain into smoothing domains is in general shown in FIGURE 2.3 with at least the minimum number of linearly independent smoothing domains. Using such a set of smoothing domains, the associated inner product is given by

$$
\begin{aligned}
(w,v)_{\mathbf{G}^1(\Omega)} &= \int_\Omega wv\mathrm{d}\Omega + \sum_{k=1}^{N_s} A_k^s \overline{\nabla w} \cdot \overline{\nabla v} \\
&= \underbrace{\int_\Omega wv\mathrm{d}\Omega}_{(w,v)_{\mathrm{L}^2(\Omega)}} + \underbrace{\sum_{k=1}^{N_s} A_k^s \left(\overline{g}_1(w)\overline{g}_1(v) + \overline{g}_2(w)\overline{g}_2(v) \right)}_{\left(\overline{\nabla w},\overline{\nabla v}\right)_{\mathrm{L}^2(\Omega)}}
\end{aligned}
\tag{2.91}
$$

where the (approximated) smoothed gradient is denoted as

$$
\begin{aligned}
\overline{\nabla w} &= \left(\frac{\overline{\partial w}}{\partial x_1} \quad \frac{\overline{\partial w}}{\partial x_2} \right) \\
&= \left(\underbrace{\frac{1}{A_k^s} \int_{\Gamma_k^s} w(s)n_1\mathrm{d}s}_{=\overline{g}_1,\ \text{constant in }\Omega_k^s} \quad \underbrace{\frac{1}{A_k^s} \int_{\Gamma_k^s} w(s)n_2\mathrm{d}s}_{=\overline{g}_2,\ \text{constant in }\Omega_k^s} \right) \\
&= \left(\overline{g}_1(w) \quad \overline{g}_2(w) \right)
\end{aligned}
\tag{2.92}
$$

where $\overline{g}_i(w)$ denotes the smoothed derivatives of w with respect to x_i. The $\mathbf{G}^1(\Omega)$ semi-norm is next defined as

$$
\left| w \right|^2_{\mathbf{G}^1(\Omega)} = \sum_{n=1}^{N_s} A_n^s \left| \overline{\nabla w} \right|^2 = \underbrace{\sum_{n=1}^{N_s} A_n^s \left(\overline{g}_1^2(w) + \overline{g}_2^2(w) \right)}_{\left(\overline{\nabla w},\overline{\nabla w}\right)}
\tag{2.93}
$$

and the $\mathbf{G}^1$ full norm becomes

$$
\left\| w \right\|^2_{\mathbf{G}^1(\Omega)} = \underbrace{\int_\Omega w^2\mathrm{d}\Omega}_{(w,w)=\|w\|^2_{L^2}} + \underbrace{\left| w \right|^2_{\mathbf{G}^1(\Omega)}}_{\left(\overline{\nabla w},\overline{\nabla w}\right)} = \left\| w \right\|^2_{L^2} + \left| w \right|^2_{\mathbf{G}^1(\Omega)}
\tag{2.94}
$$

which is induced from the inner product Equation (2.91). The definitions for 3D scalar fields are natural extensions from the 2D case and hence are omitted here.

2.3.13 G^1 norms for 2D vector fields

For vector fields, we need to use vectors of functions. The division of problem domains into smoothing domains is similar as that for the 2D scalar field discussed in the previous subsection. When the function has two components, we should have $w = \begin{pmatrix} w_1 & w_2 \end{pmatrix}$ where $w_1, w_2 \in \mathbb{G}_h^1$ are the two component functions. In this case, we have the smoothed gradient in the following form.

$$
\overline{\nabla \mathbf{w}} = \begin{pmatrix} \dfrac{\overline{\partial w_1}}{\partial x_1} & \dfrac{\overline{\partial w_1}}{\partial x_2} \\[2mm] \dfrac{\overline{\partial w_2}}{\partial x_1} & \dfrac{\overline{\partial w_2}}{\partial x_2} \end{pmatrix}
$$

$$
= \begin{pmatrix} \underbrace{\dfrac{1}{A_k^s}\int_{\Gamma_k^s} w_1(s)n_1 \mathrm{d}s}_{=\bar{g}_{11},\ \text{constant in } \Omega_k^s} & \underbrace{\dfrac{1}{A_k^s}\int_{\Gamma_k^s} w_1(s)n_2 \mathrm{d}s}_{=\bar{g}_{12},\ \text{constant in } \Omega_k^s} \\[4mm] \underbrace{\dfrac{1}{A_k^s}\int_{\Gamma_k^s} w_2(s)n_1 \mathrm{d}s}_{=\bar{g}_{21},\ \text{constant in } \Omega_k^s} & \underbrace{\dfrac{1}{A_k^s}\int_{\Gamma_k^s} w_2(s)n_2 \mathrm{d}s}_{=\bar{g}_{22},\ \text{constant in } \Omega_k^s} \end{pmatrix} \tag{2.95}
$$

$$
= \begin{pmatrix} \bar{g}_{11}(w_1) & \bar{g}_{12}(w_1) \\[2mm] \bar{g}_{21}(w_2) & \bar{g}_{22}(w_2) \end{pmatrix}
$$

where $\bar{g}_{ij}(w)$ denotes the smoothed derivatives of w_i with respect to x_j. We notice here that the (smoothed) gradient is now a matrix, and hence there can be many *equivalent* ways to define the associated inner product. In this work, we decide to have the definition associated with the type of physical problems to be studied for convenience of proving necessary theories for that type of the problems. Considering 2D solid mechanics problems, we define the associated inner product in the form of

$$
(\mathbf{w}, \mathbf{v})_{G^1(\Omega)} = \underbrace{\int_{\Omega}(w_1 v_1 + w_2 v_2)\,\mathrm{d}\Omega}_{(w,v)}
$$

$$
+ \underbrace{\sum_{k=1}^{N_s} A_k^s \big[\bar{g}_{11}(w_1)\bar{g}_{11}(v_1) + \bar{g}_{22}(w_2)\bar{g}_{22}(v_2)}_{} \tag{2.96}
$$

$$
\underbrace{+ \big(\bar{g}_{12}(w_1) + \bar{g}_{21}(w_2)\big)\big(\bar{g}_{12}(v_1) + \bar{g}_{21}(v_2)\big)\big]}_{(\nabla \mathbf{w}, \nabla \mathbf{v})}
$$

The induced $G^1(\Omega)$ semi-norm is first defined as

$$\left|\mathbf{w}\right|^2_{G^1(\Omega)} = \underbrace{\sum_{k=1}^{N_s} A_k^s \left(\overline{g}_{11}^2(w_1) + \overline{g}_{22}^2(w_2) + \left(\overline{g}_{12}(w_1) + \overline{g}_{21}(w_2) \right)^2 \right)}_{(\nabla\mathbf{w},\nabla\mathbf{w})} \tag{2.97}$$

It is clear that in our definition of the inner product and hence the induced semi-norm we have intentionally related to the strain components, and hence the L^2 norm of the vector of strains.

The associated G^1 full norm can now be defined as

$$\left\|\mathbf{w}\right\|^2_{G^1(\Omega)} = \underbrace{\int_\Omega \left(w_1^2 + w_2^2 \right) d\Omega}_{(\mathbf{w},\mathbf{w})=\|\mathbf{w}\|^2_{L^2}} + \underbrace{\left|\mathbf{w}\right|^2_{G^1(\Omega)}}_{(\nabla\mathbf{w},\nabla\mathbf{w})} = \left\|\mathbf{w}\right\|^2_{L^2} + \left|\mathbf{w}\right|^2_{G^1(\Omega)} \tag{2.98}$$

2.3.14 G^1 norms for 3D vector fields

For 3D problems, the division of problem domain into smoothing domains is similar to those in 2D but with one more dimension extension. For vector fields with three-component functions in 3D space, such as the 3D solid mechanics problems, we shall have $w = \begin{pmatrix} w_1 & w_2 & w_3 \end{pmatrix}$ where w_1, w_2, $w_3 \in \mathbb{G}_h^1$. In this case we define, naturally, the associated inner product as

$$\begin{aligned}
(\mathbf{w},\mathbf{v})_{G^1(\Omega)} = &\int_\Omega \left(w_1 v_1 + w_2 v_2 + w_3 v_3 \right) d\Omega \\
&+ \sum_{k=1}^{N_s} A_k^s \big[\overline{g}_{11}(w_1)\overline{g}_{11}(v_1) + \overline{g}_{22}(w_2)\overline{g}_{22}(v_2) + \overline{g}_{33}(w_3)\overline{g}_{33}(v_3) \\
&\quad + \left(\overline{g}_{12}(w_1) + \overline{g}_{21}(w_2) \right)\left(\overline{g}_{12}(v_1) + \overline{g}_{21}(v_2) \right) \\
&\quad + \left(\overline{g}_{13}(w_1) + \overline{g}_{31}(w_3) \right)\left(\overline{g}_{13}(v_1) + \overline{g}_{31}(v_3) \right) \\
&\quad + \left(\overline{g}_{23}(w_2) + \overline{g}_{32}(w_3) \right)\left(\overline{g}_{23}(v_2) + \overline{g}_{32}(v_3) \right) \big]
\end{aligned} \tag{2.99}$$

The associated G^1 semi-norm can be defined as

$$\left|\mathbf{w}\right|^2_{\mathbb{G}^1(\Omega)} = \sum_{k=1}^{N_s} A_k^s \Big[\bar{g}_{11}^2(w_1) + \bar{g}_{22}^2(w_2) + \bar{g}_{33}^2(w_3)$$

$$+ \Big(\bar{g}_{12}(w_1) + \bar{g}_{21}(w_2)\Big)^2 + \Big(\bar{g}_{13}(w_1) + \bar{g}_{31}(w_3)\Big)^2 \qquad (2.100)$$

$$+ \Big(\bar{g}_{23}(w_2) + \bar{g}_{32}(w_3)\Big)^2 \Big]$$

and the $\mathbb{G}^1$ full norm is given by

$$\left\|\mathbf{w}\right\|^2_{\mathbb{G}^1(\Omega)} = \underbrace{\int_\Omega \Big(w_1^2 + w_2^2 + w_3^2\Big)\mathrm{d}\Omega}_{(\mathbf{w},\mathbf{w})=\|\mathbf{w}\|_{L^2}^2} + \left|\mathbf{w}\right|^2_{\mathbb{G}^1(\Omega)} \qquad (2.101)$$

We finally denote for d-dimensional problems, a space for functions that are fixed on the Dirichlet boundaries and hence the functions cannot "float".

$$(\mathbb{G}^1_{h,0}(\Omega))^d = \Big\{ v \in (\mathbb{G}^1_h(\Omega))^d \,\big|\, v_i = 0 \quad \text{on} \ \Gamma_u \Big\} \qquad (2.102)$$

2.4 $\mathbb{G}^1_h$ space: basic properties

Because the normed $\mathbb{G}^1_h$ spaces are defined in the above-mentioned "unusual" manner, we have to show that they possess all the necessary basic properties. Here we discuss only normed $\mathbb{G}^1_h$ spaces.

2.4.1 Linearity

Remark 2.9 A $\mathbb{G}^1_h$ space is a linear space.

Consider any two functions $w, v \in \mathbb{G}^1_h$, we shall have

$$w(\mathbf{x}) = \mathbf{\Phi}(\mathbf{x})\mathbf{d}_w, \quad \mathbf{d}_w \in \mathbb{R}^{N_n}$$
$$v(\mathbf{x}) = \mathbf{\Phi}(\mathbf{x})\mathbf{d}_v, \quad \mathbf{d}_v \in \mathbb{R}^{N_n} \qquad (2.103)$$

The addition of w and v becomes

$$w(\mathbf{x}) + v(\mathbf{x}) = \mathbf{\Phi}(\mathbf{x})\underbrace{(\mathbf{d}_w + \mathbf{d}_v)}_{\in \mathbb{R}^{N_n}} \qquad (2.104)$$

which must also be in $\mathbb{G}_h^1$, because $\mathbb{R}^{N_n}$ is a linear space: $(\mathbf{d}_w + \mathbf{d}_v) \in \mathbb{R}^{N_n}$. Following exactly the same argument, we shall have that for $\forall w \in \mathbb{G}_h^1$ and $\forall \alpha \in \mathbb{R}$, $\alpha w \in \mathbb{G}_h^1$. $\qquad\qquad\square$

2.4.2 Positivity

A function in a $\mathbb{G}_h^1$ space is always non-negative in norm:

$$\left\| w \right\|_{\mathbb{G}^1} \geq 0, \qquad \forall w \in \mathbb{G}_h^1 \tag{2.105}$$

and a nonzero function in a $\mathbb{G}_h^1$ space is always strictly positive in norm:

$$\left\| w \right\|_{\mathbb{G}^1} > 0, \qquad \forall w \in \mathbb{G}_h^1, w \neq 0 \tag{2.106}$$

It is very easy to show Equations (2.105) and (2.106) from these norm definitions and their relations. Using Equation (2.101), we have

$$\left\| w \right\|_{\mathbb{G}^1(\Omega)}^2 = \left\| w \right\|_{\mathbf{L}^2}^2 + \left| w \right|_{\mathbb{G}^1(\Omega)}^2, \quad \forall w \in \mathbb{G}_h^1 \tag{2.107}$$

Since the $\mathbf{L}^2$ norm is positive and the $\mathbb{G}^1$ semi-norm is semi-positive for any nonzero $w \in \mathbb{G}_h^1$, we shall always have Equation (2.106). When w is zero, all these norms become zero, and hence Equation (2.105) holds.

2.4.3 Scalar modification

A function in a $\mathbb{G}_h^1$ space is scalable in norm:

$$\left\| \alpha w \right\|_{\mathbb{G}^1} = \left| \alpha \right| \left\| w \right\|_{\mathbb{G}^1}, \qquad \forall \alpha \in \mathbb{R}, \forall w \in \mathbb{G}_h^1 \tag{2.108}$$

which can be observed simply from the definition, say for example Equation (2.101).

2.4.4 Completeness

A $\mathbb{G}_h^1$ is a complete metric (inner product induced norm) linear space, and is a Banach space [17], meaning that for every Cauchy sequence w_j in $\mathbb{G}_h^1$ has a limit $v \in \mathbb{G}_h^1$. A Cauchy sequence is one such that $\left\| w_j - w_k \right\|_{G^1(\Omega)} \to 0$, $j,k \to \infty$, and the completeness requires that $\left\| v - w_j \right\|_{G^1(\Omega)} \to 0$ as $j \to \infty$. The completeness property can be understood easily because a $\mathbb{G}_h^1$ space is a sub-space of $\mathbb{L}^2$ space that is a Banach space [17]. Note also that the norm of $\mathbb{G}_h^1$ is a linear combination of $\mathbb{L}^2$ norms. Another way may be using the 1st inequality of equivalence of the G^1 norm of a function $w \in \mathbb{G}_h^1$ and the L^2 norm of $\mathbf{d} \in \mathbb{R}^{N_n}$, and the fact that $\mathbb{R}^{N_n}$ is known being a Banach space (that is complete).

2.4.5 Cauchy-Schwarz inequality

Because the $\mathbb{G}_h^1$ inner product defined in Equation (2.87) is a qualified bilinear form, we shall have the Cauchy-Schwarz inequality:

$$\left| \left(\overline{w'}, \overline{v'} \right)_{\mathbb{L}^2(\Omega)} \right| \leq \left| w \right|_{G^1(\Omega)} \cdot \left| v \right|_{G^1(\Omega)} \tag{2.109}$$

and

$$\left| (w, v)_{G^1(\Omega)} \right| \leq \left\| w \right\|_{G^1(\Omega)} \cdot \left\| v \right\|_{G^1(\Omega)} \tag{2.110}$$

To show this, we first observe the symmetric, because swapping places for w and v will not change the value of the inner product. Second, it is positive definite, because of the positivity of the $(w, w)_{\mathbb{L}^2(\Omega)}$ and semi-positivity of $\left(\overline{w'}, \overline{w'} \right)_{\mathbb{L}^2(\Omega)}$. Finally, it is bilinear, because of the bilinear property of $(w, v)_{\mathbb{L}^2(\Omega)}$ and $\left(\overline{w'}, \overline{v'} \right)_{\mathbb{L}^2(\Omega)}$. Equations (2.109) and (2.110) are fundamentally important for the functional analysis of G spaces.

2.4.6 Triangular inequality

We now prove the *triangular inequality* for $\mathbb{G}_h^1$ norm:

$$\left\| w + v \right\|_{G^1} \leq \left\| w \right\|_{G^1} + \left\| v \right\|_{G^1}, \quad \forall w \in \mathbb{G}_h^1, \forall v \in \mathbb{G}_h^1 \tag{2.111}$$

We first proof this for 2D scalar functions:

$$
\begin{aligned}
\|w+v\|_{\mathbb{G}^1} &= \left[\int_\Omega (w+v)^2 \, d\Omega + \sum_{k=1}^{N_s} A_k^s \left| \nabla(w+v) \right|^2 \right]^{1/2} \\
&= \left\{ \int_\Omega w^2 d\Omega + \int_\Omega v^2 d\Omega + 2\int_\Omega wv \, d\Omega \right. \\
&\quad \left. + \sum_{k=1}^{N_s} A_k^s \left[\left(\overline{g}_1(w) + \overline{g}_1(v) \right)^2 + \left(\overline{g}_2(w) + \overline{g}_2(v) \right)^2 \right] \right\}^{1/2} \\
&= \left\{ \int_\Omega w^2 d\Omega + \int_\Omega v^2 d\Omega + 2\int_\Omega wv \, d\Omega \right. \\
&\quad + \sum_{k=1}^{N_s} A_k^s \left[\overline{g}_1^2(w) + 2\overline{g}_1(w)\overline{g}_1(v) + \overline{g}_1^2(v) \right. \\
&\quad \left. \left. + \overline{g}_2^2(w) + 2\overline{g}_2(w)\overline{g}_2(v) + \overline{g}_2^2(v) \right] \right\}^{1/2} \\
&= \left\{ \underbrace{ \int_\Omega w^2 d\Omega + \sum_{k=1}^{N_s} A_k^s (\overline{g}_1^2(w) + \overline{g}_2^2(w)) }_{\|w\|_{G^1}^2} \right. \\
&\quad + \underbrace{ \int_\Omega v^2 d\Omega + \sum_{k=1}^{N_s} A_k^s (\overline{g}_1^2(v) + \overline{g}_2^2(v)) }_{\|v\|_{G^1}^2} \\
&\quad \left. + 2\left[\underbrace{ \int_\Omega wv \, d\Omega + \sum_{k=1}^{N_s} A_k^s \left(\overline{g}_1(w)\overline{g}_1(v) + \overline{g}_2(w)\overline{g}_2(v) \right) }_{(w,v)_{G^1} \le \|w\|_{G^1} \|v\|_{G^1}} \right] \right\}^{1/2} \\
&\le \left[\|w\|_{G^1}^2 + 2\|w\|_{G^1}\|v\|_{G^1} + \|v\|_{G^1}^2 \right]^{1/2} = \|w\|_{G^1} + \|v\|_{G^1}, \\
&\qquad \forall w \in \mathbb{G}_h^1, \ \forall v \in \mathbb{G}_h^1
\end{aligned}
\tag{2.112}
$$

In the above proof process, we have used the Cauchy-Schwarz inequality for our inner product induced norms.

The exact same procedure can be applied to prove the triangular inequality for vector functions, but it will be a little lengthy. We show the process here only for the 2D cases, by examining first the semi-norm of the sum of two functions $w, v \in \mathbb{G}^1$ based on the definition Equation (2.97)

$$\begin{aligned}
\left| w + v \right|^2_{\mathbb{G}^1(\Omega)} &= \sum_{k=1}^{N_s} A_k^s \left\{ \bar{g}_{11}^2 (w_1 + v_1) + \bar{g}_{22}^2 (w_2 + v_2) \right. \\
&\quad \left. + \left[\bar{g}_{12}(w_1 + v_1) + \bar{g}_{21}(w_2 + v_2) \right]^2 \right\} \\
&= \sum_{k=1}^{N_s} A_k^s \left\{ \bar{g}_{11}^2 (w_1) + \bar{g}_{11}^2 (v_1) + 2\bar{g}_{11}(w_1)\bar{g}_{11}(v_1) \right. \\
&\quad + \bar{g}_{22}^2 (w_2) + \bar{g}_{22}^2 (v_2) + 2\bar{g}_{22}(w_2)\bar{g}_{22}(v_2) \\
&\quad \left. + \left[\left(\bar{g}_{12}(w_1) + \bar{g}_{21}(w_2) \right) + \left(\bar{g}_{12}(v_1) + \bar{g}_{21}(v_2) \right) \right]^2 \right\} \\
&= \underbrace{\sum_{k=1}^{N_s} A_k^s \left[\bar{g}_{11}^2 (w_1) + \bar{g}_{22}^2 (w_2) + \left(\bar{g}_{12}(w_1) + \bar{g}_{21}(w_2) \right)^2 \right]}_{\left| w \right|^2_{\mathbb{G}^1(\Omega)}} \\
&\quad + \underbrace{\sum_{k=1}^{N_s} A_k^s \left[\bar{g}_{11}^2 (v_1) + \bar{g}_{22}^2 (v_2) + \left(\bar{g}_{12}(v_1) + \bar{g}_{21}(v_2) \right)^2 \right]}_{\left| v \right|^2_{\mathbb{G}^1(\Omega)}} \\
&\quad + 2\sum_{k=1}^{N_s} A_k^s \left[\bar{g}_{11}(w_1)\bar{g}_{11}(v_1) + \bar{g}_{22}(w_2)\bar{g}_{22}(v_2) \right. \\
&\quad \left. + \left(\bar{g}_{12}(w_1) + \bar{g}_{21}(w_2) \right)\left(\bar{g}_{12}(v_1) + \bar{g}_{21}(v_2) \right) \right]
\end{aligned} \tag{2.113}$$

We then examine the full norm of the sum of two functions $w, v \in \mathbb{G}_h^1$ based on the definition Equation (2.101):

$$\begin{aligned}
\left\| w + v \right\|^2_{\mathbb{G}^1(\Omega)} &= \int_\Omega \left((w_1 + v_1)^2 + (w_2 + v_2)^2 \right) d\Omega + \left| w + v \right|^2_{\mathbb{G}^1(\Omega)} \\
&= \int_\Omega \left(w_1^2 + v_1^2 + w_2^2 + v_2^2 + 2w_1 v_1 + 2w_2 v_2 \right) d\Omega \\
&\quad + \left| w + v \right|^2_{\mathbb{G}^1(\Omega)} \\
&= \int_\Omega \left(w_1^2 + v_1^2 \right) d\Omega + \int_\Omega \left(w_2^2 + v_2^2 \right) d\Omega \\
&\quad + 2\int_\Omega \left(w_1 v_1 + w_2 v_2 \right) d\Omega + \left| w + v \right|^2_{\mathbb{G}^1(\Omega)}
\end{aligned} \tag{2.114}$$

Substituting Equation (2.113) into (2.114) gives

$$
\begin{aligned}
\|w+v\|_{\mathbb{G}^1}^2 &= \int_\Omega \left(w_1^2 + v_1^2 \right) d\Omega + \int_\Omega \left(w_2^2 + v_2^2 \right) d\Omega \\
&\quad + 2\int_\Omega \left(w_1 v_1 + w_2 v_2 \right) d\Omega + |w|_{\mathbb{G}^1}^2 + |v|_{\mathbb{G}^1}^2 \\
&\quad + 2\sum_{k=1}^{N_s} A_k^s \Big[\overline{g}_{11}(w_1)\overline{g}_{11}(v_1) + \overline{g}_{22}(w_2)\overline{g}_{22}(v_2) \\
&\qquad\qquad\qquad + \left(\overline{g}_{12}(w_1) + \overline{g}_{21}(w_2) \right)\left(\overline{g}_{12}(v_1) + \overline{g}_{21}(v_2) \right) \Big] \\
&= \|w\|_{\mathbb{G}^1}^2 + \|v\|_{\mathbb{G}^1}^2 \\
&\quad + 2\Bigg\{ \int_\Omega \left(w_1 v_1 + w_2 v_2 \right) d\Omega \\
&\qquad + \underbrace{\sum_{k=1}^{N_s} A_k^s \begin{bmatrix} \overline{g}_{11}(w_1)\overline{g}_{11}(v_1) + \overline{g}_{22}(w_2)\overline{g}_{22}(v_2) \\ + \left(\overline{g}_{12}(w_1) + \overline{g}_{21}(w_2) \right)\left(\overline{g}_{12}(v_1) + \overline{g}_{21}(v_2) \right) \end{bmatrix} \Bigg\}}_{(w,v) \le \|w\|_{\mathbb{G}^1(\Omega)} \|v\|_{\mathbb{G}^1(\Omega)}} \\
&\le \|w\|_{\mathbb{G}^1}^2 + \|v\|_{\mathbb{G}^1}^2 + 2\|w\|_{\mathbb{G}^1}\|v\|_{\mathbb{G}^1} = \left(\|w\|_{\mathbb{G}^1} + \|v\|_{\mathbb{G}^1} \right)^2
\end{aligned}
\tag{2.115}
$$

which is Equation (2.111). Note here we used again the useful Cauchy-Schwarz inequality for inner product induced norms.

Comparing Equations (2.100) with (2.101), we obtain

$$
|w|_{\mathbb{G}^1(\Omega)} \le \|w\|_{\mathbb{G}^1(\Omega)}, \ \forall w \in \mathbb{G}_h^1
\tag{2.116}
$$

meaning that the $\mathbb{G}_h^1$ full norm is always larger than the $\mathbb{G}_h^1$ semi-norm. This is obvious.

2.5 $\mathbb{G}_h^1$ space: other properties

2.5.1 Convergence property

Remark 2.10 Convergence property

For $w, v \in \mathbb{H}^1$, when $N_s \to \infty$ and all $\Omega_i^s \to 0$, $\overline{W}$ becomes Delta functions and the integral representation becomes exact. At such a limit, we have $\overline{\nabla w} \to \nabla w$, $(w,v)_{\mathbb{G}^1(\Omega)} \to (w,v)_{\mathbb{H}^1(\Omega)}$, $\|w\|_{\mathbb{G}^1(\Omega)} \to \|w\|_{\mathbb{H}^1(\Omega)}$, $|w|_{\mathbb{G}^1(\Omega)} \to |w|_{\mathbb{H}^1(\Omega)}$, $\|w\|_{\mathbb{G}^2(\Omega)} \to \|w\|_{\mathbb{H}^2(\Omega)}$, and $|w|_{\mathbb{G}^2(\Omega)} \to |w|_{\mathbb{H}^2(\Omega)}$.

Remark 2.10 ensures that all the bound properties for G^1 norms converge to the corresponding H^1 norms defined in the same manner at the limit of $N_s \to \infty$ and all $\Omega_i^s \to 0$ for all functions in an H^1 space. We, however, need the inequalities for finite smoothing domains and for all functions in G^1 spaces, which are termed as G inequalities to be derived in the next sections.

2.5.2 First inequality

The first inequality relates the (full) G^1 norm of a function to L^2 norm of the nodal values of the function when the function is approximated based on an approximation method using local nodes scattered in the problem domain. For functions in a $\mathbb{G}_h^1$ space, there exist nonzero positive constants c_{dw}^f and c_{wd}^f, such that the 1st inequality

$$\left\|\mathbf{d}\right\|_2 \geq c_{dw}^f \left\|w\right\|_{\mathbb{G}^1(\Omega)}, \quad \forall w \in \mathbb{G}_h^1 \tag{2.117}$$

and

$$\left\|w\right\|_{\mathbb{G}^1(\Omega)} \geq c_{wd}^f \left\|\mathbf{d}\right\|_2, \quad \forall w \in \mathbb{G}_h^1 \tag{2.118}$$

hold. Equations (2.117) and (2.118) mean that the full G^1 norm of a function in a $\mathbb{G}_h^1$ space is *equivalent* (different only with a factor that is independent of $w \in \mathbb{G}_h^1$) to the L^2 norm of the nodal values of the function.

Proof

The proof of Equation (2.117) or (2.118) is straightforward. It uses the definition of the $\mathbb{G}_h^1$ space. Because any function w in $\mathbb{G}_h^1$ space is constructed using $w(\mathbf{x}) = \mathbf{\Phi}(\mathbf{x})\mathbf{d}$, and because the nodal shape functions $\mathbf{\Phi}(\mathbf{x})$ are linearly independent, the full G^1 norm of w must be equivalent to the L^2 norm of the nodal values of the function $\mathbf{d}$. $\qquad\square$

A more detailed analysis of Equation (2.117) or (2.118) for 1D cases and their extension to higher dimensions have been given in [9, 10].

2.5.3 Second inequality

The second inequality relates the G^1 semi-norm of a function to the L^2 norm of the nodal values of the function in a G^1 with a set of at least a minimum number of linearly independent smoothing domains [9, 10] created properly for evaluating the G^1 semi-norm. We now state that the 2nd inequality

$$\left|w\right|_{\mathbb{G}^1(\Omega)} \ge c^s_{wd}\left\|\mathbf{d}\right\|_2, \quad \forall w \in \mathbb{G}^1_{h,0}(\Omega) \tag{2.119}$$

and

$$\left\|\mathbf{d}\right\|_2 \ge c^s_{dw}\left|w\right|_{\mathbb{G}^1(\Omega)}, \quad \forall w \in \mathbb{G}^1_{h,0}(\Omega) \tag{2.120}$$

hold meaning that the semi G^1 norm of a function in a $\mathbb{G}^1_h$ space is equivalent to the L^2 norm of the nodal values of the function.

Proof

The proof of the 2nd inequality is also straightforward. It uses again the definition of the $\mathbb{G}^1_h$ space. Because any function w in $\mathbb{G}^1_h$ space is constructed using $w(\mathbf{x}) = \boldsymbol{\Phi}(\mathbf{x})\mathbf{d}$, the nodal shape functions $\boldsymbol{\Phi}(\mathbf{x})$ are linearly independent, and the positivity condition that $\displaystyle\sum_{k=1}^{N_s}\left(\int_{\Gamma_k^s} v(s)n_i \mathrm{d}s\right)^2 > 0$, for all $v \ne c \in \mathbb{R}$, $\forall \mathbf{d} \in \mathbb{R}^{N_n}$ and $i = 1,\cdots,d$, the semi G^1 norm of w must be equivalent to the L^2 norm of the nodal values of the function $\mathbf{d}$. $\qquad\square$

A more detailed analysis of Equation (2.119) or (2.120) for 1D cases and their extension to higher dimensions can be found in using the concept of "positivity relay" [9].

2.5.4 Third inequality

We now ready to present the 3rd inequality stated in the following theorem.

Theorem 2.1 Equivalence of G norms

Functions in a $\mathbb{G}^1_h$ space, there exists a positive nonzero constant c_G such that

$$c_G\left\|w\right\|_{\mathbb{G}^1(\Omega)} \le \left|w\right|_{\mathbb{G}^1(\Omega)}, \quad \forall w \in \mathbb{G}^1_{h,0} \tag{2.121}$$

meaning that the G^1 full norm and the G^1 semi-norm of any function in a $\mathbb{G}^1_{h,0}$ space are equivalent.

Proof

The combination of the first inequality Equation (2.117), and the second inequality (2.119) gives

$$\begin{aligned}
\left|w\right|_{\mathbb{G}^1(\Omega)} &\geq c_{wd}^s \left\|\mathbf{d}\right\|_2 \\
&\geq \underbrace{c_{wd}^s c_{dw}^f}_{c_G} \left\|w\right\|_{\mathbb{G}^1(\Omega)} \\
&\geq c_G \left\|w\right\|_{\mathbb{G}^1(\Omega)}, \quad \forall w \in \mathbb{G}_{h,0}^1(\Omega)
\end{aligned} \tag{2.122}$$

which is the third inequality. $\qquad\square$

Combination of Equation (2.116) and Equation (2.122), we arrived at the following chain inequalities:

$$c_G \left\|w\right\|_{\mathbb{G}^1(\Omega)} \leq \left|w\right|_{\mathbb{G}^1(\Omega)} \leq \left\|w\right\|_{\mathbb{G}^1(\Omega)}, \quad \forall w \in \mathbb{G}_{h,0}^1 \tag{2.123}$$

The 3rd inequality Equation (2.121) is a generalized version of the well-known Poincare-Friedrichs inequality. It is the foundation of the W^2 formulation, ensuring the stability of the solution. Equation (2.123) is essential to ensure both the uniqueness and convergence of a W^2 formulation of a *physically* stable problem. For solid mechanic problems, for example, we need the material being stable (see, Remark 1.1).

2.5.5 Softening effects

We further examine some of the important properties of functions in G spaces. For a function in an H^1 space, the G^1 semi-norm of the function is no-larger than the H^1 semi-norm (of same type) of the function.

$$\left|w\right|_{\mathbb{G}^1(\Omega)} \leq \left|w\right|_{\mathbb{H}^1(\Omega)}, \quad \forall w \in \mathbb{H}^1 \tag{2.124}$$

meaning that the smoothing operation results in a smaller semi-norm measure. This is the fundamental inequality for the so-called softening effects [18]. A proof on this remark can be found in [9].

Because of Equation (2.124) we immediately have for a function in an H^1 space, the G^1 full norm of the function is no-larger than the H^1 full norm of the function:

$$\left\|w\right\|_{\mathbb{G}^1(\Omega)} \leq \left\|w\right\|_{\mathbb{H}^1(\Omega)}, \quad \forall w \in \mathbb{H}^1 \tag{2.125}$$

meaning that the smoothing operation results in a smaller (full) norm measure. This is because the first term in the RHS of Equation (2.41) and that of Equation (2.90) are exactly the same.

Example 2.5.1 A piecewise linear and continuous function

A piecewise linear continuous function v is defined over the finite 1D domain $\Omega = (-1,1)$ that is discretized using two cells with three nodes, as shown in FIGURE 2.6. Three node-based smoothing domains are used. Show that the function v is in a $\mathbb{G}_h^1(\Omega)$ space.

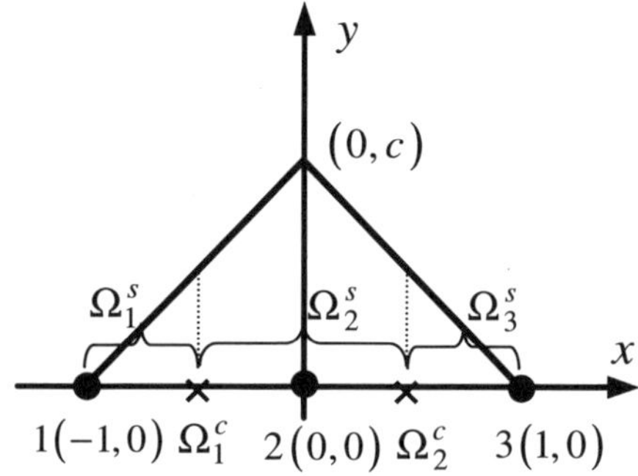

FIGURE 2.6 A piecewise linear and continuous function defined in $\Omega = (-1, 1)$. It is divided into two cells $\Omega_1^c = (-1,0)$ and $\Omega_2^c = (0,1)$ and three node-based smoothing domains $\Omega_1^s = (-1,-0.5)$, $\Omega_2^s = (-0.5,0.5)$ and $\Omega_3^s = (0.5,1)$.

To show this, we follow the definition given in Equation (2.86). First, we have in this case:

$$\mathbf{d} = \{u_1 \quad u_2 \quad u_3\} = \{0 \quad c \quad 0\} \tag{2.126}$$

These three independent nodal shape functions are

$$\phi_1 = -x; \quad \phi_2 = \begin{cases} 1+x & \text{when } -1 \le x \le 0 \\ 1-x & \text{when } 0 < x < 1 \end{cases}; \quad \phi_3 = x \tag{2.127}$$

The function v is given by

$$v(x) = \sum_{n=1}^{3} \phi_n(x) d_n \tag{2.128}$$

It is clear that $v \in \mathbb{L}^2(\Omega)$, and hence the function is bounded. In addition, using Equation (2.76), we have

$$\sum_{k=1}^{3}\left(\int_{\Gamma_k^s} v(s)n_i \mathrm{d}s\right)^2 = \left(\underbrace{v(\frac{-1}{2}) - v(-1)}_{c/2}\right)^2 + \left(\underbrace{v(\frac{1}{2}) - v(\frac{-1}{2})}_{=0}\right)^2$$
$$+ \left(\underbrace{v(\frac{1}{2}) - v(0)}_{-c/2}\right)^2 \tag{2.129}$$
$$= \frac{c^2}{4} + \frac{c^2}{4} = \frac{c^2}{2} > 0$$

Therefore, it is in a $\mathbb{G}_h^1(\Omega)$ space.

This example shows that linear PIM shape functions (see, Example 2.2.1) are in a $\mathbb{G}_h^1(\Omega)$ space.

Example 2.5.2 Heaviside type function

A Heaviside type function is defined in Equation (2.28). The problem domain $\Omega = (-1,1)$ is discretized using two cells. Three node-based smoothing domains are used. It is clear that the function v is not in a $\mathbb{G}_h^1(\Omega)$ space, because it cannot be expressed using any set of linearly independent shape functions. The only way to produce this Heaviside function is to use two duplicated nodes at $x=0$, which is, however, prohibited (see, Section 1.6.1).

In Chapter 3 we should see PIM shape functions (and the function constructed using linear combination of these shape functions) can be in a $\mathbb{G}_h^1(\Omega)$ space, despite the discontinuities.

2.6   Concluding remarks

The standard weak formulation used in FEM is applicable and has applied to many meshfree settings. As long as the meshfree shape functions created is in a proper H space, the field function constructed will also be in the same (linear) space, and thus the stability and convergence of the meshfree methods is ensured for physically well-posed problems.

Based on the G space theory presented in this chapter, a weakened weak formulation is also applicable and has applied to meshfree settings as well as weak formulation with FEM settings (because an $\mathbb{H}_h^1$ space is in a $\mathbb{G}_h^1$ space). As long as the nodal shape functions created is in a proper $\mathbb{G}_h^1$ space, the stability of the meshfree methods is ensured by these key inequalities for physically well-posed problems. Chapter 3 will thus focus on who to create nodal shape functions in $\mathbb{G}_h^1$ spaces for S-PIMs.

Finally, we note that the study on G space is still in the very early stage. What we have presented in this chapter is only the minimum for the formulation of S-PIM and other meshfree methods for problems governed by 2nd order PDEs. Much more intensive study on G space for other types of problems is required.

2.7 References

1. Hughes, T. J. R. *The Finite Element Method: Linear Static and Dynamic Finite Element Analysis*, Prentice-Hall, 1987.

2. Zienkiewicz, O. C. and Taylor R. L., The Finite Element Method, 5th ed., Butterworth Heimemann, Oxford, 2000.

3. Liu, G. R. and Quek, S. S. *The finite element method: a practical course.* Butterworth Heinemann: Oxford, 2002.

4. Liu, G. R., *Meshfree Methods: Moving beyond the Finite Element Method*, 2nd ed., CRC press, Boca Taton, USA, 2009.

5. Naylor, A. W. and Sell, G. R., *Linear operator Theory in Engineering and Science*, Springer-Verlag, 1982.

6. Liu, G. R., A generalized Gradient smoothing technique and the smoothed bilinear form for Galerkin formulation of a wide class of computational methods. *International Journal of Computational Methods*, 5(2): 199-236, 2008.

7. Liu, G. R., On a G space theory. *International Journal of Computational Methods*, 6(2): 257-289, 2009.

8. Liu, G. R. and Zhang, G. Y., A normed G space and weakened weak (W^2) formulation of cell-based smoothed point interpolation method. *International Journal of Computational Methods*, 6(1): 147-179, 2009.

9. Liu, G. R., A G space theory and a weakened weak (W^2) form for a unified formulation of compatible and incompatible methods: Part I theory. *International Journal for Numerical Methods in Engineering*, 81: 1093-1126, 2010.

10. Liu, G. R., A G space theory and a weakened weak (W^2) form for a unified formulation of compatible and incompatible methods: Part II applications to solid mechanics problems. *International Journal for Numerical Methods in Engineering*, 81: 1127-1156, 2010.

11. Liu, G. R. and Liu, M. B., *Smoothed Particle Hydrodynamics—A Meshfree Practical Method*. World Scientific: Singapore, 2003.

12. Liu, G. R. and Gu, Y. T., *An introduction to meshfree method methods and their programming*, Springer, 2005.

13. Liu, G. R. and Kee, B. B. T., A stabilized least-squares radial point collocation method (LS-RPCM) for adaptive analysis, *Computer Method in Applied Mechanics and Engineering*, 195: 4843-4861, 2006.

14. Kee, B. B. T, Liu, G. R. and Lu. C., A regularized least-squares radial point collocation method (RLS-RPCM) for adaptive analysis. *Computational Mechanics*, 40: 837-853, 2007.

15. Liu, G. R. and Xu, G. X., A gradient smoothing method (GSM) for fluid dynamics problems. *International Journal for Numerical Methods in Fluids*, 58(10): 1101-1133, 2008.

16. Xu, G. X. Liu, G. R. and Tani, A., An adaptive gradient smoothing method (GSM) for fluid dynamics problems. *International Journal for Numerical Methods in Fluids*, 62(5): 499-529, 2010.

17. Susanne, C., Brenner, S. C. and Scott, L. R., *The Mathematical Theory of Finite Element Methods*, 3rd Edition, Springer, 2008.

18. Liu, G. R. and Zhang, G. Y., Upper bound solution to elasticity problems: A unique property of the linearly conforming point interpolation method (LC-PIM). *International Journal for Numerical Methods in Engineering*, 74: 1128-1161, 2008.

Chapter 3

PIM Shape Function Creation

Creation of *shape functions* is one of the most essential and important issues in any discrete numerical method using local approximations, including the S-PIM. Therefore, the development of effective methods/techniques for creating shape functions has always been one of the most active areas in developing new numerical methods. The challenge is how to efficiently create shape functions with preferable properties using nodes *scattered* in the problem domain. This chapter discusses the theory and procedures for shape function creation, with the focus on the *point interpolation method* (PIM) that is the simplest and the most straightforward method and hence particularly preferred in the S-PIM models.

For the discussions on theory and properties, we will concentrate on the polynomial PIM shape functions. These theory and properties are largely applicable to all other variations of PIM shape functions.

3.1   Requirements on shape functions

On the theoretical aspects, a favorable set of shape functions for a discrete numerical method should satisfy the following requirements.

1) **Linear independence**: the set of nodal shape functions for all the nodes in the problem domain must be linearly independent and hence qualifies as a basis for displacement function construction.

2) **Partitions of unity**: the sum of the nodal shape functions must be *unity* at any point in the problem domain.

3) **Consistency:** the shape functions should possess a certain order of consistency to ensure the certain order of polynomials can be reproduced exactly. This book requires a minimum of 1st order or *linear consistency*.

4) **Delta function property**: preferably, the nodal shape functions should possess the Kronecker Delta function property.

5) **Compatibility**: the nodal shape functions must be compatible throughout the problem domain, if weak formulation is used. This is, however, not required for weakened weak (W^2) formulations.

On the aspects of practical implementation, a favorable method for shape function creation should satisfy the following requirements.

1) **Arbitrary node distribution:** the nodal distribution can be arbitrary within reason, and at least more flexible than that used in the FEM. The algorithm must be stable with respect to irregularity of the node distribution.

2) **Compact support:** only a small number of local *support nodes* should be used for sparseness in the resultant system equations.

3) **Efficiency:** the computational efficiency of the algorithm should be at least comparable to the FEM.

4) **Automation:** the entire creation process can be made automatic.

The PIM shape function satisfies all these requirements, except item 5) of compatibility when higher order PIM shape functions are used. The lower order PIM shape functions are ideal for establishing S-PIM models using W^2 formulations.

3.1.1 Linear independence

Because the displacement functions are constructed using the nodal shape functions of all the nodes in the problem domain Ω, these nodal shape functions *must* be linearly independent, in order to qualify as a *basis*. For a finite discrete model with N_n nodes, we need N_n *nodal shape functions* for each displacement components. The linear independence of these nodal shape functions ϕ_n requires mathematically that

$$\sum_{n=1}^{N_n} \alpha_n \phi_n(\mathbf{x}) = 0 \iff \alpha_n = 0, \quad n = 1, 2, \cdots, N_n \tag{3.1}$$

where $\mathbf{x}$ is any point inside the problem domain, and ϕ_n is the nodal shape function for the nth node. Note here that our summation is over all the nodes in the problem domain. In usual situations, such a "complete summation" is not needed as many of the nodal shape functions for a given $\mathbf{x}$ will be zero. More precise expression should use a summation over only the set of local nodes as shown in Equation (1.29). However, the use of the complete summation can make the expressions and derivation much easier, and is often used in this book. We only need to bear in mind that many of the shape functions in the completed summation can be zero.

In the FEM, these nodal shape functions are created based on elements using mostly *polynomial basis functions*, and the linear independence is ensured by element based interpolation (nonoverlapping), element topology and properly controlled *coordinate mapping* [1].

In the S-PIM, no mapping (no Jacobian matrix) is needed, and the linear independence is ensured by 1) the use of proper basis functions that can be *polynomial* and *radial*; 2) proper sets of local support nodes selection for interpolation that may overlap and 3) proper means to ensure a nonsingular *moment matrix*.

The linear independence of shape functions is often overlooked. It is, however, essential to ensure the nodal values of a function being a qualified "representative" of the whole approximated function in a norm measure. To emphasize this, we note the following theorems.

Theorem 3.1a Outer product of linearly independent shape functions: SPD

For a discrete model with a set of nodes, if the set of all these nodal shape functions are linearly independent, the matrix of the outer product of the column-matrix of all these nodal shape functions is SPD.

Proof

Consider a discrete model for a problem occupying a domain Ω with a set of N_n nodes. A field function defined in Ω can be approximated via interpolation using Equation (1.29), using all the nodal values of the field function and the shape functions that are linearly independent:

$$u^h(\mathbf{x}) = \sum_{n=1}^{N_n} \phi_n(\mathbf{x})u_n = \mathbf{\Phi}(\mathbf{x})\mathbf{d} \tag{3.2}$$

where $\mathbf{\Phi}$ denotes the row-matrix with entries of all the N_n nodal shape functions, and $\mathbf{d}$ is a vector with all the N_n nodal values u_n of the field function.

We now examine the L^2 norm of the approximated field function:

$$\begin{aligned}
\left\| u^h \right\|_{\mathbf{L}^2(\Omega)}^2 &= \int_\Omega \left(u^h(\mathbf{x}) \right)^2 \mathrm{d}\Omega = \int_\Omega \left(\sum_{n=1}^{N_n} \phi_n(\mathbf{x})u_n \right)^2 \mathrm{d}\Omega \\
&= \int_\Omega \left(\mathbf{\Phi}(\mathbf{x})\mathbf{d} \right)^{\mathrm{T}} \left(\mathbf{\Phi}(\mathbf{x})\mathbf{d} \right) \mathrm{d}\Omega = \int_\Omega \mathbf{d}^{\mathrm{T}} \left(\mathbf{\Phi}^{\mathrm{T}}(\mathbf{x})\mathbf{\Phi}(\mathbf{x}) \right) \mathbf{d} \, \mathrm{d}\Omega \\
&= \mathbf{d}^{\mathrm{T}} \underbrace{\left(\int_\Omega \mathbf{\Phi}^{\mathrm{T}}(\mathbf{x})\mathbf{\Phi}(\mathbf{x}) \, \mathrm{d}\Omega \right)}_{\mathbf{M}_s} \mathbf{d} = \mathbf{d}^{\mathrm{T}}\mathbf{M}_s\mathbf{d}
\end{aligned} \tag{3.3}$$

where

$$\mathbf{M}_s = \int_\Omega \mathbf{\Phi}^{\mathrm{T}}(\mathbf{x})\mathbf{\Phi}(\mathbf{x}) \, \mathrm{d}\Omega \tag{3.4}$$

is a $N_n \times N_n$ square matrix that is the outer product of the column-matrix of the these shape functions $\mathbf{\Phi}^{\mathrm{T}}(\mathbf{x})$ (integrated over the problem domain). Clearly, $\mathbf{M}_s$ is symmetric, because

$$\begin{aligned}
\mathbf{M}_s^{\mathrm{T}} &= \left(\int_\Omega \mathbf{\Phi}^{\mathrm{T}}(\mathbf{x}) \, \mathbf{\Phi}(\mathbf{x})\mathrm{d}\Omega \right)^{\mathrm{T}} = \int_\Omega \left(\mathbf{\Phi}^{\mathrm{T}}(\mathbf{x})\mathbf{\Phi}(\mathbf{x}) \right)^{\mathrm{T}} \mathrm{d}\Omega \\
&= \int_\Omega \left(\mathbf{\Phi}^{\mathrm{T}}(\mathbf{x})\mathbf{\Phi}(\mathbf{x}) \right) \mathrm{d}\Omega = \mathbf{M}_s
\end{aligned} \tag{3.5}$$

Now, in Equation (3.2) if all u_n ($n=1,\cdots,N_n$) are zero or $\mathbf{d}(\in \mathbb{R}^{N_n})=\mathbf{0}$, $u^h(\mathbf{x})$ must be zero, because of the linearly independence of these nodal shape functions. This leads to $\left\| u^h \right\|_{\mathbf{L}^2(\Omega)}^2 = 0$. If any of these u_n is nonzero or $\forall \mathbf{d}(\in \mathbb{R}^{N_n}) \neq \mathbf{0}$, $u^h(\mathbf{x})$ must be nonzero, because of (again) the linearly independence of the nodal shape functions. This leads to strictly $\int_\Omega \left(u^h(\mathbf{x}) \right)^2 \mathrm{d}\Omega > 0$ and hence $\left\| u^h \right\|_{\mathbf{L}^2(\Omega)}^2 > 0$. This means that $\mathbf{d}^{\mathrm{T}}\mathbf{M}_s\mathbf{d} > 0$, $\forall \mathbf{d} \in \mathbb{R}^{N_n} \neq 0$ which implies that $\mathbf{M}_s$ is positive definite and thus SPD. $\qquad \square$

This SPD property of $\mathbf{M}_s$ ensures the SPD property of mass matrices of a numerical models for solids and structures of materials of positive densities. It is thus very important for dynamics problems when mass and damping matrix are involved (see, e.g., Chapter 5). We now state the following theorem.

Theorem 3.1b SPD property of mass and damping matrices

For a discrete model with a set of nodes, if the set of all these nodal shape functions are linearly independent, the discretized mass matrix for solids of positive density is SPD. In addition, the discretized damping matrix is also SPD for solids of positive damping coefficients.

Proof

For a discretized model of a solid, the discretized mass matrix is obtained using

$$\mathbf{M} = \rho \int_\Omega \mathbf{\Phi}^{\mathrm{T}}(\mathbf{x})\mathbf{\Phi}(\mathbf{x}) \, \mathrm{d}\Omega = \rho \mathbf{M}_s \tag{3.6}$$

where ρ is the density of the solid material. Since, $\mathbf{M}_s$ is SPD and ρ is positive, $\mathbf{M}$ is SPD. In addition, the discretized damping matrix is defined as

$$\mathbf{C} = c_{dp} \int_\Omega \mathbf{\Phi}^{\mathrm{T}}(\mathbf{x})\mathbf{\Phi}(\mathbf{x}) \, \mathrm{d}\Omega = c_{dp} \mathbf{M}_s \tag{3.7}$$

where c_{dp} is the damping coefficient of the solid material. Since, $\mathbf{M}_s$ is SPD and c_{dp} is positive, $\mathbf{C}$ is SPD. $\qquad\square$

Theorem 3.2 Norm equivalence: nodal representatives of a field function

For a discrete model with a set of nodes, if the set of all these nodal shape functions are linearly independent, the set of all the nodal values of a field function is equivalent in L^2 norm to the field function interpolated using the same set of nodal shape functions.

Proof

Because of Theorem 3.1, we know that $\mathbf{M}_s$ is SPD, thus eigenvalue decomposition can always be performed:

$$\mathbf{M}_s = \mathbf{V}^\mathrm{T}\boldsymbol{\Lambda}\mathbf{V} \tag{3.8}$$

where $\boldsymbol{\Lambda}$ is a diagonal matrix with entries of a set of strictly nonzero positive eigenvalues that can be arranged in the ascending order:

$$0 < \lambda_1 \cdots \le \lambda_i \le \lambda_{i+1} \cdots \le \lambda_{N_n} < \infty \tag{3.9}$$

Matrix $\mathbf{V}$ is a unitary matrix with the eigenvectors corresponding to these eigenvalues. Using Equation (3.3), we have

$$\begin{aligned}
\left\| u^h \right\|^2_{\mathbb{L}^2(\Omega)} &= \mathbf{d}^\mathrm{T}\mathbf{M}_s\mathbf{d} = \mathbf{d}^\mathrm{T}\left[\mathbf{V}^\mathrm{T}\boldsymbol{\Lambda}\mathbf{V}\right]\mathbf{d} = (\mathbf{V}\mathbf{d})^\mathrm{T}\boldsymbol{\Lambda}(\mathbf{V}\mathbf{d}) \\
&\ge (\mathbf{V}\mathbf{d})^\mathrm{T}\lambda_1(\mathbf{V}\mathbf{d}) = \lambda_1(\mathbf{V}\mathbf{d})^\mathrm{T}(\mathbf{V}\mathbf{d}) = \lambda_1\mathbf{d}^\mathrm{T}\underbrace{\left(\mathbf{V}^\mathrm{T}\mathbf{V}\right)}_{\because\,\text{unitary},\,\therefore=\mathbf{I}}\mathbf{d} \\
&= \lambda_1\mathbf{d}^\mathrm{T}\mathbf{d} = \lambda_1\left\|\mathbf{d}\right\|^2_{\mathbb{L}^2}
\end{aligned} \tag{3.10}$$

Similarly we have

$$\begin{aligned}
\left\| u^h \right\|^2_{\mathbb{L}^2(\Omega)} &= \mathbf{d}^\mathrm{T}\mathbf{M}_s\mathbf{d} = \mathbf{d}^\mathrm{T}\left[\mathbf{V}^\mathrm{T}\boldsymbol{\Lambda}\mathbf{V}\right]\mathbf{d} = (\mathbf{V}\mathbf{d})^\mathrm{T}\boldsymbol{\Lambda}(\mathbf{V}\mathbf{d}) \\
&\le (\mathbf{V}\mathbf{d})^\mathrm{T}\lambda_{N_n}(\mathbf{V}\mathbf{d}) = \lambda_{N_n}(\mathbf{V}\mathbf{d})^\mathrm{T}(\mathbf{V}\mathbf{d}) = \lambda_{N_n}\mathbf{d}^\mathrm{T}\underbrace{\left(\mathbf{V}^\mathrm{T}\mathbf{V}\right)}_{=\mathbf{I}}\mathbf{d} \\
&= \lambda_{N_n}\mathbf{d}^\mathrm{T}\mathbf{d} = \lambda_{N_n}\left\|\mathbf{d}\right\|^2_{\mathbb{L}^2}
\end{aligned} \tag{3.11}$$

Combining inequalities (3.10) and (3.11), we obtain

$$\lambda_1\left\|\mathbf{d}\right\|^2_{\mathbb{L}^2} \le \left\| u^h \right\|^2_{\mathbb{L}^2(\Omega)} \le \lambda_{N_n}\left\|\mathbf{d}\right\|^2_{\mathbb{L}^2} \tag{3.12}$$

This means that the L^2 norm of these nodal values of the discrete model is equivalent to that of the field function interpolated using nodal shape functions.

$\square$

Theorem 3.2 states that linear independence of the nodal shape functions is a very important necessary condition for nodal values as the unknowns of a *discrete model* (of finite dimension) to represent the approximated field function in the problem domain (of infinite dimension), without losing the bound property. In other words, if $\mathbf{d} \in \mathbb{R}^{N_n}$ is bounded, so is $u^h(\mathbf{x})$ over Ω, vice versa, as long as the set of nodal shape functions are linearly independent. Note that such norm equivalence does not guarantee the stability of the solution to a PDE. It ensures only the proper approximation of the field function itself. The approximation of the derivatives (strains) of the field function is entirely another story (to be told in Chapter 4). We have to device a proper formulation procedure to solve the PDE and to make sure that the nodal unknowns obtained is at least stable (bounded) so that solution of the field function is stable. Properly discretizing the field function is only an initial but a very important step.

Due to the essential importance stated in Theorem 3.1 and Theorem 3.2, we require all the nodal shape functions used in this book being linearly independent. Such a linear independence is achieved in FEM settings through the use of element-based interpolation (with proper mapping), and in meshfree settings through the use of node-based interpolation (such as PIM with a proper T-scheme for node selection).

3.1.2 Partitions of unity

The nodal shape functions must be a part of the unit at any point in the entire problem domain Ω:

$$\sum_{i=1}^{N_n} \phi_i(\mathbf{x}) = 1, \quad \forall \mathbf{x} \in \Omega \tag{3.13}$$

This condition is essential to ensure the correct representation of the *rigid body movements* of the solid.

3.1.3 Consistency

The solution of any numerical method must *converge*, meaning that the numerical solution obtained using the method must approach the exact solution when the characteristic nodal spacing h approaches zero (see, Remark 1.10). For such a convergence, the shape functions used have to satisfy a certain degree of consistency. The degree of consistency of shape functions is measured here by "the order of the polynomial functions can be exactly reproduced in the *local domains*". If the approximation is capable of producing a constant field function exactly, the approximation is then said to have zero-order consistency, or C^0 consistency. In general, if the approximation can produce a polynomial of up to kth order exactly, the approximation is said to have kth-order consistency, or C^k consistency. In this book, when we require C^k consistency, we imply also all the lower orders of consistencies from C^0 to C^{k-1}. In addition, we require shape functions have at least C^1 or the *linear consistency*.

The requirement on the order of consistency depends also on the formulation procedures. For example, in solving any PDE of symmetric operators of order $2k$ based on the *Galerkin weak form*, the minimum consistency requirement is C^k [1, 2]. Note the consistency is required only "locally" within the elements or cells. In the linear FEM, the consistence is C^1 inside the *linear elements*, but C^0 on the interfaces of the elements. In the S-PIM based on the *GS-Galerkin weak form*, however, we allow lower order of consistency inside the smoothing domains: it can be C^{k-1} (but minimum C^1) on the boundary of the smoothing domains, but only C^{-1} inside the smoothing domains.

3.1.4 Delta function property

The Delta function property of shape functions is defined in the form of Kronecker Delta:

$$\phi_i(\mathbf{x}_j) = \begin{cases} 1 & \text{when } i = j \\ 0 & \text{when } i \neq j \end{cases} \tag{3.14}$$

where $\mathbf{x}_j$ is the coordinates of the jth node. Nodal shape functions that satisfy the Delta function property allow easy treatments for *essential boundary conditions* at the boundary nodes, without resort to special measures. The PIM shape functions possess the Delta function property and hence can easily impose the point essential boundary conditions. However, shape functions created using other methods introduced in Section 3.8 may or may not have the Kronecker Delta property [3, 4]. Shape functions that do not have the Delta function property will not be used in this book.

3.1.5 Compatibility

The term "compatibility" is very important in the standard weak formulations, such as the standard FEM. It refers essentially the consistence of the field function on the element interface. For PDEs of order $2k$, the compatibility requirement is C^{k-1} on the element interfaces to ensure the stability and convergence of the solution based on the Hilbert space theory.

In the S-PIM, because the weakened weak (W^2) formulation is used, such a compatibility requirement of C^{k-1} is not necessarily required on the cell interfaces (depending on how the smoothing domain is constructed). The stability and convergence is ensured by the G space theory [5-8].

Consistency is usually much easier to achieve in a numerical method. The compatibility is, however, often a "headache" for *weak formulations* with local approximations. The removal of the compatibility requirement in the W^2 formulation provides much freedom in establishing numerical models in more general settings, as in the S-PIM. It allows also a lot more innovative and bold approaches in creating shape functions. We are seeing a flourished situation in developing effective ways to create shape functions in both elements [9] and meshfree settings [4].

3.1.6 Basis: an essential role of shape functions

In any numerical method, a field function needs to be approximated over the problem domain using a set of nodal values of the function together with a *basis*. Given a linear space S (see Chapter 2) of dimension N_n, a set of N_n members of functions $\phi_n \in S, n = 1, 2, \cdots, N_n$ is a basis for space S if and only if $\forall w \in S$, $\exists$ unique $\alpha_n \in \mathbb{R}$, such that

$$w = \sum_{n=1}^{N_n} \alpha_n \phi_n \tag{3.15}$$

Function ϕ_n in the basis is usually given in the form of nodal shape functions in the context of FEM and S-PIM, and hence the basis is often termed as *nodal basis*. Equation (3.15) implies that these nodal shape functions must be linearly independent (Section 3.1.1). The dimension of the space S created using ϕ_n ($n = 1, 2, \cdots, N_n$) is N_n or dim(S) = N_n. Hence we can also note $S \in \mathbb{R}^{N_n}$. For d-dimensional solids, the total dimension of the space will be $d \times N_n$, because each node carries d DOFs. In this case, we have $S \in \mathbb{R}^{dN_n}$. Such a discrete space is often called FEM space or S-PIM space.

3.1.7 Interpolant

We now define *interpolant*. Given a function $w \in S$ where S is a linear space, the interpolant $\mathcal{I}_h w$ creates a new function that lives in a finite subspace $S_h \subset S$ with N_n dimensions:

$$\mathcal{I}_h w(\mathbf{x}) = \sum_{n=1}^{N_n} w(\mathbf{x}_n)\phi_n(\mathbf{x}), \quad \mathcal{I}_h w \in S_h \subset S \tag{3.16}$$

that satisfies

$$\mathcal{I}_h w(\mathbf{x}_n) = w(\mathbf{x}_n), \quad n = 1, 2, \cdots, N_n \tag{3.17}$$

where $\mathbf{x}$ is any point inside the problem domain Ω, $\mathbf{x}_n$ is the coordinates of the nth node, and ϕ_n is the nodal shape function for the nth node. To satisfy Equation (3.17), we require the shape functions satisfy the Delta property Equation (3.14).

Note a general interpolation defined in Equation (1.29) is not necessarily qualified as an interpolant: because 1) the function u^h constructed by an interpolation may not be in a subspace of that for function u to be interpolated; or 2) the nodal shape functions may not have the Kronecker Delta property. Whether or not a type of shape functions can form an interpolant of a space has implication in the error estimation to relate the interpolation error to the solution error. The PIM shape functions can be used as an interplant in a proper G^1 space, and generally not in H^1 space, except for the linear PIM shape functions (see, Chapter 2).

3.2 PIM shape functions

The point interpolation method (PIM) is a general procedure to obtain an approximation of a function at a point by letting the interpolation function pass through the function values at each local support node. It can be categorized as a series representation method [4] that uses various *basis functions*. PIM using *polynomial basis functions* and local support nodes was originally attempted in [4, 10-12]. PIM using *radial basis functions* and local scattered nodes was suggested in [4, 12-14]. The basic procedure for creating polynomial PIM shape functions is given as follows.

3.2.1 Procedure of creating shape functions

Consider a field function $u(\mathbf{x})$ (e.g., a component of the displacements) defined in the problem domain Ω with a number of scattered field nodes. For a point of interest $\mathbf{x}_Q$ (that is usually a quadrature point), the field function $u(\mathbf{x})$ is approximated using the following series representation:

$$u^h(\mathbf{x}) = \sum_{i=1}^{n} p_i(\mathbf{x})a_i = \mathbf{p}^{\mathrm{T}}(\mathbf{x})\mathbf{a} \tag{3.18}$$

where $p_i(\mathbf{x})$ is the polynomial basis functions defined in the Cartesian coordinate space $\mathbf{x}^{\mathrm{T}} = \{x, y, z\}$, n is the number of the local *support nodes* (in set S_n) selected using a T-scheme (Section 1.6.3), and a_i is the coefficient for the monomial term $p_i(\mathbf{x})$ that forms the vector $\mathbf{a}$:

$$\mathbf{a} = \left\{ a_1, a_2, \cdots, a_n \right\}^{\mathrm{T}} \tag{3.19}$$

Note that a_i is (unknown) constant in the vicinity of point of interest $\mathbf{x}_Q$, and is updated only when the support nodes associated with $\mathbf{x}_Q$ are updated. Therefore, in any finite discretization of the problem domain with non-duplicated nodes, $u^h(\mathbf{x})$ is *consistent* in finite local domains where these support nodes do not change. The order of the (local) consistence depends on the polynomial basis functions used.

The monomial $p_i(\mathbf{x})$ in Equation (3.18) is, in general, chosen in a top-down approach from the Pascal triangle, so that the basis is complete to a desired order [4]. In S-PIM, we prefer to use lower order interpolations. For 1D problems, we use

$$\mathbf{p}^{\mathrm{T}}(\mathbf{x}) = \mathbf{p}^{\mathrm{T}}(x) = \left\{1, x, x^2, \cdots \right\}_{1 \times n} \tag{3.20}$$

For 2D problems, we may use

$$\mathbf{p}^{\mathrm{T}}(\mathbf{x}) = \mathbf{p}^{\mathrm{T}}(x, y) = \left\{1, x, y, xy, x^2, y^2, \cdots\right\}_{1 \times n} \tag{3.21}$$

For 3D problems we may use

$$\mathbf{p}^{\mathrm{T}}(\mathbf{x}) = \mathbf{p}^{\mathrm{T}}(x, y, z) = \left\{1, x, y, z, xyz, \cdots\right\}_{1 \times n} \tag{3.22}$$

The coefficients a_i in Equation (3.18) can be determined by enforcing Equation (3.18) to be satisfied at these n local support nodes:

$$u_i = \mathbf{P}^{\mathrm{T}}(\mathbf{x}_i)\mathbf{a} \quad i = 1, 2, \cdots, n \tag{3.23}$$

where u_i is the nodal function value of $u(\mathbf{x}_i)$. Equation (3.23) can be rewritten in the matrix form of

$$\mathbf{d}_s = \mathbf{P}_Q \mathbf{a} \tag{3.24}$$

where $\mathbf{d}_s$ is the vector collecting the values of the field function at all the n support nodes:

$$\mathbf{d}_s = \left\{u_1 \quad u_2 \quad \cdots \quad u_n\right\}^{\mathrm{T}} \tag{3.25}$$

and $\mathbf{P}_Q$ is called the *moment matrix* written in a concise form of

$$\mathbf{P}_Q = \begin{bmatrix} \mathbf{p}^{\mathrm{T}}(\mathbf{x}_1) \\ \mathbf{p}^{\mathrm{T}}(\mathbf{x}_2) \\ \vdots \\ \mathbf{p}^{\mathrm{T}}(\mathbf{x}_n) \end{bmatrix} \tag{3.26}$$

or in more detail as (for 2D cases):

$$\mathbf{P}_Q = \begin{bmatrix} 1 & x_1 & y_1 & x_1 y_1 & x_1^2 & y_1^2 & x_1^2 y_1 & x_1 y_1^2 & x_1^3 & \cdots \\ 1 & x_2 & y_2 & x_2 y_2 & x_2^2 & y_2^2 & x_2^2 y_2 & x_2 y_2^2 & x_2^3 & \cdots \\ \vdots & \vdots & \vdots & \vdots & & \vdots & \vdots & \vdots & \vdots & \vdots \\ 1 & x_n & y_n & x_n y_n & x_n^2 & y_n^2 & x_n^2 y_n & x_n y_n^2 & x_n^3 & \cdots \end{bmatrix}_{n \times n} \tag{3.27}$$

It is clear that the moment matrix $\mathbf{P}_Q$ is asymmetric (suggesting possible instability). Assuming that the inverse of the moment matrix exists, the coefficients $\mathbf{a}$ can then be obtained from Equation (3.24)

$$\mathbf{a} = \mathbf{P}_Q^{-1} \mathbf{d}_s \tag{3.28}$$

By substituting Equation (3.28) into Equation (3.18), we obtain

$$u^h(\mathbf{x}) = \sum_{i=1}^{n} \phi_i(\mathbf{x}) u_i \tag{3.29}$$

or in the matrix form of

$$u^h(\mathbf{x}) = \mathbf{\Phi}_s(\mathbf{x}) \mathbf{d}_s \tag{3.30}$$

where $\mathbf{\Phi}_s(\mathbf{x})$ is the row-matrix of PIM shape functions:

$$\mathbf{\Phi}_s(\mathbf{x}) = \mathbf{p}^{\mathrm{T}}(\mathbf{x}) \mathbf{P}_Q^{-1} = \begin{bmatrix} \phi_1(\mathbf{x}) & \phi_2(\mathbf{x}) & \cdots & \phi_n(\mathbf{x}) \end{bmatrix} \tag{3.31}$$

Note that it is possible that the moment matrix $\mathbf{P}_Q$ is singular and the PIM procedure breaks down, which will be discussed in detail in Section 3.2.3. For now, we assume that the moment matrix is invertible.

The derivatives of the PIM shape functions can be obtained very easily when needed (W^2 models like S-PIM do not need, but weak-form models like FEM do), as all the functions involved are polynomials. The lth derivatives of the shape functions are simply given by

$$\mathbf{\Phi}_s^{(l)}(\mathbf{x}) = [\mathbf{p}^{(l)}(\mathbf{x})]^{\mathrm{T}} \mathbf{P}_Q^{-1} \tag{3.32}$$

Remark 3.1 Linear FEM shape functions: special case of PIM

We note that the above produce works also for creating linear FEM shape function, as long as the node selection is based on the triangular/tetrahedral elements. Therefore, it is clear that the linear FEM shape function is a special case of the PIM shape functions.

3.2.2 Properties of PIM shape functions

As long as the moment matrix is not singular, the PIM shape functions $\phi_i(\mathbf{x})$ can be uniquely created, and they possess the following properties [4, 12]:

Remark 3.2 PIM shape functions: linear independence

First, we argue that the PIM shape functions created for the local support nodes are linearly independent. This is because the polynomial basis functions are chosen linearly independent and the inverse of the moment matrix ($\mathbf{P}_Q^{-1}$) is assumed to exist. The existence of $\mathbf{P}_Q^{-1}$ implies that the shape functions are equivalent to the basis functions (monomials) in function space, as shown in Equation (3.31), and hence are linearly independent. Second, the T-schemes (Section 1.6.3) will be used to select the support nodes ensuring the independence of the sets of local support nodes, and hence the sets of the locally created shape functions. Therefore, the nodal PIM shape functions for all the nodes in the problem domain are linearly independent.

Remark 3.3 PIM shape functions: partitions of unity

If the constant term is included in the basis, which is always true in the PIM method, the shape functions are the partitions of unity Equation (3.13).

This property can be proven easily by the *reproduction* feature of PIM shape functions. Suppose function $u(\mathbf{x})=c$ defined in the entire problem domain, where c is an arbitrary real constant, we then have for all the nodal values defined in Equation (3.25) as:

$$\mathbf{d}_s = c\{1, \ 1, \cdots, \ 1\}^{\mathrm{T}} \tag{3.33}$$

Substituting $\mathbf{d}_s$ into Equation (3.30), we have at any point $\mathbf{x}$:

$$u(\mathbf{x}) = c = \boldsymbol{\Phi}_s(\mathbf{x})\mathbf{d}_s = c\sum_{i=1}^{n}\phi_i(\mathbf{x}) \tag{3.34}$$

which gives Equation (3.13). $\qquad\qquad\qquad\qquad\qquad\qquad\qquad\qquad\square$

It is clear now that the property of partitions of unity of PIM shape functions ensures a constant field or rigid body movement to be reproduced.

Remark 3.4 PIM shape functions: consistency

The consistency of PIM shape functions depends on the complete order of the monomials $p_i(\mathbf{x})$ used in Equation (3.18), and thus also on the number of support nodes. If the complete order of the monomials is n, the shape functions will possess C^n consistency locally, as long as the moment matrix is invertible.

To demonstrate, we consider a function that has the form of

$$f(\mathbf{x}) = \sum_{j=1}^{k} p_j(\mathbf{x})\alpha_j, \quad k \le n \tag{3.35}$$

where $p_j(\mathbf{x})$ are the jth monomial term included in Equation (3.18) and α_j is the corresponding coefficient. Such a function can always be written using Equation (3.18) with n basis terms including those in Equation (3.35):

$$f(\mathbf{x}) = \sum_{j=1}^{n} p_j(\mathbf{x})\alpha_j = \mathbf{p}^{\mathrm{T}}(\mathbf{x})\boldsymbol{\alpha} \tag{3.36}$$

where $\boldsymbol{\alpha}$ is the vector collecting all these n coefficients

$$\boldsymbol{\alpha}^{\mathrm{T}} = [\alpha_1, \alpha_2, \cdots, \alpha_k, 0, \cdots, 0] \tag{3.37}$$

Using these n nodes in a local support domain, the vector of nodal function values ($\mathbf{d}_s$) can be obtained as

$$\mathbf{d}_s = \begin{Bmatrix} f_1 \\ f_2 \\ \vdots \\ f_k \\ f_{k+1} \\ \vdots \\ f_n \end{Bmatrix} = \underbrace{\begin{bmatrix} p_1(\mathbf{x}_1) & \cdots & p_k(\mathbf{x}_1) & \cdots & p_n(\mathbf{x}_1) \\ p_1(\mathbf{x}_2) & \cdots & p_k(\mathbf{x}_2) & \cdots & p_n(\mathbf{x}_2) \\ \vdots & \cdots & \vdots & \cdots & \vdots \\ p_1(\mathbf{x}_k) & \cdots & p_k(\mathbf{x}_k) & \cdots & p_n(\mathbf{x}_k) \\ p_1(\mathbf{x}_{k+1}) & \cdots & p_k(\mathbf{x}_{k+1}) & \cdots & p_n(\mathbf{x}_{k+1}) \\ \vdots & \cdots & \vdots & \cdots & \vdots \\ p_1(\mathbf{x}_n) & \cdots & p_k(\mathbf{x}_n) & \cdots & p_n(\mathbf{x}_n) \end{bmatrix}}_{\mathbf{P}_Q} \underbrace{\begin{Bmatrix} \alpha_1 \\ \alpha_2 \\ \vdots \\ \alpha_k \\ 0 \\ \vdots \\ 0 \end{Bmatrix}}_{\alpha} = \mathbf{P}_Q \alpha \qquad (3.38)$$

Substituting Equation (3.38) into Equation (3.30), we have the approximation as

$$u^h(\mathbf{x}) = \mathbf{p}^\mathrm{T}(\mathbf{x})\mathbf{P}_Q^{-1}\mathbf{d}_s = \mathbf{p}^\mathrm{T}(\mathbf{x})\mathbf{P}_Q^{-1}\mathbf{P}_Q\alpha = \mathbf{p}^\mathrm{T}(\mathbf{x})\alpha = \sum_{j=1}^{k} p_j(\mathbf{x})\alpha_j \qquad (3.39)$$

which is exactly Equation (3.35). This shows that PIM shape functions is capable of producing exactly any field function given by Equation (3.35), as long as the given function is included in the basis functions used to create these PIM shape functions. In the working in Equation (3.39), we also require that the moment matrix $\mathbf{P}_Q$ is invertible.

Remark 3.5 PIM shape functions: reproducibility

The proof of the consistency of PIM implies another important feature of PIM: any function that appears in the basis can be reproduced exactly. This reproducibility feature of PIM can be used to improve the solution accuracy in S-PIM models by including terms in the basis functions that are good approximations of the solution of the problem. This property can also be useful for creating fields of special features, such as the singular stress field at the crack tip [9, 15-17]. This argument leads naturally the following remark.

Remark 3.6 PIM shape functions: linear reproducibility

If the first order monomial is included in the basis, the shape functions possess the linear reproducing property.

$$\sum_{i}^{n} \phi_i(\mathbf{x})x_i = \mathbf{x} \qquad (3.40)$$

Remark 3.6 allows the S-PIM the capability to pass the standard patch test. The linear reproducibility ensures the (minimum) linear consistency and hence is required for all the S-PIM models discussed in this book.

Remark 3.7 PIM shape functions: Delta function property

The PIM shape functions possess the Kronecker Delta function property Equation (3.14).

This property can be proved easily. Because PIM shape functions $\phi_i(\mathbf{x})$ created are linearly independent, any vector of length n should be uniquely produced by linear combination of n PIM shape functions. Let

$$\mathbf{d}_s = \{0, 0, \cdots, u_i, \cdots, 0\}^{\mathrm{T}} \tag{3.41}$$

Substitute the above equation into Equation (3.30), we have at $\mathbf{x}=\mathbf{x}_j$

$$u^h\left(\mathbf{x}_j\right) = \mathbf{\Phi}_s\left(\mathbf{x}_j\right)\mathbf{d}_s = \phi_i(\mathbf{x}_j)u_i \tag{3.42}$$

When $i=j$, we obtain

$$u_i = \phi_i(\mathbf{x}_i)u_i \tag{3.43}$$

which gives

$$\phi_i(\mathbf{x}_i) = 1 \tag{3.44}$$

When $i \neq j$, we have

$$u_j = 0 = \phi_i(\mathbf{x}_j)u_i \tag{3.45}$$

which leads to

$$\phi_i(\mathbf{x}_j) = 0 \tag{3.46}$$

Equations (3.44) and (3.46) give (3.14). The Kronecker Delta function property is important for easy handling essential boundary conditions.

Remark 3.8 PIM shape functions: no mapping and simplicity

Obviously, the (usual) procedure of creating PIM shape functions given in Section 3.2.1 is very straightforward. No mapping (Jacobian matrix) is needed. It is also known as "direct interpolation".

Remark 3.9 PIM shape functions: incompatibility

The PIM shape functions created in Section 3.2.1 is performed for the point of interest ($\mathbf{x}_Q$) using local support nodes. When $\mathbf{x}_Q$ changes, the local support nodes change accordingly. The created nodal shape function for a node will be consisted of different pieces that may not be connected at locations where the support nodes are updated. Therefore, functions created using these nodal PIM shape functions can be discontinuous. They will not be, in general, in an H^1 space, but a G^1 space (see, Example 3.2.3). Only when the PIM is applied in linear FEM settings with triangular/tetrahedron elements and only the nodes in the element are used as local support nodes, the shape functions are continuous. This means that when the cell-based T3-scheme is used for 2D problems, the linear PIM shape functions obtained are compatible. In fact it is the same as the FEM using linear triangular elements (see, Remark 3.1). For general polygonal elements, linear and compatible shape functions can also be created with ease as shown in [9]. A very detailed discussion on the incompatibility of PIM shape functions can be found in [4].

Remark 3.10 PIM shape functions: compact support

The shape functions are of compact support as long as they are created using a set of local support nodes. This property leads to sparse discretized system matrices making the numerical method more efficient.

Remark 3.11 PIM shape functions: interpolant in G^1 space

PIM shape functions can be used to construct an interpolant for functions in G^1 spaces, but may not be in an H^1 space. This is because the interpolant $\mathcal{I}_h w$ for a w in an H^1 space may not still be in any H^1 subspace due to the incompatibility (see, Section 3.1.7). For linear interpolations using T3-scheme, PIM shape functions are compatible and can always be used to create an interpolant in an H^1 space. For arbitrary polygonal mesh, we can use simple point interpolation techniques given in [9, 18, 19].

The PIM shape function is ideal for meshfree methods in many ways, as it possesses the above-mentioned excellent properties.

Example 3.2.1 Linear 1D PIM shape functions

A simple 1D example is used to demonstrate the properties of PIM shape functions. Consider a 1D space defined in domain [0, 4]. Five evenly distributed nodes are placed in the 1D domain, and nodal PIM shape functions are to be constructed.

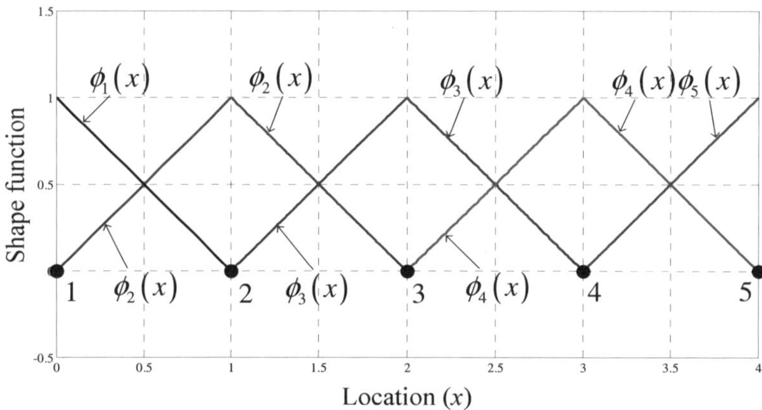

FIGURE 3.1 Linear nodal PIM shape functions for the five nodes evenly distributed in a 1D domain of [0, 4]. Note that the linear PIM shape function possesses the Kronecker Delta function property and is continuous over the problem domain.

Linear nodal PIM shape functions are created and plotted in FIGURE 3.1. It is clearly seen that when linear interpolation is used everywhere in the domain, the PIM shape functions are exactly the same as those in the standard linear FEM [1], and are continuous over the entire problem domain.

It is clear that any of these linear nodal PIM shape function is in a $\mathbb{G}_h^1(\Omega)$ space when node-based smoothing domains are use, as shown in Example 2.5.1. It is quite trivial to show that this is also true when cell-based smoothing domains are used.

We further argue that any function constructed using the linear nodal PIM shape functions will also be in a $\mathbb{G}_h^1(\Omega)$ space, simply because a $\mathbb{G}_h^1(\Omega)$ space is a linear space: a linear combination of functions from a space is still in the space.

Example 3.2.2 Mixed linear-quadratic PIM shape functions

Higher order PIM shape functions can be created using the same set of evenly distributed nodes in domain [0, 4], as in Example 3.2.1. The order of the shape functions depends on the number of the nodes used in the interpolation. In this example, we study a case of mixed linear-quadratic interpolation. The order of interpolation, support nodes selection, and polynomial basis functions used for creating this set of 1D PIM shape functions are listed in TABLE 3.1. For the point of interest ($\mathbf{x}_Q$) located in the two boundary node-based domains, i.e. in [0.0, 0.5] and [3.5, 4.0], we select two support nodes and use linear polynomial basis functions for creating the PIM shape functions; for the ($\mathbf{x}_Q$) located in the interior node-based domains, i.e. [0.5, 1.5], [1.5, 2.5] and [2.5, 3.5], three support nodes and quadratic polynomial basis functions are used.

TABLE 3.1 Support nodes selection and polynomial basis functions used for mixed linear-quadratic interpolation at different points of interest

Location of point of interest $\mathbf{x}_Q$	Local support nodes	Polynomial basis functions
[0.0, 0.5]	{1, 2}	Linear polynomial
[0.5, 1.5]	{1, 2, 3}	Quadratic polynomial
[1.5, 2.5]	{2, 3, 4}	Quadratic polynomial
[2.5, 3.5]	{3, 4, 5}	Quadratic polynomial
[3.5, 4.0]	{4, 5}	Linear polynomial

Following the procedure given in Section 3.2.1, nodal PIM shape functions can be easily created. FIGURE 3.2a shows nodal PIM shape functions for the five field nodes. It is clearly seen that the nodal PIM shape function possesses the Kronecker Delta function property. That is, $\phi_i(\mathbf{x} = \mathbf{x}_i) = 1.0$ and $\phi_i(\mathbf{x} \neq \mathbf{x}_i) = 0.0$ for $i = 1, \cdots, 5$. We note that the PIM shape functions are discontinuous at the mid points of the cells. For example, the nodal PIM shape functions for node 2 shown in FIGURE 3.2b have three discontinuity (jump) points at $x=0.5$, $x=1.5$ and $x=2.5$ in [0, 4.0], where the supports nodes are updated in the interpolation process from {1, 2} to {1, 2, 3}, {1, 2, 3} to {2, 3, 4}, and {2, 3, 4} to {3, 4, 5}, respectively.

Due to the discontinuity of the nodal PIM shape functions, the field functions constructed using these shape functions will also be discontinuous. Because the discontinuities are at the mid-cell points, this set of nodal PIM shape functions can only be in a $\mathbb{G}_h^1(\Omega)$ space, if cell-based smoothing domains are used, due to the "no-sharing rule" discussed in Section 2.3.1.

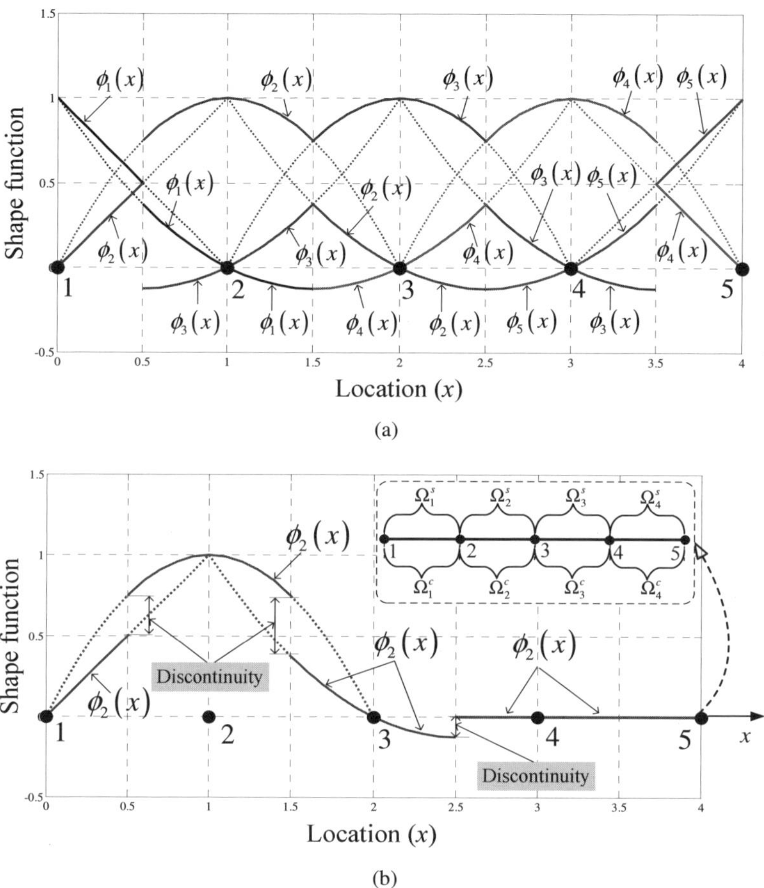

FIGURE 3.2 Nodal PIM shape functions for the five nodes evenly distributed in a 1D domain of [0, 4], created by the mixed linear-quadratic interpolation. Note that the PIM shape function possesses the Kronecker Delta function property and is discontinuous at the mid-cell points. (a) Distribution of all the five nodal PIM shape functions; (b) the nodal PIM shape function for node 2.

Using now a set of four cell-based smoothing domains $\Omega_i^s = \Omega_i^c$ (i=1, 2, 3 and 4) as illustrated in the top right corner of FIGURE 3.2b, we can now easily prove $\phi_2(\mathbf{x})$ shown in this figure is in a $\mathbb{G}_h^1(\Omega)$ space, despite the presence of these discontinuities.

Proof

First, we note that $\phi_2(\mathbf{x}) \in \mathbb{L}^2(\Omega)$, because the nodal shape function created using PIM are finite, as proven in Example 2.2.3.

Second, follow the definition given in Equation (2.86), we have

$$\begin{aligned}
\mathbf{d} &= \{u_1 \quad u_2 \quad u_3 \quad u_4 \quad u_5\} \\
&= \{0 \quad 1.0 \quad 0 \quad 0 \quad 0\}
\end{aligned} \tag{3.47}$$

Next, using Equation (2.76) for the cell-based smoothing domains, we obtain

$$\begin{aligned}
\sum_{k=1}^{5}\left(\int_{\Gamma_k^s} \phi_2(\mathbf{x})(s)n_i \mathrm{d}s\right)^2 &= \left(\phi_2(1)-\phi_2(0)\right)^2 + \left(\phi_2(2)-\phi_2(1)\right)^2 \\
&\quad + \left(\phi_2(3)-\phi_2(2)\right)^2 + \left(\phi_2(4)-\phi_2(3)\right)^2 \\
&= \left(u_2-u_1\right)^2 + \left(u_3-u_2\right)^2 \\
&\quad + \left(u_4-u_3\right)^2 + \left(u_5-u_4\right)^2 \\
&= \left(1.0-0\right)^2 + \left(0-0\right)^2 + \left(0-0\right)^2 + \left(0-0\right)^2 \\
&= 1.0 > 0
\end{aligned} \tag{3.48}$$

Therefore, the discontinuous nodal shape function $\phi_2(\mathbf{x})$ is in a $\mathbb{G}_h^1(\Omega)$ space. The same proof procedure works for any other nodal shape function shown in FIGURE 3.2. Note that any of these nodal shape functions is not in any $\mathbb{H}_h^1(\Omega)$ space, as proven in Example 2.2.5.

We further argue that any function constructed using the mixed linear-quadratic nodal PIM shape functions will also be in a $\mathbb{G}_h^1(\Omega)$ space, simply because a $\mathbb{G}_h^1(\Omega)$ space is a linear space. The following example confirms this argument.

Example 3.2.3 A discontinuous function constructed using the mixed linear-quadratic PIM shape functions

We now examine a function constructed using the discontinuous PIM shape functions studied in Example 3.2.2. The problem domain $\Omega = [1,4]$ is discretized using four cells with five evenly distributed nodes. A sinusoidal function $v(x)=\sin[(x-0.2)\pi/2]$ is numerically approximated using the mixed linear-quadratic PIM shape functions. FIGURE 3.3 plots the exact sinusoidal function and the approximated one. We can find that the PIM approximation passes through the field function values at the five field nodes due to the Delta function property of the PIM shape functions. Because the mixed linear-quadratic PIM shape functions are discontinuous at the mid points of the cells, the approximated functions are also discontinuous at these locations, as shown in FIGURE 3.3.

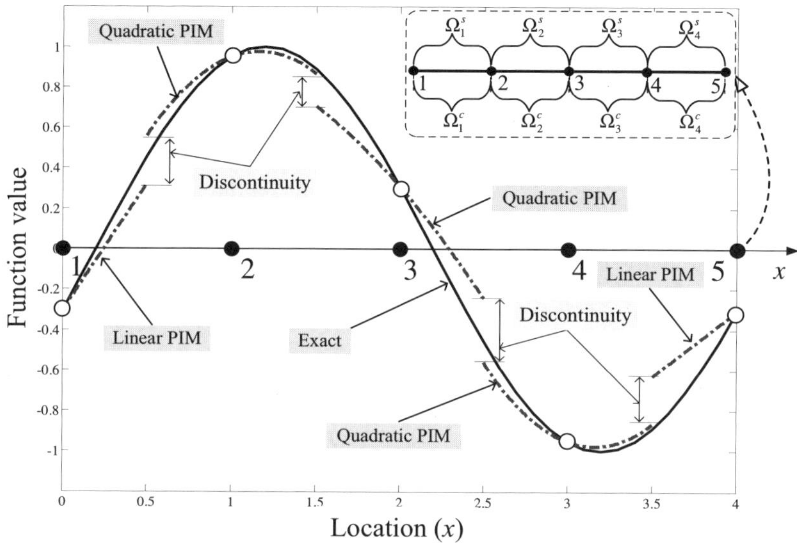

FIGURE 3.3 Sinusoidal function $v(x)=\sin[(x-0.2)\pi/2]$ is approximated using a mixed linear-quadratic PIM shape functions with five evenly distributed nodes in domain [0,4].

According to the "no-sharing rule", a set of four cell-based smoothing domains $\Omega_i^s = \Omega_i^c$ (i=1, 2, 3 and 4), is again used as illustrated in the top right corner of FIGURE 3.3, and we can show that this approximated (discontinuous) function is in a $\mathbb{G}_h^1(\Omega)$ space. First, we have for the sinusoidal function case:

$$\mathbf{d} = \{u_1 \quad u_2 \quad u_3 \quad u_4 \quad u_5\}$$
$$= \{-0.309 \quad 0.951 \quad 0.309 \quad -0.951 \quad -0.309\} \tag{3.49}$$

The approximated function $v^h(x)$ can be expressed by

$$v^h(x) = \begin{cases} \displaystyle\sum_{n=1}^{2} \phi_n(x)u_n, & \forall x \in [0.0, 0.5] \\[2mm] \displaystyle\sum_{n=1}^{3} \phi_n(x)u_n, & \forall x \in (0.5, 1.5] \\[2mm] \displaystyle\sum_{n=2}^{4} \phi_n(x)u_n, & \forall x \in (1.5, 2.5] \\[2mm] \displaystyle\sum_{n=3}^{5} \phi_n(x)u_n, & \forall x \in (2.5, 3.5] \\[2mm] \displaystyle\sum_{n=4}^{5} \phi_n(x)u_n, & \forall x \in (3.5, 4.0] \end{cases} = \sum_{n=1}^{5} \phi_n(x)u_n, \quad \forall x \in [0,4] \tag{3.50}$$

It is clear that $v^h(x) \in \mathbb{L}^2(\Omega)$, because all the nodal shape functions $\phi_n(x)$ created using PIM are finite. Hence the approximated function $v^h(x)$ is bounded. In addition, using Equation (2.76) for the cell-based smoothing domains, we have

$$\begin{aligned} \sum_{k=1}^{5} \left(\int_{\Gamma_k^s} v^h(s)n_i\,ds \right)^2 &= \left(v^h(1) - v^h(0) \right)^2 + \left(v^h(2) - v^h(1) \right)^2 \\ &\quad + \left(v^h(3) - v^h(2) \right)^2 + \left(v^h(4) - v^h(3) \right)^2 \\ &= \left(u_2 - u_1 \right)^2 + \left(u_3 - u_2 \right)^2 + \left(u_4 - u_3 \right)^2 \\ &\quad + \left(u_5 - u_4 \right)^2 \end{aligned} \tag{3.51}$$

Using Equation (3.49), we obtain for the sinusoidal function:

$$\sum_{k=1}^{5} \left(\int_{\Gamma_k^s} v^h(s)n_i\,ds \right)^2 = 3.999528 > 0 \tag{3.52}$$

Therefore, the discontinuous function $v^h(x)$ (dashed line in FIGURE 3.3) is in a $\mathbb{G}_h^1(\Omega)$ space. We note that the function shown in FIGURE 3.3 is clearly not in any $\mathbb{H}_h^1(\Omega)$ space, as discussed in Example 2.2.5.

Based on the discussion in Example 3.2.3, we now note the following remark.

Remark 3.12 Linear-quadratic approximation of a function

Any general function $v^h(x)$ defined in Equation (3.50) with the mixed linear-quadratic PIM shape function and arbitrary d_i is in a $\mathbb{G}_h^1(\Omega)$ space, when cell-based smoothing domains are used.

Proof

Let's consider first a constant function, for which we have $u_1 = u_2 = u_3 = u_4 = u_5 = c$, because our PIM shape functions have the partition of unity property (see, Remark 3.3), we shall have $v^h(x) = c$. In this case, we clearly have

$$\sum_{k=1}^{5} \left(\int_{\Gamma_k^s} v^h(s) n_i \, ds \right)^2 = \left(u_2 - u_1 \right)^2 + \left(u_3 - u_2 \right)^2$$
$$+ \left(u_4 - u_3 \right)^2 + \left(u_5 - u_4 \right)^2$$
$$= 0 \tag{3.53}$$

This fact shows that an arbitrary constant function c is in a $\mathbb{G}_h^1(\Omega)$ space, and hence we can represent the rigid body motions.

Now, for any $v^h(x) \neq c$, we have at least one $(u_{i+1} - u_i) \neq 0$ ($i = 1, 2, 3, 4$) is nonzero. From Equation (3.51), we must have strictly $\sum_{k=1}^{5} \left(\int_{\Gamma_k^s} v^h(s) n_i \, ds \right)^2 > 0$.

In other words, if $\sum_{k=1}^{5} \left(\int_{\Gamma_k^s} v^h(s) n_i \, ds \right)^2 = 0$, $v^h(x)$ must be a constant function. If $v^h(x)$ is not a constant function, we shall always have $\sum_{k=1}^{5} \left(\int_{\Gamma_k^s} v^h(s) n_i \, ds \right)^2 > 0$. Hence, any general function defined in Equation (3.50) with the mixed linear-quadratic PIM shape function and arbitrary nodal values is in a $\mathbb{G}_h^1(\Omega)$ space.

We also note that if one would like to create a node-based smoothed model for 1D problems, a set of purely linear or mixed linear-cubic (using either 2 or 4 nodes) PIM shape functions should be used. A proof on purely linear case for node-based models was given in [6].

Example 3.2.2 and Example 3.2.3 show that the basis functions used in creating PIM shape functions for a support domain can be different from those for other support domains. This important property provides a lot of flexibilities in the local enrichments, without worrying about the discontinuities as long as a proper set of smoothing domains can be constructed (see Chapter 12).

We now summarize the above discussions to the following general theorem.

Theorem 3.3 PIM shape functions: in a $\mathbf{G^1}$ space

PIM Shape functions and field functions constructed using the PIM shape functions can always be in a $\mathbb{G}_h^1(\Omega)$ space, as long as a proper set of smoothing domains can be created to ensure the positivity condition.

Theorem 3.3 is true for all the PIM shape functions created using various methods presented in this chapter. With this theorem in place, we can have a piece of mind concentrating on how creating PIM shape function successful with necessary properties for general meshfree settings.

3.2.3 Methods to avoid singular moment matrix

As presented previously, PIM shape functions possess many excellent properties that are very useful for S-PIM and other methods based on SC-Galerkin weak forms, local weak forms, and strong forms. However, the process of creating PIM shape functions for 2D and 3D problems can break down, due to the singularity of the moment matrix $\mathbf{P}_Q$. The following gives an example.

Example 3.2.4 Interpolation using four nodes at the vertices of a diamond

FIGURE 3.4 shows a typical example of four nodes in the support domain of a point of interest $\mathbf{x}_Q$. When the following bilinear polynomial basis

$$\mathbf{P}^{\mathrm{T}}(\mathbf{x}) = \{1,\, x,\, y,\, xy\} \tag{3.54}$$

is used, the moment matrix becomes

$$\mathbf{P}_Q = \begin{bmatrix} 1 & 0 & -1 & 0 \\ 1 & 1 & 0 & 0 \\ 1 & 0 & 1 & 0 \\ 1 & -1 & 0 & 0 \end{bmatrix} \tag{3.55}$$

It is obvious that the moment matrix will be singular. It has only a rank of 3.

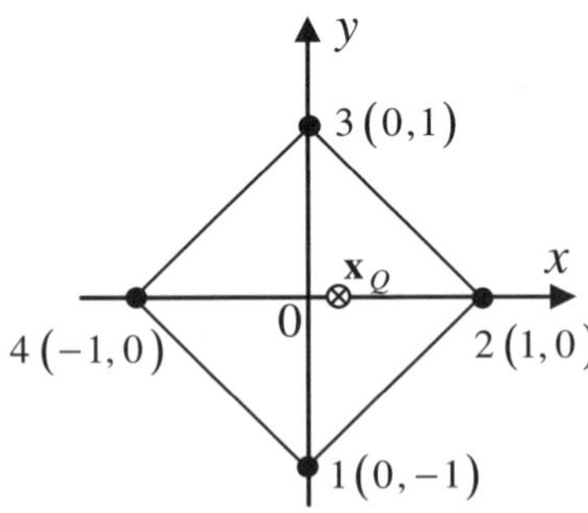

FIGURE 3.4 Four support nodes for creating shape functions for a point of interest $\mathbf{x}_Q$ inside the diamond.

In general, the existence of $\mathbf{P}_Q^{-1}$ depends on the basis used, node distribution as well as on the coordinate system. Using polynomial basis functions is known to be difficult to guarantee the existence of $\mathbf{P}_Q^{-1}$ for a set of arbitrarily scattered nodes. However, the excellent properties of PIM shape functions warrant the efforts to overcome the singular moment matrix problem. A number of methods for handling the singular moment matrix have been attempted for locally scattered nodes.

1) Performing rotational coordinate transformation to produce an invertible moment matrix $\mathbf{P}_Q$. This method makes the use of the fact that the rank of $\mathbf{P}_Q$ depends also on the coordinate system. It is not a full-proof, but simple and quite effective, when a small number of nodes are used. Because S-PIM prefers to use small number of nodes, this method suits S-PIM well. The details are given in Section 3.4.

2) The use of radial basis functions (RBFs) in creating PIM shape functions is a robust method that guarantees the existence of $\mathbf{P}_Q^{-1}$ for practically randomly distributed nodes, if certain guidelines for choosing shape parameters of the RBFs are followed. The drawback is that it uses more nodes and hence more expensive. If the efficiency is not a big concern, and the nodes are extremely irregular, this method is a good choice. The details are given in Section 3.3, where RBFs augmented with polynomial terms are used.

3) Triangular/tetrahedral-mesh-based node selection scheme (T-scheme). The T-scheme uses a triangular type background mesh to select nodes (see, Section 1.6.3). It was used for node selection in the MFree2D$^{©}$ for both element free Galerkin (EFG) and S-PIMs processors. For all the S-PIM models discussed in this book, our default choice of node selection is the T-schemes, unless specified otherwise.

4) The matrix triangularization method (MTA) is very efficient and works well for many situations, when removal of nodes in the local support domain is permitted. It can be a good alternative. Details can be found in [4, 12, 23].

5) The simplest method proposed to obtain a nonsingular moment matrix is to move or shift the nodes in the support domain by a small distance randomly in terms of both direction and the amount of shift. The method is simple and effective for some problems. However, there is still a chance that the moment matrix will be singular, which may sometimes lead to a badly

conditioned $\mathbf{P}_Q$. In addition, there are cases in which we are not allowed to move the nodes.

In this book, we will use the first three methods in our S-PIM models. The above analysis has also shown that PIM process always works in the following situations.

Remark 3.13 PIM shape functions: always work situations

- For 1D domains, the PIM process always works, as long as there is no duplicated nodes.

- For 2D domains, the PIM process always works when $n=3$, as long as these 3 nodes are not inline.

- For 3D domains, the PIM process always works when $n=4$, as long as these 4 nodes are not on-plane.

Remark 3.14 Property of PIM shape functions: a summary

Successfully created PIM shape functions are, in general, incompatible, consistent to the order of polynomials included in the formulation and possess the Kronecker Delta function property.

3.3 RPIM shape functions

The advantage of using a polynomial basis is its simplicity and high accuracy [4]. The major drawback of polynomial PIM is that the moment matrix $\mathbf{P}_Q$ may be singular and the process can break down. To overcome the singular moment matrix problem, radial basis functions (RBFs) are introduced in the PIM formulation for creating shape functions using local nodes [13-14, 20-22]. PIM using radial basis function (RBF) is termed radial PIM (RPIM).

3.3.1 Rationale for using RBFs and polynomials

RBFs are often used in the mathematics community for solving PDEs using global nodes scattered in the problem domain [24-31]. The purpose of using RBFs here in creating PIM shape functions is mainly to make use of the nonsingular moment matrix associated with RBFs, so that PIM process can always follow through. Because local nodes are used for field function approximation, we cannot rely on RBFs to improve the accuracy of the approximation. In addition, because RBFs do not have linear consistence, RPIM with pure radial basis functions has a problem passing the standard patch tests, meaning that it fails to reconstruct the linear (polynomial) field [12]. This implies also that the approximation using pure RBFs may not converge by means of nodal refinement. Therefore, our strategy is to use RBFs for stability, and to use polynomial basis for consistence (and hence convergence) for our PIM shape functions. Our study has found that, in general, adding polynomials can also improve the accuracy of the results [4]. Another additional bonus of using polynomial with RBFs is that we have much more freedom in choosing shape parameters used in the RBFs, because the sensitivity of the shape parameters on the solution accuracy is reduced.

Because of these reasons, the RPIM shape functions used in this book is always with polynomials, unless specified otherwise.

3.3.2 Formulation of polynomial augmented RPIM

Using the n local nodes, RPIM augmented with polynomial basis functions approximates the field function in the form of

$$u^h(\mathbf{x}) = \sum_{i=1}^{n} R_i(\mathbf{x})a_i + \sum_{j=1}^{m} p_i(\mathbf{x})b_j = \mathbf{R}^{\mathrm{T}}(\mathbf{x})\mathbf{a} + \mathbf{p}^{\mathrm{T}}(\mathbf{x})\mathbf{b} \tag{3.56}$$

where a_i is the coefficient for the radial basis $R_i(\mathbf{x})$, and b_j is the coefficient for the polynomial basis $p_j(\mathbf{x})$ as in the polynomial PIM (see, Section 3.2). The number of radial basis functions n is determined by the number of the support nodes, and the number of polynomial basis m can be chosen based on the reproduction requirement. The number of polynomial bases is sufficiently smaller than the terms of radial basis ($m<n$) for stability with respect to the irregularity of node distribution. If we need only pure RPIM shape functions without polynomial augmentation, we simply chose $m=0$. For 2D problems, if we want to make the S-PIM to have linear consistency (that is the minimum requirement), one needs only three terms of polynomial basis ($m=3$); to improve the accuracy, we may use bilinear terms ($m=4$), and even quadratic terms ($m=5$ or 6). For the RBF terms, the minimum number is m, and the largest may be about 20 depending on the local node distribution. We note

Remark 3.15   RPIM shape functions (for 2D cases): $m=3\sim6$; $n=m\sim20$, by default $m=3$.

Vectors **a** and **b** in Equation (3.56) collecting the coefficients are defined as

$$\mathbf{a}^{\mathrm{T}} = \{a_1 \quad a_2 \quad \cdots \quad a_n\} \tag{3.57}$$

$$\mathbf{b}^{\mathrm{T}} = \{b_1 \quad b_2 \quad \cdots \quad b_m\} \tag{3.58}$$

The polynomial basis vector **p** in Equation (3.56) has the same form of Equations (3.20)-(3.22), and can be written in the following general form

$$\mathbf{p}^{\mathrm{T}}(\mathbf{x}) = \left[p_1(\mathbf{x}), p_2(\mathbf{x}), \cdots, p_m(\mathbf{x}) \right] \tag{3.59}$$

The radial basis vector **R** has the form of

$$\mathbf{R}^{\mathrm{T}}(\mathbf{x}) = \left[R_1(\mathbf{x}), R_2(\mathbf{x}), \cdots, R_n(\mathbf{x}) \right] \tag{3.60}$$

The coefficients a_i and b_j in Equation (3.56) are determined by enforcing the interpolation passing through all these n local support nodes. The interpolation at the kth point has the following form

$$u_k = u(\mathbf{x}_k) = \sum_{i=1}^{n} R_i(\mathbf{x}_k) a_i + \sum_{j=1}^{m} P_j(\mathbf{x}_k) b_j, \quad k = 1,2,\cdots,n \tag{3.61}$$

or in the matrix form:

$$\mathbf{d}_s = \mathbf{R}_Q \mathbf{a} + \mathbf{P}_m \mathbf{b} \tag{3.62}$$

where $\mathbf{d}_s$ is the vector collecting all the nodal values of the field functions at the n support nodes, $\mathbf{R}_Q$ is the *moment matrix* of RBFs given by

$$\mathbf{R}_Q = \begin{bmatrix} R_1(r_1) & R_2(r_1) & \cdots & R_n(r_1) \\ R_1(r_2) & R_2(r_2) & \cdots & R_n(r_2) \\ \vdots & \vdots & \vdots & \vdots \\ R_1(r_n) & R_2(r_n) & \cdots & R_n(r_n) \end{bmatrix} \tag{3.63}$$

in which

$$r_k = \begin{cases} \left[(x_k - x_i)^2 + (y_k - y_i)^2 \right]^{\frac{1}{2}}, & \text{for 2D} \\ \left[(x_k - x_i)^2 + (y_k - y_i)^2 + (z_k - z_i)^2 \right]^{\frac{1}{2}}, & \text{for 3D} \end{cases} \tag{3.64}$$

As the distance r is directionless, we have

$$R_i(r_j) = R_j(r_i) \tag{3.65}$$

Therefore, matrix $\mathbf{R}_Q$ is symmetric.

There are a number of radial basis functions used in the mathematics community. TABLE 3.2 lists the four most often used RBFs with some shape parameters. A classical form is the multiquadric (MQ) basis proposed by Hardy [24]. This form has been used in surface fitting and in constructing approximate solutions for PDEs [25-31] The MQ basis function shown in TABLE 3.2 is a general form of MQ RBF with arbitrary real shape parameters that was suggested in [4, 13, 14, 32]. The general form of the MQ radial function has two shape parameters, C_m and q, which control the shape of the functions. When $q=\pm0.5$, it reduces to the original MQ RBF proposed by Hardy [24]. When $q=0.5$, it reduces to the reciprocal MQ RBF. These parameters can be tuned for better performance. The second form of radial function given in TABLE 3.2 is called the Gaussian RBF, or EXP, as it is an exponential function of the distance [27]. The EXP RBF has only one shape parameter C_e, which controls the decay rate of the function. The third radial function in TABLE 3.2 is called the thin plate spline (TPS) function. The TPS is a special case of the MQ RBF. The fourth form of radial basis function is the logarithmic RBF. This book will use the MQ RBF, unless specified otherwise. Studies on the use of other RBFs in PIMs can be found in [4].

TABLE 3.2 Typical radial basis functions (RBFs)

Item	Name	Expression	Shape parameters
1	Multiquadrics (MQ)	$R_i(\mathbf{x}) = (r_i^2 + C_m^{\ 2})^q$	C_m, q
2	Gaussian (EXP)	$R_i(\mathbf{x}) = \exp(-C_e r_i^2)$	C_e
3	Thin pleat spline (TPS)	$R_i(\mathbf{x}) = r_i^{\psi}$	ψ
4	Logarithmic RBF	$R_i(r_i) = r_i^{\psi} \log r_i$	ψ

Some radial basis functions with dimensionless shape parameters are listed in TABLE 3.3. Both sets of RBFs in TABLE 3.2 and TABLE 3.3 can be used. In this book, we use the RBFs with dimensionless shape parameters.

TABLE 3.3 Radial basis functions with dimensionless shape parameters

Item	Name	Expression	Shape parameters	Parameter relations[*]
1	Multiquadrics (MQ)	$R_i(\mathbf{x}) = \left[r_i^2 + (\alpha_c d_c)^2 \right]^q$	α_c, q	$\alpha_c = C_m / d_c \geq 0$ $q = q$
2	Gaussian (EXP)	$R_i(\mathbf{x}) = \exp[-\beta_c (\frac{r_i}{d_c})^2]$	β_c	$\beta_c = C_e d_c$

[*] The last column gives the relationship between the original parameters given in TABLE 3.2 and the dimensionless parameters given in TABLE 3.3. For MQ and EXP RBFs, d_c is the local average nodal spacing and can generally be assigned to equal to the characteristic length h (see Section 1.6.2) for a regular mesh.

In Equation (3.62), $\mathbf{P}_m$ is the moment matrix corresponding to polynomial basis given by characteristic length of cells

$$\mathbf{P}_m = \begin{bmatrix} P_1(\mathbf{x}_1) & P_2(\mathbf{x}_1) & \cdots & P_m(\mathbf{x}_1) \\ P_1(\mathbf{x}_2) & P_2(\mathbf{x}_2) & \cdots & P_m(\mathbf{x}_2) \\ \vdots & \vdots & \vdots & \vdots \\ P_1(\mathbf{x}_n) & P_2(\mathbf{x}_n) & \cdots & P_m(\mathbf{x}_n) \end{bmatrix}_{n \times m} \tag{3.66}$$

In Equation (3.62), it should be noted that the polynomial terms have to satisfy extra requirements to guarantee a unique approximation [33], which is stated as a set of homogeneous equations

$$\sum_{i=1}^{n} p_j(\mathbf{x}_i) a_i = 0 \quad j = 1, 2, \cdots, m \tag{3.67}$$

or in the following matrix form of

$$\mathbf{P}_m^{\mathrm{T}} \mathbf{a} = \mathbf{0} \tag{3.68}$$

By combining Equations (3.62) and (3.68), we have

$$\begin{bmatrix} \mathbf{R}_Q & \mathbf{P}_m \\ \mathbf{P}_m^{\mathrm{T}} & \mathbf{0} \end{bmatrix} \begin{Bmatrix} \mathbf{a} \\ \mathbf{b} \end{Bmatrix} = \begin{Bmatrix} \mathbf{d}_s \\ \mathbf{0} \end{Bmatrix} \tag{3.69}$$

or in a simpler form

$$\mathbf{G} \begin{Bmatrix} \mathbf{a} \\ \mathbf{b} \end{Bmatrix} = \begin{Bmatrix} \mathbf{d}_s \\ \mathbf{0} \end{Bmatrix} \tag{3.70}$$

where the matrix $\mathbf{G}$ is

$$\mathbf{G} = \begin{bmatrix} \mathbf{R}_Q & \mathbf{P}_m \\ \mathbf{P}_m^{\mathrm{T}} & \mathbf{0} \end{bmatrix} \tag{3.71}$$

In the above definition, because the matrix $\mathbf{R}_Q$ is symmetric, matrix $\mathbf{G}$ will also be symmetric. A unique solution for these coefficients can be obtained if the inverse of $\mathbf{G}$ exists, and we shall have

$$\begin{Bmatrix} \mathbf{a} \\ \mathbf{b} \end{Bmatrix} = \mathbf{G}^{-1} \begin{Bmatrix} \mathbf{d}_s \\ 0 \end{Bmatrix} \tag{3.72}$$

Making use of the fact that Equation (3.68) is homogeneous, Equation (3.69) can be solved using the following more efficient procedure. Starting from Equation (3.62) and using the nonsingular property of $\mathbf{R}_Q$, we have

$$\mathbf{a} = \mathbf{R}_Q^{-1}\mathbf{d}_s - \mathbf{R}_Q^{-1}\mathbf{P}_m\mathbf{b} \tag{3.73}$$

Substituting the above equation into Equation (3.68) leads to

$$\mathbf{b} = \mathbf{S}_b \mathbf{d}_s \tag{3.74}$$

where

$$\mathbf{S}_b = [\mathbf{P}_m^{\mathrm{T}}\mathbf{R}_Q^{-1}\mathbf{P}_m]^{-1}\mathbf{P}_m^{\mathrm{T}}\mathbf{R}_Q^{-1} \tag{3.75}$$

in which $\mathbf{P}_m^{\mathrm{T}}\mathbf{R}_Q^{-1}\mathbf{P}_m$ is termed as a transformed moment matrix. Substituting Equation (3.74) back into Equation (3.73), gives

$$\mathbf{a} = \mathbf{S}_a \mathbf{d}_s \tag{3.76}$$

where $\mathbf{S}_a$ is given by

$$\mathbf{S}_a = \mathbf{R}_Q^{-1}[1 - \mathbf{P}_m \mathbf{S}_b] = \mathbf{R}_Q^{-1} - \mathbf{R}_Q^{-1} \mathbf{P}_m \mathbf{S}_b \tag{3.77}$$

Substituting Equations (3.74) and (3.76) into Equation (3.56), the interpolation can now be expressed as

$$u(\mathbf{x}) = \left[\mathbf{R}^{\mathrm{T}}(\mathbf{x}) \mathbf{S}_a + \mathbf{p}^{\mathrm{T}}(\mathbf{x}) \mathbf{S}_b \right] \mathbf{d}_s = \mathbf{\Phi}_s(\mathbf{x}) \mathbf{d}_s \tag{3.78}$$

where $\mathbf{\Phi}_s(\mathbf{x})$ is the matrix containing n RPIM nodal shape functions

$$\mathbf{\Phi}_s(\mathbf{x}) = \left[\mathbf{R}^{\mathrm{T}}(\mathbf{x}) \mathbf{S}_a + \mathbf{p}^{\mathrm{T}}(\mathbf{x}) \mathbf{S}_b \right] = [\phi_1(\mathbf{x}), \phi_2(\mathbf{x}), \cdots, \phi_k(\mathbf{x}), \cdots, \phi_n(\mathbf{x})] \tag{3.79}$$

in which $\phi_k(\mathbf{x})$ is the shape function for the kth node give as

$$\phi_k(\mathbf{x}) = \sum_{i=1}^{n} R_i(\mathbf{x}) S_{ik}^a + \sum_{j=1}^{m} p_j(\mathbf{x}) S_{jk}^b \tag{3.80}$$

In the forgoing equation, S_{ik}^a is the (i, k) element of matrix $\mathbf{S}_a$ and S_{jk}^b denotes the (j, k) element of matrix $\mathbf{S}_b$, which are constant matrices for given locations of the n nodes in the support domain.

3.3.3 RPIM shape functions with pure RBFs

When pure RBFs are used, we simply drop the terms related to the polynomial basis. We shall then have matrix of shape functions $\mathbf{\Phi}(\mathbf{x})$ has the form of

$$\begin{aligned} \mathbf{\Phi}_s(\mathbf{x}) &= [R_1(\mathbf{x}), \ R_2(\mathbf{x}), \cdots, R_k(\mathbf{x}), \cdots, R_n(\mathbf{x})] \mathbf{R}_Q^{-1} \\ &= \left[\phi_1(\mathbf{x}), \phi_2(\mathbf{x}), \cdots, \phi_k(\mathbf{x}), \cdots, \phi_n(\mathbf{x}) \right] \end{aligned} \tag{3.81}$$

where

$$\phi_k(\mathbf{x}) = \sum_{i=1}^{n} R_i(\mathbf{x}) S_{ik}^a \tag{3.82}$$

in which S_{ik}^a is the (i, k) entry in the matrix $\mathbf{R}_Q^{-1}$.

According to Remark 3.15, we know that n is usually much large than m. Therefore the process of computing RPIM shape functions, the time is usually dominated by the time for computing $\mathbf{R}_Q^{-1}$. Thus, including polynomial terms have only a small effect on the computation time, especially when n is large.

3.3.4 Singularity of the moment matrix

It has been proved that the radial moment matrix $\mathbf{R}_Q$ is always invertible for scattered nodes [27, 31, 34], as long as we avoid using some shape parameters. Therefore, $\mathbf{R}_Q$ is SPD (symmetric positive definite). The general existence of $\mathbf{R}_Q^{-1}$ is the major advantage of using the radial basis over the polynomial basis. Although $\mathbf{R}_Q$ is invertible, it can be ill-conditioned, especially when too many nodes are used [25, 26]. Fortunately, in creating RPIM shape functions for S-PIM, only very few local nodes (~20 at maximum for 2D cases) are used that is much smaller compared with those (can be as much as thousands) used in [25, 26] (where all the nodes in the problem domain are used). Therefore, the conditioning in $\mathbf{R}_Q$ for creating RPIM shape functions is not much a concern.

We now examine the positivity of the transformed moment matrix $\mathbf{P}_m^T \mathbf{R}_Q^{-1} \mathbf{P}_m$ in Equation (3.75). Because $\mathbf{R}_Q$ is SPD, matrix $\mathbf{P}_m^T \mathbf{R}_Q^{-1} \mathbf{P}_m$ in is at least symmetric. If there is at least m columns in $\mathbf{P}_m$ are independent (with a rank of m), it can be easily verified that that $\mathbf{P}_m^T \mathbf{R}_Q^{-1} \mathbf{P}_m$ is invertible by invoking the full rank property of $\mathbf{R}_Q$ and $\mathbf{P}_m$. To have at least m columns of $\mathbf{P}_m$ being independent could be a problem in theory, but it is quite easy to achieve in practice. We simply try to use more nodes, so that $n \geq m$, and hence there are always at least m columns (among all these n columns) in $\mathbf{P}_m$ being independent. For example, when $m=3$, all we need is at least there are 3 support nodes in all the n support nodes that are not in-line. Of course, we do not allow duplicated nodes (see, Section 1.6.1).

However, if we deliberately arrange all the support nodes in one straight line (for 2D and 3D domains), the process will break down. This kind of situation can happen in theory, but it will not happen when the nodes are generated by triangulation defined in Section 1.6.1 and a T-scheme (see, Section 1.6.3) is used for support node selection. The purpose of defining the triangulation and T-schemes is to exclude these kinds of extreme situations. When the T6- or T2L-schemes are used, $\mathbf{P}_m^T \mathbf{R}_Q^{-1} \mathbf{P}_m$ will always be invertible at least for RPIM augmented with up to quadratic polynomial bases, and hence are recommended. The T6-scheme will be more efficient, and T2L-scheme is relatively more robust to extremely irregularly distribution nodes.

In a usual situation, it is very rare to have a case where the rank of $\mathbf{P}_m$ is less than m, even when the T-scheme is not used. It is not a guarantee, but it is a practical strategy for usual situations. Therefore, $\mathbf{P}_m^{\mathrm{T}} \mathbf{R}_Q^{-1} \mathbf{P}_m$ can be assumed invertible in usual practical cases.

3.3.5 On the range of the shape parameters

The shape parameters in RBFs used as a collocation method using all the nodes in the problem domain for solving PDEs (e.g., [25, 26]) are usually fixed at certain values. For example, in the MQ RBF, $q=\pm0.5$. The reason maybe that the RBFs with those fixed values are parts of the fundamental solutions of typical PDEs. When (global) collocation methods are used, it can produce accurate solution and the convergence property can be proven.

For the RPIM shape functions created using local support nodes, however, the RBFs are merely used as a basis function just like monomials used in the PIM shape functions. These shape functions will be used only as a means of interpolation for field function approximation in the weak and weakened weak formulations. Therefore, we decided to allow the shape parameters change freely as a real number [4, 14]. The shape parameters can then be tuned to control the incompatibility for better accuracy and performance when a Galerkin weak form is used. Careful investigations on the effects of these shape parameters on the solution accuracy have been conducted and proper guidelines for use of shape parameters for different types of problems have been provided in [14, 32] for Galerkin weak formulations. In general it is found that the reliance of the accuracy on the shape parameters can be significantly reduced by adding polynomial basis functions. In addition, such reliance is also significantly reduced when a weakened weak form such as the generalized smoothed Galerkin weak form is applied [35, 36]. In this book, unless specified otherwise, the MQ RBF with $q=1.03$ and $\alpha_c=4.0$ is used to form RPIM shape functions.

3.3.6 Properties of RPIM shape functions

Remark 3.16 General properties of RPIM shape functions

The RPIM shape functions have the same properties as the PIM shape functions in terms of linear independence, partition of unity, Kronecker Delta function, incompatibility, compact support, and interplant. The creation of RPIM is more expensive than PIM.

Remark 3.17 RPIM shape functions: consistency and linear reproducibility

The RPIM shape function created using pure RBFs has only zero order consistence [12]. The RPIM shape function with polynomial terms has a consistence up to the complete order of polynomials used in the process of creating RPIM. This polynomial reproducing property is ensure the by the imposition of condition (3.68) in the process of creating RPIM shape functions. When linear polynomials are used, RPIM shape functions can reproduce exactly the linear field.

Remark 3.18 RPIM shape functions: continuity

Because of the high continuity of RBFs, RPIM shape functions usually possess higher continuity than the polynomial PIM shape functions.

Remark 3.19 RPIM shape funtions: robustness for irregular nodes

The RPIM shape function has been found work particularly well for higher irregularly distributed nodes [4]. If sufficient support nodes are used, it works for practically randomly distributed nodes.

Example 3.3.1 Mixed linear PIM/RPIM shape functions

Examples of a set of 1D mixed linear PIM/RPIM shape functions are computed using the formulation given above. The nodes distribution for this example is exactly the same as Example 3.2.1. Five nodes evenly distributed in the support domain of [0, 4] are used for creating the shape function for node 2 at $x=1$. The pure MQ radial basis functions listed in TABLE 3.3 are used in the computation with the parameters $\alpha_c = 4.0$ and $q=1.03$. In this case, the average nodal spacing is $d_c = 1.0$.

The support nodes selection in different regions and basis functions used for creating the 1D PIM shape functions are listed in TABLE 3.4. For the point of interest $(\mathbf{x}_Q)$ located in the two boundary domains, i.e. [0.0, 0.5] and [3.5, 4.0], we select two support nodes and use linear polynomial basis functions for creating the linear PIM shape functions; for the $(\mathbf{x}_Q)$ located in the interior node-based domains, i.e. [0.5, 1.5], [1.5, 2.5] and [2.5, 3.5], four support nodes and MQ RBF are used to create RPIM shape functions.

TABLE 3.4 Support nodes selection and basis functions used for interpolation approximation at different point of interest

Location of point of interest $\mathbf{x}_Q$	Local support nodes	Basis functions
[0.0, 0.5]	{1, 2}	Linear polynomial
[0.5, 1.5]	{1, 2, 3, 4}	MQ RBF
[1.5, 2.5]	{2, 3, 4, 5}	MQ RBF
[2.5, 3.5]	{2, 3, 4, 5}	MQ RBF
[3.5, 4.0]	{4, 5}	Linear polynomial

FIGURE 3.5 shows nodal PIM shape functions for node 2. It is clearly seen that the PIM/RPIM shape function possesses the Kronecker Delta function property. That is, $\phi_i(\mathbf{x} = \mathbf{x}_i) = 1.0$ and $\phi_i(\mathbf{x} \neq \mathbf{x}_i) = 0.0$ for $i = 1, \cdots, 5$. We note that the mixed linear PIM/RPIM shape functions are discontinuous at some points, where the local support nodes are updated. FIGURE 3.5 shows that the nodal PIM shape functions for node 2 have three discontinuity (jump) points at x=0.5, x=1.5

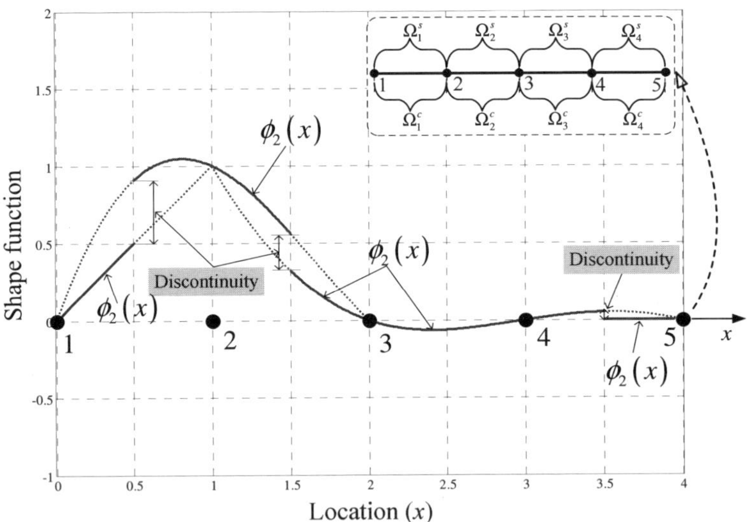

FIGURE 3.5 Nodal RPIM shape function for node 2 at x=1.0 obtained using five evenly distributed nodes in domain of [0, 4], where MQ RBF is used with $\alpha_c = 4.0$, q=1.03 and d_c=1.0. The RPIM shape function possesses the Kronecker Delta function property and is discontinuous when the support nodes change.

and $x=3.5$, where the supports nodes are updated from $\{1, 2\}$ to $\{1, 2, 3, 4\}$, $\{1, 2, 3, 4\}$ to $\{2, 3, 4, 5\}$, and $\{2, 3, 4, 5\}$ to $\{4, 5\}$, respectively. There is no discontinuity of RPM shape function at the point of $x=2.5$, because the same group of support nodes $\{2, 3, 4, 5\}$ are used for the two node-based domains of nodes 3 and 4.

Due to the discontinuity of the nodal PIM shape functions, the field functions constructed using these shape functions will also be discontinuous. Also because the discontinuities are at the mid-cell points, this set of nodal PIM/RPIM shape functions can only be in a $\mathbb{G}_h^1(\Omega)$ space, if cell-based smoothing domains are used, due to the "no-sharing rule" discussed in Section 2.3.1.

Remark 3.20 Property of RPIM shape functions augmented with polynomials: a summary

Successfully created RPIM shape functions are, in general, incompatible, consistent to the order of polynomials used in the formulation. They have the Kronecker Delta function property, and form an interpolant in a G space.

3.4 PIM-CT shape functions

Polynomial PIM with (rotational) coordinate transformation (PIM-CT) is another effective way to overcome the singularity moment matrix problem. The method was originally attempted in [14], and works well at least for 2D problems. It is devised based on the following observations [4]:

1) The linear dependence of the columns or rows in the moment matrix depends on the coordinates of the support nodes for a fixed set of monomial bases;

2) The dependence can be altered by rotating the coordinate axis, and hence a nonsingular moment matrix $\mathbf{P}_Q$ can be obtained when a proper rotation angle is found;

3) For a given set of a small number of nodes, there are only a few rotation angles that make the moment matrix singular, and hence it is quite easy to find a rotation angle leading a nonsingular moment matrix;

4) Some kinds of optimization procedure can be implemented for a set of "optimal" PIM shape functions.

The following presents the detailed procedure.

3.4.1 Coordinate transformation

We first introduce a general coordination transformation between the global coordinate system (x, y) and the local coordinate system (ξ, η), as shown in FIGURE 3.6. This transformation can be performed using

$$
\begin{aligned}
\xi &= (x - x_Q)\cos\gamma + (y - y_Q)\sin\gamma \\
\eta &= -(x - x_Q)\sin\gamma + (y - y_Q)\cos\gamma
\end{aligned}
\tag{3.83}
$$

where (x_Q, y_Q) denotes the point of interest that is also the origin of the local coordination system, and γ is the rotation angle for local coordination system with respect to the global coordinate system. The corresponding inverse transformation is given by

$$
\begin{aligned}
x &= x_Q + (\xi\cos\gamma - \eta\sin\gamma) \\
y &= y_Q + (\xi\sin\gamma + \eta\cos\gamma)
\end{aligned}
\tag{3.84}
$$

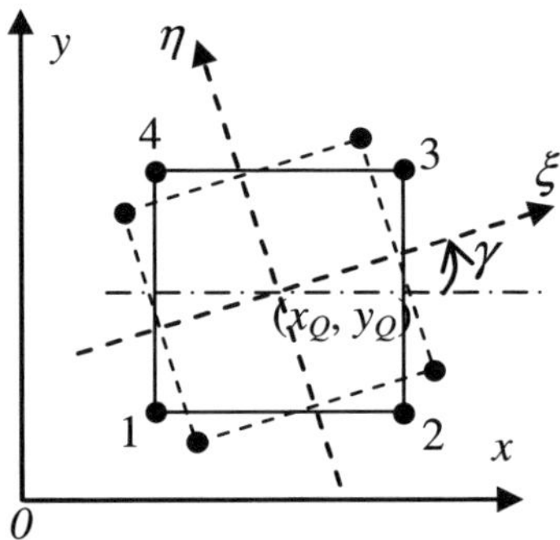

FIGURE 3.6 A local coordinate system (ξ, η) in a global coordinate system (x, y). The origin of (ξ, η) is at (x_Q, y_Q).

3.4.2 Creation of PIM-CT shape functions

In the local coordinates, n polynomial basis terms $p_i(\xi)$ $(i = 0, 1, \cdots, n-1)$ are first chosen, we then try to find a proper rotation angle γ that produces a nonsingular moment matrix $\mathbf{P}_Q^\xi$. The set of PIM shape functions in the local coordinate system can then be obtained as (see Equation (3.31))

$$
\boldsymbol{\Phi}_s(\xi) = \mathbf{p}^{\mathrm{T}}(\xi)\left(\mathbf{P}_Q^\xi\right)^{-1} = \left[\phi_1(\xi) \quad \phi_2(\xi) \quad \cdots \quad \phi_n(\xi)\right]
\tag{3.85}
$$

Using Equation (3.83), the shape functions can also be expressed as

$$\Phi_s(\mathbf{x}) = \begin{bmatrix} \phi_1(\mathbf{x}) & \phi_2(\mathbf{x}) & \cdots & \phi_n(\mathbf{x}) \end{bmatrix} \tag{3.86}$$

Example 3.4.1 Interpolation using 4 nodes selected using the edge-based T4-scheme

When the edge-based T4-scheme (see, Section 1.6.3) is used to select 4 nodes for the creation of polynomial PIM shape functions, we may encounter a possible situation shown in Example 3.2.4 The moment matrix can be singular. We now use the coordinate transformation techniques to create a set of PIM-CT shape functions.

Consider 4 nodes selected for PIM shape function creation, using bilinear polynomial basis functions. The bilinear PIM shape functions for the point of interest at (x_Q, y_Q) can be finally expressed as follows.

$$\Phi_s(\mathbf{x}_Q) = \begin{bmatrix} 1 & x_Q & y_Q & x_Q y_Q \end{bmatrix} \underbrace{\begin{bmatrix} 1 & x_1 & y_1 & x_1 y_1 \\ 1 & x_2 & y_2 & x_2 y_2 \\ 1 & x_3 & y_3 & x_3 y_3 \\ 1 & x_4 & y_4 & x_4 y_4 \end{bmatrix}^{-1}}_{\mathbf{P}_Q} \tag{3.87}$$

where $\mathbf{x}_i = (x_i, y_i)\,(i = 1, 2, 3, 4)$ are the coordinates of these four nodes. We know that the inverse of the moment matrix $\mathbf{P}_Q$ may not exist. Applying Equation (3.83) to the four support nodes, Equation (3.87) becomes

$$\Phi_s(\mathbf{x}_Q) = \begin{bmatrix} 1 & 0 & 0 & 0 \end{bmatrix} \underbrace{\begin{bmatrix} 1 & \xi_1 & \eta_1 & \xi_1 \eta_1 \\ 1 & \xi_2 & \eta_2 & \xi_2 \eta_2 \\ 1 & \xi_3 & \eta_3 & \xi_3 \eta_3 \\ 1 & \xi_4 & \eta_4 & \xi_4 \eta_4 \end{bmatrix}^{-1}}_{\mathbf{P}_Q^\xi} \tag{3.88}$$

If a γ and be found, such that $\mathbf{P}_Q^\xi$ is invertible, a set of shape function values can then be obtained successfully. We need now a proper technique to determine γ.

3.4.3 Determination of rotation angle

We first note the simple fact that when a matrix is singular, the determinant of the moment matrix will be zero. Therefore, the determinant of the matrix is a good indicator on the property of the matrix. We also know that when 4 nodes at the vertices of a square are used, the property of the moment matrix is ideal and can be surely invertible. FIGURE 3.6 shows such a four nodes located at the vertices of a square with an edge length of $2a$ in the coordinates system (x, y). Rotating the square, different moment matrixes $\mathbf{P}_Q^\xi$ can be obtained using the four nodes at these vertices. The determinant of the moment matrix, d_m, can then be obtained analytically in relation to the transformation angle. It turns out the relation has the following simple form.

$$d_m = -16a^4 \cos(2\gamma) \tag{3.89}$$

When $\gamma=0$, we shall have the "ideal" determinate

$$d_m = -16a^4 \tag{3.90}$$

Using the above relationships, we devise the following adaptive procedure to determine γ for creating bilinear PIM shape functions [8].

1) For the selected four support nodes, compute the distance between nodes 1 and 3, which is denoted as l_{13}, and then compute the determinant of the moment matrix (d_1) using these four nodes.

2) Calculated the determinant for the corresponding ideal case: $d_{m0} = -16\left(l_{13}/2\right)^4$.

3) If ($\left|d_{m1}\right| < \left|d_{m0}\right|$), we perform the following coordinate transformation:

    i.   Letting $d_m=d_{m1}$, and $a=l_{13}/2$ in Equation (3.89), solve it for the rotation angle which satisfies $\left(\gamma_1 \leq \pi/2\right)$;

    ii.   Substituting $\left(\pm \gamma_1\right)$ into Equation (3.83) and calculate determinates of these two moment matrixes $\mathbf{P}_Q^\xi$ defined in Equation (3.88);

    iii.   Select the angle from $\left(\pm \gamma_1\right)$ which gives a bigger determinant value, use it to perform the coordinate transformation, and finally obtain the shape functions using Equation (3.88).

The above CT technique is simple, can effectively overcome the singularity problem and ensures a set of bilinear PIM shape functions, as long as these four nodes are not in-line. The created bilinear PIM shape function is "optimal", in terms of largest determinant value for the moment matrix.

3.5 Isoparametric PIM shape functions

Creating PIM shape functions does not usually need coordinate mapping that are often used in the FEM to create the so-called "isoparametric" elements for the purpose of ensuring the compatibility of the FEM model [1]. However, we may want to use mimic such as isoparametric shape functions in PIM. Such an isoparametric PIM (or PIM-Iso) shape functions created using the following procedures [37].

FIGURE 3.7 shows 4 nodes selected using an edge-based T4-scheme. These 4 nodes can form a "quadrilateral element" called Q4 element (1-2-3-4). In the FEM, such a Q4 element is formulated using the isoparametric mapping to map the element in the physical coordinate (x, y) to a natural coordinate (ξ, η) defined in domain $[0,1] \times [0,1]$ for the Q4 element. Because of such a mapping the shape functions created for the Q4 element will be linear on the four edges: 1-2, 2-3, 3-4, and 4-1. Because of this "edge-linear" feature, all these element edges in any Q4 element mesh will be all linear and will be perfectly connected, leading to a compatible FEM mesh. The FEM shape functions in the natural coordinate (ξ, η) can be given as [1]:

$$\begin{cases} \phi_1(\xi,\eta) = (1-\xi)(1-\eta) \\ \phi_2(\xi,\eta) = \xi(1-\eta) \\ \phi_3(\xi,\eta) = \xi\eta \\ \phi_4(\xi,\eta) = (1-\xi)\eta \end{cases} \tag{3.91}$$

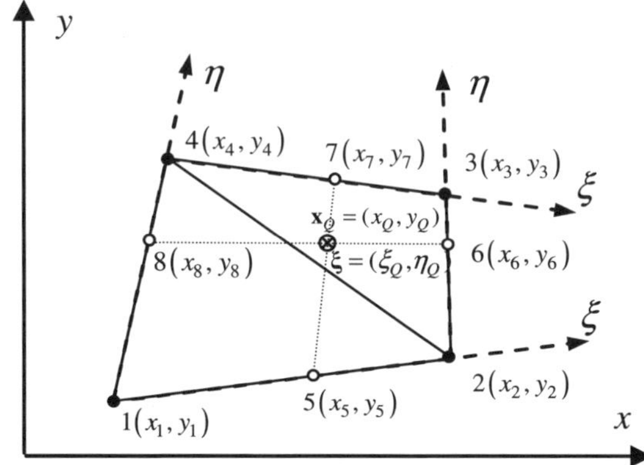

FIGURE 3.7 Four nodes selected using the edge-based T4-scheme. These 4 nodes are used to create PIM-Iso shape functions.

To mimic the same type of shape function in PIM, all we need to do is to find the value (ξ_Q, η_Q) in the natural coordinates for any given point of interest $\mathbf{x}_Q = (x_Q, y_Q)$ in the physical coordinates. We shall have two approaches to achieve this goal.

3.5.1 Approach 1: using bilinear feature

In this approach, we first define two straight lines in the natural coordinate system (ξ, η) that pass though (ξ_Q, η_Q), as shown in FIGURE 3.7. For line 5-7, we shall have

$$\xi_5 = \xi_7 = \xi_Q \tag{3.92}$$

and for line 8-6, we shall have

$$\eta_6 = \eta_8 = \eta_Q \tag{3.93}$$

Using the edge-linear feature and the bilinear mapping, we know that the shape functions will be linear along these two lines, and thus

$$\begin{cases} x_5 = x_1 + \underbrace{\left(x_2 - x_1\right)}_{x_{21}}\xi_Q \\ y_5 = y_1 + \underbrace{\left(y_2 - y_1\right)}_{y_{21}}\xi_Q \end{cases} \begin{cases} x_7 = x_4 + \underbrace{\left(x_3 - x_4\right)}_{x_{34}}\xi_Q \\ y_7 = y_4 + \underbrace{\left(y_3 - y_4\right)}_{y_{34}}\xi_Q \end{cases} \tag{3.94}$$

and

$$\begin{cases} x_6 = x_2 + \underbrace{\left(x_3 - x_2\right)}_{x_{32}}\eta_Q \\ y_6 = y_2 + \underbrace{\left(y_3 - y_2\right)}_{y_{32}}\eta_Q \end{cases} \begin{cases} x_8 = x_1 + \underbrace{\left(x_4 - x_1\right)}_{x_{41}}\eta_Q \\ y_8 = y_1 + \underbrace{\left(y_4 - y_1\right)}_{y_{41}}\eta_Q \end{cases} \tag{3.95}$$

where we used the following definition.

$$x_{IJ} = x_I - x_J, \quad y_{IJ} = y_I - y_J \quad (I, J = 1, 2, 3, 4) \tag{3.96}$$

The formulae for these two straight lines in the physical coordinate system that pass though (x_Q, y_Q) can now be expressed as follows. For line 5-7, we shall have

$$\underbrace{(x_5 - x_7)}_{x_{57}} y = \underbrace{(y_5 - y_7)}_{y_{57}} x + (x_5 - x_7)y_7 - (y_5 - y_7)x_7 \tag{3.97}$$

and for line 8-6, we shall have

$$\underbrace{(x_6 - x_8)} y = \underbrace{(y_6 - y_8)} x + (x_6 - x_8)y_8 - (y_6 - y_8)x_8 \tag{3.98}$$

Based on Equation (3.94), x_{57} and y_{57} can also be expressed as

$$\begin{aligned}
x_{57} = x_5 - x_7 = x_{14} + \underbrace{\left(x_{21} - x_{34}\right)}_{x_{2134}}\xi_Q = x_{14} + x_{2134}\xi_Q \\
y_{57} = y_5 - y_7 = y_{14} + \underbrace{\left(y_{21} - y_{34}\right)}_{y_{2134}}\xi_Q = y_{14} + y_{2134}\xi_Q
\end{aligned} \tag{3.99}$$

Substitute Equations (3.94) and (3.99) into Equation (3.97), at $\mathbf{x}_Q = (x_Q, y_Q)$, we shall have for line 5-7

$$\begin{aligned}
(x_{14} + x_{2134}\xi_Q)y_Q = {} & (y_{14} + y_{2134}\xi_Q)x_Q \\
& + (x_{14} + x_{2134}\xi_Q)(y_4 + y_{34}\xi_Q) \\
& - (y_{14} + y_{2134}\xi_Q)(x_4 + x_{34}\xi_Q)
\end{aligned} \tag{3.100}$$

or

$$\begin{aligned}
& \underbrace{x_{14}y_Q - x_Q y_{14} + x_4 y_{14} - x_{14}y_4}_{c_\xi} \\
& + \underbrace{[x_{2134}(y_Q - y_4) - y_{2134}(x_Q - x_4) + x_{34}y_{14} - x_{14}y_{34}]}_{b_\xi}\xi_Q \\
& + \underbrace{(x_{34}y_{2134} - x_{2134}y_{34})}_{a_\xi}\xi_Q^2 = 0
\end{aligned} \tag{3.101}$$

This is a quadratic equation in ξ_Q^2:

$$c_\xi + b_\xi \xi_Q + a_\xi \xi_Q^2 = 0 \tag{3.102}$$

which has a solution of

$$\xi_Q = \begin{cases} -\dfrac{c_\xi}{b_\xi}, & \text{when } a_\xi = 0 \\[2ex] \dfrac{-b_\xi \pm \sqrt{b_\xi^2 - 4a_\xi c_\xi}}{2a_\xi}, & \text{when } a_\xi \neq 0 \end{cases} \tag{3.103}$$

We shall, of course, choose the $\xi_Q \in [0,1]$ for the two multiple solutions.

Following the same procedure and using Equation (3.98) we can obtain the following equation for line 8-6:

$$\underbrace{x_{21}y_Q - x_Q y_{21} + x_1 y_{21} - x_{21}y_1}_{c_\eta}$$
$$+ \underbrace{[x_{3241}(y_Q - y_1) - y_{3241}(x_Q - x_1) + x_{41}y_{21} - x_{21}y_{41}]}_{b_\eta}\xi_Q \tag{3.104}$$
$$+ \underbrace{(x_{41}y_{3241} - x_{3241}y_{41})}_{a_\eta}\eta_Q^2 = 0$$

This is a quadratic equation in η_Q^2:

$$c_\eta + b_\eta \eta_Q + a_\eta \eta_Q^2 = 0 \tag{3.105}$$

which has a solution of

$$\eta_Q = \begin{cases} -\dfrac{c_\eta}{b_\eta}, & \text{when } a_\eta = 0 \\[2ex] \dfrac{-b_\eta \pm \sqrt{b_\eta^2 - 4a_\eta c_\eta}}{2a_\eta}, & \text{when } a_\eta \neq 0 \end{cases} \tag{3.106}$$

We shall, of course, choose the root $\eta_Q \in [0,1]$.

Finally, the PIM-Iso shape function values at $\mathbf{x}_Q = (x_Q, y_Q)$ can be obtained using

$$\begin{cases} \phi_1(x_Q, y_Q) = (1-\xi_Q)(1-\eta_Q) \\ \phi_2(x_Q, y_Q) = \xi_Q(1-\eta_Q) \\ \phi_3(x_Q, y_Q) = \xi_Q\eta_Q \\ \phi_4(x_Q, y_Q) = (1-\xi_Q)\eta_Q \end{cases} \tag{3.107}$$

Because our S-PIM models, we do not use the derivatives of the shape functions, Equation (3.107) is all we need. It is clear that the mapping in PIM-Iso is very much different from that in FEM, and no Jacobian matrix is involved.

3.5.2 Approach 2: coordinate mapping

In this approach, we perform the following coordinate mapping:

$$x = \sum_{I=1}^{4} \phi_I(\xi,\eta)x_I, \quad y = \sum_{I=1}^{4} \phi_I(\xi,\eta)y_I \tag{3.108}$$

which gives:

$$\begin{aligned} x &= (1-\xi-\eta+\xi\eta)x_1 + (\xi-\xi\eta)x_2 + \xi\eta x_3 + (\eta-\xi\eta)x_4 \\ y &= (1-\xi-\eta+\xi\eta)y_1 + (\xi-\xi\eta)y_2 + \xi\eta y_3 + (\eta-\xi\eta)y_4 \end{aligned} \tag{3.109}$$

in which we used Equation (3.91). Rearranging these terms in the forgoing equation gives

$$x = x_1 + \underbrace{(x_2-x_1)}_{x_{21}}\xi + \underbrace{(x_4-x_1)}_{x_{41}}\eta - \underbrace{(x_2-x_1+x_4-x_3)}_{x_{2134}}\xi\eta \tag{3.110}$$

or

$$x = x_1 + x_{21}\xi + x_{41}\eta - x_{2134}\xi\eta \tag{3.111}$$

Similarly, we have

$$y = y_1 + y_{21}\xi + y_{41}\eta - y_{2134}\xi\eta \tag{3.112}$$

Solving Equation (3.111) for η gives

$$\eta = \frac{(x - x_1 - x_{21}\xi)}{(x_{41} - x_{2134}\xi)} \tag{3.113}$$

Substituting the forgoing equation into Equation (3.112), we obtain

$$\begin{aligned}
y(x_{41} - x_{2134}\xi) = {}& y_1(x_{41} - x_{2134}\xi) + y_{21}(x_{41} - x_{2134}\xi)\xi \\
& + y_{41}(x - x_1 - x_{21}\xi) - y_{2134}(x - x_1 - x_{21}\xi)\xi
\end{aligned} \tag{3.114}$$

Rearranging Equation (3.114) leads to

$$\begin{aligned}
0 = {}& (y_1 - y)x_{41} + (x - x_1)\,y_{41} \\
& + [(y - y_1)x_{2134} - (x - x_1)y_{2134} - x_{21}y_{41} + x_{41}y_{21}]\xi \\
& + (x_{21}y_{2134} - y_{21}x_{2134})\xi^2
\end{aligned} \tag{3.115}$$

Now, at $\mathbf{x}_Q = (x_Q, y_Q)$, we shall have

$$\begin{aligned}
0 = {}& \underbrace{(y_1 - y_Q)x_{41} + (x_Q - x_1)\,y_{41}}_{c_Q} \\
& + \underbrace{[(y_Q - y_1)x_{2134} - (x_Q - x_1)y_{2134} - x_{21}y_{41} + x_{41}y_{21}]}_{b_Q}\xi_Q \\
& + \underbrace{(x_{21}y_{2134} - y_{21}x_{2134})}_{a_Q}\xi_Q^2
\end{aligned} \tag{3.116}$$

or

$$0 = c_Q + b_Q\xi_Q + a_Q\xi_Q^2 \tag{3.117}$$

which has solutions of

$$\xi_Q = \begin{cases}
-\dfrac{c_Q}{b_Q}, & \text{when } a_Q = 0 \\[2ex]
\dfrac{-b_Q \pm \sqrt{b_Q^2 - 4a_Q c_Q}}{2a_Q}, & \text{when } a_Q \neq 0
\end{cases} \tag{3.118}$$

We shall, of course, choose the root $\xi_Q \in [0,1]$. Using Equation (3.113) we finally obtain

$$
\eta_Q = \begin{cases} \dfrac{x_Q - x_1 - x_{21}\xi_Q}{x_{41} - x_{2134}\xi_Q}, & \text{when } a_Q = 0 \\[2em] \dfrac{x_Q - x_1 - x_{21}\xi_Q}{x_{41} - x_{2134}\xi_Q}, & \text{when } a_Q \neq 0 \end{cases}
\tag{3.119}
$$

where shall have $\eta_Q \in [0,1]$.

Approach 2 can be used for creating PIM-Iso shape functions mimicking those of any isoparametric FEM element, as long as the FEM shape functions in the natural coordinate system is available.

For 2D problems discretized using three-node triangular cells, note that the created PIM-Iso shape functions will be in general different from the PIM-CT shape functions obtained via coordinate transformation, although they all use 4 nodes. The former is edge-linear along the Q4 boundary, and the latter is linear with respect to the transformed local coordinate system that is usually different from the natural coordinate systems.

3.6 Alpha PIM shape functions

The removal of the compatibility requirement in W^2 formulations allows a lot more innovative and bold approaches in creating shape functions. Here is an example: the PIM shape functions can be combined in a "partition of unity fashion" to form a new set of shape functions. This is generally useful in establishing S-PIM models with desired properties, known as the αPIM [38]. It is particularly useful in PIM settings to quite freely control the softness of the model, so that the discrete models are capable of producing nearly "exact" solutions. The αPIM shape functions can be developed by a combination procedure given a follows.

3.6.1 Given sets of PIM shape functions

Considering two sets of nodal PIM shape functions that have already been created for point $\mathbf{x}$:

$$
\mathbf{\Phi}_s^{(I)}(\mathbf{x}) = \begin{bmatrix} \phi_{I_1}(\mathbf{x}) & \phi_{I_2}(\mathbf{x}) & \cdots & \phi_{I_n}(\mathbf{x}) \end{bmatrix}
\tag{3.120}
$$

where subscripts (I) stands for the nodal shape functions for set $S^{(I)}$. The nodes for these two sets of shape functions for the point of interest $\mathbf{x}$ do not have to be the same, but shall have some shared nodes. A displacement component u at point $\mathbf{x}$ can then be approximated using nodes in set $S^{(I)}$ as

$$u^h(\mathbf{x}) = \sum_{i=1}^{I_n} \phi_{I_i}(\mathbf{x})u_i \tag{3.121}$$

or in those in set $S^{(II)}$ as

$$u^h(\mathbf{x}) = \sum_{i=1}^{II_n} \phi_{II_i}(\mathbf{x})u_i \tag{3.122}$$

where I_n and II_n are, respectively, the number of support nodes in the two sets of $S^{(I)}$ and $S^{(II)}$.

3.6.2 Creation of αPIM shape functions

Naturally, we can approximate the displacement function by combining these two sets of PIM shape functions in a "partition of unity fashion":

$$u^h(\mathbf{x}) = \alpha\left(\sum_{i=1}^{I_n} \phi_{I_i}(\mathbf{x})u_i\right) + \underbrace{(1-\alpha)}_{\beta}\left(\sum_{i=1}^{II_n} \phi_{II_i}(\mathbf{x})u_i\right) \tag{3.123}$$

where $\alpha \in [0,1]$. Equation (3.123) can be re-arranged as

$$\begin{aligned}
u^h(\mathbf{x}) &= \alpha\left(\sum_{i=1}^{I_n} \phi_{I_i}(\mathbf{x})u_i\right) + \underbrace{(1-\alpha)}_{\beta}\left(\sum_{i=1}^{II_n} \phi_{II_i}(\mathbf{x})u_i\right) \\
&= \sum_{i \in S^{(I)} \cap S^{(II)}} \left(\alpha\phi_{I_i}(\mathbf{x}) + (1-\alpha)\phi_{II_i}(\mathbf{x})\right)u_i \\
&\quad + \sum_{i \in S^{(I)} \setminus S^{(II)}} \alpha\phi_{I_i}(\mathbf{x})u_i + \sum_{i \in S^{(II)} \setminus S^{(I)}} (1-\alpha)\phi_{II_i}(\mathbf{x})u_i
\end{aligned} \tag{3.124}$$

or

$$u^h(\mathbf{x}) = \sum_{i \in S^{(\alpha)}} \phi_i^{(\alpha)}(\mathbf{x})u_i = \mathbf{\Phi}_s^{(\alpha)}(\mathbf{x})\mathbf{d}_s \tag{3.125}$$

A new set of αPIM nodal shape functions can now be created using

$$\mathbf{\Phi}_s^{(\alpha)}(\mathbf{x}) = \begin{bmatrix} \phi_1^{(\alpha)}(\mathbf{x}) & \phi_2^{(\alpha)}(\mathbf{x}) & \cdots & \phi_i^{(\alpha)}(\mathbf{x}) & \cdots & \phi_{n^{(\alpha)}}^{(\alpha)}(\mathbf{x}) \end{bmatrix} \tag{3.126}$$

where $n^{(\alpha)}$ is the number of nodes in the set of $S^{(\alpha)} = S^{(I)} \bigcup S^{(II)}$, and

$$\phi_i^{(\alpha)}(\mathbf{x}) = \begin{cases} \alpha\phi_{I_i}(\mathbf{x}) + (1-\alpha)\phi_{II_i}(\mathbf{x}) & \forall i \in S^{(I)} \bigcap S^{(II)} \\ \alpha\phi_{I_i}(\mathbf{x}) & \forall i \in S^{(I)} \setminus S^{(II)} \\ (1-\alpha)\phi_{II_i}(\mathbf{x}) & \forall i \in S^{(II)} \setminus S^{(I)} \end{cases} \tag{3.127}$$

For the convenience in this book we term this new set of shape functions the αPIM shape functions.

3.6.3 Properties of the αPIM shape functions

It is a trivial matter to proof that the alpha shape functions preserve the partitions of unity of the original two sets of PIM shape functions:

$$\begin{aligned} \sum_{i \in S^{(\alpha)}} \phi_i^{(\alpha)}(\mathbf{x}) &= \sum_{i \in S^{(I)} \bigcap S^{(II)}} \alpha\phi_{I_i}(\mathbf{x}) + (1-\alpha)\phi_{II_i}(\mathbf{x}) \\ &\quad + \sum_{i \in S^{(I)} \setminus S^{(II)}} \alpha\phi_{I_i}(\mathbf{x}) + \sum_{i \in S^{(II)} \setminus S^{(I)}} (1-\alpha)\phi_{II_i}(\mathbf{x}) \\ &= \alpha \underbrace{\sum_{i \in S^{(I)}} \phi_{I_i}(\mathbf{x})}_{=1} + (1-\alpha) \underbrace{\sum_{i \in S^{(II)}} \phi_{II_i}(\mathbf{x})}_{=1} \\ &= \alpha + (1-\alpha) \\ &= 1 \end{aligned} \tag{3.128}$$

As for the Kronecker Delta function property, we have

$$\begin{aligned} \phi_i^{(\alpha)}(\mathbf{x}_j) &= \begin{cases} \alpha\phi_{I_i}(\mathbf{x}_j) + (1-\alpha)\phi_{II_i}(\mathbf{x}_j) & \forall i \in S^{(I)} \bigcap S^{(II)} \\ \alpha\phi_{I_i}(\mathbf{x}_j) & \forall i \in S^{(I)} \setminus S^{(II)} \\ (1-\alpha)\phi_{II_i}(\mathbf{x}_j) & \forall i \in S^{(II)} \setminus S^{(I)} \end{cases} \\ &= \begin{cases} \delta_{ij} & \forall i \in S^{(I)} \bigcap S^{(II)} \\ \alpha\delta_{ij} & \forall i \in S^{(I)} \setminus S^{(II)} \\ (1-\alpha)\delta_{ij} & \forall i \in S^{(II)} \setminus S^{(I)} \end{cases} \end{aligned} \tag{3.129}$$

which means that only when the point of interest $\mathbf{x}$ is located in the common area supported by both nodal sets $S^{(I)}$ and $S^{(II)}$, the Delta function property is satisfied. This means that the αPIM shape functions should be used in general only for the points of interest in the shared area of both nodal sets $S^{(I)}$ and $S^{(II)}$. This prevents also possible partial extrapolations in the process.

It is also easy to show that the order of consistency of the αPIM shape functions will be determined by the lowest order of consistency of these two sets shape functions. When both sets shape functions are of linear reproducibility, so will the αPIM shape functions.

The αPIM shape functions offer another a big avenue for establishing S-PIM models, and will be used to develop the CS-αRPIM in Chapter 9.

3.7 Condensed RPIM shape functions

The removal of the compatibility requirement in $\mathbf{W}^2$ formulations allows a lot more innovative and bold approaches in creating shape functions. Here is yet another example: RPIM shape functions can be condensed. This is particularly useful in creating RPIM shape functions with a small number of support nodes. It was motivated by the observation that creating RPIM shape functions tends to use too many nodes, which is clearly not desirable, because it leads eventually a less sparse stiffness matrix for S-PIM models affecting directly the computational efficiency. Ideas of condense the RPIM shape functions was then invented [37]. It works simply as follows.

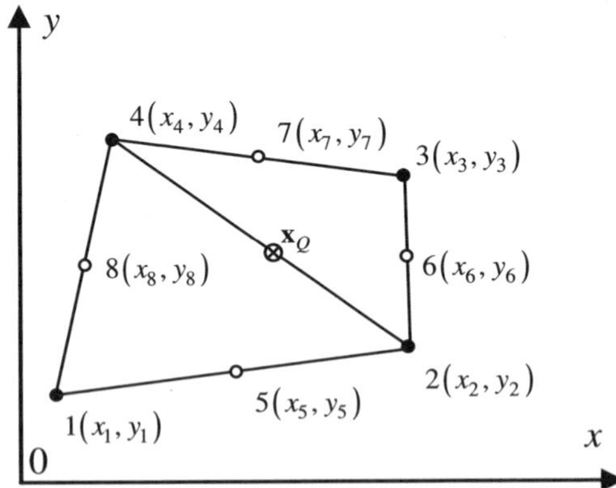

FIGURE 3.8 Four real nodes (1, 2, 3 and 4) selected using an edge-based T4-scheme plus four virtual nodes (5, 6, 7 and 8) are used to create a set of 2D condensed RPIM shape functions.

Consider a situation of using 4 support nodes to create a set of RPIM shape functions, using nodes 1, 2, 3, 4 that come from two adjacent triangular cells, as shown in FIGURE 3.8. We want to have the RPIM shape functions to contain linear, bilinear, and even quadratic polynomials. There is a number of ways to do this, and here we introduce a very simple way.

3.7.1 Introduction of virtual nodes

First, we introduce four "virtual" nodes, 5, 6, 7, and 8 at these 4 mid-edge points shown in FIGURE 3.8, and we shall have

$$\begin{cases} x_5 = \dfrac{1}{2}(x_1 + x_2) \\[2mm] y_5 = \dfrac{1}{2}(y_1 + y_2) \\[2mm] x_7 = \dfrac{1}{2}(x_3 + x_4) \\[2mm] y_7 = \dfrac{1}{2}(y_3 + y_4) \end{cases} \quad \begin{cases} x_6 = \dfrac{1}{2}(x_2 + x_3) \\[2mm] y_6 = \dfrac{1}{2}(y_2 + y_3) \\[2mm] x_8 = \dfrac{1}{2}(x_1 + x_4) \\[2mm] y_8 = \dfrac{1}{2}(y_1 + y_4) \end{cases} \quad (3.130)$$

Using these 8 nodes (4 real + 4 virtual), we create a set of RPIM shape functions using the procedure given in Section 3.3 with $m \leq 6$ polynomial terms, which gives 8 shape functions corresponding to these 8 nodes.

$$\mathbf{\Phi}_s(\mathbf{x}) = \left[\phi_1(\mathbf{x}), \phi_2(\mathbf{x}), \cdots, \phi_I(\mathbf{x}), \cdots, \phi_8(\mathbf{x})\right] \quad (3.131)$$

in which $\phi_I(\mathbf{x})$ is given by Equation (3.80). These shape functions possess the *partition of unity* and *Delta function* properties, as discussed in Section 3.3. The displacement function at the point of interest $\mathbf{x}$ can then be written as

$$u(\mathbf{x}) = \mathbf{\Phi}_s(\mathbf{x})\mathbf{d}_s = \sum_{I=1}^{8} \phi_I(\mathbf{x})u_I \quad (3.132)$$

3.7.2 Introduction of constraints

Next, we introduce the following constraints for these 4 virtual nodes:

$$\begin{aligned} \mathbf{u}_5 &= \frac{1}{2}(\mathbf{u}_1 + \mathbf{u}_2);\, \mathbf{u}_6 = \frac{1}{2}(\mathbf{u}_2 + \mathbf{u}_3) \\[2mm] \mathbf{u}_7 &= \frac{1}{2}(\mathbf{u}_3 + \mathbf{u}_4);\, \mathbf{u}_8 = \frac{1}{2}(\mathbf{u}_1 + \mathbf{u}_4) \end{aligned} \quad (3.133)$$

Substituting Equation (3.133) into (3.132) gives

$$u(\mathbf{x}) = \underbrace{\left(\phi_1 + \frac{1}{2}\phi_5 + \frac{1}{2}\phi_8 \right)}_{\phi_1^{\,c}} u_1 + \underbrace{\left(\phi_2 + \frac{1}{2}\phi_5 + \frac{1}{2}\phi_6 \right)}_{\phi_2^{\,c}} u_2$$

$$+ \underbrace{\left(\phi_3 + \frac{1}{2}\phi_6 + \frac{1}{2}\phi_7 \right)}_{\phi_3^{\,c}} u_3 + \underbrace{\left(\phi_4 + \frac{1}{2}\phi_7 + \frac{1}{2}\phi_8 \right)}_{\phi_4^{\,c}} u_4 \tag{3.134}$$

$$= \sum_{I=1}^{4} \phi_I^{\,c} u_I$$

The set of condensed RPIM shape functions for the 4 real nodes are given by

$$\phi_1^{\,c} = \phi_1 + \frac{1}{2}\phi_5 + \frac{1}{2}\phi_8 ; \quad \phi_2^{\,c} = \phi_2 + \frac{1}{2}\phi_5 + \frac{1}{2}\phi_6$$

$$\phi_3^{\,c} = \phi_3 + \frac{1}{2}\phi_6 + \frac{1}{2}\phi_7 ; \quad \phi_4^{\,c} = \phi_4 + \frac{1}{2}\phi_7 + \frac{1}{2}\phi_8 \tag{3.135}$$

This set of 4 RPIM shape functions can then be used as actual RPIM shape functions for these 4 real support nodes.

3.7.3 Properties of the condensed RPIM shape functions

It is a trivial matter to proof that these condensed RPIM shape functions satisfies the partitions:

$$\sum_{I=1}^{4} \phi_I^{\,c} = \underbrace{\left(\phi_1 + \frac{1}{2}\phi_5 + \frac{1}{2}\phi_8 \right)}_{\phi_1^{\,c}} + \underbrace{\left(\phi_2 + \frac{1}{2}\phi_5 + \frac{1}{2}\phi_6 \right)}_{\phi_2^{\,c}}$$

$$+ \underbrace{\left(\phi_3 + \frac{1}{2}\phi_6 + \frac{1}{2}\phi_7 \right)}_{\phi_3^{\,c}} + \underbrace{\left(\phi_4 + \frac{1}{2}\phi_7 + \frac{1}{2}\phi_8 \right)}_{\phi_4^{\,c}} \tag{3.136}$$

$$= \sum_{I=1}^{8} \phi_I = 1$$

Here we used the proven fact that $\phi_I(\mathbf{x})$ is of partition of unity for all these 8 nodes. The Kronecker Delta function property can also be verified easily. For example:

$$\phi_1^c(\mathbf{x}_J) = \phi_1(\mathbf{x}_J) + \frac{1}{2}\phi_5(\mathbf{x}_J) + \frac{1}{2}\phi_8(\mathbf{x}_J) = \begin{cases} 1, & \text{when } J=1 \\ 0, & \text{when } J = 2, 3, 4 \end{cases} \qquad (3.137)$$

Here we used also the proven fact that $\phi_k(\mathbf{x})$ is of Delta function property for all these 8 nodes, and hence $\phi_I(\mathbf{x})=0$ for I=5, 6, 7 and 8.

The same idea can be used for creating other types of condensed RPIM shape functions for other purpose by using proper constraints. The condensed RPIM shape functions offer yet another a big avenue for establishing S-PIM models, and will be used to develop the CS-RPIM model using only 4 local support nodes, as presented in Chapter 8. It is particularly useful for 3D S-PIM models.

3.7.4 3D condensed RPIM shape functions

Following the similar way for creating 2D cases, 3D condensed RPIM shape functions can be formulated using five field nodes selected using face-based T5-scheme plus another six mid-edge virtual nodes, as shown in FIGURE 3.9. Thus compared to the RPIM shape functions using only five field nodes, this condensed RPIM shape functions are designed to improve the performance of the results without increasing the DOFs of the model.

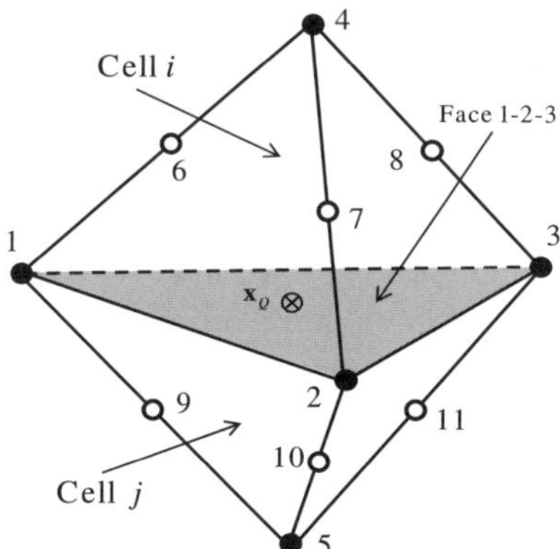

FIGURE 3.9 Five real nodes (1, 2, 3, 4 and 5) selected using the face-based T5-scheme plus six virtual nodes (6, 7, 8, 9, 10 and 11) are used to create a set of 3D condensed RPIM shape functions.

3.8   Other methods

Finally, we list some other types of shape functions in TABLE 3.5 with some important features, but without detailed elaborations. In addition, the table is not inclusive. More detailed discussions can be found in [20].

TABLE 3.5 Features of some other shape functions

Shape functions	Consistency	Global compatibility	Delta function property	Number of support nodes (2D)
SPH [39-45]	No, especially on the boundary	Yes, when sufficient particles are used	No	~25
RKPM [46, 47]	Yes	Yes, when sufficient nodes are used	No	~25
MLS [48, 49]	Yes	Yes, when sufficient nodes are used	No	~25
PIM [4, 10-12]	Yes	Yes (linear) No (higher order)	Yes	3-6 (often 3 or 4)
RPIM [4, 12-14]	Yes	Yes (linear) No (higher order)	Yes	4-18 (often 4 or 6)
Kringing (same as RPIM) [50-52]	Yes	No	Yes	4-18
Minimum length [53]	Yes (for lower orders)	No	Yes	3-6
Linear FEM	Yes	Yes	Yes	3

All these shape functions (except SPH) listed in TABLE 3.5 can be used to create a W^2 model using the GS-Galerkin weak form. Because PIM shape functions use very small number of nodes that is very close to 3 that is the smallest possible number used in the linear FEM. Our choice of PIM for S-PIM models is clear: we aim for computational efficiency. This is because the number of the local nodes used affects directly to the sparseness of the stiffness matrix of the model and hence the computational efficiency. Based on the argument, we use either the linear FEM shape function, which leads to the S-FEM [9], or the PIM shape function that leads to the S-PIM. Below is a simple intuitive analysis.

Remark 3.21 Computational efficiency: a simple analysis

Consider solving a problem using different discrete models but with the same number of say, 100K DOFs or more. These models are created using different proper types of shape functions listed in TABLE 3.5. The total number of DOFs used is the same to all these models. For such a large model, we will always use iterative solvers, and the (relative) CPU time for the equation solver should be roughly proportional to the number of the local support nodes (assuming same DOFs). In this case, an S-PIM model using 4 local support nodes can be about 6 times faster than a model using MLS shape functions created using 25 local nodes [4, 54], if an iterative solver is used (see, TABLE 1.3). If the accuracy is the same, the difference in computational efficiency will also be 6 times.

One may argue that what if we use MLS but with a small number of nodes (treat the compatibility for efficiency)? Well, this is possible in the W^2 formulations, but one needs additional treatments for the essential boundary conditions, because the MLS lacks the Delta function property. The point is that the MLS is designed for compatibility and hence suits better for the Galerkin weak form. When a W^2 form is used, we do not need MLS, and PIM or RPIM works the best, based on this analysis and the experience in the authors' group.

As for small models, the effect of the number of local nodes can be even more significant. However, we generally do not pay too much attention to small models, because the CPU time is small anyway. Letting the PC to run a little long is really not a big deal. Our concern is therefore more for large models that are much more practical, meaningful and important.

3.9 Interpolation error estimation

Error in a numerical solution is very important for reliability reasons. This section discusses the interpolation error with the understanding that the (final) solution error of a numerical model should (somehow) relate to the interpolation error, as long as the model is stable and converge. We will focus only on error bounds and convergence rates for methods using linear interpolation and node-based smoothing domains, and will not give details of proofs for these inequalities but simply list them with the reference sources. For simplicity, we consider only 1D problem in this section.

Consider now the error when a "target" function w is to be interpolated using the interpolant defined in Section 3.1.7. We first define such an *interpolation error* for the given target function w as

$$e^{I}(x) = w(x) - \mathcal{I}_h w(x), \quad \forall x \in \Omega_i^c \tag{3.138}$$

where Ω_i^c is the domain of the ith cell/element defined for a 1D domain. In the standard Galerkin weak formulation such as FEM, we should have the following bounds.

3.9.1 Error in L^2 norm

For $w \in \mathbb{H}_0^1$ and uniform mesh the interpolation error in L^2 norm is [55, 56]

$$\left| e^{I}(x) \right|_{\mathbb{L}^2} = \left| w - \mathcal{I}_h w \right|_{\mathbb{L}^2} \leq h^2 \left(\max_{i=1,\cdots,N_c} \max_{\xi \in \Omega_i^c} \left| w''(\xi) \right| \right) \tag{3.139}$$

The proof of this bound in FEM was based on the Rolle's Theorem. Equation (3.139) gives the h-dependence of the error in $\mathbb{L}^2$ norms in relation to the "strong" norm: the infinity norm of the target function to be approximated by interpolation. In the weak formulation, the h-dependence of the solution error in displacement norm is the same as that of the *interpolation error* in L^2 norm [55, 56]. Therefore, Equation (3.139) gives the essence of the h-dependence of the solution error in displacement norm.

3.9.2 Error in H^1 norm

For $w \in \mathbb{H}_0^1$ and uniform mesh the interpolation error in H^1 norm [55, 56]

$$\left| e^{I}(x) \right|_{\mathbb{H}^1} = \left| w - \mathcal{I}_h w \right|_{\mathbb{H}^1} \leq h \left(\max_{i=1,\cdots,N_c} \max_{\xi \in \Omega_i^c} \left| w''(\xi) \right| \right) \tag{3.140}$$

which gives the h-dependence of the error in $\mathbb{H}^1$ norm. In the weak formulation, the h-dependence of the solution error in energy norm is the same as that of the interpolation error in H^1 norm. Therefore, Equation (3.140) reveals the essence of the h-dependence of the solution error in energy norm. We note the following remark without proof.

Remark 3.22 The rate of convergence of linear FEM solution: a proven fact

The theoretical convergence rate of the solution error in L^2 norm (error in displacement norm) for a linear FEM (fully compatible) model is 2.0, and that in H^1 norm (error in energy norm) is 1.0.

3.9.3 Error in $\mathbf{G}^1$ norm

Theorem 3.4 Interpolation error in $\mathbb{G}_h^1$ norm

If the target function $w \in \mathbb{G}_{h,0}^1$ and $w|_{\Omega_i^c} \in \mathbb{C}^2(\Omega_i^c), i = 1, \cdots, N_c$, the linear interpolation error satisfies, in general:

$$\left| e^I \right|_{\mathbb{G}^1(\Omega)} = \left| w - \mathcal{I}_h w \right|_{\mathbb{G}^1(\Omega)} \le h_{\max} \frac{3c_{rh}}{4} \left(\max_{q=1,\cdots,N_c} \max_{\xi \in \Omega_q^c} \left| w''(\xi) \right| \right) \tag{3.141}$$

where $c_{rh} = h_{\max}/h_{\min}$. In particular, when uniform mesh is used, and the 2nd derivative of w is constant in the cells sharing the smoothing domains we further have

$$\left| e^I \right|_{\mathbb{G}^1(\Omega)} = \left| w - \mathcal{I}_h w \right|_{\mathbb{G}^1(\Omega)} = h^{1.5} \frac{1}{4} \left| w'' \right| \tag{3.142}$$

This rate of 1.5 is termed as *ideal convergence rate* for a model based on $\mathbb{G}_h^1$ space theory that is achieved for a uniform mesh and constant 2nd derivative of w.

Proof

Consider a target function $w \in \mathbb{G}_{h,0}^1$ and $w|_{\Omega_i^c} \in \mathbb{C}^2(\Omega_i^c), i = 1, \cdots, N_c$. Using the Taylor's expansion with respect to x_n, there exists a $\xi \in \Omega_n^c$ such that

$$w(x) = w(x_n) + w'(x_n)(x - x_n) + \frac{1}{2} w''(\xi(x))(x - x_n)^2, \ \forall x \in \Omega_n^c \tag{3.143}$$

In Equation (3.143), we note the fact that ξ is in fact dependent on x. We also note that we do not require the existence of w'' on the boundary of Ω_n^c : meaning that the first derivative of w can "jump" there. Using Equation (3.17), the linear interpolant has to pass through two nodes at x_n and x_{n+1}, and hence should be given as

$$\mathcal{I}_h w(x) = w(x_n) + \left[w'(x_n) + \frac{1}{2} w''(\xi(x_{n+1})) h_n \right] (x - x_n), \ \forall x \in \Omega_n^c \tag{3.144}$$

Based on the definition (2.88), for our 1D problem using node-based smoothing domains, we have the nth smoothing cell,

$$\bar{g}(w_n) = \overline{\frac{\partial w}{\partial x}}$$

$$= \frac{1}{A_n^s} \int_{\Gamma_n^s} w(s) n_1 \mathrm{d}s$$

$$= \frac{1}{A_n^s} \left(w_{n+\frac{1}{2}} - w_{n-\frac{1}{2}} \right) \tag{3.145}$$

$$= \frac{2}{h_n + h_{n-1}} \left(w_{n+\frac{1}{2}} - w_{n-\frac{1}{2}} \right)$$

where $w_n = w(x_n)$, and $h_0 = h_{N_n} = 0$. Substitute Equations (3.143) and (3.144) into (3.145), we then have for the nth smoothing cell:

$$\underbrace{\bar{g}(w_n - \mathcal{I}_h w_n)}_{e^I} = \frac{2}{h_n + h_{n-1}} \left[\frac{1}{2} w''\left(\xi(x_{n+\frac{1}{2}}) \right) (x_{n+\frac{1}{2}} - x_n)^2 \right.$$

$$- \frac{1}{2} w''\left(\xi(x_{n+1}) \right) h_n (x_{n+\frac{1}{2}} - x_n)$$

$$- \frac{1}{2} w''\left(\xi(x_{n-\frac{1}{2}}) \right) (x_{n-\frac{1}{2}} - x_{n-1})^2$$

$$\left. + \frac{1}{2} w''\left(\xi(x_n) \right) h_{n-1} (x_{n-\frac{1}{2}} - x_{n-1}) \right] \tag{3.146}$$

$$= \frac{2}{h_n + h_{n-1}} \left\{ \frac{h_n^2}{4} \left[\frac{1}{2} w''\left(\xi(x_{n+\frac{1}{2}}) \right) - w''\left(\xi(x_{n+1}) \right) \right] \right.$$

$$\left. - \frac{h_{n-1}^2}{4} \left[\frac{1}{2} w''\left(\xi(x_{n-\frac{1}{2}}) \right) - w''\left(\xi(x_n) \right) \right] \right\}$$

Let

$$w''_{\max} = \max_{i=1,\cdots,N_e} \max_{\xi \in \Omega_i^c} \left| w''(\xi) \right| \tag{3.147}$$

and hence we shall have $w''_{\max} \neq 0$, otherwise the 2nd derivative of w will be zero everywhere in the problem domain, and the interpolation will be exact: no need for error estimation. Using Equation (3.144), Equation (3.146) becomes

$$\left|\overline{g}(\underbrace{w_n - \mathcal{I}_h w_n}_{e^I})\right| = \frac{h_{\max}^2}{4}\underbrace{\frac{2}{h_n + h_{n-1}}}_{\leq \frac{1}{h_{\min}}}\left|w_{\max}''\right|$$

$$\left\|\left[\frac{1}{2}\underbrace{\frac{h_n^2}{h_{\max}^2}}_{\leq 1}\underbrace{\frac{w''\left(\xi(x_{n+\frac{1}{2}})\right)}{\left|w_{\max}''\right|}}_{-1\leq,\leq 1} - \underbrace{\frac{h_n^2}{h_{\max}^2}}_{\leq 1}\underbrace{\frac{w''\left(\xi(x_{n+1})\right)}{\left|w_{\max}''\right|}}_{-1\leq,\leq 1}\right.\right.$$

$$\left.\left.\underbrace{-\frac{1}{2}\underbrace{\frac{h_{n-1}^2}{h_{\max}^2}}_{\leq 1}\underbrace{\frac{w''\left(\xi(x_{n-\frac{1}{2}})\right)}{\left|w_{\max}''\right|}}_{-1\leq,\leq 1} + \underbrace{\frac{h_{n-1}^2}{h_{\max}^2}}_{\leq 1}\underbrace{\frac{w''\left(\xi(x_n)\right)}{\left|w_{\max}''\right|}}_{-1\leq,\leq 1}\right]}_{c_{hw}\leq 3}\right\| \tag{3.148}$$

$$\leq \frac{3h_{\max}}{4}\underbrace{\frac{h_{\max}}{h_{\min}}}_{c_{rh}<\infty}\left|w_{\max}''\right| = h_{\max}\frac{3c_{rh}}{4}\left|w_{\max}''\right|$$

The error defined in $\mathbb{G}_h^1$ norm in the global problem domain becomes

$$\left|w - \mathcal{I}_h w\right|_{\mathbb{G}^1}^2 = \sum_{n=1}^{N_s} A_n^s \left|\overline{g}(w_n - \mathcal{I}_h w_n)\right|^2 \leq \sum_{n=1}^{N_s} h_{\max}\left(h_{\max}\frac{3c_{rh}}{4}\left|w_{\max}''\right|\right)^2$$

$$= \underbrace{N_s}_{\leq \frac{1}{h_{\min}}} h_{\max}\left(h_{\max}\frac{3c_{rh}}{4}\left|w_{\max}''\right|\right)^2 \leq \left(h_{\max}\frac{3c_{rh}^{\frac{3}{2}}}{4}\left|w_{\max}''\right|\right)^2 \tag{3.149}$$

Therefore, we have Equation (3.141).

In Equation (3.148), we needed to estimate c_{hw} and gave a very sloppy bound of 3. This is in fact a very big overestimate in a usual situation. If a uniform mesh is used, and the target function has a constant 2nd derivative in the cells sharing a smoothing domain, c_{hw} should be zero for all the inner smoothing domains. In such a situation, Equation (3.148) becomes

$$\left|\overline{g}(w_n - \mathcal{I}_h w_n)\right| = \begin{cases} \dfrac{h}{4}\left|w_{\max}''\right|, & n = 1, N_n \\[2mm] 0, & n = 2, 3, \cdots, N_n - 1 \end{cases} \tag{3.150}$$

and Equation (3.149) becomes

$$
\begin{aligned}
\left| w - \mathcal{I}_h w \right|^2_{G^1} &= \sum_{n=1}^{N_s} A_n^s \left| \bar{g}(w_n - \mathcal{I}_h w_n) \right|^2 \\
&= A_1 \left| \bar{g}(w_1 - \mathcal{I}_h w_1) \right|^2 + A_{N_n} \left| \bar{g}(w_{N_n} - \mathcal{I}_h w_{N_n}) \right|^2 \\
&= \frac{h}{2}\left(\frac{h}{4}|w''_{\max}| \right)^2 + \frac{h}{2}\left(\frac{h}{4}|w''_{\max}| \right)^2 \\
&= h\left(\frac{h}{4}|w''_{\max}| \right)^2 \\
&= \frac{h^3}{4^2}|w''_{\max}|^2
\end{aligned}
\tag{3.151}
$$

which is Equation (3.142). This completes the proof. $\qquad\square$

Note in Equations (3.142) the bound is sharpest: strict equality. The proof on the above two equations were originally given in [6].

In the W^2 formulation based on G space theory, we may expect the rate of h-dependence of the solution error being the same as that of the interpolation error. However, this has not been proven theoretically. We note the following remark of expectation without proof.

Remark 3.23 The convergence rate of W^2 solution based on G^1 space: an expectation

The theoretical rate of convergence of the solution error in $\mathbb{L}^2$ norm (error in displacement norm) for W^2 model based on $\mathbb{G}_h^1$ space is expected to be 2.0, and that in $\mathbb{G}_h^1$ norm (error in energy norm) is expected in between 1.0 and the ideal rate of 1.5.

Remark 3.24 Symmetry of the smoothing domains

Note in Equation (3.150) that the symmetry of the smoothing domain plays a very important role in reducing the error. This suggests that we shall make the smoothing domains symmetric in S-PIM models. In the interior of the problem domain, we can make these smoothing domains "quite" symmetric, although not exactly symmetric. On the boundary of the problem, however, it is not possible, because these smoothing domains are always "biased", due to the cut-off of problem domain on the boundary. Equation (3.150) measures precisely the error caused by the biased smoothing domains on the boundary. This natural of the problem is quite similar to the boundary effect problem in the SPH method [45]. To reduce such an error, we need to reduce the size of the background cells on near the boundary.

3.9.4 Comparison errors in G^1 and H^1 norms

Although an exact comparison is difficult, because the space difference between $\mathbb{H}^1_{h,0}$ and $\mathbb{G}^1_{h,0}$ and hence the target function w will be different. But an indicative comparison can be useful. For errors in semi-norm measure, Equation (3.140) is quite close to Equation (3.141) for uniform mesh ($c_{rh}=1.0$): they all give a convergence rate h-dependence of 1.0. Equation (3.142) shows, however, the W^2 formulation *can* provide a convergence rate in semi-norm measure of 1.5 for cases of even division of node-based smoothing domains. Compared to Equation (3.140), the convergence rate is 50% higher.

Equation (3.142) was obtained under the conditions of 1) uniform division of elements/cells and 2) constant 2nd derivative of w in the cells sharing a smoothing domain. The first condition is essentially the same "symmetrical condition" of smoothing domains for the integral representation to produce the first gradient of a function exactly. When this condition is satisfied, all the interior smoothing domains become symmetrical, and the smoothing operation will reproduce the first derivative exactly. The only error will be on the boundary where the symmetry condition cannot be satisfied. The second condition of constant 2nd derivative of the target function seems to be very strong, but it can be rather easy to be satisfied closely, because all we need is the 2nd derivative of the target function being constant *locally* in the cells sharing a smoothing domain. When the mesh is refined, we can often expect the 2nd derivative being approximately constant locally. Therefore, the rate given in Equation (3.142) can be expected when the mesh is reasonably fine. In practical applications, on the other hand, the first condition is rather very difficult to meet, simply because it is rare to have uniform division of element/cells for practical problems of complicated geometry. However, we can in fact to expect the smoothing domains to be approximately "symmetric" locally ($h_n \approx h_{n-1}$ for node n for our 1D problems). In such cases, we can still expect Equation (3.142) holds approximately and hence a convergence rate of 1.5 in $\mathbb{G}^1_h$ semi-norm. This has been confirmed in many numerical examples and numerical rates of about 1.4 were often found, as will be shown also in later chapters.

Let's now further examine for general problem with reasonably smoothing 2nd derivative. The coefficient c_{hw} in Equation (3.148) becomes

$$
\begin{aligned}
c_{hw} &= \left\| \left[\begin{array}{c} \underbrace{\frac{1}{2}\frac{h_n^2}{h_{max}^2}}_{\leq 1}\underbrace{\frac{w''\left(\xi(x_{n+\frac{1}{2}})\right)}{\left|w''_{max}\right|}}_{-1\leq,\leq 1} - \underbrace{\frac{h_n^2}{h_{max}^2}}_{\leq 1}\underbrace{\frac{w''\left(\xi(x_{n+1})\right)}{\left|w''_{max}\right|}}_{-1\leq,\leq 1} \\[4mm] -\underbrace{\frac{1}{2}\frac{h_{n-1}^2}{h_{max}^2}}_{\leq 1}\underbrace{\frac{w''\left(\xi(x_{n-\frac{1}{2}})\right)}{\left|w''_{max}\right|}}_{-1\leq,\leq 1} + \underbrace{\frac{h_{n-1}^2}{h_{max}^2}}_{\leq 1}\underbrace{\frac{w''\left(\xi(x_{n})\right)}{\left|w''_{max}\right|}}_{-1\leq,\leq 1} \end{array} \right] \right\| \\[4mm]
&\approx \frac{1}{2}\frac{1}{h_{max}^2}\underbrace{\frac{\left|w''\left(\xi(x_n)\right)\right|}{\left|w''_{max}\right|}}_{\leq 1}\left|h_n^2 - h_{n-1}^2\right| \\[4mm]
&= \underbrace{\frac{1}{2}\frac{h_n\left(h_n+h_{n-1}\right)}{h_{max}^2}\frac{\left|w''\left(\xi(x_n)\right)\right|}{\left|w''_{max}\right|}}_{\leq 1}\left|1-\underbrace{\frac{h_{n-1}}{h_n}}_{\approx 1}\right| \\[4mm]
&\leq \left|1-\underbrace{\frac{h_{n-1}}{h_n}}_{\approx 1}\right|
\end{aligned}
\tag{3.152}
$$

If the lengths of the two neighboring cells of the interior nodes are different, the contribution of these nodes to the error norm *can* be in control (when mesh is refined), and thus the convergence rate will be about 1.0 and not 1.5. However, c_{hw} in Equation (3.148) can be a very small number. If, for example, there is a 20% length difference in the two neighboring cells of all the interior nodes ($h_{n-1}/h_n = 0.8$), we shall have $c_{hw} \approx \left|1-h_{n-1}/h_n\right| = 0.2$. This means that even the convergence rate cannot be improved, the results can be about 5 times more accurate. This was also observed in the edge-based smoothing point interpolation method (ES-PIM) where the edge-based smoothing domains are not quite symmetric, but were often found about 2~10 times more accurate in energy norm compared to the FEM using the same mesh [57, 58]. In extreme cases, where all the smoothing domain are not symmetric at all, the results of the W^2 formulation will still be better than the standard weak formulation, as shown in Equation (3.140) and Equation (3.141).

Because a finite $\mathbb{H}_h^1$ space is a sub-space of a corresponding $\mathbb{G}_h^1$ space, the convergence rates found in $\mathbb{G}_h^1$ norm measures are applicable to functions in the $\mathbb{H}_h^1$ space. This means that when a weakened weak formulation is applied to a FEM setting, we will achieve higher convergence rate and/or higher accuracy.

Finally, we note that for the weakened weak formulation, the relationship between the *interpolation error* and the *solution error* has not yet made precise. Similar relationships stated in Sections 3.9.1 and 3.9.2 should exist if the model is stable and converge, but yet proven rigorously.

3.10 Concluding remarks

We have discussed in great detail on the techniques for creating various types of PIM shape functions, the important properties, as well as the interpolation error using these shape functions. The creation of the PIM shape functions is very simple, direct interpolation, using very a small number of local nodes, no mapping is needed, and having all the necessary important properties. The linear independence of the PIM nodal shape functions is ensured using a proper T-scheme for node selection. The PIM shape functions as well as any function constructed using these shape functions are in a $\mathbb{G}_h^1$ space, as long as a proper set of smoothing domains can be established for the model. We shall use these shape functions for all our S-PIM modes in later chapters.

3.11 References

1. Liu, G. R. and Quek, S. S., *The finite element method: a practical course.* Butterworth Heinemann: Oxford. 2003.

2. Zienkiewicz, O. C. and Taylor R. L., *The Finite Element Method*, 5th ed., Butterworth Heimemann, Oxford, 2000.

3. Belytschko, T., Krongauz, Y., Organ, D., Fleming, M. and Krysl, P., Meshless method: an overview and recent developments, *Comput. Methods Appl. Mech. Eng.*, 139: 3-47, 1996.

4. Liu, G. R., *Meshfree Methods: Moving beyond the Finite Element Method*, 2nd edition, CRC press, Boca Taton, USA, 2009.

5. Liu, G. R., On the G space theory. *International Journal of Computational Methods*, 6(2): 257-289, 2009.

6. Liu, G. R., A G space theory and a weakened weak (W^2) form for a unified formulation of compatible and incompatible methods, Part I Theory, *International Journal for Numerical Methods in Engineering*, 81: 1093-1126, 2010.

7. Liu, G. R., A G space theory and a weakened weak (W^2) form for a unified formulation of compatible and incompatible methods, Part II Application to solid mechanics problems, *International Journal for Numerical Methods in Engineering*, 81: 1127-1156, 2010.

8. Liu, G. R. and Zhang, G. Y., A normed G space and weakened weak (W^2) formulation of cell-based smoothed point interpolation method. *International Journal of Computational Methods*, 6(1): 147-179, 2009.

9. Liu, G. R. and Nguyen T.T, *Smoothed Finite Element Method*, CRC press, Boca Taton, USA, 2010.

10. Liu, G. R. and Gu, Y. T., A point interpolation method, in *Proc. 4th Asia-Pacific Conference on Computational Mechanics*, December, Singapore, 1009-1014, 1999.

11. Liu, G. R. and Gu, Y. T., A point interpolation method for two-dimensional solids, *Int. J. Numer. Methods Eng.*, 50: 937-951, 2001.

12. Liu, G. R. and Gu, Y. T., *An introduction to meshfree method methods and their programming*, Springer, 2005.

13. Wang, J. G. and Liu, G. R., Radial point interpolation method for elastoplastic problems, in *Proc. 1st Int. Conf. on Structural Stability and Dynamics*, December 7–9, Taipei, China, 703–708, 2000.

14. Wang, J. G. and Liu, G. R., A point interpolation meshless method based on radial basis functions, *Int. J. Numer. Methods Eng.*, 54: 1623-1648, 2002.

15. Chen, L., Liu, G. R., Nourbakhshnia and N., Zeng, K., A singular edge-based smoothed finite element method (ES-FEM) for biomaterial interface cracks. *Computational Mechanics*, 45: 109-125, 2010.

16. Liu, G. R., Nourbakhshnia, N., Chen, L. and Zhang, Y. W., A novel general formulation of singular stress field using the ES-FEM method for the analysis of mixed-mode cracks. *International Journal of Computational Methods*, 7(1): 191-214, 2010.

17. Liu, G. R., Chen, L., Nguyen-Thoi, T., Zeng, K. and Zhang, G. Y., A novel singular node-based smoothed finite element method (NS-FEM) for upper bound solutions of fracture problems, *International Journal for Numerical Methods in Engineering*, 83(11): 1466-1497, 2010.

18. Liu, G. R., Dai, K. Y. and Nguyen-Thoi, T., A smoothed finite element method for mechanics problems, *Computational Mechanics*, 39: 859-877, 2007.

19. Dai, K. Y., Liu G. R. and Nguyen-Thoi, T., An n-sided polygonal smoothed finite element method (nSFEM) for solid mechanics. *Finite Elements in Analysis and Design*, 43: 847-860, 2007.

20. Liu, G. R., *Meshfree Methods: Moving beyond the Finite Element Method*, 1st edition, CRC press, Boca Taton, USA, 2002.

21. Liu, G. R. and Gu, Y. T., A local radial point interpolation method (LRPIM) for free vibration analyses of 2-D solids, *Journal of Sound and Vibration*, 246(1): 29-46, 2001.

22. Liu, G. R., Zhang, G. Y., Gu, Y. T. and Wang, Y. Y., A meshfree radial point interpolation method (RPIM) for three-dimensional solids, *Computational Mechanics*, 36(6): 421-430, 2005.

23. Liu, G. R. and Gu Y. T., A matrix triangularization algorithm for the polynomial point interpolation method, *Computer Methods in Applied Mechanics and Engineering*, 192(19): 2269-2295, 2003.

24. Hardy, R. L., Theory and applications of the multiquadrics—biharmonic method (20 years of discovery 1968–1988), *Comput. Math. Appl.*, 19: 163-208, 1990.

25. Kansa, E. J., Multiquadrics - A scattered data approximation scheme with application to computational fluid-dynamics .1., *Computers & Mathematics with Applications*, 19(8-9): 127-145, 1990.

26. Kansa, E. J., Multiquadrics - A scattered data approximation scheme with applications to computational fluid dynamics .2. Solutions to parabolic, hyperbolic and elliptic partial-differential equations, *Computers & Mathematics with Applications*, 19(8-9): 147-161, 1990.

27. Powell, M. J. D., The theory of radial basis function approximation in 1990, in *Advances in Numerical Analysis*, Light, F. W., Ed., Oxford University Press, Oxford, 303-322, 1992.

28. Coleman, C. J., On the use of radial basis functions in the solution of elliptic boundary value problems, *Comput. Mech.*, 17: 418-422, 1996.

29. Sharan, M., Kansa and E. J., Gupta, S., Application of the multiquadric method for numerical solution of elliptic partial differential equations, *Appl. Math. Comput.*, 84: 275-302, 1997.

30. Fasshauer, G. E., Solving partial differential equations by collocation with radial basis functions, in *Surface Fitting and Multiresolution Methods*, Mehaute, A. L., Rabut, C., and Schumaker, L. L., Eds., 131-138, 1997.

31. Wendland, H., Error estimates for interpolation by compactly supported radial basis functions of minimal degree, *J. Approx. Theor.*, 93: 258-396, 1998.

32. Wang, J. G. and Liu, G. R. On The Optimal Shape Parameters Of Radial Basis Functions Used For 2-D Meshless Methods, *Computer Methods in Applied Mechanics And Engineering*, 191: 2611-2630, 2002.

33. Golberg, M. A., Chen, C. S. and Bowman, H., Some recent results and proposals for the use of radial basis functions in the BEM, *Eng. Anal. Boundary Elements*, 23: 285-296, 1999.

34. Schaback, R., Approximation of polynomials by radial basis functions, in *Wavelets, Images and Surface Fitting*, Laurent, P. J., Mehaute, L., and Schumaker, L. L., Eds., A. K. Peters Ltd. Wellesley, MA, 459-466, 1994.

35. Liu, G. R., Li, Y. Dai, K.Y. Luan M. T. and Xue, W., A Linearly conforming radial point interpolation method for solid mechanics problems, *International Journal of Computational Methods*, 3: 401-428, 2006.

36. Li, Y., Liu, G. R., Luan, M. T., Dai, K. Y., Zhong, Z. H., Li, G. Y. and Han, X., Contact analysis for solids based on linearly conforming radial point interpolation method, *Comput. Mech.*, 39: 537-554, 2007.

37. Zhang, G. Y. and Liu, G. R., Meshfree cell-based smoothed point interpolation method using isoparametric PIM shape functions and condensed RPIM shape functions. *International Journal of Computational Methods*, 8(4): 705-730, 2011.

38. Liu, G. R., Zhang, G. Y., Zong, Z. and Li, M., Meshfree cell-based smoothed alpha radial point interpolation method (CS-αRPIM) for solid mechanics problems. *International Journal of Computational Methods*, 10(4), 31 pages, DOI: 10.1142/S0219876213500205, 2013.

39. Lucy, L., A numerical approach to testing the fission hypothesis, *Astron. J.*, 82: 1013-1024, 1977.

40. Gingold, R. A. and Monaghan, J. J., Smooth particle hydrodynamics: theory and applications to non-spherical stars, *Mon. Not. R. Astron. Soc.*, 181: 375-389, 1977.

41. Monaghan, J. J., Why particle methods work, *Siam J. Sci. Sat. Comput.*, 3(4): 423-433, 1982.

42. Monaghan, J. J., Smoothed particle hydrodynamics. *Annual Review of Astronomy and Astrophysics*, 30: 543-574, 1992.

43. Monaghan, J. J. and Lattanzio, J. C., A refined particle method for astrophysical problems. *Astronomy and Srtophsics,* 149: 135-143, 1985.

44. Libersky, L. D. and Petscheck, A. G., Smoothed particle hydrodynamics with strength of materials. In Proceeding of The Next Free Lagrange Conference, Trease, H., Fritts, J. and Crowley, W. eds., Springer-Verlag, NY, 395: 248-257, 1991.

45. Liu, G. R. and Liu, M. B., Smoothed Particle Hydrodynamics—A Meshfree Practical Method. World Scientific, Singapore, 2003.

46. Liu, W. K., Adee, J. and Jun, S., Reproducing kernel and wavelet particle methods for elastic and plastic problems, in *Advanced Computational Methods for Material Modeling*, Benson, D. J., Ed., 180/PVP 268 ASME, 175-190, 1993.

47. Liu, W. K., Jun, S. and Zhang, Y., Reproducing kernel particle methods, *Int. J. Numer. Methods Fluids*, 20: 1081-1106, 1995.

48. Nayroles, B., Touzot, G., and Villon, P., Generalizing the finite element method: diffuse approximation and diffuse elements, *Comput. Mech.*, 10: 307-318, 1992.

49. Belytschko, T., Lu, Y. Y. and Gu, L., Element-free Galerkin methods, *Int. J. Numer. Methods Eng.*, 37: 229-256, 1994.

50. Gu L, Moving Kriging interpolation and element-free Galerkin method, *International Journal for Numerical Methods in Engineering*, 56: 1-11, 2003.

51. Dai, K. Y., Liu, G. R., Lim, K. M. and Gu, Y. T., Comparison between the radial point interpolation and the Kriging based interpolation using in meshfree methods. *Computational Mechanics*, 32: 60-70, 2003.

52. Tongsuk, P. and Kanok, N. W., Further investigation of element-free Galerkin method using moving kriging interpolation, *International Journal of Computational Methods*, 1: 345-365, 2004.

53. Liu, G. R., Dai, K. Y., Han, X. and Li, Y., A meshfree minimum length method for 2-D problems. *Computational Mechanics*, 38: 533-550, 2006.

54. Chen, L., Nguyen-Thoi, T., Zeng, K. Y. and Wu, S.C., Assessment of smoothed point interpolation methods for elastic mechanics. *Communications in Numerical Methods in Engineering*, 26: 1635-1655, 2010.

55. Peraire, J., *Lecture notes on finite element methods for elliptic problems.* MIT. 1999.

56. Strang, G. and Fix, G. J., *An analysis of the finite element method*, Prentice-hall, 1973.

57. Liu, G. R. and Zhang, G. Y., Edge-based smoothed point interpolation method (ES-PIM), *International Journal of Computational Methods*. 5(4): 621-646, 2008.

58. Liu, G. R., Nguyen-Thoi, T. and Lam, K. Y., An edge-based smoothed finite element method (ES-FEM) for static, free and forced vibration analyses in solids. *Journal of Sound and Vibration*, 320: 1100-1130, 2009.

Chapter 4

Strain Field Construction

The standard finite element method (FEM) [1, 2] procedure consists of largely three steps:

1) domain discretization using elements,

2) displacement field construction (shape function creation based on elements), and

3) weak formulation for the discretized algebraic equations system.

In the S-PIM, however, we need to introduce one more additional step: *strain field construction* [3] and have

1) domain discretization by triagulation,

2) displacement field construction (shape function creation based on nodes),

3) strain field construction, and

4) weakened weak formulation for the discretized algebraic equations system.

This additional third step offers a new avenue to develop models of special properties and to improve the solution accuracy. Compared to the displacement construction, the theories and techniques for strain field construction are, however, relatively less developed, and there are still many theoretical issues need to be further studied. With Theorems 3.1, 3.2 and 3.3 in place to ensure a proper displacement construction, we are ready now to discuss in this chapter some of the techniques used for strain field construction in S-PIM models. A model using a constructed strain field is, in general, coined as the strain-constructed model or SC-model [3]. Our S-PIM models are typical SC-models that use PIM shape functions.

4.1 Why constructing strain field?

Once the displacement function in a proper H space is constructed, the compatible strain field is already available using simply the strain-displacement relation. In such a standard Galerkin FEM formulation, because the assumed displacement field and the resultant strain field are all *compatible* over the entire problem domain, the Galerkin model is said *fully compatible*. However, this type of fully compatible Galerkin formulation behaviors "overly-stiff", which can have possible consequences of

1) the so-called "locking" behavior for many problems,

2) inaccuracy in stress solutions,

3) poor solution using triangular mesh,

4) reliance on quality quadrilateral mesh and sensitive to mesh distortion, and

5) giving only lower bound.

We, engineers dealing with practical engineering problems, often prefer using lower order of triangular types of mesh as they can be created much more easily in automatic manner for complicated geometries, and minimum requirement on the solution regularity. In addition, we require often quality stress solutions. Moreover, we demand not only for solutions but *certificates* on the quality of the solutions for reliable design of structural systems.

Many efforts have been made in the FEM settings to resolve the overly-stiff phenomenon, especially in the area of hybrid or mixed FEM formulations based on three or two field principles [4, 5]. Some kinds strain modifications have also been used in FEM settings. However, these efforts are made within the framework of elements. It is clear that more effective means or out-of-box approaches beyond the standard variational principles or operations beyond the elements that brings in information from the neighboring elements are necessary, and can be much more effective.

The S-PIM works naturally well with triangular meshes and offers solutions to the above-mentioned issues without the increase of degrees of freedom (DOFs). The essential idea in the S-PIM is *to construct a new strain field*, hoping a "Galerkin" model using the newly constructed strain field can have some good properties. Such a construction is performed *beyond the element/cell*. Since the strain field is constructed anew, we can further reduce the requirement on the assumed displacement functions, and thus the strain-constructed Galerkin (SC-Galerkin, Chapter 5) weak forms can be used, and offer many important and attractive properties, such as upper bound, lower bound, superconvergence, free of locking, and ultra-accuracy. In addition, various incompatible methods can also be effectively formulated with ensured stability and convergence. The idea of strain construction opens a new window of opportunity for a new class of numerical methods outside the framework of the standard principles [6, 7].

4.2 Discrete models: a base for strain construction

Consider a d-dimensional problem, the problem domain Ω is first divided, using the triangulation procedure discussed in Section 1.6.1, into N_c nonoverlapping and seamless (NOSL) triangular/tetrahedron background cells, such that $\boxed{\Omega} = \bigcup_{i=1}^{N_c} \boxed{\Omega}_i^c$ and $\Omega_i^c \cap \Omega_j^c = \varnothing$, $\forall i \neq j$, with a set of N_n nodes. The strain construction in the S-PIM is then based on this discrete model of triangular background cells. The strain at a point or in a smoothing domain is constructed using the information from more than one local surrounding cell in a proper manner. Note that other types of background cells can also be used for strain field construction. We choose to use triangular cells, simply because of the easy creation of such cells and convenience in adaptive analyses.

4.3 General procedure for strain construction

In general, the strain component $\hat{\varepsilon}_{ij}(\mathbf{u}^h) \in \mathbb{L}^2(\Omega)$ is constructed using only the assumed displacement functions $\mathbf{u}^h \in (\mathbb{G}_h^1)^d$, without introducing additional degrees of freedom (DOFs). In the matrix form we can write

$$\hat{\boldsymbol{\varepsilon}}(\mathbf{u}^h) = \mathcal{B}\left(\mathbf{u}^h\right) \tag{4.1}$$

where $\mathcal{B}$ is a general *transformation* operator over the coordinate space for a given input $\mathbf{u}^h \in (\mathbb{G}_h^1)^d$.

For the boundedness of the strain energy, the constructed strain field $\hat{\boldsymbol{\varepsilon}}(\mathbf{u}^h)$ needs only in $\mathbb{L}^2(\Omega)$, meaning that we only require $\hat{\boldsymbol{\varepsilon}}(\mathbf{u}^h)$ being piecewise continuous in the problem domain Ω. Therefore, in an actual discrete model, the domain Ω is usually divided into a set of quadrature/integration domains $\boxed{\Omega} = \bigcup_{i=1}^{N_q} \boxed{\Omega}_i^q$, and the strain construction is done piecewise over in each Ω_i^q. In addition, $\hat{\boldsymbol{\varepsilon}}(\mathbf{u}^h)$ in Ω_i^q is usually made continuous of lower order, often constant or linear, for easy strain energy integration. We then try to obtain strains at a number of points in domain $\boxed{\Omega}_i^q$ for the given $\mathbf{u}^h \in (\mathbb{G}_h^1)^d$, and the constructed strain field $\hat{\boldsymbol{\varepsilon}}_q(\mathbf{u}^h)$ in Ω_i^q can be obtained using simply (again) the point interpolation method (PIM), or other approximation methods discussed in Chapter 3, as in the construction of the displacement field. After the strain field is constructed, the strain potential energy can then be evaluated.

For any $\mathbf{u}^h \in (\mathbb{G}_{h,0}^1)^d$ (including $\mathbf{u}^h \in (\mathbb{H}_0^1)^d$), the strain energy potential for the constructed strain field becomes:

$$\hat{U}_{PE}(\mathbf{u}^h) = \int_\Omega \frac{1}{2} \hat{\boldsymbol{\varepsilon}}^{\mathrm{T}}(\mathbf{u}^h)\, \mathbf{c}\hat{\boldsymbol{\varepsilon}}(\mathbf{u}^h) \mathrm{d}\Omega \tag{4.2}$$

Remark 4.1 SC-models: possible zero-energy modes

The potential energy computed using Equation (4.2) is bounded because $\hat{\varepsilon}_{ij}(\mathbf{u}^h) \in \mathbb{L}^2(\Omega)$. It will be surely nonnegative, but can be zero depends on how the construction is performed. This means that the resultant stiffness matrix of the SC-model can be semi symmetric positive definite (semi-SPD) even for solids of stable materials, leading to zero energy modes and hence can be spatially instable (Remark 1.9). We thus need to impose some sort of admissible conditions to the constructed strain fields.

We note in the standard FEM, the strain energy potential for the original compatible strain field becomes:

$$\tilde{U}_{PE}(\mathbf{u}^h) = \int_\Omega \frac{1}{2} \tilde{\boldsymbol{\varepsilon}}^{\mathrm{T}}(\mathbf{u}^h)\, \mathbf{c}\tilde{\boldsymbol{\varepsilon}}(\mathbf{u}^h) \mathrm{d}\Omega \tag{4.3}$$

where $\mathbf{u}^h \in (\mathbb{H}_0^1)^d$ and $\tilde{\boldsymbol{\varepsilon}}$ is the vector of compatible strains obtained using Equation (1.5).

Remark 4.2 SC-models: no increase of DOFs

Since the constructed strain field depends entirely on the assumed displacement field, there are no additional unknowns or other parameters introduced: the degrees of freedom (DOFs) of the SC-model are not increased compared to the standard linear FEM model.

4.4 Admissible conditions for constructed strain field

The condition of $\hat{\boldsymbol{\varepsilon}} \in \mathbb{L}^2$ is only a minimum condition ensuring the potential energy defined in Equation (4.2) being bounded, and the spatial stability is left uncontrolled. Liu [3] has presented three additional admissible conditions for a constructed strain field for establishing stable and convergent SC-models.

4.4.1 Condition 1: orthogonal condition

For any $\mathbf{v} \in (\mathbb{H}^1_{h,0})^d$, when the constructed strain field $\hat{\boldsymbol{\varepsilon}}(\mathbf{v})$ and the compatible strain field $\tilde{\boldsymbol{\varepsilon}}(\mathbf{v}) = \mathbf{L}_d \mathbf{v}$ satisfy

$$\int_\Omega \hat{\boldsymbol{\varepsilon}}^{\mathrm{T}}(\mathbf{v}) \mathbf{c} \hat{\boldsymbol{\varepsilon}}(\mathbf{v}) \mathrm{d}\Omega = \int_\Omega \hat{\boldsymbol{\varepsilon}}^{\mathrm{T}}(\mathbf{v}) \mathbf{c} (\mathbf{L}_d \mathbf{v}) \mathrm{d}\Omega \tag{4.4}$$

a strain-constructed Galerkin model can be derived directly from the single-field Hellinger-Reisnner principle [3, 8] and hence is variationally consistent (see, Chapter 5). Equation (4.4) is in fact the well-known orthogonal condition [4] for displacement compatible models. This condition has been used to construct hybrid FEM models based on elements, and can also be used in S-PIM when the assumed displacement functions are in a proper H^1 space.

4.4.2 Condition 2a: norm equivalence condition

The norm equivalence condition is derived based on a so-called *base* model that can be any compatible standard FEM model. In our S-PIM, we choose to use the linear FEM model (FEM-Tr3) as the base model, because it is the simplest of all proven FEM models. In this case, the compatible strain field is a constant in a triangular cell.

Consider a discrete vector displacement function $\mathbf{v} \in (\mathbb{G}_{h,0}^1)^d$ with a vector $\mathbf{d} \in \mathbb{R}_0^{dN_n}$, in which $\mathbb{R}_0^{dN_n}$ is a sub-space of $\mathbb{R}^{dN_n}$ where the displacements at the nodes on the essential boundary are imposed. The nodal displacements arranged in the form of

$$
\mathbf{d} = \left\{ \underbrace{u_{1(1)} \quad u_{2(1)}}_{\mathbf{d}_1 \text{ for node } (1)} \quad \underbrace{u_{1(2)} \quad u_{2(2)}}_{\mathbf{d}_2 \text{ for node } (2)} \quad \cdots \quad \underbrace{u_{1(n)} \quad u_{2(n)}}_{\mathbf{d}_n \text{ for node } (n)} \right.
$$

$$
\left. \cdots \quad \underbrace{u_{1(N_n)} \quad u_{2(N_n)}}_{\mathbf{d}_{Nn} \text{ for node } (N_n)} \right\}^{\mathrm{T}} , \quad \text{for 2D}
$$

$$
\mathbf{d} = \left\{ \underbrace{u_{1(1)} \quad u_{2(1)} \quad u_{3(1)}}_{\mathbf{d}_1 \text{ for node } (1)} \quad \underbrace{u_{1(2)} \quad u_{2(2)} \quad u_{3(2)}}_{\mathbf{d}_2 \text{ for node } (2)} \quad \cdots \quad \underbrace{u_{1(n)} \quad u_{2(n)} \quad u_{3(n)}}_{\mathbf{d}_n \text{ for node } (n)} \right.
$$

$$
\left. \cdots \quad \underbrace{u_{1(N_n)} \quad u_{2(N_n)} \quad u_{3(N_n)}}_{\mathbf{d}_{Nn} \text{ for node } (N_n)} \right\}^{\mathrm{T}} , \quad \text{for 3D}
$$

(4.5)

where $\mathbf{d}_I = \mathbf{u}(\mathbf{x}_I), I = 1, \cdots, N_n$ is the nodal displacement vector. The compatible constant base strain field $\tilde{\boldsymbol{\varepsilon}}_b(\mathbf{d})$ can be obtained using an FEM-Tr3 model but with the same set of nodal displacements $\mathbf{d} \in \mathbb{R}_0^{dN_n}$:

$$
\tilde{\boldsymbol{\varepsilon}}_b(\mathbf{d}) = \underbrace{\mathbf{L}_d \boldsymbol{\Phi}}_{\mathbf{B}_b} \mathbf{d}
\tag{4.6}
$$

where $\boldsymbol{\Phi}$ is the row-matrix of shape functions for all the nodes and $\mathbf{B}_b$ is the *global* strain-displacement matrix has entries assembled from the elemental strain-displacement matrix defined by

$$
\mathbf{B}_b = \begin{bmatrix} \mathbf{L}_d \boldsymbol{\Phi}_1 & \mathbf{L}_d \boldsymbol{\Phi}_2 & \cdots & \mathbf{L}_d \boldsymbol{\Phi}_e & \cdots & \mathbf{L}_d \boldsymbol{\Phi}_{N_n} \end{bmatrix}
\tag{4.7}
$$

in which the matrix of shape functions of the base model has the form of

$$\mathbf{\Phi}_e = \left[\begin{array}{cc|cc|cc} \phi_{1e}^b & 0 & \phi_{2e}^b & 0 & \phi_{3e}^b & 0 \\ 0 & \phi_{1e}^b & 0 & \phi_{2e}^b & 0 & \phi_{3e}^b \\ \underbrace{\qquad}_{\text{node 1, element } e} & & \underbrace{\qquad}_{\text{node 2, element } e} & & \underbrace{\qquad}_{\text{node 3, element } e} & \end{array} \right], \quad \text{for 2D}$$

$$\mathbf{\Phi}_e = \left[\begin{array}{ccc} \phi_{1e}^b & 0 & 0 \\ 0 & \phi_{1e}^b & 0 \\ 0 & 0 & \phi_{1e}^b \\ \underbrace{\qquad}_{\text{node 1, element } e} & & \end{array} \right.$$

$$\left. \begin{array}{ccc|ccc} \phi_{2e}^b & 0 & 0 & \phi_{3e}^b & 0 & 0 \\ 0 & \phi_{2e}^b & 0 & 0 & \phi_{3e}^b & 0 \\ 0 & 0 & \phi_{2e}^b & 0 & 0 & \phi_{3e}^b \\ \underbrace{\qquad}_{\text{node 2, element } e} & & & \underbrace{\qquad}_{\text{node 3, element } e} & & \end{array} \right.$$

$$\left. \begin{array}{ccc} \phi_{4e}^b & 0 & 0 \\ 0 & \phi_{4e}^b & 0 \\ 0 & 0 & \phi_{4e}^b \\ \underbrace{\qquad}_{\text{node 4, element } e} & & \end{array} \right], \quad \text{for 3D} \tag{4.8}$$

The *global* strain-displacement matrix $\mathbf{B}_b$ is a very sparse matrix with dimension of $3N_c \times 2N_n$ for 2D and $6N_c \times 3N_n$ for 3D models. Because all the nodal shape functions of the (proven stable) base model are linearly independent, $\mathbf{B}_b$ has $2N_n$ (for 2D) and $3N_n$ (for 3D) linear independent columns (here we consider constrained solids and N_n is the number of unconstrained nodes). The (global) strain vector $\tilde{\boldsymbol{\varepsilon}}_b(\mathbf{d})$ can be written in the form of

$$\tilde{\boldsymbol{\varepsilon}}_b(\mathbf{d}) = \left\{ \underbrace{\tilde{\boldsymbol{\varepsilon}}_{b(1)}^{\mathrm{T}}}_{\text{element 1}} \quad \underbrace{\tilde{\boldsymbol{\varepsilon}}_{b(2)}^{\mathrm{T}}}_{\text{element 2}} \quad \cdots \quad \underbrace{\tilde{\boldsymbol{\varepsilon}}_{b(N_c)}^{\mathrm{T}}}_{\text{element } N_c} \right\}^{\mathrm{T}} \tag{4.9}$$

where N_c is the total number of the triangular elements of the base model that are the same as the background cells for the SC-model, and

$$\tilde{\boldsymbol{\varepsilon}}_{b(i)} = \left\{ \tilde{\varepsilon}_{11} \quad \tilde{\varepsilon}_{22} \quad 2\tilde{\varepsilon}_{12} \right\}_{(i)}^{\mathrm{T}},$$

$$(i = 1, 2, \cdots, N_c) \text{ for 2D},$$

$$\tilde{\boldsymbol{\varepsilon}}_{b(i)} = \left\{ \tilde{\varepsilon}_{11} \quad \tilde{\varepsilon}_{22} \quad \tilde{\varepsilon}_{33} \quad 2\tilde{\varepsilon}_{23} \quad 2\tilde{\varepsilon}_{13} \quad 2\tilde{\varepsilon}_{12} \right\}_{(i)}^{\mathrm{T}},$$

$$(i = 1, 2, \cdots, N_c) \text{ for 3D} \tag{4.10}$$

Because linear triangular elements are used in the base model, these strain components $\tilde{\varepsilon}_{ij}$ are constants in these triangular background cells.

On the other hand, the constructed strain field $\hat{\boldsymbol{\varepsilon}}(\mathbf{d})$ for the SC-model can also be written in the similar form of

$$\hat{\boldsymbol{\varepsilon}}(\mathbf{d}) = \left\{ \underbrace{\hat{\boldsymbol{\varepsilon}}_{(1)}^{\mathrm{T}}}_{\text{cell 1}} \quad \underbrace{\hat{\boldsymbol{\varepsilon}}_{(2)}^{\mathrm{T}}}_{\text{cell 2}} \quad \cdots \quad \underbrace{\hat{\boldsymbol{\varepsilon}}_{(N_q)}^{\mathrm{T}}}_{\text{cell } N_q} \right\}^{\mathrm{T}} \tag{4.11}$$

where N_q is the total number of the *quadrature* (or integration) *cells* used in the SC-model, and

$$\hat{\boldsymbol{\varepsilon}}_{(i)} = \left\{ \hat{\varepsilon}_{11} \quad \hat{\varepsilon}_{22} \quad 2\hat{\varepsilon}_{12} \right\}_{(i)}^{\mathrm{T}},$$
$$(i = 1, 2, \cdots, N_q) \text{ for 2D},$$
$$\hat{\boldsymbol{\varepsilon}}_{(i)} = \left\{ \hat{\varepsilon}_{11} \quad \hat{\varepsilon}_{22} \quad \hat{\varepsilon}_{33} \quad 2\hat{\varepsilon}_{23} \quad 2\hat{\varepsilon}_{13} \quad 2\hat{\varepsilon}_{12} \right\}_{(i)}^{\mathrm{T}},$$
$$(i = 1, 2, \cdots, N_q) \text{ for 3D} \tag{4.12}$$

These constructed strain components $\hat{\varepsilon}_{ij}$ are, in general, not constant in an integration cell.

To ensure the stability and convergence of the SC-model, the constructed strain field $\hat{\boldsymbol{\varepsilon}}(\mathbf{d})$ needs to be *equivalent in a norm* to the strain field $\tilde{\boldsymbol{\varepsilon}}_b(\mathbf{d})$ of the base model, by which we mean that there exist nonzero positive constants c_{ac} and c_{ca} such that

$$\left\| \hat{\boldsymbol{\varepsilon}}(\mathbf{d}) \right\|_{\mathbb{L}^2(\Omega)} \le c_{ac} \left\| \tilde{\boldsymbol{\varepsilon}}_b(\mathbf{d}) \right\|_{\mathbb{L}^2(\Omega)} \text{ and } \left\| \tilde{\boldsymbol{\varepsilon}}_b(\mathbf{d}) \right\|_{\mathbb{L}^2(\Omega)} \le c_{ca} \left\| \hat{\boldsymbol{\varepsilon}}(\mathbf{d}) \right\|_{\mathbb{L}^2(\Omega)}$$
$$\forall \mathbf{d} \in \mathbb{R}_0^{dN_n} \tag{4.13}$$

where c_{ac} and c_{ca} are general constants independent of $\mathbf{d}\left(\in \mathbb{R}_0^{dN_n}\right)$. Since our strain field is defined in a vector form, measuring in the L^2 norm is handy, and workable for both $\tilde{\boldsymbol{\varepsilon}}_b(\mathbf{d})$ and $\hat{\boldsymbol{\varepsilon}}(\mathbf{d})$, although they can be different in length. Clearly, to satisfy conditions (4.13), the *minimum* requirement should be

$$N_q \ge N_c \tag{4.14}$$

which means that we should "sample" more locations for the strain energy in the SC-model. Equation (4.14) is a very important condition for SC-models. The similar condition is practiced in the S-FEM [9-13], and proved for general settings based on the argument of ensuring positivity [6, 7].

The norm equivalence ensures only the stability and hence convergence, but we do not know where it converges to. To ensure the solution converges to the exact solution, we need the strain convergence condition.

4.4.3 Condition 2b: strain convergence condition

The convergence condition for the constructed strain field is defined as [14]:

$$\lim_{\substack{h \to 0 \\ N_n \to \infty}} \hat{\boldsymbol{\varepsilon}}(\mathbf{x},\mathbf{d}) \to \tilde{\boldsymbol{\varepsilon}}_b(\mathbf{x},\mathbf{d}), \quad \forall \mathbf{x} \in \Omega, \forall \mathbf{d} \in \mathbb{R}_0^{dN_n} \tag{4.15}$$

When the constructed stain field satisfies both conditions in Equations (4.13) and (4.15), the resultant SC-Galerkin model (such as our S-PIM model) is stable and converges to the exact solution of the original strong form defined in Section 1.1. In this case, the formulation may not be variationally consistent in the conventional sense, depending on 1) whether or not $\mathbf{v}$ is in a proper H^1 space, and 2) how the strain field is constructed.

4.4.4 Condition 3: zero-sum condition

For any $\mathbf{v} \in \left(\mathbb{H}^1_{h,0}\right)^d$, we may construct the strain in the following form

$$\hat{\boldsymbol{\varepsilon}}(\mathbf{v}) = \tilde{\boldsymbol{\varepsilon}}_b(\mathbf{v}) + \alpha \hat{\boldsymbol{\varepsilon}}_m(\mathbf{v}) \tag{4.16}$$

where $\tilde{\boldsymbol{\varepsilon}}_b$ is the (compatible) strain field of a base model, $\hat{\boldsymbol{\varepsilon}}_m$ is the constructed portion of the strain field, and $\alpha \in \mathbb{R}$ is a parameter regularizing the amount of modification. The *zero-sum condition* is then defined as

$$\int_\Omega \hat{\boldsymbol{\varepsilon}}_m(\mathbf{v}) \mathrm{d}\Omega = 0, \quad \forall \mathbf{v} \in \left(\mathbb{H}^1_{h,0}\right)^d \tag{4.17}$$

When both Equations (4.16) and (4.17) are satisfied, a Galerkin-like weak form is then applicable (see, Chapter 5), and the formulation is *variationally consistent*. The zero-sum condition is similar to orthogonal condition used in FEM for the stabilization of quadrilateral elements using reduced integration [15, 16]. In our S-PIM, however, we operate beyond the elements/cells. We now observe that when α is set to zero, we have the FEM setting.

In summarizing the above discussions on admissible conditions for contracted strain fields, we note the following theorem.

Theorem 4.1 Admissible conditions for strain field: basis for $\mathbf{W}^2$ formulation

For any $\mathbf{u}^h \in \left(\mathbb{G}^1_h\right)^d$, if the constructed stain field satisfies one of these three sets of admissible conditions, a stable and convergent $\mathbf{W}^2$ formulation can then be performed for elasticity problems (or problems of that nature).

With Theorem 4.1 in place, we can now concentrate our efforts on how to construct proper strain fields and the resultant properties.

4.5   Strain construction techniques

We now present some practical techniques for constructing strain fields that can be used in SC-models. These techniques are listed in TABLE 4.1. It is clear that the first two techniques use the generalized smoothing technique, and are used in the S-PIM models.

4.5.1   At a glance

TABLE 4.1 Some techniques for the construction of strain fields

Item	Schemes	Brief description	Examples/Sources
1	Generalized smoothing	The constructed strains in each smoothing domain are assumed to be constant, obtained using the generalized smoothing technique over the smoothing domains that can be cell-based, node-based, edge-based, and face-based.	S-FEM [9-12, 17] NS-PIM [17-23] ES-PIM [17, 24, 25] CS-PIM [17, 26-28]
2	Point interpolation method	At points in the problem domain, the strains are obtained using the generalized smoothing technique over a proper smoothing domain. The strain field is then constructed using a point interpolation method.	SC-PIM [17, 29, 30] PIM-CS [14, 31]
3	Least square approximation	At points in the problem domain, the strains are obtained using the generalized smoothing technique over a proper smoothing domain. The strain field is then constructed using a least square approximation.	PIM-LSS [32]
4	Strain smoothing	The constructed strains in each smoothing domain are assumed to be constant, obtained via smoothing the compatible strain field over the smoothing domains that can be cell-based, node-based, edge-based and face-based.	SCNI [8] NS-FEM [13, 33] ES-FEM [13, 34] FS-FEM [13, 35]
5	Strain gradient scaling	Using a factor to scale the gradient of the compatible strain field.	αFEM [36] VCαFEM [37]

Note: this table may not be exclusive.

4.5.2 Strain construction by generalized smoothing

4.5.2.1 Smoothing domain construction

In the S-PIM, on top of the background cells, the problem domain is divided into smoothing domains in nonoverlapping and no-gap manner such that $\boxed{\Omega} = \bigcup_{i=1}^{N_s}\boxed{\Omega}_i^s$ and $\Omega_i^s \cap \Omega_j^s = \varnothing$, $\forall i \neq j$, as shown in FIGURE 2.3, where N_s is the number of smoothing domains and the smoothing domain Ω_i^s bounded by Γ_i^s is associated with the point at $\mathbf{x}_i$. In constructing the smoothing domains, the important "no-sharing rule" presented in Section 2.3.1 must be observed: Γ_i^s does not share any *finite* portion of discontinuous lines of the assumed displacement field. Γ_i^s can only go across these segments. It is clear that such a setting is for creating functions in a G^1 space (see, Chapter 2). In this setting, the smoothing domains are also the integration domains, and hence we have $\Omega_i^s = \Omega_i^q$ and $N_s = N_q$.

The constructed strain in Ω_i^s is assumed to be constant $\hat{\boldsymbol{\varepsilon}}_i$ and equals the smoothed strain:

$$\hat{\boldsymbol{\varepsilon}}_i = \bar{\boldsymbol{\varepsilon}}_i = \bar{\boldsymbol{\varepsilon}}(\mathbf{x}_i), \quad \forall \mathbf{x} \in \Omega_i^s \tag{4.18}$$

where

$$\bar{\boldsymbol{\varepsilon}}(\mathbf{x}_i) = \int_{\Gamma_i^s} \mathbf{L}_n \mathbf{u}^h(\boldsymbol{\xi})\widehat{\mathbf{W}}(\mathbf{x}_i - \boldsymbol{\xi})\,\mathrm{d}\Gamma - \int_{\Omega_i^s} \mathbf{u}^h(\boldsymbol{\xi})\mathbf{L}_d^T\left(\widehat{\mathbf{W}}(\mathbf{x}_i - \boldsymbol{\xi})\right)\mathrm{d}\boldsymbol{\xi} \tag{4.19}$$

in which $\widehat{\mathbf{W}}$ is a diagonal matrix of smoothing functions $\widehat{W}$. In Equation (4.19), $\mathbf{u}^h(\boldsymbol{\xi}) \in \mathbb{C}^{-1}(\Omega_i^s)$ and the smoothing functions $\widehat{\mathbf{W}}$ satisfies the unity condition over the smoothing domain Ω_i^s. We require here that $\widehat{W}$ is at least 1st order differentiable in (open) Ω_i^s. When $\mathbf{u}^h$ is continuous, Equation (4.19) can be derived by using the Green's divergence theorem; when $\mathbf{u}^h$ is discontinuous, the Green's divergence theorem is not applicable, and hence the equation is only an "approximation".

Remark 4.3 "Smoothed strain" by boundary-flux-approximation

The smoothed strain defined in Equation (4.19) is a generalized terminology. It is generally NOT "the strain obtained by smoothing the compatible strain field", because such a compatible strain field does not in general exist! Rigorously speaking, the smoothed strain is the outward flux of the assumed displacement field across the smoothing domain boundary Γ_i^s. Such a boundary-flux-approximation preserves the overall conservation of the strain-displacement relation over the smoothing domain, which is important to ensure the energy balance in energy principles using such (generalized) smoothed strains. It is essential for the GS-Galerkin weak forms workable for assumed functions in a $\mathbb{G}_h^1$ space. Observing Equation (4.19), we note that the differentiation on the assumed displacement has now been transferred to on the smoothing functions. Therefore, the continuity requirement on the assumed displacement function is reduced by one order, if only the smoothed strains are required.

Remark 4.4 Smoothed strain: convergence property

For any $\mathbf{u}^h \in (\mathbb{H}_h^1)^d$, when the dimension of Ω_i^s approaches to zero, the smoothing function $\widehat{W}$ approaches to the Delta function. At such a limit $\widehat{\boldsymbol{\varepsilon}} \to \tilde{\boldsymbol{\varepsilon}}$, and the constructed strain field approaches to the compatible strain field. For any $\mathbf{u}^h \in (\mathbb{G}_h^1)^d$, the compatible strain for $\mathbf{u}^h$ may not exist at locations in the problem domain, the convergence needs to be assessed against to the base model, as discussed in Section 4.4.3.

Note that the minimum number of smoothing domains needs to be determined based on TABLE 2.2. Violating of TABLE 2.2, the discretized system equations established using the GS-Gakerkin weak form will be singular, and no unique solution can be obtained. Generally, a finer division of smoothing domains leads to a stiffer model (see Theorem 4.5).

When using the Heaviside type of smoothing function $\bar{W}$ given in Equation (2. 74) for the smoothing domain Ω_i^s, the last domain integral term in Equation (4.19) vanishes due to the constant smoothing functions in Ω_i^s. We shall have for $\mathbf{u}^h(\xi) \in \mathbb{C}^{-1}(\Omega_i^s)$

$$\bar{\boldsymbol{\varepsilon}}(\mathbf{x}_i) = \frac{1}{A_i^s} \int_{\Gamma_i^s} \mathbf{L}_n \mathbf{u}^h \mathrm{d}\Gamma \tag{4.20}$$

Since the strain is assumed constant in any entire smoothing domain Ω_i^s,

$$\hat{\boldsymbol{\varepsilon}}_i = \bar{\boldsymbol{\varepsilon}}_i = \bar{\boldsymbol{\varepsilon}}(\mathbf{x}_i), \quad \forall \mathbf{x} \in \Omega_i^s \tag{4.21}$$

a piecewise constant strain field is constructed for the entire problem domain. The potential energy for the smoothed strain field becomes:

$$\bar{U}_{PE}^{D} = \int_{\Omega} \frac{1}{2} \bar{\boldsymbol{\varepsilon}}^{\mathrm{T}} \, \mathbf{c} \bar{\boldsymbol{\varepsilon}} \, \mathrm{d}\xi = \frac{1}{2} \sum_{i=1}^{N_s} A_i^s \bar{\boldsymbol{\varepsilon}}_i^{\mathrm{T}} \mathbf{c} \bar{\boldsymbol{\varepsilon}}_i \tag{4.22}$$

4.5.2.2 Properties of the smoothed strain field

We now discuss briefly the properties of the *smoothed* strain field and the strain energy of the smoothed strain field. More detailed discussions and proofs on these properties can be found in [3, 6, 7, 17-21].

Theorem 4.2 Orthogonal projection

For any $\mathbf{u}^h \in \left(\mathbb{H}_h^1 \right)^d$, the smoothed strain defined in Equation (4.18) satisfies the orthogonal condition (4.4), and the smoothed strain field is an orthogonal projection of the compatible strain field on to the coordinate space of the smoothed strain field.

The proof for the orthogonal projection Theorem 4.2 can be found in [3]. Theorem 4.2 essentially states that a Galerkin formulation using the smoothed strain will be variationally consistent (see, Chapter 5), as long as $\mathbf{u}^h \in \mathbb{H}_h^1$. We also note that the orthogonality is in fact "uniform", meaning that for each of the smoothing domains we have [3]

$$\int_{\Omega_i^s} \bar{\boldsymbol{\varepsilon}}^{\mathrm{T}}(\mathbf{v}) \mathbf{c} \bar{\boldsymbol{\varepsilon}}(\mathbf{v}) \mathrm{d}\Omega = \int_{\Omega_i^s} \bar{\boldsymbol{\varepsilon}}^{\mathrm{T}}(\mathbf{v}) \mathbf{c} (\mathbf{L}_d \mathbf{v}) \mathrm{d}\Omega, \quad \forall \mathbf{v} \in \left(\mathbb{H}_h^1 \right)^d \tag{4.23}$$

which is in fact a quite strong condition.

Theorem 4.3 Softening effects

For any assumed admissible displacement function $\mathbf{u}^h \in \left(\mathbb{H}_h^1 \right)^d$ and linear elastic solids of stable materials, and the smoothed strain is obtained using Equation (4.18), we have

$$\bar{U}_{PE}^{D}(\mathbf{u}^h) \leq \tilde{U}_{PE}(\mathbf{u}^h) \tag{4.24}$$

meaning that the potential energy for the smoothed strain field is always smaller than that of the compatible strain field, for any same assumed displacement field in a proper Hilbert space.

The proof and further discussion for the softening effect theorem can be found in [3, 17, 20]. Theorem 4.3 offers a theoretical foundation for creating a discrete model that is softer than the FEM model. Because of Equation (4.23), we should have a "stronger" theorem.

Theorem 4.4 Uniform softening effects

For any assumed admissible displacement function $\mathbf{u}^h \in \mathbb{H}_h^1$ and linear elastic solids of stable materials, and the smoothed strain is obtained using Equation (4.21), we have for each smoothing domain that

$$\underbrace{\int_{\Omega_i^s} \frac{1}{2}\overline{\boldsymbol{\varepsilon}}^{\mathrm{T}}\mathbf{c}\overline{\boldsymbol{\varepsilon}}\,\mathrm{d}\Omega}_{\overline{U}_{PE(i)}^D(\mathbf{u}^h)} \le \underbrace{\int_{\Omega_i^s} \frac{1}{2}\tilde{\boldsymbol{\varepsilon}}^{\mathrm{T}}\mathbf{c}\tilde{\boldsymbol{\varepsilon}}\,\mathrm{d}\Omega}_{\tilde{U}_{PE(i)}(\mathbf{u}^h)} \tag{4.25}$$

meaning that the potential energy for the smoothed strain field in each smoothing domain is always smaller than that of the compatible strain field, for any same assumed displacement field in a proper Hilbert space.

Proof

We first examine the strain energy in each smoothing domain resulted from the difference of the compatible and smoothed strain fields:

$$\int_{\Omega_i^s} \frac{1}{2}\left(\overline{\boldsymbol{\varepsilon}}-\tilde{\boldsymbol{\varepsilon}}\right)^{\mathrm{T}}\mathbf{c}\left(\overline{\boldsymbol{\varepsilon}}-\tilde{\boldsymbol{\varepsilon}}\right)\mathrm{d}\Omega$$

$$= \int_{\Omega_i^s} \frac{1}{2}\overline{\boldsymbol{\varepsilon}}^{\mathrm{T}}\mathbf{c}\overline{\boldsymbol{\varepsilon}}\,\mathrm{d}\Omega - \int_{\Omega_i^s} \frac{1}{2}\overline{\boldsymbol{\varepsilon}}^{\mathrm{T}}\mathbf{c}\tilde{\boldsymbol{\varepsilon}}\,\mathrm{d}\Omega - \int_{\Omega_i^s} \frac{1}{2}\tilde{\boldsymbol{\varepsilon}}^{\mathrm{T}}\mathbf{c}\overline{\boldsymbol{\varepsilon}}\,\mathrm{d}\Omega + \int_{\Omega_i^s} \frac{1}{2}\tilde{\boldsymbol{\varepsilon}}^{\mathrm{T}}\mathbf{c}\tilde{\boldsymbol{\varepsilon}}\,\mathrm{d}\Omega \tag{4.26}$$

$$= \int_{\Omega_i^s} \frac{1}{2}\overline{\boldsymbol{\varepsilon}}^{\mathrm{T}}\mathbf{c}\overline{\boldsymbol{\varepsilon}}\,\mathrm{d}\Omega - \int_{\Omega_i^s} \overline{\boldsymbol{\varepsilon}}^{\mathrm{T}}\mathbf{c}\tilde{\boldsymbol{\varepsilon}}\,\mathrm{d}\Omega + \int_{\Omega_i^s} \frac{1}{2}\tilde{\boldsymbol{\varepsilon}}^{\mathrm{T}}\mathbf{c}\tilde{\boldsymbol{\varepsilon}}\,\mathrm{d}\Omega$$

where $\tilde{\boldsymbol{\varepsilon}} = \mathbf{L}_d\mathbf{u}^h$ is the compatible strain field. From Equation (4.23), and SPD property of $\mathbf{c}$ for stable materials, we have

$$\underbrace{\int_{\Omega} \frac{1}{2}\left(\overline{\boldsymbol{\varepsilon}}-\tilde{\boldsymbol{\varepsilon}}\right)^{\mathrm{T}}\mathbf{c}\left(\overline{\boldsymbol{\varepsilon}}-\tilde{\boldsymbol{\varepsilon}}\right)\mathrm{d}\Omega}_{\ge 0} = \underbrace{\int_{\Omega_i^s} \frac{1}{2}\tilde{\boldsymbol{\varepsilon}}^{\mathrm{T}}\mathbf{c}\tilde{\boldsymbol{\varepsilon}}\,\mathrm{d}\Omega}_{\tilde{U}_{PE(i)}(\mathbf{u}^h)} - \underbrace{\int_{\Omega_i^s} \frac{1}{2}\overline{\boldsymbol{\varepsilon}}^{\mathrm{T}}\mathbf{c}\overline{\boldsymbol{\varepsilon}}\,d\Omega}_{\overline{U}_{PE(i)}^D(\mathbf{u}^h)} \ge 0 \tag{4.27}$$

which gives Equation (4.25). $\qquad\qquad\square$

Theorem 4.4 indicates that the SC-model is softer than the FEM model uniformly "everywhere" (in each and very smoothing domains) in the problem domain.

Theorem 4.5 Monotonic convergence

In a given division D_1 of domain Ω into a set of smoothing domains $[\Omega] = \bigcup_{i=1}^{N_s} [\Omega]_i^s$, if a new division D_2 is performed by sub-dividing one of the smoothing domains in D_1 into n_{sd} sub-smoothing-domains: $[\Omega]_i^s = \bigcup_{j=1}^{n_{sd}} [\Omega]_{i,j}^s$, then the following inequality stands

$$\overline{U}_{PE}^{D_1}(\mathbf{u}^h) \le \overline{U}_{PE}^{D_2}(\mathbf{u}^h) \tag{4.28}$$

This implies that the "softening" effect provided by the smoothing operation will be *monotonically* reduced with the increase of the smoothing domains in a nested manner. A simple proof can be found in [17].

4.5.3 Strain construction by point interpolation

In this more general approach, the strain field $\hat{\boldsymbol{\varepsilon}}$ is constructed using the point interpolation method with the following steps.

Step 1: Determination of critical points

Select a proper set of points for strain interpolation based on the triangular background cells. The density of the points should usually be higher than the field nodes to satisfy the norm equivalence condition. We shall also try to have these points coincide with the nodes so that the strain convergence condition (4.15) can be satisfied easily. We usually choose all the field nodes first, and then may add in the middle points of cell edges and the centroidals of the triangular cells as the interpolation points. Such selection of strain interpolation points are used in PIM-CS [14, 31], SαFEM [38], and SC-PIM [29, 30].

Step 2: Compute strains at the critical points

For any assumed admissible displacement $\mathbf{u}^h$, at a strain interpolation point, the strain is assigned as either the compatible strain obtained using $\mathbf{u}^h \in \mathbb{H}_h^1$, or a smoothed strain using a proper local domain for that point using Equation (4.19) or (4.20). We can choose node-based smoothing domains for points at the nodes, and edge-based smoothing domains for points at the mid-edge. For points at the centroidals, cell-based smoothing domain may be used. This is practiced in PIM-CS [14, 31], SC-PIM [29, 30], and SαFEM [38]. A new set of (preferably triangular) cells for the set of strain interpolation points is then created for the strain field interpolation. This set of cells is used also for integration of the weak form, and are called quadrature or integration cells. The new triangular quadrature cells are hosted by the original triangles, and typically a triangular cell is divided into 1, 3, 4 or 6 integration cells, as in SC-PIM [29]. Chapter 10 gives more details.

Step 3: Strain field construction over cells

The strain field in each of the quadrature cells is then constructed using the point interpolation method (Chapter 3) based on these strains at the set of points, using shape functions created for these strain interpolation points. Because we use triangular integration cells, the integration of the weak form can be performed analytically using the well-known *area coordinates* [2], and no numerical integration is required. The strain field within each quadrature cell can then be given by:

$$\hat{\boldsymbol{\varepsilon}}(\mathbf{x}) = \sum_{i=1}^{3} \boldsymbol{\Phi}_i(\mathbf{x}) \breve{\boldsymbol{\varepsilon}}_i \tag{4.29}$$

where $\hat{\boldsymbol{\varepsilon}}$ is the vector of constructed strains over the triangular quadrature cell, and $\breve{\boldsymbol{\varepsilon}}$ is the vector of strains at each vertex cell obtained at step 1. In Equation (4.29), $\boldsymbol{\Phi}_i$ is a diagonal matrix of linear PIM shape functions:

$$\boldsymbol{\Phi}_i(\mathbf{x}) = \begin{bmatrix} L_i(\mathbf{x}) & 0 & 0 \\ 0 & L_i(\mathbf{x}) & 0 \\ 0 & 0 & L_i(\mathbf{x}) \end{bmatrix} \tag{4.30}$$

where L_i is the area coordinates for node i of the quadrature cell. Equation (4.29) can be written in detail as

$$\begin{Bmatrix} \hat{\varepsilon}_{11}(\mathbf{x}) \\ \hat{\varepsilon}_{22}(\mathbf{x}) \\ 2\hat{\varepsilon}_{12}(\mathbf{x}) \end{Bmatrix} = \begin{Bmatrix} L_1(\mathbf{x})\breve{\varepsilon}_{11}(\mathbf{x}_1) + L_2(\mathbf{x})\breve{\varepsilon}_{11}(\mathbf{x}_2) + L_3(\mathbf{x})\breve{\varepsilon}_{11}(\mathbf{x}_3) \\ L_1(\mathbf{x})\breve{\varepsilon}_{22}(\mathbf{x}_1) + L_2(\mathbf{x})\breve{\varepsilon}_{22}(\mathbf{x}_2) + L_3(\mathbf{x})\breve{\varepsilon}_{22}(\mathbf{x}_3) \\ 2\left(L_1(\mathbf{x})\breve{\varepsilon}_{12}(\mathbf{x}_1) + L_2(\mathbf{x})\breve{\varepsilon}_{12}(\mathbf{x}_2) + L_3(\mathbf{x})\breve{\varepsilon}_{12}(\mathbf{x}_3)\right) \end{Bmatrix} \tag{4.31}$$

Step 4: Evaluation of strain energy

Using Equation (4.2), the strain energy potential can be obtained as

$$
\hat{U}_{PE} = \underbrace{\int_{\Omega} \frac{1}{2} \hat{\boldsymbol{\varepsilon}}^{\mathrm{T}}(\mathbf{x}) \mathbf{c} \hat{\boldsymbol{\varepsilon}}(\mathbf{x}) \mathrm{d}\Omega}_{\text{finite, } \because \hat{\varepsilon}_{ij} \in L^2(\Omega)}
$$

$$
= \frac{1}{2} \sum_{j=1}^{N_q} \int_{\Omega_j^q} \left(\sum_{i=1}^{3} \boldsymbol{\Phi}_i(\mathbf{x}) \breve{\boldsymbol{\varepsilon}}_i \right)^{\mathrm{T}} \mathbf{c} \left(\sum_{i=1}^{3} \boldsymbol{\Phi}_i(\mathbf{x}) \breve{\boldsymbol{\varepsilon}}_i \right) \mathrm{d}\Omega
$$

$$
= \frac{1}{2} \sum_{j=1}^{N_q} \int_{\Omega_j^q} \left(\breve{\boldsymbol{\varepsilon}}_1^{\mathrm{T}} \boldsymbol{\Phi}_1^{\mathrm{T}} \mathbf{c} \boldsymbol{\Phi}_1 \breve{\boldsymbol{\varepsilon}}_1 + \breve{\boldsymbol{\varepsilon}}_2^{\mathrm{T}} \boldsymbol{\Phi}_2^{\mathrm{T}} \mathbf{c} \boldsymbol{\Phi}_2 \breve{\boldsymbol{\varepsilon}}_2 + \breve{\boldsymbol{\varepsilon}}_3^{\mathrm{T}} \boldsymbol{\Phi}_3^{\mathrm{T}} \mathbf{c} \boldsymbol{\Phi}_3 \breve{\boldsymbol{\varepsilon}}_3 \right.
$$

$$
\left. + 2 \breve{\boldsymbol{\varepsilon}}_1^{\mathrm{T}} \boldsymbol{\Phi}_1^{\mathrm{T}} \mathbf{c} \boldsymbol{\Phi}_2 \breve{\boldsymbol{\varepsilon}}_2 + 2 \breve{\boldsymbol{\varepsilon}}_1^{\mathrm{T}} \boldsymbol{\Phi}_1^{\mathrm{T}} \mathbf{c} \boldsymbol{\Phi}_3 \breve{\boldsymbol{\varepsilon}}_3 + 2 \breve{\boldsymbol{\varepsilon}}_2^{\mathrm{T}} \boldsymbol{\Phi}_2^{\mathrm{T}} \mathbf{c} \boldsymbol{\Phi}_3 \breve{\boldsymbol{\varepsilon}}_3 \right) \mathrm{d}\Omega \tag{4.32}
$$

$$
= \frac{1}{2} \sum_{j=1}^{N_q} \left(\breve{\boldsymbol{\varepsilon}}_1^{\mathrm{T}} \mathbf{c} \breve{\boldsymbol{\varepsilon}}_1 \int_{\Omega_j^q} L_1^2 \mathrm{d}\Omega + \breve{\boldsymbol{\varepsilon}}_2^{\mathrm{T}} \mathbf{c} \breve{\boldsymbol{\varepsilon}}_2 \int_{\Omega_j^q} L_2^2 \mathrm{d}\Omega \right.
$$

$$
+ \breve{\boldsymbol{\varepsilon}}_3^{\mathrm{T}} \mathbf{c} \breve{\boldsymbol{\varepsilon}}_3 \int_{\Omega_j^q} L_3^2 \mathrm{d}\Omega + 2 \breve{\boldsymbol{\varepsilon}}_1^{\mathrm{T}} \mathbf{c} \breve{\boldsymbol{\varepsilon}}_2 \int_{\Omega_j^q} L_1 L_2 \mathrm{d}\Omega
$$

$$
\left. + 2 \breve{\boldsymbol{\varepsilon}}_1^{\mathrm{T}} \mathbf{c} \breve{\boldsymbol{\varepsilon}}_3 \int_{\Omega_j^q} L_1 L_3 \mathrm{d}\Omega + 2 \breve{\boldsymbol{\varepsilon}}_2^{\mathrm{T}} \mathbf{c} \breve{\boldsymbol{\varepsilon}}_3 \int_{\Omega_j^q} L_2 L_3 \mathrm{d}\Omega \right)
$$

Using the Eisenberg-Malvern formula for any general triangle with an area A [2]:

$$
\int_A L_1^p L_2^q L_3^r \mathrm{d}\Omega = \frac{p! \, q! \, r!}{(p+q+r+2)!} 2A \tag{4.33}
$$

which gives for a quadrature cell:

$$
\int_{\Omega_j^q} L_1^2 \mathrm{d}\Omega = \int_{\Omega_j^q} L_2^2 \mathrm{d}\Omega = \int_{\Omega_j^q} L_2^2 \mathrm{d}\Omega = \frac{2!}{4!} 2A_j^q = \frac{1}{6} A_j^q
$$

$$
\int_{\Omega_j^q} L_1 L_2 \mathrm{d}\Omega = \int_{\Omega_j^q} L_1 L_3 \mathrm{d}\Omega = \int_{\Omega_j^q} L_2 L_3 \mathrm{d}\Omega = \frac{1!}{4!} 2A_j^q = \frac{1}{12} A_j^q \tag{4.34}
$$

where A^q is the area of the quadrature cell. Finally, Equation (4.32) becomes

$$
\hat{U}_{PE} = \frac{1}{2} \sum_{j=1}^{N_q} \frac{A_j^q}{6} \left(\breve{\boldsymbol{\varepsilon}}_1^{\mathrm{T}} \mathbf{c} \breve{\boldsymbol{\varepsilon}}_1 + \breve{\boldsymbol{\varepsilon}}_2^{\mathrm{T}} \mathbf{c} \breve{\boldsymbol{\varepsilon}}_2 + \breve{\boldsymbol{\varepsilon}}_3^{\mathrm{T}} \mathbf{c} \breve{\boldsymbol{\varepsilon}}_3 + \breve{\boldsymbol{\varepsilon}}_1^{\mathrm{T}} \mathbf{c} \breve{\boldsymbol{\varepsilon}}_2 + \breve{\boldsymbol{\varepsilon}}_1^{\mathrm{T}} \mathbf{c} \breve{\boldsymbol{\varepsilon}}_3 + \breve{\boldsymbol{\varepsilon}}_2^{\mathrm{T}} \mathbf{c} \breve{\boldsymbol{\varepsilon}}_3 \right) \tag{4.35}
$$

It is clear that the strain energy potential for the constructed strain field is the average of the all the six components of energy combinations of strains at these three vertices of the quadrature cell.

The properties of the S-PIM depend on how the strain field is constructed. We often introduce some parameters controlling the constructed strain field for desired properties. The above mentioned steps were used in PIM-CS [14, 31], SC-PIM [29, 30], and SαFEM [38].

4.5.4 Strain construction by least square approximation

Least square approximation is a type of orthogonal projection, found useful in strain field construction. It was implemented in the least square point interpolation methods (PIM-LSS) [32]. The constructed strain field can easily satisfy the norm equivalence condition Equation (4.13). As long as the strain convergence condition (4.15) can be satisfied, we can establish a convergent SC-model. The procedure for constructing strain field using least square approximation is quite similar to the interpolation method. The four steps procedure given in Section 4.5.3 stand, except to replace the interpolation by a least square approximation in Step 3. When least square approximation is used, we can have large quadrature cells, and use more strain points for approximation. We can also have wider choice of polynomial basis terms for creating models of desired properties. More details are given in [32].

4.6   A brief historical note

In this section, we frequently used the smoothing techniques. In the past, various smoothing techniques have been used for different purposes, including in the nonlocal continuum mechanics [39] to model the size effects, and in the smoothed particle hydrodynamics [40-42] to approximate field functions and their derivatives. The gradient smoothing was widely used in finite volume method (FVM) [43]. It was also used in formulating the so-called quasi-conforming elements [44], and for the discretization of differential operators based on nodes [45]. The strain smoothing was used to deal with the material instabilities [46] and the spatial instability in the nodal integrated meshfree methods [47]. It was later used in so-called gradient smoothing method (GSM) to approximate the derivatives using various types of properly nested smoothing domains to construct strong-form models for solid mechanics problems [48, 49], compressible fluids [50, 51], and incompressible fluids [52]. It was found out that as long as the smoothing operation is *strictly* used in all the gradient approximations in a *properly nested manner, stable* strong-form numerical models can be formulated [3]. More detailed discussions can be found in [3].

Since 2005 [18], the strain smoothing technique was used as standard numerical tool in the developing a family of S-PIM models as a class of general numerical methods, as listed in TABLE 4.1. Intensive theoretical studies have been conducted to examine the stability, convergence, and various properties of SC-models, leading to various ways of constructing novel smoothing domains: cell-, node-, edge-, and face-based smoothing domains. With these theoretical developments, we now know exactly how to make an S-PIM model stable and convergent, and how to tailor the model for special properties, as listed in TABLEs 1.5 and 1.6.

4.7 Concluding remarks

Remark 4.5 Solution to displacement incompatibility

Strain construction technique is an effective way to solve some of the displacement incompatibility problem in the standard FEM modeling [2]. The essential issue for the incompatibility is the possible unbounded strain energy resulted from the incompatible assumed displacements. Our idea of constructing the strain field is a very effective means to avoid the use of the compatible strain field, and hence the strain energy can always be bounded. For the solid mechanics problems, the assumed displacement fields do not have to be continuous!

Remark 4.6 Stability and convergence

When we open the door to construct the strain field anew, possible issues related to instability may arise. In addition, the solution of such a SC-model may not converge to the exact solution, even if the model is made stable. Hence admissible conditions to the constructed strain field shall also be imposed, as listed in Section 4.4.

Remark 4.7 Minimum conditions for stable and convergent models

It was asserted [4] that as long as a set of linearly independent nodal shape functions with at least linear consistency can be created, we can always establish a stable model that produces solutions converging to the exact solution of the problem defined in Section 1.1, with proper ways to construct the strain field.

Remark 4.8 Efficiency issues

To construct the strain field in S-PIM, additional efforts or alternative computational treatments are required. From the experience of the authors' group, such efforts/ treatments used in the S-PIM do not contribute significantly to the overall computational time. The CPU time is still heavily dominated by the equation solver for a usual size model [3]. In addition, some of the operations are simpler compared to the models that use directly compatible strain fields. For example, in the S-PIM models, we do not have to compute the derivatives of the shape functions; we do not need to perform domain Gauss integrations; energy integration becomes simple summation; etc. It is the authors' opinion that a strain-constructed model does not necessarily increase the computation cost compared to the corresponding fully compatible model, as long as proper techniques are used.

Remark 4.9 Accuracy issues

A properly constructed S-PIM model, can improve the solution accuracy significantly, as will be demonstrated in these models in later chapters. One can practically establish models that give "close-to-exact" solutions (at least in a norm), as shown in PIM-CS [14, 31], SC-PIM [29, 30], SαFEM [38], and CS-αRPIM [53]. This is made possible to feature in a "knob" to tune S-PIM model for desired properties.

Remark 4.10 Other properties

Other astrictive properties, such as upper bound, lower bound, and tight bounds can also be achieved in S-PIM, by devising properly a strain construction scheme. More discussions will be given in later chapters.

4.8 References

1. Zienkiewicz, O. C., Taylor R. L., *The Finite Element Method*, 5th ed., Butterworth Heimemann, Oxford, 2000.

2. Liu, G. R. and Quek, S. S., *The finite element method: a practical course.* Butterworth Heinemann: Oxford, 2003.

3.   Liu, G. R., *Meshfree Methods: Moving beyond the Finite Element Method*, 2nd Edition, CRC press, Boca Taton, USA, 2009.

4.   Simo, J. C. and Hughes, T. J. R., *Computational Inelasticity*. Springer-Verlag, New York, 1998.

5.   Pian, T. H. H. and Wu, C. C., *Hybrid and Incompatible finite element methods*. CRC Press, Boca Raton, 2006.

6.   Liu, G. R., A G space theory and a weakened weak (W^2) form for a unified formulation of compatible and incompatible methods, Part I Theory, *International Journal for Numerical Methods in Engineering*, 81: 1093-1126, 2009.

7.   Liu, G. R., A G space theory and a weakened weak (W^2) form for a unified formulation of compatible and incompatible methods, Part II Application to solid mechanics problems, *International Journal for Numerical Methods in Engineering*, 81: 1127-1156, 2009.

8.   Wu, H. C., *Variational Principle in Elasticity and Applications*. Scientific Press, Beijing, 1982.

9.   Liu, G. R., Dai, K. Y. and Nguyen-Thoi, T., A smoothed finite element method for mechanics problems, *Computational Mechanics* 39: 859-877. 2007.

10.   Dai, K. Y. and Liu, G. R., Free and forced vibration analysis using the smoothed finite element method (SFEM), *Journal of Sound and Vibration*, 301: 803-820, 2007.

11.   Dai, K. Y., Liu, G. R. and Nguyen-Thoi, T., An n-sided polygonal smoothed finite element method (nSFEM) for solid mechanics. *Finite Elements in Analysis and Design*, 43: 847-860, 2007.

12.   Liu, G. R., Nguyen-Thoi, T., Dai, K. Y. and Lam, K. Y., Theoretical aspects of the smoothed finite element method (SFEM), *International Journal for Numerical Methods in Engineering*, 71: 902-930, 2007.

13.   Liu, G. R. and Nguyen-Thoi, T., *Smoothed Finite Element Method*, CRC press, Boca Taton, USA, 2010.

14.   Liu, G. R., Xu, X., Zhang, G. Y. and Gu, Y. T., An extended Galerkin weak form and a point interpolation method with continuous strain field and superconvergence using triangular mesh. *Computational Mechanics*, 43: 651-673, 2009.

15.   Belytschko, T. and Bachrach, W. E., Efficient implementation of quadrilaterals with high coarse-mesh accuracy. *Computer Methods in Applied Mechanics and Engineering*, 54: 279-301, 1986.

16. Belytschko, T. and Bindeman, L. P., Assumed strain stabilization of the 4-node quadrilateral with 1-point quadrature for nonlinear problems. *Computer Methods in Applied Mechanics and Engineering*, 88: 311-340, 1993.

17. Liu, G. R., A generalized Gradient smoothing technique and the smoothed bilinear form for Galerkin formulation of a wide class of computational methods. *International Journal of Computational Methods*, 5(2): 199-236, 2008.

18. Liu, G. R., Zhang, G. Y., Dai, K. Y., Wang, Y. Y., Zhong, Z. H., Li, G. Y. and Han, X., A linearly conforming point interpolation method (LC-PIM) for 2D solid mechanics problems, *International Journal of Computational Methods*, 2(4): 645-665, 2005.

19. Zhang, G.Y., Liu, G. R., Wang, Y. Y., Huang, H. T., Zhong, Z. H., Li, G. Y. and Han, X., A linearly conforming point interpolation method (LC-PIM) for three-dimensional elasticity problems, *International Journal for Numerical Methods in Engineering*, 72: 1524-1543, 2007.

20. Liu, G. R. and Zhang, G. Y., Upper bound solution to elasticity problems: A unique property of the linearly conforming point interpolation method (LC-PIM). *International Journal for Numerical Methods in Engineering*, 72: 1524-1543, 2007.

21. Zhang, G. Y., Liu, G. R., Nguyen-Thoi, T., Song, C. X., Han, X., Zhong, Z. H. and Li, G. Y., The upper bound property for solid mechanics of the linearly conforming radial point interpolation method (LC-RPIM). *International Journal of Computational Methods*, 4(3): 521-541, 2007.

22. Liu, G. R, Li, Y., Dai, K. Y., Luan, M. T. and Xue, W., A linearly conforming radial point interpolation method for solid mechanics problems. *International Journal of Computational Methods,* 3(4): 401-428, 2006.

23. Li, Y., Liu, G. R., Luan, M. T., Dai, K. Y., Zhong, Z. H., Li, G. Y. and Han, X., Contact analysis for solids based on linearly conforming radial point interpolation method, *Computational Mechanics*, 39: 537-554, 2007.

24. Liu, G. R. and Zhang, G. Y., Edge-based smoothed point interpolation methods, *International Journal of Computational Methods*, 5(4): 621-646, 2008.

25. Liu, G. R., Wang, Z., Zhang, G. Y., Zong, Z. and Wang, S., An edge-based smoothed point interpolation method for material discontinuity. *Mechanics of Advanced Materials and Structures*, 19(1-3): 3-17, 2012.

26. Liu, G. R. and Zhang, G. Y. A normed G space and weakened weak (W^2) formulation of a cell-based smoothed point interpolation method. *International Journal of Computational Methods*, 6(1): 147-179, 2009.

27. Zhang, G. Y. and Liu, G. R., Meshfree cell-based smoothed point interpolation method using isoparametric PIM shape functions and condensed RPIM shape functions. *International Journal of Computational Methods*, 8(4): 705-730, 2011.

28. Liu, G. R., Jiang, Y., Chen, L., Zhang, G. Y. and Zhang, Y. W., A singular cell-based smoothed radial point interpolation method for fracture problems. *Computers & Structures*, 89(13-14): 1378-1396, 2011.

29. Zhang, G. Y., Liu, G. R. and Xu, X., A strain-constructed point interpolation method (SC-PIM) and strain field construction schemes for solid mechanics problems using triangular mesh. *Applied Mathematics and Computation,* 219: 2067-2086, 2012.

30. Liu, G. R. and Zhang, G. Y., A novel scheme of strain-constructed point interpolation method for static and dynamic mechanics problems. *International Journal of Applied Mechanics,* 1(1): 233-258, 2009.

31. Liu G. R., Xu X., Zhang G. Y. and Nguyen T. T., A superconvergent point interpolation method (SC-PIM) with piecewise linear strain field using triangular mesh, *International Journal for Numerical Method in Engineering*, 77: 1439-1467, 2009.

32. Xu, X., Liu, G. R. and Zhang, G. Y., A point interpolation method with least square strain field (PIM-LSS) for solution bounds and ultra-accurate solutions using triangular mesh. *Computer Methods in Applied Mechanics and Engineering*, 198: 1486-1499, 2009.

33. Liu, G. R., Nguyen-Thoi, T., Nguyen-Xuan, H. and Lam, K. Y., A node-based smoothed finite element method (NS-FEM) for upper bound solutions to solid mechanics problems, *Computers and Structures*, 87: 14-26, 2009.

34. Liu, G. R., Nguyen-Thoi, T. and Lam, K. Y. An edge-based smoothed finite element method (ES-FEM) for static, free and forced vibration analyses in solids. *Journal of Sound and Vibration*, 320: 1100-1130, 2009.

35. Nguyen-Thoi T, Liu, G. R., Lam, K. Y. and Zhang, G. Y., A Face-based Smoothed Finite Element Method (FS-FEM) for 3D linear and nonlinear solid mechanics problems using 4-node tetrahedral elements. *International Journal for Numerical Methods in Engineering*, 78: 324-353, 2009.

36. Liu, G. R., Nguyen-Thoi, T. and Lam, K. Y., A novel FEM by scaling the gradient of strains with factor α (αFEM). *Computational Mechanics*, 43: 369-391, 2009.

37. Liu, G. R., Nguyen-Xuan, H. and Nguyen-Thoi, T., A variationally consistent αFEM (VCαFEM) for solution bounds and nearly exact solution to solid

mechanics problems using quadrilateral elements, *International Journal for Numerical Methods in Engineering*, 85(4): 461-497, 2011.

38.   Liu, G. R., Nguyen-Xuan, H., Nguyen-Thoi, T. and Xu, X., A novel Galerkin-like weak form and a superconvergent alpha finite element method (SαFEM) for mechanics problems using triangular meshes, *Journal of Computational Physics*, 228(11): 4055-4087, 2009.

39.   Eringen, A. C., Nonlocal polar elastic continua, *International Journal of Engineering Science*, 10: 1-16, 1972.

40.   Lucy, L., A numerical approach to testing the fission hypothesis, *Astron. J.*, 82: 1013-1024, 1977.

41.   Monaghan, J. J., Why particle methods work, *Siam J. Sci. Sat. Comput.*, 3(4): 423-433, 1982.

42.   Liu, G. R. and Liu, M. B., *Smoothed Particle Hydrodynamics - A Meshfree Practical Method*. World Scientific, Singapore, 2003.

43.   LeVeque R. J., *Finite volume methods for hyperbolic problems*, Cambridge University Press, New York, 2002.

44.   Tang, L. M., Chen, W. J. and Liu, Y., Formulation of quasi-conforming element and Hu-Washizu Principle, *Computers and Structures*, 19: 247-250, 1984.

45.   Tang, L. M., Zhang, Y. Z., Wu, J. X., Jie, M. Y., Lu, H. X. and Chen, W. J., On discretization of differential operator for solids, (in Chinese), J. Dalian Instit. Tech., pp: 7-30, 1973.

46.   Chen, J. S., Wu, C. T. and Belytschko, T., Regularization of material instabilities by meshfree approximations with intrinsic length scales. *International Journal for Numerical Methods in Engineering*, 47: 1303-1322, 2000.

47.   Chen, J. S., Wu, C. T. and Yoon, S. Y., A stabilized conforming nodal integration for Galerkin mesh-free methods. *International Journal for Numerical Methods in Engineering*, 50: 435-466, 2001.

48.   Liu, G. R., Zhang J. and Lam, K. Y., A gradient smoothing method (GSM) with directional correction for solid mechanics problems. *Computational Mechanics*, 41: 457-472, 2008.

49.   Zhang, J., Liu, G. R., Lam, K.Y., Li, H. and Xu, George X., A gradient smoothing method (GSM) based on strong form governing equation for adaptive analysis of solid mechanics problems. *Finite Elements in Analysis and Design*, 44: 889-909, 2008.

50. Liu, G. R. and Xu, George X., A gradient smoothing method (GSM) for fluid dynamics problems, *International Journal for Numerical Methods in Fluids*, 58(10):1101-1133, 2008.

51. Xu, George X., Liu, G. R. and Tani, A., An adaptive gradient smoothing method (GSM) for fluid dynamics problems. *International Journal for Numerical Methods in Fluids*, 62(5): 499-529, 2010.

52. Xu, George X., Li, E., Tan, V. and Liu, G. R., Simulation of steady and unsteady incompressible flow using gradient smoothing method (GSM), *Computers & Structures,* 90-91: 131-144, 2012.

53. Liu, G. R., Zhang, G. Y., Zong, Z. and Li, M., Meshfree cell-based smoothed alpha radial point interpolation method (CS-αRPIM) for solid mechanics problems. *International Journal of Computational Methods*, 10(4), 31 pages, DOI: 10.1142/S0219876213500205, 2013.

Chapter 5

Weak and Weakened Weak Formulations

Strong-form equations are those given in the form of partial differential equations (PDEs), such as those given in Section 1.1 for solid mechanics problems (see Remark 1.7). Obtaining the exact solution for such strong-form system equations is ideal, but is usually not possible for practical engineering problems with complicated geometry and settings. We, therefore, have to search for approximate solutions numerically. A typical *strong-form method* for approximated solutions is the well-known *collocation* method. In such a method, when nodal distribution is arbitrary and only local nodes are used for function approximation, special techniques are needed to stabilize the numerical solution [1-6]. In addition, the discretized system equations are generally *asymmetric* for irregularly distributed nodes, even for PDEs with symmetric operators. Therefore, strong-form methods are not widely used currently in solving practical problems with complicated geometries.

In Chapters 3 and 4, we understand how to properly create displacement fields and how to construct strain fields. With Theorems 3.1-3.3, and 4.1 in place, we need now put two together to provide various detailed formulations for stable and convergent numerical methods: weak and weakened weak (W^2) forms.

This chapter describes first the standard *weak* formulation based on the H space theory for establishing the standard FEM models, and then the relatively new *weakened weak* (W^2) formulations based on G space theory for establishing S-PIM models. Some important properties of the weak forms and W^2 forms will be examined. This book gives high preference to the so-called *Galerkin* weak form for reasons of simplicity, symmetry, and hence efficiency for problems with symmetric operators, such as the ones defined in Section 1.1. When we perform all sorts of numerical advances and manipulations in S-PIM models, we always try to keep the *form* of "Galerkin" (only in the sense that the trail and test functions are from the same space), even though our formulation has gone far beyond the standard Galerkin weak form, in terms of implementation, solution/function spaces and properties. This preference is particularly important for PDEs of symmetric operators: we certainly do not want to destroy the important symmetry property of the original PDEs, but to make full use of it. After all the efficiency is eventually a crucial factor for any successful numerical method to survive.

This chapter will focus on the discussion of 1) the standard Galerkin weak form, 2) general SC-Galerkin weak form, and 3) a particular GS-Galerkin weak form. A more detailed general discussion on various strong and weak formulations can be found in [4].

5.1 Briefing on weak forms

5.1.1 Weak forms

Weak form is in integral form that is an alternative representation of the strong form. It requires a *weaker* consistency on the assumed field functions, and the foundation is on the Sobolev (or H) space theory (see Chapter 2). The order of differentiation on the field variables in a weak form is reduced by half, compared to the strong form. Therefore, the consistency requirements on the assumed displacement functions are also reduced by half. The most widely used is the Galerkin weak form, which is essentially the same as the physical principle for solid mechanics problems known as the minimum potential energy principle. Formulation based on the Galerkin weak forms can produce a stable set of algebraic system equations for well-posed problems, because the positivity of the bilinear form (or energy in physical term) is ensured using functions from proper spaces. The discretized system equations are *symmetric* for irregularly distributed nodes for problems of symmetric operators. Therefore, it preserves the symmetry property and hence has good stability, accuracy and efficiency. The FEM is established using the Galerkin weak form.

5.1.2 Weakened weak forms

Weakened weak (W^2) forms [7, 8, 11] require a further weakened consistency on the displacement functions. It can have various versions. The first W^2 form was the generalized smoothed Galerkin or GS-Galerkin weak form built using the generalized gradient smoothing technique [9]. The foundation is on the normed G space theory [10, 11]. The second W^2 form was the strain-constructed Galerkin or SC-Galerkin weak form that is a norm general W^2 form [7, 8, 12] using functions from basically unnormed G spaces. These W^2 forms can further reduce the consistency requirement on the assumed functions, compared to the weak forms. The W^2 formulations are relatively new but have been used for a number of meshfree methods for important properties including upper bound property, ultra-accuracy and superconvergence. The discrete system equations of a W^2 formulation are also *symmetric* for irregularly distributed nodes for problems of symmetric operators. Therefore, it preserves the symmetry property and hence has good stability, accuracy and efficiency. The S-PIM models are established using the W^2 forms.

5.2 Galerkin weak form

5.2.1 Bilinear form

Consider the solid mechanics problems of *d*-dimension defined in Section 1.1. Using the weighted residual method [4], we have

$$\int_\Omega \widehat{\mathbf{W}} \underbrace{\left(\mathbf{L}_d^{\mathrm{T}}\mathbf{c}\mathbf{L}_d\mathbf{u}+\mathbf{b}\right)}_{\text{residual}}\mathrm{d}\Omega = \int_\Omega \widehat{\mathbf{W}}\left(\mathbf{L}_d^{\mathrm{T}}\mathbf{c}\mathbf{L}_d\mathbf{u}\right)\mathrm{d}\Omega + \int_\Omega \widehat{\mathbf{W}}\mathbf{b}\mathrm{d}\Omega = \mathbf{0} \tag{5.1}$$

where $\widehat{\mathbf{W}}$ is a vector or a diagonal matrix of *weight* or *test functions* defined in the problem domain Ω bounded by Γ. We hope that the weighted residual resulted from substituting a particular *trial* function $\mathbf{u}(\mathbf{x})\in\left(\mathbb{H}^1(\Omega)\right)^d$ into the strong-form Equation (1.21) can vanish in the integral form of Equation (5.1). When Equation (5.1) can be satisfied for a set of properly selected $\widehat{\mathbf{W}}$, a good approximation of the exact solution can be expected. If the weight or test functions are chosen differentiable, we can then perform integration by parts (or the Gauss divergence theorem) to the first term in Equation (5.1):

$$\int_\Omega \left(\mathbf{L}_d\widehat{\mathbf{W}}\right)^{\mathrm{T}}\mathbf{c}\left(\mathbf{L}_d\mathbf{u}\right)\mathrm{d}\Omega - \int_\Gamma \left(\mathbf{L}_n\widehat{\mathbf{W}}\right)^{\mathrm{T}}\left(\mathbf{c}\mathbf{L}_d\mathbf{u}\right)\mathrm{d}\Omega - \int_\Omega \widehat{\mathbf{W}}^{\mathrm{T}}\mathbf{b}\mathrm{d}\Omega = 0 \tag{5.2}$$

It is clear now that we have only first order derivatives to either the field function $\mathbf{u}$ or test function $\hat{\mathbf{W}}$: a *weak form*.

Next, we choose with *test* functions in the vector form of $\mathbf{v} \in \left(\mathbb{H}_0^1(\Omega)\right)^d \subset \left(\mathbb{H}^1(\Omega)\right)^d$, we obtain:

$$\int_\Omega \left(\mathbf{L}_d \mathbf{v}\right)^{\mathrm{T}} \mathbf{c}\left(\mathbf{L}_d \mathbf{u}\right) d\Omega - \underbrace{\int_{\Gamma_u} \left(\mathbf{L}_n \mathbf{v}\right)^{\mathrm{T}} \left(\mathbf{c}\mathbf{L}_d \mathbf{u}\right) d\Omega}_{=0,\, \because \mathbf{v}=0 \text{ on } \Gamma_u}$$
$$- \int_{\Gamma_t} \mathbf{v}^{\mathrm{T}} \mathbf{L}_n^{\mathrm{T}} \underbrace{\underbrace{\left(\mathbf{c}\mathbf{L}_d \mathbf{u}\right)}_{\sigma} d\Omega}_{\mathbf{t}_\Gamma} - \int_\Omega \mathbf{v}^{\mathrm{T}} \mathbf{b} d\Omega = 0 \tag{5.3}$$

Because $\mathbf{u}$ and $\mathbf{v}$ are all from the same space, the weak form given in (5.3) becomes the well-known Galerkin weak form.

In Equation (5.3), since $\mathbf{v} \in \left(\mathbb{H}_0^1(\Omega)\right)^d$, we shall have $\mathbf{v}=0$ on Γ_u, and thus the 2nd term becomes zero. Using the natural boundary condition Equation 1.23 on Γ_t, we shall have

$$\underbrace{\int_\Omega \left(\mathbf{L}_d \mathbf{v}\right)^{\mathrm{T}} \mathbf{c}\left(\mathbf{L}_d \mathbf{u}\right) d\Omega}_{a(\mathbf{u},\mathbf{v})} = \underbrace{\left(\int_{\Gamma_t} \mathbf{v}^{\mathrm{T}} \mathbf{t}_\Gamma d\Omega + \int_\Omega \mathbf{v}^{\mathrm{T}} \mathbf{b} d\Omega\right)}_{f(\mathbf{v})} \tag{5.4}$$

We now have the widely-used *bilinear form* for the solid mechanics problems:

$$a(\mathbf{u},\mathbf{v}) = \int_\Omega \underbrace{\left(\mathbf{L}_d \mathbf{v}\right)^{\mathrm{T}}}_{\varepsilon(\mathbf{v})} \mathbf{c}\underbrace{\left(\mathbf{L}_d \mathbf{u}\right)}_{\varepsilon(\mathbf{u})} d\Omega = \int_\Omega \varepsilon(\mathbf{v})^{\mathrm{T}} \mathbf{c}\varepsilon(\mathbf{u}) d\Omega \tag{5.5}$$

The basic properties of the bilinear form are:
Symmetry

$$a(\mathbf{u},\mathbf{v}) = a(\mathbf{v},\mathbf{u}), \quad \forall \mathbf{u} \in \left(\mathbb{H}^1(\Omega)\right)^d, \forall \mathbf{v} \in \left(\mathbb{H}^1(\Omega)\right)^d \tag{5.6}$$

Ellipticity

$$a(\mathbf{v},\mathbf{v}) \geq C_p \|\mathbf{v}\|_{H^1(\Omega)}^2, \quad \forall \mathbf{v} \in \left(\mathbb{H}_0^1(\Omega)\right)^d \tag{5.7}$$

Continuity

$$a(\mathbf{u}, \mathbf{v}) \le C_c \|\mathbf{u}\|_{H^1(\Omega)}^2 \|\mathbf{v}\|_{H^1(\Omega)}^2, \quad \forall \mathbf{u} \in \left(\mathbb{H}^1(\Omega)\right)^d, \forall \mathbf{v} \in \left(\mathbb{H}^1(\Omega)\right)^d \tag{5.8}$$

where C_p and C_c are constants that are independent of $\mathbf{v}$ and $\mathbf{u}$. The symmetry is obvious; the ellipticity is the consequence of the 2nd Korn's inequality applied to solids of stable materials [4]:

$$C_k \|\mathbf{v}\|_{\mathbb{H}^1(\Omega)} \le \|\boldsymbol{\varepsilon}(\mathbf{v})\|_{L^2}, \quad \forall \mathbf{v} \in \left(\mathbb{H}_0^1(\Omega)\right)^d \tag{5.9}$$

where C_k is a general constant independent of $\mathbf{v}$. The 2nd Korn's inequality is in turn rooted at the Poincare-Friedrichs inequality Equation (2.44). The continuity is resulted from the Cauchy-Schwarz inequality Equation (2.15) for bilinear forms for stable materials (see, Remark 1.1).

Using Equation (5.5), we have

$$U_{PE}\left(\boldsymbol{\varepsilon}(\mathbf{v})\right) = \frac{1}{2} \int_\Omega \boldsymbol{\varepsilon}^T(\mathbf{v}) \mathbf{c} \boldsymbol{\varepsilon}(\mathbf{v}) d\Omega = \frac{1}{2} a(\mathbf{v}, \mathbf{v}) \tag{5.10}$$

where $U_{PE}\left(\boldsymbol{\varepsilon}(\mathbf{v})\right)$ is the strain potential energy in the entire solid for a given displacement field $\mathbf{v} \in \left(\mathbb{H}_0^1(\Omega)\right)^d$. We see clearly here the relationship between the mathematical term of bilinear form and the engineering term of strain energy.

The *linear functional* for the solid mechanics problem is defined as

$$f(\mathbf{v}) = \int_{\Gamma_t} \mathbf{v}^T \mathbf{t}_\Gamma d\Omega + \int_\Omega \mathbf{v}^T \mathbf{b} d\Omega \tag{5.11}$$

This is the work done by the external forces (body force $\mathbf{b}$ and forces on the natural boundary $\mathbf{t}_\Gamma$) under displacement field $\mathbf{v}$.

5.2.2 Weak statement

Following Equation (5.3) we now give the *weak statement*: the exact solution of the displacement $\mathbf{u} \in \mathbb{S} \subset \left(\mathbb{H}_0^1(\Omega)\right)^d$ of the strong-form equations given in Section 1.1 satisfies

$$a(\mathbf{u}, \mathbf{v}) = f(\mathbf{v}), \quad \forall \mathbf{v} \in \mathbb{S} \tag{5.12}$$

This weak statement is also known as the Galerkin weak form, because $\mathbf{u}$ and $\mathbf{v}$ are in the same space.

5.3 Galerkin weak forms: alternative expressions

The above mathematical approach is very handy in proving the theories and properties for weak and W^2 formulations. In the engineering community, we often use the energy principles such as the principle of virtual work, and the Galerkin weak form can also be established directly from such a physical principle. The principle of virtual work states that "if a solid is in an equilibrium status, the total virtual work performed by all the *internal* stresses in the solid and all the *external* forces applied on the solid should vanish, when the solid is subjected to an arbitrary *virtual field of displacement* $\delta\mathbf{u}$. Mathematically, it sates (see, e.g., [4]):

$$\underbrace{\int_\Omega \delta(\boldsymbol{\varepsilon}(\delta\mathbf{u}))^{\mathrm{T}} \underbrace{\mathbf{c}\boldsymbol{\varepsilon}}_{\boldsymbol{\sigma}(\delta\mathbf{u})} \, \mathrm{d}\Omega}_{\text{Virtual work by internal stress } \boldsymbol{\sigma}} - \underbrace{\left(\int_\Omega \delta\mathbf{u}^{\mathrm{T}}\mathbf{b}\mathrm{d}\Omega + \int_{\Gamma_t} \delta\mathbf{u}^{\mathrm{T}}\mathbf{t}_\Gamma\mathrm{d}\Gamma \right)}_{\text{Virtual work by external forces } \mathbf{b} \text{ and } \mathbf{t}} = 0 \tag{5.13}$$

which is the Galerkin weak form for static solid mechanics problems.

Using the displacement strain relation $\boldsymbol{\varepsilon} = \mathbf{L}_d\mathbf{u}$, Equation (5.13) can be written explicitly in terms of displacements.

$$\underbrace{\int_\Omega (\mathbf{L}_d\delta\mathbf{u})^{\mathrm{T}}\mathbf{c}(\mathbf{L}_d\mathbf{u})\mathrm{d}\Omega}_{a(\delta\mathbf{u},\mathbf{u})} - \underbrace{\left(\int_\Omega \delta\mathbf{u}^{\mathrm{T}}\mathbf{b}\mathrm{d}\Omega + \int_{\Gamma_t} \delta\mathbf{u}^{\mathrm{T}}\mathbf{t}_\Gamma\mathrm{d}\Gamma \right)}_{f(\delta\mathbf{u})} = 0 \tag{5.14}$$

We can now observe the relationship between the physical energy principle and the weak-form statement given in Equation (5.12): they are essentially the same. In Equation (5.12) we require the equation to be satisfied for all functions $\mathbf{v} \in \mathbb{H}_0^1$, while in Equation (5.14) we require the same for all *arbitrary variation* of admissible displacements $\mathbf{u}$. The Galerkin weak form is then looking for a displacement field in $\mathbf{u} \in \left(\mathbb{H}_0^1\right)^d$ that satisfies Equation (5.14) for any arbitrary $\delta\mathbf{u} \in \left(\mathbb{H}_0^1\right)^d$. Physically, at such a displacement field the solid stays stable (in equilibrium).

For dynamic problems, the Galerkin weak form can be directly derived using Hamilton's principle (see, e.g., [4]). In general, we have

$$\int_{\Omega} \delta\boldsymbol{\varepsilon}^{\mathrm{T}} \boldsymbol{\sigma} \, \mathrm{d}\Omega - \int_{\Omega} \delta\mathbf{u}^{\mathrm{T}} \mathbf{b} \mathrm{d}\Omega - \int_{\Gamma_t} \delta\mathbf{u}^{\mathrm{T}} \mathbf{t}_\Gamma \mathrm{d}\Gamma + \int_{\Omega} \rho \delta\mathbf{u}^{\mathrm{T}} \ddot{\mathbf{u}} \mathrm{d}\Omega = 0 \qquad (5.15)$$

Removing the dynamic term Equation (5.15) reduces to Equation (5.13). Using the stress-strain relation, and then the strain-displacement relation, Equation (5.15) can be expressed in terms of displacements:

$$\int_{\Omega} \left(\mathbf{L}_d \delta\mathbf{u} \right)^{\mathrm{T}} \mathbf{c} \left(\mathbf{L}_d \mathbf{u} \right) \mathrm{d}\Omega - \int_{\Omega} \delta\mathbf{u}^{\mathrm{T}} \mathbf{b} \mathrm{d}\Omega - \int_{\Gamma_t} \delta\mathbf{u}^{\mathrm{T}} \mathbf{t}_\Gamma \mathrm{d}\Gamma + \int_{\Omega} \rho \delta\mathbf{u}^{\mathrm{T}} \ddot{\mathbf{u}} \mathrm{d}\Omega = 0 \qquad (5.16)$$

This is the Galerkin weak form for dynamic problems written in terms of displacement, which is widely used in the displacement methods such as the FEM and some of the meshfree methods.

Remark 5.1 Galerkin weak formulation: proven facts

The statement Equation (5.12) has a unique and stable solution for a well-posed physical problem, which is ensured mathematically by the well-known Lax-Milgram theorem, attributed to the ellipticity and continuity for the bilinear form Equations (5.7) and (5.8). From Equation (5.4), it is seen that we need only the first derivatives for all functions involved in the formulation. This is because a half of the 2nd order derivatives on $\mathbf{u}$ has been "transferred" to the so-called test function $\mathbf{v}$. As results, the continuity requirement on function $\mathbf{u}$ is weakened: it needs to be only 1st order differentiable, compared with that of 2nd order differentiable requirement in the strong formulation Equation (1.21). Equation (5.12) is thus termed as the weak formulation, which is the foundation for the well-known and widely used FEM. Creating a function space $\mathbb{S} \subset \left(\mathbb{H}_0^1(\Omega) \right)^d$, however, requires special considerations (see, e.g., [13])

5.4 FEM: a typical Galerkin weak formulation

In practice, it is generally not possible to solve the governing equations either in strong or weak forms in analytical means for the *exact* solution, except a very few simple cases. We thus resort to numerical methods to obtain *approximate* solutions, of which the finite element method (FEM) is the most popular. We present now briefly the formulation of the widely used FEM that is formulated using the Galerkin weak formulation and a mesh of elements.

Consider solid mechanics problems presented in Section 1.1. The problem domain Ω is discretized into a mesh of elements, such that $\boxed{\Omega} = \bigcup_{i=1}^{N_e} \boxed{\Omega}_i^e$ and

$\Omega_i^e \bigcap \Omega_j^e = \varnothing$, $\forall i \neq j$ where N_e is the total number of elements. Each of the elements is bounded by Γ_i^e, as shown in FIGURE 5.1. The element can have various shapes but the compatibility conditions have to be satisfied (see, e.g., [13]). For good accuracy we prefer using quality quadrilateral (Q4) types of elements, but generating such an element mesh can be a problem for complicated geometries. For easy and automatically mesh generation, triangular (Tr3) types of meshes are preferred, but the FEM solutions using such a mesh are usually poor.

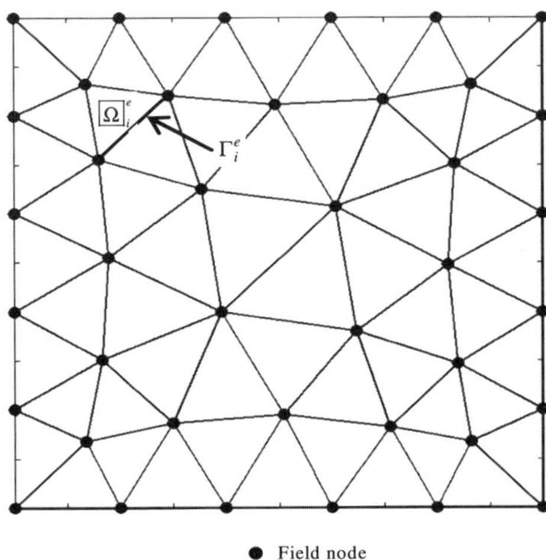

FIGURE 5.1 Illustration of three-node triangular (Tr3) elements for 2D domain.

5.4.1 Weak statement for FEM models

The FEM formulation is conveniently done using the weak-form statement, as long as we can create a discrete (hence finite) $\mathbb{H}_h^1 \subset \mathbb{H}$ space based on a mesh of elements, known as the FEM space. When the mesh is refined and $N_e \to \infty$ leading to $\mathbb{H}_h^1 \to \mathbb{H}$ (that is a Banach space: complete metric), and hence we can expect that $\tilde{\mathbf{u}} \to \mathbf{u}$, meaning that the approximate FEM solution will approach to the exact solution when the sizes of all the elements approach to zero.

The finite element displacement solution $\tilde{\mathbf{u}} \in \left(\mathbb{H}_{h,0}^1(\Omega)\right)^d \subset \left(\mathbb{H}_0^1(\Omega)\right)^d$ is an approximation to the solution to the problem defined in Section 1.1 satisfies the following weak statement

$$a\left(\tilde{\mathbf{u}},\mathbf{v}\right)=f\left(\mathbf{v}\right),\quad\forall\mathbf{v}\in\left(\mathbb{H}^1_{h,0}(\Omega)\right)^d \tag{5.17}$$

We need now to create the discrete FEM solution space.

5.4.2 $\mathbb{H}^1_h$: A space of FEM displacements

In constructing the FEM space, we use a set of linearly independent nodal shape functions $\phi^H_i\in\mathbb{H}^1_h$ obtained based on the elements with proper mapping. Because $\left(\mathbb{H}^1(\Omega)\right)^d$ is a linear space, the displacement function $\tilde{\mathbf{u}}\in\left(\mathbb{H}^1_h(\Omega)\right)^d$ can then be constructed using the following linear combination of the nodal shape functions:

$$\tilde{\mathbf{u}}\left(\mathbf{x}\right)=\sum_{I\in S_e}\mathbf{\Phi}^H_I\left(\mathbf{x}\right)\tilde{\mathbf{d}}_I \tag{5.18}$$

where $\mathbf{x}=\left\{x_1,\cdots,x_d\right\}^{\mathrm{T}}$, S_e is the set of the nodes in the element that hosts $\mathbf{x}$, $\tilde{\mathbf{d}}_I=\tilde{\mathbf{u}}\left(\mathbf{x}_I\right)$ is a nodal displacement in the FEM model (see, Equation (4.5)), and

$$\mathbf{\Phi}^H_I\left(\mathbf{x}\right)=\begin{bmatrix}\phi^H_I\left(\mathbf{x}\right) & 0\\ 0 & \phi^H_I\left(\mathbf{x}\right)\end{bmatrix}\qquad\text{for 2D: } d=2$$

$$\mathbf{\Phi}^H_I\left(\mathbf{x}\right)=\begin{bmatrix}\phi^H_I\left(\mathbf{x}\right) & 0 & 0\\ 0 & \phi^H_I\left(\mathbf{x}\right) & 0\\ 0 & 0 & \phi^H_I\left(\mathbf{x}\right)\end{bmatrix}\qquad\text{for 3D: } d=3 \tag{5.19}$$

Because the nodal basis functions are the Kronecker Delta: $\phi^H_i\left(\mathbf{x}_j\right)=\delta_{ij}$, we shall have

$$\tilde{\mathbf{u}}\left(\mathbf{x}_I\right)=\tilde{\mathbf{u}}_I \tag{5.20}$$

This means that the essential boundary condition at the nodes can be imposed directly (the same is not true when the MLS shape functions are used in some meshfree settings (see, e.g., [4]).

Equation (5.18) implies that the finite element solution space is spanned by these nodal shape functions:

$$\left(\mathbb{H}^1_h(\Omega)\right)^d = \text{span}\left\{\mathbf{\Phi}^H_I(\mathbf{x})\right\}^{N_n}_{I=1} \tag{5.21}$$

For convenience in later derivations, Equation (5.18) is rewritten in the following form

$$\tilde{\mathbf{u}}(\mathbf{x}) = \sum_{I \in S_e} \mathbf{\Phi}^H_I(\mathbf{x})\tilde{\mathbf{d}}_I = \sum_{I=1}^{N_n} \mathbf{\Phi}^H_I(\mathbf{x})\tilde{\mathbf{d}}_I$$

$$= \underbrace{\left[\mathbf{\Phi}^H_1(\mathbf{x}) \quad \mathbf{\Phi}^H_2(\mathbf{x}) \quad \cdots \quad \mathbf{\Phi}^H_I(\mathbf{x}) \quad \cdots \quad \mathbf{\Phi}^H_{N_n}(\mathbf{x})\right]}_{\mathbf{\Phi}^H(\mathbf{x})} \underbrace{\left\{\begin{array}{c} \tilde{\mathbf{d}}_1 \\ \tilde{\mathbf{d}}_2 \\ \vdots \\ \tilde{\mathbf{d}}_{N_n} \end{array}\right\}}_{\tilde{\mathbf{d}}} \tag{5.22}$$

$$= \mathbf{\Phi}^H(\mathbf{x})\tilde{\mathbf{d}}$$

where $\mathbf{x} = \{x_1, \cdots, x_d\}^T$, $\mathbf{\Phi}^H$ is the row-matrix of the nodal shape function matrix, and $\tilde{\mathbf{d}}$ is the nodal displacements for all the nodes in the entire problem domain:

$$\tilde{\mathbf{d}} = \left\{\tilde{\mathbf{d}}_1(\mathbf{x}) \quad \tilde{\mathbf{d}}_2(\mathbf{x}) \quad \cdots \quad \tilde{\mathbf{d}}_I(\mathbf{x}) \quad \cdots \quad \tilde{\mathbf{d}}_{N_n}(\mathbf{x})\right\}^T \tag{5.23}$$

We need to note the following remark.

Remark 5.2 Sparseness of the shape function matrix

Because $\mathbf{\Phi}_I$ for node I is locally supported by the nodes in the elements connected to node I, it is zero beyond the support domain. Any given $\mathbf{x} \in \Omega$ will surely fall in an element that contains a very small number of nodes (for example, 3 nodes for a Tr3 element). Therefore, $\tilde{\mathbf{u}}(\mathbf{x})$ in Equation (5.22) relates only to a very small number of nodal shape functions, and the shape function matrix $\mathbf{\Phi}(\mathbf{x})$ will be an extremely sparse matrix with lots of zeros for an actual model having large number of nodes. The expression of $\tilde{\mathbf{u}}(\mathbf{x})$ in Equation (5.22) using all the field nodes is only for convenience in the formulation. In actual computation, we need only Equation (5.18).

5.4.3 Compatible strain field

Using the strain-displacement relation Equation (1.19) and Equation (5.18), we obtain the compatible strain field for the FEM model as

$$\tilde{\boldsymbol{\varepsilon}} = \mathbf{L}_d \tilde{\mathbf{u}} = \sum_{I \in S_e} \underbrace{\mathbf{L}_d \boldsymbol{\Phi}_I^H (\mathbf{x})}_{\tilde{\mathbf{B}}_I(\mathbf{x})} \tilde{\mathbf{d}}_I = \sum_{I \in S_e} \tilde{\mathbf{B}}_I (\mathbf{x}) \tilde{\mathbf{d}}_I \tag{5.24}$$

For the convenience in the later derivations, Equation (5.24) is rewritten as

$$\tilde{\boldsymbol{\varepsilon}} = \mathbf{L}_d \tilde{\mathbf{u}} = \sum_{I \in S_e} \underbrace{\mathbf{L}_d \boldsymbol{\Phi}_I^H (\mathbf{x})}_{\tilde{\mathbf{B}}_I(\mathbf{x})} \tilde{\mathbf{d}}_I = \sum_{I=1}^{N_n} \underbrace{\mathbf{L}_d \boldsymbol{\Phi}_I^H (\mathbf{x})}_{\tilde{\mathbf{B}}_I(\mathbf{x})} \tilde{\mathbf{d}}_I$$

$$= \underbrace{\left[\tilde{\mathbf{B}}_1 (\mathbf{x}) \quad \tilde{\mathbf{B}}_2 (\mathbf{x}) \quad \cdots \quad \tilde{\mathbf{B}}_{N_n} (\mathbf{x}) \right]}_{\tilde{\mathbf{B}}(\mathbf{x})} \underbrace{\begin{Bmatrix} \tilde{\mathbf{d}}_1 \\ \tilde{\mathbf{d}}_2 \\ \vdots \\ \tilde{\mathbf{d}}_{N_n} \end{Bmatrix}}_{\tilde{\mathbf{d}}} \tag{5.25}$$

$$= \tilde{\mathbf{B}} (\mathbf{x}) \tilde{\mathbf{d}}$$

where $\tilde{\mathbf{B}}_I (\mathbf{x})$ is the "strain-displacement matrix" for node I, which contains differentiations of shape functions given by

$$\tilde{\mathbf{B}}_I (\mathbf{x}) = \mathbf{L}_d \boldsymbol{\Phi}_I^H (\mathbf{x})$$

$$= \begin{cases} \begin{bmatrix} \partial \phi_I (\mathbf{x})/\partial x & 0 \\ 0 & \partial \phi_I (\mathbf{x})/\partial y \\ \partial \phi_I (\mathbf{x})/\partial y & \partial \phi_I (\mathbf{x})/\partial x \end{bmatrix} & \text{for 2D: } d=2 \\[2em] \begin{bmatrix} \partial \phi_I (\mathbf{x})/\partial x & 0 & 0 \\ 0 & \partial \phi_I (\mathbf{x})/\partial y & 0 \\ 0 & 0 & \partial \phi_I (\mathbf{x})/\partial z \\ 0 & \partial \phi_I (\mathbf{x})/\partial z & \partial \phi_I (\mathbf{x})/\partial y \\ \partial \phi_I (\mathbf{x})/\partial z & 0 & \partial \phi_I (\mathbf{x})/\partial x \\ \partial \phi_I (\mathbf{x})/\partial y & \partial \phi_I (\mathbf{x})/\partial x & 0 \end{bmatrix} & \text{for 3D: } d=3 \end{cases} \tag{5.26}$$

Matrix $\tilde{\mathbf{B}}_I(\mathbf{x})$ is often called "strain matrix" for short, because the use of $\tilde{\mathbf{B}}_I(\mathbf{x})$ in Equation (5.24) gives the strains. The global strain-displacement matrix can be given by

$$\tilde{\mathbf{B}}(\mathbf{x}) = \mathbf{L}_d \boldsymbol{\Phi}^H(\mathbf{x}) \tag{5.27}$$

It is clear that $\tilde{\mathbf{B}}(\mathbf{x})$ will also be very sparse because of the sparseness of the shape function matrix $\boldsymbol{\Phi}^H(\mathbf{x})$.

5.4.4 Discrete system equations

We now substitute Equations (5.22) and (5.25) into Equation (5.13), which gives

$$\int_\Omega \delta\left(\tilde{\mathbf{B}}\tilde{\mathbf{d}}\right)^{\mathrm{T}}\left(\mathbf{c}\tilde{\mathbf{B}}\tilde{\mathbf{d}}\right)\mathrm{d}\Omega - \int_\Omega \delta\left(\boldsymbol{\Phi}^H\tilde{\mathbf{d}}\right)^{\mathrm{T}}\mathbf{b}\mathrm{d}\Omega - \int_{\Gamma_t}\delta\left(\boldsymbol{\Phi}^H\tilde{\mathbf{d}}\right)^{\mathrm{T}}\mathbf{t}_\Gamma\mathrm{d}\Gamma = 0 \tag{5.28}$$

After simple manipulation, and factoring out $\delta\tilde{\mathbf{d}}^{\mathrm{T}}$, Equation (5.28) can be rewritten as

$$\delta\tilde{\mathbf{d}}^{\mathrm{T}}\left(\int_\Omega \tilde{\mathbf{B}}^{\mathrm{T}}\mathbf{c}\tilde{\mathbf{B}}\tilde{\mathbf{d}}\mathrm{d}\Omega - \int_\Omega (\boldsymbol{\Phi}^H)^{\mathrm{T}}\mathbf{b}\mathrm{d}\Omega - \int_{\Gamma_t}(\boldsymbol{\Phi}^H)^{\mathrm{T}}\mathbf{t}_\Gamma\mathrm{d}\Gamma\right) = 0 \tag{5.29}$$

To have the above equation satisfied for arbitrary $\delta\tilde{\mathbf{d}}^{\mathrm{T}}$, the terms in the bracket must vanish:

$$\underbrace{\int_\Omega \tilde{\mathbf{B}}^{\mathrm{T}}\mathbf{c}\tilde{\mathbf{B}}\mathrm{d}\Omega}_{\tilde{\mathbf{K}}}\tilde{\mathbf{d}} - \left(\underbrace{\int_\Omega (\boldsymbol{\Phi}^H)^{\mathrm{T}}\mathbf{b}\mathrm{d}\Omega + \int_{\Gamma_t}(\boldsymbol{\Phi}^H)^{\mathrm{T}}\mathbf{t}_\Gamma\mathrm{d}\Gamma}_{\tilde{\mathbf{F}}}\right) = 0 \tag{5.30}$$

We finally have the following set of discretized algebraic linear system equations of FEM models for static problems.

$$\tilde{\mathbf{K}}\tilde{\mathbf{d}} = \tilde{\mathbf{f}} \tag{5.31}$$

where

$$\tilde{\mathbf{K}} = \int_\Omega \tilde{\mathbf{B}}^{\mathrm{T}}\mathbf{c}\tilde{\mathbf{B}}\mathrm{d}\Omega \tag{5.32}$$

is the (global) stiffness matrix, and

$$\tilde{\mathbf{f}} = \int_{\Omega} (\mathbf{\Phi}^H)^{\mathrm{T}} \mathbf{b} \mathrm{d}\Omega + \int_{\Gamma_t} (\mathbf{\Phi}^H)^{\mathrm{T}} \mathbf{t}_\Gamma \mathrm{d}\Gamma \tag{5.33}$$

is the (global) nodal force vector for all the external forces applied at all the nodes.

Because of the sparseness of the strain-displacement matrix $\tilde{\mathbf{B}}$, we will not actually use Equation (5.32) in the computation of the stiffness matrix. Instead, we evaluate the entries of $\tilde{\mathbf{K}}$ using

$$\tilde{\mathbf{K}}_{IJ} = \int_{\Omega} \tilde{\mathbf{B}}_I^{\mathrm{T}} \mathbf{c} \tilde{\mathbf{B}}_J \mathrm{d}\Omega = \sum_{i=1}^{N_e} \underbrace{\int_{\Omega_i^e} \tilde{\mathbf{B}}_I^{\mathrm{T}} \mathbf{c} \tilde{\mathbf{B}}_J \mathrm{d}\Omega}_{\tilde{\mathbf{K}}_{IJ,i}^e} \tag{5.34}$$

where Ω_i^e is the domain of the ith element in the FEM model, and

$$\tilde{\mathbf{K}}_{IJ,i}^e = \int_{\Omega_i^e} \tilde{\mathbf{B}}_I^{\mathrm{T}} \mathbf{c} \tilde{\mathbf{B}}_J \mathrm{d}\Omega \tag{5.35}$$

This conversion of problem domain integration to the sum of elemental integrations is possible, because $\tilde{\mathbf{u}} \in \left(\mathbb{H}_h^1(\Omega) \right)^d$ and hence no energy loss on the interfaces of the elements. Because of the sparseness of the $\tilde{\mathbf{B}}$ matrix, leads to $\tilde{\mathbf{K}}_{IJ} = 0$ for node I and J that are not attached to the same Ω_i^e, which means that the stiffness matrix $\tilde{\mathbf{K}}$ will be sparse. Therefore, in the actual evaluation of Equation (5.32) becomes that of Equation (5.35) for each element, and assemble (a summation associated with nodes) it to the global matrix $\tilde{\mathbf{K}}$, which avoids the computations of zero entries. The integration in Equation (5.35) can be performed effectively using the standard Gauss integration technique.

Likewise, the evaluation of the vector $\tilde{\mathbf{f}}$ is also performed via an assembly process as follows.

$$\begin{aligned} \tilde{\mathbf{f}}_I &= \int_{\Omega} (\mathbf{\Phi}_I^H(\mathbf{x}))^{\mathrm{T}} \mathbf{b} \mathrm{d}\Omega + \int_{\Gamma_t} (\mathbf{\Phi}_I^H(\mathbf{x}))^{\mathrm{T}} \mathbf{t}_\Gamma \mathrm{d}\Gamma \\ &= \sum_{i=1}^{N_e} \int_{\Omega_i^e} (\mathbf{\Phi}_I^H(\mathbf{x}))^{\mathrm{T}} \mathbf{b} \mathrm{d}\Omega + \sum_{i=1}^{N_e} \int_{\Gamma_{t,i}^e} (\mathbf{\Phi}_I^H(\mathbf{x}))^{\mathrm{T}} \mathbf{t}_\Gamma \mathrm{d}\Gamma \end{aligned} \tag{5.36}$$

which can be evaluated in the similar way as for the stiffness matrix.

For dynamics problems, the FEM equations become

$$\tilde{\mathbf{K}}\tilde{\mathbf{d}} + \mathbf{M}\ddot{\tilde{\mathbf{d}}} = \tilde{\mathbf{f}} \tag{5.37}$$

where the global mass matrix is given by

$$\mathbf{M} = \int_{\Omega} \rho \left(\mathbf{\Phi}^{H}\right)^{\mathrm{T}} \mathbf{\Phi}^{H} \mathrm{d}\Omega \tag{5.38}$$

For all $\rho > 0$, it is clear that $\mathbf{M}$ is SPD, due to Theorem 3.1.

5.4.5 Imposition of the essential boundary conditions

From Equation (5.32) it is clear that $\tilde{\mathbf{K}}$ is symmetric. However, at this stage $\tilde{\mathbf{K}}$ is usually singular, because we used $\tilde{\mathbf{u}} \in \left(\mathbb{H}_{h}^{1}(\Omega)\right)^{d}$ not $\tilde{\mathbf{u}} \in \left(\mathbb{H}_{h,0}^{1}(\Omega)\right)^{d}$ meaning that the essential boundary conditions given in Equation (1.22) are not yet satisfied and the solid can have rigid body movements. Therefore, we have to impose these essential boundary conditions. This is done via simply removing (or modifying) the corresponding rows and columns in the stiffness matrix $\tilde{\mathbf{K}}$ leading to a condensed (or modified) $\tilde{\mathbf{K}}$ matrix [13]. Such a simple treatments are possible because the FEM shape functions possess the Delta function property. The condensed $\tilde{\mathbf{K}}$ matrix will become nonsingular, and hence will be symmetric positive definite (SPD).

5.4.6 FEM solutions

Equation (5.31) can now be solved with ease using the standard equation solver routines for the nodal displacements $\tilde{\mathbf{d}}$. Because of the SPD property of the $\tilde{\mathbf{K}}$ matrix, the displacement solution will be stable and unique. Next, the solutions in terms of strains in each element can be finally retrieved, using Equation (5.24) with the shape functions and the nodal displacements obtained. Finally, the stress solutions in each element are computed using the constitutive equations (1.2). When needed, the solution in terms of strain energy of the solid can be computed using

$$\tilde{U}_{PE}\left(\tilde{\boldsymbol{\varepsilon}}(\tilde{\mathbf{u}})\right) = \frac{1}{2}\int_{\Omega} \tilde{\boldsymbol{\varepsilon}}^{\mathrm{T}}(\tilde{\mathbf{u}})\mathbf{c}\tilde{\boldsymbol{\varepsilon}}(\tilde{\mathbf{u}})\mathrm{d}\Omega = \frac{1}{2}\tilde{\mathbf{d}}^{\mathrm{T}}\tilde{\mathbf{K}}\tilde{\mathbf{d}} = \frac{1}{2}a\left(\tilde{\mathbf{u}}, \tilde{\mathbf{u}}\right) \tag{5.39}$$

As presented above, the formulation of the standard FEM is in fact quite straightforward for solid mechanics problems. Implementations of FEM to practical problems, formulations for mechanics of structures, such as beams, plates, shells, crack-tip elements, heat transfer and acoustic problems can also be performed in similar manners [13]. Issues related practical modeling techniques, and application of FEM results to actual design problems can be quite intensive. Readers may refer to other dedicated books [13] for more details. We now summarize some of the important properties of the FEM models and solutions.

Remark 5.3 Full compatibility

A fully compatible FEM model satisfies three conditions: 1) the strain-displacement relation; 2) the essential (displacement) boundary conditions; 3) the nodal shape functions ϕ_i^H are compatible. To ensure the shape functions being compatible in an FEM setting, we practice mainly two tricks: a) using only nodes of the element to create ϕ_i^H in the natural coordinate system; b) use proper mapping for the element to ensure the continuity of ϕ_i^H on the interfaces of the elements.

Remark 5.4 Lower bound property

The strain energy of the fully compatible FEM solution is a lower bound of the strain energy of the exact solution.

$$\tilde{U}_{PE}(\tilde{\boldsymbol{\varepsilon}}) = \frac{1}{2}a(\tilde{\mathbf{u}},\ \tilde{\mathbf{u}}) \le \frac{1}{2}a(\mathbf{u},\ \mathbf{u}) = U_{PE}(\boldsymbol{\varepsilon}) \tag{5.40}$$

The proof of the lower bound property can be found in a number of references, for example [14] in variational formulation and [15] in matrix formulation based on the energy principle. The lower bound property implies the well-known fact that the FEM solution always underestimates the strain energy. This property of FEM solution provides a good global measure of the lower bound solution with respect to the exact solution. We will discuss more in Chapter 6.

Remark 5.5 Monotonic convergence property

When a sequence of n_m nested element meshes $M_1, M_2, \cdots, M_{n_m}$ with the corresponding solution spaces of $\mathbb{H}^1_{M_1} \subset \mathbb{H}^1_{M_2} \cdots \subset \mathbb{H}^1_{M_{n_m}} \subset \mathbb{H}^1$ are used to solve force-driving solid mechanics problems, then the solution satisfies

$$\tilde{U}\left(\tilde{\boldsymbol{\varepsilon}}_{M_1}\right) \leq \tilde{U}\left(\tilde{\boldsymbol{\varepsilon}}_{M_2}\right) \leq \cdots \leq \tilde{U}\left(\tilde{\boldsymbol{\varepsilon}}_{M_{n_m}}\right) \leq U\left(\boldsymbol{\varepsilon}\right) \tag{5.41}$$

where $\tilde{\boldsymbol{\varepsilon}}_{M_i}$ is the FEM compatible solution of strains obtained using mesh M_i. This property can be shown easily using the arguments given by Oliveira [16].

Remark 5.6 Reproducibility of FEM

If the exact solution $u \in \mathbb{H}^1_{h,0}$, then a fully compatible FEM model created using the standard Galerkin weak form with functions in $\mathbb{H}^1_{h,0}$ reproduces the exact solution.

This property can be proven [14, 17, 18]. Intuitively, it can also be understood easily. Because an fully compatible FEM model guarantees producing a unique, stable and the best possible solution from the given $\mathbb{H}^1_{h,0}$ (in energy norm), if the exact solution is in its solution space, it will surely produce it as it must be the best one.

5.5 SC-Galerkin: a general W^2 formulation

The strain-constructed Galerkin (SC-Galerkin) weak form is a general W^2 form that uses functions from basically a proper unnormed G space, and has been applied to create a number of SC-PIM models [19, 20]. The stability and convergence of the SC-Galerkin models is controlled by the admissible conditions imposed on the constructed strain fields discussed in Chapter 4.

For a vector displacement field $\mathbf{v} \in \left(\mathbb{G}^1_{h,0}\right)^d$, we first construct a strain field $\hat{\boldsymbol{\varepsilon}}(\mathbf{v})$ in the form of

$$\hat{\boldsymbol{\varepsilon}}(\mathbf{v}) = \mathscr{B}\left(\mathbf{v}\right) \tag{5.42}$$

The strain energy potential for the constructed strain field becomes

$$\hat{U}_{PE}(\mathbf{v}) = \int_{\Omega} \frac{1}{2} \hat{\boldsymbol{\varepsilon}}^{\mathrm{T}}(\mathbf{v}) \, \mathbf{c}\hat{\boldsymbol{\varepsilon}}(\mathbf{v}) \mathrm{d}\Omega \tag{5.43}$$

The solution of an SC-model shall satisfy the SC-Galerkin weak form

$$\int_{\Omega} \delta\hat{\boldsymbol{\varepsilon}}^{\mathrm{T}}(\hat{\mathbf{u}})\mathbf{c}\hat{\boldsymbol{\varepsilon}}(\hat{\mathbf{u}})\,\mathrm{d}\Omega - \int_{\Omega} \delta\hat{\mathbf{u}}^{\mathrm{T}}\mathbf{b}\,\mathrm{d}\Omega - \int_{\Gamma_{t}} \delta\hat{\mathbf{u}}^{\mathrm{T}}\mathbf{t}_{\Gamma}\mathrm{d}\Gamma = 0 \tag{5.44}$$

where $\hat{\mathbf{u}} \in \left(\mathbb{G}_{h,0}^{1}\right)^{d}$, $\hat{\boldsymbol{\varepsilon}}(\hat{\mathbf{u}}) \in \left(\mathbb{L}^{2}\right)^{d}$ and satisfies 1) the norm equivalence condition (4.13) and 2) strain convergence condition (4.15).

Theorem 5.1 SC-Galerkin: a stable and convergent $\mathbf{W}^{2}$ formulation

The SC-Galerkin weak form defined in Equation (5.44) using displacement functions in a $\mathbb{G}_{h,0}^{1}$ space and constructed strain fields that satisfy 1) the norm equivalence condition (4.13) and 2) strain convergence condition (4.15) produces a stable and unique solution that convergences to the exact solution for stable solids when $h \rightarrow 0$ (and hence $\Omega_{i}^{s} \rightarrow 0$).

A proof of Theorem 5.1 can be found in [4]. We note here the following remarks.

Remark 5.7 SC-Galerkin: variational consistence

An SC-Galerkin model is, in general, not variationally consistent in the usual sense. It is a general $\mathbf{W}^{2}$ formulation using assumed displacement functions in a proper $\mathbb{G}_{h}^{1}$ space that can be discontinuous.

Remark 5.8 SC-Galerkin: temporal stability

An SC-Galerkin model is spatially stable and will not have zero-energy modes, but it does not necessarily ensure a temporally stability. The model may have spurious modes at a higher energy level depending on how the strain field is constructed, and care must be taken in creating such an SC-Galerkin model.

Remark 5.9 SC-Galerkin: no additional DOFs

In an SC-Galerkin model, there are only nodal displacement variables, and no additional unknowns compared to the standard Galerkin model.

5.6 GS-Galerkin: a widely used $\mathbf{W}^2$ form

5.6.1 Bilinear forms in $\mathbb{G}_h^1$ spaces

For general S-PIM settings, we need to form a proper bilinear form using functions in a proper $\mathbb{G}_h^1$ space defined in Chapter 2. Such a bilinear form in a $\mathbb{G}_h^1$ space is also called the *generalized smoothed* (GS) bilinear form because of the use of the generalized smoothing operations allowing the use of discontinuous functions [9].

Consider solid mechanics problems governed by strong-form PDEs given in Section 1.1, the problem domain Ω is discretized with a set of background cells using the triangulation procedure defined in Section 1.6.1. On top of the background cells a set of smoothing domains are created following the procedure detailed in Section 2.3.1. We require also that at least the minimum number of linearly independent smoothing domains is used (see Section 2.3.10). The GS bilinear form can be defined as

$$\bar{a}_D\left(\mathbf{w}, \mathbf{v}\right) = \sum_{i=1}^{N_s} A_i^s \bar{\boldsymbol{\varepsilon}}_i^{\mathrm{T}}(\mathbf{w})\mathbf{c}\bar{\boldsymbol{\varepsilon}}_i\left(\mathbf{v}\right) \tag{5.45}$$

where $\mathbf{w}, \mathbf{v} \in (\mathbb{G}_h^1)^d$, and $\bar{\boldsymbol{\varepsilon}}_i$ is the *generalized smoothed strain* for the smoothed domain Ω_i^s (see, Remark 4.3). Note that in Equation (5.45) $N_s = N_q$ for GS-Galerkin based models using the generalized smoothed strains (see Section 4.5.2).

Because $\mathbf{w}, \mathbf{v} \in (\mathbb{G}_h^1)^d$, the displacement functions can be discontinuous. Therefore, the no-sharing rule is important: the boundaries of Ω_i^s should not share any *finite* portion of lines where the displacements are discontinuous. In other words, the no-sharing rule ensures that the displacement functions are continuous on the smoothing domain boundaries. Therefore, the summation used in Equation (5.45) is possible because there will be no loss of energy along these smoothing domain boundaries. This concept is also known as *energy consistency*.

For *any* given displacement $\mathbf{w} = \{w_1 \quad w_2\}^{\mathrm{T}}$ with $\mathbf{w} \in (\mathbb{G}_h^1)^2$, $\bar{\boldsymbol{\varepsilon}}_i$ can be written as (for 2D problems as an example)

$$\overline{\boldsymbol{\varepsilon}}_i\left(\mathbf{w}\right)=\frac{1}{A_i^s}\int_{\Gamma_i^s}\mathbf{L}_n\mathbf{w}\left(\mathbf{x}\right)\mathrm{d}s=\left\{\overline{\varepsilon}_{11}\quad\overline{\varepsilon}_{22}\quad 2\overline{\varepsilon}_{12}\right\}_i^{\mathrm{T}}$$

$$=\left\{\underbrace{\frac{\partial w_1}{\partial x_1}}_{\overline{g}_{11}}\quad\underbrace{\frac{\partial w_2}{\partial x_2}}_{\overline{g}_{22}}\quad\left(\underbrace{\frac{\partial w_1}{\partial x_2}}_{\overline{g}_{12}}+\underbrace{\frac{\partial w_2}{\partial x_1}}_{\overline{g}_{21}}\right)\right\}_i^{\mathrm{T}} \tag{5.46}$$

$$=\left\{\overline{g}_{11}\quad\overline{g}_{22}\quad\left(\overline{g}_{12}+\overline{g}_{21}\right)\right\}_i^{\mathrm{T}}$$

The L^2 norm of the strain vector $\left\Vert\overline{\boldsymbol{\varepsilon}}\right\Vert_{L^2}^2$ can be written as

$$\begin{aligned}\left\Vert\overline{\boldsymbol{\varepsilon}}\right\Vert_{L^2}^2&=\sum_{i=1}^{N_s}A_i^s\left(\overline{\varepsilon}_{11}^2+\overline{\varepsilon}_{22}^2+4\overline{\varepsilon}_{12}^2\right)\\&=\sum_{i=1}^{N_s}A_i^s\left(\overline{g}_{11}^2(w_1)+\overline{g}_{22}^2(w_2)+\left(\overline{g}_{12}(w_1)+\overline{g}_{21}(w_2)\right)^2\right)\end{aligned} \tag{5.47}$$

The L^2 norm of the strain vector is the same as the $\mathbb{G}_h^1$ semi-norm (see, Equation (2.100)

$$\left\vert\mathbf{w}\right\vert_{\mathbb{G}^1(\Omega)}=\left\Vert\overline{\boldsymbol{\varepsilon}}\right\Vert_{L^2},\quad\forall\mathbf{w}\in\left(\mathbb{G}_{h,0}^1\right)^2 \tag{5.48}$$

Combining Equations (2.121) and (5.48) leads to

Remark 5.10 Fourth inequality:

$$c_G\left\Vert\mathbf{w}\right\Vert_{\mathbb{G}^1(\Omega)}\le\left\Vert\overline{\boldsymbol{\varepsilon}}\right\Vert_{\mathbf{L}^2},\quad\forall\mathbf{w}\in\left(\mathbb{G}_{h,0}^1\right)^2 \tag{5.49}$$

which is equivalent to the 2nd Korn's inequality for functions in an $\mathbb{H}_h^1$ space (see, Equation (5.9)).

Combining Equations (2.116), (5.48) and (5.49), we have the following chain inequality for functions in $\mathbb{G}_h^1$ space

$$c_G\left\Vert\mathbf{w}\right\Vert_{\mathbb{G}^1(\Omega)}\le\left\Vert\overline{\boldsymbol{\varepsilon}}\right\Vert_{L^2}\le\left\Vert\mathbf{w}\right\Vert_{\mathbb{G}^1(\Omega)},\quad\forall\mathbf{w}\in\left(\mathbb{G}_{h,0}^1\right)^2 \tag{5.50}$$

5.6.2 Properties of GS bilinear form

Remark 5.11 Symmetry and semi-positivity

By simple observation, it is found that the GS bilinear form has the basic property of symmetry

$$\bar{a}_D\left(\mathbf{w},\mathbf{v}\right)=\bar{a}_D\left(\mathbf{v},\mathbf{w}\right),\quad \forall \mathbf{w},\mathbf{v}\in\left(\mathbb{G}_h^1\right)^d \tag{5.51}$$

Because of the symmetry and the SPD property of the elastic material constants, the GS bilinear form is semi-positive definite.

Remark 5.12 Convergence property for functions in an H space

If $\mathbf{w},\mathbf{v}\in\mathbb{H}_h^1$, when $N_s\to\infty$ and all $\Omega_k^s\to 0$, $\overline{W}$ becomes the Delta function and the integral representation of the strain field will be exact (see, Remark 4.4). At such a limit $\bar{a}_D\left(\mathbf{w},\mathbf{v}\right)\to a\left(\mathbf{w},\mathbf{v}\right)$. Based on the known property of $a\left(\mathbf{w},\mathbf{w}\right)$ given in Equation (5.7), we should have the ellipticity property: there exists a nonzero positive constant c such that

$$\lim_{\substack{N_s\to\infty\\ \text{all }\Omega_k\to 0}}\bar{a}_D\left(\mathbf{w},\mathbf{w}\right)\geq c\left\|\mathbf{w}\right\|_{\mathbb{H}^1(\Omega)}^2,\quad \forall \mathbf{w}\in\left(\mathbb{H}_{h,0}^1\right)^d \tag{5.52}$$

The ellipticity ensures that when the $\mathbb{H}_h^1$ space is enriched by element/cell refinement (and hence the smoothing domains) the GS-Galerkin solution approaches to that of the standard Galerkin and hence the exact solution.

Theorem 5.2 Ellipticity with respect to G semi-norm

For solids of stable materials, there exists a nonzero positive constant c_{aw}^s such that

$$\bar{a}_D\left(\mathbf{w},\mathbf{w}\right)\geq c_{aw}^s\left|\mathbf{w}\right|_{\mathbb{G}^1(\Omega)}^2,\quad \forall \mathbf{w}\in\left(\mathbb{G}_{h,0}^1\right)^d \tag{5.53}$$

where c_{aw}^s is independent of $\mathbf{w}$ (related to positivity constants of the material). A proof of Theorem 5.2 can be found in [4].

Theorem 5.3 Ellipticity (coercivity): fifth inequality

For solids of stable materials, there exists a nonzero positive constant c_{aw}^{f} such that

$$\overline{a}_{D}\left(\mathbf{w},\mathbf{w}\right) \geq c_{aw}^{f}\left\|\mathbf{w}\right\|_{\mathbb{G}^{1}(\Omega)}^{2}, \forall \mathbf{w} \in \left(\mathbb{G}_{h,0}^{1}\right)^{d} \tag{5.54}$$

where c_{aw}^{f} is independent of $\mathbf{w}$. The inequality (5.54) implies the *ellipticity* or *coercivity* of the GS bilinear forms.

A proof Theorem 5.3 can be found in [4]. Theorem 5.3 is important because it ensures the existence and hence the uniqueness, and consequently stability of the solution of the GS-Galerkin formulation.

Theorem 5.4 Continuity: sixth inequality

For solids of stable materials, there exists a nonzero positive constant c_{awv}^{f} such that

$$\overline{a}_{D}\left(\mathbf{w},\mathbf{v}\right) \leq c_{awv}^{f}\left\|\mathbf{w}\right\|_{\mathbb{G}^{1}(\Omega)}\left\|\mathbf{v}\right\|_{\mathbb{G}^{1}(\Omega)}, \quad \forall \mathbf{w} \in \left(\mathbb{G}_{h}^{1}\right)^{d}, \forall \mathbf{v} \in \left(\mathbb{G}_{h}^{1}\right)^{d} \tag{5.55}$$

where c_{awv}^{f} is independent of both $\mathbf{w}$ and $\mathbf{v}$.

A proof Theorem 5.4 can be found in [4]. Theorem 5.4 is important because it ensures that the GS bilinear from is continuous. Together with the ellipticity, it ensures the stability of the solution of the GS-Galerkin formulation [4].

Remark 5.13 Softened model: seventh inequality

For solids of stable materials and for any $\mathbf{w} \in \left(\mathbb{H}_{h}^{1}\right)^{d}$ the GS bilinear form is smaller than the bilinear form:

$$\overline{a}_{D}\left(\mathbf{w},\mathbf{w}\right) \leq a\left(\mathbf{w},\mathbf{w}\right), \quad \forall \mathbf{w} \in \left(\mathbb{H}_{h}^{1}\right)^{d} \tag{5.56}$$

Remark 5.13 is the same as the Theorem 4.3 stated in terms of strain energy. A more general inequality than Equation (5.56) can be found in [9]. Remark 5.13 implies that a model established based on the GS bilinear form will be "softer" than that of the standard bilinear form. This so-called softening effect was discovered in [15].

Remark 5.14 Monotonic convergence property: eighth inequality

In a given smoothing domain division D_1, $\boxed{\Omega} = \bigcup_{i=1}^{N_s} \boxed{\Omega}_i^s$, if a further division D_2 is conducted by sub-dividing a smoothing domain in D_1 into n_{sd} sub-smoothing-domains: $\boxed{\Omega}_i^s = \bigcup_{j=1}^{n_{sd}} \boxed{\Omega}_{,j}^s$, we then have the following inequality

$$\bar{a}_{D_1}(\mathbf{w},\mathbf{w}) \leq \bar{a}_{D_2}(\mathbf{w},\mathbf{w}) \tag{5.57}$$

which was found in [15].

Remark 5.14 is the same as the Theorem 4.5 stated in terms of strain energy. It implies that the "softening" effect provided by the GS-Galerkin formulation will be *monotonically* reduced with the increase of the number of smoothing domain in a nested manner. This provides a theoretical base to reduce or increase the stiffness or softness of the GS-Galerkin model.

5.6.3 GS-Galerkin statement

For solid mechanics problems given in the strong statements in Section 1.1, our W^2 statement using the GS-Galerkin form becomes: An approximate solution $\bar{\mathbf{u}} \in \left(\mathbb{G}_{h,0}^1\right)^d$ associated with the strong statements satisfies

$$\bar{a}_D(\bar{\mathbf{u}},\mathbf{v}) = f(\mathbf{v}), \quad \forall \mathbf{v} \in \left(\mathbb{G}_{h,0}^1\right)^d \tag{5.58}$$

Note here we make no changes to the linear functional $f(\mathbf{v})$ and Equation (5.11) stands. The properties of the W^2 statement are expressed in the following theorems.

Theorem 5.5 W^2 solution in $\mathbb{H}_h^1$ spaces: variationally consistent

If the GS-Galerkin solution is sought from an $\mathbb{H}_h^1$ space, the W^2 statement Equation (5.58) is then variationally consistent, and hence the solution will be stable, unique and converge to the exact solution of the strong statement when $h \to 0$ (and hence $\Omega_i^s \to 0$).

The proof can be found in [4].

Theorem 5.6 W^2 solution in $\mathbb{G}_h^1$ spaces: convergence

For solids of stable materials, the solution of the W^2 statement Equation (5.58) is stable, unique and convergent to the exact solution of the strong statement when $h \to 0$ (and hence $\Omega_i^s \to 0$).

The proof of Theorem 5.6 can be found in [4].

Theorem 5.7 Upper bound to the Galerkin weak-form solution

The strain energy of the GS-Galerkin solution $\overline{\mathbf{u}} \in \left(\mathbb{H}_{h,0}^1 \right)^d$ is no-less than that of the Galerkin solution $\tilde{\mathbf{u}} \in \left(\mathbb{H}_{h,0}^1 \right)^d$, when the same mesh is used for creating the numerical model.

$$\tilde{U}_{PE}\left(\tilde{\mathbf{u}}\right) \leq \overline{U}_{PE}^D\left(\overline{\mathbf{u}}\right) \tag{5.59}$$

The proof is a little lengthy and can be found in [10] in variational formulation in [15] in energy principle formulation. It can be explained intuitively based on the soft effect Theorems 4.3 and 4.4 (or Remark 5.13). We know that the GS-Galerkin model is always (for any assumed $\mathbf{u}^h \in \left(\mathbb{H}_h^1 \right)^d$) softer than the Galerkin counterpart. Therefore, the displacement field obtained from the GS-Galerkin model should be "larger" in a uniform manner and so the strain field. The strain energy obtained using such a uniformly "larger" strain field should also be "larger" compared to that of the Galerkin model. In other words, Equation (5.59) is the consequence of the uniformly softening effect of the GS-Galerkin model.

Remark 5.15 Upper bound to the exact solution: special cases

The strain energy of the GS-Galerkin solution $\overline{\mathbf{u}}$ is no-less than that of the exact solution $\mathbf{u}$, if $\overline{\mathbf{u}}$ is found from an $\mathbb{H}^1$ space that contains the exact solution:

$$U_{PE}\left(\mathbf{u}\right) \leq \overline{U}_{PE}^D\left(\overline{\mathbf{u}}\right) \tag{5.60}$$

This remark can be easily understood intuitively: the $\mathbb{H}^1$ space contains the exact solution, and Equation (5.59) becomes (5.60).

Remark 5.16 Reproducibility of S-PIM

Note that when an FEM model created using the Galerkin weak form is used to compute a solution from a space that contains the exact solution, it will reproduce the exact solution (see, Remark 5.6). An S-PIM model created using the GS-Galerkin weak form, however, will not necessarily reproduce the exact solution, even if the solution space contains the exact solution. It produces the exact solution only when all the smoothing domains are refined. This means that such an S-PIM model has the reproducibility property only at the limit of all $\Omega_i^s \rightarrow 0$.

Remark 5.16 has only theoretical significance, because 1) it is difficult to assume a space that contains the exact solution, unless the exact solution has a very simple polynomial form; 2) even if it can be done, one can then simply use the FEM to reproduce the exact solution, and hence there is no need for any other form of solution! The following two remarks are of more practical importance.

Remark 5.17 Upper bound to the exact solution: existence

If a discrete GS-Galerkin model is sufficiently refined and hence the "stiffness" of the model can change reasonably smoothly with the change of the number of linearly independent smoothing domains $N_s \geq N_s^{\min}$, the GS-Galerkin model can produce an upper bound solution by reducing the number of smoothing domains N_s.

Remark 5.17 implies that we can, in theory, make a GS-Galerkin model as soft as we want to by reducing N_s. Therefore, we can always obtain an upper bound solution, when the model becomes sufficinetly soft via the change of N_s [4]. Remark 5.17 is important because it ensures (with conditions) that a GS-Galerkin model can always be built for an upper bound to the exact solution at least in energy norm. This provides a theoretical foundation for establishing S-PIM models for upper bound solutions.

In actual practice, the S-PIM models created are discrete finite, the "stiffness" of the model may not change very smoothly with the change of N_s, especially when the number of nodes in the model is too small. In addition, the division of smoothing domain is always somehow tied together with the cell/element mesh for reasons of

1)   convenience in implementation,

2)   efficiency in computation,

3) ensuring smoothing domains are created properly, and

4) tighter upper bounds. ·

Therefore, when the background cell is refined, the smoothing domains are also refined accordingly. In this case, the smoothing effects will be reduced with the refinement of the cells. Therefore, proper techniques need to be designed for a GS-Galerkin model for upper bound solutions. By the same token, one can also build, by design, a GS-Galerkin model for lower bound solutions.

Note that the upper bound property of an S-PIM depends on the "battle between the softening and hardening effects" [15]. In an S-PIM setting, we have one more efficient way to make the model softer (or stiffer), by increasing (reducing) the order of PIM shape functions.

Remark 5.18 Upper bound to the exact solution: usual occurrences

The strain energy of the solution $\bar{\mathbf{u}} \in \left(\mathbb{G}_h^1 \right)^d$ of a GS-Galerkin model is no-less than that of the exact solution $\mathbf{u}$, when the smoothing domains are properly created for sufficient smoothing effects.

$$U_{PE}(\mathbf{u}) \le \bar{U}_{PE}^D(\bar{\mathbf{u}}) \tag{5.61}$$

Precise proof for Remark 5.18 is difficult due to the difficulty in quantifying these conditions. The inequality (5.61) was found mostly true when node-based smoothing domains are used as in the NS-PIMs, as will be shown Chapter 6. Intuitive explanation and arguments can be found in [10, 15, 21, 22]. Remark 5.18 is of practical importance, because it offers a means by properly chosen the smoothing domains to build a GS-Galerkin model can provide upper bound to the exact solution in energy norm. It offers an alternative way to obtain upper (or lower) bound solutions by properly constructing the smoothing domain rather than how many smoothing domains are used in creating a GS-Galerkin model.

Remark 5.19 The "form" of Galerkin

Equation (5.58) has a form of Galerkin, in the sense that the trial and test functions are all from the same space. It is much more general framework, but still in the simple form of Galerkin, for establishing a wide class of numerical methods. It offers plenty freedoms to create shape functions (that do not have to be continuous), and the smoothing domains based on settings of both FEM and S-PIM.

5.6.4 GS-Galerkin: a special case of SC-Galerkin

The GS-Galerkin weak form is a special case of the SC-Galerkin forms when the strain is constructed using the generalized smoothing technique with a minimum number of linearly independent smoothing domains created in the way described in Section 2.3.1.

$$\hat{\boldsymbol{\varepsilon}}(\overline{\mathbf{u}}) = \overline{\boldsymbol{\varepsilon}}(\overline{\mathbf{u}}), \quad \overline{\mathbf{u}} \in (\mathbb{G}_{h,0}^1)^d \tag{5.62}$$

The discretized form of the potential energy for the generalized smoothed strains becomes:

$$\overline{U}_{PE}^{D} = \int_{\Omega} \frac{1}{2} \overline{\boldsymbol{\varepsilon}}^{\mathrm{T}} \mathbf{c} \overline{\boldsymbol{\varepsilon}} \, \mathrm{d}\xi = \frac{1}{2} \sum_{i=1}^{N_q} A_i^s \overline{\boldsymbol{\varepsilon}}_i^{\mathrm{T}} \mathbf{c} \overline{\boldsymbol{\varepsilon}}_i = \frac{1}{2} \sum_{i=1}^{N_s} A_i^s \overline{\boldsymbol{\varepsilon}}_i^{\mathrm{T}} \mathbf{c} \overline{\boldsymbol{\varepsilon}}_i \tag{5.63}$$

The GS-Galerkin weak form can be written as

$$\underbrace{\sum_{i=1}^{N_s} A_i^s \delta \overline{\boldsymbol{\varepsilon}}_i^{\mathrm{T}}(\overline{\mathbf{u}}) \mathbf{c} \overline{\boldsymbol{\varepsilon}}_i(\overline{\mathbf{u}})}_{\overline{a}_D(\overline{\mathbf{u}}, \delta \overline{\mathbf{u}})} - \underbrace{\left(\int_{\Omega} \delta \overline{\mathbf{u}}^{\mathrm{T}} \mathbf{b} \mathrm{d}\Omega + \int_{\Gamma_t} \delta \overline{\mathbf{u}}^{\mathrm{T}} \mathbf{t} \mathrm{d}\Gamma \right)}_{f(\delta \overline{\mathbf{u}})} = 0 \tag{5.64}$$

which is exactly the same as Equation (5.58). Because the smoothed strains are constant in the smoothing domain, and the fact that the displacement is continuous on the smoothing domain interfaces, the summation in Equation (5.64) is made possible. In GS-Galerkin based models, because the smoothing domains are also used as integration/quadrature domains, the summations in Equations (5.63) and (5.64) are conducted based on all the smoothing domains.

For dynamic problems, we have

$$\sum_{i=1}^{N_s} A_i^s \delta \overline{\boldsymbol{\varepsilon}}_i^{\mathrm{T}}(\overline{\mathbf{u}}) \mathbf{c} \overline{\boldsymbol{\varepsilon}}_i(\overline{\mathbf{u}}) - \int_{\Omega} \delta \overline{\mathbf{u}}^{\mathrm{T}} \mathbf{b} \mathrm{d}\Omega - \int_{\Gamma_t} \delta \overline{\mathbf{u}}^{\mathrm{T}} \mathbf{t} \mathrm{d}\Gamma + \int_{\Omega} \rho \delta \overline{\mathbf{u}}^{\mathrm{T}} \ddot{\overline{\mathbf{u}}} \mathrm{d}\Omega = 0 \tag{5.65}$$

The proof of the GS-Galerkin being a special case of SC-Galerkin can be found in [4].

5.7 S-PIM: a typical $\mathbf{W}^2$ formulation

We now present the general formulation of the smoothed point interpolation method (S-PIM) that is a typical W^2 formulation using PIM shape functions. The major advantages of PIM shape functions, as discussed in Chapter 3, are 1) excellent accuracy, 2) efficiency, and 3) Kronecker Delta function property that allows simple impositions of essential boundary conditions as in the conventional finite element method (FEM). In the lengthy process of developing the PIM method, the battle has been on two major fronts: issues of singular moment matrix and suitable weak forms that can always ensure stable and convergent solutions. On the first battle front, three effective approaches have been developed: use of radial basis functions (RBFs), coordinate transformation, and T-schemes for local node selection, as discussed in Section 3.2.3.

On the second battle front, the local Petrov-Galerkin approach was often used together with the PIM shape functions in the earlier stage [3]. It has been found recently that the weakened weak (W^2) formulation is much more effective with the PIM shape functions. In particular, we have now two important W^2 formulations: the generalized smoothed Galerkin (GS-Galerkin) weak form, and the strain-constructed Galerkin (SC-Galerkin) weak forms. The W^2 formulation allows the use of discontinuous PIM shape functions without worrying about the compatibility issues. It offers variety ways of implement S-PIM models with excellent properties, including the upper bound, lower bound, superconvergence, accurate stress solutions, free of locking and working well with triangular/tetrahedral meshes. Typical S-PIM models include the node-based smoothed point interpolation method (NS-PIM), the edge-based smoothed point interpolation method (ES-PIM), the cell-based smoothed point interpolation method (CS-PIM) and the strain-constructed point interpolation method (SC-PIM). The NS-PIM, ES-PIM and CS-PIM models are constructed based on the GS-Galerkin weak form and the SC-PIM is built based on the SC-Galerkin weak form. As we discussed in Section 5.6.4, the GS-Galerkin weak form is a special case of SC-Galerkin weak form and is widely used in our S-PIM models. In this section, we provide a unified formulation for these S-PIM models formulated using the GS-Galerkin formulation and PIM shape functions.

Consider solid mechanics problems presented in Section 1.1. The problem domain Ω is discretized into a set of background cells, such that $\boxed{\Omega} = \bigcup_{i=1}^{N_c} \boxed{\Omega}_i^c$ and $\Omega_i^c \bigcap \Omega_j^c = \varnothing$, $\forall i \neq j$ where N_c is the total number of background cells. Each of the cells is bounded by Γ_i^c. We shall have a total of N_n nodes each of which coincides with the vertices of the cells, and N_{eg} edges of the cells (for 2D) or N_f faces of the cells (for 3D). In general, any FEM mesh can be used as the background cells for S-PIM. We, however, prefer to use triangular cells constructed in Section 1.6.1 for the simple reason of easy generation automatically.

5.7.1 PIM displacement field

Based on the background cells, a set of local nodes can be selected using a T-scheme presented in Section 1.6.3. PIM shape functions can then be created using the techniques detailed in Section 3.2. The solution of the displacement function for an S-PIM model at any point $\mathbf{x}$ in the problem domain can be constructed as

$$\bar{\mathbf{u}}(\mathbf{x}) = \left\{ \begin{matrix} \bar{u}(\mathbf{x}) \\ \bar{v}(\mathbf{x}) \end{matrix} \right\}^h = \sum_{I \in S_n} \underbrace{\left[\begin{matrix} \phi_I(\mathbf{x}) & 0 \\ 0 & \phi_I(\mathbf{x}) \end{matrix} \right]}_{\boldsymbol{\Phi}_I(\mathbf{x})} \underbrace{\left\{ \begin{matrix} \bar{u}_I \\ \bar{v}_I \end{matrix} \right\}}_{\bar{\mathbf{d}}_I} = \sum_{I \in S_n} \boldsymbol{\Phi}_I(\mathbf{x}) \bar{\mathbf{d}}_I \tag{5.66}$$

where S_n is the set of these local support nodes selected using a T-scheme for a cell hosting $\mathbf{x}$, $\boldsymbol{\Phi}_I$ is the matrix of the nodal PIM shape functions, and $\bar{\mathbf{d}}_I$ is the vector of the nodal displacements for the Ith support node.

Note that in the S-PIM formulation, the derivatives of the PIM shape functions are not required, and no mapping is needed, due to the use of the W^2 formulation.

For easy derivation, we rewrite Equation (5.66) in following expanded form.

$$\begin{aligned} \bar{\mathbf{u}}(\mathbf{x}) &= \sum_{I \in S_n} \boldsymbol{\Phi}_I(\mathbf{x}) \bar{\mathbf{d}}_I = \sum_{I=1}^{N_n} \boldsymbol{\Phi}_I(\mathbf{x}) \bar{\mathbf{d}}_I \\ &= \underbrace{\left[\begin{matrix} \boldsymbol{\Phi}_1(\mathbf{x}) & \boldsymbol{\Phi}_2(\mathbf{x}) & \cdots & \boldsymbol{\Phi}_I(\mathbf{x}) & \cdots & \boldsymbol{\Phi}_{N_n}(\mathbf{x}) \end{matrix} \right]}_{\boldsymbol{\Phi}(\mathbf{x})} \bar{\mathbf{d}} = \boldsymbol{\Phi}(\mathbf{x}) \bar{\mathbf{d}} \end{aligned} \tag{5.67}$$

where $\boldsymbol{\Phi}$ is the matrix of all the nodal shape function matrices, and $\bar{\mathbf{d}}$ is the nodal displacements for all the nodes in the entire problem domain:

$$\bar{\mathbf{d}} = \left\{ \begin{matrix} \bar{\mathbf{d}}_1(\mathbf{x}) & \bar{\mathbf{d}}_2(\mathbf{x}) & \cdots & \bar{\mathbf{d}}_I(\mathbf{x}) & \cdots & \bar{\mathbf{d}}_{N_n}(\mathbf{x}) \end{matrix} \right\}^{\mathrm{T}} \tag{5.68}$$

We note the following remark.

Remark 5.20 Sparseness of shape function matrix

Because $\mathbf{\Phi}_I$ for node I is locally supported by the nodes in the support domain defined by a node selections scheme, it is zero beyond the support domain. Any given $\mathbf{x} \in \Omega$ will surely fall in a support domain that contains a very small number of nodes (for example, 3 nodes for the cell-based T3-scheme). Therefore, $\bar{\mathbf{u}}(\mathbf{x})$ in Equation (5.67) relates only to a very small number of nodal shape functions, and the shape function matrix $\mathbf{\Phi}(\mathbf{x})$ will be an extremely sparse matrix with lots of zeros for an actual model with large number of nodes. The expanded expression of $\bar{\mathbf{u}}(\mathbf{x})$ in Equation (5.67) using all the field nodes is only for the convenience in the formulation. In actual computation, we need only Equation (5.66). This sparseness in an S-PIM is largely the same as in the FEM.

5.7.2 Smoothing domain creation in S-PIM

TABLE 5.1 Types of smoothing domains and S-PIM models

Smoothing domains	S-PIM models	Formulation	Major features
Node-based $N_s=N_n$ (2D & 3D)	Node-based smoothed point interpolation method: NS-PIM & NS-RPIM	GS-Galerkin; PIM or RPIM shape function	Linearly conforming; Volumetric locking free; Upper bound; Superconvergence
Edge-based $N_s=N_{eg}$ (2D)	Edge-based smoothed point interpolation method: ES-PIM & ES-RPIM	GS-Galerkin; PIM or RPIM shape function	Linearly conforming; Ultra-accuracy; Very efficient; Superconvergence
Face-based $N_s=N_f$ (3D)	Face-based smoothed point interpolation method: FS-PIM & FS-RPIM	GS-Galerkin; PIM or RPIM shape functions	Linearly conforming; Ultra-accuracy; Very efficient; Superconvergence
Cell-based $N_s=N_c$ (2D & 3D)	Cell-based smoothed point interpolation method: CS-PIM & CS-RPIM	GS-Galerkin; PIM or RPIM shape function	Linearly conforming; Ultra-accuracy; Very efficient; Superconvergence
Various smoothing domains listed above (2D)	Strain-constructed point interpolation method: SC-PIM & SC-RPIM	SC-Galerkin; PIM or RPIM shape function; Strain construction	Linearly conforming; Ultra-accuracy; Lower & upper bounds; Superconvergence

The "smoothed" strain used in the GS-Galerkin form is constructed using only the PIM shape function values and a set of smoothing domains. In the S-PIM, the problem domain is divided, on top of the background cells, into N_s NOSL smoothing domains in a proper manner, such that $\boxed{\Omega} = \bigcup_{k=1}^{N_s} \boxed{\Omega}_k^s$ and $\Omega_k^s \cap \Omega_l^s = \varnothing$, $\forall k \neq l$. This is to ensure the PIM shape functions can be in a $\mathbb{G}_h^1$ space. The creation of such smoothing domains can be arbitrary as long as two conditions are satisfied: 1) The "no-sharing" rules given in Section 2.3.1, 2) at least the minimum number of linearly independent smoothing domains has to be used (see TABLE 2.2). The proven methods of smoothing domain divisions are given in TABLE 5.1, together with the S-PIM models and their major features.

Remark 5.21 Smoothing domains and integration/quadrature domains

- For S-PIM models based on the SC-Galerkin weak form presented in this book, such as NS-PIM, ES-PIM/FS-PIM and CS-PIM methods, the smoothing domains also serve as quadrature domains and hence $N_q = N_s$.

- For S-PIM models based on the SC-Galerkin weak form, such as SC-PIM, the smoothing domains are generally not the quadrature domains and hence $N_q \neq N_s$.

5.7.3 Smoothed strain field

Within each smoothing domain, the "smoothed" strains can now be obtained using the generalized gradient smoothing operation, as detailed in Section 4.5.2. When the Heaviside type of smoothing domains are used, the smoothed strains over the smoothing domain for node i can be given as

$$\bar{\boldsymbol{\varepsilon}}(\mathbf{x}_i) = \frac{1}{A_i^s} \int_{\Gamma_i^s} \mathbf{L}_n \bar{\mathbf{u}} \, d\Gamma \tag{5.69}$$

where $\bar{\mathbf{u}} \in \mathbb{G}_h^1(\Omega)$, $A_i^s = \int_{\Omega_i^s} d\Omega$ is the area of the smoothing domain Ω_i^s, $\mathbf{L}_n$ is the matrix of outwards normal components given in Equation (1.24), and $\bar{\mathbf{u}}$ is the assumed displacement given in Equation (5.66).

Substituting Equation (5.66) into (5.69), the smoothed strain can be written in the following matrix form

$$\bar{\boldsymbol{\varepsilon}}(\mathbf{x}_i) = \sum_{I \in S_s} \bar{\mathbf{B}}_I(\mathbf{x}) \bar{\mathbf{d}}_I \tag{5.70}$$

where S_s is the set of the support nodes for the smoothing domain Ω_i^s that contains all the nodes involved in the interpolation for all the quadrature points located on the boundary of this smoothing domain Γ_i^s. For example, the node-based smoothing domain Ω_i^s shown in FIGURE 6.2 consists of 6 triangular cells, which is also plotted in FIGURE 5.2 particularly. If linear interpolation using cell-based T3-scheme is used in the NS-PIM, the number of support nodes for the smoothing domain is 7. The smoothed strain-displacement matrix $\bar{\mathbf{B}}$ has the form of

$$\bar{\mathbf{B}}_I(\mathbf{x}) = \frac{1}{A_i^s}\int_{\Gamma_i^s}\mathbf{L}_n(\mathbf{x})\mathbf{\Phi}_I(\mathbf{x})\,\mathrm{d}\Gamma = \begin{bmatrix} \bar{\phi}_{Ix}(\mathbf{x}) & 0 \\ 0 & \bar{\phi}_{Iy}(\mathbf{x}) \\ \bar{\phi}_{Iy}(\mathbf{x}) & \bar{\phi}_{Ix}(\mathbf{x}) \end{bmatrix} \tag{5.71}$$

in which

$$\bar{\phi}_{Il} = \frac{1}{A_i^s}\int_{\Gamma_i^s}\phi_I(\mathbf{x})n_l(\mathbf{x})\,\mathrm{d}\Gamma, \quad (l = x,\,y) \tag{5.72}$$

where no differentiations on the shape functions.

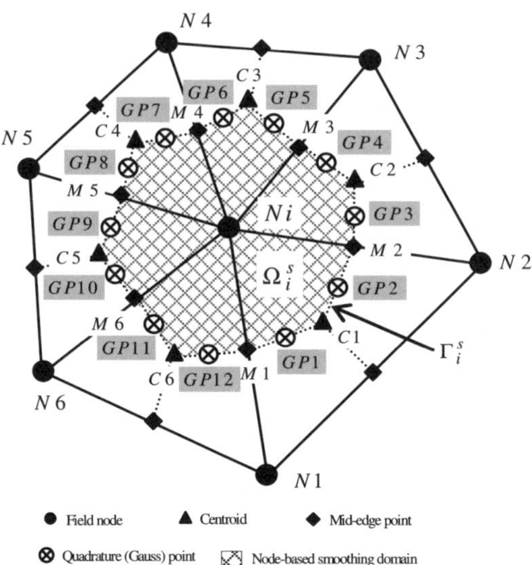

FIGURE 5.2 An example of a node-based equally-shared smoothing domain Ω_i^s bounded by Γ_i^s for node i.

The integration in Equation (5.72) is curve integration. It can be performed easily using the standard Gauss integration scheme looping around Γ_i^s (by the standard convention). Because the smoothing domain boundary is usually consists of straight lines segments. Such a numerical integration can be simply performed for each line-segment and then summations.

$$\bar{\phi}_{ll} = \frac{1}{A_i^s} \sum_{m=1}^{n_{seg}} \left[\sum_{n=1}^{n_G} W_n^G \left(\phi_I \left(\mathbf{x}_{m,n} \right) n_{l,m} \right) \right], \quad (l = x, y) \tag{5.73}$$

where n_{seg} is the number of line-segments of the boundary Γ_i^s, $n_{l,m}$ is the component of the unit outwards normal on the mth segment of Γ_i^s, n_G is the number of Gauss points distributed in each segment, $\mathbf{x}_{m,n}$ is the nth Gauss point located on the mth segment of Γ_i^s, and W_n^G is the corresponding Gauss weight.

For example, for the node-based smoothing domain of node i shown in FIGURE 5.2, Γ_i^s consists of 12 line-segments, i.e. $n_{seg} = 12$. If linear interpolation is used in the NS-PIM, the numerical integration along each line-segment needs only one Gauss point ($n_G = 1$), and therefore, there are a total of 12 ($= n_{seg} \times n_G$) Gauss points (Gp1 to Gp12) to be used in Equation (5.73) for the entire smoothing domain of Ω_i^s. In this simple linear interpolation case, the shape function values at all these 12 Gauss points can be easily tabulated in TABLE 5.2 by inspection.

When higher order PIM shape functions are used, the shape function values need to be computed numerically, which can be easily programmed following the procedure presented in Section 3.2. Note from Equations (5.70)-(5.73) that in computing the smoothed strain-displacement matrix, we do not need to perform differentiation for the shape functions. In the weak-form methods like in FEM, we need to perform 1st order differentiations to the shape functions, as shown in Equation (5.26). This shows explicitly that the consistence requirement to the shape functions is further reduced: a weakened weak (W^2) formulation. Because of this, whether or not the shape functions are continuous within (strictly) the smoothing domains is not an issue.

We now further assume that the constructed strain in each of the smoothing domains is the constant, and it is the same as that given in Equation (5.69):

TABLE 5.2 Shape functions values at different sites on the node-based smoothing domain boundary for node i

Sites		N_i	N_1	N_2	N_3	N_4	N_5	N_6
Field node	N_i	1.0	0.0	0.0	0.0	0.0	0.0	0.0
Field node	N_1	0.0	1.0	0.0	0.0	0.0	0.0	0.0
Field node	N_2	0.0	0.0	1.0	0.0	0.0	0.0	0.0
Field node	N_3	0.0	0.0	0.0	1.0	0.0	0.0	0.0
Field node	N_4	0.0	0.0	0.0	0.0	1.0	0.0	0.0
Field node	N_5	0.0	0.0	0.0	0.0	0.0	1.0	0.0
Field node	N_6	0.0	0.0	0.0	0.0	0.0	0.0	1.0
Mid-edge point	M_1	1/2	1/2	0.0	0.0	0.0	0.0	0.0
Centroid	C_1	1/3	1/3	1/3	0.0	0.0	0.0	0.0
Mid-edge point	M_2	1/2	0.0	1/2	0.0	0.0	0.0	0.0
Centroid	C_2	1/3	0.0	1/3	1/3	0.0	0.0	0.0
Mid-edge point	M_3	1/2	0.0	0.0	1/2	0.0	0.0	0.0
Centroid	C_3	1/3	0.0	0.0	1/3	1/3	0.0	0.0
Mid-edge point	M_4	1/2	0.0	0.0	0.0	1/2	0.0	0.0
Centroid	C_4	1/3	0.0	0.0	0.0	1/3	1/3	0.0
Mid-edge point	M_5	1/2	0.0	0.0	0.0	0.0	1/2	0.0
Centroid	C_5	1/3	0.0	0.0	0.0	0.0	1/3	1/3
Mid-edge point	M_6	1/2	0.0	0.0	0.0	0.0	0.0	1/2
Centroid	C_6	1/3	1/3	0.0	0.0	0.0	0.0	1/3
Mid-segment	GP_1	5/12	5/12	1/6	0.0	0.0	0.0	0.0
Mid-segment	GP_2	5/12	1/6	5/12	0.0	0.0	0.0	0.0
Mid-segment	GP_3	5/12	0.0	5/12	1/6	0.0	0.0	0.0
Mid-segment	GP_4	5/12	0.0	1/6	5/12	0.0	0.0	0.0
Mid-segment	GP_5	5/12	0.0	0.0	5/12	1/6	0.0	0.0
Mid-segment	GP_6	5/12	0.0	0.0	1/6	5/12	0.0	0.0
Mid-segment	GP_7	5/12	0.0	0.0	0.0	5/12	1/6	0.0
Mid-segment	GP_8	5/12	0.0	0.0	0.0	1/6	5/12	0.0
Mid-segment	GP_9	5/12	0.0	0.0	0.0	0.0	5/12	1/6
Mid-segment	GP_{10}	5/12	0.0	0.0	0.0	0.0	1/6	5/12
Mid-segment	GP_{11}	5/12	1/6	0.0	0.0	0.0	0.0	5/12
Mid-segment	GP_{12}	5/12	5/12	0.0	0.0	0.0	0.0	1/6

$$\overline{\boldsymbol{\varepsilon}}_i(\mathbf{x}) = \overline{\boldsymbol{\varepsilon}}(\mathbf{x}_i), \quad \forall \mathbf{x} \in \Omega_i^s \tag{5.74}$$

For easy derivation, we rewrite Equation (5.70) in the following form

$$\begin{aligned}
\overline{\boldsymbol{\varepsilon}}_i(\mathbf{x}) &= \sum_{I \in S_s} \overline{\mathbf{B}}_I(\mathbf{x}) \overline{\mathbf{d}}_I = \sum_{I=1}^{N_n} \overline{\mathbf{B}}_I(\mathbf{x}) \overline{\mathbf{d}}_I \\
&= \underbrace{\left[\, \overline{\mathbf{B}}_1(\mathbf{x}) \quad \overline{\mathbf{B}}_2(\mathbf{x}) \quad \cdots \quad \overline{\mathbf{B}}_{N_n}(\mathbf{x}) \,\right]}_{\overline{\mathbf{B}}(\mathbf{x})} \overline{\mathbf{d}} \\
&= \overline{\mathbf{B}}(\mathbf{x}) \overline{\mathbf{d}}
\end{aligned} \tag{5.75}$$

where $\overline{\mathbf{B}}$ is the *global* "smoothed strain-displacement" matrix. It is clear that due to the sparseness of shape function matrix (see Remark 5.2), $\overline{\mathbf{B}}$ is also extremely sparse, because it is obtained using the PIM shape functions. In an S-PIM model, $\overline{\mathbf{B}}_I(\mathbf{x})$ is nonzero only for the set of nodes "supporting" the smoothing domain Ω_k^s that hosts $\mathbf{x}$. This set of nodes consists of all the nodes of the cells contributing to the smoothing domain Ω_k^s. For example, the support nodes for Ω_i^s shown in FIGURE 5.2, there are 7 support nodes: $Ni, N1, \cdots, N6$. The expression of $\overline{\boldsymbol{\varepsilon}}_i(\mathbf{x})$ in Equation (5.75) using the global smoothed strain-displacement matrix is only for the convenience in the formulation. In actual computation, we need only Equation (5.70).

5.7.4   Discretized equations for S-PIM

Substituting the assumed displacement Equation (5.67) and the smoothed strains Equation (5.75) into Equation (5.64), we shall have

$$\sum_{i=1}^{N_s} A_i^s \delta\left(\overline{\mathbf{B}}\overline{\mathbf{d}}\right)^{\mathrm{T}} \left(\mathbf{c}\overline{\mathbf{B}}\overline{\mathbf{d}}\right) - \int_{\Omega} \delta\left(\boldsymbol{\Phi}\overline{\mathbf{d}}\right)^{\mathrm{T}} \mathbf{b}\,\mathrm{d}\Omega - \int_{\Gamma_t} \delta\left(\boldsymbol{\Phi}\overline{\mathbf{d}}\right)^{\mathrm{T}} \mathbf{t}\,\mathrm{d}\Gamma = 0 \tag{5.76}$$

After simple manipulation, and factoring out $\delta\overline{\mathbf{d}}^{\mathrm{T}}$, Equation (5.76) becomes

$$\delta\overline{\mathbf{d}}^{\mathrm{T}} \left(\sum_{i=1}^{N_s} A_i^s \overline{\mathbf{B}}^{\mathrm{T}} \mathbf{c}\overline{\mathbf{B}}\overline{\mathbf{d}} - \int_{\Omega} \boldsymbol{\Phi}^{\mathrm{T}} \mathbf{b}\,\mathrm{d}\Omega - \int_{\Gamma_t} \boldsymbol{\Phi}^{\mathrm{T}} \mathbf{t}\,\mathrm{d}\Gamma \right) = 0 \tag{5.77}$$

To have the above equation satisfied for arbitrary $\delta\overline{\mathbf{d}}^{\mathrm{T}}$, the terms in the bracket must vanish:

$$\underbrace{\sum_{i=1}^{N_s} A_i^s \overline{\mathbf{B}}^{\mathrm{T}} \mathbf{c} \overline{\mathbf{B}} \, \overline{\mathbf{d}}}_{\overline{\mathbf{K}}} - \underbrace{\left(\int_{\Omega} \mathbf{\Phi}^{\mathrm{T}} \mathbf{b} \, d\Omega + \int_{\Gamma_t} \mathbf{\Phi}^{\mathrm{T}} \mathbf{t}_{\Gamma} d\Gamma \right)}_{\tilde{\mathbf{F}}} = 0 \tag{5.78}$$

We obtain the following discretized algebraic system of equations for the S-PIM model.

$$\overline{\mathbf{K}} \overline{\mathbf{d}} = \tilde{\mathbf{f}} \tag{5.79}$$

where

$$\overline{\mathbf{K}} = \sum_{i=1}^{N_s} A_i^s \overline{\mathbf{B}}^{\mathrm{T}} \mathbf{c} \overline{\mathbf{B}} \tag{5.80}$$

is the global smoothed stiffness matrix for the S-PIM model. For constrained stable solids, $\overline{\mathbf{K}}$ is SPD due to the coercivity of the W^2 bilinear form $\overline{a}_D$ (see, Theorem 5.3). In Equation (5.78), the global nodal force vector is given as

$$\tilde{\mathbf{f}} = \int_{\Omega} \mathbf{\Phi}^{\mathrm{T}} \mathbf{b} \, d\Omega + \int_{\Gamma_t} \mathbf{\Phi}^{\mathrm{T}} \mathbf{t}_{\Gamma} d\Gamma \tag{5.81}$$

which contains nodal force vectors for all the nodes in the problem domain. Equation (5.81) is essentially the same as that given in Equation (5.33), except the possible difference in the shape function used. This is because the S-PIM models do not change the linear forms. Therefore, the treatment of the force vectors is exactly the same as those in the FEM model, and $\tilde{\mathbf{f}}$ is assembled using entries of

$$\tilde{\mathbf{f}}_I = \int_{\Omega} \mathbf{\Phi}_I^{\mathrm{T}} \mathbf{b} \, d\Omega + \int_{\Gamma_t} \mathbf{\Phi}_I^{\mathrm{T}} \mathbf{t}_{\Gamma} d\Gamma \tag{5.82}$$

Because of the sparseness of the smoothed strain-displacement matrix, the stiffness matrix $\overline{\mathbf{K}}$ given in Equation (5.80) will be very sparse. To avoid computations of zero entries, we will not actually use Equation (5.80) in the computation. Instead, we evaluate the entries of $\overline{\mathbf{K}}$ using

$$\overline{\mathbf{K}}_{IJ} = \sum_{i=1}^{N_s} \underbrace{A_i^s \overline{\mathbf{B}}_I^{\mathrm{T}} \mathbf{c} \overline{\mathbf{B}}_J}_{\overline{\mathbf{K}}_{IJ,i}^s} \tag{5.83}$$

where A_i^s is the area of the ith smoothing domain used in the S-PIM model, and

$$\overline{\mathbf{K}}_{IJ,i}^s = A_i^s \overline{\mathbf{B}}_I^{\mathrm{T}} \mathbf{c} \overline{\mathbf{B}}_J \tag{5.84}$$

Compared to FEM models, we note that in the computation, 1) we do not need numerical integration, and 2) the assembly is performed over smoothing domains in S-PIM models.

Because $\overline{\mathbf{K}}$ is SPD, Equation (5.79) has a unique solution and can be solved efficiently using a standard linear equation solvers. For dynamics problems, the S-PIM equations become

$$\overline{\mathbf{K}}\overline{\mathbf{d}} + \mathbf{M}\ddot{\overline{\mathbf{d}}} = \tilde{\overline{\mathbf{f}}} \tag{5.85}$$

where the global mass matrix is

$$\mathbf{M} = \int_\Omega \rho \mathbf{\Phi}^{\mathrm{T}} \mathbf{\Phi} \mathrm{d}\Omega \tag{5.86}$$

which is essentially the same as that given in Equation (5.38), except the possible difference in the shape function used. For all $\rho > 0$, it is clear that $\mathbf{M}$ is SPD, due to Theorem 3.1. Because $\overline{\mathbf{K}}$ and $\mathbf{M}$ are all SPD, Equation (5.85) has a unique solution and can be solved efficiently using a standard time integration solver for the linear equation system.

5.7.5 Imposition of essential boundary conditions

From Equation (5.80) it is clear that $\overline{\mathbf{K}}$ is symmetric. However, at this stage it is usually singular, because we used $\overline{\mathbf{u}} \in \left(\mathbb{G}_h^1(\Omega)\right)^d$ not $\overline{\mathbf{u}} \in \left(\mathbb{G}_{h,0}^1(\Omega)\right)^d$ meaning that the essential boundary conditions given in Equation (1.22) are not yet satisfied and the solid can have rigid body movements. Therefore, we have to impose these essential boundary conditions, which can be done exactly in the same way as in the FEM, because the PIM shape functions used in S-PIM models possess also the Delta function property (see, Chapter 3). We simply remove (or modify) the corresponding rows and columns in the stiffness matrix $\overline{\mathbf{K}}$ leading to a condensed (or modified) $\overline{\mathbf{K}}$ matrix [13]. The condensed $\overline{\mathbf{K}}$

matrix will become nonsingular, and hence will be symmetric positive definite (SPD), according to Theorem 5.3.

5.7.6 S-PIM solutions

Equation (5.79) can now be solved with ease using standard equation solver routines for the nodal displacements $\bar{\mathbf{d}}$. Because of the SPD property of the $\bar{\mathbf{K}}$ matrix after the imposition of the essential boundary conditions, the displacement solution will be stable and unique. When the cells are refined, the solution will approach to the exact solution according to Theorem 5.6. Next, the solutions in terms of strains in each smoothing domain can be finally retrieved, using Equation (5.70) with the shape functions and the nodal displacements obtained. Finally, the stress solutions in each element are computed using the constitutive equations (1.2). When needed, the solution in terms of strain energy of the solid can be computed using Equation (5.63).

Remark 5.22 Incompatibility

An S-PIM model will not be, in general, compatible. First, the displacement functions used in S-PIM models are not generally compatible. Second, the smoothed strain field is also not compatible in terms of the strain-displacement relations. Even if compatible displacements are used in creating an S-PIM model, it is still not compatible because of the violation of the strain-displacement relations. Only when the linear S-PIM is used with cell-based smoothing domains, the linear CS-PIM-Tr2 is identical to the linear FEM-Tr3, and it is compatible. It is this incompatibility that gives the S-PIM models properties presented in Sections 5.6.2 and 5.6.3.

5.8 Error assessment in S-PIM and FEM models

To examine the accuracy and efficiency of our S-PIM models, many benchmarking studies will also be conducted against the analytical or reference solutions. Comparisons will also be made against the standard linear FEM that is used as the common base of comparison. For quantitative study of the error and convergence rate of all these discrete methods, two types of error norms are used in this book, *i.e.*, *displacement norm* and *energy norm* [24].

5.8.1 Error in displacement norm

The standard definition of error in displacements is a measure in L^2 norm:

$$e_d = \left\| \mathbf{u}^{ref} - \mathbf{u}^{num} \right\|_{\mathbb{L}^2} = \left[\int_\Omega \left(\mathbf{u}^{ref} - \mathbf{u}^{num} \right)^{\mathrm{T}} \left(\mathbf{u}^{ref} - \mathbf{u}^{num} \right) d\Omega \right]^{\frac{1}{2}} \tag{5.87}$$

where $\mathbf{u}^{ref}$ is the reference (or exact or analytical) solution for the displacements, and $\mathbf{u}^{num}$ is the numerical solution obtained using an S-PIM or FEM model. To avoid the integrations in Equation (5.87), we use the following simpler summation form with nodal displacements

$$e_d = \left[\frac{\sum\limits_{i=1}^{N_n} (\mathbf{u}_i^{ref} - \mathbf{u}_i^{num})^2}{\sum\limits_{i=1}^{N_n} (\mathbf{u}_i^{ref})^2} \right]^{\frac{1}{2}} \tag{5.88}$$

where $\mathbf{u}_i^{ref}$ is the reference solution for the displacement at node i, $\mathbf{u}_i^{num}$ is the numerical solution for the displacement at node i, and N_n the total number of the field nodes used in the problem domain.

These two definitions given in Equations (5.87) and (5.88) will give different values of the errors. However, they should be equivalent in terms of the convergence rate. In addition, Equation (5.88) is much more convenient and cheap to use, compared to Equation (5.87), and can be implemented in the exactly the same way for all the discrete models. Therefore, Equation (5.88) will be used in this book for all the S-PIM and FEM models. The comparison based on this error indicator is fair and rigorous for all there models.

5.8.2 Error in energy norm

The definition of error in energy norm (or energy norm error) can have different ways. For weak-form models, we shall have the original H^1 (semi) norm defined as

$$e_{\mathrm{H}^1} = \left| \mathbf{u}^{ref} - \mathbf{u}^{num} \right|_{\mathbb{H}^1(\Omega)}$$

$$= \left[\int_\Omega \left((\mathbf{u}')^{ref} - (\mathbf{u}')^{num} \right)^{\mathrm{T}} \left((\mathbf{u}')^{ref} - (\mathbf{u}')^{num} \right) d\Omega \right]^{\frac{1}{2}} \tag{5.89}$$

where $(\mathbf{u}')^{ref}$ and $(\mathbf{u}')^{num}$ are the reference and numerical solutions of the derivatives of displacements, respectively. For elasticity problems the *error in energy norm* is defined using the H^1 norm as follows

$$e_e = \left[\int_\Omega \left(\boldsymbol{\varepsilon}^{ref} - \boldsymbol{\varepsilon}^{num} \right)^{\mathrm{T}} \mathbf{c} \left(\boldsymbol{\varepsilon}^{ref} - \boldsymbol{\varepsilon}^{num} \right) d\Omega \right]^{\frac{1}{2}} \tag{5.90}$$

where $\boldsymbol{\varepsilon}^{ref}$ is the reference (or exact or analytical) solution for the strains, and $\boldsymbol{\varepsilon}^{num}$ is the numerical solution for the strains obtained using a numerical method. The error defined by the energy norm is a physical measure of the energy caused by the distributed strain errors (It is not the error in energy solution!).

To calculate the integral in the above equation, the following summation form is generally used in the FEM models.

$$e_e^H = \left[\sum_{i=1}^{N_e} \int_{\Omega_i^e} \left(\boldsymbol{\varepsilon}^{ref} - \tilde{\boldsymbol{\varepsilon}}^{num} \right)^{\mathrm{T}} \mathbf{c} \left(\boldsymbol{\varepsilon}^{ref} - \tilde{\boldsymbol{\varepsilon}}^{num} \right) d\Omega \right]^{\frac{1}{2}} \tag{5.91}$$

where N_e is the number of elements and $\tilde{\boldsymbol{\varepsilon}}^{num}$ is the FEM solution for the strains. The integral over each element is performed using a mapping procedure and the standard Gauss integration, with proper number of Gauss points chosen depending on the order of the integrand, which can be quite involved when the analytic solution is in a complicated form.

For W^2 models based on the GS-Galerkin weak form with constant smoothed strains, the original G^1 (semi) norm is defined as

$$e_{G^1} = \left| \mathbf{u}^{ref} - \mathbf{u}^{num} \right|_{\mathbb{G}^1(\Omega)}$$

$$= \left[\sum_{i=1}^{N_s} A_i^s \left(\left(\mathbf{u}'(\mathbf{x}_c) \right)^{ref} - \left(\overline{\mathbf{u}'} \right)^{num} \right)^{\mathrm{T}} \left(\left(\mathbf{u}'(\mathbf{x}_c) \right)^{ref} - \left(\overline{\mathbf{u}'} \right)^{num} \right) \right]^{\frac{1}{2}} \tag{5.92}$$

where $\mathbf{x}_c$ is the "center" of the smoothing domain, $\overline{\mathbf{u}'}$ is the value of the smoothed derivative over the ith smoothing domain Ω_i^s, A_i^s is the area of smoothing domain, and N_s is the total number of smoothing domains. For elasticity problems the *error in energy norm* can also be defined based on the G^1 norm as follows.

$$e_e^G = \left[\sum_{i=1}^{N_s} A_i^s \left(\boldsymbol{\varepsilon}^{ref}(\mathbf{x}_c) - \overline{\boldsymbol{\varepsilon}}_i^{num} \right)^{\mathrm{T}} \mathbf{c} \left(\boldsymbol{\varepsilon}^{ref}(\mathbf{x}_c) - \overline{\boldsymbol{\varepsilon}}_i^{num} \right) \right]^{\frac{1}{2}} \tag{5.93}$$

where $\overline{\boldsymbol{\varepsilon}}^{num}$ is the strains solutions obtained using W^2 models based on the GS-Galerkin formulation. These two definitions given in Equations (5.91) and (5.93) are different, due to the difference in the space theory used. For later comparison studies, this book uses the following "unified definition" for the error in energy norm for all the discrete models with piecewise constant strain field.

$$e_e = \begin{cases} \left[\sum\limits_{i=1}^{N_s} A_i^s \left(\boldsymbol{\varepsilon}^{ref}(\mathbf{x}_c) - \overline{\boldsymbol{\varepsilon}}^{num}(\mathbf{x}_c) \right)^{\mathrm{T}} \mathbf{c} \left(\boldsymbol{\varepsilon}^{ref}(\mathbf{x}_c) - \overline{\boldsymbol{\varepsilon}}^{num}(\mathbf{x}_c) \right) \right]^{\frac{1}{2}} \\ \qquad\qquad\qquad \text{for GS-Galerkin based S-PIM} \\[2ex] \left[\sum\limits_{i=1}^{N_e} A_i^e \left(\boldsymbol{\varepsilon}^{ref}(\mathbf{x}_c) - \tilde{\boldsymbol{\varepsilon}}^{num}(\mathbf{x}_c) \right)^{\mathrm{T}} \mathbf{c} \left(\boldsymbol{\varepsilon}^{ref}(\mathbf{x}_c) - \tilde{\boldsymbol{\varepsilon}}^{num}(\mathbf{x}_c) \right) \right]^{\frac{1}{2}} \\ \qquad\qquad\qquad \text{for linear FEM} \end{cases} \tag{5.94}$$

where A^s is the area for one smoothing domain, A^e is the area for one element, and $\mathbf{x}_c$ is the "center" of the smoothing domain or that of the element.

For S-PIM models using the SC-Galerkin formulation with constructed strain field that is linear in each quadrature domain, for example the SC-PIM introduced in Chapter 10, we use the following definition for the error in energy norm

$$e_e = \left[\sum_{i=1}^{N_q} \sum_{j=1}^{N_G} W_j^G \left(\boldsymbol{\varepsilon}^{ref}(\mathbf{x}_{i,j}) - \hat{\boldsymbol{\varepsilon}}_i^{num}(\mathbf{x}_{i,j}) \right)^{\mathrm{T}} \mathbf{c} \left(\boldsymbol{\varepsilon}^{ref}(\mathbf{x}_{i,j}) - \hat{\boldsymbol{\varepsilon}}_i^{num}(\mathbf{x}_{i,j}) \right) \right]^{\frac{1}{2}} \tag{5.95}$$

where $\hat{\boldsymbol{\varepsilon}}^{num}$ is the strains solutions obtained using W^2 models based on the SC-Galerkin formulation, N_G is the number of Gauss points in each quadrature domain and W_j^G is the Gauss weight corresponding to the Gauss point j. For the SC-PIM models with linear strain field in each triangular quadrature domain, $N_G=3$ is used (see Chapter 10).

For linear FEM models, the 2nd equation in (5.94) is the same as Equation (5.91) evaluated using one Gauss point. It is not the full evaluation of the errors, but an easily computable error measure. The similar can be said for S-PIM models. We note also that Equation (5.94) uses raw data without any post-processing. It is very efficient to use and can be implemented for all these discrete S-PIM models and lower order (linear and bilinear) FEM models.

Remark 5.23 On the comparison of errors in energy norms

In the FEM practice, we can obtain more accurate solutions for the strain and stress fields by performing a post-processing called recovery operation. The error in energy norm can be evaluated using Equation (5.91) with the recovered strains. The same can be done for S-PIM models. The error analyses based on the recovery strain solution are quite complicated and very subjective to the ways the strain field is recovered. Therefore, the outcome of the comparison may depend on a number of issues. When only displacement functions in H^1 space are used to create the discrete models, such a recovery strain based error analyses have been performed in great detail in [26]. In such cases, the S-PIM becomes its special case of smoothed finite element method (S-FEM), and its superiority over the standard FEM was also demonstrated in detail in [26]. In this book, we go for simplicity, and use Equation (5.94) with raw data for evaluating the error in energy norm for all the discrete models. The comparison is on a common footing for all the models, but the fairness may be subjective to possible ways of post-processing. Readers are advised to keep this in mind in the interpretation of these comparison studies for the error in energy norm.

5.9 Concluding remarks

Before leaving this chapter, we make the following remarks:

Remark 5.24 S-PIM models: operations beyond elements

The procedures used in S-PIMs are in general different from those in FEM, because of 1) the differences in the shape functions: FEM uses compatible FEM shape functions but S-PIM uses PIM shape functions that may not be compatible; and 2) strain field used: FEM uses compatible strain fields but S-PIM uses constructed strain fields. Displacement interpolation, strain field construction and integration in S-PIM are all beyond the elements/cells.

Remark 5.25 S-PIM models: larger bandwidth

The stiffness matrix $\bar{\mathbf{K}}$ will also be banded if the nodes are properly numbered, as that in the FEM. Because of the beyond-elements/cells operations used in S-PIM, the bandwidth of the stiffness matrix will be in general larger than that of FEM counterpart. Therefore, for a given model the S-PIM will take more CPU time when a bandwidth solver or iterative solver is used to solve the system equations, even though the S-PIM has the same DOFs as the FEM counterpart. However, in terms of computational efficiency (CPU time for the solutions of same accuracy), S-PIM can be more effective. In addition, S-PIM models will have many other properties (see Sections 5.6.2 and 5.6.3), in contrary to the FEM counterpart.

Remark 5.26 S-PIM models: smaller condition number

For a given model, the stiffness matrix of an S-PIM model will generally have a smaller condition number than that of the FEM counterpart. Therefore, the computation cost of an S-PIM can be compatible to the fastest linear FEM model with the same set nodes, if the system equations are solved using the iterative solver which depends on the condition number of the stiffness matrix. Detailed comparison on CPU time and computational efficiency between S-PIM models and FEM is conducted in the following chapters (See Chapters 6, 7, 8, 9 and 10).

Remark 5.27 FEM: a special case of S-PIM

The FEM may be regarded as a special case of S-PIM: 1) when the linear S-PIM is used with cell-based smoothing domains, the linear CS-PIM-Tr2 is identical to the linear FEM-Tr3; 2) when FEM shape functions are used and all the smoothing domains approach to zero, S-PIM solution approaches to the FEM solution.

Remark 5.28 S-FEM: a special case of S-PIM

The S-FEM [26] is a special case of S-PIM using compatible shape functions.

Remark 5.29 W^2 form: for the next generation of numerical methods

The concept of the W^2 formulation has profound impact on the development of the next generation of numerical methods. The family of S-PIM methods to be detailed in the following chapters is only a beginning of the development in this direction.

Remark 5.30 Stiffer and softer models

We finally defined two intuitive engineering terms often used in this book: stiffer model and softer model. A stiffer model has a "stiffness" that is larger than that of the exact model (that usually cannot be established). A more precise definition is that "all the eigenvalues of the stiffness matrix of the discrete stiffer model should be larger than the corresponding exact eigenvalues of the original (continuous) problem". Such a stiffer model should provide lower bound solutions for force driving problems. A stiffer model is both spatially and temporally stable. A fully compatible FEM model is a typical stiffer model.

On the other hand, A softer model has a "stiffness" that is smaller than that of the exact model. A more precise definition is that "all the eigenvalues of the stiffness matrix of the discrete softer model should be smaller than the corresponding exact eigenvalues of the original problem". Such a softer model should provide upper bound solutions for force driving problems. A softer model is spatially stable but can be temporally instable. A properly established NS-PIM model [15, 22, 23] is a typical softer model.

Remark 5.31 W^2 models: close-to-exact stiffness

W^2 formulations can produce models that have very close-to-exact stiffness. A more precise definition is that "the important lowest eigenvalues of the stiffness matrix of the discrete W^2 model should be very close to the corresponding exact eigenvalues of the original problem". A properly established ES-PIM model [25] is a typical stiffer model with close-to-exact stiffness.

Remark 5.32 W^2 forms: with two important knobs

The W^2 formulation offers two knobs for us to establish a model for special properties. The major knob is the ways of constructing smoothing domains that make the model softer or stiffer in a relatively decisive way. The other knob is the different ways to construct PIM shape functions that can fine tune the stiffness of the mode. With these two important knobs, one can practically tailor our W^2 model for desired properties. The following chapters are essentially about the "art of tuning" these two knobs, based on the theory described in this chapter.

In following chapters, we use the GS-Galerkin weak form, the more general SC-Galerkin weak form, to formulate these S-PIM methods for mechanics problems of solids and structures. The standard Galerkin weak form is only used in the FEM models involved in the comparison studies.

Remark 5.33 W^2 form: a long way to go

Theoretical issues related to W^2 formulation have not fully been resolved such as the G dual spaces, solution regularity, allowable linear functional, convergence rate of the W^2 solutions, etc. In other words, at this moment, we only know it works very well, but do not yet know much theoretically on how-well. Helps from mathematical communities in this kind of theoretical work is needed.

5.10 References

1.   Belytschko, T., Krongauz, Y., Organ, D. Fleming, M. and Krysl, P., Meshless methods: An overview and recent developments. *Computer Methods in Applied Mechanics and Engineering*, 139(1-4): 3-47, 1996.

2.   Fairweather, G. and Karageorghis, A., The method of fundamental solutions for elliptic boundary value problems. *Advances in Computational Mathematics*, 9(1-2): 69-95, 1998.

3.   Liu, G. R. and Gu, Y. T., *An introduction to meshfree methods and their programming*, Springer, 2005.

4.   Liu, G. R., *Meshfree Methods: Moving beyond the Finite Element Method*, 2nd Edition, CRC press, Boca Taton, USA, 2009.

5.   Liu, G. R. and Kee, B. B. T., A stabilized least-squares radial point collocation method (LS-RPCM) for adaptive analysis, *Computer Method in Applied Mechanics and Engineering*, 195: 4843-4861, 2006.

6.   Kee, B. B. T., Liu, G. R. and Lu. C., A regularized least-squares radial point collocation method (RLS-RPCM) for adaptive analysis. *Computational Mechanics*, 40: 837-853, 2007.

7.   Liu, G. R., A G space theory and a weakened weak (W^2) form for a unified formulation of compatible and incompatible methods, Part I Theory, *International Journal for Numerical Methods in Engineering*, 81: 1093-1126, 2009.

8. Liu, G. R., A G space theory and a weakened weak (W^2) form for a unified formulation of compatible and incompatible methods, Part II Application to solid mechanics problems, *International Journal for Numerical Methods in Engineering*, 81: 1127-1156, 2009.

9. Liu, G. R., A generalized Gradient smoothing technique and the smoothed bilinear form for Galerkin formulation of a wide class of computational methods. *International Journal of Computational Methods*, 5(2): 199-236, 2008.

10. Liu, G. R., On a G space theory. *International Journal of Computational Methods*, 6(2): 257-289, 2009.

11. Liu, G. R. and Zhang, G. Y. A normed G space and weakened weak (W^2) formulation of a cell-based smoothed point interpolation method. *International Journal of Computational Methods*, 6(1): 147-179, 2009.

12. Liu, G. R., Xu, X., Zhang, G. Y. and Gu, Y. T., An extended Galerkin weak form and a point interpolation method with continuous strain field and superconvergence (PIM-CS) using triangular mesh, *Computational Mechanics*, 43: 651-673, 2009.

13. Liu, G. R. and Quek, S. S. *The finite element method: a practical course.* Butterworth Heinemann: Oxford, 2003.

14. Hughes, T. J. R. *The Finite Element Method: Linear Static and Dynamic Finite Element Analysis*, Prentice-Hall, 1987.

15. Liu, G. R. and Zhang, G. Y., Upper bound solution to elasticity problems: A unique property of the linearly conforming point interpolation method (LC-PIM). *International Journal for Numerical Methods in Engineering*, 72: 1524-1543, 2007.

16. Oliveira Eduardo R De Arantes E., Theoretical Foundations of the Finite Element Method. *International Journal of Solids and Structures*. 4: 929-952, 1968.

17. Peraire, J., *Lecture notes on finite element methods for elliptic problems.* MIT, 1999.

18. Strang, G. and Fix, G. J., *An analysis of the finite element method*, Prentice-hall, 1973.

19. Zhang, G. Y., Liu, G. R. and Xu, X., A strain-constructed point interpolation method (SC-PIM) and strain field construction schemes for solid mechanics problems using triangular mesh. *Applied Mathematics and Computation*, 219: 2067-2086, 2012.

20. Liu, G. R. and Zhang, G. Y., A novel scheme of strain-constructed point interpolation method for static and dynamic mechanics problems. *International Journal of Applied Mechanics*, 1(1): 233-258, 2009.

21. Zhang, G.Y., Liu, G. R., Wang, Y. Y., Huang, H. T., Zhong, Z. H., Li, G. Y. and Han, X., A linearly conforming point interpolation method (LC-PIM) for three-dimensional elasticity problems, *International Journal for Numerical Methods in Engineering*, 72: 1524-1543, 2007.

22. Zhang, G. Y., Liu, G. R., Nguyen-Thoi, T., Song, C. X., Han, X., Zhong, Z. H. and Li, G. Y., The upper bound property for solid mechanics of the linearly conforming radial point interpolation method (LC-RPIM). *International Journal of Computational Methods*, 4(3): 521-541, 2007.

23. Liu, G. R., Zhang, G. Y., Dai, K. Y., Wang, Y. Y., Zhong, Z. H., Li, G. Y. and Han, X., A linearly conforming point interpolation method (LC-PIM) for 2D solid mechanics problems, *International Journal of Computational Methods*, 2(4): 645-665, 2005.

24. Zienkiewicz, O. C. and Taylor R. L., *The Finite Element Method*, 5th ed., Butterworth Heimemann, Oxford, 2000.

25. Liu, G. R. and Zhang, G. Y., Edge-based smoothed point interpolation methods, *International Journal of Computational Methods*, 5(4): 621-646, 2008.

26. Liu, G. R. and Nguyen-Thoi, T., *Smoothed Finite Element Methods*, CRC press, Boca Taton, USA, 2010.

Chapter 6

Node-based Smoothed Point Interpolation Method (NS-PIM)

With the theory given in Chapters 2-5, we are now ready to present a particular S-PIM method: the node-based smoothed point interpolation method or NS-PIM. Because the NS-PIM is the first to be introduced, we shall present it as in detail as possible.

The NS-PIM was originally proposed by Liu and Zhang et al. [1-4] using the generalized gradient smoothing technique [3] and the PIM shape functions [5-13] created using only a small number of local nodes in meshfree settings. The NS-PIM is a typical method created using generalized smoothed Galerkin (GS-Galerkin) weak form based on the normed G space theory that allows the use of discontinuous displacement functions. This chapter formulates the NS-PIM for mechanics problems for 2D and 3D solids. We focus first on 2D cases, because it is much easier to describe and comprehend. We will then extend the formulations to 3D cases by simply highlighting the major differences.

The PIM shape function used in the NS-PIM was created using the T-schemes with nodes that can come from more than one cell, and hence are discontinuous in general. The NS-PIM was termed as linearly conforming point interpolation method (or LC-PIM) [1, 2, 4], because it is at least linearly conforming despite the use of discontinuous PIM shape functions. It was later termed as NS-PIM because the smoothing operation is node based. The name of NS-PIM is more convenient for distinguishing from other S-PIM models, many of which are also linearly conforming but use different types of smoothing domains for different properties and features. The NS-PIM was found possessing the following important features.

1) The T-schemes based on triangular background cells are used for node selection, which helps to minimize the number of nodes and to overcome the singular moment matrix issue (see, Chapter 3), and ensures the efficiency in computing the PIM shape functions of different orders.

2) Shape functions generated using polynomial basis functions and simple interpolation method ensure that the PIM shape functions possesses at least linearly consistence, which is essential for the convergence and the order of accuracy of the model.

3) No mapping is needed in the creation of PIM shape functions, and hence no Jacobian matrix is involved, and hence the NS-PIM is naturally insensitive to the mesh distortion.

4) Shape functions generated has the Delta function property facilitating easy implementations of essential boundary conditions.

5) The GS-Galerkin weak form is used, and it is a W^2 form that allows the use of incompatible displacement functions created using the PIM shape functions with stability guaranteed.

6) Because of the use of W^2 form, we do not need the derivatives of the PIM shape functions in the formulation of NS-PIM models.

7) The number of node-based smoothing domain is very small but sufficient to meet the minimum requirement for (spatial) stability. The use of small number of smoothing domains provides sufficient softness to the NS-PIM model. It may be the most practical model that can produce upper bound solutions for force-driving problems with domain of arbitrary complexity.

8) NS-PIM uses primarily triangular/tetrahedral background cells (although it works also for other types of cells). Both the cells and smoothing domains can all be generated automatically, and hence is suited for adaptive analysis.

Due to these excellent features, the NS-PIM is very easy to implement, guarantees (spatial) stability and convergence, accurate in stress solution, computationally can be as efficient as the FEM using the same mesh (depending on the norm measure), and most importantly it can produce upper bound solution with respect to the exact solution.

6.1   NS-PIM for 2D solids

6.1.1   Approximation of displacement field

In the NS-PIM, the problem domain is first represented by a set of properly scattered N_n field nodes, as shown in FIGURE 6.1a. It is next triangulated into N_c nonoverlapping and seamless (NOSL) triangular background cells, such that $\boxed{\Omega} = \bigcup_{i=1}^{N_c} \boxed{\Omega}_i^c$ and $\Omega_i^c \bigcap \Omega_j^c = \varnothing, \forall i \neq j$, following the procedure detailed in Section 1.6.1. The triangulation can be performed automatically for complicated geometries.

Following the procedure presented in Section 3.2.1, PIM shape functions using polynomial basis functions can be constructed with a small number of support nodes. Based on the triangular background cells, we select the support nodes using the T-schemes as presented in Section 1.6.3. When the cell-based T3-scheme (see, FIGURE 6.1b, which is the same as FIGURE 1.7a) is used we have a linear NS-PIM-Tr3 model; when the edge-based T4-scheme (see, FIGURE 6.1c) is used it produces a bilinear NS-PIM-Tr4 model, and when the cell-based T6/3-scheme (see, FIGURE 6.1d, which is the same as FIGURE 1.7b) is used we obtain a quadratic NS-PIM-Tr6/3. Note the edge-based T4-scheme shown in FIGURE 6.1c shares the same name with that described in FIGURE 1.7f, but there is difference between them. The former one selects support nodes for the point of interest located in the vicinity of an edge (diamond for interior edge and triangle for boundary edge); the latter selects support nodes for those located on the edge. Therefore the latter one can be viewed a special case of the former. However the edge-based T4-scheme shown in FIGURE 1.7f is used in various 2D cell-based S-PIM models (see Chapters 8 and 9), while the one in FIGURE 6.1c is used only for the bilinear NS-PIM-Tr4 model. We also note that the node selection is performed automatically without manual intervention: another important feature of using triangular background cells.

For the NS-PIM-Tr3 model using the cell-based T3-scheme to select support nodes, the approximated displacement is continuous over the problem domain. Consider the edge-based T4-scheme used in the bilinear NS-PIM-Tr4 model, FIGURE 6.1c shows four nodes selected based on an interior edge i_1-i_2, and two nodes selected based on a boundary edge j_1-j_2, for the quadrature point near the edge. Therefore, in the NS-PIM-Tr4, the approximated displacement field is not continuous and the discontinuity will occur particularly on the lines radiated from the node to the centroid of the triangle, as illustrated using dotted lines in the figure. For the NS-PIM-Tr6/3 model using the cell-based T6/3-scheme for node selection, the displacement field is also discontinuous and the discontinuity will occur on the interfaces of triangular cells.

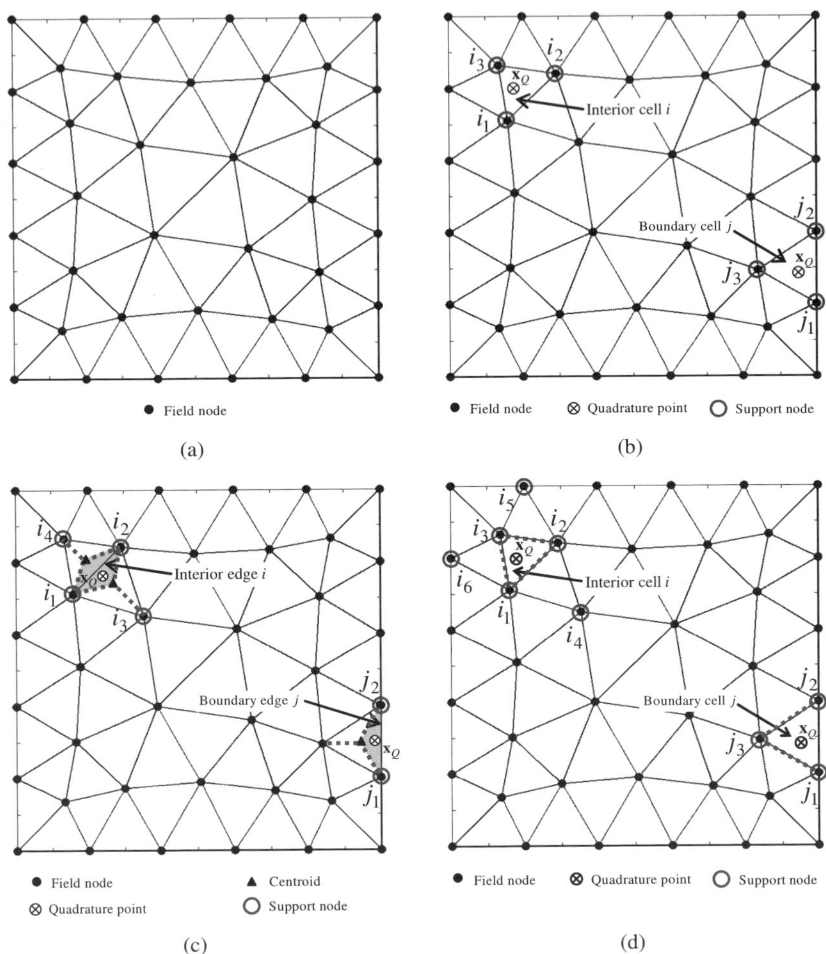

FIGURE 6.1 T-schemes for selecting support nodes for the construction of PIM shape functions: (a) triangular background cells; (b) cell-based T3-scheme; (c) edge-based T4-scheme; (d) cell-based T6/3-scheme.

The cell-based T3-scheme is the simplest one which leads to a linear NS-PIM model: NS-PIM-Tr3. In this case, the PIM shape functions are the same as those used in the linear FEM (FEM-Tr3). Therefore, such a linear NS-PIM is also called NS-FEM-Tr3 [14, 15]. However, all the other major numerical operations and the solution properties of the NS-PIM-Tr3 will be very much different from that of the FEM-Tr3 model, even if exactly the same mesh is used. The FEM-Tr3 is known behaving "overly-stiff" and produces a lower bound solution in strain energy norm for "force-driving" problems (see Remark 5.4). The NS-PIM-Tr3 model is much "softer" than the FEM-Tr3 model, and produces an upper bound solution. In fact, the NS-PIM-Tr3 is often found "overly-soft", in contrary to the "overly-stiff" FEM-Tr3.

Once a proper set of small number of local support nodes are selected using a T-scheme, PIM shape functions can then be created using the techniques detailed in Section 3.2. We can then

1) assume the solution of the displacement function for the S-PIM model at any point of interest in the problem domain using Equation (5.66);

2) compute the generalized node-based smoothed strains following the procedure given in Section 6.1.2;

3) compute the force vector using Equation (5.82) and

4) for dynamics problems, compute the mass matrix using Equation (5.86).

6.1.2 Evaluation of node-based smoothed strains

The "smoothed" strain used in the NS-PIM is constructed using only the PIM shape function values with the help of a proper set of smoothing domains. In the NS-PIM, the problem domain is divided, on top of the triangular background cells, into N_s NOSL smoothing domains *associated with nodes* following the "no-sharing" rule given in Section 2.3.1, such that $\left|\Omega\right| = \bigcup_{k=1}^{N_s}\left|\Omega\right|_k^s$ and $\Omega_k^s \cap \Omega_l^s = \varnothing$, $\forall k \neq l$. Each node-based smoothing domain contains only one node and covers portions of cells sharing the node. Thus, the number of smoothing domains is the same as that of the field nodes, i.e. $N_s = N_n$. Therefore, the number of smoothing domain used satisfies the minimum number given in TABLE 2.2: a condition to ensure the spatial stability (Remark 1.9) of the NS-PIM model. These smoothing domains serve also as the basis for the summation in the W^2 formulation of system matrices, and hence we have for the NS-PIM models $N_q = N_s = N_n$.

As discussed in Chapter 5, because the GS-Galerkin weak form that allows the use of discontinuous displacement functions is used to create the discretized system equations, the no-sharing rule is important: the boundaries of any

Ω_k^s should not share any *finite* portion of lines where the displacements are discontinuous. This implies that we need to be careful in choosing the T-schemes and in constructing smoothing domains. Obeying the no-sharing rule ensures the continuity (compatibility) of the displacements along the boundaries of all these smoothing domains. In the case of NS-PIM, displacement fields are approximated using the PIM shape functions and local support nodes selected using a proper T-scheme detailed in Section 6.1.1. The displacement field discontinuity in the vicinity of a smoothing domain can possibly occur on the interfaces of background cells supporting the cell or along the lines radiated from the node to the centroid of the triangle. Therefore, we require that the boundaries of node-based smoothing domains should not share any finite portion of these interfaces of the triangular cells, and the lines radiated from the node to the centroid of the triangle. The following are two proven ways for constructing node-based smoothing domains.

6.1.3 Equally-shared smoothing domains

The most commonly used node-based smoothing domains, is the so-called *equally-shared* smoothing domains crated based on the triangular (for 2D) and tetrahedral (for 3D) background cells. FIGURE 6.2 shows a typical 2D case. The problem domain Ω is divided into smoothing domains each of which contains only one node. For example Ω_i^s bounded by Γ_i^s contains node i, as shown in FIGURE 6.2. The smoothing domain Ω_i^s is constructed by connecting sequentially the mid-edge points to the centroids of the surrounding triangular cells sharing node i. The union of all these smoothing domains Ω_i^s forms exactly the problem Ω: no overlap or gap is allowed. Because the area of each triangular cell is equally shared by its three nodes and the length of the edge of a cell is also equally shared by these two nodes at the edge-ends, the smoothing domain created is termed as "equally-shared smoothing domain".

It is clear that the boundary of an equally-shared smoothing domain will not share any finite portion of these interfaces of the triangular cells and the lines radiated from the node to the centroid of the triangle. It goes across them, shares with them only at (infinitely small) point, and hence the no-sharing rule is well observed. It is also clear that these equally-shared smoothing domains are independent each other (see, Section 2.3.2), and hence can be used for our NS-PIM models.

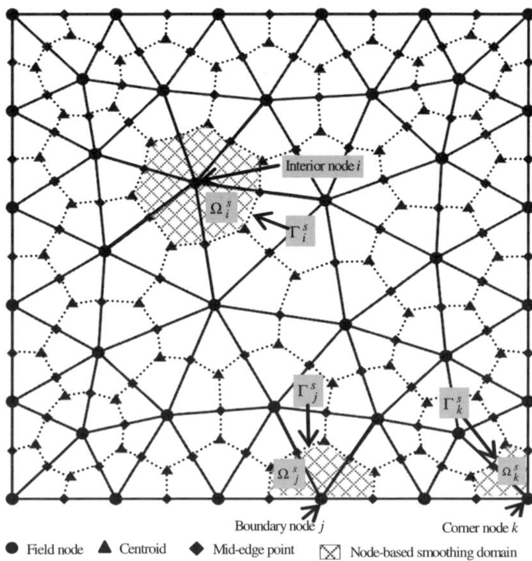

FIGURE 6.2 Node-based equally-shared smoothing domain Ω_i^s bounded by Γ_i^s for node i. The smoothing domains for the boundary and corner nodes are biased, such as Ω_j^s for the boundary node j and Ω_k^s for the corner node k.

6.1.4 Voronoi smoothing domains

Alternatively, we can also use the so-called the Voronoi smoothing domain that was used in [16] for stabilizing the nodal integrated meshfree method. The Voronoi smoothing domain is constructed using the standard Voronoi diagram, which results in smoothing domains of different shapes associated with nodes, as shown in FIGURE 6.3. It is clear that the no-sharing rule is also observed, and these Voronoi smoothing domains are also independent, and hence can be used for our NS-PIM models.

We note that the Voronoi smoothing domains are convex, but the equally-shared smoothing domains are not. Whether or not a smoothing domain is convex is immaterial, because our NS-PIM formulation does not require convexity. One of the biggest advantages of using the equally-shared smoothing domains is the convenience in constructing various T-schemes for local node selection and the evaluation of the generalized smoothed strains. The authors' group uses mainly the equally-shared smoothing domain, but the use of the Voronoi smoothing domain can also be found in [5, 17, 18].

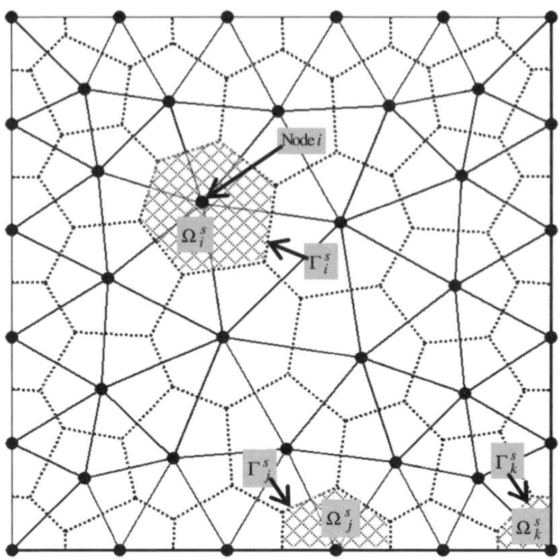

FIGURE 6.3 Voronoi smoothing domain. The smoothing domains for the boundary and corner nodes are biased.

Within each constructed node-based smoothing domain, the "smoothed" strains can now be obtained using the generalized gradient smoothing operation, as detailed in Section 4.5.2. When the Heaviside type of smoothing functions are used, the smoothed strains over the smoothing domain for node i can be given in Equation (4.20).

As an special case, when the linear interpolation is used in our NS-PIM, the numerical integration along each line-segment needs only one Gauss point ($n_G = 1$), and the shape function values at all these Gauss points can be easily tabulated in TABLE 5.2 by simple inspection.

Note from FIGURE 6.2 and FIGURE 6.3 that the smoothing domains for the boundary and corner nodes are biased, which can be a source of the numerical error (see, Remark 3.24). To reduce such an error, it is preferable to reduce the size of the background cells near the problem boundary for NS-PIM models, whenever possible.

6.1.5　Stiffness matrix for NS-PIM

The stiffness matrix for an NS-PIM model follows Equation (5.79) and the linear system of equations of the NS-PIM has the form of

$$\overline{\mathbf{K}}^{\text{NS-PIM}}\overline{\mathbf{d}} = \tilde{\mathbf{f}} \tag{6.1}$$

where $\overline{\mathbf{K}}^{\text{NS-PIM}}$ is the *smoothed* stiffness matrix whose entries are given by

$$\overline{\mathbf{K}}_{IJ}^{\text{NS-PIM}} = \int_{\Omega} \overline{\mathbf{B}}_I^{\text{T}} \mathbf{c} \overline{\mathbf{B}}_J \, \mathrm{d}\Omega = \sum_{k=1}^{N_n} \int_{\Omega_k^s} \overline{\mathbf{B}}_I^{\text{T}} \mathbf{c} \underbrace{\overline{\mathbf{B}}_J}_{\text{constant in } \Omega_k^s} \mathrm{d}\Omega = \sum_{k=1}^{N_n} A_k^s \overline{\mathbf{B}}_I^{\text{T}} \mathbf{c} \overline{\mathbf{B}}_J \tag{6.2}$$

in which $A_k^s = \int_{\Omega_k^s} \mathrm{d}\Omega$ is the area of node-based smoothing domain Ω_k^s, and the smoothed strain-displacement matrix $\overline{\mathbf{B}}_I$ is computed using Equation (5.71).

Note that the sub-matrix of stiffness $\overline{\mathbf{K}}_{IJ}^{\text{NS-PIM}}$ defined in Equation (6.2) needs to be computed only when nodes I and J share a same smoothing domain. Otherwise, it is zero. Hence, the global stiffness matrix in an NS-PIM model will be very sparse, in addition to the obvious symmetry property. The sparseness depends on the node selections scheme used. It is determined by the difference of node numbers of the nodes supporting the smoothing domain. For 2D cases with linear interpolations, a node-based smoothing domain is usually supported by 4-8 nodes (cover 2 rows of nodes), depending on the background cells created. Based on the discussion given in Section 5.7.6, the global stiffness matrix $\overline{\mathbf{K}}^{\text{NS-PIM}}$ of a NS-PIM model is SPD. The NS-PIM solution will be stable, unique and converge to the exact solution when the cells are refined (together with the smoothing domains).

Remark 6.1   Features of the global stiffness matrix of NS-PIM

The global stiffness matrix of an NS-PIM model is SPD and sparse, but the sparseness of the linear NS-PIM is larger than that of the linear FEM, due to more nodes used in the interpolation. It is also banded if the field nodes are properly numbered. The bandwidth of the linear NS-PIM is usually larger than that of FEM.

6.1.6   Comparison of NS-PIM, NS-FEM and FEM

6.1.6.1   NS-PIM vs. FEM

When the same triangular mesh is used, we mention the following points by comparing the NS-PIM with the traditional FEM.

1) The interpolation in the NS-PIM is based on a group of support nodes selected using a T-scheme. The support nodes can generally be selected from more than one background cell, and the interpolation area may overlap. No mapping is needed in NS-PIM, and hence no Jacobian matrix is involved. The interpolation in the FEM is strictly based on elements and there is no overlap. In addition, a proper mapping in FEM is a *must* to insure the compatibility on the element interfaces, except for linear triangular elements.

2) Shape functions used in both FEM and NS-PIM have the property of the Kronecker Delta function. The imposition of essential boundary conditions in NS-PIM is as simple as that in the standard FEM.

3) The FEM uses the compatible displacement over the problem domain and compatible strains within the elements, and hence it is a fully compatible mode. NS-PIM can use incompatible displacements and the constructed strain field which is also incompatible.

4) The numerical operations are element based in the FEM, while those in the NS-PIM are beyond the cells/elements.

5) Both FEM and NS-PIM are at least linearly conforming: linear displacement field can be produced exactly and hence can pass the standard patch test (to machine accuracy).

6) The FEM solution does not in general satisfy the equilibrium conditions locally. The NS-PIM solution satisfies the (body force free) equilibrium equations at any point within the smoothing domains. Therefore, the NS-PIM behaviors somewhat like an equilibrium model.

7) The standard FEM model is compatible everywhere: inside the elements, on the element interfaces, and on the essential boundary. It is thus said "fully compatible". The NS-PIM models are in general not compatible, because the displacement field is continuous only on the interfaces of the smoothing domains and the essential boundary but may be discontinuous within the smoothing domains. Because the equilibrium status is met inside the smoothing domains, a "complementary" situation has been established: where the compatibility condition is violated the equilibrium is satisfied, and where the equilibrium is violated the compatibility is ensured. Such a complimentary situation prevents any energy loss in the NS-PIM model, and hence the NS-PIM model can be said *energy consistent*. It is also the essential reason for being convergent.

8) The FEM model using Tr3 elements behaviors very "stiffly", and stress result is in general not accurate. The NS-PIM model using exactly the same triangular cells behaviors much softer than the FEM, and stress solution is in general more accurate.

9) The linear FEM and NS-PIM models have the same set of nodes, nodal displacements, the same size in the discrete system equations and the number of DOFs. The stiffness matrices obtained using both FEM and NS-PIM are symmetric positive definite (SPD) for solids of stable material, if sufficient constraints are applied to eliminate the rigid body movement. The proof on the SPD for FEM is based on the theory of weak formulation, and that for the NS-PIM is based on those of the W^2 formulation [19-21].

10) For a force-driving problem (with homogeneous essential boundary conditions), the fully compatible FEM provides a lower bound for the solution in energy norm, and the NS-PIM can provide an upper bound of the solution which will be discussed intensively in Section 6.4.

6.1.6.2 NS-PIM vs. NS-FEM

1) If the same set of compatible shape functions is used, the NS-PIM formulation can lead to the node-based smoothed FEM (NS-FEM) [14, 15]. When only triangular elements and linear interpolation are used, the NS-PIM-Tr3 is identical to the NS-FEM-Tr3, and it gives the same results as the NIFEM proposed by Dohrmann et al. [22].

2) The NS-PIM is much more general than the NS-FEM. The NS-PIM is basically conceived from the meshfree procedures: shape functions are constructed using nodes beyond the cells (may overlap). In the case of NS-RPIM (see Section 6.2), the nodes can practically randomly distribute. The NS-FEM can be considered as a special case of NS-PIM.

6.1.7 Macro flowchart of the NS-PIM

The macro flowchart for constructing the global matrices and vectors of NS-PIM models can be briefly summarized as follows.

1) Loop over the field nodes (or node-based smoothing domains).

2) Loop over surrounding cells directly connected to the node.

3) Loop over the Gauss points located along the (two) boundary segments of the smoothing domain located in the cell:

   (a) Select support nodes using a proper T-scheme and create PIM shape functions.

   (b) Compute the components of smoothed strain-displacement matrix and form the matrix.

   (c) Compute nodal stiffness matrix and force vector.

   (d) Assemble the nodal contributions to the global matrices and vectors.

4) End the Gauss points loop.

5) End the cell loop.

6) End the smoothing domain loop.

6.1.8 Possible NS-PIM models

Proper T-schemes (see, Section 1.6.3) have to be chosen for the compliance to the no-sharing rule in the NS-PIM settings. Based on the schemes used in the construction of displacement field, we can have the following proven NS-PIM models.

NS-PIM-Tr3

The NS-PIM-Tr3 is the simplest NS-PIM model. It uses linear polynomial PIM shape functions and the support nodes selected using the cell-based T3-scheme, as shown in FIGURE 6.1b. The displacement field is *compatible* over the problem domain and the corresponding NS-PIM-Tr3 formulation is variationally consistent. In this case the NS-PIM-Tr3 is exactly as same as the NS-FEM-Tr3.

For the NS-PIM models, it is clear that more support nodes will be used to construct the stiffness matrix according to each node-based smoothing domain than the FEM-Tr3 model. To have a rough picture on this issue, the mesh with 248 irregularly distributed nodes shown in FIGURE 6.4 is used to find out the average number of support nodes for node-based smoothing domains. For the NS-PIM-Tr3 model, the support nodes for a node-based smoothing domain varies from 4 to 8 and the average number is 6.4 for the mesh with 248 nodes, which is about 2.1 times as large as the number (3) of the support nodes for an element in the FEM-Tr3 model. Therefore, the average number of the nonzero entries n_{ze} in a row of the stiffness matrix of an NS-PIM-Tr3 model is about 2.1

times that of the FEM-Tr3 counterpart using the same mesh, as listed in TABLE 6.1.

Note that the FEM-Tr3 model is the fastest (accuracy aside) model among all the possible 2D FEM (and meshfree) models with the same number of nodes.

NS-PIM-Tr4-CT

The NS-PIM-Tr4-CT model uses polynomial PIM shape functions and support nodes selected using the edge-based T4-scheme, as shown in FIGURE 6.1c. Thus in this case we have a bilinear interpolation for the displacement field, and the displacement is *incompatible*. The discontinuity of displacement field will occur on the lines connecting vertices and the centroid of a triangular cell. Note the adaptive coordinates transformation (CT) technique presented in Section 3.4 will be used to overcome the possible singularity problem in the process of constructing bilinear PIM shape functions. Note also that linear displacements are always used along the problem boundaries.

For the NS-PIM-Tr4-CT model, the number of the support nodes for a smoothing domain is in fact the same as the NS-PIM-Tr3 that is about 4-8 with the average value of 6.4, and hence the average number of the nonzero entries in a row of the stiffness matrix will be roughly the same, as shown in TABLE 6.1.

NS-PIM-Tr6/3

In the NS-PIM-Tr6/3, the displacement field is constructed using polynomial PIM shape functions and support nodes selected by the cell-based Tr6/3-scheme, as shown in FIGURE 6.1d. The displacement field located in an interior cell will be approximated using quadratic PIM shape functions; while displacement in a boundary cell will be approximated using linear PIM shape functions. The displacement field in the NS-PIM-Tr6/3 is incompatible. The discontinuity of displacement field will occur on the interfaces of triangular background cells.

Due to using the cell-based T6/3-scheme, it is clear that the NS-PIM-Tr6/3 model will use more nodes for each node-based smoothing domain. For the second mesh shown in FIGURE 6.4, the number of the support nodes for a smoothing domain is from 4 to 15 and the average number is 10.6, that is about 3.5 times that of the NS-PIM-Tr3. Therefore, n_{ze} for an NS-PIM-Tr6/3 model is about 3.5 times that of the FEM-Tr3 counterpart using the same mesh, as listed in TABLE 6.1.

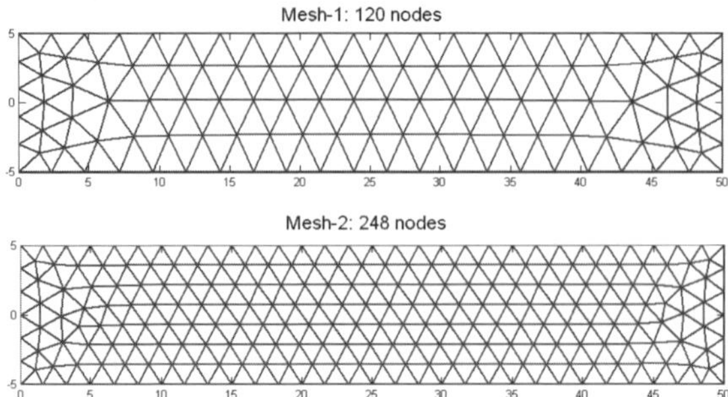

FIGURE 6.4 Two irregular meshes, with 120 and 248 distributed nodes respectively, for the 2D rectangular cantilever problem described in Example 6.1.2.

TABLE 6.1 NS-PIM models and overall characteristics in relation to the FEM-Tr3 model using the same mesh

Numerical method	Support nodes in an integral cell/element	Estimated average number of nonzero entries	Estimated Solver CPU time
FEM-Tr3	3	n_{ze}	t_{CPU}
NS-PIM-Tr3	4-8 (6.4)	$2.1\,n_{ze}$	$2.1\,\bar{n}_{iter}t_{CPU}$
NS-PIM-Tr4-CT	4-8 (6.4)	$2.1\,n_{ze}$	$2.1\,\bar{n}_{iter}t_{CPU}$
NS-PIM-Tr6/3	4-15 (10.6)	$3.5\,n_{ze}$	$3.5\,\bar{n}_{iter}t_{CPU}$

In TABLE 6.1, we assume that an iterative solver is used and the CPU time of such a solver can be estimated (see Section 1.2.2) using

$$t_{CPU} \propto n_{iter} N_{DOF} n_{ze} \tag{6.3}$$

where n_{iter} is the number of iterations needed to get a converged solution in the iterative solver, N_{DOF} is the number of degree of freedoms, and n_{ze} is the (average) nonzero entries in the system matrix. In TABLE 6.1, $\bar{n}_{iter}$ is the relative iteration number that is the ratio of the iteration number n_{iter} of a model to the counterpart of the FEM. Here we use the linear FEM as the base model for the purpose of comparison.

6.1.9 Condition number of NS-PIM models

The condition number of the global stiffness matrix, $cond(\mathbf{K})$, is an important indicator for a numerical method, as discussed in Section 1.2.2. When an iteration solver is used to solve the algebraic system equation, it affects directly the number of iterations, because $n_{iter} \propto \sqrt{cond(\mathbf{K})}$ (see, Equation (1.28)).

TABLE 6.2 lists the condition numbers of the global stiffness matrixes for various NS-PIM models in relation to the FEM-Tr3 using the same triangular meshes for the cantilever beam problem shown in FIGURE 6.4. We found that all these three NS-PIM models have smaller condition numbers than the linear FEM-Tr3, although they use higher order interpolation and more support nodes as shown in TABLE 6.1. Compared to the NS-PIM-Tr6/3 model whose condition number is about one half of that of linear FEM, the NS-PIM-Tr3 and NS-PIM-Tr4-CT have even smaller conditions numbers which are about one third of the counterpart of the FEM-Tr3. With the increase of DOFs, we found the condition number will increase accordingly for all the models as expected. However, the ratio of the condition numbers of a model with respect to the FEM-Tr3 model is almost constant regardless of the DOFs, as shown in TABLE 6.2. In the following analysis, we will use the relative $\overline{n}_{iter}$ given in the last column in TABLE 6.2.

TABLE 6.2 Condition number of the global stiffness matrixes for various NS-PIM models for the cantilever beam in relation to the FEM-Tr3 using the same mesh

Numerical method	Mesh-1 of 120 nodes		Mesh-2 of 248 nodes	
	$cond(\mathbf{K})$	$\overline{n}_{iter}$	$cond(\mathbf{K})$	$\overline{n}_{iter}$
FEM-Tr3	1.645e+08	1.00	2.399e+08	1.00
NS-PIM-Tr3	5.280e+07	0.57	7.794e+07	0.57
NS-PIM-Tr4-CT	5.296e+07	0.57	7.786e+07	0.57
NS-PIM-Tr6/3	7.402e+07	0.67	1.109e+08	0.68

6.1.10 Estimation of computational cost

Combining TABLE 6.1 and TABLE 6.2 we shall have a rough estimation of computational cost for the NS-PIM models in relation to the standard FEM-Tr3 using the same mesh as shown in TABLE 6.3. Here an iterative solver is supposed to be used and we estimate the complexity of the solver using approximately $n_{iter} N_{DOF} n_{ze}$, as discussed in Section 1.2.2. We choose the linear FEM-Tr3 for comparison, because it is the simplest and fastest FEM model for a given set of nodes. It is seen that the first two NS-PIM models, i.e. NS-PIM-Tr3 and NS-PIM-Tr4-CT only cost about 20% more time than the FEM-Tr3 owing to the smaller condition numbers. However, the NS-PIM-Tr6/3 will cost about 2.4 times that of the FEM-Tr3 model for the same set of nodes.

TABLE 6.3 Estimation of computational cost for NS-PIM models in relation to the FEM-Tr3 model using the same mesh

Numerical method	Support nodes in an integral cell/element	Estimated Solver CPU time
FEM-Tr3	3	t_{CPU}
NS-PIM-Tr3	4-8 (6.4)	$1.2\, t_{CPU}$
NS-PIM-Tr4-CT	4-8 (6.4)	$1.2\, t_{CPU}$
NS-PIM-Tr6/3	4-15 (10.6)	$2.4\, t_{CPU}$

Note that the total CPU time to obtain the solution of a numerical model should also contain the overhead time needed to establish Equation (6.1). Detailed analysis on this is too complicated. Fortunately, for a not-too-small model, such an overhead time is usually very small compared to the solver time, as examined in [5]. Therefore, the solver CPU time can be used as a good gauge on the computational cost of a model in relation to the simplest and fastest model: FEM-Tr3.

From TABLE 6.3, it is clear that an NS-PIM model will always be slower than the FEM-Tr3 model for the same set of nodes. However, the NS-PIM model can give more accurate results than the FEM-Tr3 using the same mesh, as will be shown later. Therefore, a fair comparison must also take the accuracy into consideration. In the following sections, we will investigate the *efficiency* which takes into consideration of both the computation cost and the accuracy of the numerical results.

6.1.11 Issues on the treatments along boundaries

For smoothing domains on the boundary of the problem domain, cares need to be taken for the following two issues.

Remark 6.2 Boundary effect

For a field node located on the boundary of the problem domain, for example point j shown in FIGURE 6.2, some boundary segments of the node-based smoothing domain will share some parts of the problem boundaries. Such a smoothing domain becomes "biased" leading to some *boundary effects* similar to those discussed in [23]. The numerical integration for the evaluation of the strain-displacement matrix should still be performed using Equation (5.69) as per normal: loop around the boundary of the biased smoothing domains. This practice will result in a less accurate "smoothed" strain due to such a boundary effects, but we do not have much choice. To mitigate such a boundary effects, smaller triangular cells should be used on the problem boundary (this is also somewhat supported by the mesh pattern generated from an adaptive analysis, as shown in Chapter 13).

Remark 6.3 Linear interpolation along boundaries

Regardless of the types of PIM shape functions and T-schemes used, a linear interpolation will *always* be used in an S-PIM model for those Gauss points located on the boundaries of the problem domain, when we perform the numerical integration using Equation (5.73) along boundaries to compute the strain-displacement matrix. This is to ensure the exact enforcement of linear essential boundary conditions and hence the linear conformability of the model.

6.1.12 Evaluation of nodal strain (stress)

In the scheme of NS-PIM models, the node-based smoothed strains are corresponding to the field nodes. Therefore, once the nodal displacement results have been obtained by solving Equation (6.1), the nodal strain results can be evaluated directly using Equation (5.70) and the nodal stress can be further obtained using Equation (1.10).

Note that the evaluation of nodal strain (stress) results is conducted directly based on the nodal displacements and there is no any post-processing needed. In the FEM-Tr3, the nodal strains (stresses) are usually obtained using the area weighted recovering process based on the elemental strain (stress) results.

6.1.13 Rank test of NS-PIM

Based on the theory presented in Chapter 5, the NS-PIM model should possess only the "legal" zero-energy modes representing the physical rigid motions, and there exist no spurious unphysical zero-energy modes. This means that NS-PIM models are spatially stable. In other words, for a constrained solid, all the eigenvalues of the stiffness matrix must be strictly positive. Such stability is ensured in a discrete model by the points summarized as follows.

1) In an NS-PIM model, the number of smoothing domains equals to the number of field nodes, i.e. $N_s=N_n$. It thus satisfies the minimum number of smoothing domains for problems of all dimensions as required in TABLE 2.2.

2) PIM shape functions used in an NS-PIM model are of partitions of unity, which ensures a proper representation of the rigid movements.

3) The nodal PIM shape functions created are linearly independent, as long as a T-scheme is used in node selection. Together with the independent node-based smoothing domains, the smoothed strain matrix will have sufficient number of independent columns, providing the basic condition for the stability of the model.

4) The stiffness matrix of an NS-PIM model is SPD for constrained solids of stable materials, after all the rigid motions are fixed.

5) The use of the GS-Galerkin formulation ensures a proper measure in evaluating the strain energy in the model.

Therefore, NS-PIM models will have proper number of zero eigenmodes representing rigid body movements, and will not have any spurious zero-energy models. It means that any finite deformation (except the rigid motions) in an NS-PIM model will result in a finite amount of strain energy in the model: spatially stable.

However, an NS-PIM model can have *spurious* nonzero-energy models, and can be temporally unstable, and hence cannot be applied directly to solve some dynamic and nonlinear problems. Some special stabilization techniques have been developed for NS-PIM models to solve dynamic problems [24, 25]. We will not discuss this matter in detail in this book, and interested reader may refer to [24, 25].

Remark 6.4 Stability property of NS-PIM models

NS-PIM models are spatially stable but may be temporally instable. An NS-PIM model possesses only "legal" zero-energy modes that represents the physic rigid motions, there exist no spurious zero-energy modes, but may have spurious nonzero-energy modes.

6.1.14 Numerical examples for 2D solids

Several numerical examples of 2D solids are studied in this section using the NS-PIM models. The materials for the solids are all linear elastic with Young's modulus $E=3.0\times10^7$ Pa and passion's radio $v=0.3$ and the units used are based on the international standard unit system, unless specially mentioned. Numerical results are evaluated using the error indicators in both the displacement and energy norms defined in Equation (5.88) and Equations (5.94), respectively.

Example 6.1.1 2D linear patch test

For a numerical method working for solid mechanics problems, the sufficient requirement for the convergence is passing the standard patch test [26]. Therefore, we present the first example: the standard patch test using the NS-PIM models. The patch is a square domain with the dimension of 10×10, and is represented using regular and irregular nodes as shown in FIGURE 6.5. The displacements are prescribed on entire boundary by the following linear function.

$$\begin{cases} u_i = 0.6x_i \\ v_i = 0.6y_i \end{cases} \tag{6.4}$$

where u_i and v_i are, respectively, the displacement components in x and y directions for node i, and (x_i, y_i) are the coordinates of node i.

Passing the patch test requires that the numerical displacements obtained for all the interior nodes in the patch follow "exactly" (to machine precision) the same linear function of the imposed displacement on the boundary. TABLE 6.4 lists the displacement norms of the solution error which are calculated using Equation (5.88). It can be clearly found that all the NS-PIM models, including NS-PIM-Tr3, NS-PIM-Tr4-CT and NS-PIM-Tr6/3, can pass the standard patch test with regularly and irregularly distributed nodes. Note that when quadratic PIM shape functions are used with support nodes selected using the cell-based Tr6/3-scheme, the discontinuity of displacements will occur along the interfaces of background cells and the displacements are incompatible. Similar discontinuity of displacements also exists for the NS-PIM-Tr4-CT model, where the displacements are discontinuous along the lines radiated from the node to the centroid of the cells.

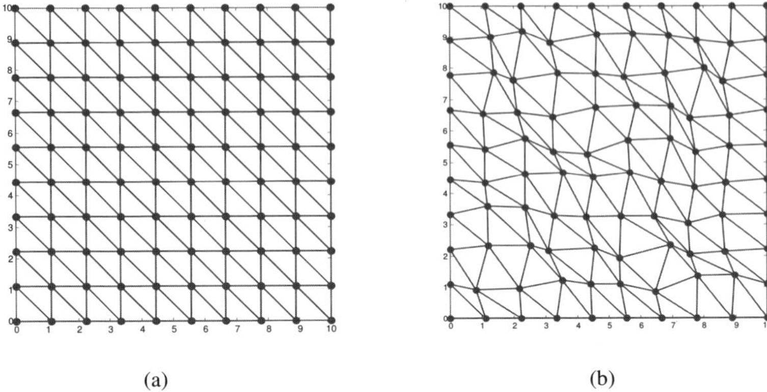

(a)   (b)

FIGURE 6.5 A 2D patch with regularly and irregularly distributed nodes and triangular meshes: (a) regularly distributed nodes and (b) irregularly distributed nodes.

TABLE 6.4 Error norm in displacements of numerical results for the standard patch test obtained using three NS-PIM models

NS-PIM models	Regular mesh	Irregular mesh
NS-PIM-Tr3 (displacement compatible)	2.015E-15	2.099E-15
NS-PIM-Tr4-CT (displacement incompatible)	1.238E-14	5.555E-15
NS-PIM-Tr6/3 (displacement incompatible)	2.303E-14	1.794E-14

Remark 6.5   NS-PIM: linearly conforming and 2nd order accuracy

This example demonstrates numerically that the NS-PIM can reproduce linear fields exactly, regardless of the incompatible displacement field used: *linearly conforming*. Together with the stability (Theorem 5.1), the NS-PIM solution will converge to the exact solution of any well-posed linear elasticity problem. This implies also that the model is at least 2nd order accuracy: the solution error in displacements is at the terms of 2nd order and above. This supports Remark 3.23.

Remark 6.6   Requirements for a GS-Galerkin model to pass the patch test

To pass the standard patch test, a stable numerical method based on the GS-Galerkin weak form needs to satisfy the following requirements.

- The shape functions are at least linearly consistent, which implies that the shape function must be created using at least complete linear polynomial terms (see Chapter 3).

- The essential boundary conditions (displacement constraints on the boundary of the patch) have to be accurately imposed (see Remark 6.3).

Of course, we require also that all the numerical operations are accurate.

As discussed in Chapter 3, PIM shape functions created using linear, bilinear and quadratic polynomial basis are at least linearly consistent. Using cell-based T-schemes (T3- or T6/3-schemes) or edge-based T4-scheme, linear PIM shape functions with Delta function property are generated for the points of interests located along domain boundaries, which guarantees the accurate imposition of essential boundary conditions of the patch. These PIM shape functions may be incompatible (such as the cell-based T6/3-scheme and the edge-based T4-scheme), but they all pass the standard patch test, as shown in TABLE 6.4. In contrast to the numerical method based on the standard Galerkin weak form, the condition of displacement compatibility is not required for a numerical method built based on the GS-Galerkin weak form, such as the present NS-PIM models.

Example 6.1.2 Rectangular cantilever

A benchmark problem of a rectangular cantilever is now studied using our NS-PIM models. The cantilever is of length L, height D, and subjected to a parabolic traction of P along the right edge, as shown in FIGURE 6.6. The analytical solution for this problem is available and can be found in the textbook by Timoshenko and Goodier [27].

The analytical solutions of displacement components are

$$u_x = -\frac{Py}{6EI}\left[(6L-3x)x+(2+v)\left(y^2-\frac{D^2}{4}\right)\right] \tag{6.5}$$

$$u_y = \frac{P}{6EI}\left[3vy^2(L-x)+(4+5v)\frac{D^2x}{4}+(3L-x)x^2\right] \tag{6.6}$$

where I is the moment of inertia for the cantilever with rectangular cross section and unit thickness given by $I = D^3/12$.

The analytical solutions of the stress components are

$$\sigma_{xx} = -\frac{P(L-x)y}{I} \tag{6.7}$$

$$\sigma_{yy} = 0 \tag{6.8}$$

$$\tau_{xy} = \frac{p}{2I}\left[\frac{D^2}{4} - y^2\right] \tag{6.9}$$

In the present study, the beam model is supported on the left edge by prescribing the displacements obtained the analytical formula (Equations (6.5) and (6.6)) with $x = 0$. The external parabolic traction (P) at the right edge is computed using Equation (6.9) with $P = -1000$. This is to remove the modeling error for the boundary conditions. The geometric parameters are taken as: $L = 50$ and $D = 10$.

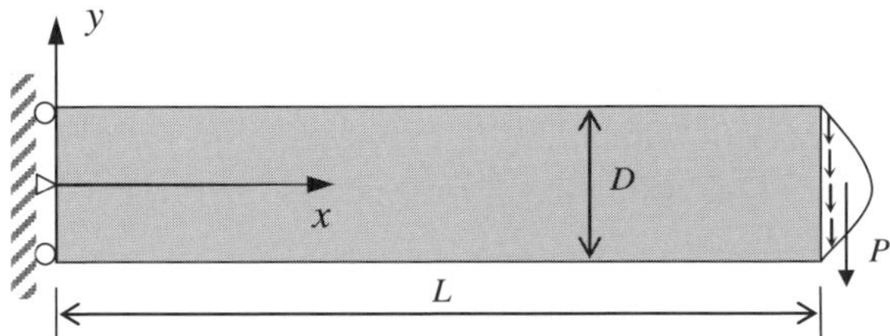

FIGURE 6.6 Rectangular cantilever subjected to traction P in a parabolic fashion at the right end.

First, to investigate the effect of the irregularity of nodal distribution, four meshes of 369 distributed nodes with different irregularity (shown in FIGURE 6.7) are used to examine the NS-PIM. Mesh-1 shown in FIGURE 6.7 is presented using regular triangular mesh and other three meshes of irregular meshes are created by altering the coordinates of the interior nodes using

$$\begin{aligned}
x_i' &= x_i + \Delta x \cdot r_c \cdot \alpha_{ir} \\
y_i' &= y_i + \Delta y \cdot r_c \cdot \alpha_{ir}
\end{aligned} \tag{6.10}$$

where $\left(x'_i, y'_i\right)$ is the coordinates of irregular interior nodes generated based on the original coordinates $\left(x_i, y_i\right)$, Δx and Δy are the initial regular mesh sizes in the x and y directions, respectively; $r_c \in [-1,1]$ is a computer-generated random number, and α_{ir} is a prescribed factor controlling irregularity. When $\alpha_{ir} = 0.3$ is used the mesh is extremely irregular as Mesh-4 shown in FIGURE 6.7.

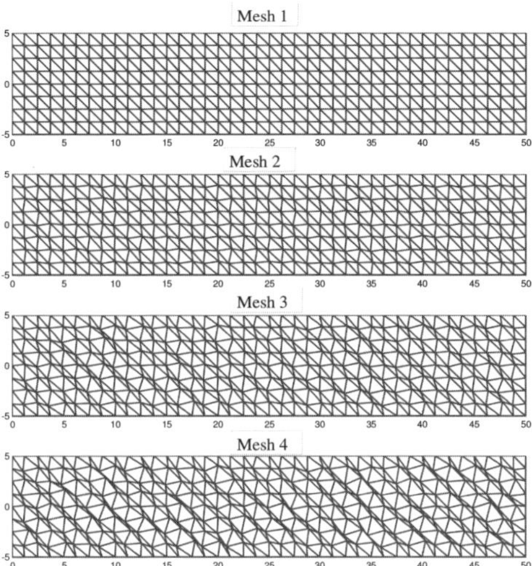

FIGURE 6.7 Four meshes of triangular mesh presentation of the rectangular cantilever with different irregularity. The values of irregularity factors for these four models are 0.0, 0.1, 0.2 and 0.3 respectively.

The numerical results of deflection along the neutral line and shear stress along the line of $x = L/2$ of the beam are plotted in FIGURE 6.8a and FIGURE 6.8b, respectively, together with the analytical solutions. It can be found that the numerical results, including the deflections and shear stress components, obtained using NS-PIM-Tr6/3 with these four meshes are all in good agreement with the analytical ones. The irregularity of the nodal distribution has little effect on the numerical results. The NS-PIM is insensitive to the mesh distortion.

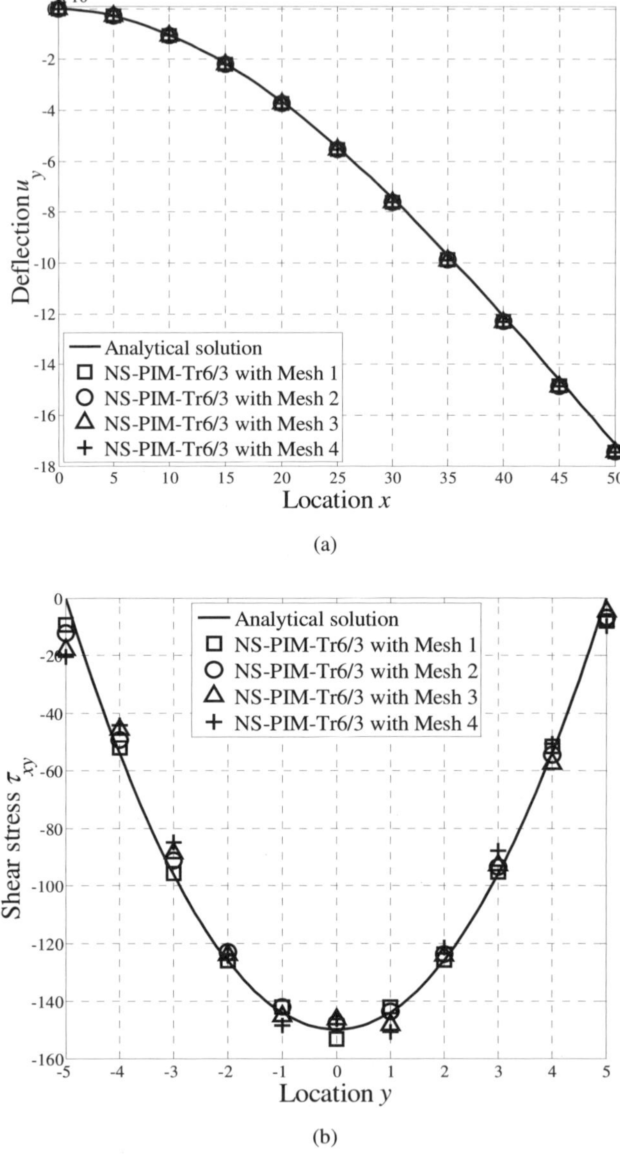

(a)

(b)

FIGURE 6.8 Comparison of the numerical results of NS-PIM-Tr6/3 with triangular meshes of different irregularity with the analytical solutions: (a) deflection u_y along the neutral line of the cantilever; (b) distribution of shear stress along the line of $x=L/2$ of the cantilever.

Remark 6.7   NS-PIM: insensitive to mesh distortion

Numerical results of NS-PIM models are insensitive to mesh distortion. This implies NS-PIM works well with triangular mesh and does not even demand for quality of the mesh.

To investigate the rate of the NS-PIM solution converging to the exact one, four mesh models with different nodal density are used to compute the errors in both displacement and energy norms. To remove possible "coincidence" in the numerical tests, we use irregularly distributed nodes (shown in FIGURE 6.9) which can be automatically generated for any 2D domain with complicated shape. Note that this set of meshes will be used to examine all the S-PIM models presented in this book for the problem of rectangular cantilever to provide a common ground of comparison.

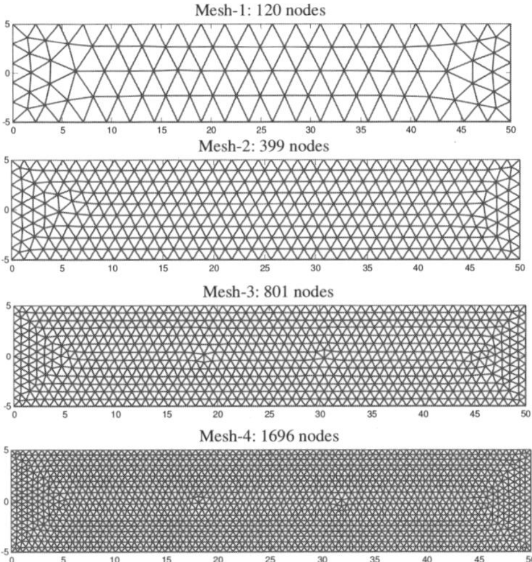

FIGURE 6.9 Domain discretization of the 2D rectangular cantilever using irregular triangular meshes.

Four numerical models, including linear FEM (FEM-Tr3), linear NS-PIM with cell-based T3-scheme (NS-PIM-Tr3), bilinear NS-PIM with edge-based T4-scheme (NS-PIM-Tr4-CT) and quadratic NS-PIM with cell-based T6/3-scheme (NS-PIM-Tr6/3) are used to study this cantilever problem with the same set of meshes. Using a particular mesh (Mesh-2 of 399 field nodes), we first study the problem using three NS-PIM models. The numerical results of both displacements and shear stress at those nodes along two particular lines, i.e. the neutral line and the middle line of $x = L/2$, are plotted in FIGURE 6.10. It is seen that the numerical results are all in good agreement with the analytical ones.

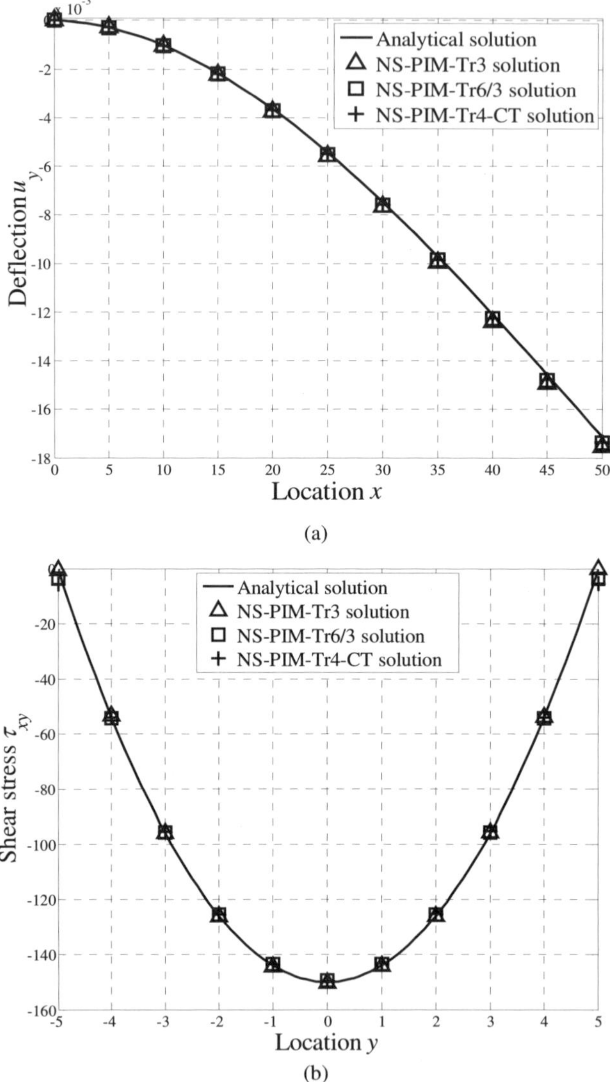

(a)

(b)

FIGURE 6.10 Comparison of numerical results obtained using NS-PIM models for the cantilever with the analytical ones: (a) deflection along the neutral line of the cantilever; (b) shear stress along the line of $x = L/2$ of the cantilever.

The convergence of the solutions errors for the NS-PIM models are computed using Equations (5.88) and (5.94) with the four meshes shown in FIGURE 6.9. FIGURE 6.11 plots in the logarithm scale the solution error in displacement norm with different density of the mesh measured by the characteristic length h as described in Section 1.6.2. The errors in displacement norm for the numerical results obtained using the same set of triangular mesh (Mesh-4 of 1,696 nodes) are listed in TABLE 6.5. Considering the computational cost given in TABLE 6.3 for these methods, the estimated computational efficiency for each method is listed in TABLE 6.5, which is measured by calculating the reciprocal of the product of solution error and computational cost (see Section 1.2.2). Following points can be observed clearly for this case.

1) The error decreases almost linearly with the decrease of h, showing a nice monotonic convergence for all these models.

2) In terms of convergence rate, we estimated the numerical convergence rates (CR) from the slop of the plot. It is found that the linear FEM achieves a numerical rate of 2.0 that is the same as the theoretical rate for linear displacement weak-form methods (see Section 3.9). The linear NS-PIM achieves a rate of 2.04, the bilinear NS-PIM achieved 1.98 and the quadratic NS-PIM achieved 2.09. These rates for the NS-PIM models are all around the theoretical value of 2.0.

3) The convergence rates for the NS-PIM-Tr3 and NS-PIM-Tr6/3 models is a little above the theoretical value of 2.0 showing a weak *superconvergence*[*] even in displacement norm.

4) In terms of accuracy, the linear NS-PIM-Tr3 has almost exactly the same accuracy as the linear FEM in displacement norm for this problem. The bilinear and quadratic NS-PIM models are about twice more accurate than the linear FEM, when Mesh-4 is used.

5) In terms of computational efficiency (see TABLE 6.5), the linear NS-PIM-Tr3 is about 20%, and the NS-PIM-Tr6/3 will be about 30% less efficient than the linear FEM-Tr3 model, in displacement norm measure. The bilinear NS-PIM-Tr4-CT stands out clearly for this problem, which is about 1.4 times more efficient than the linear FEM and 1.7 times more efficient than the linear NS-PIM.

---

[*] The superconvergence is defined in this book loosely as a rate that exceeds the theoretical rate.

TABLE 6.5 Estimated computational efficiency of different NS-PIM models measured in *displacement norm* error for the numerical results of the cantilever beam problem with the same set of triangular mesh (Mesh-4 of 1,696 nodes)

Numerical method	Solution error	Error ratio to FEM-Tr3	Efficiency
FEM-Tr3	5.2334E-03	1.00	1.0
NS-PIM-Tr3	5.5340E-03	1.06	0.8
NS-PIM-Tr4-CT	3.0198E-03	0.58	1.4
NS-PIM-Tr6/3	3.2376E-03	0.62	0.7

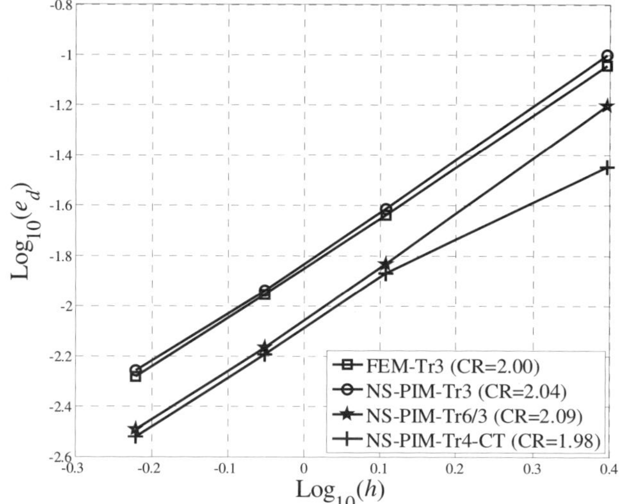

FIGURE 6.11 Comparison of convergence rates and accuracy of the numerical results in displacement norm obtained using linear FEM and three NS-PIM models for the rectangular cantilever problem.

FIGURE 6.12 plots also in the logarithm scale the solution error in energy norm with respect to the characteristic nodal spacing h. Considering both the computational cost given in TABLE 6.3 and the errors in energy norm for the numerical results listed in TABLE 6.6 which are obtained using the same set of triangular mesh (Mesh-4), the estimated computational efficiency measured in energy norm is calculated and given in the last column in TABLE 6.6. It is observed again for this case we have the following points.

1) The error decreases almost linearly with the decrease of the nodal spacing, showing a nice monotonic convergence also in energy norm for all these models.

2) The convergence rates (CR) are also estimated numerically from the slop of the plot for these models. The linear FEM achieves a rate of 0.99 that is very close to the theoretical rate of 1.0 (see Section 3.9) for linear displacement weak-form methods. The rates for these three NS-PIM models are found as 1.38, 1.46 and 1.35, respectively. They are all above the theoretical value of 1.0 showing a quite strong *superconvergence* in energy norm. It is found that these rates are quite close to the ideal theoretical convergence rate of 1.5 for a GS-Galerkin model (see Remark 3.23). The rate of convergence for the quadratic NS-PIM is slightly higher than the linear and bilinear NS-PIMs.

3) In terms of accuracy, the three NS-PIM models are about five times more accurate than the linear FEM in energy norm for this problem, when Mesh-4 is used. The accuracy of the bilinear NS-PIM is slightly higher than the other two NS-PIM models.

4) In terms of computational efficiency (see TABLE 6.6), the linear NS-PIM-Tr3 will be about 4.0 times, the bilinear NS-PIM-Tr4-CT will be 4.6 times, and the quadratic NS-PIM-Tr6/3 will be 2.2 times more efficient than the linear FEM. Compared to the performance in terms of displacement results, we may find that these NS-PIM models have much better performance about stress results.

5) Because of the much higher convergence rate, the NS-PIM models are expected to become even much more efficient than the FEM-Tr3, when a finer mesh is used; and hence we can conclude that the NS-PIM will be more efficient than the FEM-Tr3, when fine meshes are used and the measure is in energy norm. Note that the energy norm measures in weak form and W^2 form are not exactly the same, as detailed in Section 5.8.2.

TABLE 6.6 Estimated computational efficiency of different NS-PIM models measured in *energy norm* error for the numerical results for the cantilever beam problem with the same set of triangular mesh (Mesh-4 of 1,696 nodes)

Numerical method	Solution error	Error ratio to FEM-Tr3	Efficiency
FEM-Tr3	2.4406E-01	1.00	1.0
NS-PIM-Tr3	5.0076E-02	0.21	4.0
NS-PIM-Tr4-CT	4.3389E-02	0.18	4.6
NS-PIM-Tr6/3	4.6204E-02	0.19	2.2

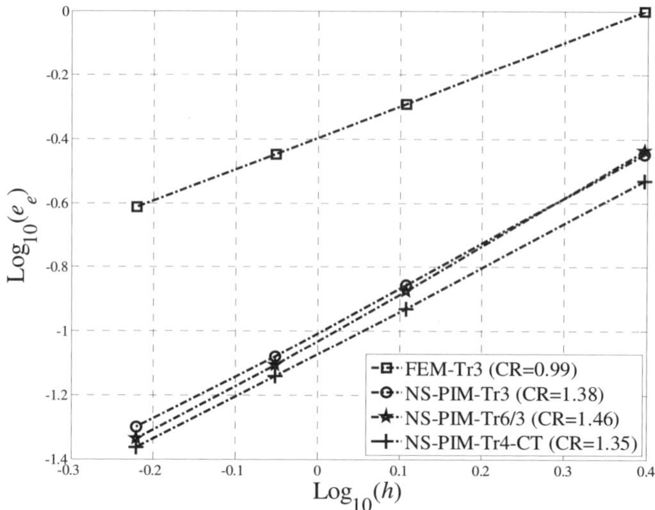

FIGURE 6.12 Comparison of convergence rates and accuracy of the numerical results in energy norm obtained using linear FEM and three NS-PIM models for the rectangular cantilever problem.

Remark 6.8　Bilinear NS-PIM-Tr4-CT: the best NS-PIM model

We conclude that the bilinear NS-PIM-Tr4-CT is the best in terms of overall performance among all these three NS-PIM models found so far. Compared to the linear FEM-Tr3, it is about 1.4 times more efficient in displacement norm, and about 4.6 times more efficient in energy norm.

The outstanding performance of the NS-PIM-Tr4-CT in efficiency is partially because the fact that it has the same sparseness as the NS-PIM-Tr3 and hence roughly the same cost, but it gives much more accurate solutions.

Remark 6.9　Superconvergence in both displacement and energy norms

The so-called *superconvergence* refers to a phenomenon in FEM that solution in energy norm at some points in the elements converges faster than the predicted theoretical value [26]. In a weak formulation, such a superconvergence will be observed in energy norm, but not in displacement norm. In a weakened weak formulation, such as the NS-PIM models, it can be observed in both displacement and energy norms.

Example 6.1.3 Infinite solid with a circular hole

Another often used benchmark problem is an infinite solid with a central circular hole ($r=a$) and subjected to a unidirectional tensile (T_x), as shown in FIGURE

6.13a. Due to the two-fold symmetry, only one quarter of the hole centralized square with side length b is modeled, as shown in **FIGURE 6.13b**, and symmetric conditions are imposed on the left and bottom edges as follows:

$$
\begin{aligned}
u_x &= 0 \quad \text{when } x = 0 \\
u_y &= 0 \quad \text{when } y = 0
\end{aligned}
\tag{6.11}
$$

The analytical solutions of this problem are available in the textbook by Timoshenko and Goodier [27]. The exact solution for the stresses is

$$
\sigma_{xx} = T_x \left\{ 1 - \frac{a^2}{r^2} \left[\frac{3}{2}\cos(2\theta) + \cos(4\theta) \right] + \frac{3a^4}{2r^4}\cos(4\theta) \right\}
\tag{6.12}
$$

$$
\sigma_{yy} = -T_x \left\{ \frac{a^2}{r^2} \left[\frac{1}{2}\cos(2\theta) - \cos(4\theta) \right] + \frac{3a^4}{2r^4}\cos(4\theta) \right\}
\tag{6.13}
$$

$$
\tau_{xy} = -T_x \left\{ \frac{a^2}{r^2} \left[\frac{1}{2}\sin(2\theta) + \sin(4\theta) \right] - \frac{3a^4}{2r^4}\sin(4\theta) \right\}
\tag{6.14}
$$

where (r,θ) are the polar coordinates and θ is measured counterclockwise from the positive x-axis. Traction boundary conditions are imposed on the right and top edges using the values of the exact solutions calculated using Equations (6.12)-(6.14). This is to make sure that the exact and numerical solutions are under the same boundary conditions and hence comparable. The analytical solutions of displacements corresponding to the stresses are

$$
u_x = \frac{T_x a}{8\mu} \left[\frac{r}{a}(\kappa+1)\cos\theta + 2\frac{a}{r}\left((1+\kappa)\cos\theta + \cos(3\theta)\right) - 2\frac{a^3}{r^3}\cos(3\theta) \right]
\tag{6.15}
$$

$$
u_y = \frac{T_x a}{8\mu} \left[\frac{r}{a}(\kappa-3)\sin\theta + 2\frac{a}{r}\left((1-\kappa)\sin\theta + \sin(3\theta)\right) - 2\frac{a^3}{r^3}\sin(3\theta) \right]
\tag{6.16}
$$

where μ is the shear modulus and κ is the Kolosov constant defined as

$$
\mu = \frac{E}{2(1+v)}
\tag{6.17}
$$

$$
\kappa = \begin{cases} 3-4v & (\text{Plane strain}) \\ \dfrac{3-v}{1+v} & (\text{Plane stress}) \end{cases}
\tag{6.18}
$$

When the condition $b/a=5$ is satisfied, the solution of this finite square panel should be very close to that of an infinite 2D solid. Therefore we use $a=1$, $b=5$ and $T_x = 10$ in all numerical models.

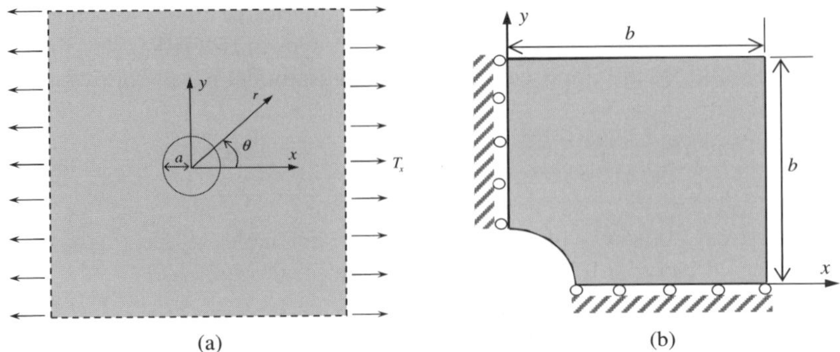

(a)   (b)

FIGURE 6.13 Infinite solid with a circular hole subjected to a tensile load in the horizontal direction: (a) problem setting; (b) a quarter model.

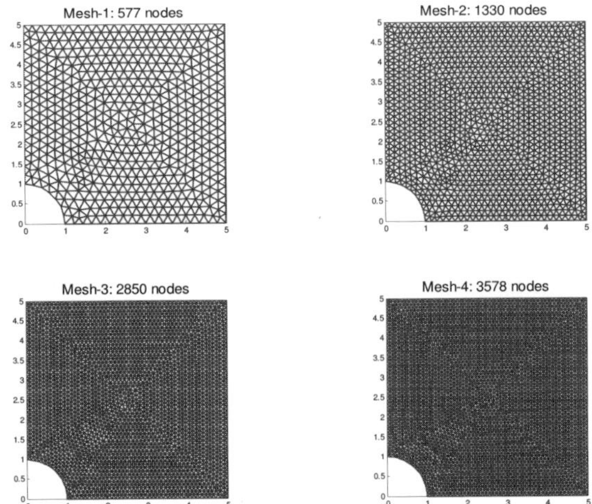

FIGURE 6.14 Domain discretization of the infinite solid with circular hole using a set of irregular triangular meshes.

A set of triangular meshes of the quarter model have been used in the present numerical study, as shown in FIGURE 6.14. First, plane stress problem is considered and Mesh-2 in FIGURE 6.14 is used. The displacement components for nodes located along the bottom and left edges are calculated and plotted in FIGURE 6.15a and FIGURE 6.15b, respectively. FIGURE 6.16 shows the distribution of normal stresses at the nodes located along the bottom and left edges respectively. These figures show that all the numerical results, including displacement and stress components, obtained using the NS-PIM models agree well with the analytical ones.

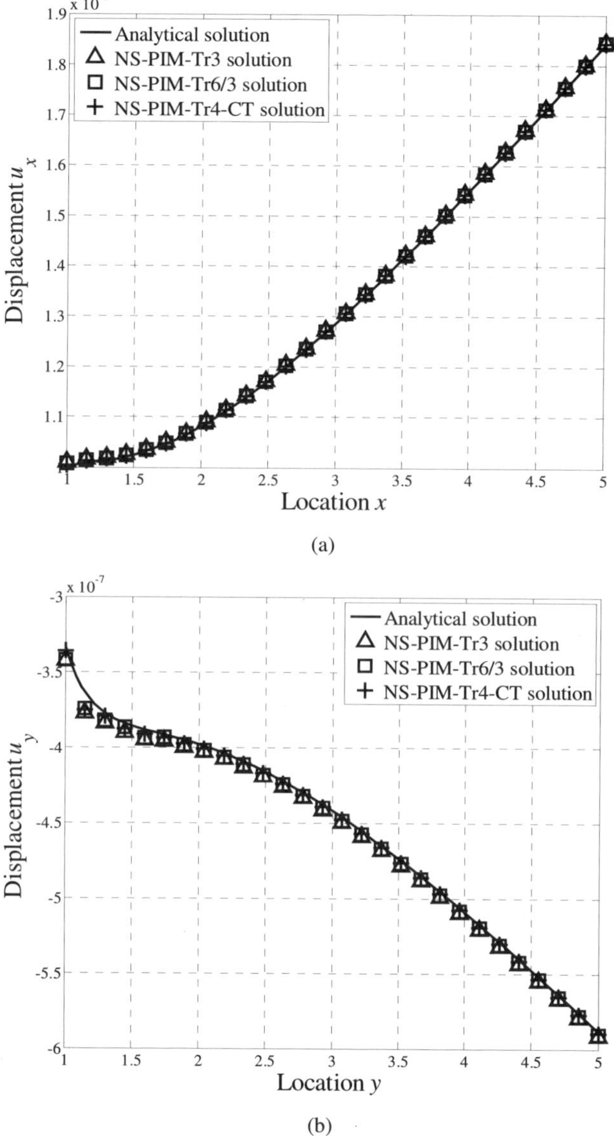

(a)

(b)

FIGURE 6.15 Displacements distribution along the bottom and the left edges for the quarter model of the infinite solid with hole: (a) displacement (u_x) for nodes located along the bottom edge; (b) displacement (u_y) for nodes located along the left edge.

 Smoothed Point Interpolation Methods

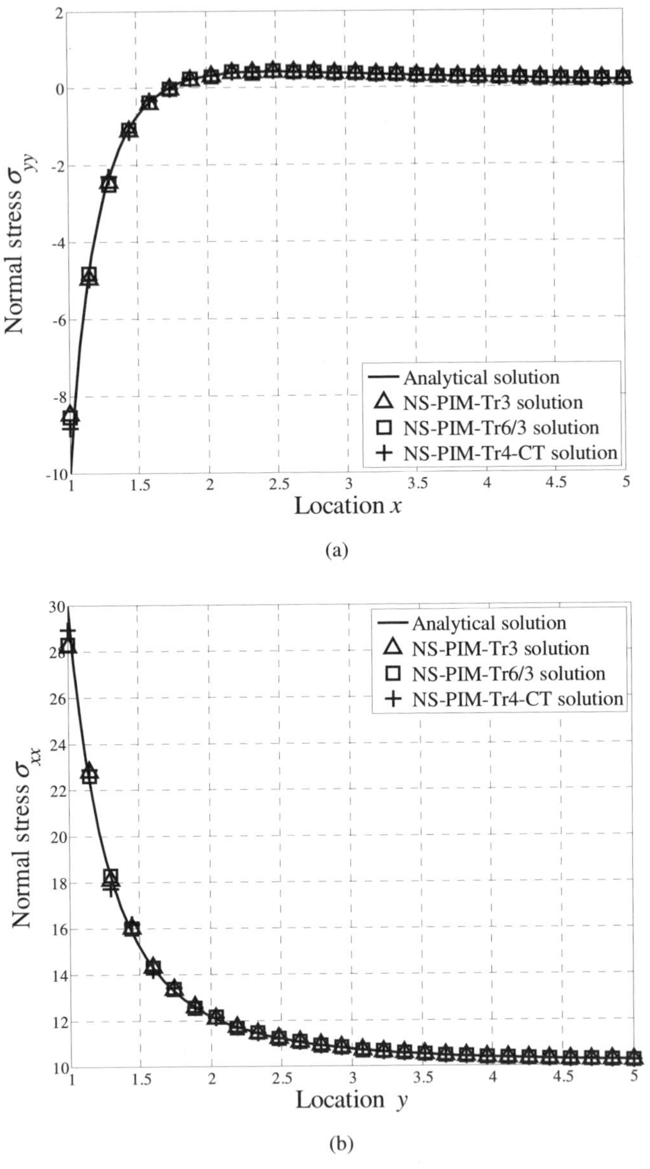

FIGURE 6.16 Stresses distribution along bottom and left edges for the quarter model of the infinite solid with hole: (a) normal stresses in y-direction for nodes located along the bottom edge; (b) normal stresses in x-direction for nodes located along the left edge.

To investigate the convergence property, four numerical models, FEM-Tr3, NS-PIM-Tr3, NS-PIM-Tr4-CT and NS-PIM-Tr6/3 are used to analyze this problem with same set of meshes shown in FIGURE 6.14. FIGURE 6.17 plots in the logarithm scale the solution error in displacement norm with different density of the mesh measured by the characteristic nodal spacing h. Considering the computational cost given in TABLE 6.3 and the displacement norm errors of the numerical results obtained using the same triangular mesh (Mesh-4 of 3,578 nodes), the estimated computational efficiency measured in displacement norm are calculated and listed in the last column in TABLE 6.7. From these results, we have the following points.

1) Both the FEM and NS-PIM results have shown a nice monotonic convergence for this problem.
2) In terms of convergence rate, the linear FEM achieves 1.92, which is very close to the theoretical one of 2.0. The linear NS-PIM-Tr3 achieves the same convergence rate as the FEM for this problem. The convergence rate for the bilinear NS-PIM-Tr4-CT is 1.86, which is a little lower than the linear FEM, but is also close to the theoretical value. The quadratic NS-PIM-Tr6/3 model achieves 2.2 of the convergence rate, which is above the theoretical value of 2.0 due to the higher order interpolation.
3) In terms of accuracy, the linear NS-PIM-Tr3 is about 20% more accurate than the linear FEM; while the bilinear and quadratic NS-PIM models are about twice more accurate than the FEM-Tr3, when Mesh-4 is used, as listed in TABLE 6.7.
4) In terms of efficiency (see TABLE 6.7), the linear NS-PIM-Tr3 has the same efficiency as the linear FEM, and the NS-PIM-Tr6/3 is 10% less efficient than the linear FEM-Tr3 model, in displacement norm measure. The NS-PIM-Tr4-CT stands out clearly again which is about 1.9 times more efficient than the FEM-Tr3.

TABLE 6.7 Estimated computational efficiency of different NS-PIM models measured in *displacement norm* error for the numerical results for the problem of infinite solid with a circular hole using the same triangular mesh (Mesh-4 of 3,578 nodes)

Numerical method	Solution error	Error ratio to FEM-Tr3	Efficiency
FEM-Tr3	1.4435E-03	1.00	1.0
NS-PIM-Tr3	1.1520E-03	0.80	1.0
NS-PIM-Tr4-CT	6.3699E-04	0.44	1.9
NS-PIM-Tr6/3	7.0986E-04	0.49	0.9

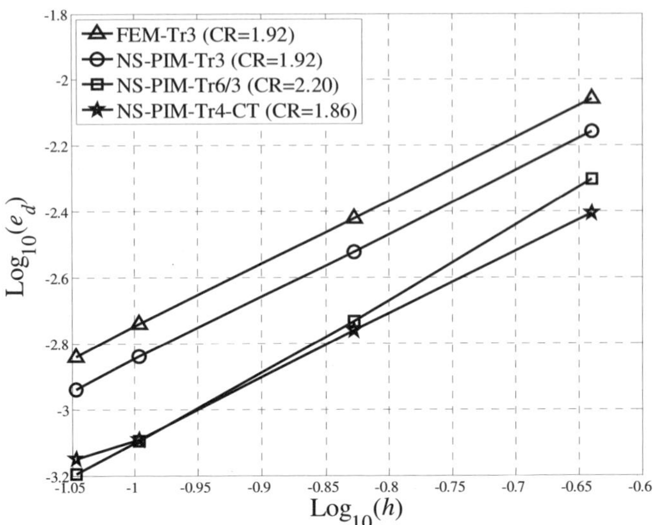

FIGURE 6.17 Comparison of convergence rates and accuracy of the numerical results in displacement norm obtained using linear FEM and three NS-PIM models for the infinite solid with circular hole problem.

FIGURE 6.18 plots the solution error of the numerical results in energy norm for meshes of various node densities. The errors in the energy norm for the numerical results obtained using the same set of triangular mesh (Mesh-4 of 3,578 nodes) are listed in TABLE 6.8. Considering the computational cost given in TABLE 6.3 for these methods, the estimated computational efficiency measured in energy norm measure are calculated and given in the last column in TABLE 6.8. The following points may be noted for this example.

1) All the three NS-PIM models have clearly better performance in energy norm measure than the linear FEM in terms of both accuracy and convergence rate.

2) In terms of convergence rate, the linear FEM achieves 0.95, which is very close to the theoretical value of 2.0. The convergence rates for the linear, bilinear and quadratic NS-PIM models are 1.4, 1.36 and 1.47, respectively, which are very close to the ideal rate for W^2 models and much higher than the linear FEM showing quite strong superconvergence.

3) In terms of accuracy, the three NS-PIM models are about 2.7 times more accurate than the linear FEM, when the finest Mesh-4 is used.

4) Considering the computational cost in TABLE 6.3, the three NS-PIMs are about 2.3, 2.3 and 1.2 times more efficient than the linear FEM.

TABLE 6.8 Estimated computational efficiency of different NS-PIM models measured in *energy norm* error for the numerical results for the problem of infinite solid with circular hole using the same triangular mesh (Mesh-4 of 3,578 nodes)

Numerical method	Solution error	Error ratio to FEM-Tr3	Efficiency
FEM-Tr3	1.3831E-04	1.00	1.0
NS-PIM-Tr3	5.0516E-05	0.37	2.3
NS-PIM-Tr4-CT	4.9617E-05	0.36	2.3
NS-PIM-Tr6/3	5.0352E-05	0.36	1.2

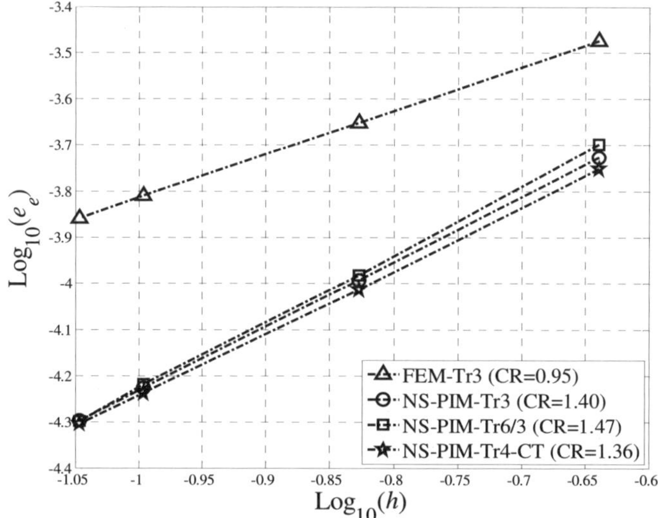

FIGURE 6.18 Comparison of convergence rates and accuracy of the numerical results in energy norm obtained using linear FEM and three NS-PIM models for the infinite solid with circular hole problem.

We observe again the outstanding performance of the NS-PIM-Tr4-CT: it has the same sparseness as the NS-PIM-Tr3 and hence roughly the same cost but gives much more accurate solutions for this problem. In terms of the computational efficiency, the NS-PIM-Tr4-CT shall be the best among all these three NS-PIM models.

Finally, we study the volumetric locking issue (see Remark 1.2) using this example. Plane strain problem is now considered and the Poisson's ratio is set to vary from 0.4 to 0.4999999. The NS-PIM models are used to solve this problem together with the linear FEM for comparison. Mesh-2 shown in FIGURE 6.14 is used for all the methods.

FIGURE 6.19a plots the error of numerical solution in displacement norm against Poisson's ration varying from 0.4 to 0.4999999 and TABLE 6.9 gives the detailed values. FIGURE 6.19b plots the error of numerical solution in energy norm. It can be clearly found that the linear FEM suffers from the volumetric locking: when the Poisson's ratio is larger than 0.4, the solution error increases drastically, as predicated in Remark 1.2. Special treatments are needed in FEM for this type of problems. However, the results show that all these three NS-PIM models are naturally immune from the volumetric locking for nearly incompressible materials: the accuracy of the numerical results has not been affected by the increasing incompressibility without any additional treatment.

Remark 6.10 NS-PIM: Immune from volumetric locking

NS-PIM is naturally immune from the volumetric locking, and no special treatments are needed for solving mechanics problem of solids with nearly incompressible materials. It behaves like an equilibrium model in this regard.

TABLE 6.9 Error in displacement norm of numerical solutions for the 2D plane strain problem with Poisson's ratio approaching to 0.5

Poisson's ratio	FEM solution	NS-PIM-Tr3 solution	NS-PIM-Tr6/3 solution	NS-PIM-Tr4-CT solution
0.4	0.45740639E-02	0.29932681E-02	0.18795295E-02	0.17600807E-02
0.49	0.17083257E-01	0.29433390E-02	0.19133028E-02	0.17876324E-02
0.499	0.52619287E-01	0.29460681E-02	0.19377587E-02	0.18067138E-02
0.4999	0.81990377E-01	0.29466914E-02	0.19410219E-02	0.18091744E-02
0.49999	0.88546719E-01	0.29467582E-02	0.19413586E-02	0.18094273E-02
0.499999	0.89356594E-01	0.29467643E-02	0.19413909E-02	0.18094527E-02
0.4999999	0.89440415E-01	0.29467622E-02	0.19413994E-02	0.18094546E-02

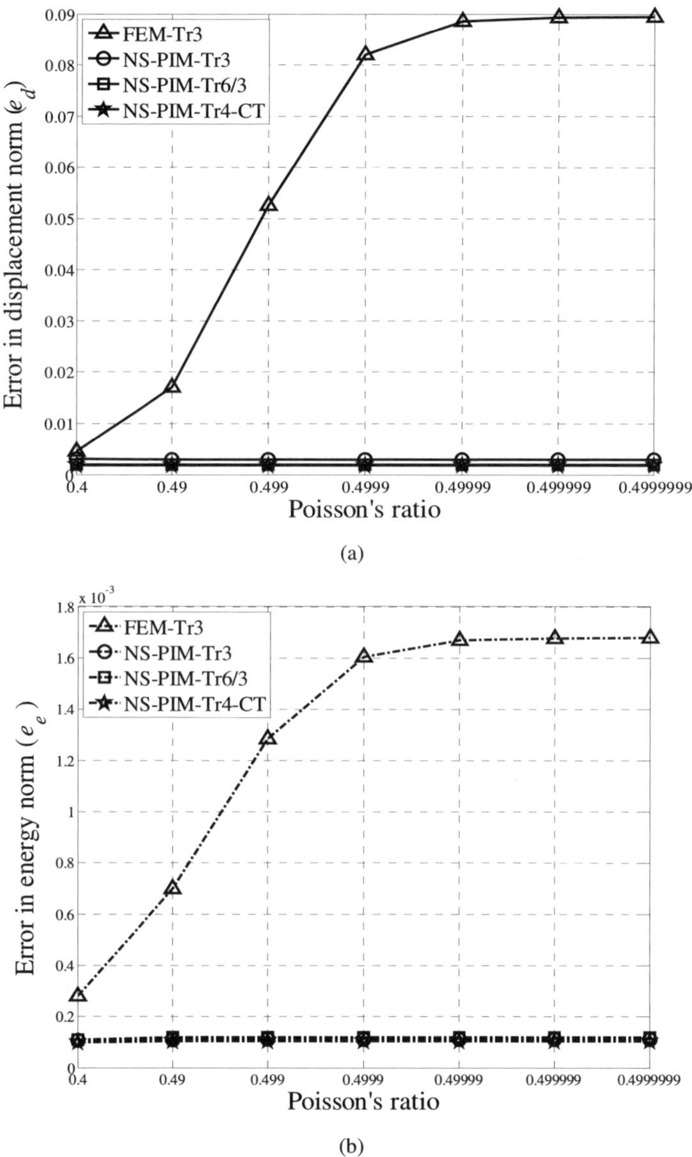

FIGURE 6.19 Error of solution against Poisson's ratio changing from 0.4 to 0.4999999. FEM solution is locked when Poisson's ration larger than 0.4, but NS-PIM models are naturally immune from the volumetric locking: (a) solution error in displacement norm; (b) solution error in energy norm.

Example 6.1.4 A mechanical part: 2D rim

As an application of NS-PIMs to practical design problem, a typical rim model of automotive component with a complicated shape is now studied using NS-PIM models. As shown in FIGURE 6.20a, the rim model is fixed at the nodes located along the inner circle and a pressure of 100 units is applied along a portion of the lower arc edge in the range of 60°.

The rim is considered as plane stress problem and meshed using triangular cells with 2,608 nodes, as shown in FIGURE 6.20b. As analytical solutions are not available for this problem, it is very difficult to examine the solution error in either the displacement and energy norm. We therefore examine the distribution of the solutions in displacements and stresses. For comparison purpose, a reference solution is obtained using the FEM with a very fine mesh of six-node triangular elements with total 341,188 elements. Displacement and stress results at the nodes located along the lower half circle of the rim termed as *M-N*, are computed using the bilinear NS-PIM-Tr4-CT model and plotted in FIGURE 6.21 and FIGURE 6.22, together with the reference solutions. For this practical problem with complicated shape, we find that the bilinear NS-PIM solutions in both displacements and stresses are in good agreement with the references.

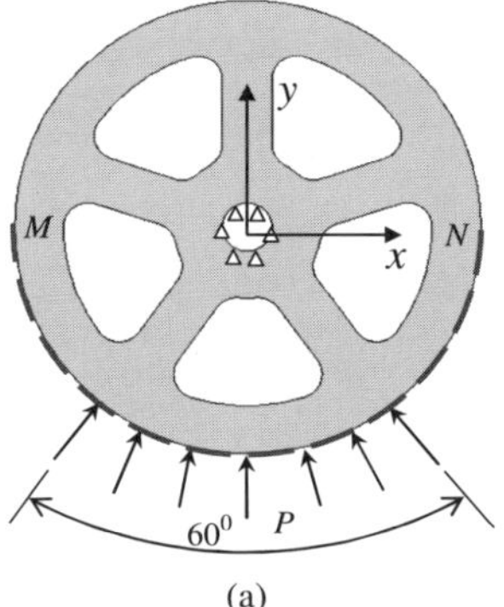

(a)

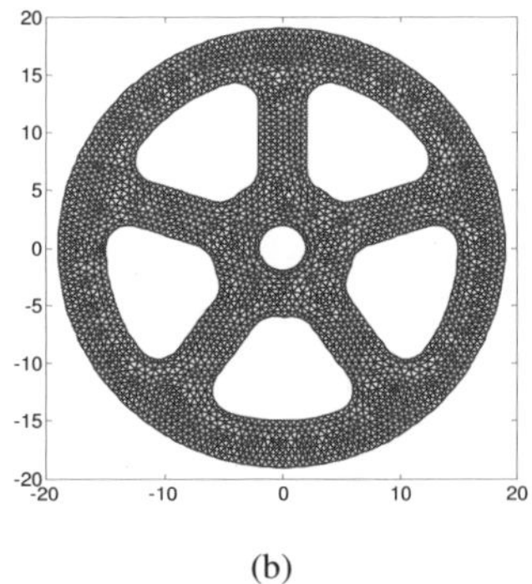

(b)

FIGURE 6.20 An automotive rim subjected to pressure along a portion of the lower arc edge of 1/3 circumferential length: (a) problem setting; (b) domain discretization with triangular mesh of 2,608 nodes.

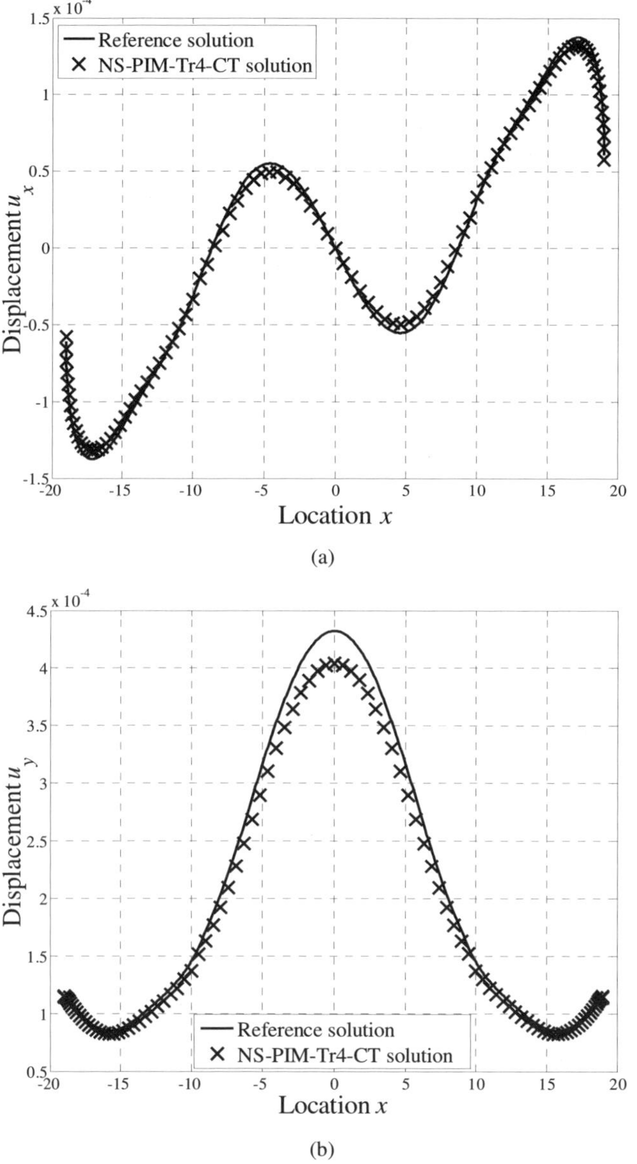

FIGURE 6.21 Displacements distribution along the *M-N* line of the rim: (a) displacements in *x* direction; (b) displacements in *y* direction.

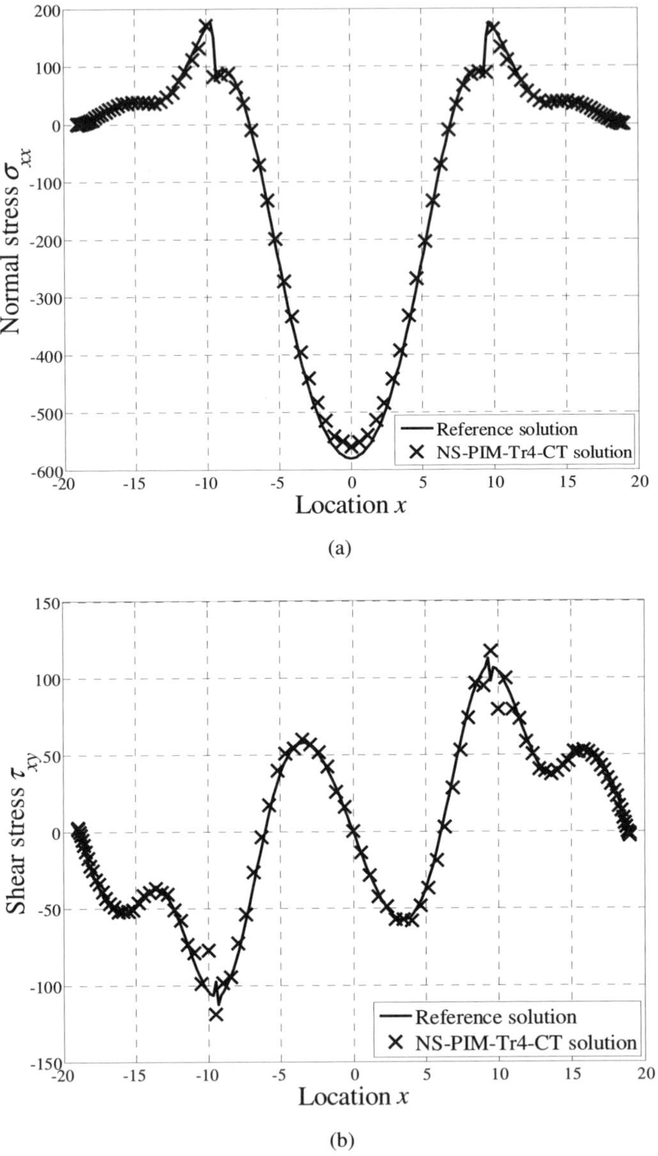

(a)

(b)

FIGURE 6.22 Stresses distribution along the *M-N* line of the rim: (a) distribution of normal stresses σ_{xx} and (b) shear stresses τ_{xy}.

6.2 NS-RPIM for 2D solids

6.2.1 Considerations

Chapter 3 demonstrated that the use of radial functions as basis functions (with proper shape parameters) can guarantee a nonsingular moment matrix, and shape functions with Delta function property can be created via simple point interpolation method for virtually randomly distributed nodes. In this section, we introduce the so-called node-based smoothed radial point interpolation method or NS-RPIM for short, which uses node-based strain smoothing operation and PIM shape functions constructed using radial basis functions instead of polynomial basis functions. The method was called originally the linearly conforming radial point interpolation method (LC-RPIM) [17, 18, 28], because it is at least linearly conforming. We need now to rename it as NS-RPIM, because the other methods developed later, such as the ES-PIM, CS-PIM, etc, are all linearly conforming but very much different in the use of smoothing domains, and properties.

6.2.2 Support node selection

The material of this section is mostly based on the work of Refs. [17, 18, 28]. The formulation of NS-RPIM is largely the same as that of the NS-PIM, except the procedure of the support node selection and shape function creation. So the NS-RPIM follows the same flowchart as the NS-PIM for both 2D and 3D problems and all we need to do is to replace the polynomial PIM shape functions with the radial PIM shape functions. In this work, we use RPIM shape functions with linear polynomial basis to restore the (polynomial) reproducibility, which has been presented in Section 3.3.

As presented in Chapter 1, we use the triangular cell based node selection schemes (short as T-schemes and detailed in Section 1.6.3) to select support nodes for constructing PIM shape functions for 2D problems. Following the procedure presented in Section 3.3, RPIM shape functions (with proper parameters) can always be created using a small number of support nodes. In this work, we use the cell-based T6-scheme and T2L-scheme to select support nodes for constructing 2D RPIM shape functions.

When we use the cell-based T6-scheme, as shown in FIGURE 6.23a which is the same as that summarized in FIGURE 1.7c, to select support nodes for constructing RPIM shape functions, the approximated displacement field will be discontinuous and the discontinuity will occur on the interfaces of triangular background cells, which are marked with dished lines in the figure. When the cell-based T2L-scheme is used to select support nodes for constructing RPIM shape functions, the approximated displacement field has the same discontinuity feature, as shown FIGURE 6.23b which is the same as that in FIGURE 1.7d.

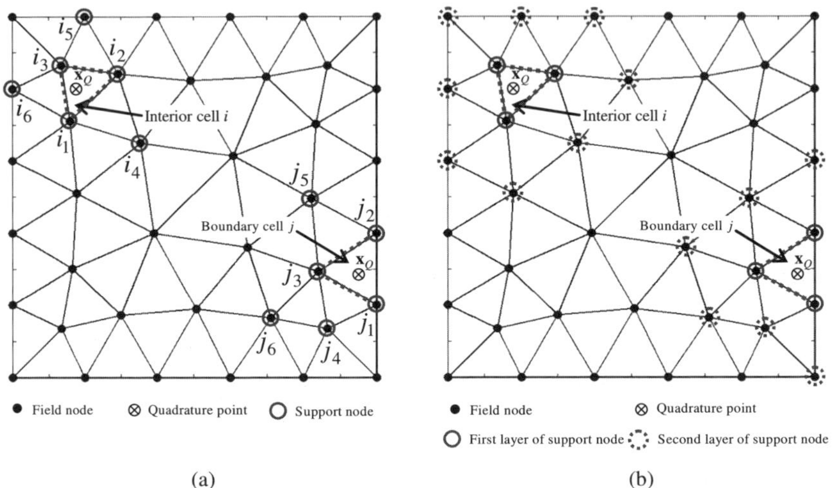

FIGURE 6.23 T-schemes used for selecting support nodes for the construction of 2D RPIM shape functions: (a) cell-based T6-scheme; (b) cell-based T2L-scheme.

6.2.3 Possible 2D NS-RPIM models

According to different cell-based T-schemes using for those quadrature points located within background cells, the NS-RPIM has the following two models for 2D problems.

NS-RPIM-Tr6

A NS-RPIM-Tr6 uses RPIM shape functions and support nodes selected by the cell-based T6-scheme, as shown in FIGURE 6.23a. The cell-based T6-scheme is well controlled, and hence is very robust and efficient. The equally-shared smoothing domain is preferred for this scheme, for better efficiency and the convenience in implementation. Note that the cell-based T3-scheme used in NS-PIM can also be used in the present NS-RPIM. However, the NS-RPIM model so constructed produces the same results as the linear NS-PIM-Tr3, and hence we do not use the cell-based T3-scheme for NS-RPIM, except for debugging purpose. The displacement field in the NS-RPIM-Tr6 is incompatible.

To find out the general support nodes for each node-based smoothing domain, two meshes of the cantilever beam shown in FIGURE 6.4 are also used. The number of the support nodes for a smoothing domain of the NS-RPIM-Tr6 model is about 6-15 and the average number is 11.5, that is about 3.8 times that of the FEM-Tr3. Therefore, n_{ze} for an NS-RPIM-Tr6 model is about 3.8 times that of the FEM-Tr3 counterpart using the same mesh, as shown in TABLE 6.10.

NS-RPIM-Tr2L

A NS-RPIM-Tr2L uses RPIM shape functions with the support nodes selected by the cell-based T2L-scheme, as shown in FIGURE 6.23b. This type of T-scheme is also well controlled, and with a lot of freedom. This is a very robust scheme for node selection which works well for virtually randomly distributed nodes, but less efficient than the cell-based T6-scheme. The displacement field in the NS-RPIM-Tr2L is incompatible.

For this model, the number of the support nodes for a smoothing domain is about 8-22 and the average number is 16, which is about 5.3 times that of the FEM-Tr3. Therefore, n_{ze} for an NS-RPIM-Tr2L model is about 5.3 times that of the FEM-Tr3 counterpart using the same mesh, as listed in TABLE 6.10.

TABLE 6.10 NS-RPIM models and overall characteristics in relation to the FEM-Tr3 model using the same mesh

Numerical method	Support nodes in an integral cell/element	Estimated average number of nonzero entries	Estimated Solver CPU time
FEM-Tr3	3	n_{ze}	t_{CPU}
NS-RPIM-Tr6	6-15 (11.5)	$3.8\,n_{ze}$	$3.8\,\bar{n}_{iter}t_{CPU}$
NS-RPIM-Tr2L	8-22 (16)	$5.3\,n_{ze}$	$5.3\,\bar{n}_{iter}t_{CPU}$

6.2.4   Condition number of NS-RPIM models

TABLE 6.11 lists the condition numbers of the global stiffness matrixes for two NS-RPIM models in relation to the FEM-Tr3 using the same triangular meshes for the cantilever beam problem. We may find that both NS-RPIM models have smaller condition numbers than the linear FEM-Tr3, although they use RPIM shape functions with more support nodes as shown in TABLE 6.10.

Similar to previous studies, the ratio between the condition numbers of different methods has little change with the increase of the DOFs. In the following analysis, we will use the relative $\bar{n}_{iter}$ given in the last column in TABLE 6.11.

TABLE 6.11 Condition number of the global stiffness matrixes for various NS-RPIM models for the 2D cantilever beam in relation to the FEM-Tr3 model using the same mesh

Numerical method	Mesh-1 of 120 nodes		Mesh-2 of 248 nodes	
	$cond(\mathbf{K})$	$\bar{n}_{iter}$	$cond(\mathbf{K})$	$\bar{n}_{iter}$
FEM-Tr3	1.645e+08	1.00	2.399e+08	1.00
NS-RPIM-Tr6	1.169e+08	0.84	1.613e+08	0.82
NS-RPIM-Tr2L	1.043e+08	0.80	1.535e+08	0.80

6.2.5 Estimation of computational cost for 2D NS-RPIM

Combining TABLE 6.10 and TABLE 6.11 we shall have a rough estimation of computational cost for the 2D NS-RPIM models in relation to the standard FEM-Tr3 model using the same mesh, as shown in TABLE 6.12. It is seen that these two NS-RPIM models, i.e. NS-RPIM-Tr6 and NS-RPIM-Tr2L, will cost about 3.1 and 4.2 time compared to the FEM-Tr3, respectively.

TABLE 6.12 Estimation of computational cost for 2D NS-RPIM models in relation to the FEM-Tr3 model using the same mesh

Numerical methods	Support nodes in an integral cell/element	Estimated Solver CPU time
FEM-Tr3	3	t_{CPU}
NS-RPIM-Tr6	6-15 (11.5)	$3.1\, t_{CPU}$
NS-RPIM-Tr2L	8-22 (16)	$4.2\, t_{CPU}$

From TABLE 6.12, it is clear that the proposed NS-RPIM models are both much slower than the FEM-Tr3 model using the same mesh. Then for these two NS-RPIM models, they must obtain more accurate results for being more efficient than the linear FEM model.

6.2.6 Numerical examples for 2D solids

Example 6.2.1 A 2D linear patch test

The linear patch test conducted in Example 6.1.1 is again studied using the NS-RPIM models with the meshes presented in FIGURE 6.5. Displacement errors of the numerical results are listed in TABLE 6.13. The results show that NS-RPIR models can pass the linear patch test within machine precision, which demonstrates numerically that the NS-RPIMs can reproduce linear fields exactly, and hence the NS-RPIM solutions will converge to the exact ones.

TABLE 6.13 Error norm in displacements of numerical results for the standard patch test obtained using NS-RPIM models

NS-RPIM models	Regular mesh	Irregular mesh
NS-RPIM-Tr6	1.015E-15	1.099E-15
NS-RPIM-Tr2L	2.303E-15	1.794E-15

Example 6.2.2 Rectangular cantilever

The benchmark rectangular cantilever problem described in Example 6.1.2 is again studied using present NS-RPIM models. FIGURE 6.24 plots the distribution of numerical solution along two particular lines of the problem, which are obtained using Mesh-2 of 399 nodes as shown in FIGURE 6.9. We can find that numerical results of both displacement and stress components agree well with the analytical ones.

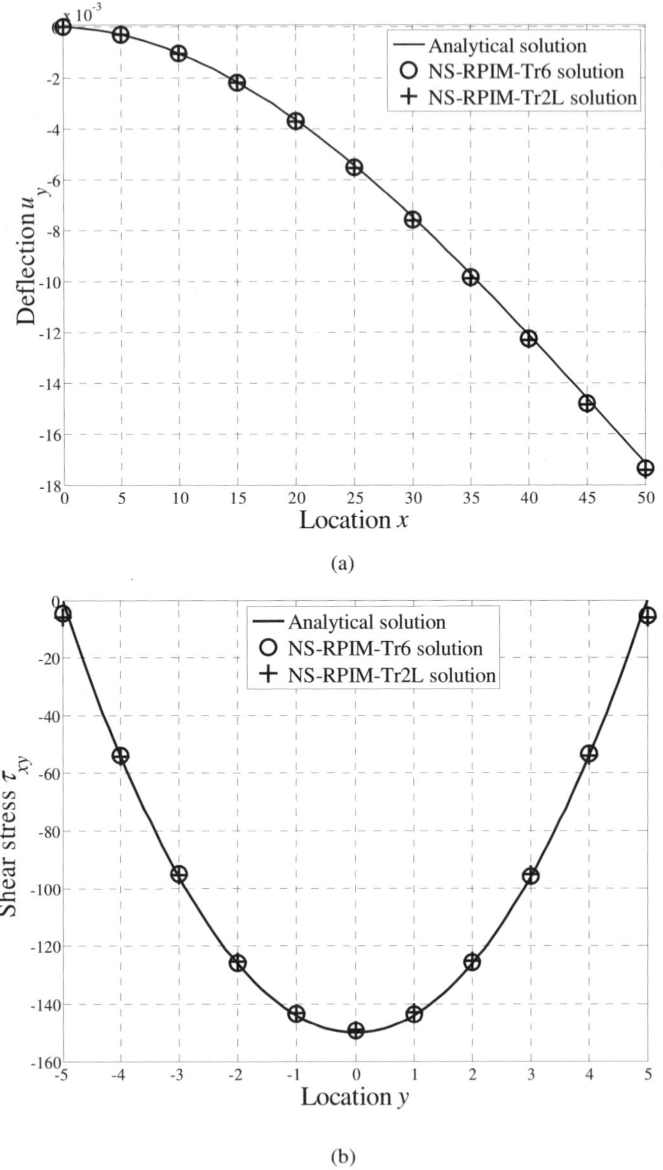

(a)

(b)

FIGURE 6.24 Comparison of numerical results obtained using NS-RPIM models for the rectangular cantilever problem with analytical ones: (a) deflection along the neutral line of the cantilever; (b) shear stress along the line of $x=L/2$ of the cantilever.

To investigate convergence of the NS-RPIM models, study is also conducted using the irregular triangular meshes shown in FIGURE 6.9. FIGURE 6.25 plots in the logarithm scale the solution error in displacement norm with different density of the mesh measured by the characteristic nodal spacing h. TABLE 6.14 lists the displacement norm error for the numerical results obtained using Mesh-4, together with the estimated computational efficiency by considering the computational cost given in TABLE 6.12. We can conclude the following points.

1) These two NS-RPIM models have shown a nice monotonic convergence measured in the displacement norm.

2) In terms of convergence rate, the NS-RPIM-Tr6 achieves 2.00 in displacement norm. The NS-RPIM-Tr2L achieves 1.98, which is very close to the theoretical value.

3) In terms of accuracy, the NS-RPIM-Tr6 is about 1.6 times, and the NS-RPIM-Tr2L is about 1.8 times more accurate than the FEM-Tr3, when Mesh-4 is used.

4) In terms of efficiency, the NS-RPIM models are about twice less efficient than the FEM-Tr3, due to the use of larger number of support nodes in computing the RPIM shape functions.

5) For comparison purpose, TABLE 6.14 lists also the results for the NS-PIM models obtained in Example 6.1.2. It is seen that NS-RPIM-Tr6 has about the same accuracy as the NS-PIM-Tr6/3, but is ~30% less efficient. NS-RPIM-Tr2L is the most accurate NS- model, but the least efficient, due to the use of many local support nodes. We note also that the NS-RPIM-Tr2L is expected most robust for irregular nodes.

TABLE 6.14 Estimated computational efficiency of different NS-PIM/NS-RPIM models measured in *displacement norm* error for the numerical results for the cantilever beam problem with the same set of triangular mesh (Mesh-4 of 1,696 nodes)

Numerical method	Solution error	Error ratio to FEM-Tr3	Efficiency
FEM-Tr3	5.2334E-03	1.00	1.0
NS-PIM-Tr3	5.5340E-03	1.06	0.8
NS-PIM-Tr4-CT	3.0198E-03	0.58	1.4
NS-PIM-Tr6/3	3.2376E-03	0.62	0.7
NS-RPIM-Tr6	3.2163E-03	0.61	0.5
NS-RPIM-Tr2L	2.9699E-03	0.57	0.4

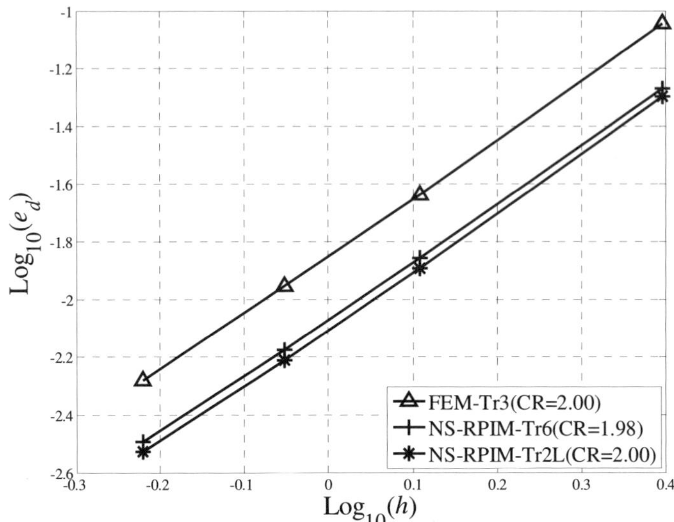

FIGURE 6.25 Comparison of convergence rates and accuracy of the numerical results in displacement norm obtained using linear FEM and two NS-RPIM models for the rectangular cantilever problem.

FIGURE 6.26 plots in the logarithm scale the solution error in energy norm with respect to the characteristic nodal spacing h. TABLE 6.15 lists the energy norm error for the numerical results obtained using Mesh-4 and the estimated computational efficiency by considering the computational cost given in TABLE 6.12. The following points may be noted for this example.

1) The energy norms error of the NS-RPIM solutions decreases linearly showing a nice monotonic convergence property.

2) The two NS-RPIM models achieve similar convergence rates of 1.44 and 1.42, which are much higher than the theoretical value of 1.0 showing a strong superconvergence in energy norm. These rates are very close to the ideal value of 1.5 for a GS-Galerkin model (see Remark 3.23).

3) These two NS-RPIM models obtain much more accurate results than the linear FEM in terms of energy norm. When Mesh-4 is used, they will be about 6 times more accurate than the FEM-Tr3, as listed in TABLE 6.15.

4) Owing to the superiority in accuracy, the NS-RIPM-Tr6 and NS-RPIM-Tr2L are about 1.9 and 1.5 times more efficient than the linear FEM, despite the much higher cost listed in TABLE 6.12.

5) Compared to the measure in displacement norm, NS-RPIM models show much better performance in the measure of energy norm.

6) For comparison purpose, TABLE 6.15 lists also the results for the NS-PIM models obtained in Example 6.1.2. It is seen that NS-RPIM-Tr6 is about 10% more accurate than NS-PIM-Tr6/3, but ~14% less efficient. NS-RPIM-Tr2L is the most accurate NS- model, but the least efficient, due to the use of many local nodes.

TABLE 6.15 Estimated computational efficiency of different NS-PIM/NS-RPIM models measured in *energy norm* error for the numerical results for the cantilever beam problem with the same set of triangular mesh (Mesh-4 of 1,696 nodes)

Numerical method	Solution error	Error ratio to FEM-Tr3	Efficiency
FEM-Tr3	2.4406E-01	1.00	1.0
NS-PIM-Tr3	5.0076E-02	0.21	4.0
NS-PIM-Tr4-CT	4.3389E-02	0.18	4.6
NS-PIM-Tr6/3	4.6204E-02	0.19	2.2
NS-RPIM-Tr6	4.0940E-02	0.17	1.9
NS-RPIM-Tr2L	3.9730E-02	0.16	1.5

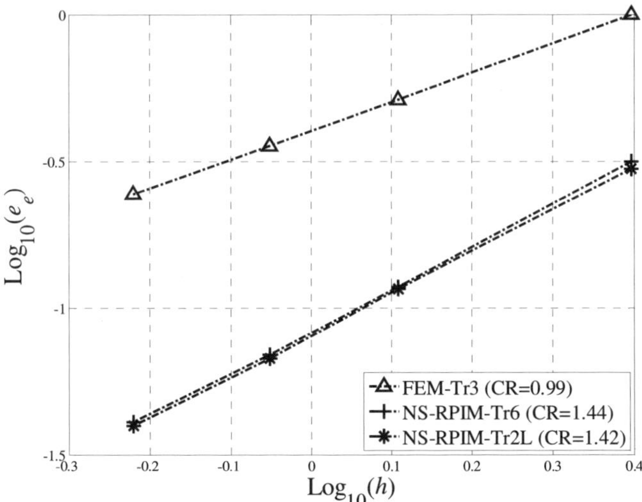

FIGURE 6.26 Comparison of convergence rates and accuracy of the numerical results in energy norm obtained using linear FEM and two NS-RPIM models for the rectangular cantilever problem.

Example 6.2.3 Infinite solid with a circular hole

The infinite solid with a central circular hole problem described in Example 6.1.3 is studied again using the NS-RPIM models. For comparison, the linear FEM is also adopted using the same irregular triangular meshes shown in FIGURE 6.14.

The convergence process of the numerical results in displacement norm is shown in FIGURE 6.27. TABLE 6.16 lists the displacement norm error of numerical results computed using Mesh-4 and the estimated efficiency considering the computational cost in TABLE 6.12. We have the following points.

1) The NS-RPIM models show monotonic convergence also for this problem. The NS-RPIM-Tr6 achieves the convergence rate of 2.23 showing superconvergence. The convergence rate of the NS-RPIM-Tr2L results is 1.99, which is very close to the theoretical one.

2) In terms of accuracy, these two NS-RPIM models are about three times more accurate than the linear FEM, when Mesh-4 is used.

3) In terms of efficiency, the NS-RPIM-Tr6 and NS-RPIM-Tr2L are, respectively, about 10% and 30% less efficient than the linear FEM in displacement norm measure, due to higher computational cost.

4) For comparison purpose, TABLE 6.16 lists also the results for the NS-PIM models obtained in Example 6.1.3. It is seen that NS-RPIM-Tr6 is about 30% more accurate than NS-PIM-Tr6/3, but their efficiencies are the same. NS-RPIM-Tr2L is the most accurate NS- model, but the least efficient, due to the use of many local nodes. We note also that the NS-RPIM-Tr2L is the most robust for irregular notes.

TABLE 6.16 Estimated computational efficiency of different NS-PIM/NS-RPIM models measured in *displacement norm* error for the numerical results for the infinite solid with circular hole problem with the same set of triangular mesh (Mesh-4 of 3,578 nodes)

Numerical method	Solution error	Error ratio to FEM-Tr3	Efficiency
FEM-Tr3	1.4435E-03	1.00	1.0
NS-PIM-Tr3	1.1520E-03	0.80	1.0
NS-PIM-Tr4-CT	6.3699E-04	0.44	1.9
NS-PIM-Tr6/3	7.0986E-04	0.49	0.9
NS-RPIM-Tr6	5.1289E-04	0.36	0.9
NS-RPIM-Tr2L	4.8237E-04	0.33	0.7

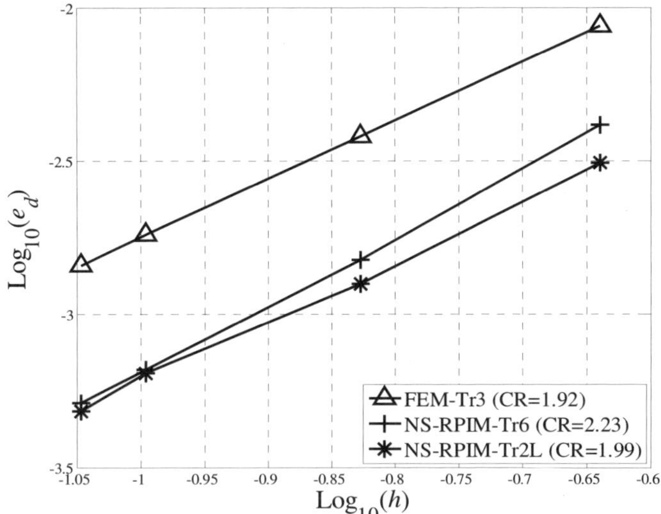

FIGURE 6.27 Comparison of convergence rates and accuracy of the numerical results in displacement norm obtained using linear FEM and two NS-RPIM models for the infinite solid with circular hole problem.

Measuring the numerical results in energy norm, FIGURE 6.28 plots the solution error of the computed results obtained using different methods. TABLE 6.17 lists the solution error in energy norm obtained using Mesh-4, together with the estimated efficiency by considering the computational cost listed in TABLE 6.12. The following points can be concluded for this example.

1) All the numerical models show monotonic convergence. The two NS-RPIM models achieve much higher convergence rates (1.53 and 1.51 respectively) than the linear FEM measured in energy norm, showing strong superconvergence property. The convergence rate of the NS-RPIM for this problem is even slightly higher than the ideal value of 1.5 (see Remark 3.23).

2) In terms of accuracy, the NS-RPIM models are about three times more accurate than the linear FEM, when Mesh-4 is used.

3) In terms of efficiency, the NS-RPIM-Tr6 has the similar efficiency, and NS-RPIM-Tr2L is 30% less efficient compared to the FEM-Tr3 model.

4) In comparing with the NS-PIM models obtained in Example 6.1.3, it is seen that NS-RPIM-Tr6 is about 10% more accurate than NS-PIM-Tr6/3, but

~17% less efficient. NS-RPIM-Tr2L is the most accurate NS- model, but the least efficient.

TABLE 6.17 Estimated computational efficiency of different NS-PIM/NS-RPIM models measured in *energy norm* error for the numerical results for the infinite solid with circular hole problem with the same set of triangular mesh (Mesh-4 of 3,578 nodes)

Numerical method	Solution error	Error ratio to FEM-Tr3	Efficiency
FEM-Tr3	1.3831E-04	1.00	1.0
NS-PIM-Tr3	5.0516E-05	0.37	2.3
NS-PIM-Tr4-CT	4.9617E-05	0.36	2.3
NS-PIM-Tr6/3	5.0352E-05	0.36	1.2
NS-RPIM-Tr6	4.4355E-05	0.32	1.0
NS-RPIM-Tr2L	4.4501E-05	0.32	0.7

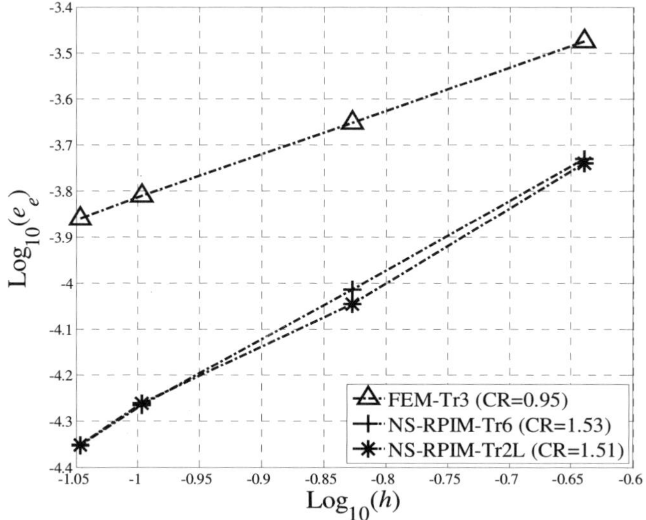

FIGURE 6.28 Comparison of convergence rates and accuracy of the numerical results in energy norm obtained using linear FEM and two NS-RPIM models for the infinite solid with circular hole problem.

We study now the volumetric locking issue using this example. Plane strain condition is now considered and the Poisson's ratio varies from 0.4 to 0.4999999. The NS-RPIM models are used to solve this problem together with the linear FEM for comparison by using Mesh-2 shown in FIGURE 6.14.

TABLE 6.18 gives the error of numerical solution in displacement norm against Poisson's ration varying from 0.4 to 0.4999999. It can be clearly found that the FEM-Tr3 suffers from the volumetric locking, while NS-RPIM models are naturally immune from the volumetric locking for nearly incompressible materials.

TABLE 6.18 Error of solution in displacement norm for 2D plane strain problem with Poisson's ration changing from 0.4 to 0.4999999

Poisson's ratio	FEM-Tr3	NS-RPIM-Tr6	NS-RPIM-Tr2L
0.4	0.45740639E-02	0.15035416E-02	0.12763577E-02
0.49	0.17083257E-01	0.14847734E-02	0.12796987E-02
0.499	0.52619287E-01	0.14908740E-02	0.12859326E-02
0.4999	0.81990377E-01	0.14918365E-02	0.12868340E-02
0.49999	0.88546719E-01	0.14919372E-02	0.12869277E-02
0.499999	0.89356594E-01	0.14919483E-02	0.12869344E-02
0.4999999	0.89440415E-01	0.14919521E-02	0.12869600E-02

6.3 NS-PIM/NS-RPIM for 3D solids

The formulation of NS-PIM/NS-RPIM for 3D solid mechanics problems is almost the same as that for 2D, except that all the operations have to be extended to one more dimension. Most of the formulae for 3D have the same forms, but the following major changes are needed. In this section, we extend the NS-PIM/NS-RPIM for 3D problems by simply mention the major difference of formulations between the 2D and 3D. More detailed formulation can be found in [2].

6.3.1 Approximation of displacement

For 3D problems, the problem domain is discretized using four-node tetrahedron cells (Te4), which is simplest and can be generated automatically for complicated 3D geometries. Based on the tetrahedral background cells, the support nodes can be selected using cell-based T-schemes and PIM shape functions can be created as previously presented in Chapter 3. For the linear NS-PIM, we use the cell-based T4-scheme (see Section 1.6.3) to select four support nodes and the linear PIM shape functions so constructed are exactly same as those of linear FEM (FEM-Te4). Hence the displacement field approximated using linear PIM shape functions is continuous over the whole problem domain. For higher order NS-PIM models in three dimensions, more nodes should be selected, and it is a little tricky in node selection, due to the possible singular moment matrix. Therefore for higher order approximation, we suggest using the RPIM shape functions which selects the support nodes by the cell-based T2L-scheme (see Section 1.6.3) and work well for very irregularly distributed nodes without much special treatments. Note that in this case, the constructed displacement is discontinuous and the discontinuity occurs on the interface triangles of the tetrahedral background cells.

6.3.2 Computation of node-based smoothed strains

In the scheme of NS-PIM/NS-RPIM for 3D problems, the problem domain discretized using tetrahedral cells will be further divided into N_s NOSL smoothing domains associated with nodes following the rules given in Section 2.3.1. Each node-based smoothing domain contains a node, thus the number of smoothing domains is the same as that of the field nodes and also equals to the number of quadrature cells, i.e. $N_s = N_n = N_q$.

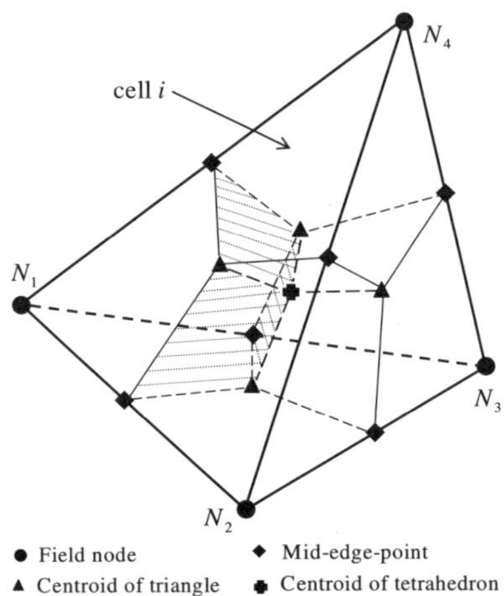

FIGURE 6.29 Formation of node-based smoothing domain based on a set of tetrahedral background cells. For node N_1 interested, the portion of the smoothing domain contributed from cell i is created by connecting the centroid of tetrahedron cell i, the centroids of the three surface triangles containing node N_1, the three mid-edge points and node N_1 itself.

We use the equally-shared smoothing domains for the 3D problems in this book and FIGURE 6.29 shows the formation of smoothing domain for a particular node N_1. The node-based smoothing domain for node N_1 consists of portions from all tetrahedral cells sharing the node. As shown in FIGURE 6.29, the tetrahedron cell i, which contains the node N_1, will contribute equally to four smoothing domains associated with nodes N_1, N_2, N_3, and N_4. For node N_1 interested, the portion of the smoothing domain contributed from cell i is created by connecting the centroid of tetrahedron cell i, the centroids of the three surface triangles containing node N_1, the three mid-edge points and node N_1 itself. Hence the portion of a 3D node-based smoothing domain contributed by one cell is a hexahedron. Similarly, other portions of the smoothing domain of node N_1 can also be generated and the whole node-based smoothing domain can be finally formed by union all these portions.

Within each node-based smoothing domain, the smoothed strains can be calculated using the generalized gradient smoothing operation following the similar way presented in Section 4.5.2. Note that the area in 2D becomes volume in 3D, the area integration in 2D now becomes volume integration in 3D, and the curve integration in 2D becomes surface integration in 3D.

6.3.3 Stiffness matrix of 3D NS-PIM

The discretized system equations have the same form as Equation (6.1), and the sub-stiffness matrix can be written as

$$\overline{\mathbf{K}}_{IJ} = \sum_{i=1}^{N_s} V_i^s \overline{\mathbf{B}}_I^{\mathrm{T}} \mathbf{c} \overline{\mathbf{B}}_J \tag{6.19}$$

where $V_i^s = \int_{\Omega_i^s} d\Omega$ is the volume of the smoothing domain i, and

$$\overline{\mathbf{B}}_I(\mathbf{x}_i) = \begin{bmatrix} \overline{\phi}_{Ix}(\mathbf{x}_i) & 0 & 0 \\ 0 & \overline{\phi}_{Iy}(\mathbf{x}_i) & 0 \\ 0 & 0 & \overline{\phi}_{Iz}(\mathbf{x}_i) \\ \overline{\phi}_{Iy}(\mathbf{x}_i) & \overline{\phi}_{Ix}(\mathbf{x}_i) & 0 \\ 0 & \overline{\phi}_{Iz}(\mathbf{x}_i) & \overline{\phi}_{Iy}(\mathbf{x}_i) \\ \overline{\phi}_{Iz}(\mathbf{x}_i) & 0 & \overline{\phi}_{Ix}(\mathbf{x}_i) \end{bmatrix} \tag{6.20}$$

$$\overline{\phi}_{Il} = \frac{1}{V_i^s} \sum_{m=1}^{n_{seg}} \left[\sum_{n=1}^{n_G} W_n^G \left(\phi_I(\mathbf{x}_{m,n}) n_{l,m} \right) \right] \quad (l = x,\, y,\, z) \tag{6.21}$$

in which n_{seg} refers to the number of surface quadrilaterals of the smoothing domain and n_G is the number of Gauss points located within each surface quadrilateral. Note that to perform the numerical integration exactly, a proper number of n_G should be used. In this study, $n_G = 1$ is used for the model using linear PIM shape functions and $n_G = 2 \times 2$ is used when RPIM shape functions are adopted.

6.3.4 Possible 3D NS-PIM/NS-RPIM models

NS-PIM-Te4

The NS-PIM-Te4 is the simplest NS-PIM model for 3D problems. It uses linear polynomial PIM shape functions and the support nodes selected by the cell-based T4-scheme. The displacement field is *compatible* over the problem domain and the corresponding NS-PIM-Te4 formulation is variationally consistent. In this case the NS-PIM-Te4 is exactly same as the NS-FEM-Te4 [14].

To find out the number of support nodes for each node-based smoothing domain, the model of the 3D Lame (Example 6.3.2) problem with irregular mesh of 173 nodes is used. For this case, we found that the number of the support nodes for a node-based smoothing domain varies from 6 to 20 and the average number is 10.9, which is about 2.7 times as large as that of the linear FEM model. Therefore, the average number of the nonzero entries n_{ze} in a row of the stiffness matrix of an NS-PIM-Te4 model is about 2.7 times that of the FEM-Te4 counterpart using the same mesh, as listed in TABLE 6.19.

NS-RPIM-Te2L

The NS-RPIM-Te2L uses the RPIM shape functions and support nodes selected by the cell-based T2L-scheme. The displacement field in the NS-RPIM-Te2L model is incompatible and the discontinuity will occur on the interface triangles between tetrahedral background cells.

As listed in TABLE 6.19, the number of support nodes for the NS-RPIM-Te2L model varies from 17 to 85 and the average number is 38, which is about 9.5 times as large as that of the linear FEM model. Therefore, the average number of the nonzero entries n_{ze} in a row of the stiffness matrix of an NS-RPIM-Te2L model is about 9.5 times that of the FEM-Te4 counterpart using the same mesh, as listed in TABLE 6.19.

TABLE 6.19 3D NS-PIM/NS-RPIM models and overall characteristics in relation to the FEM-Te4 model using the same mesh

Numerical method	Support nodes in an integral cell/element	Estimated average number of nonzero entries	Estimated Solver CPU time
FEM-Te4	4	n_{ze}	t_{CPU}
NS-PIM-Te4	6-20 (10.9)	$2.7\,n_{ze}$	$2.7\,\bar{n}_{iter}t_{CPU}$
NS-RPIM-Te2L	17-85 (38.0)	$9.5\,n_{ze}$	$9.5\,\bar{n}_{iter}t_{CPU}$

6.3.5 Condition number of 3D NS-PIM/NS-RPIM models

TABLE 6.20 lists the condition numbers of the global stiffness matrixes for these two NS-PIM models in relation to the FEM-Te4 using the same tetrahedral meshes for the 3D Lame problem. We may find that the NS-PIM-Te4 model has smaller condition numbers than the linear FEM-Te4, although it uses more support nodes as shown in TABLE 6.19. However, the NS-RPIM-Te2L model has bigger condition numbers than the FEM-Te4. We also find that the ratio between condition numbers of different methods keeps almost a constant with the refinement of the model. In the following analysis, we will use the relative $\bar{n}_{iter}$ given in the last column in TABLE 6.20.

TABLE 6.20 Condition number of the global stiffness matrixes for various 3D NS-PIM/NS-RPIM models for the Lame problem in relation to the FEM-Te4 model using the same mesh

Numerical method	Mesh of 96 nodes		Mesh of 173 nodes	
	$cond(\mathbf{K})$	$\bar{n}_{iter}$	$cond(\mathbf{K})$	$\bar{n}_{iter}$
FEM-Te4	3.284e+02	1.00	5.274e+02	1.00
NS-PIM-Te4	1.074e+02	0.57	1.638e+02	0.56
NS-RPIM-Te2L	8.753e+02	1.63	1.327e+03	1.59

6.3.6 Estimation of computational cost

Combining TABLE 6.19 and TABLE 6.20 we shall have a rough estimation of computational cost for the 3D NS-PIM/NS-RPIM models in relation to the standard FEM-Te4 model using the same four-node tetrahedral meshes, as shown in TABLE 6.21. Although the NS-PIM-Te4 model uses more support nodes for each smoothing domain, we can find that it costs only 50% more time than the FEM-Te4 model owing to smaller condition number. However, the NS-

RPIM-Te2L model will cost much more time, about 15.1 times, than the FEM-Te4 model for this particular problem. Therefore, we only use the NS-PIM-Te4 model to study 3D numerical examples in the following sections.

From TABLE 6.21, it is clear that a 3D NS-PIM model may be in the same level of computational cost compared to the linear FEM-Te4 model for the same mesh. In the following Sections, we will investigate the efficiency issue which is also related to the accuracy of the numerical results.

TABLE 6.21 Estimation of computational cost for 3D NS-PIM/NS-RPIM models in relation to the FEM-Te4 model using the same mesh

Numerical method	Support nodes in an integral cell/element	Estimated Solver CPU time
FEM-Te4	4	t_{CPU}
NS-PIM-Te4	6-20 (10.9)	$1.5\, t_{CPU}$
NS-RPIM-Te2L	17-85 (38.0)	$15.1\, t_{CPU}$

6.3.7 Numerical examples for 3D solids

A 3D code of linear NS-PIM-Te4 has been developed, and it is examined using the following examples. The error indicators defined in Equations (5.88) and (5.94) are also used to study the numerical results, with the only change that area A has been replaced with volume V.

Example 6.3.1 Linear patch test

The first case is the linear patch test with a cubic patch of $10\times10\times10$. The displacements are prescribed on all outside boundary surfaces by the following linear function.

$$\begin{cases} u_i = 0.6x_i \\ v_i = 0.6y_i \\ w_i = 0.6z_i \end{cases} \tag{6.22}$$

where u_i, v_i and w_i are, respectively, the displacement components in x, y and z directions for node i, and (x_i, y_i, z_i) are the coordinates of node i.

Two patches are represented using four-node tetrahedral cells with 125 regularly and 166 irregularly distributed nodes, as shown in FIGURE 6.30. The errors of linear NS-PIM-Te4 results in displacement norm as defined in Equation (5.88) are found to be 1.2837×10^{-15} for the regular and 1.2036×10^{-15} for the irregular patch respectively, which are almost the lever of the machine precision. The results show that the displacements of all the interior nodes follow "exactly" the same function of the imposed displacement. This example demonstrates numerically that the 3D NS-PIM solution will converge owing to its ability to reproduce linear fields.

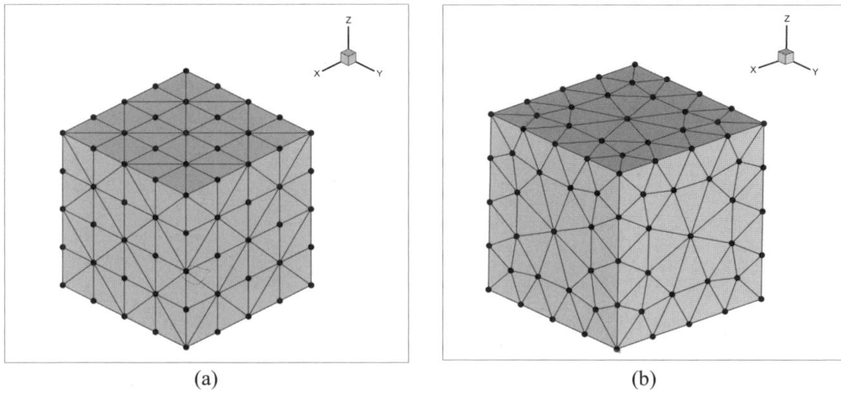

(a)   (b)

FIGURE 6.30 A 3D patch with regularly and irregularly distributed nodes of tetrahedral mesh: (a) tetrahedral mesh with 125 regularly distributed nodes; (b) tetrahedral mesh with 166 irregularly distributed nodes.

Example 6.3.2 Lame problem

The 3D Lame problem is a hollow sphere with inner radius (a) and outer radius (b) and subjected to internal pressure P, as shown in FIGURE 6.31a. For this benchmark problem, the analytical solution is available in spherical coordinate system as follows [27]

$$u_r = \frac{pa^3 r}{E\left(b^3 - a^3\right)}\left[\left(1 - 2v\right) + \left(1 + v\right)\frac{b^3}{2r^3}\right] \tag{6.23}$$

$$\sigma_r = \frac{pa^3\left(b^3 - r^3\right)}{r^3\left(a^3 - b^3\right)} \tag{6.24}$$

$$\sigma_\theta = \sigma_\varphi = \frac{pa^3\left(b^3 + 2r^3\right)}{2r^3\left(b^3 - a^3\right)} \tag{6.25}$$

where r is the radial distance from the centroid of the sphere to the point of interest in the sphere. As the problem is spherically symmetrical, only one-eighth of the sphere is modeled, as shown in FIGURE 6.31b, and symmetry conditions are imposed on the three planes of symmetry. The parameters used in the present study are taken as: $E = 1$, $v = 0.3$, $a = 1$, $b = 2$ and $P = 1$.

The problem model is presented using four-node tetrahedral cells with total 1,138 irregularly distributed nodes, as shown in FIGURE 6.31b, and the computed results of displacement and stress components along the x-axis are plotted together with the analytical solutions in FIGURE 6.32. It can be clearly seen that the numerical results agree well with the analytical ones.

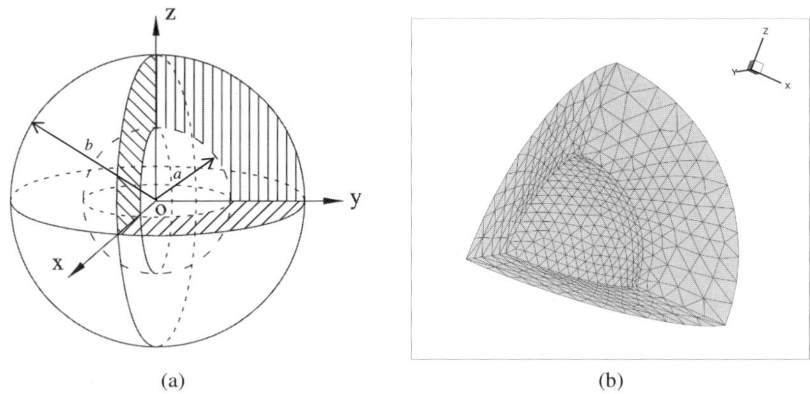

(a)   (b)

FIGURE 6.31 3D Lame problem of a hollow sphere under international pressure and its one-eighth model: (a) the 3D Lame problem; (b) domain discretization of the one-eighth model with tetrahedral mesh of 1,138 nodes.

To investigate the convergence and efficiency of the NS-PIM, four irregular meshes of tetrahedron cells with 173, 317, 729, and 1,304 nodes are employed. The convergence process of the numerical results measured in displacement norm is plotted FIGURE 6.33. TABLE 6.22 lists the solution error in displacement norm obtained using Mesh-4 of 1,304 nodes and the computation efficiency considering the cost in TABLE 6.21. The following points can be found.

1) The displacement norm error decreases almost linearly with the decrease of h, showing a nice monotonic convergence.

2) In terms of convergence rate, the linear FEM-Te4 achieves 1.88, which is a little lower than the theoretical value of 2.0. For the linear NS-PIM-Te4, the

computed convergence rate is 1.98, which is higher than the counterpart of the FEM and is very close to the theoretical one.

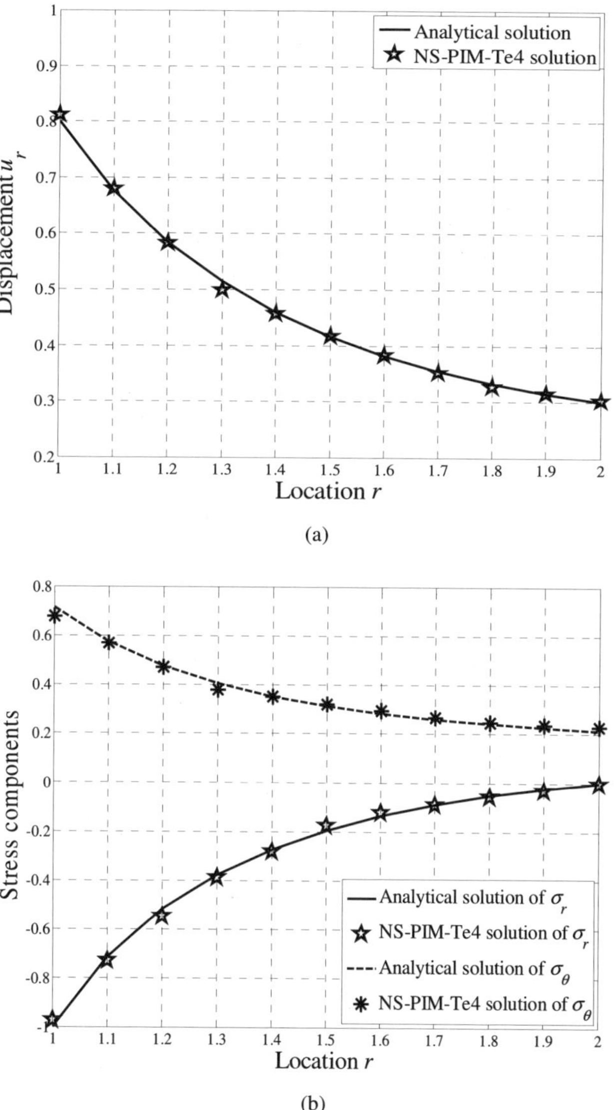

(a)

(b)

FIGURE 6.32 Displacement and stress distribution along the *x*-axis of the one-eighth model of the 3D Lame problem: (a) distribution of the radial displacement; (b) distribution of radial and tangential stress components.

3) In terms of accuracy, the FEM-Te4 is about twice more accurate than the NS-PIM-Te4 for this problem, when Mesh-4 is used.

4) In terms of efficiency, the linear NS-PIM-Te4 is about 3 times less efficient than the FEM-Te4 measured in displacement norm.

TABLE 6.22 Estimated computational efficiency of different methods measured in *displacement error* for the numerical results for the Lame problem with the same set of tetrahedral mesh (Mesh-4 of 1,304 nodes)

Numerical method	Solution error	Error ratio to the FEM-Te4	Efficiency
FEM-Te4	1.8381E-02	1.00	1.00
NS-PIM-Te4	3.5123E-02	1.91	0.35

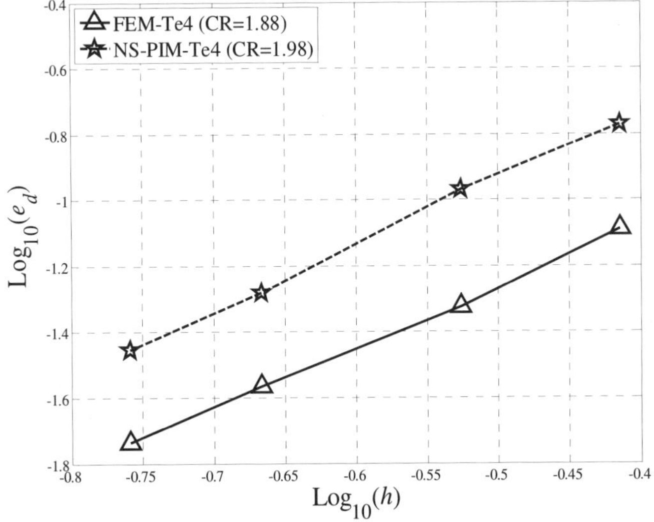

FIGURE 6.33 Comparison of convergence rates and accuracy of the numerical results in displacement norm obtained using linear FEM-Te4 and NS-PIM-Te4 models for the 3D Lame problem.

FIGURE 6.34 plots in the logarithm scale the solution error in energy norm with respect to the characteristic nodal spacing h. TABLE 6.23 lists the solution error in energy norm obtained using Mesh-4 and the estimated efficiency considering the computational cost in TABLE 6.21. Following points can be observed.

1) The energy norm errors decrease almost linearly with the decrease of the nodal spacing, showing a nice monotonic convergence also in energy norm.

2) In terms of convergence rate, the linear FEM-Te4 achieves the value of 0.92, which is very close to the theoretical one of 1.0 for standard weak formulation. The convergence rate of the linear NS-PIM-Te4 is 1.44, which is much higher than the theoretical one showing strong superconvergence and very close to the ideal value of 1.5 for W^2 formulation (see Remark 3.23).

TABLE 6.23 Estimated computational efficiency of different methods measured in *energy error* for the numerical results for the Lame problem with the same set of tetrahedral mesh (Mesh-4 of 1,304 nodes)

Numerical method	Solution error	Error ratio to the FEM-Te4	Efficiency
FEM-Te4	1.5008E-01	1.00	1.0
NS-PIM-Te4	7.7785E-02	0.52	1.3

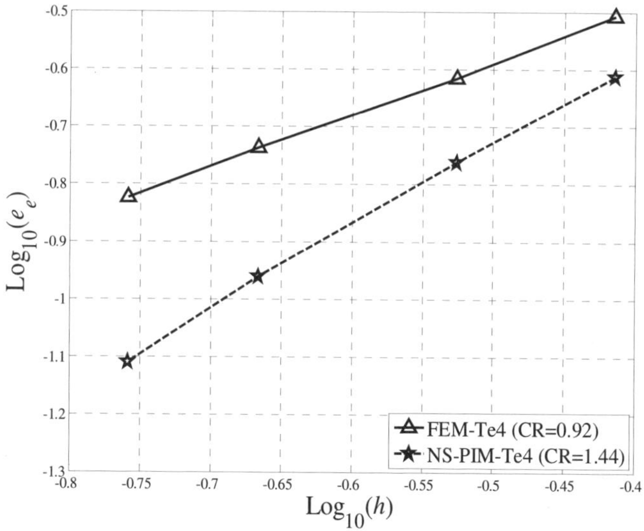

FIGURE 6.34 Comparison of convergence rates and accuracy of the numerical results in energy norm obtained using linear FEM-Te4 and NS-PIM-Te4 models for the 3D Lame problem.

3) In terms of accuracy, the NS-PIM-Te4 is about twice more accurate than the linear FEM, when the same Mesh-4 is used.

4) In terms of computational efficiency, the NS-PIM-Te4 is about 1.3 times more efficient than the linear FEM measured in energy norm.

Example 6.3.3 A mechanical part: 3D rim

A typical 2D rim model of automotive component with a complicated shape has been studied in Example 6.1.4 as a plane stress problem. A 3D model of the rim is now considered with a thickness of 3, as shown in FIGURE 6.35a. The 3D model is also fixed at the nodes located along the inner circle face and a pressure of 100 units is applied along a portion of the lower arc edge.

The rim is meshed using tetrahedral cells with tetrahedral mesh of 7,972 nodes, as shown in FIGURE 6.35b. A reference solution is obtained using the standard FEM with a very fine mesh of total 607,684 elements and 115,486 nodes. The computed results of stress components at the nodes on the plane of $z=0$ are plotted in the contour form. The pictures in FIGURE 6.36 show the comparison of the stress contours between the reference solutions and the NS-PIM-Te4 solutions for σ_{xx}, σ_{yy} and σ_{xy}, respectively. It can be clearly seen that the solutions of the NS-PIM-Te4 agree well with the reference ones.

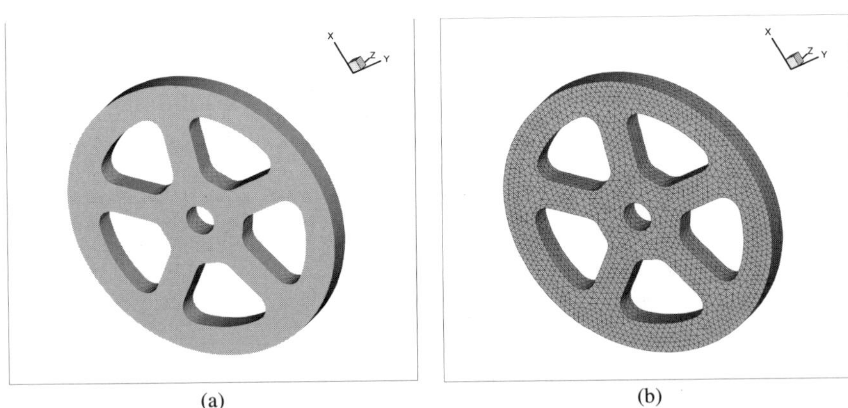

FIGURE 6.35 A 3D model of an automotive rim component: (a) problem setting; (b) domain discretization with tetrahedral mesh of 7,972 nodes.

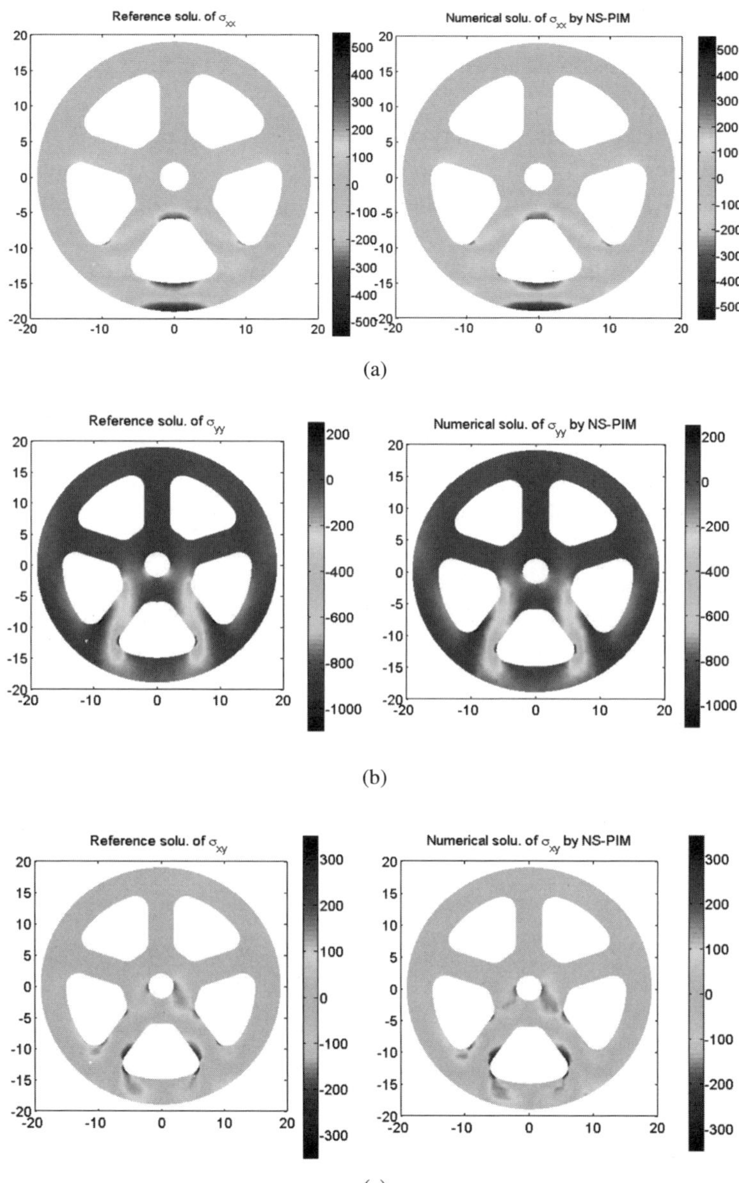

(a)

(b)

(c)

FIGURE 6.36 Comparison of contours of stress components between the reference and NS-PIM-Te4 solutions: (a) contour of stress σ_{xx} on the plane of $z=0$ in the rim component; (b) contour of stress σ_{yy} on the plane of $z=0$ in the rim component; (c) contour of stress σ_{xy} on the plane of $z=0$ in the rim component.

Example 6.3.4 A 3D riser connector

This example comes from a real offshore project of a Floating Production and Storage Unit (FPSO). Fluid of oil-gas-water mixture is transferred between the FPSO and subsea pipeline. This process is carried out through a kind of flexible pipe called riser, which is attached to the FPSO shipside by a riser connector, as shown in FIGURE 6.37a, which is the cross section drawing of the problem. The problem is simplified with a 3D model as shown in FIGURE 6.37b. The load is applied on the top flange of the model. The boundary conditions are defined at the end of I-beams where riser connector is supported by other structures. Due to the complexity of the actual structure, we omit the detailed specification of the model.

Reference solution of this problem is obtained using the FEM model with very fine mesh of total 27,072 nodes, and the contour of Von Mises stress is plotted in FIGURE 6.38a together with the deformed shape of the riser connector. The problem is studied using the present NS-PIM-Te4 with a tetrahedron mesh of 2,228 nodes, and the computed results of Von Mises stress are plotted in counter form in FIGURE 6.38b. It can be found that, although the riser connector is presented with less than one tenth of the number of nodes of the reference model, the linear NS-PIM solution matches with the reference ones very well.

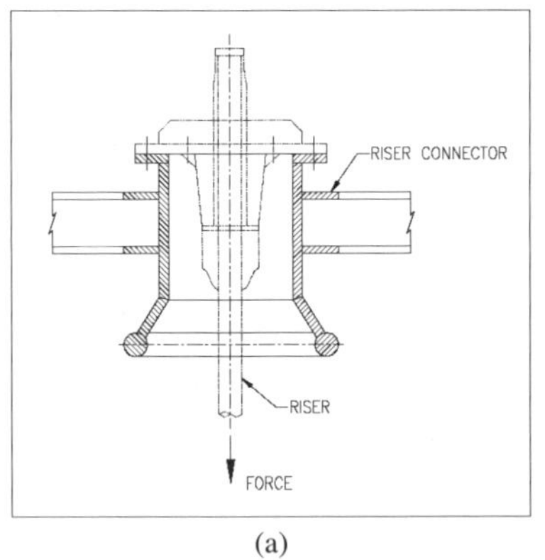

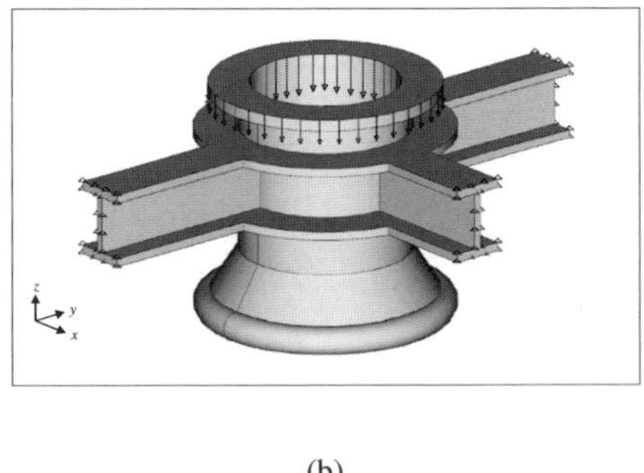

(a)   (b)

FIGURE 6.37 Simplified model of the three-dimensional riser connector in an offshore platform: (a) cross section drawing of the problem setting; (b) simplified 3D model of the riser connector.

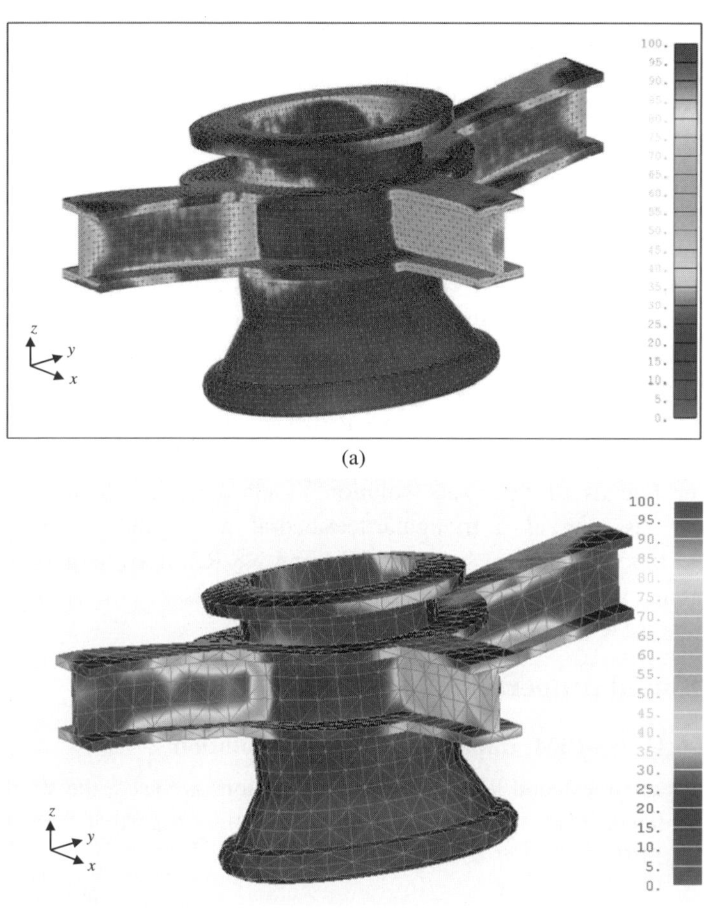

(a)

(b)

FIGURE 6.38 Comparison of contours of Von Mises stress between the reference and NS-PIM-Te4 solutions for the 3D riser connector: (a) reference solution obtained using FEM with fine mesh of 27,072 nodes; (b) numerical solutions obtained using NS-PIM-Te4 with a coarse mesh of 2,228 nodes.

6.4 Upper bound properties of NS-PIM/NS-RPIM

6.4.1 Background

It is well known that the fully compatible FEM [26, 29] provides a lower bound in energy norm for the exact solution to force-driving elasticity problems (see,

Remark 5.4). It is, however, much more difficult to bound the solution from above for complicated practical problems. It has been a dream of many to find a general systematical way to obtain an upper bound of the exact solution for practical problems, so that we can then bound the solution from both sides and our numerical solution can be properly "certified". It has been discovered [4, 15, 25, 28, 30] recently that the NS-PIM and the NS-RPIM with sufficient *softening effects* can provide an upper bound solution in energy norm for such problems. The theory on the softening effects and upper bounds has been presented in Chapter 5. This section discusses practical matters on obtaining an upper bound using NS-PIM or NS-RPIM models, and demonstrates the upper bound property of NS-PIM and NS-RPIM through a number of numerical examples. Using a properly constructed NS-PIM or NS-RPIM model together with the FEM counterpart, we now have a systematically way to numerically obtain both upper and lower bounds of the exact solution to elasticity problems of arbitrary complexity, as long as a triangular/tetrahedral mesh can be built. As the theoretical fundamentals for both NS-PIM and NS-RPIM are largely the same, our discussion will focus mainly on the NS-PIM models.

6.4.2 Bound properties of NS-PIM models

Remark 6.11 NS-PIM: upper bound to FEM solution

When the same mesh and linear PIM shape functions are used, the strain energy solution obtained from the NS-PIM results is no-less than that from the fully compatible FEM.

$$\underbrace{\frac{1}{2}\bar{\mathbf{d}}^{\mathrm{T}}\bar{\mathbf{K}}\bar{\mathbf{d}}}_{\bar{U}_{PE}^{D}(\bar{\mathbf{d}})} \geq \underbrace{\frac{1}{2}\tilde{\mathbf{d}}^{\mathrm{T}}\tilde{\mathbf{K}}\tilde{\mathbf{d}}}_{\tilde{U}_{PE}(\tilde{\mathbf{d}})} \tag{6.26}$$

This inequality is essentially the same as that given in Theorem 5.7, except that here is expressed in solution of discrete nodal displacements. It was first presented and proven in [4]. An alternative proof based on variational formulation can be found in [31]. The equality is true when both NS-PIM and FEM produce the exact solution, or the smoothing operation is performed independently for each individual part of the triangular/tetrahedral elements connecting to the node and hence the node-based smoothing has no effect.

Remark 6.12 NS-PIM: upper bound to exact solution

Via examples of force-driving problems, it has been found that $\overline{U}_{PE}^{D}(\overline{\mathbf{d}}) \geq U_{PE}(\mathbf{d}) \geq \tilde{U}_{PE}(\tilde{\mathbf{d}})$ stands, except for a few trivial cases. This means that the solution of an NS-PIM can give an upper bound of the exact solution in energy norm. This is discussed in Remark 5.17. Here we further discuss these exceptional cases for NS-PIM (or NS-RPIM) models, based on the argument of "the battle of softening and stiffening effects" [4].

Remark 6.13 The battle of softening and stiffening effects

The NS-PIM is a typical GS-Galerkin model based on the G space theory. Remark 5.17 states that a GS-Galerkin model *can* produce an upper bound solution by performing proper smoothing operations. The NS-PIM model performs smoothing operation based on nodes and is often found capable of producing sufficient softening effects leading to an upper bound solution.

Remark 6.13 can be understood intuitively (not very precise) as follows. According to Theorem 5.7 and Remark 5.15, we know that an NS-PIM model can always provide an upper bound for the exact solution in energy norm due to the softening effects induced by the "smoothing" operation, when the shape functions corresponding to the exact solution are used (so that the Galerkin model will be exact). For a general problem, however, finding the exact shape functions is not possible. Therefore, the NS-PIM can only use the usual discrete PIM shape functions (or the FEM shape functions). The use of any shape functions in the place of the exact shape functions in the construction of displacements will, on the other hand, causes some stiffening effects to the model. The battle between the softening effects caused by the smoothing operation and the stiffening effects caused by the displacement construction will determine whether or not an NS-PIM model can in fact provide an upper bound solution to the problem.

Remark 6.14 Factors affecting the softening effects

In an actual discrete NS-PIM model, there could be a number of factors affecting the softening effects. For easy discussion we consider first NS-PIM-Tr3 model. The following factors may affect the softening effect.

1) The number of cells connected to a node of a smoothing domain: the more the cells, the more the smoothing effects. In an extreme case, if the smoothing domain is not based on nodes but defined for each cell to

perform the smoothing operation, there will no softening effect at all. In this case the NS-PIM and FEM produce the same solution, and the NS-PIM will not provide an upper bound, but a lower bound solution.

2) The number of node-based smoothing domains used in the problem domain. In theory, one does not have to perform the smoothing operation for all the node-based smoothing domains. The softening effect will proportionally depend on the number of smoothing operations performed.

3) The dimension of the smoothing domain. In NS-PIM, the smoothing domains are usually "seamless", meaning that there are no gaps and overlaps between these neighboring smoothing domains. If however, one chooses to use a smaller smoothing domain, the method will still work (such as the *a*FEM [32]) but the softening effects will be reduced.

4) The number of nodes used in the model. When a small number of nodes are used, the displacement constructed using the PIM shape functions in a smoothing domain is far from the exact solution. When a smoothing operation is performed, it gives a heavy smoothing to the strain field, and hence a strong softening effect. On the other hand, when a large number of nodes are used, the displacement constructed using the PIM shape functions is closer to the exact solution, and hence less softening effect. At the extreme of infinitely small elements are used, the smoothing effects diminish and the NS-PIM solution (also the FEM solution) will approach to the exact solution. This reveals, from another view point, why the NS-PIM models will converge to the exact solution.

Remark 6.15 Factors affecting the stiffening effect

1) The stiffening effect depends on the order of the PIM shape functions used in the displacement construction. When high order PIM shape functions are used, the displacement constructed using the PIM shape functions in a smoothing domain is usually closer to the exact solution, which reduces the stiffening effect for "stiff" models and softening effect for "soft" models. Therefore, we can expect, in general, a tighter upper bound, when higher order PIM shape functions are used.

2) The stiffening effect depends also on the number of nodes used in the model. When a small number of nodes are used, the displacement constructed using the PIM shape functions is far from the exact solution, the stiffening effect is thus small, and vice versa. At the extreme of infinitely

small cells are used, the stiffening effects diminish and the NS-PIM solution (also the FEM solution) will approach to the exact solution.

The softening effect provided by the smoothing operation is found more significant than the stiffening effects in the NS-PIM setting, due mainly to the use of quite small number of smoothing domains in comparison to other S-PIM models. Therefore, the NS-PIM always produces an upper bound solution for 1D, 2D and 3D solids, except the following few special cases.

Remark 6.16 Upper bound by NS-PIM: a few exceptions

1) Too few cells are used. In an extreme case, when only one cell is used with linear interpolations, there is only one cell with constant strain participating in the smoothing. In this case there is no smoothing effect at all, the solutions of NS-PIM and FEM are the same, and NS-PIM-Tr3 gives only a lower bound solution. In order to obtain sufficient smoothing effects to produce upper bound solutions, the number of cells should not be too small. For 1D case, the number of elements should be at least 2, as will be shown in Section 6.4.3.

2) Hanging elements are used in an NS-PIM model with small number of cells. FIGURE 6.39 shows a simple model with three hanging triangular cells attached to the domain. At the corner nodes of these three cells, only one cell for each node can participate in the smoothing operation. In such a case, there is no smoothing effect at all for these three node-based smoothing domains. Note that such hanging elements are not supposed to use even in the FEM models, because the stress there are always zero when the nodes are free. Thus it is equivalent to remove entire the corner elements! The similar situations can occur for cells on the two ends of a 1D domain and on the corners of 3D domains.

In our numerical study conducted so far, we found that NS-PIM can produce upper bound solutions for all the problems we have studied, except the very special cases mentioned above. Examples given in Section 6.4.3 will support these facts.

Note that the above discussions on the NS-PIM are largely applicable to the NS-RPIM, as they all share the same theoretical background. The difference is only in the shape functions used. Therefore, we will omit the detailed discussions on NS-RPIM, and refer the readers to the paper [28] for more discussions. In general, the NS-RPIM gives tighter upper bounds due to higher order shape functions used.

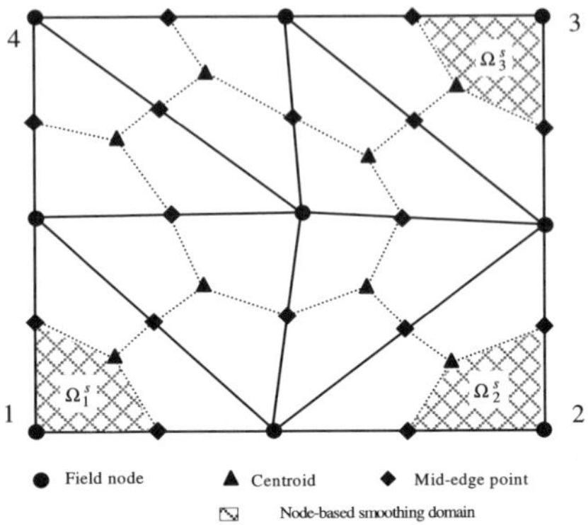

FIGURE 6.39 Hanging elements in 2D domains: smoothing operation on nodes 1, 2 and 3 has no smoothing effects.

6.4.3 Upper bound solutions: numerical examples

Example 6.4.1 A 1D bar problem

Consider first a very simple 1D bar of length L and of uniform cross-sectional area A, as shown in FIGURE 6.40. The bar is fixed at the left end and subjected to a uniform body force b in the x-direction. The parameters used in the analysis are $L=1$, $A=1$, $b=1$, and $E=1$. The governing equation and boundary conditions for this problem are:

$$E\frac{d^2u}{dx^2}+1=0 \tag{6.27}$$

$$u\left(x=0\right)=0$$

$$\sigma\left(x=L\right)=0 \tag{6.28}$$

The exact solution that satisfies the above equations can be easily obtained as

$$u\left(x\right)=-\frac{1}{2E}x^2+\frac{1}{E}x \tag{6.29}$$

The exact strain energy solution can be calculated using

$$U_{PE}\left(\mathbf{u}\right)=\frac{1}{2}\int_{L}\boldsymbol{\varepsilon}^{\mathrm{T}}E\boldsymbol{\varepsilon}\,\mathrm{d}x=\frac{1}{6E} \tag{6.30}$$

Although the problem is very simple, it is very useful to show explicitly some of the important properties of the NS-PIM.

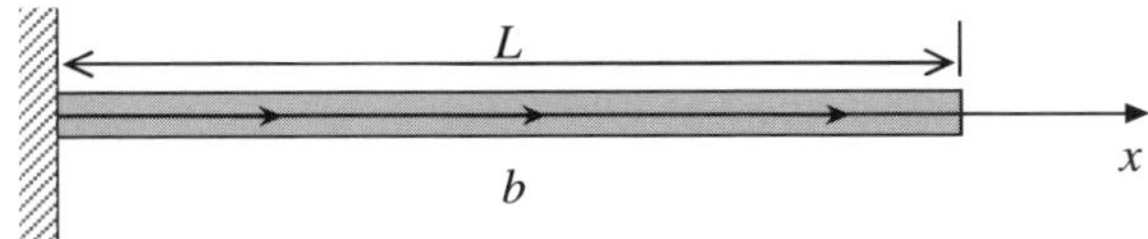

FIGURE 6.40 One-dimensional bar of uniform cross-sectional area *A* subjected to a uniformly distributed body force *b* along the *x*-axis.

To start, the effect of the size of the smoothing domain is studied using this simple 1D problem. As shown in FIGURE 6.41, the problem domain of the bar is presented using two uniform background cells and three nodes: node 1 locates at the left end, node 2 locates at the midpoint, and node 3 locates at the right end. In the usual NS-PIM settings using the equally-shared node-based smoothing domains, the smoothing domain length for node 2 is obtained by connecting these two midpoints of cell 1 and cell 2, which is $L/2$. The length of the two smoothing domains for these two end-nodes is $L/4$. Since we use linear interpolation based on each of these two cells, the compatible strain is constant in each of these cells. The smoothing operations for nodes 1 and 3 have no effect, and hence no need to perform. We can now intentionally change the size of the smoothing domain length for node 2 by allowing the smoothing domain to shrink or stretch beyond the midpoints of these two cells, so that we can study the effect of the smoothing domain length on the linear NS-PIM solution.

FIGURE 6.41 plots the strain energy of the NS-PIM solution using different smoothing domain lengths. It is found that the strain energy increases monotonically with the increase of the smoothing domain length that is measured as the ratio between the smoothing domain length of node 2 and the length L of the entire problem domain. When the smoothing length reduces to zero, the NS-PIM model becomes exactly the FEM model of two linear elements with 3 nodes, which gives a lower bound solution. When the ratio of the smoothing domain length increases to about 0.43, the NS-PIM solution of strain energy is larger than that of the exact solution, and an upper bound solution is attained. Increasing the ratio of the smoothing domain further to 0.9, the NS-PIM model becomes very soft and the strain energy becomes much larger than the exact one. This implies that one can make the NS-PIM model as softer as desired by reducing the number but increase the dimension of smoothing domain. These findings from this simple example also support the discussions given in Remark 5.17. If we stretch the smoothing domain to the entire problem, meaning that we use only one smoothing domain, the stiffness matrix will become singular and the NS-PIM model becomes unstable, which confirms our theory on the minimum number of smoothing domains (see Section 2.3.10). From TABLE 2.2, we know that we need at least two smoothing domains for this 1D problem with two nodal DOFs for a stable solution.

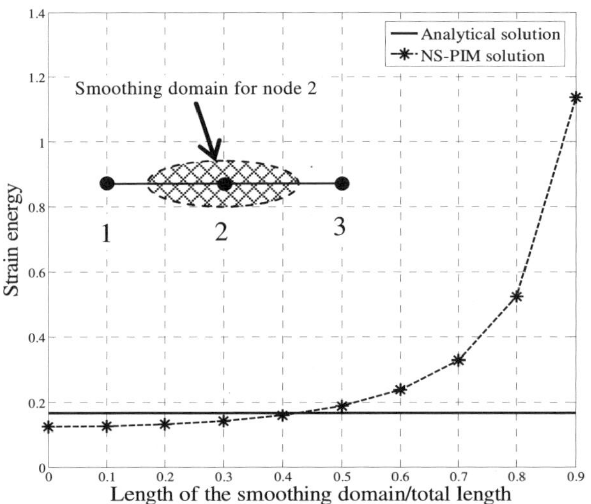

FIGURE 6.41 The effect of the length of the smoothing domain for 1D problem.

Next, for the same 1D problem we study further the convergence process, by increasing the number of nodes and examine the properties of the solutions of linear NS-PIM and NS-RPIM, together with that of the linear FEM using exactly the same meshes. Six models of different numbers of uniformly distributed nodes are used with equally-shared node-based smoothing domains. Note with these node-based smoothing domains, PIM shape functions are created using the same set of support nodes for the point of interest located in one cell. Then the discontinuities of PIM shape functions (except using linear

PIM) are at the field nodes, this set of nodal PIM shape functions can only be in a $\mathbb{G}_h^1(\Omega)$ space, if node-based smoothing domains are used, due to the "no-sharing rule" discussed in Section 2.3.1.

The computed solutions of strain energy are plotted in FIGURE 6.42 against the number of DOFs (or nodes) used, together with the exact solution. It can be found that, NS-PIM and NS-RPIM produce same results as FEM when only two nodes are used. This is because in this case two smoothing domains are used for, respectively, the two field nodes at the two ends of the bar, hence the smoothing has no effect at all to the entire problem, and the solution is the same as the FEM giving a lower bound (see, Remark 5.4). When the number of nodes is more than 2, we then have internal nodes, the smoothing takes effect for these nodes, and the NS-PIM provides an upper bound solution. With the increase of the number of nodes, the FEM solution approaches monotonically to the exact solution from bellow. The NS-PIM and NS-RPIM solutions, however, approach to the exact solution monotonically from above. These findings confirm Remark 6.12 on NS-PIMs models. This simple 1D example shows clearly the very important fact that we now can bound the exact solution from both sides. It is also observed that the upper bound provided by the NS-RPIM is much tighter

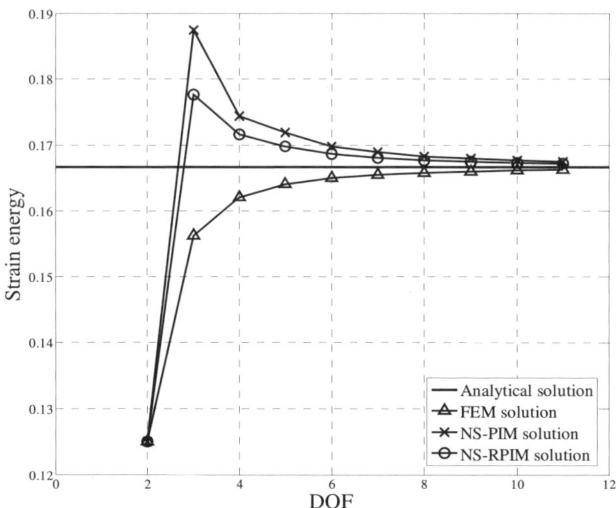

FIGURE 6.42 Upper bound solutions obtained using the linear NS-PIM and NS-RPIM for the 1D bar problem; the lower bound solution is obtained using the FEM with linear elements of the same mesh.

than that provided by the NS-PIM. This is due to the higher order RPIM shape functions used in the NS-RPIM, which reduces both the softening and stiffening effects pushing the solution closer to the exact one.

Example 6.4.2 A 2D rectangular cantilever

In this example, we revisit the cantilever problem but focus on the examination of the upper bound property of the NS-PIM models, in comparison with the FEM model using the same meshes. The settings and parameters used are exactly the same as those given in Example 6.1.2. In this study, we use equally-shared smoothing domains for all the NS-PIM models.

FIGURE 6.43 plots the solution convergence process to the exact solution with the increase of the DOFs for the NS-PIM/NS-RPIM models and the linear FEM with the same mesh shown in FIGURE 6.9. The errors in the strain energy solution for the numerical results obtained using the same set of triangular mesh (Mesh-4) are listed in TABLE 6.24. Considering the computational cost given in TABLE 6.3 and TABLE 6.12 for these methods, the estimated computational efficiency measured in strain energy solution is given in the last column of TABLE 6.24. We find for this case the following points.

1) With the increase of DOFs, the solutions of all the models converge to the exact solution.

2) The linear FEM-Tr3 gives a lower bound solution, and all the NS-PIM/NS-RPIM models provide upper bound solutions to the exact one.

3) Compared to the linear NS-PIM-Tr3, the other four NS-PIM/NS-RPIM models provide much tighter upper bounds to the exact solution. This is because that higher order PIM or RPIM shape functions are used in these models, which reduce both the softening and stiffening effects and push the solution closer to the exact one.

4) In terms of relative error measured in strain energy, the linear NS-PIM-Tr3 has similar accuracy as the linear FEM-Tr3, and the other four models of NS-PIM/NS-RPIM are about two times more accurate than the FEM.

5) In terms of computational efficiency, except that the bilinear NS-PIM-Tr4-CT is 1.5 times more efficient, other four NS-PIM/NS-RPIM models are all less efficient than the FEM-Tr3.

6) Among all these models, the NS-PIM-Tr4-CT stands out for this problem, in terms of both accuracy and efficiency measured in strain energy.

TABLE 6.24 Estimated computational efficiency of different methods measured in the error in *strain energy solution* of the numerical results for the cantilever beam problem with the same set of triangular mesh (Mesh-4 of 1,696 nodes)

Numerical method	Strain energy Solution	Error[*] (%)	Error ratio to FEM-Tr3	Efficiency
FEM-Tr3	8.5474	-0.5345	1.00	1.0
NS-PIM-Tr3	8.6398	0.5407	1.01	0.8
NS-PIM-Tr4-CT	8.6184	0.2917	0.55	1.5
NS-PIM-Tr6/3	8.6201	0.3115	0.58	0.7
NS-RPIM-Tr6	8.6199	0.3092	0.58	0.6
NS-RPIM-Tr2L	8.6178	0.2847	0.53	0.4

[*] The analytical value of the strain energy for the cantilever beam problem is 8.59333333. Negative value of error indicates lower bound, and positive indicates upper bound.

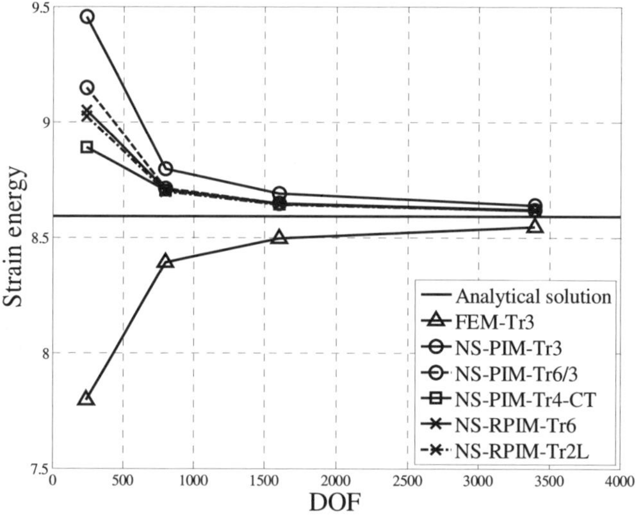

FIGURE 6.43 Upper bound solution obtained using the NS-PIM and NS-RPIM models for the 2D rectangular beam problem; the lower bound solution is obtained using the FEM with linear elements of the same meshes.

Example 6.4.3 A 2D Infinite solid with a circular hole

We now revisit the problem of a circular hole in infinite solid but with the focus on the examination of the upper bound properties of the NS-PIM and NS-RPIM, in comparison with the linear FEM. The settings and the parameters for this example problem are the same as those given in Example 6.1.3.

Using the same set of triangular meshes shown in FIGURE 6.14, we studied this problem and plotted the convergence process of strain energy solutions against the increase of DOFs in FIGURE 6.44. TABLE 6.25 lists the relative errors of the computed strain energy for the numerical results obtained using the same set of triangular mesh (Mesh-4). Considering the computational cost given in TABLE 6.3 and TABLE 6.12 for these methods, the estimated computational efficiency measured in strain energy solution is give in the last column in TABLE 6.25. The following points may be noted for this example.

1) The solutions of all the models converge to the exact solution with the increase of DOFs.

2) The linear FEM-Tr3 provides a lower bound solution, and all the NS-PIM/NS-RPIM models provide upper bound solutions to the exact one. Compared to the linear NS-PIM-Tr3, the other four NS-PIM/NS-RPIM models provide much tighter upper bounds to the exact solution.

3) In terms of the relative error of strain energy and computational efficiency, all the NS-PIM/NS-RPIM models show better performance than the linear FEM-Tr3.

4) Among all these models, the NS-PIM-Tr4-CT stands out clearly for this problem, in terms of efficiency measured in strain energy. It is about 3 times more efficient than the linear FEM-Tr3.

5) Other four NS-PIM/NS-RPIM models are also more efficient (10% to 30%) than the linear FEM.

TABLE 6.25 Estimated computational efficiency of different methods measured in the error in *strain energy solution* of the numerical results for the infinite solid with circular hole problem with the same set of triangular mesh (Mesh-4 of 3,578 nodes)

Numerical method	Strain energy Solution	Error[*] (%)	Error ratio to FEM-Tr3	Efficiency
FEM-Tr3	4.3230E-05	-0.0495	1.00	1.0
NS-PIM-Tr3	4.3265E-05	0.0314	0.63	1.3
NS-PIM-Tr4-CT	4.3257E-05	0.0130	0.26	3.2
NS-PIM-Tr6/3	4.3258E-05	0.0153	0.31	1.3
NS-RPIM-Tr6	4.3257E-05	0.0130	0.26	1.2
NS-RPIM-Tr2L	4.3256E-05	0.0106	0.21	1.1

[*] The analytical value of the strain energy for the infinite solid with hole problem is 4.32513989E-05.

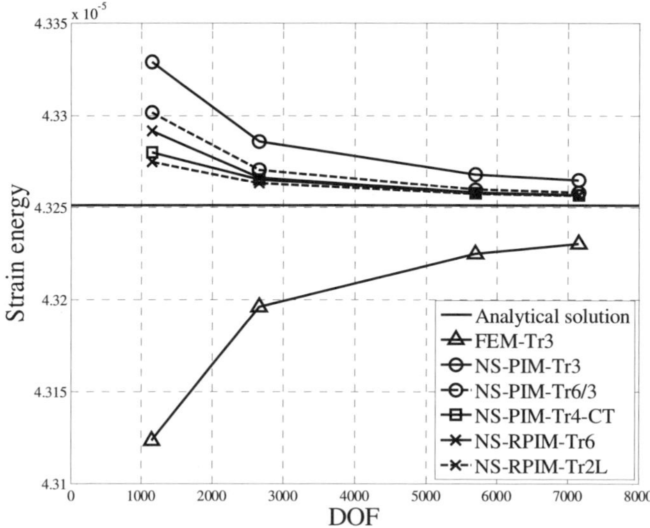

FIGURE 6.44 Upper bound solution obtained using the NS-PIM and NS-RPIM models for the 2D infinite solid with hole problem; the lower bound solution is obtained using the FEM with linear elements of the same meshes.

Example 6.4.4 A 2D connecting rod problem

A practical problem of typical connecting rod used in automobiles is studied focusing on the upper bound property of various NS-PIM/NS-RPIM models. The rod is constrained along the left circle and subjected to a uniform radial pressure $P=100$ along the right circle, as shown in FIGURE 6.45. Four triangular meshes with 319, 671, 1,050 and 2,155 nodes are created and used in the computation, as shown in FIGURE 6.46.

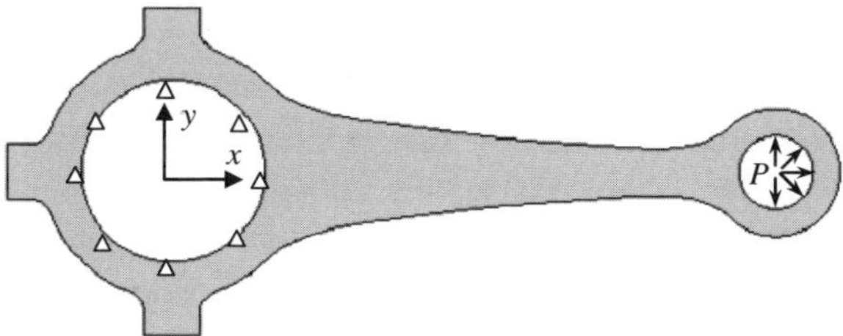

FIGURE 6.45 Model of the 2D connecting rod problem.

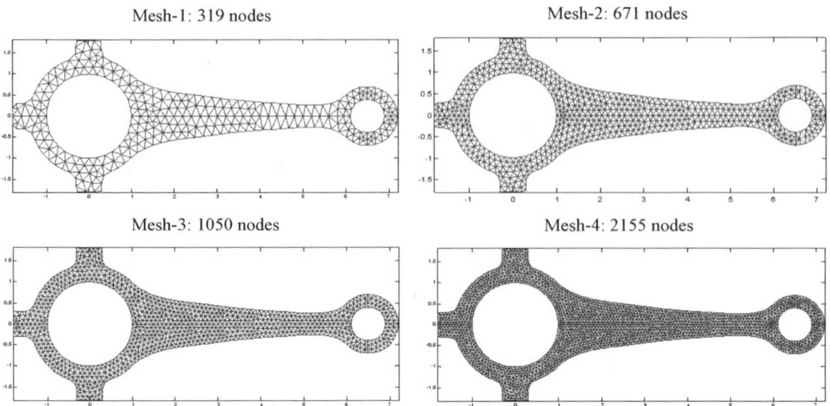

FIGURE 6.46 Domain discretization of the 2D connecting rod using irregular triangular meshes.

Solutions of strain energy for both FEM and NS-PIM models are plotted against the DOFs. FIGURE 6.47 shows the convergence process of the strain energy solution for this practical problem with complicated shape. TABLE 6.26 lists the relative errors of the computed strain energy for the numerical results obtained using the same set of triangular mesh (Mesh-4). Considering the computational cost for these methods, the estimated computational efficiency measured in strain energy solution is give in the last column in TABLE 6.26. We have the following points.

1) All the numerical solutions converge to the exact solution with the increase of DOFs.

2) The linear FEM-Tr3 provides a lower bound solution, and all the NS-PIM/NS-RPIM models provide upper bound solutions to the reference one, when fine meshes are used.

3) The NS-PIM-Tr4-CT, NS-RPIM-Tr6 and NS-RPIM-Tr2L models show non-monotonic-convergence for this problem, as shown in FIGURE 6.47. When the mesh is very coarse, the softening effects are small, and the model is stiff and may give lower bound solutions, such as the NS-RPIM-Tr2L model using Mesh-1 does. When the mesh is refined, the softening effects increase leading to upper bounds. Compared to other numerical methods, these three models provide much tighter bound solutions for this problem.

4) Among all these models, the NS-PIM-Tr4-CT stands out clearly for this practical problem, in terms computational efficiency measured in strain energy. It is about 7 times more efficient than the linear FEM-Tr3.

TABLE 6.26 Estimated computational efficiency of different methods measured in the error in *strain energy solution* of the numerical results for the connecting rod problem with the same set of triangular mesh (Mesh-4 of 2,155 nodes)

Numerical method	Strain energy Solution	Error[*] (%)	Error ratio to FEM-Tr3	Efficiency
FEM-Tr3	1.0814E-03	-2.105	1.00	1.0
NS-PIM-Tr3	1.1165E-03	1.073	0.51	1.6
NS-PIM-Tr4-CT	1.1074E-03	0.249	0.12	6.9
NS-PIM-Tr6/3	1.1110E-03	0.575	0.27	1.5
NS-RPIM-Tr6	1.1110E-03	0.575	0.27	1.2
NS-RPIM-Tr2L	1.1108E-03	0.557	0.26	0.9

[*] The reference value of the strain energy for the connecting rod problem is 1.1046479E-03, which is obtained using FEM with a very fine mesh (six-node triangular mesh of 192,420 elements and 387,977 nodes).

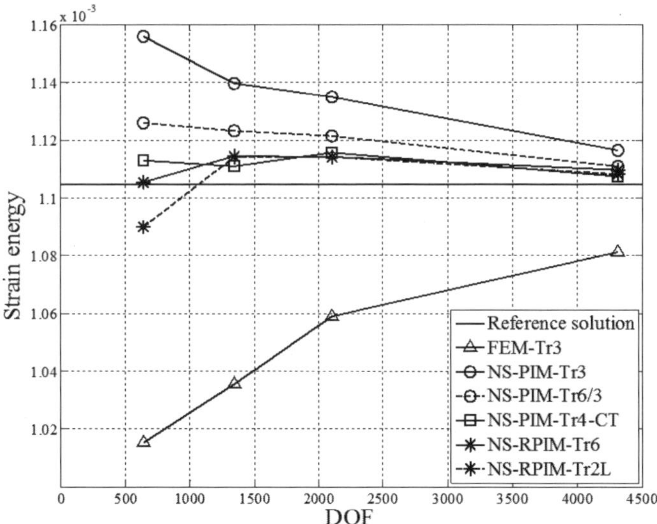

FIGURE 6.47 Upper bound solution obtained using the NS-PIM and NS-RPIM models for the 2D connecting rod problem; the lower bound solution is obtained using the FEM with linear elements of the same meshes.

Example 6.4.5 A 3D Lame problem

The 3D Lame problem described in Example 6.3.2 is revisited with focusing on the examination of the upper bound properties of the linear NS-PIM-Te4, in comparison with the linear FEM-Te4. The settings and the parameters for this example problem are the same as those given in Example 6.3.2.

Using the same set of tetrahedral meshes used in Example 6.3.2, we studied this problem and plotted the convergence process of strain energy solutions against the increase of DOF, as shown in FIGURE 6.48. Using the same Mesh-4 of 1,304 nodes, the relative errors of the computed strain energy for the numerical results are listed in TABLE 6.27, together with the estimated computational efficiency in the last column by considering the cost in TABLE 6.21. It can be found that

TABLE 6.27 Estimated computational efficiency of different methods measured in the error in strain energy solution of the numerical results for the Lame problem with the same set of tetrahedral mesh (Mesh-4 of 1,304 nodes)

Numerical method	Strain energy Solution	Error (%)	Error ratio to FEM-Te4	Efficiency
FEM-Te4	6.1701E-01	-1.86	1.00	1.0
NS-PIM-Te4	6.5268E-01	3.81	2.00	0.3

[*] The reference value of the strain energy for the Lame problem is 6.2869698E-01.

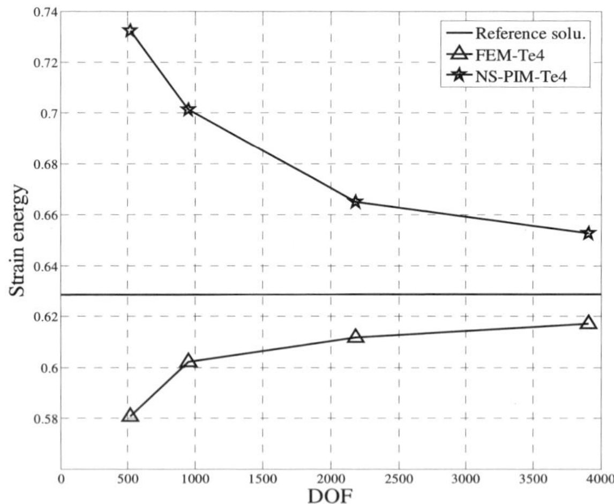

FIGURE 6.48 Upper bound solution obtained using the linear NS-PIM-Te4 for the 3D Lame problem; the lower bound solution is obtained using the FEM-Te4 with linear elements of the same meshes.

1) the numerical solutions converge to the exact one with the increase of DOF;

2) the FEM-Te4 and NS-PIM-Te4 give lower and upper bounds to the exact solution, respectively; and

3) for this problem, the linear NS-PIM-Te4 has worse accuracy and lower efficiency than the linear FEM-Te4.

Example 6.4.6 A 3D axletree base

Finally, linear NS-PIM and FEM are used to solve a 3D practical problem of a mechanical component of axletree base to examine the solution bounds. As shown in FIGURE 6.49, the axletree base is symmetric with respect to the $y-z$ plane. It is fixed at the locations of four lower cylindrical holes and subjected to a uniform pressure of $P=100$ on the concave annulus. The material constants used in analysis are $E=3.0\times10^7$ and $v=0.3$.

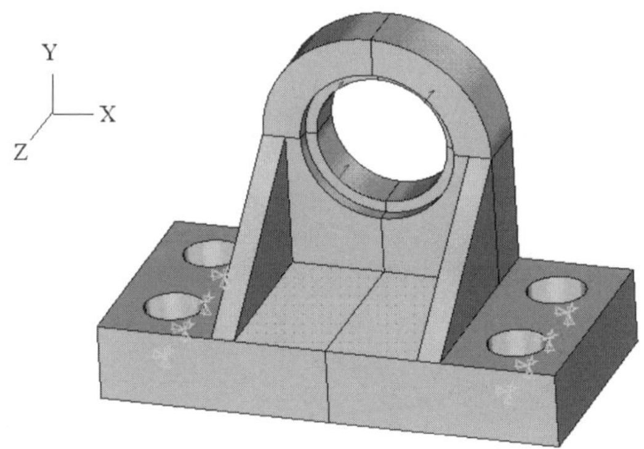

FIGURE 6.49 A mechanical component of 3D axletree base.

Four tetrahedral meshes with 781, 1828, 2,566 and 3,675 nodes are created and used in the computation. Solutions of strain energy for both FEM and NS-PIM are plotted against the DOFs. FIGURE 6.50 shows the convergence process of the strain energy solution for this practical problem with complicated shape.

TABLE 6.28 lists the relative errors of the computed strain energy for the numerical results obtained using the same set of triangular mesh (Mesh-4). Considering the computational cost given in TABLE 6.21 for these methods, the estimated computational efficiency measured in strain energy solution is give in the last column in TABLE 6.28. The following points may be noted.

1) The computed strain energy converges to the reference one with the increase of DOF.

2) The FEM-Te4 and NS-PIM-Te4 give lower and upper bounds to the reference solution, respectively.

3) For this problem, the linear NS-PIM-Te4 has similar accuracy and lower efficiency than the linear FEM-Te4.

TABLE 6.28 Estimated computational efficiency of different methods measured in the error in strain energy solution of the numerical results for the 3D axletree base with the same set of tetrahedral mesh (Mesh-4 of 3,675 nodes)

Numerical method	Strain energy Solution	Error (%)	Error ratio to FEM-Te4	Efficiency
FEM-Te4	1.7479E-03	-16.3	1.00	1.0
NS-PIM-Te4	2.4388E-03	16.8	1.03	0.65

[*] The reference value of the strain energy for the 3D axletree base is 2.0878887E-03, which is obtained using FEM with a very fine mesh (ten-node tetrahedral mesh of 238,041 elements).

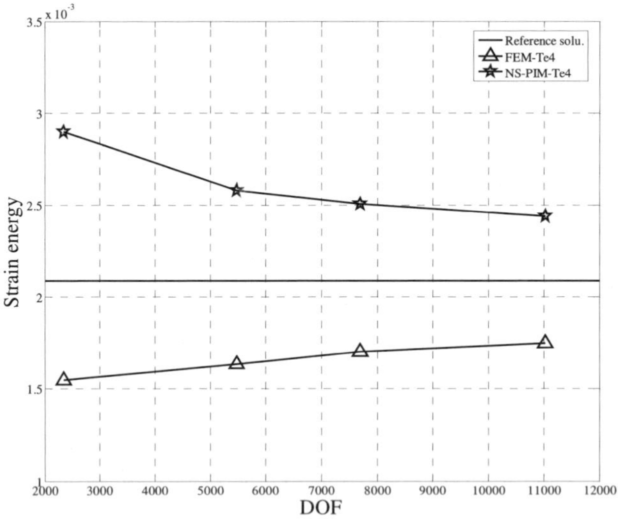

FIGURE 6.50 Upper and lower bounds for solution in strain energy obtained using the linear NS-PIM and FEM for the 3D axletree base problem.

Note in this example, the reference solution is computed using the FEM, and hence itself is a lower bound solution to the exact solution that we do not know. Therefore, the exact solution plotted in FIGURE 6.50 should be higher than the reference solution, meaning that it should be closer to the NS-PIM solution. The similar issues exist for all the problems using reference solution obtained via an FEM with a fine mesh.

6.5   Concluding remarks

This chapter presents in detail the NS-PIM with various models. It is a W^2 formulation based on the GS-Galerkin weak form. It has a number of important properties in contrary to the standard FEM. Following is the summary of these properties.

Remark 6.17 NS-PIM: accuracy

In terms of accuracy in displacement norm, the best performer NS-PIM-Tr4-CT is about 2 times more accurate than that of the FEM-Tr3 (see, e.g., TABLE 6.5 and TABLE 6.7). Using the iterative solver, the NS-PIM-Tr4-CT will cost about 1.2 times the CPU time compared to the linear FEM and hence be about 2 times more efficient.

In terms of accuracy in energy norm, the NS-PIM-Tr4-CT is about 3 times more accurate than the FEM-Tr3 (see, e.g., TABLE 6.6 and TABLE 6.8). Considering the computational cost listed in TABLE 6.3, the NS-PIM-Tr4-CT can be more than 3 times efficient than the linear FEM model.

Remark 6.18 NS-PIM: upper bound property

In this section, we studied the upper bound property of the node-based smoothed point interpolation methods (NS-PIMs) for 1D, 2D and 3D problems. In a summary, some concluding remarks may be made as follows.

1) The smoothing operation provides softening effects, while the construction of displacement introduces stiffening effects to the NS-PIM models.

2) Because node-based smoothing operations are performed in NS-PIM models, the softening effects are found sufficiently strong, which results in upper bound solutions to the exact one.

3) Compared to other NS-PIM and NS-RPIM models, the linear NS-PIM-Tr3 provides the loosest upper bound. The bilinear NS-PIM-Tr4-CT can provide

a very much tighter upper bound solution, although both models have the same sparseness. The NS-RPIM-T2L provides the tightest upper bound, but it uses a lot more nodes (about 8-22) compared to NS-PIM-Tr4-CT (about 4-8). Therefore, NS-PIM-Tr4-CT is recommended.

4) Together with the FEM model, NS-PIM can bound the solution from both sides for problems of arbitrary complexity as long as a triangular/tetrahedral mesh can be built.

Remark 6.19 NS-PIM: a quasi-equilibrium model

The NS-PIM is in general an incompatible model, it is found free of volumetric locking, superconvergence in displacement and energy norms, produces upper bound solutions, and there exist spurious modes at high energy level. This behavior is quite similar to that of an equilibrium model. In fact, at any point in all these smoothing domains, the body-force-free equilibrium equations are satisfied in an NS-PIM model. It is, however, not a fully equilibrium model because the stresses on these interfaces of the smoothing domains are not in equilibrium. It is said a *quasi-equilibrium model.*

Remark 6.20 Solution bound for general problems

In this section, we focus on force-driving problems. For displacement driving problems: zero external forces but nonzero prescribed displacements on the essential boundary, we can expect that the FEM and NS-PIM to swap their roles. The NS-PIM gives the lower bound and FEM gives the upper bound. For general problems with mixed force and displacement boundary conditions, we will not know which plays what role before the analysis. However, we can still expect these two models bound the exact solution from both sides. Which model is on which side will be problem dependent.

Remark 6.21 On solution bounds

With the development of NS-PIM, we now have a practical means to obtain both upper and lower bounds for problems with arbitrary complexity as long as a triangular/tetrahedral mesh can be built. We do not really worry too much about where the exact solution is, because we know the error of the approximate solution. In addition, such a "certified" solution can help to determine how fine the mesh should be. As soon as the error is acceptable for the design purpose, we can then stop further refining the model. This know-where-to-stop is very important because it gives us confidence for the solution as well as preventing using unnecessarily large models in the analysis and design. This translates to savings of computational and manpower resources.

The development of practical methods for producing certified solutions will become more and more important to engineering analysis and design. Techniques that can provide upper bound solutions like NS-PIMs are very much in demand. Even we have very powerful computer and can use hundreds of millions of DOFs to solve a problem with extremely high accuracy, if we cannot quantify the error, it is practically useless! On the other hand, when one uses a coarse model of only thousand DOFs but obtains a certified solution of 10% error, it is in fact much more useful in design.

Remark 6.22 Extension to general n-sided polygonal cells/elements

In the above discussions on NS-PIM, we focused on the use of triangular/ tetrahedral background cells/elements. However, the same idea can be applied for general n-sided cells/elements, as practiced in the S-FEM [14, 15].

6.6   Computer program

For the convenience of the readers who are interested in using and further develop S-PIM methods, we provide for free a set of source codes in FORTRAN 90 for solving 2D linear elastic solids problems. The codes are design for easy to read, but not for optimal efficiency.

Because the similarity of the program structure for different S-PIM models, most of the subroutines are designed to be commonly shareable for all the S-PIMs. All these *common* subroutines together with global parameters and array variables are provided with explanations in the Appendix. In this section, only the main code and a few *particular* subroutines for S-PIM models are provided. This section focuses the discussion on the source codes particularly for various NS-PIM/NS-RPIM models.

6.6.1   On the structure of the source codes

For easy comprehension, we use Example 6.1.2 to introduce the code structure. In solving Example 6.1.2, we use Mesh-1 with 120 nodes shown in FIGURE 6.9. TABLE 6.29 lists the major subroutines which are called in the main code NSPIM.F90 (to be detailed in Section 6.6.2). The functions of these subroutines are also given in TABLE 6.29.

TABLE 6.29 Major subroutines used in NSPIM.F90 and their functions

Subroutines	Type	Function explanation
Input	Common routine	Input data from an external file and find out the nodes with essential and natural boundary conditions
C_materialM	Common routine	Compute the material matrix
Cell_information	Common routine	Obtain necessary information of the background cell
StiffM_Intedomain	Common routine	Compute the stiffness matrix for an integration domain
Form_GK	Common routine	Form the global stiffness matrix by assembling the local stiffness matrixes
Natural_BC	Common routine	Enforce the nature boundary conditions
Essential_BC	Common routine	Apply the essential boundary conditions
Solver_LAE	Common routine	Solve the linear algebraic equations
Dispnorm_error	Common routine	Compute the global displacement norm error
FormB_NSPIM	Particular for NS-PIM	Form the strain-displacement matrix for a node-based smoothing domain
Nodalstress_NSPIM	Particular for NS-PIM	Obtain nodal stress and strain components for the NS-PIM models
Enernorm_error_NSPIM	Particular for NS-PIM	Compute the global energy norm error and strain energy for NS-PIM results

From TABLE 6.29, we note that except three subroutines are coded particularly for the NS-PIM models, all the other subroutines are the common routines and used by all the S-PIM models. So we only present the three particular subroutines in this chapter, and put all the common routine in the code library located in the Appendix for easy reference.

The two switches, i.e. NTscheme and NPIMSF, controlling the model types of S-PIM are described in TABLE 6.30.

TABLE 6.30 Two switches used in S-PIM models

	NTscheme	NPIMSF
Type	Integer	Integer
Usage	Input	Input
Function	Determine the type of T-schemes used for support nodes selection	Determine the type of PIM shape functions
Explanation	1: Cell-based T3-scheme	1: Normal polynomial PIM shape functions
	2: Cell-based T6/3-scheme	2: Normal radial PIM (RPIM) shape functions
	3: Cell-based T6-scheme	3: Polynomial PIM shape functions with coordinate transformation (CT)
	4: Cell-based T2L-scheme	
	5: Edge-based T4-scheme	4: Isoparametric PIM shape functions
	6: Edge-based T2L-scheme	5: Condensed RPIM shape functions

By assigning different values to these two switches of NTscheme and NPIMSF shown in TABLE 6.30, four NS-PIM/NS-RPIM models can be established using the same set of source codes, as shown in TABLE 6.31.

TABLE 6.31 Choices of variable values for different NS-PIM/NS-RPIM models

NS-PIM/NS-RPIM models	Value of NTscheme	Value of NPIMSF
NS-PIM-Tr3	1	1
NS-PIM-Tr6/3	2	1
NS-RPIM-Tr6	3	2
NS-RPIM-Tr2L	4	2

For the reader, the complete source code for a particular NS-PIM model can be "assembled" in the following straightforward way:

1) using the main program and the three particular subroutines described in Section 6.6.2;

2) assembling the input file "DATAINPUT" and two modules ("Parameters" and "Variables") detailed in Appendix 2 and Appendix 3, respectively;

3) assembling the associated common subroutines (see Appendix 4) called in the main code of NSPIM.F90; and

4) finally assigning the values of two switches of NTscheme and NPIMSF for the particular NS-PIM model.

Similarly, source codes for various ES- and CS-PIM models can be assembled as described in Chapters 7 and 8, respectively.

6.6.2 Source codes in FORTRAN 90

Program 6.1 Main code of "NSPIM.F90"

```fortran
PROGRAM NSPIM
!================================================================
! The main program for conducting the 2D NS-PIM/NS-RPIM methods
!================================================================
Use Parameters
```

```
Use Variables
Implicit real*8 (a-h, o-z)
Character*20 DATAINPUT, DATAOUTPUT
Open (5, FILE=' DATAINPUT ')
Open (10, FILE=' DATAOUTPUT ', status='unknown')

Call Input
Call C_materialM
Call Cell_information
Gk=0.d0; force=0.d0
Loop_on_nodes: do ia=1, numnode
    Bmat=0.d0; ndexnow=0; nvlist=0
    Call FormB_NSPIM(ia)
    area_inte=area_node(ia)
    Call StiffM_Intedomain(ia,area_inte,gpk)
    Call Form_GK
End do Loop_on_nodes
Call Natural_BC
Call Essential_BC
Call Solver_LAE(Gk,force,nx*numnode,nx*numn)
Call Dispnorm_error
Call Nodalstress_NSPIM
Call Enernorm_error_NSPIM

End PROGRAM NSPIM
```

Program 6.2 Source code of "FormB_NSPIM"

```
Subroutine FormB_NSPIM (ia)
!=================================================================
! Form the strain-displacement matrix of the ia_th node-based smoothing domain
!
! Input:     x, nxcell, area_node, kcell, numneig, neighbour and nodesix.
! Output:    Bmat, nvlist and ndexnow.
!=================================================================
```

```fortran
use Parameters
use Variables
implicit real*8 (a-h,o-z)
! work array
dimension xg(nx,2),xxg(nx,ngauss),wei(ngauss)
dimension xgg(2,3),onxy(2,2),nv(ndom),gpos(nx),phi(ndom)

ss=area_node(ia)
! Loop on the neighbouring cells of node_ia
LooP_on_neighbors: do i1=1,numneig(ia)
  kcc=neighbour(i1,ia); kcb=kcell(kcc)
  n1=nxcell(1,kcc); n2=nxcell(2,kcc); n3=nxcell(3,kcc)
  xx1=x(1,n1); yy1=x(2,n1); xx2=x(1,n2); yy2=x(2,n2)
  xx3=x(1,n3); yy3=x(2,n3)
  xm1=(x(1,n1)+x(1,n2))/2.; ym1=(x(2,n1)+x(2,n2))/2.
  xm2=(x(1,n2)+x(1,n3))/2.; ym2=(x(2,n2)+x(2,n3))/2.
  xm3=(x(1,n3)+x(1,n1))/2.; ym3=(x(2,n3)+x(2,n1))/2.
  xm5=center(1,kcc); ym5=center(2,kcc)
  xL1=((xm1-xm5)**2+(ym1-ym5)**2)**0.5
  onx11=(ym5-ym1)/xL1; ony11=(xm1-xm5)/xL1; onx12=-onx11; ony12=-ony11
  xL2=((xm2-xm5)**2+(ym2-ym5)**2)**0.5
  onx22=(ym5-ym2)/xL2; ony22=(xm2-xm5)/xL2; onx23=-onx22; ony23=-ony22
  xL3=((xm3-xm5)**2+(ym3-ym5)**2)**0.5
  onx33=(ym5-ym3)/xL3; ony33=(xm3-xm5)/xL3; onx31=-onx33; ony31=-ony33
  if (ia.eq.n1) then
      onxy(1,1)=onx11; onxy(2,1)=ony11; onxy(1,2)=onx31; onxy(2,2)=ony31
      xgg(1,1)=xm1; xgg(2,1)=ym1; xgg(1,2)=xm5; xgg(2,2)=ym5
      xgg(1,3)=xm3; xgg(2,3)=ym3
  else if (ia.eq.n2) then
      onxy(1,1)=onx22; onxy(2,1)=ony22; onxy(1,2)=onx12; onxy(2,2)=ony12
      xgg(1,1)=xm2; xgg(2,1)=ym2; xgg(1,2)=xm5; xgg(2,2)=ym5
      xgg(1,3)=xm1; xgg(2,3)=ym1
  else
      onxy(1,1)=onx33; onxy(2,1)=ony33; onxy(1,2)=onx23; onxy(2,2)=ony23
      xgg(1,1)=xm3; xgg(2,1)=ym3; xgg(1,2)=xm5; xgg(2,2)=ym5
      xgg(1,3)=xm2; xgg(2,3)=ym2
  endif
  ndex=0; nv=0
```

```fortran
select case(NTscheme)
case(1)
    ndex=3
    do kka=1,ndex
    nv(kka)=nxcell(kka,kcc)
    enddo
case(2)
    select case(kcb)
    case(0); ndex=6; case(1); ndex=3; case(2); ndex=3
    endselect
    do ik=1,ndex
    nv(ik)=nodesix(ik,kcc)
    enddo
case(3)
    ndex=6
    do kka=1,ndex
    nv(kka)=nodesix(kka,kcc)
    enddo
case(4)
    call cell_basedT2L(kcc,ndex,nv)
endselect
! Loop on two segments of smoothing domain boundaries located in cell_kcc
Loop_on_segments: do i2=1,2
    onx=onxy(1,i2); ony=onxy(2,i2)
    xg=0.d0; xxg=0.d0; wei=0.d0
    select case(i2)
    case(1)
    xg(1,1)=xgg(1,1); xg(2,1)=xgg(2,1); xg(1,2)=xgg(1,2); xg(2,2)=xgg(2,2)
    case(2)
    xg(1,1)=xgg(1,3); xg(2,1)=xgg(2,3); xg(1,2)=xgg(1,2); xg(2,2)=xgg(2,2)
    endselect
    call Line_gauss(xg,xxg,wei,det,Ngauss)
    ! Loop on the Gauss points located on the i2_th segment
    Loop_on_GP: do i3=1,ngauss
      gpos(1)=xxg(1,i3); gpos(2)=xxg(2,i3); weight=wei(i3)
      padx=onx*weight*det/ss; pady=ony*weight*det/ss; phi=0.d0
      select case(NPIMSF)
      case(1); call PPIM_SF2D(gpos,nv,ndex,phi)
```

```fortran
        case(2); call RPIM_SF2D(gpos,nv,ndex,phi)
        endselect
        kp=i1+i2+i3
        if (kp.eq.3) then
        ndexnow=ndex
        Inner1_1: do j2=1,ndex
            nvlist(j2)=nv(j2); k1=nx*j2-1; k2=nx*j2-0
            bmat(1,k1)=bmat(1,k1)+padx*phi(j2); bmat(2,k2)=bmat(2,k2)+pady*phi(j2)
            bmat(3,k1)=bmat(3,k1)+pady*phi(j2); bmat(3,k2)=bmat(3,k2)+padx*phi(j2)
        enddo Inner1_1
        else
        ndexlast=ndexnow
        Inner1_2: do j3=1,ndex
          ka=nv(j3); kindex=0
          Inner1_2_1: do j4=1,ndexlast
            kb=nvlist(j4)
            if (ka.eq.kb) then
            kindex=1; kc=nx*j4-1; kd=nx*j4-0
            bmat(1,kc)=bmat(1,kc)+padx*phi(j3)
            bmat(2,kd)=bmat(2,kd)+pady*phi(j3)
            bmat(3,kc)=bmat(3,kc)+pady*phi(j3)
            bmat(3,kd)=bmat(3,kd)+padx*phi(j3)
            endif
          enddo Inner1_2_1
          if (kindex.eq.0) then
          ndexnow=ndexnow+1;nvlist(ndexnow)=ka;          kc=nx*ndexnow-1; kd=2*ndexnow
            bmat(1,kc)=bmat(1,kc)+padx*phi(j3); bmat(2,kd)=bmat(2,kd)+pady*phi(j3)
            bmat(3,kc)=bmat(3,kc)+pady*phi(j3); bmat(3,kd)=bmat(3,kd)+padx*phi(j3)
          endif
        enddo Inner1_2
        endif
      enddo Loop_on_GP
     enddo Loop_on_segments
    if (ia.eq.n1) then
    k1=n1; k2=n2; k3=n3
    else if (ia.eq.n2) then
    k1=n2; k2=n3; k3=n1
```

```fortran
else
k1=n3; k2=n1; k3=n2
endif
kk1=nodein(k1); kk2=nodein(k2); kk3=nodein(k3); kk4=kk1+kk2+kk3
! Linear interpolation for the Gauss Points located on domain boundaries
if (kk1.eq.1 .and. kk4>=2) then
    Inner1_3: do m1=1,2
    select case(m1)
    case(1)
    kkm1=kk2; kkp=k2; xg=0.d0
    xg(1,1)=x(1,k1); xg(2,1)=x(2,k1)
    xg(1,2)=(x(1,k1)+x(1,kkp))/2.; xg(2,2)=(x(2,k1)+x(2,kkp))/2.
    case(2)
    kkm1=kk3; kkp=k3; xg=0.d0
    xg(1,1)=(x(1,k1)+x(1,kkp))/2.;xg(2,1)=(x(2,k1)+x(2,kkp))/2.
    xg(1,2)=x(1,k1); xg(2,2)=x(2,k1)
    endselect
    if (kkm1.eq.1) then
    xL12=((xg(1,1)-xg(1,2))**2+(xg(2,1)-xg(2,2))**2)**0.5
    onx=(xg(2,2)-xg(2,1))/xL12; ony=(xg(1,1)-xg(1,2))/xL12
    ndex=2; nv=0; nv(1)=k1; nv(2)=kkp
    ma=k1; mb=kkp
    call Line_gauss(xg,xxg,wei,det,Ngauss)
    Inner1_3_1: do i5=1,ngauss
      gpos(1)=xxg(1,i5); gpos(2)=xxg(2,i5)
      ss1=((gpos(1)-x(1,ma))**2+(gpos(2)-x(2,ma))**2)**0.5
      ss2=((gpos(1)-x(1,mb))**2+(gpos(2)-x(2,mb))**2)**0.5
      f1=ss2/(ss1+ss2); f2=ss1/(ss1+ss2); phi(1)=f1; phi(2)=f2
      weight=wei(i5); padx=onx*weight*det/ss; pady=ony*weight*det/ss
      do 10 i7=1,ndex
         ka=nv(i7)
         do 20 i6=1,ndexnow
         kb=nvlist(i6)
         if (kb.eq.ka) then
         kc=nx*i6-1; kd=nx*i6-0
         bmat(1,kc)=bmat(1,kc)+padx*phi(i7); bmat(2,kd)=bmat(2,kd)+pady*phi(i7)
         bmat(3,kc)=bmat(3,kc)+pady*phi(i7); bmat(3,kd)=bmat(3,kd)+padx*phi(i7)
         goto 10
```

```
        endif
     20 continue
   10  continue
  enddo Inner1_3_1
    endif
  enddo Inner1_3
 endif
end do LooP_on_neighbors


return
end subroutine FormB_NSPIM
```

Program 6.3 Source code of "Nodalstress_NSPIM"

```
Subroutine Nodalstress_NSPIM
!==================================================================
! Obtain nodal stress and strain compenents for the NS-PIM models
!
! Input:     Cmat, area_node and force (displacement solutions).
! Output:    stress_node and strain_node.
!==================================================================
use Parameters
use Variables
implicit real*8 (a-h,o-z)
! wrok array
dimension Dinv(3,3),ne(nx*nndom)

Dinv=0.d0; Dinv=Cmat
call inversion(3,3,Dinv)
stress_node=0.d0; strain_node=0.d0
Loop_on_nodes: do ia=1,numnode
  weight=area_node(ia)
  Bmat=0.d0; ndexnow=0; nvlist=0
  call formB_NSPIM(ia)
```

```fortran
  do ih=1,ndexnow
    n1=nx*ih-1; n2=nx*ih-0
    ne(n1)=nx*nvlist(ih)-1
    ne(n2)=nx*nvlist(ih)-0
  enddo
  do ij=1,3
  do ik=1,3
  do il=1,nx*ndexnow
    mn=ne(il)
    stress_node(ij,ia)=stress_node(ij,ia)+Cmat(ij,ik)*Bmat(ik,il)*force(mn)
  enddo
  enddo
  enddo
  do iq=1,3
  do ir=1,3
      strain_node(iq,ia)=strain_node(iq,ia)+Dinv(iq,ir)*stress_node(ir,ia)
  enddo
  enddo
enddo Loop_on_nodes

!output of the nodal stress results for the first 20 nodes
write(10,11) 'Field No.','Stress Sigma_xx','Stress Sigma_yy','Stress Tau_xy'
do i=1,20
write(10,12) i,(stress_node(j,i),j=1,3)
enddo
11 format(a10,3a18)
12 format(i10,3e18.8)


Return
End subroutine Nodalstress_NSPIM
```

Program 6.4 Source code of "Enernorm_error_NSPIM"

```fortran
Subroutine Enernorm_error_NSPIM
!==============================================================
```

```fortran
! Computing the global energy norm error and strain energy for NS-PIM results
!
! Input:      area_node, stress_node and strain_node;
! Output:     Enernorm(the global displacement norm error) and
!             StrainE(the computed strain energy).
!===============================================================
use Parameters
use Variables
implicit real*8 (a-h,o-z)
! wrok array
dimension Dinv(3,3),ne(nx*nndom),gpos(nx)
dimension stressex(3),strainex(3),errstress(3),errstrain(3)

Dinv=0.d0; Dinv=Cmat
call inversion(3,3,Dinv)
StrainE=0.; Enernorm=0.
Loop_on_nodes: do ia=1,numnode
  weight=area_node(ia)
  ai=ylength**3/12.d0
  gpos(1)=x(1,ia); gpos(2)=x(2,ia)
  stressex=0.d0; strainex=0.d0
  stressex(1)=-pLoad*(xlength-gpos(1))*gpos(2)/ai
  stressex(3)=pLoad/2./ai*(ylength**2/4.-gpos(2)**2)
  do iq=1,3
  do ir=1,3
     strainex(iq)=strainex(iq)+Dinv(iq,ir)*stressex(ir)
  enddo
  enddo
  errstress=0.d0; errstrain=0.d0
  do im=1,3
     errstress(im)=stress_node(im,ia)-stressex(im)
     errstrain(im)=strain_node(im,ia)-strainex(im)
  enddo
  do ip=1,3
     StrainE=StrainE+weight*0.5*strain_node(ip,ia)*stress_node(ip,ia)
  enddo
  do ip=1,3
     Enernorm=Enernorm+weight*errstrain(ip)*errstress(ip)
```

```
    enddo
end do Loop_on_nodes

Enernorm=dsqrt(Enernorm)
write(10,11) 'Global error in energy norm:',Enernorm
write(10,12) 'Calculated strain energy:',StrainE
11 format(a29,e18.8)
12 format(a26,e21.8)


Return
End subroutine Enernorm_error_NSPIM
```

6.6.3 Computed results

Running the present program by taking NTscheme=1 and NPIMSF=1, analysis using the NS-PIM-Tr3 will be conducted and the following output file can be obtained. Note for demonstration purpose, only the results for the first 20 nodes are listed.

Output file of "DATAOUTPUT"

Field No.	Disp. U_x	Disp. U_y
1	-0.36800000E-05	-0.30000000E-05
2	0.36800000E-05	-0.30000000E-05
3	-0.73600000E-05	-0.27000000E-04
4	0.73600000E-05	-0.27000000E-04
5	0.80869917E-06	-0.44008734E-04
6	0.63591818E-04	-0.50840315E-04
7	-0.65040286E-04	-0.51601703E-04
8	0.11991704E-03	-0.71799785E-04
9	-0.11879454E-03	-0.72011376E-04
10	0.00000000E+00	-0.75000000E-04
11	0.00000000E+00	-0.75000000E-04
12	-0.97316430E-04	-0.18705557E-03
13	0.10933880E-03	-0.19445588E-03
14	-0.19805873E-03	-0.17494183E-03

15	0.20475191E-03	-0.17798958E-03
16	0.25212437E-03	-0.14337391E-03
17	-0.25237550E-03	-0.14419816E-03
18	-0.29995239E-03	-0.38564068E-03
19	0.32436281E-03	-0.39802453E-03
20	0.48685937E-03	-0.34661957E-03

Global error in displacement norm: 0.10005440E+00

Field No.	Stress Sigma_xx	Stress Sigma_yy	Stress Tau_xy
1	0.57801561E+03	0.23094914E+01	-0.14896716E+03
2	-0.58404691E+03	-0.27820076E+01	-0.14817227E+03
3	0.18498566E+04	0.39177617E+02	-0.95573915E+02
4	-0.18521429E+04	-0.42542664E+02	-0.10417259E+03
5	0.10190858E+02	0.99218472E+00	-0.11703368E+03
6	0.12026337E+04	0.94298412E+00	-0.13052978E+03
7	-0.11850624E+04	-0.11651399E+01	-0.13821381E+03
8	0.22189640E+04	-0.49511752E+01	-0.38243214E+02
9	-0.22155967E+04	0.64477254E+01	-0.42701896E+02
10	0.28186059E+04	-0.23645117E+02	-0.79931913E+02
11	-0.28076986E+04	0.30548866E+02	-0.77258376E+02
12	-0.60369265E+03	-0.14193107E+00	-0.15345653E+03
13	0.69142316E+03	0.13835419E+01	-0.14816786E+03
14	-0.18819205E+04	-0.80472940E+01	-0.11124850E+03
15	0.19285477E+04	0.12238271E+02	-0.10662486E+03
16	0.26823416E+04	0.13612592E+02	0.16109827E+01
17	-0.26770556E+04	-0.17443827E+02	-0.18514016E+01
18	-0.15560841E+04	0.66275529E+01	-0.10534061E+03
19	0.16495992E+04	-0.11471220E+02	-0.93116514E+02
20	0.26256208E+04	-0.42901197E+00	0.13923425E+02

Global error in energy norm: 0.35600820E+00

Calculated strain energy: 0.94551396E+01

6.7   References

1.   Liu, G. R., Zhang, G. Y., Dai, K. Y, Wang, Y. Y., Zhong, Z. H., Li, G. Y. and Han, X., A linearly conforming point interpolation method (LC-PIM) for 2D solid mechanics problems, *International Journal of Computational Methods*, 2(4): 645-665, 2005.

2.   Zhang, G. Y., Liu, G. R., Wang, Y. Y., Huang, H. T., Zhong, Z. H., Li, G. Y. and Han, X., A linearly conforming point interpolation method (LC-PIM) for three-dimensional elasticity problems, *International Journal for Numerical Methods in Engineering*, 72: 1524-1543, 2007.

3.   Liu, G. R., A generalized Gradient smoothing technique and the smoothed bilinear form for Galerkin formulation of a wide class of computational methods. *International Journal of Computational Methods*, 5(2): 199-236, 2008.

4.   Liu, G. R. and Zhang, G. Y., Upper bound solution to elasticity problems: A unique property of the linearly conforming point interpolation method (LC-PIM). *International Journal for Numerical Methods in Engineering*, 74: 1128-1161, 2008.

5.   Liu, G. R., *Meshfree Methods: Moving beyond the Finite Element Method*, 2nd Edition, CRC press, Boca Taton, USA, 2009.

6.   Trefethen, L. N., and Bau, D., Numerical Linear Algebra, AIAM, Philadelphia. 1997.

7.   Liu, G. R. and Gu, Y. T., A point interpolation method, in *Proc. 4th Asia-Pacific Conference on Computational Mechanics*, December, Singapore, 1009-1014, 1999.

8.   Liu, G. R. and Gu, Y. T., A point interpolation method for two-dimensional solids, *Int. J. Numer. Methods Eng.*, 50: 937-951, 2001.

9.   Liu, G. R. and Gu, Y. T., A local point interpolation method for stress analysis of two-dimensional solids, *Struct. Eng. Mech.*, 11(2): 221-236, 2001.

10.   Wang, J. G. and Liu, G. R., Radial point interpolation method for elastoplastic problems, in *Proc. 1st Int. Conf. on Structural Stability and Dynamics*, December 7-9, Taipei, China, 703-708, 2000.

11.   Wang, J. G. and Liu, G. R., A point interpolation meshless method based on radial basis functions, *International Journal for Numerical Methods in Engineering*, 54: 1623-1648, 2002.

12.   Liu, G. R. and Gu, Y. T., A local radial point interpolation method (LR-PIM) for free vibration analyses of 2-D solids, *J. Sound Vib.*, 246(1): 29-46, 2001.

13.   Liu, G. R. and Gu, Y. T., *An introduction to meshfree method methods and their programming*, Springer, 2005.

14.   Liu, G. R. and Nguyen-Thoi, T., *Smoothed Finite Element Methods*, CRC press, Boca Taton, USA, 2010.

15.   Liu, G. R., Nguyen-Thoi, T., Nguyen-Xuan, H. and Lam, K. Y., A node-based smoothed finite element method (NS-FEM) for upper bound solutions to solid mechanics problems, *Computers and Structures*, 87: 14-26, 2009.

16.   Chen, J. S., Wu C. T., Yoon, S. and You, Y., A stabilized conforming nodal integration for Galerkin mesh-free methods. *International Journal for Numerical Methods in Engineering*, 50: 435-466, 2001.

17.   Liu, G. R., Li, Y., Dai, K. Y., Luan M.T. and Xue, W., A Linearly conforming radial point interpolation method for solid mechanics problems, *International Journal of Computational Methods*, 3: 401-428, 2006.

18.   Li, Y., Liu, G. R., Luan, M. T., Dai, K. Y., Zhong, Z. H., Li, G. Y. and Han, X., Contact analysis for solids based on linearly conforming radial point interpolation method, *Computational Mechanics*, 39: 537-554, 2007.

19. Liu, G. R., A G space theory and a weakened weak (W^2) form for a unified formulation of compatible and incompatible methods, Part I Theory, *International Journal for Numerical Methods in Engineering*, 81: 1093-1126, 2009.

20. Liu, G. R., A G space theory and a weakened weak (W^2) form for a unified formulation of compatible and incompatible methods, Part II Application to solid mechanics problems, *International Journal for Numerical Methods in Engineering*, 81: 1127-1156, 2009.

21. Liu, G. R. and Zhang, G. Y. A normed G space and weakened weak (W^2) formulation of a cell-based smoothed point interpolation method. *International Journal of Computational Methods*, 6(1): 147-179, 2009.

22. Dohrmann, C. R., Heinstein, M. W., Jung, J., Key, S. W. and Witkowski, W. R., Node-based uniform strain elements for three-node triangular and four-node tetrahedral meshes. *International Journal for Numerical Methods in Engineering* 47: 1549-1568, 2000.

23. Liu, G. R. and Liu, M. B., *Smoothed particle Hydrodynamics-A Meshfree Particle Method*. World Scientific: Singapore, 2003.

24. Zhang, Z. Q. and Liu, G. R., Temporal stabilization of the node-based smoothed finite element method (NS-FEM) and solution bound of linear elastostatics and vibration problems. *Computational Mechanics*, 46(2): 229-246, 2010.

25. Zhang, Z. Q. and Liu, G. R., Upper and lower bounds for natural frequencies: a property of the smoothed finite element methods. *International Journal for Numerical Methods in Engineering*, 84(2): 149-178, 2010.

26. Zienkiewicz, O. C. and Taylor R. L., *The Finite Element Method*, 5th ed., Butterworth Heimemann, Oxford, 2000.

27. Timoshenko, S. P. and Goodier, J. N., *Theory of Elasticity*, 3rd ed., McGraw-Hill, New York, 1970.

28. Zhang, G. Y., Liu, G. R., Nguyen-Thoi, T., Song, C. X., Han, X., Zhong, Z. H. and Li, G. Y., The upper bound property for solid mechanics of the linearly conforming radial point interpolation method (LC-RPIM). *International Journal of Computational Methods*, 4(3): 521−541, 2007.

29. Liu, G. R. and Quek, S. S., *The finite element method: a practical course*. Butterworth Heinemann: Oxford, 2003.

30. Liu, G. R. and Zhang, G. Y., LETTER TO THE EDITOR: Comments on "Upper bound solution to elasticity problems: A unique property of the linearly conforming point interpolation method (LC-PIM)". *International Journal for Numerical Methods in Engineering*, 77: 1046-1050, 2009.

31. Liu, G. R., On a G space theory. *International Journal of Computational Methods*, 6(2): 257-289, 2009.

32. Liu, G. R., Nguyen-Thoi, T. and Lam, K. Y. A novel Alpha Finite Element Method (αFEM) for exact solution to mechanics problems using triangular and tetrahedral elements. *Computer Methods in Applied Mechanics and Engineering*, 197: 3883-3897, 2008.

Chapter 7

Edge-based Smoothed Point Interpolation Method (ES-PIM)

In Chapter 6, we have presented the node-based smoothed PIM (NS-PIM), and examined the properties of various NS-PIM models. These NS-PIM models are softer models, spatially stable, having nonzero-energy modes, work very well for static mechanics problems, and have very important upper bound and super-convergence properties in the strain energy [1-8]. However, the NS-PIM was found *temporally* instable [7-10], when applied to solve dynamic problems. Such instability is often observed at higher energy level as spurious (non-physical) nonzero-energy modes in the free vibration analysis of (stable and constrained) solids. Our studies have found that the cause of such instability is the "overly-soft" nature of these models introduced by the node-based smoothing operations. The authors believe that any method that has upper bound property can have spurious modes at a higher energy level and hence may suffer from the instability for solving dynamic problems. Therefore, for dynamic analysis of solid and structures, we need alternatives or special treatments to install temporal stability for these overly-soft models.

In this chapter, we introduce another important S-PIM model called edge-based smoothed point interpolation method or ES-PIM, which has been found both spatially and temporally stable, and works well for both static and dynamic problems [9, 10]. It is found also that the ES-PIM can produce much more accurate results compared to the NS-PIM and FEM using the same mesh [9-21]. The ES-PIM models are usually stiffer models with very "close-to-exact" stiffness. In particular the linear ES-PIM-Tr3 is found as one of the best linear models, and is regarded as a "star performer" among all the linear models.

This chapter focuses on presenting 2D ES-PIM models for both static and dynamic problems that were originated in [9]. For 3D models, we can apply the same idea, and to develop the so-called face-based smoothed point interpolation method (or FS-PIM) and face-based smoothed radial point interpolation method (or FS-RPIM). In such cases, the linear FS-PIM-Te4 was found particularly effective, and it is also known as the FS-FEM. Here we only give a brief introduction on 3D FS-PIM models in Section 7.4.3 and interested readers are referred to [20] for more details. Numerical results of FS-PIM for solid mechanics problems will be presented and compared with those of other models in Chapters 8 and 9.

7.1   Approximation of displacement field

Same as any S-PIM model, in the ES-PIM the problem domain is first represented by a set of N_c nonoverlapping and seamless triangular background cells with N_n field nodes, by triangulation. The selection of nodes for the construction of the shape function is performed in the similar way as in the NS-PIM, using a proper T-scheme detailed in Section 1.6.3. We can also use either PIM or RPIM shape functions following the procedure given in Chapter 3, which leads to variations of ES-PIM and edge-based smoothed radial point interpolation method (ES-RPIM) models.

When the cell-based T3-scheme, as shown in FIGURE 7.1a, is used we have a linear ES-PIM-Tr3 model that is the same as the ES-FEM-Tr3 [8, 10]; when the cell-based T6/3-scheme (see FIGURE 7.1b) is used we obtain a quadratic ES-PIM-Tr6/3. If the cell-based T6-scheme, as shown in FIGURE 7.1c, is used with RPIM shape functions we have a quadratic ES-RPIM-Tr6; and if the cell-based T2L-scheme (see, FIGURE 7.1d) is used with RPIM shape functions we obtain an ES-RPIM-Tr2L. It is also possible to use the edge-based T4-scheme, but has not yet been implemented, because we expect its performance being in between the ES-PIM-Tr3 and ES-PIM-Tr6/3. Note that the node selection is performed automatically without manual intervention.

Using PIM shape functions and support nodes selected by different T-schemes, the constructed displacement field in the scheme of ES-PIM models is not always continuous. Except for the ES-PIM-Tr3, where the displacement field is continuous over the problem domain and hence the model is variationally consistent, the displacement field in other ES-PIM models is not continuous and the discontinuity will occur on the interfaces of triangular background cells, as illustrated using dashed lines in FIGURE 7.1b-d.

Once a proper set of small number of local support nodes are selected using a cell-based T-scheme, PIM shape functions can then be created using the techniques detailed in Section 3.2. We can then

1) assume the solution of the displacement function for the ES-PIM models at any point **x** in the problem domain using Equation (5.66);

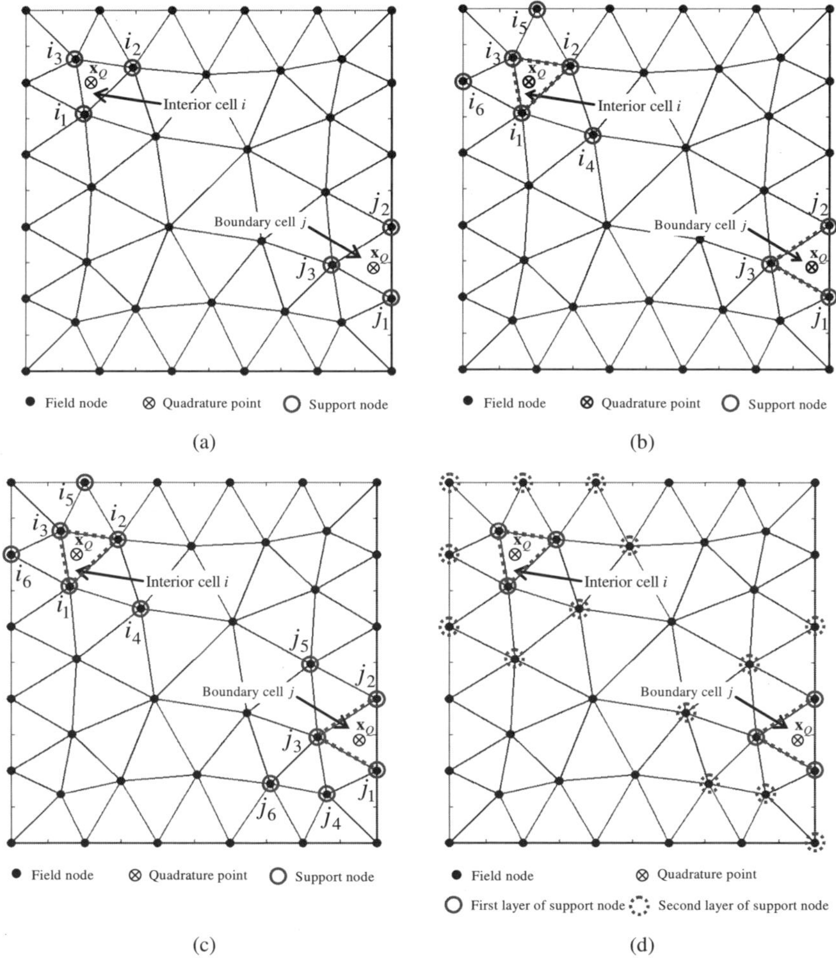

FIGURE 7.1 T-schemes for selecting support nodes for the construction of PIM/RPIM shape functions in the ES-PIM models: (a) cell-based T3-scheme used in ES-PIM-Tr3; (b) cell-based T6/3-scheme used in ES-PIM-Tr6/3; (c) cell-based T6-scheme used in ES-RPIM-Tr6; (d) cell-based T2L-scheme used in ES-RPIM-Tr2L.

2) compute the generalized smoothed strains following the procedure given in Section 2);

3) compute the force vector using Equation (5.82); and

4) for dynamics problems, compute the mass matrix using Equation (5.86).

7.2 Evaluation of edge-based smoothed strains

On top of the triangular background cells, the problem domain Ω is further divided into N_s edge-based smoothing domains associated with edges in a seamless and nonoverlapping manner, such that $\boxed{\Omega} = \bigcup_{i=1}^{N_s} \boxed{\Omega}_i^s$ and $\Omega_i^s \cap \Omega_j^s = \varnothing$, $\forall i \neq j$. The total number of smoothing domain becomes the total number of edges of triangular cells in the entire problem domain, i.e. $N_s = N_{eg}$. As the method based on the GS-Galerkin weak form, each edge-based smoothing domain also serves as quadrature/integration domain, and hence $N_q = N_s = N_{eg}$.

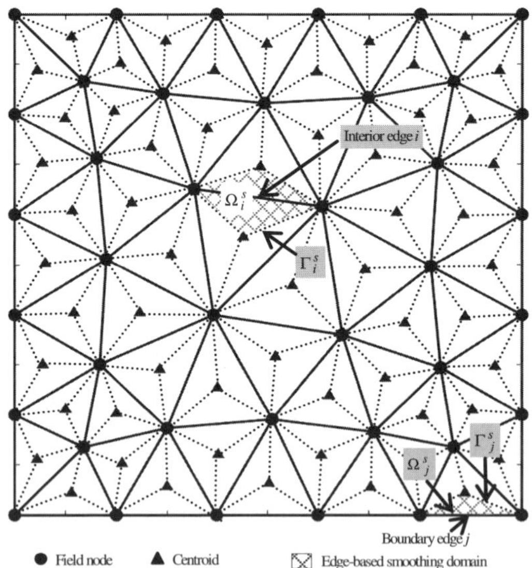

FIGURE 7.2 Triangular background cells and edge-based smoothing domains. In the problem domain the edge-based smoothing domain is a quadrilateral, such as Ω_i^s for the interior edge I; and on the problem boundary, it is a triangle and only one-sided, such as Ω_j^s for the boundary edge j.

The smoothing domain Ω_i^s associated with the edge i is created by connecting the nodes at the ends of the edge to the two centroids of the two adjacent elements sharing the edge, as shown in FIGURE 7.2. Clearly such a division of smoothing domains is "legal" and satisfies the conditions given in Section 2.3.1. In an ES-PIM model, we assume also the strain is constant in each smoothing domain which can be obtained following the strain smoothing operation described in Section 4.5.2. Hence the G space theory applies, and the GS-Galerkin weak form can be used to establish ES-PIM models, as discussed in Chapters 4 and 5.

Note from FIGURE 7.2 that the smoothing domains on the boundary edges are biased, which can be a source of the numerical error (see, Remark 3.24). To reduce such an error, it is preferable to reduce the size of the background cells near the problem boundary for ES-PIM models, whenever possible.

7.3 ES-PIM formulations

7.3.1 Dynamic analysis

Since ES-PIM models work very well for dynamic problems, our formulation shall be directly for dynamics which includes both inertial and damping terms. The discretized differential equations with respect to times follow Equation (5.85) and shall have the matrix form of

$$\overline{\mathbf{K}}^{\text{ES-PIM}}\overline{\mathbf{d}} + \mathbf{C}\dot{\overline{\mathbf{d}}} + \mathbf{M}\ddot{\overline{\mathbf{d}}} = \tilde{\mathbf{f}} \tag{7.1}$$

where a dot stands for a differentiation operation with respect to time, and $\overline{\mathbf{K}}^{\text{ES-PIM}}$ is the *smoothed* stiffness matrix whose entries are given by

$$\overline{\mathbf{K}}_{IJ}^{\text{ES-PIM}} = \int_{\Omega} \overline{\mathbf{B}}_I^{\text{T}} \mathbf{c} \overline{\mathbf{B}}_J d\Omega = \sum_{k=1}^{N_{eg}} \int_{\Omega_k^s} \overline{\mathbf{B}}_I^{\text{T}} \mathbf{c} \underbrace{\overline{\mathbf{B}}_J}_{\text{constant in } \Omega_k^s} d\Omega = \sum_{k=1}^{N_{eg}} A_k^s \overline{\mathbf{B}}_I^{\text{T}} \mathbf{c} \overline{\mathbf{B}}_J \tag{7.2}$$

in which $A_k^s = \int_{\Omega_k^s} d\Omega$ is the area of the edge-based smoothing domain Ω_k^s. The smoothed strain-displacement matrix $\overline{\mathbf{B}}_I$ is computed using Equation (5.71).

In Equation (7.1), $\mathbf{M}$ is the mass matrix that is assembled using the following entries.

$$\mathbf{M}_{IJ} = \int_{\Omega} \mathbf{\Phi}_I^{\mathrm{T}} \rho \mathbf{\Phi}_J \, \mathrm{d}\Omega \tag{7.3}$$

where ρ is the mass density of the material. $\mathbf{M}$ is always SPD, as long as ρ is positive (see Theorem 3.1b), and the nodal shape functions are linearly independent.

Matrix $\mathbf{C}$ is called the damping matrix that is assembled using the following entries.

$$\mathbf{C}_{IJ} = \int_{\Omega} \mathbf{\Phi}_I^{\mathrm{T}} c_{dp} \mathbf{\Phi}_J \, \mathrm{d}\Omega \tag{7.4}$$

where c_{dp} is the damping parameter. Similarly, $\mathbf{C}$ is also SPD, as long as c_{dp} is positive, and the nodal shape functions are linearly independent (see Theorem 3.1b).

Note that the computation of $\mathbf{M}$, $\mathbf{C}$, and $\tilde{\mathbf{f}}$ are basically the same as we do in the standard FEM, and no smoothing operation is applied.

7.3.2 Static analysis

For static problems, the system equations can be obtained by simply dropping the dynamic terms in Equation (7.1)

$$\overline{\mathbf{K}}^{\,\text{ES-PIM}}\overline{\mathbf{d}} = \tilde{\mathbf{f}} \tag{7.5}$$

The solution procedure of ES-PIM for static problems is the essentially the same as in the NS-PIM detailed in Chapter 6.

7.3.3 Free vibration analysis

Considering the free vibration of a stable solid, we shall not have the damping and external force terms but the inertial term. Equation (7.1) becomes

$$\overline{\mathbf{K}}^{\,\text{ES-PIM}}\overline{\mathbf{d}} + \mathbf{M}\ddot{\overline{\mathbf{d}}} = \mathbf{0} \tag{7.6}$$

A general solution to the foregoing equation can be given as

$$\overline{\mathbf{d}} = \overline{\mathbf{d}}_A \exp(i\omega t) \tag{7.7}$$

where t indicates time, $\bar{\mathbf{d}}_A$ is the amplitude of the nodal displacement (or the eigenvector satisfying the equation below), and ω is the natural frequency that is found by solving the following eigenvalue equation.

$$\left(\bar{\mathbf{K}}^{\text{ES-PIM}} - \omega^2 \mathbf{M}\right)\bar{\mathbf{d}}_A = \mathbf{0} \tag{7.8}$$

We know from Equations (7.2) and (7.3) that $\bar{\mathbf{K}}^{\text{ES-PIM}}$ is at least non-negative. When a proper set of essential boundary conditions are imposed, $\bar{\mathbf{K}}^{\text{ES-PIM}}$ will be all SPD, and hence the foregoing equation has a set of unique eigenvalues (may have multiplicity) and eigenvectors (when normalized). These eigenvalues correspond to the natural frequencies and the eigenvectors correspond to the vibration modes for the free vibrating solid.

7.3.4 Forced vibration analysis

For forced or transient vibration problems, Equation (7.1) can be solved by direct integration methods as in the FEM. For simplicity, the Rayleigh damping is assumed in this book, and damping matrix $\mathbf{C}$ becomes a linear combination of $\mathbf{M}$ and $\bar{\mathbf{K}}^{\text{ES-PIM}}$

$$\mathbf{C} = \alpha_R \mathbf{M} + \beta_R \bar{\mathbf{K}}^{\text{ES-PIM}} \tag{7.9}$$

where α_R and β_R are the Rayleigh damping coefficients.

Many direct integration schemes are available to solve the second-order time dependent problems, such as the Newmark method and Grank-Nicholson method. In this chapter, we use the Newmark method.

When the initial state at $t = t_0 \left(\bar{\mathbf{d}}_0, \dot{\bar{\mathbf{d}}}_0, \ddot{\bar{\mathbf{d}}}_0\right)$ for the model is known, we aim to find the next state at $t_1 = t_0 + \theta\Delta t \left(\bar{\mathbf{d}}_1, \dot{\bar{\mathbf{d}}}_1, \ddot{\bar{\mathbf{d}}}_1\right)$ where $0.5 \le \theta \le 1$, using the following time-marching formulations.

    1)   Solve first

$$\begin{aligned}
&\left[\left(\alpha_R + \frac{1}{\theta\Delta t}\right)\mathbf{M} + \left(\beta_R + \theta\Delta t\right)\bar{\mathbf{K}}^{\text{ES-PIM}}\right]\bar{\mathbf{d}}_1 = \theta\Delta t \tilde{\mathbf{f}}_1 + (1-\theta)\Delta t \tilde{\mathbf{f}}_0 \\
&+\left(\alpha_R + \frac{1}{\theta\Delta t}\right)\mathbf{M}\bar{\mathbf{d}}_0 + \frac{1}{\theta}\mathbf{M}\dot{\bar{\mathbf{d}}}_0 + \left[\beta_R - (1-\theta)\right]\bar{\mathbf{K}}^{\text{ES-PIM}}\bar{\mathbf{d}}_0
\end{aligned} \tag{7.10}$$

        for $\bar{\mathbf{d}}_1$ and then

2) compute the velocity at t_1 using

$$\dot{\bar{\mathbf{d}}}_1 = \frac{1}{\theta \Delta t}\left(\bar{\mathbf{d}}_1 - \bar{\mathbf{d}}_0\right) - \frac{1-\theta}{\theta}\dot{\bar{\mathbf{d}}}_0 \qquad (7.11)$$

3) and next the acceleration at t_1 with

$$\ddot{\bar{\mathbf{d}}}_1 = \frac{1}{\theta \Delta t}\left(\dot{\bar{\mathbf{d}}}_1 - \dot{\bar{\mathbf{d}}}_0\right) - \frac{1-\theta}{\theta}\ddot{\bar{\mathbf{d}}}_0 \qquad (7.12)$$

March forward one time step and go back to step 1), until the final time is reached. The solution will always be stable, as long as the numerical model is temporally stable, meaning that 1) $\bar{\mathbf{K}}^{\text{ES-PIM}}$ and $\mathbf{M}$ are SPD; and 2) $\bar{\mathbf{K}}^{\text{ES-PIM}}$ is not "soft".

7.4 Numerical implementation

7.4.1 Macro flowchart of the ES-PIM

The macro flowchart for constructing the global matrices and vectors of ES-PIM models can be briefly summarized as follows.

1) Loop over all the edges of the background cells (or edge-based smoothing domains).

2) Loop over surrounding cells directly connected to the edge.

3) Loop over the Gauss points located along the (two) boundary segments of the smoothing domain in the cell:

   (a) Select support nodes using a proper T-scheme and construct the PIM shape functions.

   (b) Compute the components of smoothed strain-displacement matrix and form the matrix.

   (c) Compute nodal stiffness matrices and force vectors.

   (d) Assemble the nodal contributions to the global matrices and vectors.

4) End the Gauss points loop.

5) End the cell loop.

6) End the smoothing domain loop.

7.4.2 Possible ES-PIM models

In the ES-PIM settings, a proper cell-based T-scheme (see Section 1.6.3) has to be used for the compliance to the no-sharing rule. Based on the interpolation schemes used in the construction of the displacement field, we can have the following possible variation of ES-PIM models.

ES-PIM-Tr3

This is the simplest ES-PIM model. It uses linear polynomial PIM shape functions and the support nodes are selected using the cell-based T3-scheme, as shown in FIGURE 7.1a. In this particular case, the displacement field is *compatible* over the problem domain and the corresponding ES-PIM-Tr3 formulation is variationally consistent. The ES-PIM-Tr3 is exactly the same as the ES-FEM-Tr3 [10].

To find out the rough average number of support nodes for edge-based smoothing domains, the model of the cantilever beam presented using 248 irregular nodes, as shown in FIGURE 6.4, is again used for analysis. For the ES-PIM-Tr3 model, the number of the support nodes for an edge-based smoothing domain is 3-4 and the average number is found 3.9 (for this beam model), which is about 1.3 times the number (3) of the support nodes for a linear element in the FEM-Tr3 model. Therefore, average number (n_{ze}) of nonzero entries in a row in the stiffness matrix of an ES-PIM-Tr3 model is about 1.3 times that of the FEM-Tr3 counterpart using the same mesh.

ES-PIM-Tr6/3

In ES-PIM-Tr6/3, the displacement field is constructed using polynomial PIM shape functions and support nodes selected by the cell-based T6/3-scheme, as shown in FIGURE 7.1b. The displacement field located in an interior cell will be approximated using quadratic PIM shape functions and six nodes; while displacement in a boundary cell will be approximated using linear PIM shape functions. The displacement field in the ES-PIM-Tr6/3 is incompatible.

For the ES-PIM-Tr6/3 model, the number of the support nodes for a smoothing domain is 3-8 and the average number is 7, that is about 2.3 times that of the FEM-Tr3. Therefore, the average number of nonzero entries in a row in the stiffness matrix of an ES-PIM-Tr6/3 model is about 2.3 time of that of the FEM-Tr3 counterpart using the same mesh.

ES-RPIM-Tr6

An ES-RPIM-Tr6 model uses RPIM shape functions with the support nodes selected by the cell-based T6-scheme, as shown in FIGURE 7.1c. There is no singularity problem in constructing RPIM shape functions, the cell-based T6-scheme is well controlled, and hence this model is very robust. Note that an ES-RPIM model using cell-based T3-scheme produces the same results as the ES-PIM-Tr3. The displacement field in an ES-RPIM-Tr6 model is incompatible.

For this model, the number of the support nodes for a smoothing domain varies from 6 to 8 and the average number is 7.7, which is about 2.6 times that of the FEM-Tr3. Therefore, the average number of nonzero entries in a row in the stiffness matrix of an ES-RPIM-Tr6 model is about 2.6 time of that of the FEM-Tr3 counterpart using the same mesh.

ES-RPIM-Tr2L

In ES-RPIM-Tr2L, the RPIM shape functions are used with the support nodes selected by the cell-based T2L-scheme, as shown in FIGURE 7.1d. This type of T-scheme is also well controlled and with a lot of freedom. Compared to the cell-based T6-scheme, this scheme works better for virtually randomly distributed nodes, but less efficient. The displacement field in an ES-RPIM-Tr2L model is incompatible.

For the ES-RPIM-Tr2L mode, the number of the support nodes for a smoothing domain is about 7-15 and the average number is 12.6, which is about 4.2 times the counterpart in the FEM-Tr3. Therefore, the average number of nonzero entries in a row in the stiffness matrix of an ES-RPIM-Tr2L model is about 4.2 times that of the FEM-Tr3 using the same mesh.

Note that the above analysis on the overall characteristics is just a rough estimation. Such an analysis gives us useful indications on what we can expect, before the model is actually built. For easy reference later, we summarize the above discussion in TABLE 7.1.

TABLE 7.1 ES-PIM models and overall characteristics in relation to the FEM-Tr3 model using the same mesh

Numerical method	Support nodes in an integral cell/element	Estimated average number of nonzero entries	Estimated Solver CPU time
FEM-Tr3	3	n_{ze}	t_{CPU}
ES-PIM-Tr3	3-4 (3.9)	$1.3\, n_{ze}$	$1.3\, \bar{n}_{iter} t_{CPU}$
ES-PIM-Tr6/3	3-8 (7.0)	$2.3\, n_{ze}$	$2.3\, \bar{n}_{iter} t_{CPU}$
ES-RPIM-Tr6	6-8 (7.7)	$2.6\, n_{ze}$	$2.6\, \bar{n}_{iter} t_{CPU}$
ES-RPIM-Tr2L	7-15 (12.6)	$4.2\, n_{ze}$	$4.2\, \bar{n}_{iter} t_{CPU}$

Note: for an iterative solver we assume $t_{CPU} \propto n_{iter} N_{DOF} n_{ze}$; $\bar{n}_{iter}$ is the iteration number n_{iter} of the model related to the linear FEM model.

7.4.3 FS-PIM for 3D solids

To extend the formulation of ES-PIM from 2D to 3D solids is straightforward. Most of the formulae for 3D have the same forms as those for 2D cases, and the major differences include that the displacement approximation is now conducted in 3D domains based on the tetrahedral background cells and the strain smoothing is performed based on the face-based smoothing domains. Therefore, the ES-PIM for 3D problems is usually called FS-PIM models. Following we simply brief some important issues.

For 3D problems, the problem domain is first discretized using four-node tetrahedron cells. Based on the tetrahedral background cells, the support nodes can be selected using a proper cell-based T-scheme and PIM shape functions can be constructed as presented previously in Chapter 3. For the linear FS-PIM which is exactly same as the FS-FEM [20], we use the cell-based T4-scheme to select four support nodes to construct the linear PIM shape functions. The displacement field in the scheme of linear FS-PIM is continuous over the whole problem domain. For higher order FS-PIM models in three dimensions, we suggest using the RPIM shape functions which selects the support nodes by the cell-based T2L-scheme due to the possible singularity problem using PIM shape functions. Note that in this case, the constructed displacement is discontinuous and the discontinuity occurs on the interface triangles of the tetrahedral background cells.

In the scheme of FS-PIM for 3D problems, the problem domain discretized using tetrahedral cells will be further divided into N_s NOSL smoothing domains associated with faces of tetrahedrons following the rules given in Section 2.3.1. Each face-based smoothing domain contains a face, thus the number of smoothing domains is the same as that of the faces and also equals to the number of quadrature cells, i.e. $N_q = N_s = N_f$. FIGURE 7.3 shows a particular face (triangle) N_2-N_3-N_4 which is the common surface of two neighboring tetrahedron cells, i.e. cell i and cell j. Then the smoothing domain based on face N_2-N_3-N_4 is formed by connecting two centroids of the two cells (C_i and C_j) with the three vertices of the triangle (N_2, N_3, and N_4) and this face-based smoothing domain is a hexahedron. Similarly if a face is located on the surface of the problem domain, the face-based smoothing domain will be formed by connecting the centroid of the host tetrahedron cell and three vertices of this triangle surface and the constructed smoothing domain is a tetrahedron.

Within each face-based smoothing domain, the smoothed strain can be calculated using the generalized gradient smoothing operation as presented in Section 4.5.2.

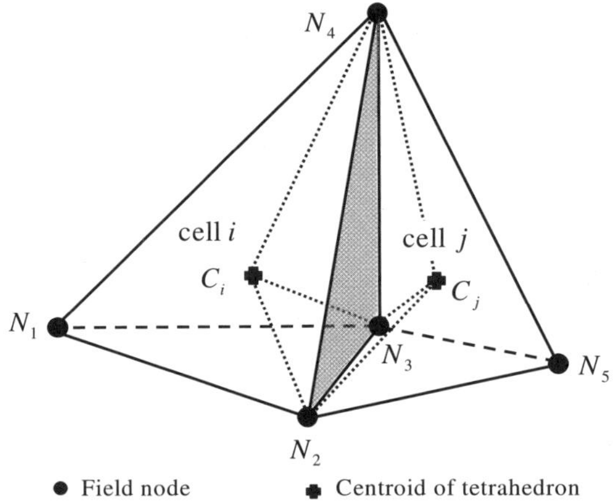

FIGURE 7.3 Formation of face-based smoothing domain based on a set of tetrahedral background cells. The smoothing domain based on face N_2-N_3-N_4 is formed by connecting two centroids of the two cells (C_i and C_j) with the three vertices of the triangle (N_2, N_3, and N_4) and this face-based smoothing domain is a hexahedron.

There are two FS-PIM models that have been developed so far by Liu's group, including the linear FS-PIM-Te4 using linear PIM shape functions and four support nodes selected using the face-based T4-scheme and the high order FS-RPIM-Te2L using RPIM shape functions and support nodes selected using the face-based T2L-scheme [7]. The FS-RPIM-Te2L is generally more expensive than other methods due to much more support nodes used in the field approximation. Therefore in this book, only the linear FS-PIM-Te4 model is used and the properties including condition number and computational cost and the performance will be examined and compared with other models in the following Chapters 8 and 9.

7.4.4 Evaluation of nodal strain (stress)

In the numerical implementation of S-PIM models, except NS-PIM models that produce directly the nodal strain (stress) values, the strains (stresses) at one node will be evaluated via (area-weighted) averaging the raw strains (stresses) in the smoothing domains around this node.

In the scheme of ES-PIM models, the edge-based smoothed strains are corresponding to each edge-based smoothing domain. Once the nodal displacement results have been obtained by solving Equation (7.5), the smoothed strains corresponding to each smoothing domain can then be calculated using Equation (5.70). For one particular field node i, the nodal strain can be obtained as

$$\overline{\boldsymbol{\varepsilon}}\left(\mathbf{x}_i\right) = \frac{1}{A_i^{ns}} \sum_{k=1}^{n_i^s} \overline{\boldsymbol{\varepsilon}}_k A_k^s \tag{7.13}$$

where n_i^s is the number of smoothing domains around node i, $A_i^{ns} = \sum_{k=1}^{n_i^s} A_k^s$ is the total area of these n_i^s smoothing domains, and $\overline{\boldsymbol{\varepsilon}}_k$ is the smoothed strain corresponding to the smoothing domain Ω_k^s of area A_k^s.

FIGURE 7.4 shows the smoothing domains used to evaluate the strains at the particular node i, where totally six smoothing domains are involved and hence $n_i^s = 6$.

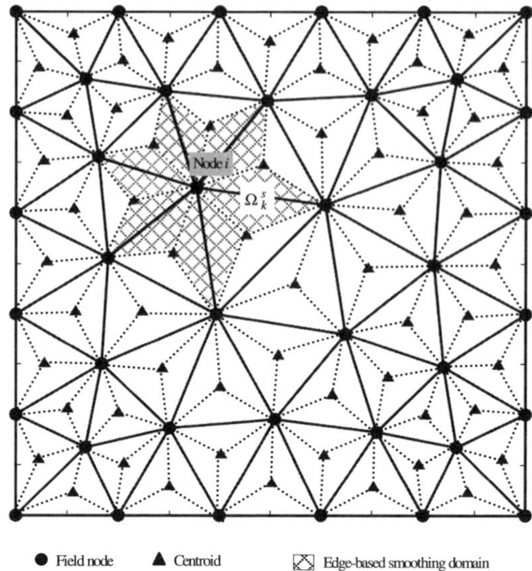

FIGURE 7.4 Smoothing domains used to evaluate the strains at node i in the scheme of ES-PIM models. For the particular node i shown in the figure, six edge-based smoothing domains are involved in the calculation, and hence $n_i^s = 6$.

7.4.5 Condition number of ES-PIM models

TABLE 7.2 lists the condition numbers of the global stiffness matrixes of various ES-PIM models in relation to the FEM-Tr3, which are obtained using the same triangular meshes shown in FIGURE 6.4 for the cantilever beam problem. We may find that the simplest ES-PIM model, i.e. the ES-PIM-Tr3 has smaller condition number than the FEM-Tr3, although the former uses more support nodes for each integration domain. The other three ES-PIM models have bigger condition number than the linear FEM model. We also find that the ES-RPIM-Tr2L model has similar conditions number as the linear FEM, even though it uses much more support nodes than other models.

Although the condition number changes when the size of the model (node number) increases, the ratio of the condition number for different methods in relation to the linear FEM-Tr3 model has little change regardless of the size of the models. In the following analysis, we will thus use the relative $\bar{n}_{iter}$ given in last column in TABLE 7.2.

TABLE 7.2 Condition number of the global stiffness matrixes for various ES-PIM models for the cantilever beam in relation to the FEM-Tr3 model using the same mesh

Numerical method	120 nodes		248 nodes	
	$cond(\mathbf{K})$	$\overline{n}_{iter}$	$cond(\mathbf{K})$	$\overline{n}_{iter}$
FEM-Tr3	1.645e+08	1.00	2.399e+08	1.00
ES-PIM-Tr3	1.094e+08	0.82	1.574e+08	0.81
ES-PIM-Tr6/3	1.972e+08	1.09	3.397e+08	1.19
ES-RPIM-Tr6	2.327e+08	1.19	3.515e+08	1.21
ES-RPIM-Tr2L	1.738e+08	1.03	2.497e+08	1.02

7.4.6 Estimation of computational cost for ES-PIM

Combining TABLE 7.1and TABLE 7.2 we shall have a rough estimation of computational cost for the ES-PIM models in relation to the standard FEM-Tr3 model using the same mesh as shown in TABLE 7.3. We find that the ES-PIM-Tr3 model only costs slightly more time than the linear FEM model, owning to the smaller condition number of the global stiffness matrix. The other three ES-PIM models will cost more (2.7 to 4.3 times) than the FEM-Tr3 when using the same triangular mesh.

TABLE 7.3 Estimation of computational cost for ES-PIM models in relation to the FEM-Tr3 model using the same mesh

Numerical methods	Support nodes in an integral cell/element	Estimated Solver CPU time
FEM-Tr3	3	t_{CPU}
ES-PIM-Tr3	3-4 (3.9)	1.1 t_{CPU}
ES-PIM-Tr6/3	3-8 (7.0)	2.7 t_{CPU}
ES-RPIM-Tr6	6-8 (7.7)	3.1 t_{CPU}
ES-RPIM-Tr2L	7-15 (12.6)	4.3 t_{CPU}

7.4.7 Rank analysis for ES-PIM

Based on the theory presented in Chapter 5, the ES-PIM models shall only possess the "legal" zero-energy modes representing the physical rigid movements of the solid, and there exist no spurious zero-energy modes. This means that ES-PIM models are spatially stable. Such stability is ensured and the following points are noted.

1) In an ES-PIM model, the number of smoothing domains equals to the number of triangular edges, i.e. $N_s=N_{eg}$, which is always equal and often much bigger than the number of field nodes for any discretization. Thus the minimum number of smoothing domains required in TABLE 2.2 can always be satisfied.

2) The edge-based smoothing domains are independent with each other. Together with the independence of the nodal PIM shape functions, the strain smoothing operation ensures linearly independent columns (or rows) in the stiffness matrix.

3) The PIM shape functions used in an ES-PIM model are of partitions of unity, which ensures a proper representation of the rigid movements.

4) The stiffness matrix of an ES-PIM model is SPD for solids of stable materials, after the rigid movements are constrained.

Therefore, there is no deformed zero-energy mode existing in an ES-PIM model, and any finite deformation in the model (except the rigid motions) will result in a finite amount of strain energy in an ES-PIM model.

7.4.8 Temporal stability analysis

We now provide a rough intuitive analysis on the temporal stability of the ES-PIM models, in comparison against the FEM-Tr3. The stability, both spatial and temporal, is directly related to the number of the "samplings" of the integrand in the weak form. In an FEM-Tr3 model, the compatible strain in each element is constant, and only one Gauss point is needed for an element in the domain integration in the weak form. This means that the number of Gauss points used in the whole domain equals to the number of elements. Such an FEM-Tr3 model is known for both spatially and temporally stable.

In an ES-PIM model, the number of the smoothing domains equals to the number of edges in the mesh, which is always larger than the number of elements. It means that the number of samplings in an ES-PIM model is always larger than that in the FEM-Tr3, which indicates that ES-PIM models will not be "soft" and should be always temporally stable and shall have no spurious nonzero-energy modes.

On the other hand, in an NS-PIM model, each smoothing domain is associated with a node and the strain over each smoothing domain is constant. For stability considerations, each smoothing domain can be viewed equivalent to one Gauss point in terms of sampling the integrand in the weak form. As the number of field nodes (also the number of smoothing domains) can be smaller than the number of the elements, there is thus a chance for spurious modes to appear at a high energy level. This phenomenon is quite well-known in the nodal integrated meshfree methods.

Remark 7.1 Stability property of ES-PIM models

ES-PIM models are both spatially and temporally stable. An ES-PIM model possesses only "legal" zero-energy modes that represents the physical rigid motions, there exist no spurious zero-energy or unphysical nonzero-energy modes, and is well suitable for solving dynamic problems.

Remark 7.2 Bound property of ES-PIM models

The stiffness of an ES-PIM model is in between those of the FEM and the NS-PIM models using the same mesh. The strain energy solution of the ES-PIM will be in between those of the FEM and NS-PIM models using the same mesh.

7.5 Numerical examples

Several numerical examples are studied in this section to examine the ES-PIM models. The materials used are linear elastic with Young's modulus $E=3.0\times10^7$ Pa and passion's ratio $v=0.3$, and the units used are based on the international standard unit system unless specified explicitly. Numerical results are assessed using the error indicators in displacement and energy norms which are defined in Equation (5.88) and Equations (5.94), respectively.

Example 7.5.1 Linear patch test

The 2D linear patch test described in Example 6.1.1 is again studied using the ES-PIM models. TABLE 7.4 lists the displacement norm errors of the ES-PIM solutions for the patch test using both the regular and irregular meshes shown in

FIGURE 6.5. We found that all the ES-PIM models can pass the test exactly (to the machine accuracy). Note that except in the ES-PIM-Tr3 model, the displacement field in other models is incompatible, and discontinuity occurs along the interfaces of background cells.

Table 7.4 Error norm in displacements of numerical results for the standard patch test obtained using ES-PIM models

ES-PIM models	Regular mesh	Irregular mesh
ES-PIM-Tr3 (displacement compatible)	1.118E-15	1.327E-15
ES-PIM-Tr6/3 (displacement incompatible)	2.117E-14	1.808E-14
ES-RPIM-Tr6 (displacement incompatible)	8.763E-16	1.080E-15
ES-RPIM-Tr2L (displacement incompatible)	8.707E-16	1.598E-15

Remark 7.3 ES-PIM: linearly conforming and 2nd order accuracy

This patch test example demonstrates numerically that the ES-PIM can reproduce linear fields exactly, regardless of the incompatible displacement field used: linearly conforming. Together with the stability (Theorem 5.1), the ES-PIM solution will converge to the exact solution of any well-posed linear elasticity problem. This implies also that the model is at least 2nd order accuracy: the solution error in displacement is on the terms of 2nd order and above. This supports Remark 3.23.

Example 7.5.2 Rectangular cantilever

The benchmark problem of rectangular cantilever described in Example 6.1.2 is studied again using the ES-PIM models. For comparison, the linear FEM and NS-PIM are also used in the analyses with the same set of triangular meshes shown in FIGURE 6.9.

Using a particular mesh (Mesh-2 of 399 nodes), we first analyze the problem using these four ES-PIM models. The numerical results at those nodes along two lines are plotted in FIGURE 7.5. It can be found that all the ES-PIM solutions, including deflections and shear stress, are all in a very good agreement with the analytical ones.

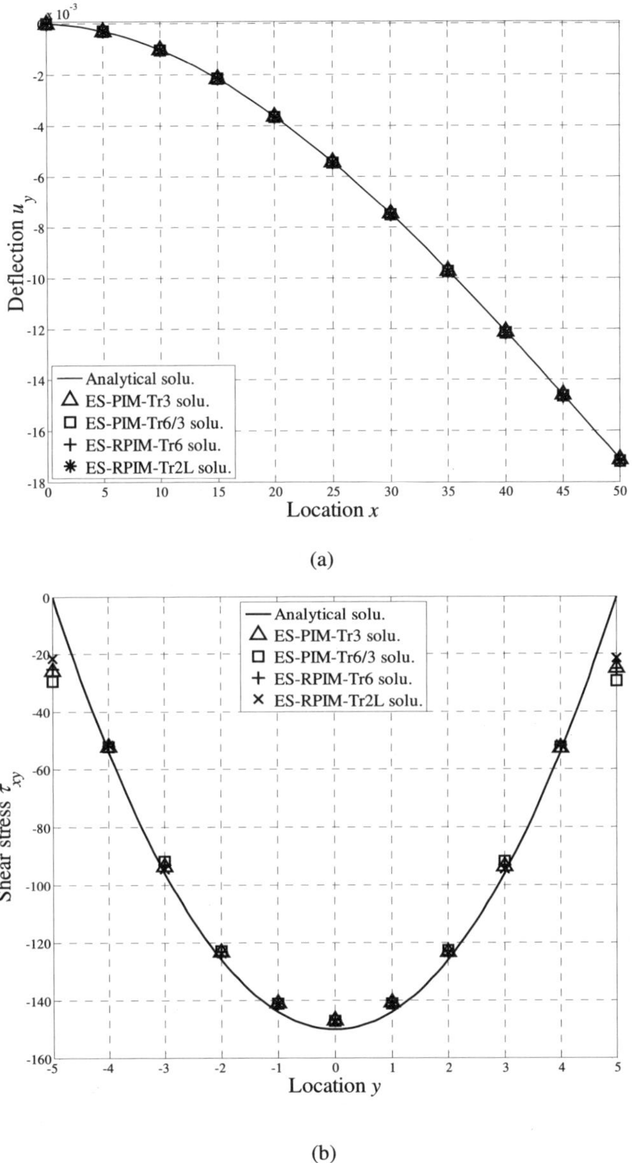

(a)

(b)

FIGURE 7.5 Comparison of numerical results obtained using ES-PIM models for the rectangular cantilever problem with analytical ones: (a) deflection along the neutral line of the cantilever; (b) shear stress along the line of $x=L/2$ of the cantilever.

The convergence of the error in displacement norm is plotted in FIGURE 7.6. TABLE 7.5 lists the solution error in displacement norm obtained using Mesh-4 (see FIGURE 6.9) and the estimated efficiency by considering the computational cost in TABLE 7.3. We observe the following.

1) All the ES-PIM models provide much better numerical results than that of FEM-Tr3, in terms of accuracy. The ES-PIM-Tr3 stands out clearly among all the ES-PIM models. Its accuracy is about 100 times that of FEM-Tr3 for Mesh-4 used. Other three ES-PIM models have similar accuracy, which is about 5 times more accurate than the linear FEM-Tr3.

2) In terms of convergence rate, the linear ES-PIM-Tr3 achieves a very high rate of 2.55 that is much higher than of FEM-Tr3 and the theoretical value of 2.0 (see Remark 3.23). We observe clearly the *superconvergence* property of the ES-PIM-Tr3 in *displacement norm*. The other three higher order ES-PIM models have numerical rates around 1.5 that is less than theoretical value of 2.0.

3) In terms of computational efficiency, the linear ES-PIM-Tr3 performs the best, which is about more than 100 times more efficient than the linear FEM using the same mesh. The ES-PIM-Tr6/3 and ES-RPIM-Tr6 are about 2 times more efficient than the FEM-Tr3. The ES-RPIM-Tr2L has almost the same efficiency as the linear FEM.

4) For comparison, we have also listed the results obtained using two NS-PIM models. It is seen that ES-PIM-Tr3 performs far better than these NS-PIM models, in terms of accuracy and efficiency.

TABLE 7.5 Estimated computational efficiency of different ES-PIM models in relation to the FEM-Tr3 measured in *displacement norm* error for the numerical results of the cantilever beam problem with the same set of triangular mesh (Mesh-4 of 1,696 nodes)

Numerical method	Solution error	Error ratio to FEM-Tr3	Efficiency
FEM-Tr3	5.2334E-03	1.00	1.0
NS-PIM-Tr3	5.5340E-03	1.06	0.8
NS-PIM-Tr4-CT	3.0198E-03	0.58	1.4
ES-PIM-Tr3	4.6937E-05	0.0090	101.0
ES-PIM-Tr6/3	1.1160E-03	0.21	1.8
ES-RPIM-Tr6	9.9995E-04	0.19	1.7
ES-RPIM-Tr2L	1.2239E-03	0.23	1.0

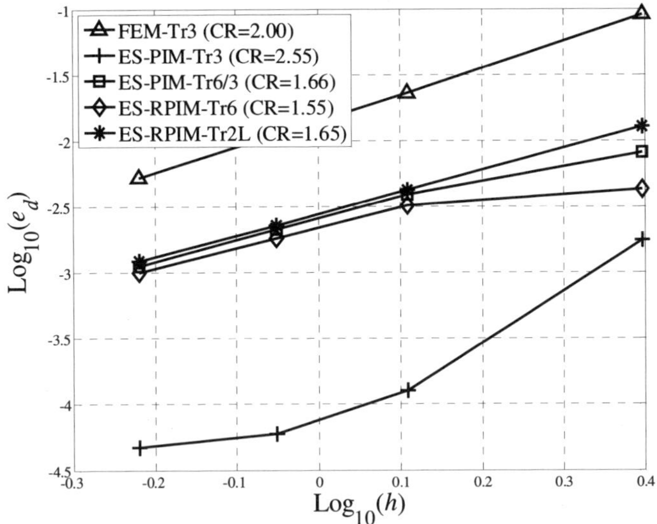

FIGURE 7.6 Comparison of convergence rates and accuracy of the numerical results in displacement norm obtained using linear FEM and three ES-PIM models for the rectangular cantilever problem.

FIGURE 7.7 shows the convergence of the error in energy norm for different ES-PIM models. TABLE 7.6 lists the detailed energy norm error obtained using Mesh-4 (see FIGURE 6.9) and the estimated computational efficiency in relation to the linear FEM. The following points can be found.

1) All the ES-PIM models converge monotonically in a "linear" fashion, and achieve much better accuracy and higher convergence rate than the linear FEM, showing superconvergence in energy norm. The rates of convergence are around 1.44-1.53 that is quite close to the ideal rate of 1.5 (see Remark 3.23).

2) We can find again that the linear ES-PIM model stands out clearly in terms of accuracy, convergence rate and computational efficiency. The ES-PIM-Tr3 is about 8 times more accurate and 7 times more efficient than the linear FEM.

3) The ES-PIM-Tr6/3 and ES-RPIM-Tr6 are more efficient than the FEM and the ES-RPIM-Tr2L is about 10% less efficient compared to the FEM-Tr3.

4) For comparison, we have also listed the results obtained using two NS-PIM models. It is seen that ES-PIM-Tr3 performs far better than these NS-PIM

models, in terms of accuracy and efficiency. These two NS-PIM models performed much better than the other ES-PIM models, in energy norm measure.

TABLE 7.6 Estimated computational efficiency of different ES-PIM models in relation to the FEM-Tr3 measured in *energy norm* error for the numerical results of the cantilever beam problem with the same set of triangular mesh (Mesh-4 of 1,696 nodes)

Numerical method	Solution error	Error ratio to FEM-Tr3	Efficiency
FEM-Tr3	2.4406E-01	1.00	1.00
NS-PIM-Tr3	5.0076E-02	0.21	4.0
NS-PIM-Tr4-CT	4.3389E-02	0.18	4.6
ES-PIM-Tr3	2.8075E-02	0.12	7.6
ES-PIM-Tr6/3	5.1029E-02	0.21	1.8
ES-RPIM-Tr6	6.1977E-02	0.25	1.3
ES-RPIM-Tr2L	6.2482E-02	0.26	0.9

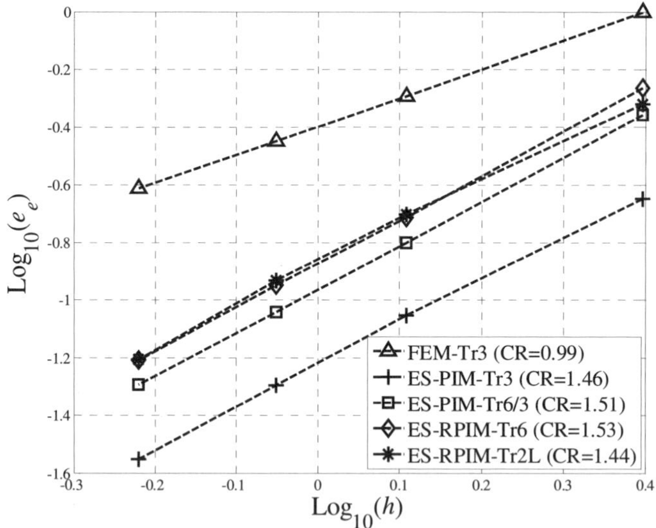

FIGURE 7.7 Comparison of convergence rates and accuracy of the numerical results in energy norm obtained using linear FEM and three ES-PIM models for the rectangular cantilever problem. In displacement norm.

The convergence process of the strain energy solution for different ES-PIM models is shown in FIGURE 7.8, together with the FEM-Tr3 and NS-PIM-Tr3 models. TABLE 7.7 lists the detailed values of computed strain energy and the efficiency in relation to the linear FEM. We observed the following points.

1) The solutions of all the models converge to the exact solution.

2) The linear FEM-Tr3 gives a lower bound, and the NS-PIM-Tr3 gives an upper bound. The solutions of all the ES-PIM models are in between, as stated in Remark 7.2.

3) The solutions of all the ES-PIM models are much more accurate than the FEM-Tr3, and they offer much tighter bounds to the exact solution.

4) The linear ES-PIM-Tr3 stands out clearly; it is more than 130 times more efficient than the linear FEM. ES-PIM-Tr6/3 and ES-RPIM-Tr6 are about twice more efficient than the linear FEM. Even the ES-RPIM-Tr2L that uses a lot of local nodes in the interpolation, the efficiency is about 10% higher than the FEM-Tr3, in the strain energy solution measure.

5) It is seen that ES-PIM-Tr3 performs far better than these NS-PIM models, in terms of both accuracy and efficiency.

TABLE 7.7 Estimated computational efficiency of different ES-PIM models in relation to the FEM-Tr3 measured in the error in *strain energy solution* of the numerical results for the cantilever beam problem with the same set of triangular mesh (Mesh-4 of 1,696 nodes)

Numerical method	Strain energy Solution	Error[*] (%)	Error ratio to FEM-Tr3	Efficiency
FEM-Tr3	8.5474	-0.5345	1.00	1.0
NS-PIM-Tr3	8.6398	0.5407	1.01	0.8
NS-PIM-Tr4-CT	8.6184	0.2917	0.55	1.5
ES-PIM-Tr3	8.5930	-0.0035	0.0066	137.7
ES-PIM-Tr6/3	8.6022	0.1032	0.19	1.9
ES-RPIM-Tr6	8.6012	0.0915	0.17	1.9
ES-RPIM-Tr2L	8.6031	0.1137	0.21	1.1

[*] The analytical value of the strain energy for the cantilever beam problem is 8.59333333.

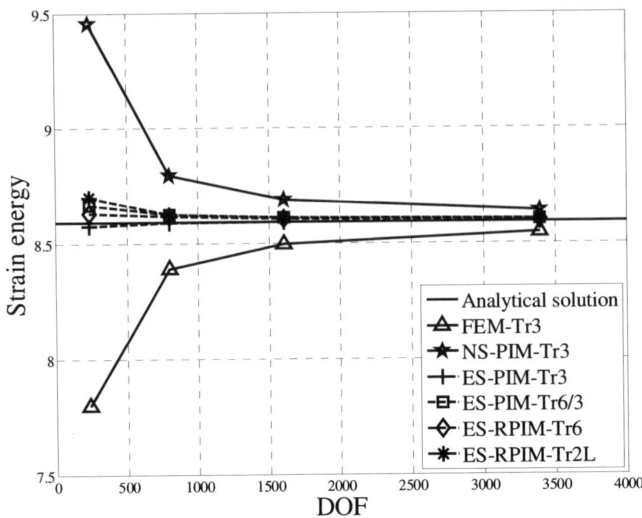

FIGURE 7.8 Converging process of the numerical results in strain energy for the problem of rectangular cantilever.

To provide a clearer view on the convergence process of the strain energy for different ES-PIM models, a close-up plot is given in FIGURE 7.9. We can find the following points.

1) The linear ES-PIM gives a very tight lower bound solution for this problem.

2) All the other three ES-PIM models all provide upper bound solutions, and the ES-RPIM-Tr6 provides the tightest upper bound for this problem;

3) As discussed in Remark 6.15, one issue affecting the softness of a model is the order of shape functions used. When higher order shape functions are used in an ES-PIM, the model becomes softer than the linear ES-PIM.

4) Compared to the linear ES-PIM, the merit of the higher order ES-PIM models is only in providing upper bound solutions. In terms of accuracy and computational efficiency the linear ES-PIM-Tr3 is recommended.

Remark 7.4 ES-PIM-Tr3 model: superconvergence in both norms

The linear ES-PIM-Tr3 is of superconvergence in both displacement and energy norms.

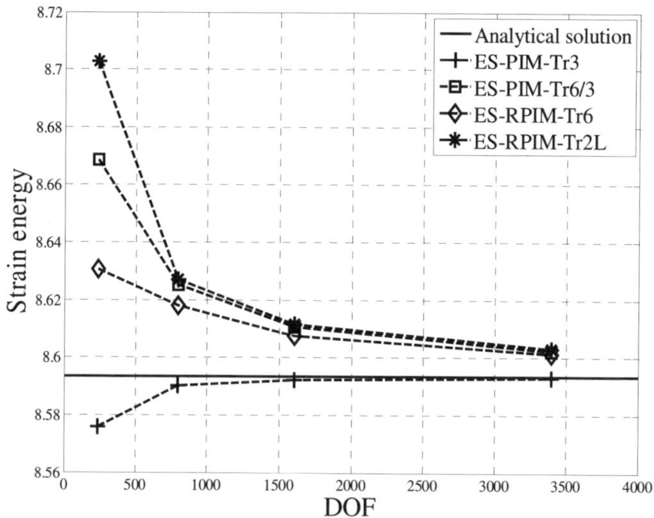

FIGURE 7.9 Converging process of the numerical results in strain energy for the cantilever problem: a close-up comparison between the ES-PIM models.

Remark 7.5 ES-PIM-Tr3 model: a star performer

In terms of accuracy, convergence rate and computational efficiency the linear ES-PIM-Tr3 is found outstanding. The computational efficiency of ES-PIM-Tr3 (with Mesh-4 of 1,696 nodes) is about 100 times that of the FEM-Tr3 in displacement norm, and 7.6 times in energy norm for this cantilever problem. In addition, it is the simplest S-PIM model, and very easy to implement. It is regarded as the "star performer".

Example 7.5.3 Infinite solid with a circular hole

The infinite 2D solid with a circular hole problem described in Example 6.1.3 is studied again using the ES-PIM models. Using Mesh-2 shown in FIGURE 6.14, the displacement components are computed and the results along two particular lines of the model (the bottom and the left edges) are plotted in FIGURE 7.10. It is seen that the displacement results obtained by all the ES-PIM models are all in a very good agreement with the analytical ones, and these results are not distinguishable from these figures.

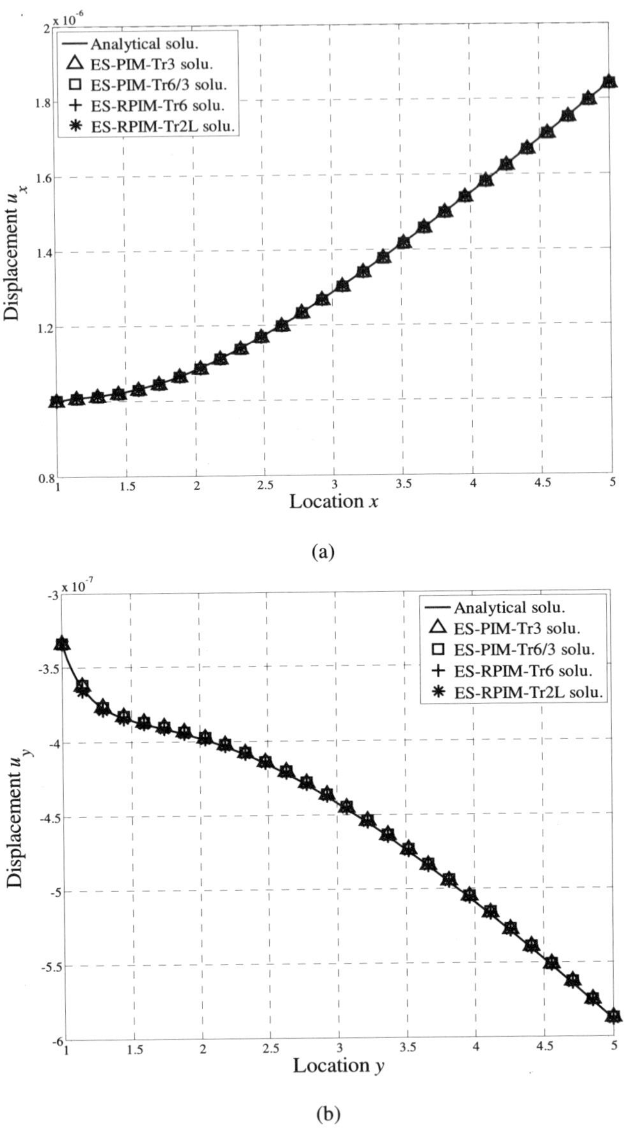

(a)

(b)

FIGURE 7.10 Displacements distribution along bottom and left edges for the quarter model of the infinite solid with hole: (a) displacement (u_x) for nodes located along the bottom edge; (b) displacement (u_y) for nodes located along the left edge.

FIGURE 7.11 plots the numerical results for the stress components distributed along the bottom and the left edges. It is also seen that the results obtained by the ES-PIM models are all in good agreement with the analytical ones, and these stress results are also not distinguishable from these figures.

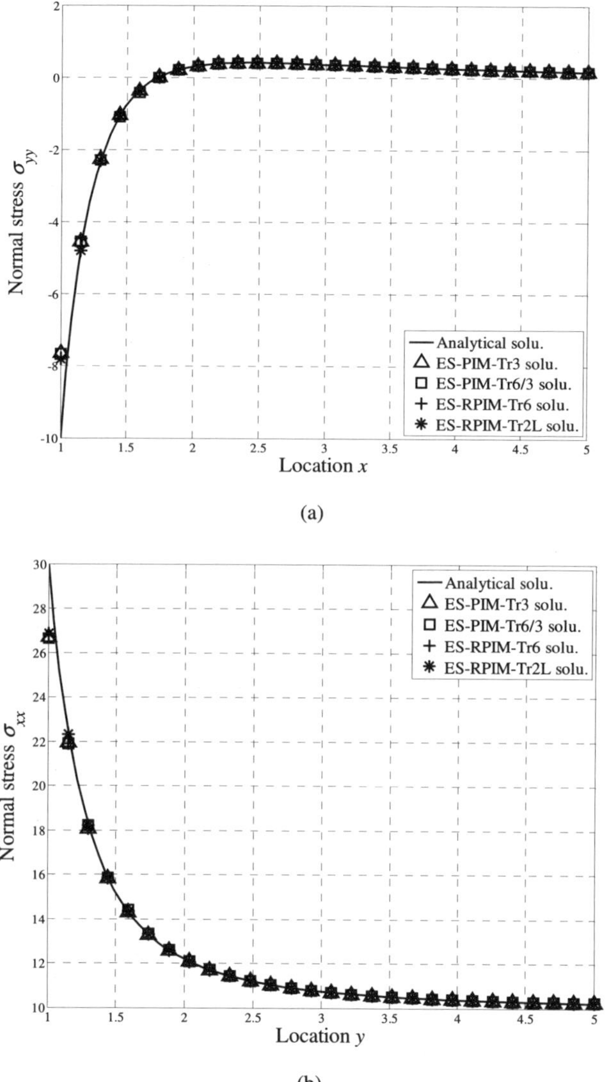

(a)

(b)

FIGURE 7.11 Stresses distribution along bottom and left edges for the quarter model of the infinite solid with hole: (a) normal stresses in y-direction for at the nodes located along the bottom edge; (b) normal stresses in x-direction at the nodes located along the left edge.

Using the meshes of different nodal density shown in FIGURE 6.14, the convergence of the solution errors for the ES-PIM models are computed using Equations (5.88) and (5.94). FIGURE 7.12 plots these results in displacement norm. Considering the computational cost in TABLE 7.3, TABLE 7.8 lists the detailed solution error in displacement norm using the same Mesh-4 (3,578 nodes) and the estimated computational efficiency. The following can be observed.

1) All the ES-PIM solutions are of much more accurate and the convergence rates are much higher than the linear FEM.

2) The convergence rates of those ES-PIM solutions are from 2.17 to 3.14 that are much higher than the theoretical value of 2.0, showing superconvergence in displacement norm.

3) The ES-RPIM-Tr2L stands out among all the ES-PIM models in terms of both accuracy and convergence rate. When the finest mesh is used (Mesh-4 of 3,578 nodes), it is about 14 times more accurate than the FEM-Tr3. The computational efficiency (see, TABLE 7.8) of ES-RPIM-Tr2L will be about 3 times more efficient than the FEM-Tr3. In addition, because of the much higher convergence rate, the ES-RPIM-Tr2L is expected to be even more efficient than the FEM-Tr3, when a finer mesh is used. This argument is application to all the ES-PIM models, and hence we can conclude that the ES-PIM will be more efficient than the FEM-Tr3, when fine mesh is used for this problem.

4) The ES-PIM-Tr3 is about 4 times more accurate than FEM-Tr3 when the Mesh-4 is used for this problem. The computational efficiency of the ES-PIM-Tr3 will still be about 3.5 times more efficient than the FEM-Tr3. Such efficiency will be further improved when a fine mesh is used. The ES-PIM-Tr3 shows superconvergence property in displacement norm with a rate of 2.17 for this problem, which supports Remark 7.4.

5) For this problem, the higher order ES-PIM models are more accurate than the linear ES-PIM-Tr3 for the same mesh used.

6) The high order ES-RPIM-Tr2L model is as efficient as the star performer ES-PIM-Tr3 for this case.

7) ES-PIM-Tr3 performs much better than these NS-PIM models, in terms of accuracy and efficiency.

TABLE 7.8 Estimated computational efficiency of different ES-PIM models in relation to the FEM-Tr3 measured in *displacement norm* error for the numerical results of the infinite solid with a circular hole problem with the same triangular mesh (Mesh-4 of 3,578 nodes)

Numerical method	Solution error	Error ratio to FEM-Tr3	Efficiency
FEM-Tr3	1.4435E-03	1.00	1.0
NS-PIM-Tr3	1.1520E-03	0.80	1.0
NS-PIM-Tr4-CT	6.3699E-04	0.44	1.9
ES-PIM-Tr3	3.7620E-04	0.26	3.5
ES-PIM-Tr6/3	2.6928E-04	0.19	1.9
ES-RPIM-Tr6	1.9453E-04	0.13	2.5
ES-RPIM-Tr2L	1.0166E-04	0.070	3.3

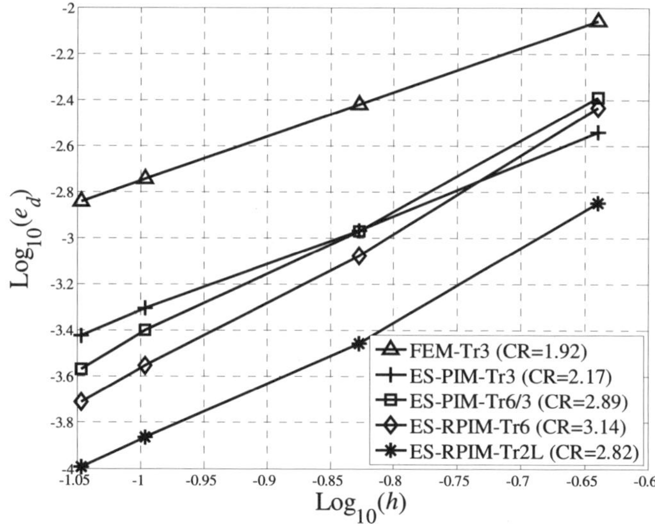

FIGURE 7.12 Comparison of convergence rates and accuracy of the numerical results in displacement norm obtained using ES-PIM models for the problem of infinite solid with circular hole.

FIGURE 7.13 plots these results in energy norm. TABLE 7.9 lists the detailed solution error in energy norm and the computational efficiency in relation to the linear FEM. The following points may be noted.

1)   All the ES-PIM solutions are of much more accurate and the convergence rates are much higher than the linear FEM.

2)   The convergence rates of those ES-PIM solutions are from 1.32 to 1.61 that are much higher than the theoretical value of 1.0 for linear weak-form models, and around the ideal rate of 1.5 for W^2 models (see, Remark 3.23).

3)   The ES-RPIM-Tr2L is still the best performer in terms of both accuracy and convergence rate. When the finest mesh (Mesh-4 of h=0.09) is used, it is about 3.5 times more accurate than the FEM-Tr3. The computational efficiency of the ES-RPIM-Tr2L (see, TABLE 7.9) will be about 20% less efficient than the FEM-Tr3. However, because of the much higher convergence rate, the ES-RPIM-Tr2L is expected to be more efficient than the FEM-Tr3, when a finer mesh is used. This argument is application to all the ES-PIM models for this problem.

4)   The ES-PIM-Tr3 is also about 3 times more accurate than FEM-Tr3 when the finest mesh (Mesh-4) is used for this problem, and hence the ES-PIM-Tr3 will be about 3 times more efficient than the FEM-Tr3. Such efficiency will be further improved when a fine mesh is used. The ES-PIM-Tr3 exhibits the superconvergence also in energy norm for this problem with a rate of 1.32, which supports Remark 7.4.

5)   For this problem, the higher order ES-PIM-Tr6/3 and ES-RPIM-Tr6 models performed worse in energy norm than the linear ES-PIM-Tr3 in terms of accuracy and efficiency.

6)   ES-PIM-Tr3 performs about 20% better than these NS-PIM models, in terms of accuracy and efficiency.

TABLE 7.9 Estimated computational efficiency of different ES-PIM models in relation to the FEM-Tr3 measured in *energy norm* error for the numerical results of the infinite solid with circular hole problem with the same triangular mesh (Mesh-4 of 3,578 nodes)

Numerical method	Solution error	Error ratio to FEM-Tr3	Efficiency
FEM-Tr3	1.3831E-04	1.00	1.0
NS-PIM-Tr3	5.0516E-05	0.37	2.3
NS-PIM-Tr4-CT	4.9617E-05	0.36	2.3
ES-PIM-Tr3	4.2694E-05	0.31	2.9
ES-PIM-Tr6/3	6.8476E-05	0.50	0.7
ES-RPIM-Tr6	6.1823E-05	0.45	0.7
ES-RPIM-Tr2L	3.8275E-05	0.28	0.8

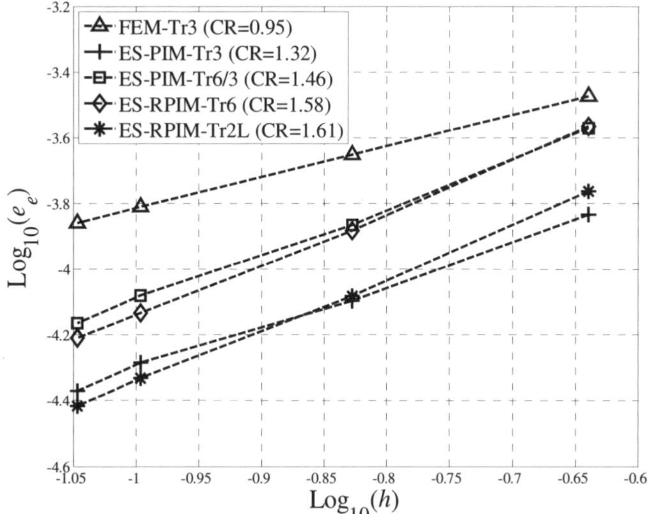

FIGURE 7.13 Comparison of convergence rates and accuracy of the numerical results in energy norm obtained using ES-PIM models for the problem of infinite solid with circular hole.

The convergence process of the strain energy solution obtained using different ES-PIM models is shown in FIGURE 7.14. TABLE 7.10 lists the computed strain energy of the numerical solutions using the same Mesh-4 and the estimated computational efficiency of ES-PIM models in relation to the linear FEM. For this particular problem, we note the following.

1) ES-PIM models behave softer than the overly-stiff FEM and stiffer than the overly-soft NS-PIM-Tr3 model. All the ES-PIM solutions are bounded by FEM-Tr3 and NS-PIM-Tr3, which supports Remark 7.2.

2) As shown in FIGURE 7.14 b, all these ES-PIM models give lower bound solution in strain energy. This is in contrary to the results in Example 7.5.2. This indicates that the bound property of the ES-PIM solution can change with the problem. Because the ES-PIM solutions are all very close to the exact solution, they can be on both sides.

3) Using high order PIM shape functions, the three higher order ES-PIM models perform softer and provide tighter lower bound solutions, compared to the linear ES-PIM. The ES-RPIM-Tr2L model gives the tightest bound solution among all these ES-PIM models.

4) We observe that the higher order ES-RPIM-Tr2L model is about 20% more efficient even than the star performer ES-PIM-Tr3.

5) ES-PIM-Tr3 performs about 10% worse than the NS-PIM-Tr4-CT model in terms of accuracy and efficiency.

TABLE 7.10 Estimated computational efficiency of different ES-PIM models in relation to the FEM-Tr3 measured in the error in *strain energy solution* of the numerical results for the infinite solid with circular hole problem with the same set of triangular mesh (Mesh-4 of 3,578 nodes)

Numerical method	Strain energy Solution	Error[*] (%)	Error ratio to FEM-Tr3	Efficiency
FEM-Tr3	4.3230E-05	-0.0495	1.00	1.0
NS-PIM-Tr3	4.3265E-05	0.0314	0.63	1.3
NS-PIM-Tr4-CT	4.3257E-05	0.0130	0.26	3.2
ES-PIM-Tr3	4.3245E-05	-0.0148	0.30	3.0
ES-PIM-Tr6/3	4.3247E-05	-0.0102	0.21	1.8
ES-RPIM-Tr6	4.3248E-05	-0.0079	0.16	2.0
ES-RPIM-Tr2L	4.3250E-05	-0.0032	0.065	3.6

[*] The analytical value of the strain energy for the infinite solid with hole problem is 4.32513989E-05.

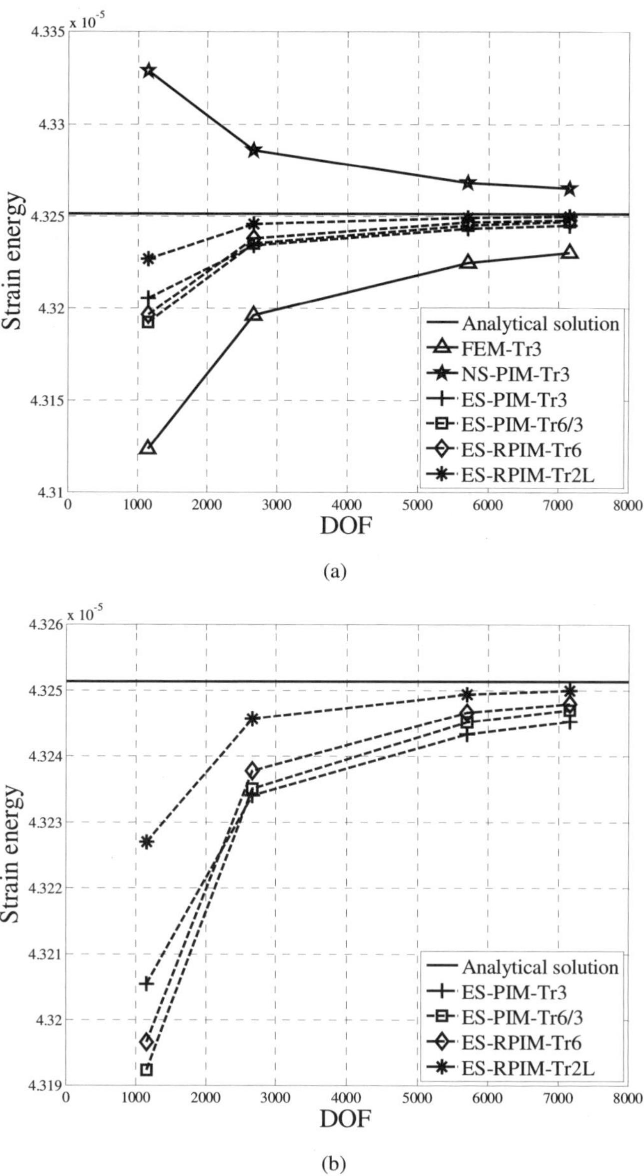

(a)

(b)

FIGURE 7.14 Converging process of the strain energy solution for the problem of infinite plate with circular hole: (a) comparison between ES-PIM models and other methods; (b) close-up comparison between the ES-PIM models.

Example 7.5.4 Free vibration of a slender cantilever

In this example, a free vibration analysis is conducted for a slender cantilever beam with length $L = 100$mm, height $H = 10$mm and thickness $t = 1.0$mm. The material properties are: Young's modulus $E = 2.1 \times 10^4$ kgf/mm^2, Poisson's ratio $v = 0.3$, and mass density $\rho = 8.0 \times 10^{-10}$ kgf s^2/mm^4. The cantilever is treated as a 2D solid in our numerical models and plane stress condition is considered. Because the slenderness, the cantilever behaves like a "thin beam", and a reference fundamental frequency $f_1 = 0.08276 \times 10^4$ Hz is obtained using the Euler-Bernoulli thin beam theory. Numerical results obtained by the FEM using four node quadrangular elements (FEM-Q4) with a very fine mesh (100×10) are also used as reference solutions for this problem.

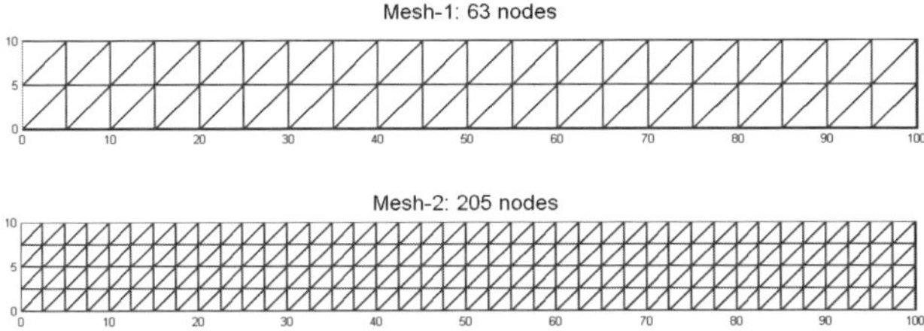

FIGURE 7.15 Meshes of three-node triangular elements for the slender cantilever.

Eigenvalue Equation (7.8) is established using our ES-PIM models and solved using standard eigenvalue (symmetric) solvers for eigenvalues that give natural frequencies and eigenvectors leading to vibration modes. Two meshes of triangular cells are used to represent the problem domain, as shown in FIGURE 7.15. TABLE 7.11 lists the first twelve natural frequencies of the cantilever obtained using different numerical methods. The first twelve corresponding vibration modes obtained using the NS-PIM-Tr3 are plotted in FIGURE 7.16 a. Those using the ES-PIM models plotted in FIGURE 7.16 b-d. The following points can be observed.

1)　Four spurious nonzero-energy modes (mode 5, 8, 11 and 12) have been found for the NS-PIM solution, due to its overly-soft behavior.

2)　The natural frequencies obtained using FEM-Tr3 are much larger than the reference ones and those using ES-PIM models, due to the overly-stiff behavior.

3)　All the ES-PIM models do not have any spurious modes.

4) The natural frequencies obtained using ES-PIM models are generally between those of NS-PIM-Tr3 and FEM-Tr3, showing that the stiffness of ES-PIM is in between those two models.

5) The natural frequencies obtained using ES-PIM models are generally closer to the reference solution than that obtained using the linear FEM.

Because the natural frequencies can be used as a good measure on the stiffness of a model, the above findings confirm again that the ES-PIM model has a very close-to-exact stiffness. Hence, the ES-PIM models are well suited for dynamic problems.

TABLE 7.11 First twelve nature frequencies (in 10kHz) of the slender cantilever

	NS-PIM-Tr3	ES-PIM-Tr3	ES-PIM-Tr6/3	ES-RPIM-Tr2L	FEM-Tr3	Reference
	0.0675	0.0853	0.0944	0.0814	0.1117	0.0824
	0.4032	0.5078	0.5628	0.4936	0.6539	0.4944
	1.0518	1.2828	1.2833	1.2832	1.2843	1.2824
	1.2810	1.3246	1.4792	1.3230	1.6748	1.3022
	<u>1.6467</u>	2.3783	2.6883	2.4541	2.9554	2.3663
Mesh 1	1.8786	3.5784	3.8368	3.8108	3.8424	3.6085
(21×3=63	<u>2.7823</u>	3.8298	4.1043	3.8450	4.3866	3.8442
nodes)	3.0926	4.8533	5.6450	5.3366	5.8836	4.9674
	3.6783	6.1527	6.3607	6.3641	6.3751	6.3960
	3.8089	6.3182	7.2557	6.9503	7.4046	6.4023
	<u>4.0543</u>	7.4419	8.7940	8.5845	8.8210	7.8853
	<u>4.1605</u>	8.6776	8.9124	8.8394	8.9411	8.9290
	0.0778	0.0827	0.0839	0.0816	0.0906	0.0824
	0.4654	0.4950	0.5032	0.4889	0.5409	0.4944
	1.2199	1.2826	1.2828	1.2827	1.2831	1.2824
	1.2818	1.3006	1.3251	1.2871	1.4161	1.3022
	<u>1.6689</u>	2.3554	2.4086	2.3381	2.5570	2.3663
Mesh 2	2.2012	3.5778	3.6754	3.5648	3.8433	3.6085
(41×5=205	<u>3.2517</u>	3.8408	3.8451	3.8448	3.8786	3.8442
nodes)	3.3270	4.9029	5.0661	4.9060	5.3087	4.9674
	3.8344	6.2867	6.3958	6.3205	6.3935	6.3960
	4.5248	6.3774	6.5421	6.3965	6.8093	6.4023
	<u>4.6406</u>	7.6987	8.0793	7.7804	8.3473	7.8853
	<u>5.3275</u>	8.8751	8.9271	8.9282	8.9183	8.9290

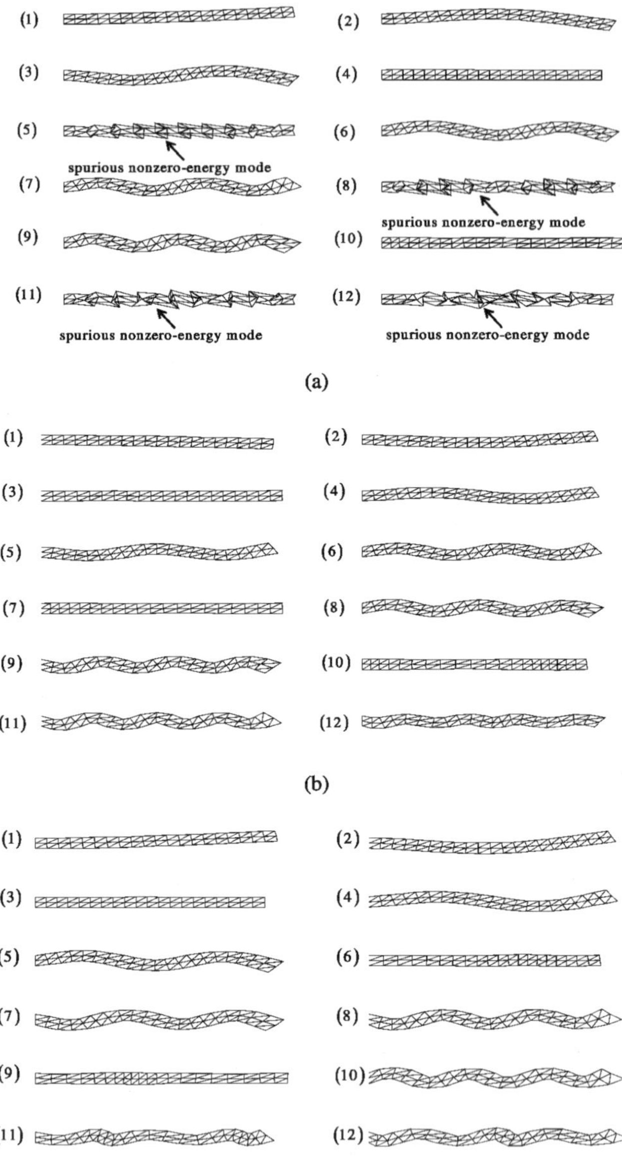

(a)

(b)

(c)

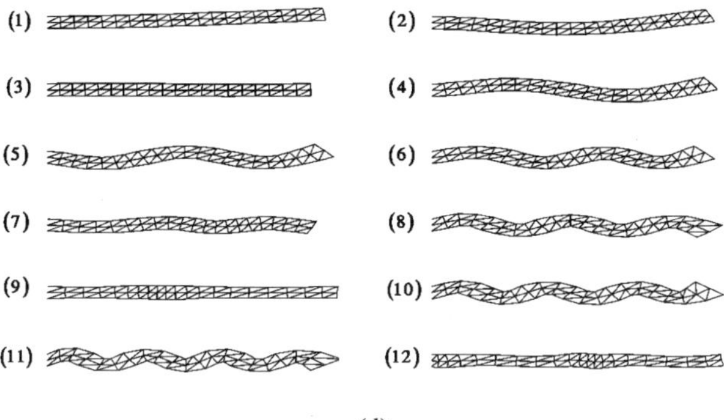

(d)

FIGURE 7.16 First twelve free vibration modes of the cantilever obtained using the NS-PIM-Tr3 and ES-PIM models with Mesh 1 of 63 nodes: (a) NS-PIM-Tr3; (b) ES-PIM-Tr3; (c) ES-PIM-Tr6/3; (d) ES-RPIM-Tr2L.

Example 7.5.5 Forced vibration of a spherical shell

Finally, a forced vibration analysis is conducted for the spherical shell shown in FIGURE 7.17 The shell is subjected to a concentrated vertical force at the apex, and is modeled as a 2D plane strain problem. Due to symmetry, a half of the shell is modeled as shown in FIGURE 7.18 and two types of meshes including triangular and quadrilateral elements are used in the FEM models. The non-dimensional parameters used in the computation: $R=12$, $t=0.1$, $\phi=10.9°$. $\theta=0.5$, $E=1.0$, $v=0.3$, and $\rho=1.0$. ES-PIM-Tr3 and ES-RPIM-Tr2L are used in the computation.

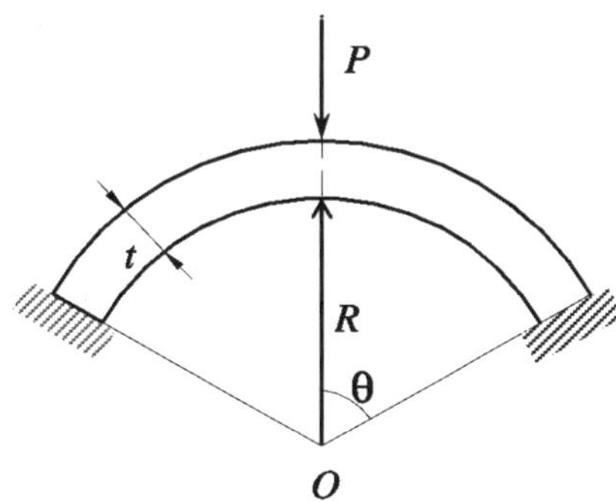

FIGURE 7.17 A spherical shell subjected to point loading at the apex.

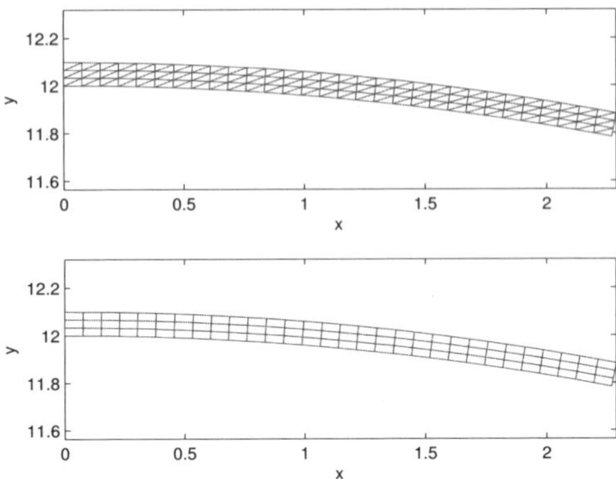

FIGURE 7.18 Meshes of three-node triangular and four-node quadrilateral (Q4) elements for half a spherical shell.

This problem is first studied for a shell subjected to a harmonic loading of $f(t) = \cos\omega_f t$, with $\omega_f = 0.05$. The time history of the deflection at the apex of the shell is computed using time step of $\Delta t = 5$, and the results are plotted in FIGURE 7.19. We can find the following points.

1) The solutions of the two ES-PIM models are much more accurate than that of FEM-T3.

2) The ES-PIM solutions are comparable to that of the FEM-Q4.

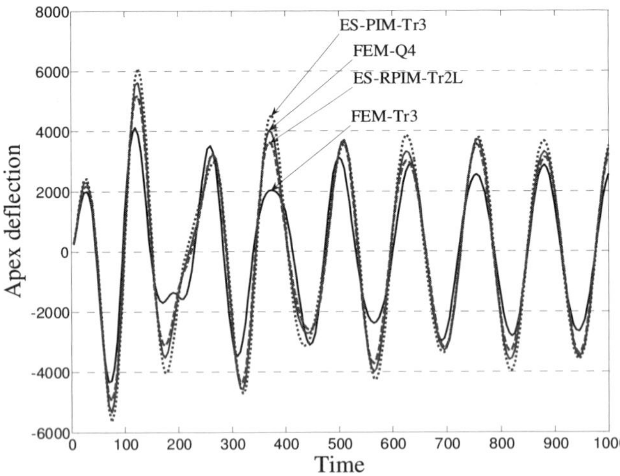

FIGURE 7.19 Transient responses of the spherical shell subjected to a harmonic loading obtained using different methods.

Further, an analysis is conducted for the shell excited by a constant step load $f(t) = 1$ starting from $t=0$, using these two ES-PIM models. The responses are plotted in FIGURE 7.20. The dash line is for the results without damping. We

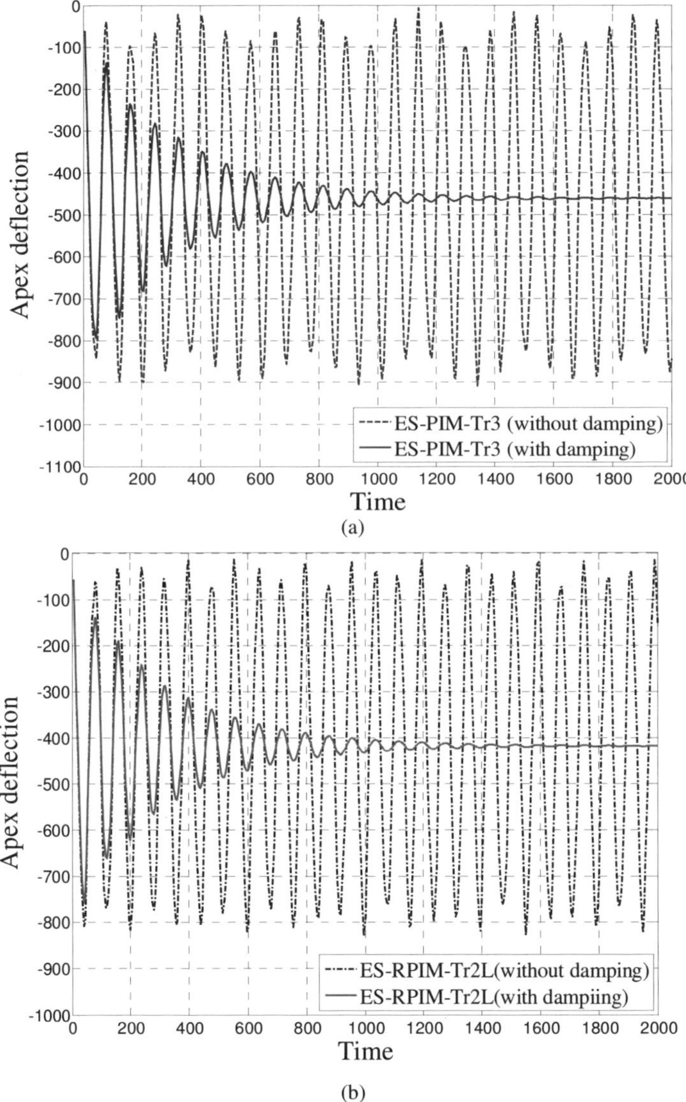

FIGURE 7.20 Transient responses obtained using ES-PIM models for the spherical shell subjected to a Heaviside step loading: (a) ES-PIM-Tr3 solution; (b) ES-RPIM-Tr2L solution.

observe that the amplitude of the deflection approach constant value with increasing time, but the vibration will never stop. When a Rayleigh damping of α_R=0.005 and β_R=0.272 are considered, the vibration is damped out, and the response converges to constant very quickly as expected. Such a constant value gives the static solution of the shell subjected to a unit force.

These dynamics examples lead to the following remark.

Remark 7.6 ES-PIM: works well for dynamic problems

The star performer ES-PIM models are usually stiffer models, temporally (and of course spatially) stable, produce close-to-exact stiffness, and work well for dynamic problems. We mentioned in the beginning of the chapter that models that produce upper bound solutions may have temporal instability problem. Some of the ES-PIM models produce upper bound solution for some problems, but no temporal instability has been observed. This may be because the stiffness of the ES-PIM models are so close to the exact model, so that even it is on the "soft" side and there may be spurious models, these models must be very higher order modes (at very higher energy level). Hence these spurious modes will not show in the numerical results for usual engineering problems with squarely integrable time excitations that do not excite high order (energy) modes.

7.6   Concluding remarks

This chapter presents in detail the ES-PIM with various models. It is a W^2 formulation based on the GS-Galerkin weak form. It has a number of important properties in comparison with the standard FEM using the same mesh. The following is a few additional issues worth to mention.

Remark 7.7 ES-PIM models: bound property

Because the ES-PIM models have very close-to-exact stiffness, their solutions are very close to the exact solution. Also because of this, these solutions can be both (often) lower and (but possibly) upper bounds, depending on the settings of the problem. Our experience shows that the linear ES-PIM-Tr3 gives usually lower bounds, and higher order models are less certain and can give both.

Remark 7.8 Extension to general n-sided polygonal cells/elements

In the above discussions on ES-PIM, we focused on the use of triangular background cells. However, the same edge-based smoothing idea can be applied for general n-sided cells, as practiced in the ES-FEM [10, 22].

7.7 Computer program

As in Chapter 6, we provide now the source codes for various ES-PIM/ES-RPIM models written in FORTRAN 90. The main program and three particular subroutines for ES-PIM models are given in Section 7.7.2, and all the other common subroutines are provided in the Appendix.

7.7.1 About the program

Consider the setting of Example 6.1.2 with the Mesh-1 of 120 nodes shown in FIGURE 6.9. TABLE 7.12 lists the three subroutines which are particularly coded for ES-PIM models.

TABLE 7.12 Particular subroutines coded for ESPIM.F90 and their functions

Subroutines	Function explanation
FormB_ESPIM	Form the strain-displacement matrix for an edge-based smoothing domain
Nodalstress_ESPIM	Obtain nodal stress and strain components for the ES-PIM models
Enernorm_error_ESPIM	Compute the global energy norm error and strain energy for the ES-PIM results

By adjusting the values of the two switches shown in shown in TABLE 7.13, four ES-PIM/ES-RPIM models can be established using the same set of source codes.

TABLE 7.13 Choices of the values of the switches for four ES-PIM/ RPIM models

ES-PIM/ES-RPIM models	NTscheme	NPIMSF
ES-PIM-Tr3	1	1
ES-PIM-Tr6/3	2	1
ES-RPIM-Tr6	3	2
ES-RPIM-Tr2L	4	2

7.7.2 Source codes in FORTRAN 90

Program 7.1 Main code of "ESPIM.F90"

```fortran
PROGRAM ESPIM
!================================================================
! The main program for conducting 2D ES-PIM/ES-RPIM methods
!================================================================
Use Parameters
Use Variables
Implicit real*8 (a-h, o-z)
Character*20 DATAINPUT, DATAOUTPUT
Open (5, FILE='DATAINPUT')
Open (10, FILE= DATAOUTPUT, status='unknown')

Call input
Call C_materialM
Call Cell_information
Gk=0.d0; force=0.d0
Loop_on_edges: do ia=1, numedge
    Bmat=0.d0; ndexnow=0; nvlist=0
    Call formB_ESPIM(ia)
    area_inte=area_edge(ia)
    Call StiffM_Intedomain(ia,area_inte,gpk)
    Call Form_GK
End do Loop_on_edges
Call Natural_BC
Call Essential_BC
Call Solver_LAE(Gk,force,nx*numnode,nx*numn)
Call Dispnorm_error
Call Nodalstress_ESPIM
Call Enernorm_error_ESPIM

End PROGRAM ESPIM
```

Program 7.2 Source code of "FormB_ESPIM"

```fortran
Subroutine FormB_ESPIM (ia)
!===============================================================
! Form the strain-displacement matrix of the ia_th edge-based smoothing domain
!
! Input:   x, nxcell, area_edge, kcell, Node_edge, Neig_edge and nodesix.
! Output: Bmat, nvlist and ndexnow.
!===============================================================
use Parameters
use Variables
implicit real*8 (a-h,o-z)
! work array
dimension xg(nx,2),xxg(nx,ngauss),wei(ngauss)
dimension xgg(nx,3),onxy(nx,2),nv(ndom),gpos(nx),phi(ndom)


ss=area_edge(ia); n1=Node_edge(1,ia); n2=Node_edge(2,ia)
! Loop on the neighbouring cells of edge_ia
LooP_on_neighbors: do i1=1,2
  kcc=Neig_edge(i1,ia)
  if (kcc/=0) then
  kcb=kcell(kcc)
  nn1=nxcell(1,kcc); nn2=nxcell(2,kcc); nn3=nxcell(3,kcc)
  xn1=x(1,nn1); yn1=x(2,nn1)
  xn2=x(1,nn2); yn2=x(2,nn2)
  xn3=x(1,nn3); yn3=x(2,nn3)
  xc=center(1,kcc); yc=center(2,kcc)
  indexn=0
  if (n1.eq.nn1 .and. n2.eq.nn2) then
      indexn=3
  else if (n2.eq.nn1 .and. n1.eq.nn2) then
      indexn=3
  else if (n1.eq.nn2 .and. n2.eq.nn3) then
      indexn=5
  else if (n2.eq.nn2 .and. n1.eq.nn3) then
      indexn=5
  else if (n1.eq.nn1 .and. n2.eq.nn3) then
```

```fortran
            indexn=4
    else if (n2.eq.nn1 .and. n1.eq.nn3) then
            indexn=4
    else
        write(*,*) 'Smothing wrong on [Neig_edge]'
        stop
    endif
    xgg=0.d0; onxy=0.d0
    select case(indexn)
    case(3)
        xgg(1,1)=xn2; xgg(2,1)=yn2; xgg(1,2)=xc; xgg(2,2)=yc
        xgg(1,3)=xn1; xgg(2,3)=yn1; xL1=((xn2-xc)**2+(yn2-yc)**2)**0.5
        onxy(1,1)=(yc-yn2)/xL1; onxy(2,1)=(xn2-xc)/xL1
        xL2=((xn1-xc)**2+(yn1-yc)**2)**0.5
        onxy(1,2)=(yn1-yc)/xL2; onxy(2,2)=(xc-xn1)/xL2
    case(5)
        xgg(1,1)=xn3; xgg(2,1)=yn3; xgg(1,2)=xc; xgg(2,2)=yc
        xgg(1,3)=xn2; xgg(2,3)=yn2; xL1=((xn3-xc)**2+(yn3-yc)**2)**0.5
        onxy(1,1)=(yc-yn3)/xL1; onxy(2,1)=(xn3-xc)/xL1
        xL2=((xn2-xc)**2+(yn2-yc)**2)**0.5
        onxy(1,2)=(yn2-yc)/xL2; onxy(2,2)=(xc-xn2)/xL2
    case(4)
        xgg(1,1)=xn1; xgg(2,1)=yn1; xgg(1,2)=xc; xgg(2,2)=yc
        xgg(1,3)=xn3; xgg(2,3)=yn3; xL1=((xn1-xc)**2+(yn1-yc)**2)**0.5
        onxy(1,1)=(yc-yn1)/xL1; onxy(2,1)=(xn1-xc)/xL1
        xL2=((xn3-xc)**2+(yn3-yc)**2)**0.5
        onxy(1,2)=(yn3-yc)/xL2; onxy(2,2)=(xc-xn3)/xL2
    endselect
    ndex=0; nv=0
    select case(NTscheme)
    case(1)
        ndex=3
        do kka=1,ndex
        nv(kka)=nxcell(kka,kcc)
        enddo
    case(2)
        select case(kcb)
        case(0); ndex=6; case(1); ndex=3; case(2); ndex=3
```

```fortran
      endselect
      do ik=1,ndex
      nv(ik)=nodesix(ik,kcc)
      enddo
case(3)
      ndex=6
      do kka=1,ndex
      nv(kka)=nodesix(kka,kcc)
      enddo
case(4)
      call cell_basedT2L(kcc,ndex,nv)
endselect
! Loop on two segments of smoothing domain boundaries located in cell_kcc
Loop_on_segments: do i2=1,2
   onx=onxy(1,i2); ony=onxy(2,i2)
   xg=0.d0; xxg=0.d0; wei=0.d0
   select case(i2)
      case(1)
      xg(1,1)=xgg(1,1); xg(2,1)=xgg(2,1); xg(1,2)=xgg(1,2); xg(2,2)=xgg(2,2)
      case(2)
      xg(1,1)=xgg(1,3); xg(2,1)=xgg(2,3); xg(1,2)=xgg(1,2); xg(2,2)=xgg(2,2)
   endselect
   call Line_gauss(xg,xxg,wei,det,Ngauss)
   ! Loop on the Gauss points located on the i2_th segment
   Loop_on_GP: do i3=1,ngauss
      gpos(1)=xxg(1,i3); gpos(2)=xxg(2,i3); weight=wei(i3)
      padx=onx*weight*det/ss; pady=ony*weight*det/ss; phi=0.d0
         select case(NPIMSF)
         case(1); call PPIM_SF2D(gpos,nv,ndex,phi)
         case(2); call RPIM_SF2D(gpos,nv,ndex,phi)
         endselect
      kp=i1+i2+i3
      if (kp.eq.3) then
      ndexnow=ndex
      Inner1_1: do j2=1,ndex
         nvlist(j2)=nv(j2); k1=nx*j2-1; k2=nx*j2-0
         bmat(1,k1)=bmat(1,k1)+padx*phi(j2); bmat(2,k2)=bmat(2,k2)+pady*phi(j2)
         bmat(3,k1)=bmat(3,k1)+pady*phi(j2); bmat(3,k2)=bmat(3,k2)+padx*phi(j2)
```

```
        enddo Inner1_1
        else
        ndexlast=ndexnow
        Inner1_2: do j3=1,ndex
          ka=nv(j3); kindex=0
          Inner1_2_1: do j4=1,ndexlast
          kb=nvlist(j4)
          if (ka.eq.kb) then
          kindex=1; kc=nx*j4-1; kd=nx*j4-0
          bmat(1,kc)=bmat(1,kc)+padx*phi(j3);bmat(2,kd)=bmat(2,kd)+pady*phi(j3)
          bmat(3,kc)=bmat(3,kc)+pady*phi(j3);bmat(3,kd)=bmat(3,kd)+padx*phi(j3)
          endif
          enddo Inner1_2_1
          if (kindex.eq.0) then
          ndexnow=ndexnow+1; nvlist(ndexnow)=ka
          kc=nx*ndexnow-1; kd=2*ndexnow-0
          bmat(1,kc)=bmat(1,kc)+padx*phi(j3);bmat(2,kd)=bmat(2,kd)+pady*phi(j3)
          bmat(3,kc)=bmat(3,kc)+pady*phi(j3);bmat(3,kd)=bmat(3,kd)+padx*phi(j3)
          endif
        enddo Inner1_2
        endif
    enddo Loop_on_GP
  enddo Loop_on_segments
  else
  ! Linear interpolation for the Gauss Points located on domain boundaries
  kcc=Neig_edge(1,ia); xg=0.d0; xxg=0.d0; wei=0.d0
  select case(indexn)
  case(3)
    xL=((xn1-xn2)**2+(yn1-yn2)**2)**0.5; onx=(yn2-yn1)/xL; ony=(xn1-xn2)/xL
    xg(1,1)=xn1; xg(2,1)=yn1; xg(1,2)=xn2; xg(2,2)=yn2
  case(5)
    xL=((xn3-xn2)**2+(yn3-yn2)**2)**0.5; onx=(yn3-yn2)/xL; ony=(xn2-xn3)/xL
    xg(1,1)=xn3; xg(2,1)=yn3; xg(1,2)=xn2; xg(2,2)=yn2
  case(4)
    xL=((xn3-xn1)**2+(yn3-yn1)**2)**0.5; onx=(yn1-yn3)/xL; ony=(xn3-xn1)/xL
    xg(1,1)=xn3; xg(2,1)=yn3; xg(1,2)=xn1; xg(2,2)=yn1
  endselect
  call Line_gauss(xg,xxg,wei,det,Ngauss)
```

```fortran
    ndex=0; nv=0; ndex=3
    do i9=1,ndex
       nv(i9)=nxcell(i9,kcc)
    enddo
    Inner1_3: do i5=1,ngauss
        gpos(1)=xxg(1,i5); gpos(2)=xxg(2,i5); weight=wei(i5)
        padx=onx*weight*det/ss; pady=ony*weight*det/ss; phi=0.d0
        call PPIM_SF2D(gpos,nv,ndex,phi)
        do 100 i7=1,ndex
           ka=nv(i7)
           do 110 i6=1,ndexnow
           kb=nvlist(i6)
             if (kb.eq.ka) then
             kc=2*i6-1; kd=2*i6-0
             bmat(1,kc)=bmat(1,kc)+padx*phi(i7);bmat(2,kd)=bmat(2,kd)+pady*phi(i7)
             bmat(3,kc)=bmat(3,kc)+pady*phi(i7);bmat(3,kd)=bmat(3,kd)+padx*phi(i7)
             goto 100
             endif
         110 continue
      100 continue
    enddo Inner1_3
    endif
end do LooP_on_neighbors


Return
End subroutine FormB_ESPIM
```

Program 7.3 Source code of "Nodalstress_ESPIM"

```fortran
Subroutine Nodalstress_ESPIM
!=============================================================
! Obtain nodal stress and strain compenents  for the ES-PIM models
!
! Input:    Cmat, area_edge, NoSmoo_node, NaSmoo_node and force(displacemets).
```

```fortran
! Output: stress_edge, strain_edge, stress_node and strain_node.
!===============================================================
use Parameters
use Variables
implicit real*8 (a-h,o-z)
! wrok array
dimension Dinv(3,3),ne(nx*nndom)

Dinv=0.d0; Dinv=Cmat
call inversion(3,3,Dinv)
stress_node=0.d0; strain_node=0.d0; stress_edge=0.d0; strain_edge=0.d0
! Computing stress and strain solutions for edge-based smoothing domain
Loop_on_edges: do ia=1,numedge
   weight=area_edge(ia)
   Bmat=0.d0; ndexnow=0; nvlist=0
   call formB_ESPIM(ia)
   do ih=1,ndexnow
      n1=nx*ih-1; n2=nx*ih-0
      ne(n1)=nx*nvlist(ih)-1
      ne(n2)=nx*nvlist(ih)-0
   enddo
   do ij=1,3
   do ik=1,3
   do il=1,nx*ndexnow
      mn=ne(il)
      stress_edge(ij,ia)=stress_edge(ij,ia)+Cmat(ij,ik)*Bmat(ik,il)*force(mn)
   enddo
   enddo
   enddo
   do iq=1,3
   do ir=1,3
      strain_edge(iq,ia)=strain_edge(iq,ia)+Dinv(iq,ir)*stress_edge(ir,ia)
   enddo
   enddo
enddo Loop_on_edges
! Computing nodal stress and strain components
Loop_on_nodes: do i=1,numnode
   ia=NoSmoo_node(i)
```

```
  ss=0.; ss1=0.; ss2=0.; ss3=0.; sn1=0.; sn2=0.; sn3=0.
  Inner: do j=1,ia
      ib=NaSmoo_node(j,i); ss=ss+area_edge(ib)
      ss1=ss1+stress_edge(1,ib)*area_edge(ib); ss2=ss2+stress_edge(2,ib)*area_edge(ib)
      ss3=ss3+stress_edge(3,ib)*area_edge(ib); sn1=sn1+strain_edge(1,ib)*area_edge(ib)
      sn2=sn2+strain_edge(2,ib)*area_edge(ib);
sn3=sn3+strain_edge(3,ib)*area_edge(ib)
  enddo Inner
  stress_node(1,i)=ss1/ss; stress_node(2,i)=ss2/ss; stress_node(3,i)=ss3/ss
  strain_node(1,i)=sn1/ss; strain_node(2,i)=sn2/ss; strain_node(3,i)=sn3/ss
enddo Loop_on_nodes

!output of the nodal stress results for the first 20 nodes
write(10,11) 'Field No.','Stress Sigma_xx','Stress Sigma_yy','Stress Tau_xy'
do i=1,20
write(10,12) i,(stress_node(j,i),j=1,3)
enddo
11 format(a10,3a18)
12 format(i10,3e18.8)


Return
End subroutine Nodalstress_ESPIM
```

Program 7.4 Source code of "Enernorm_error_ESPIM"

```
Subroutine Enernorm_error_ESPIM
!================================================================
! Computing the global energy norm error and strain energy for ES-PIM results
!
! Input:    area_edge, center, stress_edge and strain_edge;
! Output: Enernorm(the global displacement norm error) and
!            StrainE(the computed strain energy).
!================================================================
use Parameters
```

```fortran
use Variables
implicit real*8 (a-h,o-z)
! wrok array
dimension Dinv(3,3),ne(nx*nndom),gpos(nx)
dimension stressex(3),strainex(3),errstress(3),errstrain(3)

ai=ylength**3/12.d0
Dinv=0.d0; Dinv=Cmat
call inversion(3,3,Dinv)
StrainE=0.; Enernorm=0.
Loop_on_edges: do ia=1,numedge
  weight=area_edge(ia)
  nn1=Node_edge(1,ia); nn2=Node_edge(2,ia)
  ne1=Neig_edge(1,ia); ne2=Neig_edge(2,ia)
  gpos=0.d0
  if (ne2/=0) then
  gpos(1)=(x(1,nn1)+x(1,nn2))/2.; gpos(2)=(x(2,nn1)+x(2,nn2))/2.
  else
  gpos(1)=(x(1,nn1)+x(1,nn2)+center(1,ne1))/3.
   gpos(2)=(x(2,nn1)+x(2,nn2)+center(2,ne1))/3.
  endif
  stressex=0.d0; strainex=0.d0
  stressex(1)=-pLoad*(xlength-gpos(1))*gpos(2)/ai
  stressex(3)=pLoad/2./ai*(ylength**2/4.-gpos(2)**2)
  do iq=1,3
  do ir=1,3
      strainex(iq)=strainex(iq)+Dinv(iq,ir)*stressex(ir)
  enddo
  enddo
  errstress=0.d0; errstrain=0.d0
  do im=1,3
      errstress(im)=stress_edge(im,ia)-stressex(im)
      errstrain(im)=strain_edge(im,ia)-strainex(im)
  enddo
  do ip=1,3
     StrainE=StrainE+weight*0.5*strain_edge(ip,ia)*stress_edge(ip,ia)
  enddo
  do ip=1,3
```

```
    Enernorm=Enernorm+weight*errstrain(ip)*errstress(ip)
  enddo
end do Loop_on_edges

Enernorm=dsqrt(Enernorm)
write(10,11) 'Global error in energy norm:',Enernorm
write(10,12) 'Calculated strain energy:',StrainE
11 format(a29,e18.8)
12 format(a26,e21.8)


Return
End subroutine Enernorm_error_ESPIM
```

7.7.3　Computed results

Running the present program by taking NTscheme=1 and NPIMSF=1, the ES-PIM-Tr3 will be established and the following output file can be obtained.

Output file of "DATAOUTPUT"

Field No.	Disp. U_x	Disp. U_y
1	-0.36800000E-05	-0.30000000E-05
2	0.36800000E-05	-0.30000000E-05
3	-0.73600000E-05	-0.27000000E-04
4	0.73600000E-05	-0.27000000E-04
5	0.10833994E-05	-0.46670088E-04
6	0.62104288E-04	-0.51821450E-04
7	-0.62237532E-04	-0.52047751E-04
8	0.10576088E-03	-0.73441372E-04
9	-0.10583292E-03	-0.73435465E-04
10	0.00000000E+00	-0.75000000E-04
11	0.00000000E+00	-0.75000000E-04
12	-0.77752346E-04	-0.17538265E-03
13	0.91539773E-04	-0.18356409E-03
14	-0.19951809E-03	-0.16294205E-03

15	0.20608814E-03	-0.16673635E-03
16	0.23344330E-03	-0.14149516E-03
17	-0.23375317E-03	-0.14166129E-03
18	-0.28350844E-03	-0.36922598E-03
19	0.30523709E-03	-0.37847812E-03
20	0.45404712E-03	-0.32507782E-03

Global error in displacement norm: 0.17645926E-02

Field No.	Stress Sigma_xx	Stress Sigma_yy	Stress Tau_xy
1	0.56967452E+03	0.73901161E+01	-0.16407240E+03
2	-0.55937834E+03	-0.56301153E+01	-0.16306953E+03
3	0.16805649E+04	-0.16021231E+02	-0.13326585E+03
4	-0.16830206E+04	0.14073806E+02	-0.13408853E+03
5	0.20811599E+02	-0.17242869E+01	-0.12905948E+03
6	0.11295968E+04	0.23253353E+01	-0.10712029E+03
7	-0.11014904E+04	-0.20777726E+00	-0.10618984E+03
8	0.21572035E+04	-0.18734162E+02	-0.55531781E+02
9	-0.21510747E+04	0.21405070E+02	-0.53074992E+02
10	0.25489276E+04	-0.72873418E+02	-0.72119185E+02
11	-0.25514380E+04	0.73447413E+02	-0.71936105E+02
12	-0.54965522E+03	0.47221964E+01	-0.14259541E+03
13	0.63697711E+03	-0.45514976E+01	-0.14136317E+03
14	-0.17833599E+04	0.18513092E+01	-0.84627031E+02
15	0.18154417E+04	0.33555130E+01	-0.83984658E+02
16	0.25854973E+04	-0.17861306E+02	-0.23904864E+02
17	-0.25860610E+04	0.16570882E+02	-0.26607392E+02
18	-0.14236509E+04	0.12038368E+01	-0.87551740E+02
19	0.15066403E+04	-0.86395298E+00	-0.83294770E+02
20	0.23815246E+04	-0.24497132E+01	-0.45556866E+02

Global error in energy norm: 0.22500975E+00
Calculated strain energy: 0.85759038E+01

7.8 References

1.   Liu, G. R., Zhang, G. Y., Dai, K. Y, Wang, Y. Y., Zhong, Z. H., Li, G. Y. and Han, X., A linearly conforming point interpolation method (LC-PIM) for 2D solid mechanics problems, *International Journal of Computational Methods*, 2(4): 645-665, 2005.

2. Zhang, G. Y., Liu, G. R., Wang, Y. Y., Huang, H. T. Zhong, Z. H., Li, G. Y. and Han, X., A linearly conforming point interpolation method (LC-PIM) for three-dimensional elasticity problems, *International Journal for Numerical Methods in Engineering*, 72: 1524-1543, 2007.

3. Liu, G. R. and Zhang, G. Y., Upper bound solution to elasticity problems: A unique property of the linearly conforming point interpolation method (LC-PIM). *International Journal for Numerical Methods in Engineering*, 74: 1128-1161, 2008.

4. Zhang, G. Y., Liu, G. R., Nguyen-Thoi, T., Song, C. X., Han, X., Zhong, Z. H. and Li, G. Y., The upper bound property for solid mechanics of the linearly conforming radial point interpolation method (LC-RPIM). *International Journal of Computational Methods*, 4(3): 521-541, 2007.

5. Liu, G. R., A G space theory and a weakened weak (W^2) form for a unified formulation of compatible and incompatible methods: Part I theory. *International Journal for Numerical Methods in Engineering*, 81: 1093-1126, 2010.

6. Liu, G. R., A G space theory and a weakened weak (W^2) form for a unified formulation of compatible and incompatible methods: Part II applications to solid mechanics problems. *International Journal for Numerical Methods in Engineering*, 81: 1127-1156, 2010.

7. Liu, G. R., *Meshfree Methods: Moving beyond the Finite Element Method*, 2nd Edition, CRC press, Boca Taton, USA, 2009.

8. Liu, G. R. and Nguyen-Thoi, T., *Smoothed Finite Element Methods*, CRC press, Boca Taton, USA, June, 2010.

9. Liu, G. R. and Zhang, G. Y., Edge-based smoothed point interpolation methods, *International Journal of Computational Methods*, 5(4): 621-646, 2008.

10. Liu, G. R., Nguyen-Thoi, T. and Lam, K. Y. An edge-based smoothed finite element method (ES-FEM) for static, free and forced vibration analyses in solids. *Journal of Sound and Vibration*, 320: 1100-1130, 2009.

11. Tang, Q., Zhang, G. Y., Liu, G. R., Zhong, Z. H. and He, Z. C., An efficient adaptive analysis procedure using the edge-based smoothed point interpolation method (ES-PIM) for 2D and 3D problems. *Engineering Analysis with Boundary Elements,* 36: 1424-1443, 2012.

12. Liu, G. R., Wang, Z., Zhang, G. Y., Zong, Z. and Wang, S., An edge-based smoothed point interpolation method for material discontinuity. *Mechanics of Advanced Materials and Structures*, 19(1-3): 3-17, 2012.

13. He, Z. C., Cheng, A. G., Zhang, G. Y., Zhong, Z. H. and Liu, G. R., Dispersion error reduction for acoustic problems using the edge-based smoothed finite element method (ES-FEM). *International Journal for Numerical Methods in Engineering*, 86: 1322-1338, 2011.

14. He, Z. C., Liu, G. R., Zhong, Z. H., Cui, X. Y., Zhang, G. Y. and Cheng, A. G., A coupled edge-/face-based smoothed finite element method for structural-acoustic problems. *Applied Acoustics,* 71(10): 955-964, 2010.

15. Cui, X. Y., Liu, G. R., Li, G. Y., Zhang, G. Y., Zheng, G., Analysis of plates and shells using an edge-based smoothed finite element method, *Computational Mechanics*, 45: 141-156, 2010.

16. Wu, S. C., Liu, G. R., Cui, X. Y., Nguyen-Thoi, T. and Zhang, G. Y., An edge-based smoothed point interpolation method (ES-PIM) for heat transfer analysis of rapid manufacturing system. *International Journal of Heat and Mass Transfer*, 53: 1938-1950, 2010.

17. Cui, X. Y., Liu, G. R., Li, G. Y., Zhang, G. Y. and Sun, G. Y., Analysis of elastic-plastic problems using edge-based smoothed finite element method (ES-FEM). *International Journal of Pressure Vessels and Piping*, 86(10): 711-718, 2009.

18. Nguyen-Xuan, H., Liu, G. R., Nguyen-Thoi, T. and Nguyen-Tran, C., An edge-based smoothed finite element method for analysis of two-dimensional piezoelectric structures, *Smart Materials & Structures*, 18(6): 065015, 2009.

19. Nguyen-Thoi, T., Liu, G. R. and Vu-Do, H. C., Nguyen-Xuan, H., An edge-based smoothed finite element method (ES-FEM) for visco-elastoplastic analyses of 2D solids using triangular mesh. *Computational Mechanics*, 45: 23-44, 2009.

20. Nguyen-Thoi, T., Liu, G. R., Lam, K. Y. and Zhang, G. Y., A Face-based Smoothed Finite Element Method (FS-FEM) for 3D linear and nonlinear solid mechanics problems using 4-node tetrahedral elements. *International Journal for Numerical Methods in Engineering*, 78: 324-353, 2009.

21. Nguyen-Thoi, T., Liu, G. R., Vu-Do, H. C. and Nguyen-Xuan, H., A face-based smoothed finite element method (FS-FEM) for visco-elastoplastic analyses of 3D solids using tetrahedral mesh. *Computer Methods in Applied Mechanics and Engineering*, 198: 3479-3498, 2009.

22. Dai, K. Y., Liu, G. R. and Nguyen-Thoi, T., An n-sided polygonal smoothed finite element method (nSFEM) for solid mechanics. *Finite elements in analysis and design*, 43: 847-860, 2007.

Chapter 8

Cell-based Smoothed Point Interpolation Method (CS-PIM)

In Chapters 6 and 7, we have presented various models of NS-PIM and ES-PIM and examined their properties. The NS-PIMs are "softer models", spatially stable, work well for static problems, with a very important upper bound property, but instable temporally due to the "overly-soft" behavior [1-6]. The ES-PIM models, on the other hand, have very "close-to-exact" stiffness, and can thus provide very accurate and superconvergent solutions in both displacement and energy norms. The ES-PIM models are usually "stiffer models", both spatially and temporally stable, and hence work well for both static and dynamic problems [1, 7-11].

In this chapter, we introduce yet another powerful method called cell-based smoothed point interpolation method or CS-PIM that was originated in [12-14]. Different from NS-PIM and ES-PIM, the CS-PIM models use the triangular (2D) and tetrahedral (3D) background cells as smoothing domains. The CS-PIM models can be made both spatially and temporally stable and hence work well for both static and dynamic problems. They are stiffer than the overly-soft NS-PIMs, but generally softer than the ES-PIMs, and hence possess properties of both. A CS-PIM model can have a very close-to-exact stiffness, and can be made softer or stiffer as desired. Examples will be presented to examine in detail the accuracy and computational efficiency of the CS-PIM models in comparison with FEM and other S-PIM models.

8.1 CS-PIM for 2D solids

8.1.1 Approximation of displacement field

The problem domain is first represented by a set of N_c nonoverlapping and seamless (NOSL) triangular background cells with N_n field nodes, as in any S-PIM model, such that $\overline{\Omega} = \bigcup_{i=1}^{N_c} \overline{\Omega}_i^c$. In the CS-PIM, the points of interest (quadrature points) are always located on the edges of triangles, because of the cell-based strain smoothing. The displacement field will be approximated using the PIM shape functions (both polynomial PIM and RPIM [1, 15-17]) with the support nodes selected using the edge-based T-schemes as described in Section 1.6.3.

When the edge-based T2-scheme is used, as shown in FIGURE 8.1a which is the same as FIGURE 1.7e, the approximated displacement field is compatible over the problem domain, and the constructed CS-PIM-Tr2 model is exactly as same as the linear FEM. When the edge-based T4-scheme (shown in FIGURE 8.1b and is the same as FIGURE 1.7f) is used to select support nodes, we have bilinear displacement field within the problem domain and linear displacement field along the domain boundaries. The approximated displacement field is not compatible and the discontinuity will occur along the lines radiated from the node to the centroid of the triangle, as illustrated using dashed line in FIGURE 8.1b. Note in this case, singularity problem of the moment matrix may be encountered for forming polynomial PIM shape functions, and effective measures have been proposed in Chapter 3. When we use the edge-based T2L-scheme (illustrated in FIGURE 8.1c and is the same as FIGURE 1.7g) to select support nodes for constructing RPIM shape functions, we have the CS-RPIM-Tr2L model. In this case, the displacement field approximated is not continuous and the discontinuity will occur along the lines radiated from the node to the centroid of the triangle, as indicated by the dashed line in FIGURE 8.1c.

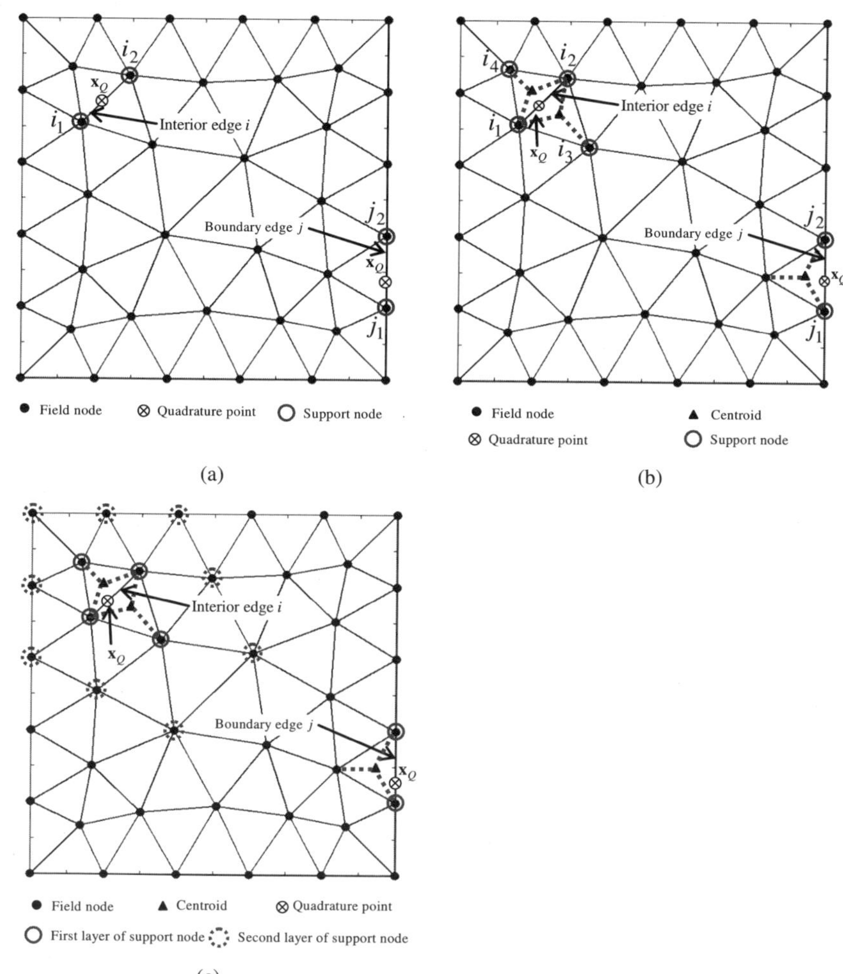

(a)

(b)

(c)

FIGURE 8.1 T-schemes used for selecting support nodes for the construction of PIM/RPIM shape functions used in various CS-PIM models: (a) edge-based T2-scheme; (b) edge-based T4-scheme; (c) edge-based T2L-scheme.

8.1.2 Evaluation of cell-based smoothed strains

In the CS-PIM, the gradient smoothing domains are exactly the background cells and also used as quadrature domains. It is hence termed as cell-based smoothing domains, and no additional operation is needed to construct the smoothing domains. We thus have $\boxed{\Omega} = \bigcup_{k=1}^{N_s} \boxed{\Omega}_k^s$ where $\Omega_i^s \bigcap \Omega_j^s = \varnothing$, $\forall i \neq j$, $\Omega_k^s = \Omega_k^c$, and $N_q = N_s = N_c$. FIGURE 8.2 shows a 2D domain with triangular cells, each of which also serves as a cell-based smoothing domain for CS-PIM models. It can be clearly seen that such a division of smoothing domains is "legal" and satisfies the conditions described in Section 2.3.1, because the node selection for creating PIM shape functions is based on the edges of these cells. We also assume the smoothed strain is constant in each smoothing domain which can be obtained by following the strain smoothing operation described in Section 4.5.2.

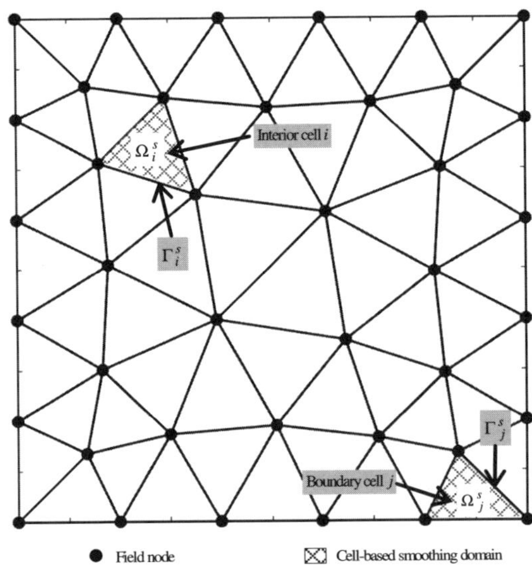

FIGURE 8.2 A CS-PIM model: triangular background cells for a 2D domain, each of which serves also as a smoothing domain. The "smoothed" strain is constant in a cell, and is obtained using the generalized smoothing operation over the cell.

8.1.3 CS-PIM formulations

For dynamic problems, the discretized algebraic system equation follows Equation (5.85) and has the form of

$$\bar{\mathbf{K}}^{\text{CS-PIM}}\bar{\mathbf{d}} + \mathbf{C}\dot{\bar{\mathbf{d}}} + \mathbf{M}\ddot{\bar{\mathbf{d}}} = \tilde{\mathbf{f}} \tag{8.1}$$

where $\overline{\mathbf{K}}^{\text{CS-PIM}}$ is the global *smoothed* stiffness matrix whose entries are given by

$$\overline{\mathbf{K}}_{IJ}^{\text{CS-PIM}} = \int_{\Omega} \overline{\mathbf{B}}_I^{\text{T}} \mathbf{c} \overline{\mathbf{B}}_J \, d\Omega = \sum_{k=1}^{N_c} \int_{\Omega_k^s} \overline{\mathbf{B}}_I^{\text{T}} \mathbf{c} \underbrace{\overline{\mathbf{B}}_J}_{\text{constant in } \Omega_k^s} d\Omega = \sum_{k=1}^{N_c} A_k^s \overline{\mathbf{B}}_I^{\text{T}} \mathbf{c} \overline{\mathbf{B}}_J \tag{8.2}$$

in which $A_k^s = \int_{\Omega_k^s} d\Omega$ is the area of the cell-based smoothing domain Ω_k^s, and

the smoothed strain-displacement matrix $\overline{\mathbf{B}}_I$ is computed using Equation (5.71).

In Equation (8.1), $\mathbf{M}$ is the global mass matrix and $\mathbf{C}$ is the global damping matrix that is assembled using the following entries, respectively.

$$\mathbf{M}_{IJ} = \int_{\Omega} \mathbf{\Phi}_I^{\text{T}} \rho \mathbf{\Phi}_J \, d\Omega \tag{8.3}$$

$$\mathbf{C}_{IJ} = \int_{\Omega} \mathbf{\Phi}_I^{\text{T}} c_{dp} \mathbf{\Phi}_J \, d\Omega \tag{8.4}$$

where ρ is the mass density and c_{dp} is the damping coefficient. Note that there is no smoothing operation for $\mathbf{M}$, $\mathbf{C}$, and $\tilde{\mathbf{f}}$, and hence these matrices are obtained basically in the same way as in the standard FEM.

8.1.3.1   Static analysis

For static problems, the system equation can be obtained by dropping the dynamic terms in Equation (8.1):

$$\overline{\mathbf{K}}^{\text{CS-PIM}} \overline{\mathbf{d}} = \tilde{\mathbf{f}} \tag{8.5}$$

The solution procedure of CS-PIM for static problems is exactly the same as in other S-PIMs or FEM.

8.1.3.2   Free vibration analysis

Considering free vibration analysis, we shall not have the damping and the external force terms, and Equation (8.1) becomes

$$\overline{\mathbf{K}}^{\text{CS-PIM}} \overline{\mathbf{d}} + \mathbf{M} \ddot{\overline{\mathbf{d}}} = \mathbf{0} \tag{8.6}$$

A general solution to the foregoing homogeneous equation can be given as

$$\overline{\mathbf{d}} = \overline{\mathbf{d}}_A \exp\left(i\omega t\right) \tag{8.7}$$

where t stands for time, $\overline{\mathbf{d}}_A$ is the amplitude of the nodal displacement or the eigenvector and ω is natural frequency that is found from

$$\left(\overline{\mathbf{K}}^{\text{CS-PIM}} - \omega^2 \mathbf{M}\right)\overline{\mathbf{d}}_A = \mathbf{0} \tag{8.8}$$

8.1.3.3 Forced vibration analysis

For forced or transient vibration problems, Equation (8.1) can be solved by direct integration methods as in the FEM. For simplicity, the Rayleigh damping is assumed in this book, and damping matrix $\mathbf{C}$ is assumed to be a linear combination of $\mathbf{M}$ and $\overline{\mathbf{K}}^{\text{CS-PIM}}$.

$$\mathbf{C} = \alpha_R \mathbf{M} + \beta_R \overline{\mathbf{K}}^{\text{CS-PIM}} \tag{8.9}$$

where α_R and β_R are the Rayleigh damping coefficients.

In this chapter, we also use the Newmark method to solve the forced vibration problems as presented in Section 7.3.4.

8.1.4 Numerical implementation

8.1.4.1 Macro flowchart of the CS-PIM

The macro flowchart for constructing the global matrices and vectors of CS-PIM models can be briefly summarized as follows.

1) Loop over all the cell-based smoothing domains, i.e. the background cells.

2) Loop over the Gauss points located along boundary edges of each cell:

    (a) Select support nodes using a proper edge-based T-scheme and create PIM shape functions.

    (b) Compute the components of smoothed strain-displacement matrix and form the smoothed strain-displacement matrix.

(c) Compute nodal stiffness matrices and force vectors.

(d) Assemble the nodal contributions to the global matrices and vectors.

3) End the Gauss points loop.

4) End the smoothing domain loop.

8.1.4.2 Possible 2D CS-PIM models

In the CS-PIM settings, various edge-based T-scheme (see Section 1.6.3) have been proposed to select support nodes for constructing PIM or RPIM shape functions. Based on different PIM shape functions and the associated T-schemes used, we have the following CS-PIM models developed so far.

CS-PIM-Tr2: same as FEM-Tr3

A CS-PIM-Tr2 model uses linear PIM shape functions and the two support nodes at the two ends of the edge (selected by the edge-based T2-scheme shown in FIGURE 8.1a). This is the simplest CS-PIM model and is exactly the same as the FEM-Tr3, and hence it is a fully compatible model. Therefore, the CS-PIM-Tr2 model will not be discussed any further in this book.

CS-PIM-Tr4-CT

In a CS-PIM-Tr4-CT model, the displacement field is constructed using polynomial PIM shape functions and support nodes selected by the edge-based T4-scheme with the adaptive coordinate transformation (CT) technique [12]. In this case, we have bilinear displacement field within the problem domain and linear displacements along the problem boundaries. Note the displacement field in the CS-PIM-Tr4-CT model is incompatible.

Note that the support nodes for interior and boundary cells are different. To find the rough average number of support nodes for cell-based smoothing domains, the discrete model for the cantilever beam presented by 248 irregular nodes, shown in FIGURE 6.4, is used. For the CS-PIM-Tr4-CT model, the number of the support nodes for a smoothing domain is found 5-6 and the average number is found about 5.8, that is about 1.9 times that of the FEM-Tr3. Therefore, the average number of nonzero entries in a row in the (global) stiffness matrix of a CS-PIM-Tr4-CT model (n_{ze}) is about 1.9 time that of the FEM-Tr3 counterpart using the same mesh, as listed in TABLE 8.1.

CS-PIM-Tr4-Iso

A CS-PIM-Tr4-Iso model uses isoparametric PIM shape functions (see, Section 3.5) to construct displacement fields, where the support nodes are selected by the edge-based T4-scheme [13]. Compared to the previous bilinear PIM shape functions using coordinates transformation technique, the isoparametric PIM shape functions can be obtained in a very straightforward way, no extra operation is needed, and the singularity issue can also been effectively solved. The displacement field in the CS-PIM-Tr4-Iso model is incompatible.

For the CS-PIM-Tr4-Iso model, the estimation of the number of the support nodes for a smoothing domain and the average number of nonzero entries in a row in the stiffness matrix are as same as that in the CS-PIM-Tr4-CT model.

CS-RPIM-Tr4

The displacement field in the CS-RPIM-Tr4 model is constructed by the RPIM shape functions with the support nodes selected using the edge-based T4-scheme. Since the RPIM shape functions can always be created without the singularity problem, there is no need to perform the coordinate transformation or other techniques in the process of shape function construction. The displacement field in the CS-RPIM-Tr4 model is incompatible.

For this model, the estimation of the number of the support nodes for a smoothing domain, and the average number of nonzero entries in a row in the stiffness matrix are the same as the CS-PIM-Tr4-CT model.

CS-RPIM-Tr4-Cd

The CS-RPIM-Tr4-Cd uses the condensed RPIM shape functions to construct displacement field, where the support nodes are selected by the edge-based T4-scheme [13]. As described in Section 3.7, the condensed RPIM shape functions are formulated by using four real field nodes selected using the edge-based T4-scheme plus four virtual nodes. Thus compared to the RPIM shape functions using only four real nodes (as used in the CS-RPIM-Tr4 model), this condensed RPIM shape functions are supposed to improve the performance of the results but do not increase the DOFs of the problem. The displacement field in the CS-RPIM-Tr4-Cd is incompatible.

Note that compared to the CS-RPIM-Tr4 model, the CS-RPIM-Tr4-Cd uses the condensed RPIM shape functions, and there is no increase for either the support nodes of a smoothing domain or the DOFs of the problem. Therefore for the CS-RPIM-Tr4-Cd model, the estimation of the number of the support nodes for a smoothing domain and the average number of nonzero entries in a row in the stiffness matrix are as same as that of the CS-PIM-Tr4-CT model.

CS-RPIM-Tr2L

In a CS-RPIM-Tr2L model, the RPIM shape functions are used with the support nodes selected by the edge-based T2L-scheme. This type of T-scheme is well controlled and works well for virtually randomly distributed nodes, but can be less efficient, because many nodes are usually used resulting in a less sparse system equation. The displacement field in the CS-RPIM-Tr2L model is incompatible.

The number of the support nodes for a smoothing domain is 10-19 and the average number is about 16, that is about 5.4 times that of the FEM-Tr3. Therefore, the average number of nonzero entries in a row in the stiffness matrix of a CS-RPIM-Tr2L model is about 5.4 times that of the FEM-Tr3 counterpart using the same mesh, as listed in TABLE 8.1.

TABLE 8.1 CS-PIM models and overall characteristics in relation to the FEM-Tr3 model using the same mesh

Numerical method	Support nodes in an integral cell/element	Estimated average number of nonzero entries	Estimated Solver CPU time
FEM-Tr3	3	n_{ze}	t_{CPU}
CS-PIM-Tr4-CT	5-6 (5.8)	$1.9\, n_{ze}$	$1.9\, \bar{n}_{iter} t_{CPU}$
CS-PIM-Tr4-Iso	5-6 (5.8)	$1.9\, n_{ze}$	$1.9\, \bar{n}_{iter} t_{CPU}$
CS-RPIM-Tr4-Cd	5-6 (5.8)	$1.9\, n_{ze}$	$1.9\, \bar{n}_{iter} t_{CPU}$
CS-RPIM-Tr4	5-6 (5.8)	$1.9\, n_{ze}$	$1.9\, \bar{n}_{iter} t_{CPU}$
CS-RPIM-Tr2L	10-19 (16.1)	$5.4\, n_{ze}$	$5.4\, \bar{n}_{iter} t_{CPU}$

8.1.4.3 Condition number of CS-PIM models

The condition number of the global stiffness matrix, $cond(\mathbf{K})$, is an important indicator for the numerical property of a numerical method, as discussed in Section 1.2.2. When an iteration solver is used to solve the algebraic system equation, it affects directly the number of iterations needed to obtain a converged solution in the manner of $n_{iter} \propto \sqrt{cond(\mathbf{K})}$ (see, Equation 1.28).

TABLE 8.2 lists the condition numbers of the global stiffness matrixes computed for various CS-PIM models in relation to the FEM-Tr3, where we use the same triangular meshes for the cantilever beam problem shown in FIGRUE 6.4. We can find that except the CS-RPIM-Tr2L using RPIM shape functions with much more support nodes, other CS-PIM models have a smaller condition numbers than the linear FEM-Tr3, although higher order interpolation with more support nodes are used as shown in TABLE 8.1. The condition number of the CS-RPIM-Tr4 model is a little smaller than the linear FEM, and other three CS-PIM models, i.e., CS-PIM-Tr4-CT, CS-PIM-Tr4-Iso and CS-RPIM-Tr4-Cd, all have a smaller condition number with is about 60% of the counterpart of the linear FEM. We also find that using the condensed RPIM shape functions without increasing DOFs and support nodes, the condition number of the numerical model can be reduced significantly: the condition number of the CS-RPIM-Tr4-Cd model is about 60% of the counterpart of the CS-RPIM-Tr4 model.

We can also find that the ratio between the condition numbers of different methods with respect to the linear FEM-Tr3 model has little change when the field nodes are more than doubled, as shown in TABLE 8.2. In the following estimation of computation cost, we will use the relative number given in last column in TABLE 8.2.

TABLE 8.2 Condition number of the global stiffness matrixes for various CS-PIM models for the cantilever in relation to the FEM-Tr3 model using the same mesh

Numerical method	Mesh-1 of 120 nodes		Mesh-2 of 248 nodes	
	$cond(\mathbf{K})$	$\bar{n}_{iter}$	$cond(\mathbf{K})$	$\bar{n}_{iter}$
FEM-Tr3	1.645e+08	1.00	2.399e+08	1.00
CS-PIM-Tr4-CT	6.097e+07	0.61	9.222e+07	0.62
CS-PIM-Tr4-Iso	5.420e+07	0.57	8.352e+07	0.59
CS-RPIM-Tr4-Cd	5.415e+07	0.57	8.362e+07	0.59
CS-RPIM-Tr4	1.529e+08	0.96	2.257e+08	0.97
CS-RPIM-Tr2L	2.279e+08	1.18	3.064e+08	1.13

8.1.4.4 Estimation of computational cost for 2D CS-PIM

Combining TABLE 8.1 and TABLE 8.2 we shall have a rough estimation of computational cost for the CS-PIM models in relation to the standard FEM-Tr3 model using the same mesh, as shown in TABLE 8.3. It is seen that the first three CS-PIM models including the CS-PIM-Tr4-CT, the CS-PIM-Tr4-Iso and the CS-RPIM-Tr4-Cd, only cost slightly more time than the linear FEM model for the same mesh. If these three CS-PIM models can be 20% more accurate than the FEM model, they can be more computationally *efficient*. The CS-RPIM-Tr4 will cost about 1.8 times, and the CS-RPIM-Tr2L will cost about 6 times that of the FEM for the same problem using the same set of nodes.

TABLE 8.3 Estimation of computational cost for 2D CS-PIM models in relation to the FEM-Tr3 model using the same mesh

Numerical method	Support nodes in an integral cell/element	Estimated Solver CPU time
FEM-Tr3	3	t_{CPU}
CS-PIM-Tr4-CT	5-6 (5.8)	$1.2\,t_{CPU}$
CS-PIM-Tr4-Iso	5-6 (5.8)	$1.1\,t_{CPU}$
CS-RPIM-Tr4-Cd	5-6 (5.8)	$1.1\,t_{CPU}$
CS-RPIM-Tr4	5-6 (5.8)	$1.8\,t_{CPU}$
CS-RPIM-Tr2L	10-19 (16.1)	$6.1\,t_{CPU}$

8.1.4.5 Evaluation of nodal strain (stress)

In the scheme of CS-PIM models, the cell-based smoothed strains (constant) are corresponding to each triangular background cell. Once the nodal displacement results have been obtained by solving Equation (8.5), the smoothed strains corresponding to each smoothing domain can then be calculated using Equation (5.70). The "nodal strain" can then be further obtained using the area-weighted averaging the smoothed strains, as discussed in Section 7.4.4. For node i, the nodal strain becomes

$$\overline{\varepsilon}\left(\mathbf{x}_i\right) = \frac{1}{A_i^{ns}} \sum_{k=1}^{n_i^s} \overline{\varepsilon}_k A_k^s \tag{8.10}$$

where n_i^s is the number of smoothing domains around node i, $A_i^{ns} = \sum_{k=1}^{ni^s} A_k^s$ is the total area of these n_i^s smoothing domains, and $\overline{\varepsilon}_k$ is the smoothed strain corresponding to the smoothing domain Ω_k^s of area A_k^s.

FIGURE 8.3 shows the smoothing domains used to evaluate the strains at the particular node i, where totally six smoothing domains are involved and hence $n_i^s = 6$.

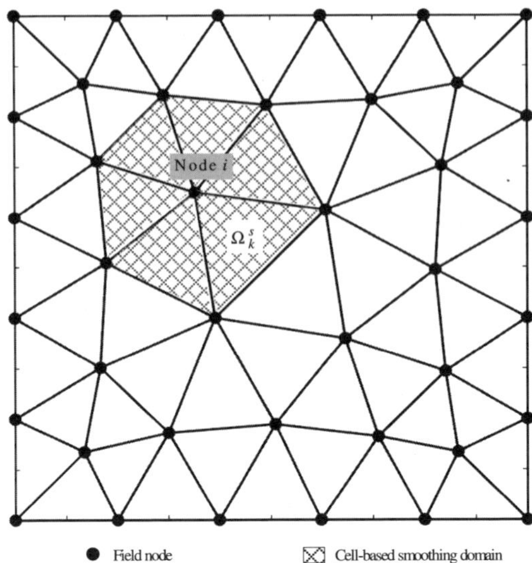

FIGURE 8.3 Smoothing domains used to evaluate the strains at node i in the scheme of CS-PIM models. For the particular node i shown in the figure, six cell-based smoothing domains around node i are involved in the calculation, and hence $n_i^s = 6$.

8.1.4.6 Rank analysis for CS-PIM

Same as other two S-PIM models presented in the previous chapters, CS-PIM models possess only the "legal" zero-energy modes representing the physical rigid motions, and there exist no spurious zero-energy modes. The spatial stability of CS-PIM models is ensured by the following points.

1) In a 2D CS-PIM model, the number of smoothing domains equals to the number of triangular background cells, i.e. $N_s = N_c$. Thus the minimum number of smoothing domains required in TABLE 2.2 can always be satisfied. The worst case is a mesh with only one triangular cell. Even in such a case, we have $N_s = N_s^{\min}$. When more nodes are used, the N_c/N_n ratio should in general increase, CS-PIM models should thus become stiffer and hence have better stability.

2) The cell-based smoothing domains are independent with each other. Together with the independence of the nodal PIM shape functions, the strain smoothing operation ensures linearly independent columns (or rows) in the stiffness matrix.

3) The PIM shape functions used in a CS-PIM model are of partitions of unity, which ensures a proper representation of the rigid movements.

4) The stiffness matrix of a CS-PIM model is SPD for constrained solids of stable materials.

Remark 8.1 CS-PIM models: spatially stable

CS-PIM models are spatially stable. A CS-PIM model possesses only "legal" zero-energy modes that represent the physic rigid motions, and there exist no spurious unphysical zero-energy modes. Therefore, there is no deformed zero-energy mode existing in a CS-PIM model, and any finite deformation in the model (except the rigid motions) will results in a finite amount of strain energy in a CS-PIM model.

8.1.4.7   Temporal stability analysis

Based on the discussion in Section 7.4.8, we know both spatial and temporal stability are directly related to the number of the samplings of the integrand in the weak form. To study the temporal stability of S-PIMs, we have used the linear FEM that is known as a stiffer model (in fact, it is overly stiff) and hence stable both spatially and temporally as the base model for comparison.

For the present CS-PIM models, the smoothing domains, which have constant smoothed strains and serve as quadrature domains, are exactly the same as the background cells. Thus a CS-PIM model has the same number of the "samplings" of the integrand as the linear FEM-Tr3. When the edge-based T2-scheme is used, the CS-PIM is exactly the same as the FEM-Tr3, and hence is both spatially and temporally stable. The higher order CS-PIM models, however, can be temporally instable, due to the softening effect introduced by 1) the higher order PIM interpolation, and 2) the cell-based gradient smoothing operation. In our previous studies [12], we have found that CS-PIM models possess very "close-to-exact" stiffness, which is even closer to the exact one than the ES-PIM models. This feature also indicates that some CS-PIM models may behave temporally instable. At this stage, we have the following intuitive understanding.

1) The CS-PIM-Tr2 model is *always* both spatially and temporally stable.

2) Higher order CS-PIM models are always spatially stable and should be in general softer than FEM-Tr3 (or CS-PIM-Tr2).

3) When the mesh is refined, the N_c/N_n rate should in general increase, CS-PIM models should thus become more stable temporally. This implies that CS-PIM models should be temporally stable when mesh is not too coarse, at least for lower modes.

4) The temporal stability for a CS-PIM model can *always* be restored by introducing the edge-based T2-scheme (two-node) for PIM interpolations for more cell edges.

5) In CS-PIM setting, the PIM interpolation is always performed for nodes on the cell edges, and hence replacing a higher order interpolation by two-node interpolation can be done very easily.

Remark 8.2 CS-PIM models: temporally stable

The above-mentioned points ensure essentially that we can always make a CS-PIM model temporally stable. This can be achieved by either using not-too-coarse mesh, or using the edge-based T2-scheme for more edges in the PIM interpolation.

The temporal stability property of CS-PIM models will be further discussed in great detail in Example 8.1.5.

8.1.5 Numerical examples for 2D solids

Several numerical examples are studied in this section to examine the CS-PIM models. The materials used are all linear elastic with Young's modulus $E=3.0\times10^7$ Pa and passion's ratio $v=0.3$, and the units used are based on the international standard unit system unless specified explicitly. Numerical results are assessed using the error indicators in displacement and energy norms which are defined in Equation (5.88) and Equations (5.94), respectively.

Example 8.1.1 A 2D standard patch test

The standard patch test described in Example 6.1.1 is first studied using the CS-PIM models. TABLE 8.4 lists the displacement norm errors (defined in Equation 5.88) of the CS-PIM solutions for the patch test using both the regular and irregular meshes shown in FIGURE 6.5. All the CS-PIM models can pass these tests exactly (to the machine accuracy). Note that the displacement field in all the studied CS-PIM models is incompatible.

TABLE 8.4 Error norm in displacements of numerical results for the standard patch test obtained using various CS-PIM models

CS-PIM models	Regular mesh	Irregular mesh
CS-PIM-Tr4-CT (displacement incompatible)	5.157E-15	3.793E-15
CS-PIM-Tr4-Iso (displacement incompatible)	4.132E-14	1.286E-15
CS-RPIM-Tr4-Cd (displacement incompatible)	2.652E-12	3.668E-11
CS-RPIM-Tr4 (displacement incompatible)	5.744E-15	1.679E-14
CS-RPIM-Tr2L (displacement incompatible)	1.068E-12	2.259E-12

Remark 8.3 CS-PIM: linearly conforming and 2nd order accuracy

This standard patch test example demonstrates numerically that the CS-PIM is linearly conforming: it can reproduce linear fields exactly regardless of the incompatible displacement field used. Together with the stability (Theorem 5.1), the CS-PIM solution will converge to the exact solution of any well-posed linear elasticity problem. This implies also that the model is at least 2nd order accuracy: the solution error in displacement is on the terms of 2nd order and above. This is consistent with Remark 3.23.

Example 8.1.2 Rectangular cantilever

The benchmark rectangular cantilever problem described in Example 6.1.2 is again studied using various CS-PIM models. For comparison, the linear FEM, NS-PIM and ES-PIM are also adopted in this analysis with the same set of triangular meshes shown in FIGURE 6.9.

Using Mesh-2 (with 399 nodes) in FIGURE 6.9, this cantilever problem is first studied using various CS-PIM models and the calculated numerical results along two particular lines are plotted in FIGURE 8.4. It shows that both the displacements and stress results are all in good agreement with the analytical ones.

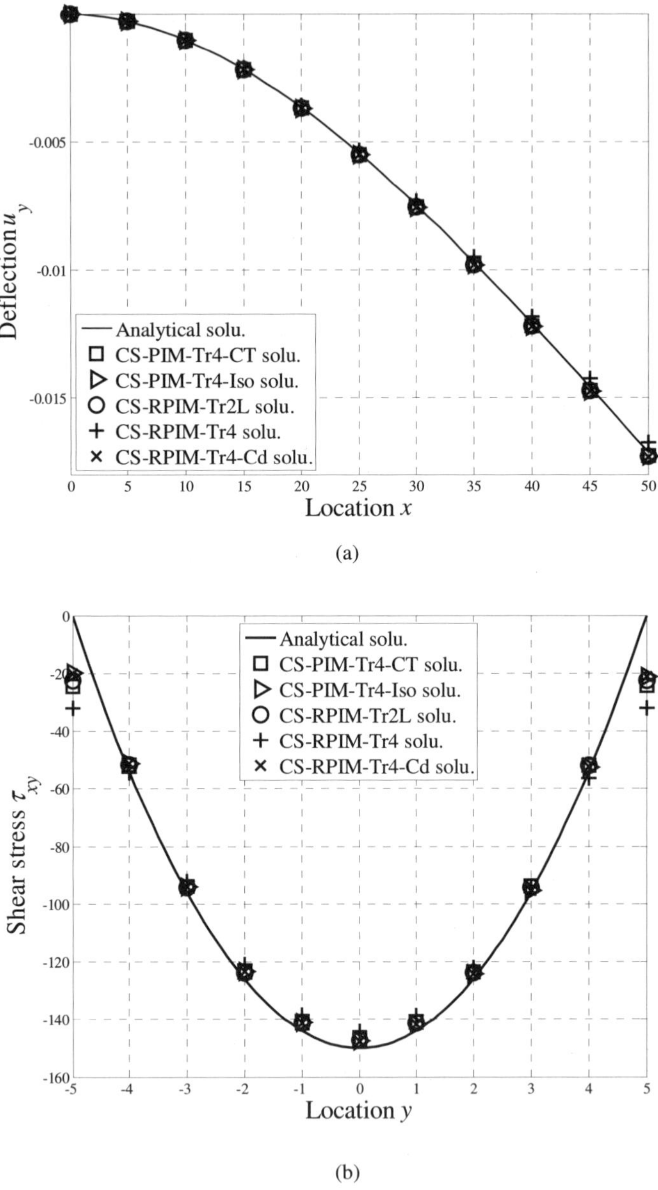

(a)

(b)

FIGURE 8.4 Comparison of numerical results obtained using CS-PIM/CS-RPIM models for the rectangular cantilever problem with analytical ones: (a) deflection along the neutral line of the cantilever; (b) shear stress along the line of $x=L/2$ of the cantilever.

The convergence of the solution error in displacement norm for the cantilever problem is plotted in FIGURE 8.5. The errors in the displacement norm for the numerical results obtained using the same set of triangular mesh (Mesh-4) are listed in TABLE 8.5. Taking the computational cost given in TABLE 8.3 into consideration for these methods, the estimated computational efficiency measured in displacement error is calculated and listed in the last column in TABLE 8.5. We can now clearly observe the following.

1) Except the CS-RPIM-Tr4 model, all the other CS-PIM models provide much better numerical results than that of FEM-Tr3 and NS-PIM-Tr3 models, in terms of both accuracy and convergence rate.

2) For the CS-RPIM-Tr4 model, where only four nodes are selected for the construction of RPIM shape functions, the calculated results in displacement norm are of similar accuracy and convergence rate compared to the FEM-Tr3 and NS-PIM-Tr3. This model has been found performing worst among all these CS-PIM models.

3) For the CS-RPIM-Tr4-Cd model, where the condensed RPIM shape functions are used with four support nodes plus four virtual nodes, the performance of the numerical results has been greatly improved, compared to the CS-RPIM-Tr4 model using RPIM shape functions with only four support nodes.

4) The three models, including the CS-PIM-Tr4-CT, CS-PIM-Tr4-Iso and CS-RPIM-Tr4-Cd, show similar performance for this problem. In terms of accuracy, their results are about more than 2 times that of the FEM-Tr3. Considering the estimated CPU time shown in TABLE 8.3, the computational *efficiency* of these three models will be about 2 times that of the FEM-Tr3, as shown in the last column in TABLE 8.5. For the CS-RPIM-Tr4 model, estimated CPU time is about 1.8 times that of the FEM-Tr3, and hence the computational efficiency will be lower, about 0.6 times that of the FEM-Tr3. For the CS-RPIM-Tr2L model, estimated CPU time is about 6.1 times that of the FEM-Tr3, and hence the computational efficiency will also be lower, about 0.4 times that of the FEM-Tr3.

5) In terms of convergence rate, we found that all the CS-PIM models, except the CS-RPIM-Tr4, have higher numerical rates than the theoretical value of 2.0, showing a weak *superconvergence* in displacement norm.

6) The ES-PIM-Tr3 is still clearly the best performer in the displacement norm measure for this problem.

TABLE 8.5 Estimated computational efficiency of different methods measured in *displacement norm* error for the numerical results for the cantilever beam problem with the same set of triangular mesh (Mesh-4 of 1,696 nodes)

Numerical method	Solution error	Error ratio to FEM-Tr3	Efficiency
FEM-Tr3	5.2334E-03	1.00	1.0
NS-PIM-Tr3	5.5340E-03	1.06	0.8
ES-PIM-Tr3	4.6937E-05	0.0090	101.0
CS-PIM-Tr4-CT	1.9940E-03	0.38	2.2
CS-PIM-Tr4-Iso	2.2586E-03	0.43	2.1
CS-RPIM-Tr4-Cd	2.3951E-03	0.46	2.0
CS-RPIM-Tr4	5.0426E-03	0.96	0.58
CS-RPIM-Tr2L	2.1737E-03	0.42	0.39

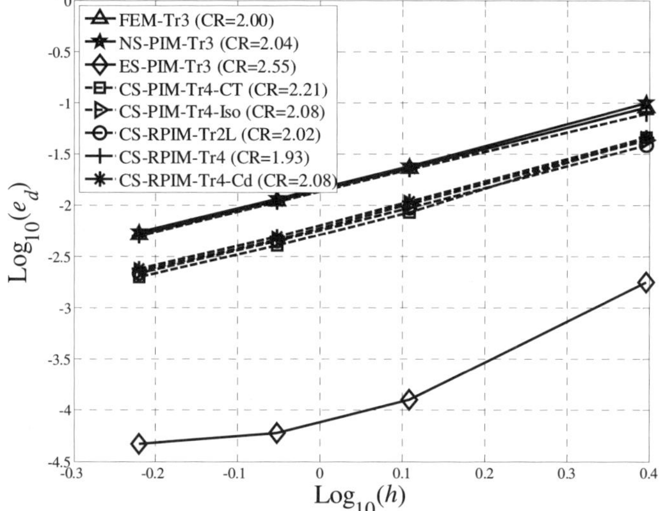

FIGURE 8.5 Comparison of convergence rates and accuracy of the numerical results in displacement norm obtained using FEM-Tr3, NS-PIM-Tr3, ES-PIM-Tr3 and five CS-PIM models for the rectangular cantilever problem.

FIGURE 8.6 shows the convergence of the solution error in energy norm (defined in Equation 5.94) for different methods. The errors in the energy norm for the numerical results obtained using the same set of triangular mesh (Mesh-4) are listed in TABLE 8.6. Considering the computational cost given in TABLE 8.3 for these methods, the estimated computational efficiency measured in energy norm measure is calculated and given in the last column in TABLE 8.6. The following can be found.

1) The three CS-PIM models, including the CS-PIM-Tr4-Iso, CS-RPIM-Tr4-Cd, and CS-RPIM-Tr4, have similar performance as the linear FEM in terms of accuracy and convergence rate. Therefore, in terms of computational efficiency, the CS-PIM-Tr4-Iso is about 8%, CS-RPIM-Tr4-Cd is about 4%, and CS-RPIM-Tr4 is about 43% less than the linear FEM model, in the energy norm measure.

2) The CS-PIM-Tr4-CT and CS-RPIM-Tr2L models, together with NS-PIM-Tr3 and ES-PIM-Tr3 models, achieve much better accuracy and higher convergence rate than the linear FEM, showing *superconvergence* in energy norm. The rates of convergence are 1.38 and 1.51 for the two CS-PIM models, respectively, which are very close to the ideal rate of 1.5 (see Remark 3.23).

3) Among all the S-PIM models, we found the CS-PIM-Tr4-CT stands out clearly in terms of both accuracy and convergence rate. When the finest Mesh-4 is used, the CS-PIM-Tr4-CT result is about 15 times more accurate than that of FEM-Tr3. The estimated CPU time (see TABLE 8.3) for the CS-RPIM-Tr4-CT, is about 1.2 times that of the FEM-Tr3, and hence the computational efficiency of the CS-PIM-Tr4-CT will be about 12 times that of the FEM-Tr3. Compared to the star performer ES-PIM-Tr3, the CS-PIM-Tr4-CT is about 1.8 more accurate. In terms of the computational efficiency, the CS-PIM-Tr4-CT is about 60% higher even than the star performer in energy norm measure for this problem.

TABLE 8.6 Estimated computational efficiency of different methods measured in *energy norm* error for the numerical results for the cantilever beam problem with the same set of triangular mesh (Mesh-4 of 1,696 nodes)

Numerical method	Solution error	Error ratio to FEM-Tr3	Efficiency
FEM-Tr3	2.4406E-01	1.00	1.0
NS-PIM-Tr3	5.0076E-02	0.21	4.0
ES-PIM-Tr3	2.8075E-02	0.12	7.6
CS-PIM-Tr4-CT	1.6635E-02	0.068	12.0
CS-PIM-Tr4-Iso	2.4256E-01	0.99	0.92
CS-RPIM-Tr4-Cd	2.3207E-01	0.95	0.96
CS-RPIM-Tr4	2.3950E-01	0.98	0.57
CS-RPIM-Tr2L	3.1301E-02	0.13	1.3

4)	The CS-RPIM-Tr2L delivers quite accurate results in energy norm that is about 7.7 times more accurate than the FEM-Tr3. Although the CS-RPIM-Tr2L cost 6.1 times more, the computational efficiency is still 1.3 times that of the FEM-Tr3. Therefore, the CS-RPIM-Tr2L is in fact quite a good model, if our interest is in the accuracy of the stress solution. In addition, the CS-RPIM-Tr2L model is very robust against extreme node distributions.

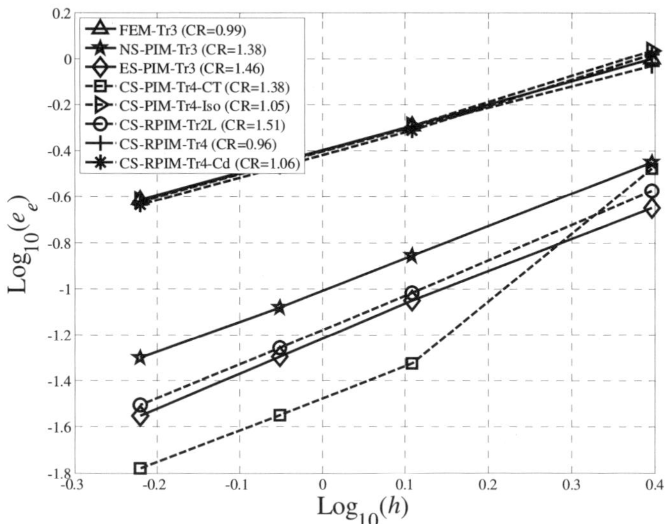

FIGURE 8.6 Comparison of convergence rates and accuracy of the numerical results in energy norm obtained using FEM-Tr3, NS-PIM-Tr3, ES-PIM-Tr3 and five CS-PIM models for the rectangular cantilever problem.

The convergence process of the strain energy solution for different CS-PIM models is shown in FIGURE 8.7, together with the FEM-Tr3, NS-PIM-Tr3 and ES-PIM-Tr3 models. The errors in the strain energy solution for the numerical results obtained using the same set of triangular mesh (Mesh-4) are listed in TABLE 8.7. Taking into the computational cost given in TABLE 8.3 for these methods, the estimated computational efficiency measured in strain energy solution are given in the last column in TABLE 8.7. We may find the following.

1)	With the increase of DOF, the solutions of all the models converge to the exact solution.

2) The linear FEM-Tr3 gives a lower bound, and the NS-PIM-Tr3 gives an upper bound. The solutions of all the CS-PIM and ES-PIM-Tr3 models are in between.

3) The CS-RPIM-Tr4 model performs similarly as the linear FEM and gives a lower bound solution; this indicates that the RPIM-Tr4 shape functions behave very similarly to the PIM-Tr2 shape functions.

4) All the other CS-PIM models give much more accurate results than both FEM-Tr3 and NS-PIM-Tr3, and provide much tighter upper bounds for this problem. They have quite close-to-exact stiffness, but on the softer side.

5) The solutions of all the CS-PIM models are in between the NS-PIM-Tr3 and ES-PIM-Tr3.

6) In terms of computational efficiency, the three models, CS-PIM-Tr4-CT, CS-PIM-Tr4-Iso and CS-RPIM-Tr4-Cd show similar performance for this problem and they are all about 2.2 times more efficient than the FEM-Tr3.

7) Among all these models, the ES-PIM-Tr3 stands out very clearly for this problem, in terms of computational efficiency measured in strain energy norm. It is massive 137 times more efficient even than the FEM-Tr3, and at least about 60 time more efficient than the CS-PIM models.

TABLE 8.7 Estimated computational efficiency of different methods measured in the error in *strain energy solution* of the numerical results for the cantilever beam problem with the same set of triangular mesh (Mesh-4 of 1,696 nodes)

Numerical method	Strain energy Solution	Error[*] (%)	Error ratio to FEM-Tr3	Efficiency
FEM-Tr3	8.5474	-0.5341	1.00	1.0
NS-PIM-Tr3	8.6398	0.5411	1.01	0.8
ES-PIM-Tr3	8.5930	-0.0035	0.0066	137.7
CS-PIM-Tr4-CT	8.6097	0.1908	0.36	2.3
CS-PIM-Tr4-Iso	8.6114	0.2106	0.39	2.3
CS-RPIM-Tr4-Cd	8.6127	0.2258	0.42	2.2
CS-RPIM-Tr4	8.5490	-0.5155	0.97	0.57
CS-RPIM-Tr2L	8.6111	0.2071	0.39	0.42

[*] The analytical value of the strain energy for the cantilever beam problem is 8.59333333.

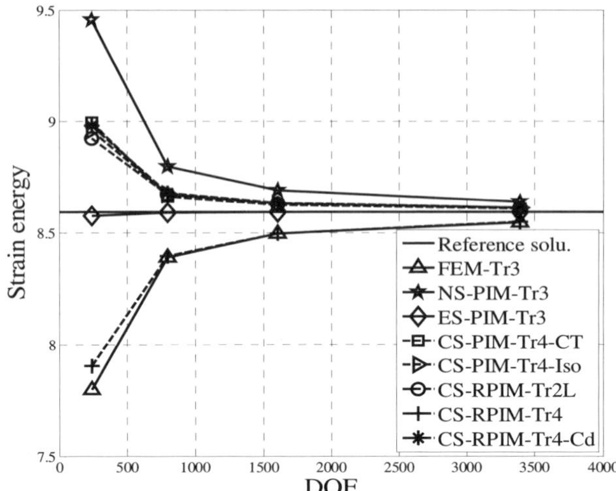

FIGURE 8.7 Converging process of the numerical results (in strain energy norm) for the problem of rectangular cantilever.

To provide a clearer view on the convergence process of the strain energy for the CS-PIM models providing upper bound solutions for this case, a close-up plot is given in FIGURE 8.8. We can find that among these four CS-PIM models, the CS-PIM-Tr4-CT provides the tightest upper bound to the exact solution.

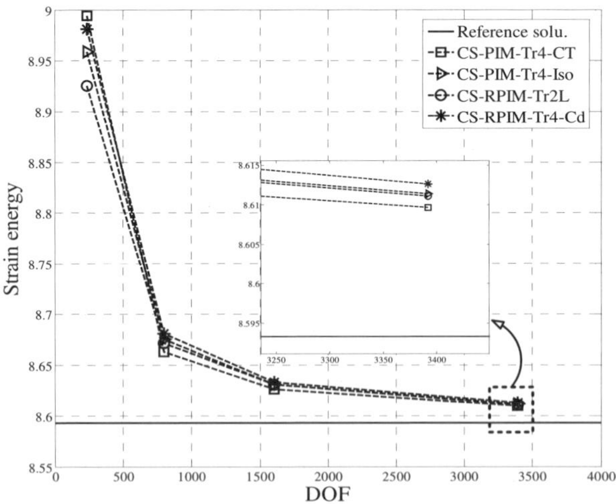

FIGURE 8.8 Converging process of the numerical results (in strain energy norm) for the problem of rectangular cantilever. A close-up comparison between the CS-PIM models.

Example 8.1.3 Infinite solid with a circular hole

The infinite 2D solid with a circular hole problem described in Example 6.1.3 is again studied using the CS-PIM models. Using Mesh-2 shown in FIGURE 6.14, the displacement components are computed and the results along two particular lines of the model (the bottom and the left edges) are plotted in FIGURE 8.9.

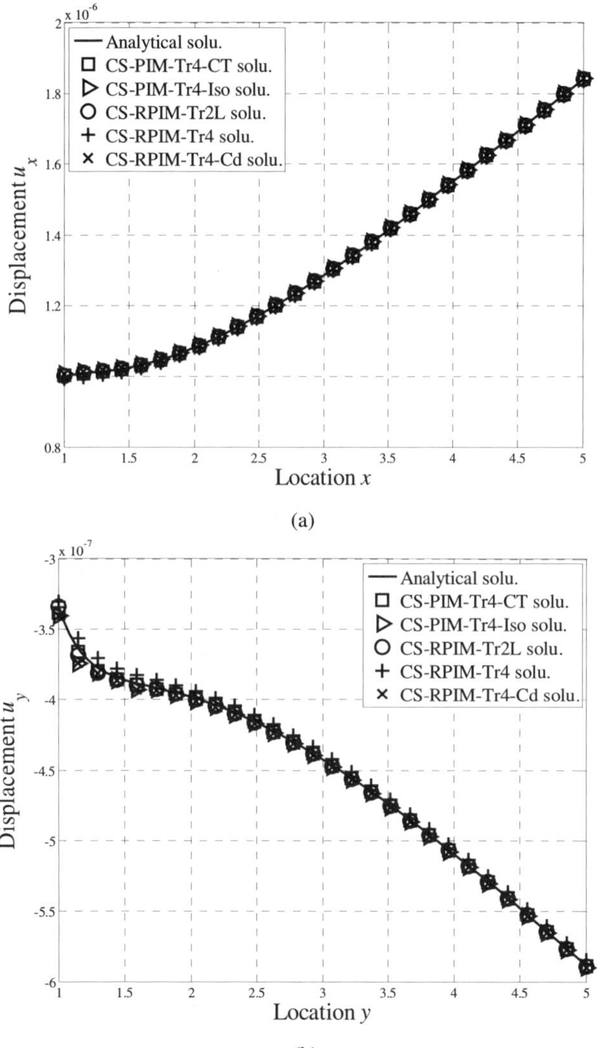

(a)

(b)

FIGURE 8.9 Displacements distribution along bottom and left edges for the quarter model of the infinite solid with hole: (a) displacement (u_x) for nodes located along the bottom edge; (b) displacement (u_y) for nodes located along the left edge.

It is seen that the displacement results obtained by all the CS-PIM models are all in good agreement with the analytical ones.

FIGURE 8.10 plots the numerical results for the stress components distributed along the bottom and the left edges. It is also seen that the results obtained by

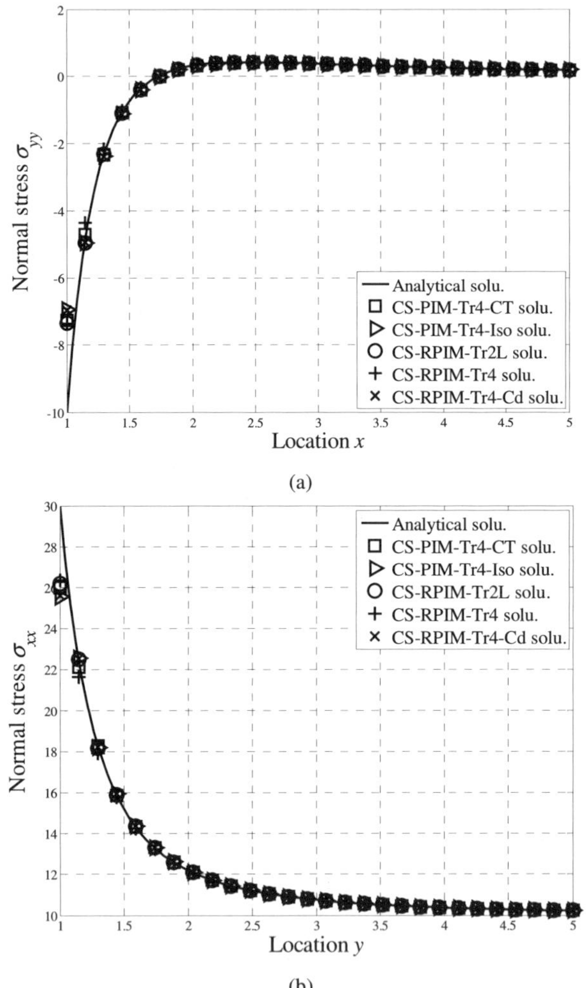

FIGURE 8.10 Stresses distribution along bottom and left edges for the quarter model of the infinite solid with hole: (a) normal stresses in the y-direction for nodes located along the bottom edge; (b) normal stresses in the x-direction for nodes located along the left edge.

the CS-PIM models are all in good agreement with the analytical ones, and these results are not distinguishable from these figures.

Using the meshes of different nodal density shown in FIGURE 6.14, the convergence of the solution errors for the CS-PIM models are computed using Equations (5.88) and (5.94). The convergence of the solution error in displacement norm is plotted in FIGURE 8.11. The errors in the displacement norm for the numerical results obtained using the same set of triangular mesh (Mesh-4 of 3,578 nodes) are listed in TABLE 8.8. Taking the computational cost given in TABLE 8.3 into consideration for these methods, the estimated computational efficiency measured in displacement error are calculated and listed in the last column in TABLE 8.8. We can note the following observations.

1) Similar to the results of the cantilever case, the CS-RPIM-Tr4 model performs worst among all the CS-PIM models. It has similar accuracy and convergence rate as that of FEM-Tr3 and NS-PIM-Tr3.

2) In terms of accuracy, all the other CS-PIM models provide much more accurate results than the FEM-Tr3. The CS-PIM-Tr4-CT and CS-RPIM-Tr2L stand out, achieving even better results than the ES-PIM-Tr3 for this problem.

3) In terms of convergence rate, the CS-PIM models generally achieve higher convergence rates than the FEM-Tr3 which are around the theoretical value of 2.0.

4) When the finest Mesh-4 is used for this problem, the CS-PIM-Tr4-CT is about 6 times more accurate than FEM-Tr3. The computational efficiency of the CS-PIM-Tr4-CT is about 5 times that of the FEM-Tr3. Compared to the star performer ES-PIM-Tr3, the CS-PIM-Tr4-CT is about 1.7 more accurate. In terms of the computational efficiency, the CS-PIM-Tr4-CT is about 60% higher even than the star performer in displacement norm measure for this problem.

5) Among all these models, the CS-PIM-Tr4-CT stands out for this problem, in terms of both accuracy and computational efficiency measured in displacement norm.

TABLE 8.8 Estimated computational efficiency of different methods measured in *displacement norm* error for the numerical results for the infinite solid with hole problem with the same set of triangular mesh (Mesh-4 of 3,578 nodes)

Numerical method	Solution error	Error ratio to the FEM-Tr3	Efficiency
FEM-Tr3	1.4435E-03	1.00	1.0
NS-PIM-Tr3	1.1520E-03	0.80	1.0
ES-PIM-Tr3	3.7620E-04	0.26	3.5
CS-PIM-Tr4-CT	2.1172E-04	0.15	5.6
CS-PIM-Tr4-Iso	5.0191E-04	0.35	2.6
CS-RPIM-Tr4-Cd	5.1579E-04	0.36	2.5
CS-RPIM-Tr4	1.2694E-03	0.88	0.63
CS-RPIM-Tr2L	2.1513E-04	0.15	1.1

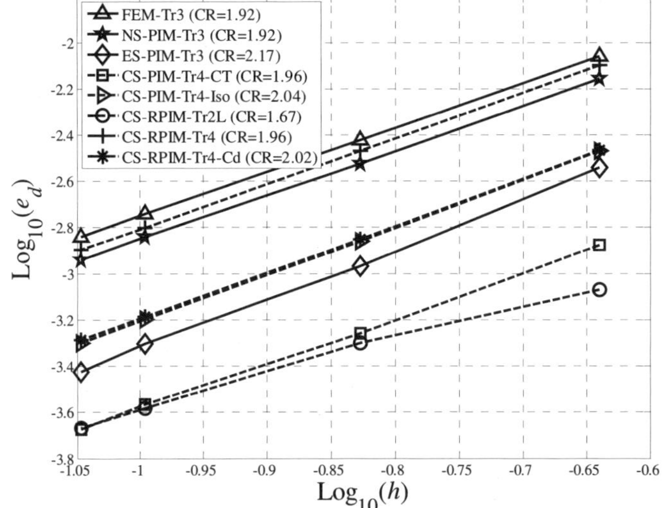

FIGURE 8.11 Comparison of convergence rates of the numerical results in displacement norm obtained using CS-PIM models for the problem of infinite solid with circular hole.

FIGURE 8.12 plots the solution error of the numerical results in energy norm for meshes of various node densities. The errors in the energy norm for the numerical results obtained using the same set of triangular mesh (Mesh-4) are listed in TABLE 8.9. Considering the computational cost given in TABLE 8.3 for these methods, the estimated computational efficiency measured in energy norm measure are calculated and given in the last column in TABLE 8.9. The following points may be noted for this example.

1) Three CS-PIM models (CS-PIM-Tr4-Iso, CS-RPIM-Tr4 and CS-RPIM-Tr4-Cd) have slightly better performance than the linear FEM in terms of accuracy and convergence rate.

2) In terms of accuracy, the CS-PIM-Tr4-CT and CS-RPIM-Tr2L provide much more accurate results than the FEM-Tr3, which are even better than the NS-PIM-Tr3 and ES-PIM-Tr3.

3) In terms of convergence rate, the CS-PIM-Tr4-CT and CS-RPIM-Tr2L models achieve much higher convergence rate than the linear FEM, showing *superconvergence* in energy norm. The rates of convergence for these two models are 1.14 and 1.56 respectively, which are much higher than the theoretical value of 1.0 for the linear Galerkin weak-form models, and around the ideal rate of 1.5 for W^2 models (see, Remark 3.23).

4) When the finest Mesh-4 is used for this problem, the CS-PIM-Tr4-CT is about 4.8 times more accurate, and hence about 4 times more efficient than the FEM-Tr3. Compared to the star performer ES-PIM-Tr3, the CS-PIM-Tr4-CT is about 1.5 more accurate. In terms of the computational efficiency, the CS-PIM-Tr4-CT is about 40% higher even than the star performer in energy norm measure for this problem.

5) Among all these models, the CS-PIM-Tr4-CT stands out for this problem, in terms of both accuracy and computational efficiency measured in energy norm.

TABLE 8.9 Estimated computational efficiency of different methods measured in *energy norm* error for the numerical results for the infinite solid with hole problem with the same set of triangular mesh (Mesh-4 of 3,578 nodes)

Numerical method	Solution error	Error ratio to FEM-Tr3	Efficiency
FEM-Tr3	1.3831E-04	1.00	1.0
NS-PIM-Tr3	5.0516E-05	0.37	2.3
ES-PIM-Tr3	4.2694E-05	0.31	2.9
CS-PIM-Tr4-CT	2.8684E-05	0.21	4.0
CS-PIM-Tr4-Iso	1.3229E-04	0.96	0.95
CS-RPIM-Tr4-Cd	1.2637E-04	0.91	1.00
CS-RPIM-Tr4	1.2701E-04	0.92	0.60
CS-RPIM-Tr2L	2.7543E-05	0.20	0.82

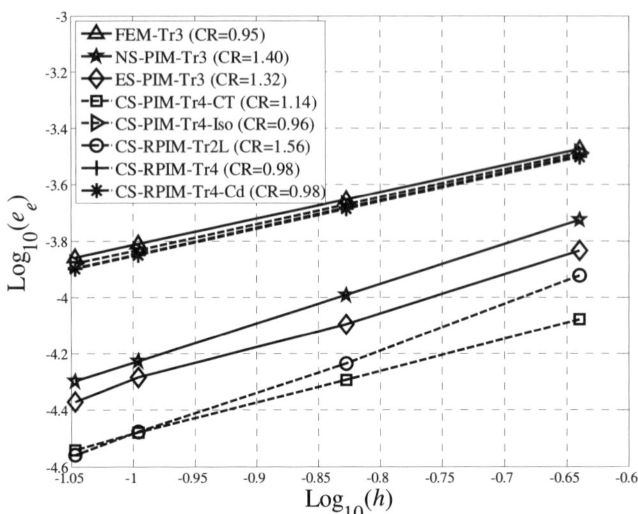

FIGURE 8.12 Comparison of convergence rates of the numerical results in energy norm obtained using CS-PIM models for the problem of infinite solid with circular hole.

The convergence process of the strain energy solution for different CS-PIM models is shown in FIGURE 8.13, together with the FEM-Tr3, NS-PIM-Tr3 and ES-PIM-Tr3 models. The errors in the strain energy solution for the numerical results obtained using the same set of triangular mesh (Mesh-4) are listed in TABLE 8.10. Considering the computational cost given in TABLE 8.3 for these methods, the estimated computational efficiency measured in strain energy solution are calculated and given in the last column in TABLE 8.10. We may find the following.

1)	CS-PIM models are softer than the overly-stiff FEM model giving lower bounds and stiffer than the overly-soft NS-PIM-Tr3 giving upper bounds. All the CS-PIM models are bounded by the FEM-Tr3 and NS-PIM-Tr3 solutions.

2)	The CS-RPIM-Tr4 provides lower bound solutions to this problem, and performs similarly as the FEM-Tr3.

3)	The CS-PIM-Tr4-Iso and the CS-RPIM-Tr4-Cd models provide very tight upper bound solutions for this problem, compared to the NS-PIM-Tr3 model.

4) The CS-PIM-Tr4-CT and the CS-RPIM-Tr2L models show non-momotonic convergence for this problem, as shown in FIGURE 8.13. When the mesh is too coarse, the softening effects are small, and the model is stiff giving lower bound solutions. When the mesh is refined, the softening effects increase leading to upper bounds. Compared to other numerical methods, these two models provide much tighter bound solutions and possess close-to-exact stiffness for this problem.

5) In terms of computational efficiency, all these S-PIM models (except CS-RPIM-Tr4) perform much (2.2-30 times) better than the FEM-Tr3.

6) Among all these models, the CS-PIM-Tr4-CT stands out very clearly for this problem, in terms of computational efficiency measured in strain energy solution. It is about 10 times more efficiency even than the star performer ES-PIM-Tr3.

TABLE 8.10 Estimated computational efficiency of different methods measured in the error in *strain energy solution* of the numerical results for the infinite solid with hole problem with the same set of triangular mesh (Mesh-4 of 3,578 nodes)

Numerical method	Strain energy Solution	Error[*] (%)	Error ratio to FEM-Tr3	Efficiency
FEM-Tr3	4.3230E-05	-0.0495	1.00	1.0
NS-PIM-Tr3	4.3265E-05	0.0314	0.63	1.3
ES-PIM-Tr3	4.3245E-05	-0.0148	0.30	3.0
CS-PIM-Tr4-CT	4.3252E-05	0.0014	0.028	30.0
CS-PIM-Tr4-Iso	4.3256E-05	0.0106	0.21	4.3
CS-RPIM-Tr4-Cd	4.3257E-05	0.0130	0.26	3.5
CS-RPIM-Tr4	4.3233E-05	-0.0425	0.86	0.65
CS-RPIM-Tr2L	4.3253E-05	0.0037	0.075	2.2

[*]The analytical value of the strain energy for the infinite solid with hole problem is 4.32513989E-05.

Remark 8.4 CS-PIM-Tr4-CT: another star performer

The CS-PIM-Tr4-CT model is of superconvergence in both displacement and energy norms. In terms of accuracy and convergence rate, this model performs even better than the star performer ES-PIM-Tr3 for many cases.

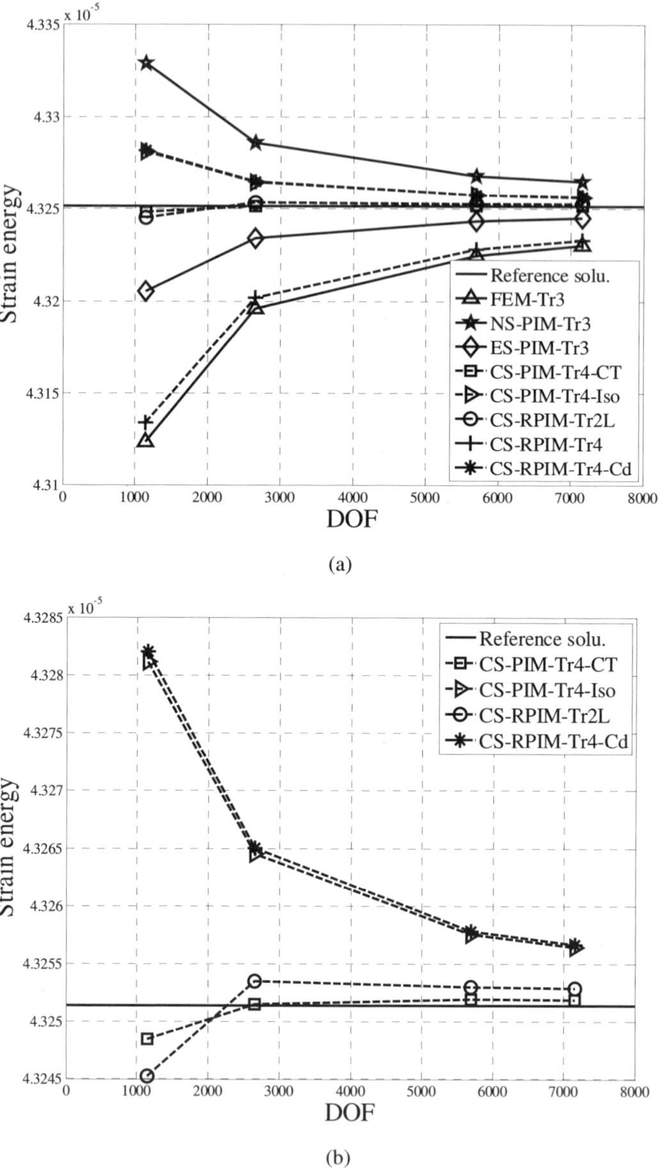

(a)

(b)

FIGURE 8.13 Converging process of the numerical results (in strain energy norm) for the problem of infinite plate with circular hole: (a) comparison between CS-PIM models and other methods; (b) close-up comparison between the CS-PIM models.

Example 8.1.4 A mechanical part: 2D connecting rod

The connecting rod problem described in Example 6.4.4 is studied here using the CS-PIM models. Using the same sets of triangular meshes (four meshes of 319, 671, 1,050 and 2,155 nodes respectively shown in FIGURE 6.46), this problem is studied using different methods and the convergence processes of the numerical results in strain energy norm are plotted in FIGURE 8.14. Considering the computational cost given in TABLE 8.3 for these methods, the estimated computational efficiency measured in strain energy solution are calculated and given in the last column in TABLE 8.11. For this mechanical model with complicated shape, we can find the followings.

1) All the numerical methods converge to the reference solution with the increase of DOFs.

2) The three CS-PIM models (CS-PIM-Tr4-CT, CS-RPIM-Tr2L and CS-RPIM-Tr4-Cd) provide very tight bound solutions, which are bounded by the NS-PIM-Tr3 and FEM-Tr3 solutions.

3) The CS-RPIM-Tr4-Cd and the CS-RPIM-Tr2L provide much tighter upper and lower bound solutions respectively.

4) In terms of computational efficiency and accuracy, all the S-PIM models perform better than the FEM-Tr3 for this problem. The CS-RPIM-Tr4-Cd stands out clearly among these methods.

5) Note that the reference solution is obtained using the FEM with a very fine mesh and is a lower bound of the exact solution. Therefore, the exact solution should be higher than the reference one used in this study.

TABLE 8.11 Estimated computational efficiency of different methods measured in the error in *strain energy solution* of the numerical results for the connecting rod problem with the same set of triangular mesh (Mesh-4 of 2,155 nodes)

Numerical method	Strain energy Solution	Error[*] (%)	Error ratio to FEM-Tr3	Efficiency
FEM-Tr3	1.0814E-03	-2.105	1.00	1.0
NS-PIM-Tr3	1.1165E-03	1.073	0.51	1.6
ES-PIM-Tr3	1.0955E-03	-0.828	0.39	2.3
CS-PIM-Tr4-CT	1.0928E-03	-1.073	0.51	1.6
CS-RPIM-Tr4-Cd	1.1070E-03	0.213	0.10	9.1
CS-RPIM-Tr2L	1.1026E-03	-0.185	0.09	1.8

[*] The reference value of the strain energy for the connecting rod problem is 1.1046479E-03, which is obtained using FEM with a very fine mesh (six-node triangular mesh of 192,420 elements and 387,977 nodes).

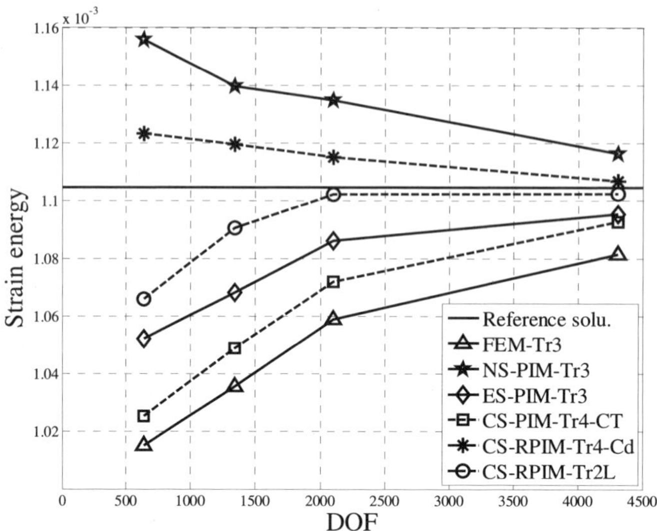

FIGURE 8.14 Converging process of the numerical results (in strain energy norm) for the connecting rod problem.

Example 8.1.5 Free vibration of a slender cantilever

The free vibration analysis for a slender cantilever beam described in Example 7.5.4 is again studied using the CS-PIM models. Numerical results obtained by the FEM using four-node quadrangular elements (FEM-Q4) with a very fine mesh (100×10) are used as reference solutions for this problem. For the purpose of comparison, the standard FEM-Tr3 and the linear ES-PIM-Tr3 are also used to study this problem using the same set of triangular meshes shown in FIGURE 7.15.

TABLE 8.12 lists the first twelve natural frequencies of the cantilever obtained using the different numerical methods. The first twelve corresponding vibration modes obtained using the five CS-PIM models are plotted in FIGURE 8.15. We observe the following.

1) All the CS-PIM models produce smaller natural frequencies for all the modes, compared to the FEM-Tr3 model. This means that they are softer than the overly-stiff FEM-Tr3 model.

2) Spurious nonzero-energy modes are found for the CS-PIM-Tr4-Iso and the CS-RPIM-Tr4-Cd models, showing that these models are softer models and can be temporally instable. However, there are no zero-energy modes for all the CS-PIM models. This shows that all these CS-PIM models are spatially stable, which confirms numerically Remark 8.1.

3) When Mesh-1 (two layers of cells with 63 nodes) is used, only the highest three 9th-12th modes obtained using the CS-PIM-Tr4-Iso and CS-RPIM-Tr4-Cd are spurious, and no spurious modes are found for all the other three CS-PIM models, as shown in FIGURE 8.15. When the finer Mesh-2 (4 layers of cells with 205 nodes) is used, no spurious modes are found for all the CS-PIM models within the first 11 modes, and only the 12th modes obtained using the CS-PIM-Tr4-Iso and CS-RPIM-Tr4-Cd are spurious. Note that use of 4 layer cells and 205 nodes is considered very small model in actual practice of modeling. Therefore, we can conclude that in actual practice, the CS-PIM models are all temporally stable, when a fine mesh is used to represent the problem domain. This finding confirms numerically Remark 8.2.

4) When the Mesh-2 is used, all the CS-PIM models can generally provide closer natural frequencies to the reference ones, compared to the stiffer FEM-Tr3 model and even the star performer ES-PIM-Tr3.

TABLE 8.12 First twelve nature frequencies (in 10kHz) of the slender cantilever

	CS-PIM-Tr4-CT	CS-RPIM-Tr4	CS-RPIM-Tr2L	CS-PIM-Tr4-Iso	CS-RPIM-Tr4-Cd	ES-PIM-Tr3	FEM-Tr3	Reference solution
	0.0769	0.0804	0.0754	0.0823	0.0803	0.0853	0.1117	0.0824
	0.4623	0.4833	0.4578	0.4889	0.4781	0.5078	0.6539	0.4944
	1.2227	1.2772	1.2266	1.2689	1.2422	1.2828	1.2843	1.2824
	1.2823	1.2841	1.2827	1.2815	1.2812	1.3246	1.6748	1.3022
Mesh 1	2.2325	2.3360	2.2734	2.2572	2.2116	2.3783	2.9554	2.3663
(21×3=63	3.4165	3.5774	3.5298	3.3447	3.2795	3.5784	3.8424	3.6085
nodes)	3.8350	3.8411	3.8353	3.8119	3.8122	3.8298	4.3866	3.8442
	4.7133	4.9426	4.9287	4.4315	4.3475	4.8533	5.8836	4.9674
	6.0720	6.3374	6.3429	4.9845	5.0341	6.1527	6.3751	6.3960
	6.3512	6.4129	6.4111	5.4110	5.3744	6.3182	7.4046	6.4023
	7.4630	7.8432	7.9040	5.4664	5.5963	7.4419	8.8210	7.8853
	8.7693	8.8507	8.7874	5.6634	5.7264	8.6776	8.9411	8.9290

Mesh 2 (41×5=205 nodes)							
0.0809	0.0819	0.0805	0.0831	0.0825	0.0827	0.0906	0.0824
0.4858	0.4918	0.4825	0.4958	0.4924	0.4950	0.5409	0.4944
1.2810	1.2825	1.2699	1.2819	1.2812	1.2826	1.2831	1.2824
1.2825	1.2978	1.2825	1.2961	1.2881	1.3006	1.4161	1.3022
2.3308	2.3631	2.3054	2.3337	2.3206	2.3554	2.5570	2.3663
3.5572	3.6107	3.5112	3.5224	3.5043	3.5778	3.8433	3.6085
3.8421	3.8437	3.8436	3.8362	3.8362	3.8408	3.8786	3.8442
4.8981	4.9791	4.8258	4.7929	4.7701	4.9029	5.3087	4.9674
6.3102	6.3896	6.2071	6.0966	6.0694	6.2867	6.3935	6.3960
6.3862	6.4283	6.3922	6.3605	6.3607	6.3774	6.8093	6.4023
7.7646	7.9212	7.6261	7.3963	7.3652	7.6987	8.3473	7.8853
8.8996	8.9164	8.9177	<u>7.8005</u>	<u>8.2072</u>	8.8751	8.9183	8.9290

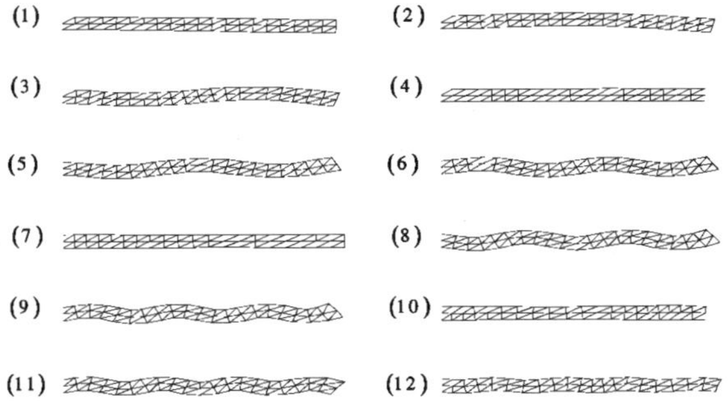

(a)

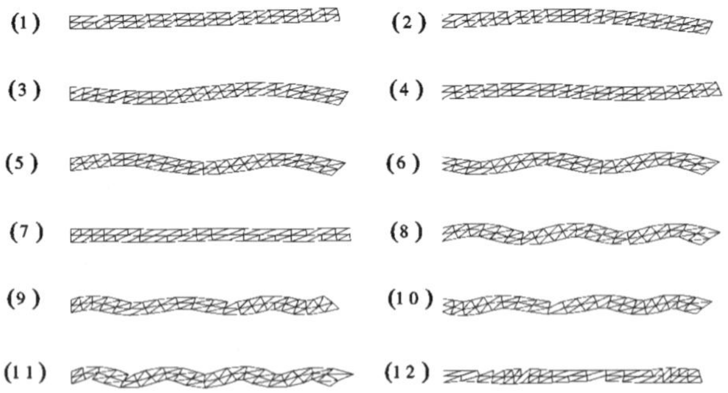

(b)

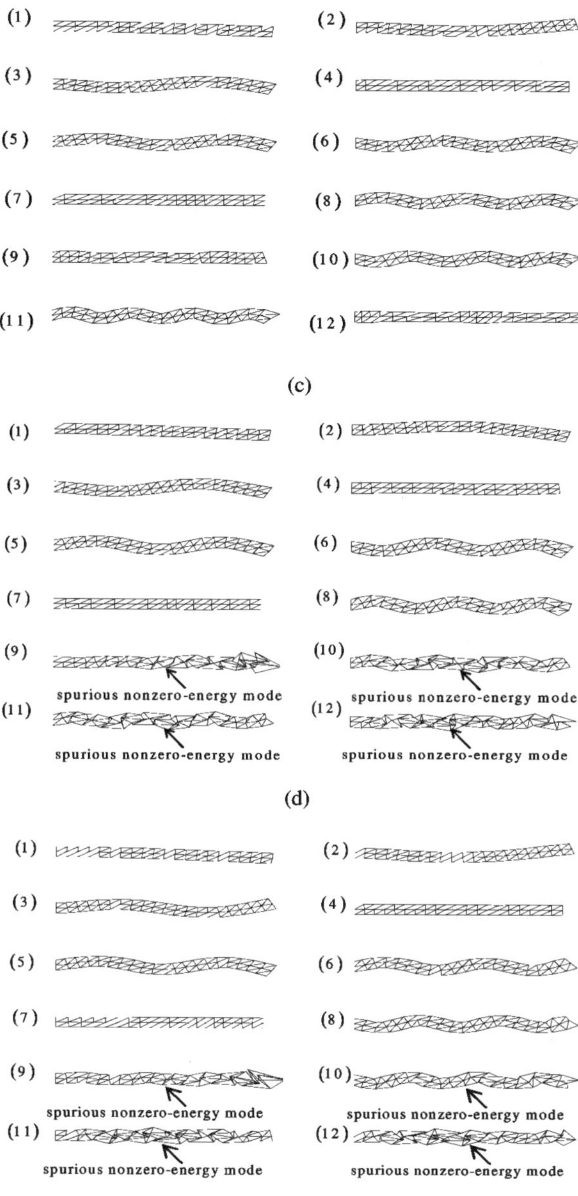

FIGURE 8.15 First twelve free vibration modes of the slender cantilever by CS-PIM models: (a) solutions of CS-PIM-Tr4-CT; (b) solutions of CS-RPIM-Tr4; (c) solutions of CS-RPIM-Tr2L; (d) solutions of CS-PIM-Tr4-Iso; (e) solutions of CS-RPIM-Tr4-Cd.

Example 8.1.6 Forced vibration of a spherical shell

Finally, the forced vibration described in Example 7.5.5 is again studied using three CS-PIM models. This problem is first studied for a shell subjected to a harmonic loading of $f(t) = \cos\omega_f t$, with $\omega_f = 0.05$. The time history of the

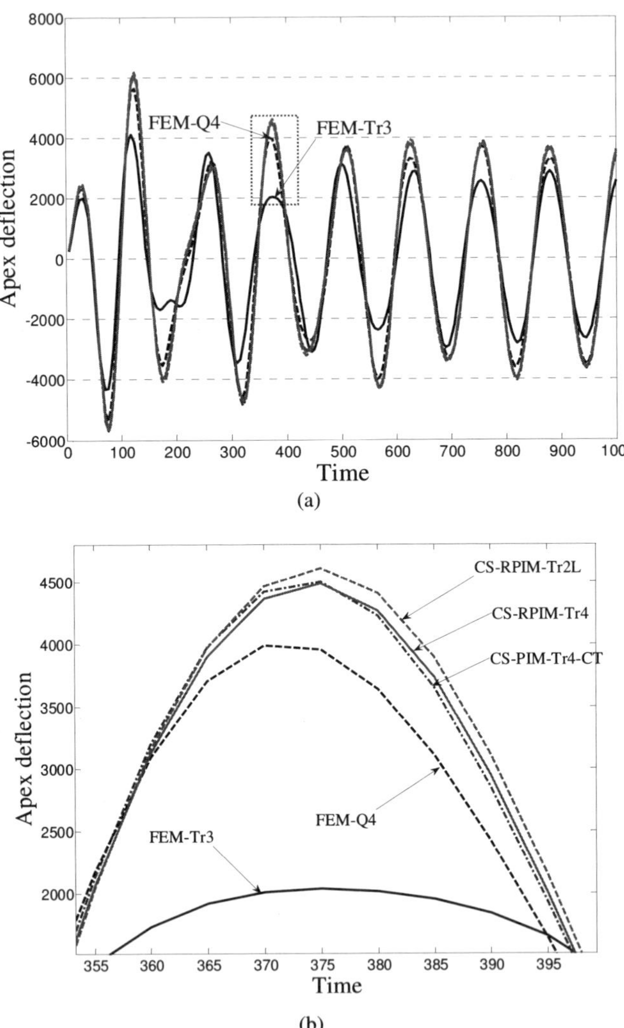

FIGURE 8.16 Transient responses of the spherical shell subjected to a harmonic loading obtained using different methods: (a) comparison between CS-PIM models and other methods; (b) zooming in for the local part located in the dashed frame.

deflection at the apex of the shell is computed using time steps of $\Delta t = 5$, and the results are plotted in FIGURE 8.16. We noted the following points.

1) The solutions of the three CS-PIM models (CS-PIM-Tr4-CT, CS-RPIM-Tr2L and CS-RPIM-Tr4), are much more accurate than that of FEM-T3.

2) The CS-PIM solutions are comparable to those of the FEM-Q4.

Further, an analysis is conducted for the shell excited by a constant step load $f(t) = 1$ starting from $t=0$, using these three CS-PIM models. The responses are plotted in FIGURE 8.17. The dash line is for the results without damping. We observed that the amplitude of the deflection approach constant value with increasing time, but the vibration will never stop. When a Rayleigh damping of $\alpha_R=0.005$ and $\beta_R=0.272$ are considered, the vibration is damped out, and the response converges to constant very quickly as expected. Such a constant value gives the static solution of the shell subjected to a unit force.

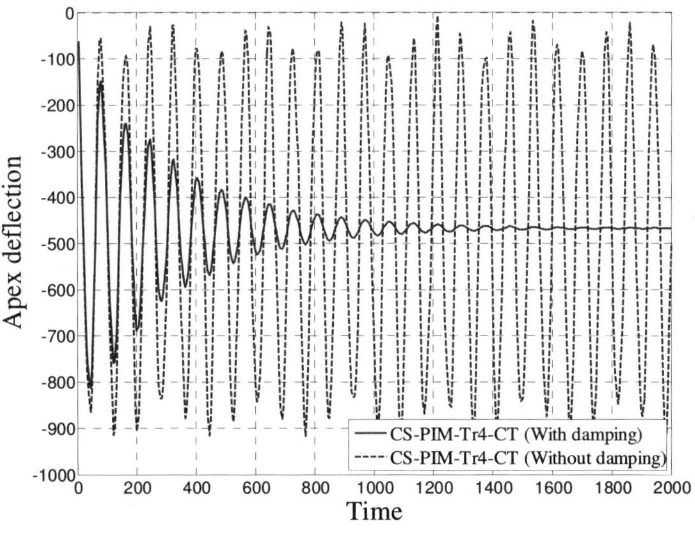

(a)

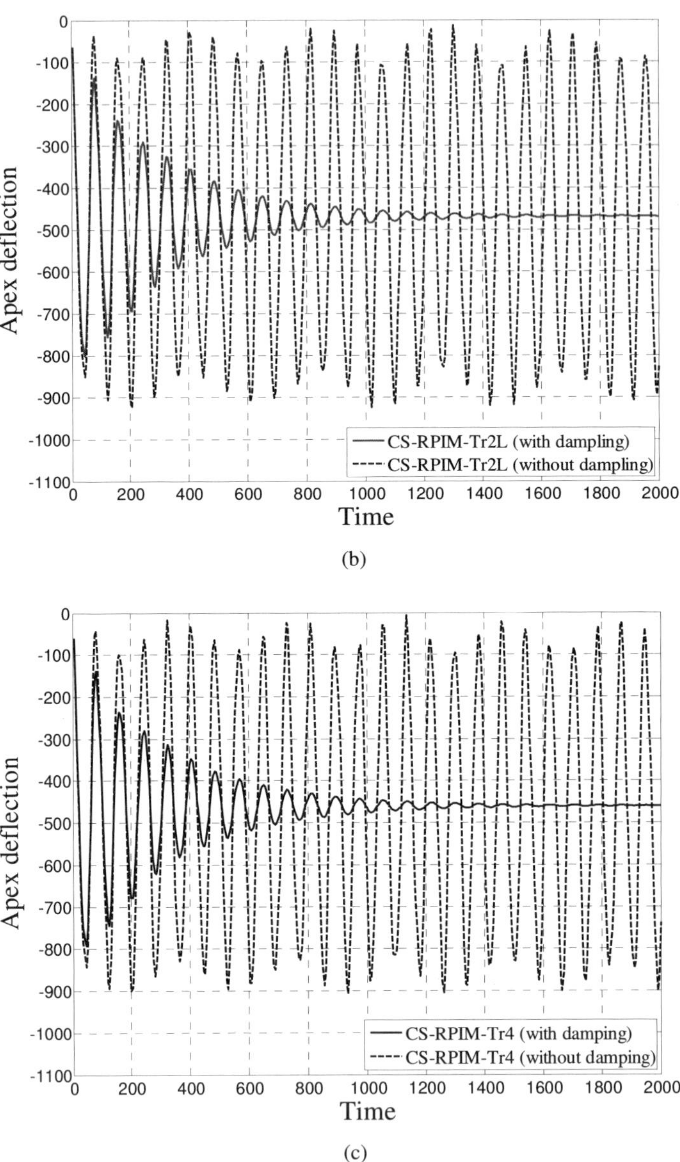

FIGURE 8.17 Transient responses obtained using CS-PIM models for the spherical shell subjected to a Heaviside step loading: (a) transient responses by the CS-PIM-Tr4-CT; (b) transient responses by the CS-RPIM-Tr2L; (c) transient responses by the CS-RPIM-Tr4.

8.2 CS-PIM for 3D solids

In this section, we extend the CS-PIM for three-dimensional (3D) solid problems, by simply mentioning the major difference of formulations between the 2D and 3D models.

8.2.1 Approximation of displacement

In this book, we use four-node tetrahedron (Te4) cells to discretize the 3D problem domain, which is essentially the same mesh as in the standard FEM using linear tetrahedron elements (FEM-Te4). The Te4 elements are known as the simplest and can be generated automatically for complicated geometries. In the FEM-Te4, linear tetrahedron elements are used for the displacement functions approximation. In our CS-PIM, however, we will use more nodes that may beyond the cell for the interpolation of displacement field, as seen in the 2D cases.

Based on the tetrahedral background cells, the support nodes used in the CS-PIM can be selected using face-based T-schemes, and PIM shape functions can be constructed using the procedure presented in Chapter 3. If we use the face-based T3-scheme to select the support nodes, as shown in FIGURE 8.18a, only 3 nodes of the vertices of the face are selected, and the linear PIM shape functions will be created. In such a case, the constructed CS-PIM model will be *exactly* the same as that of linear FEM-Te4. For higher order 3D CS-PIM models, more nodes should be selected, and it is quite tricky in the node selection, due to the possible singular moment matrix as discussed in Chapter 3. Therefore for higher order approximation, we suggest using the RPIM shape functions which selects the support nodes by the face-based T5-scheme (shown in FIGURE 8.18b) or T2L-scheme and work well for very irregularly distributed nodes without special treatments. Note that when we use these two T-schemes to select support nodes, the approximated displacement field will be discontinuous and the discontinuity will occur on the triangular surfaces formed by connecting the centroid and two vertices of the tetrahedron. As presented in Section 3.7, we can also use the condensed RPIM shape functions to construct the displacement fields for 3D CS-PIM models. The RPIM shape functions can be, in general, created for virtually randomly distributed nodes, if more nodes are used. In practice, however, we try to use as few nodes as possible, for efficiency reason, which is particularly critical to 3D problems. Therefore, the face-based T5-scheme selecting only five nodes for an interior face (shown in FIGURE 8.18b) is recommended for constructing RPIM shape functions in the scheme of the present 3D CS-PIM models.

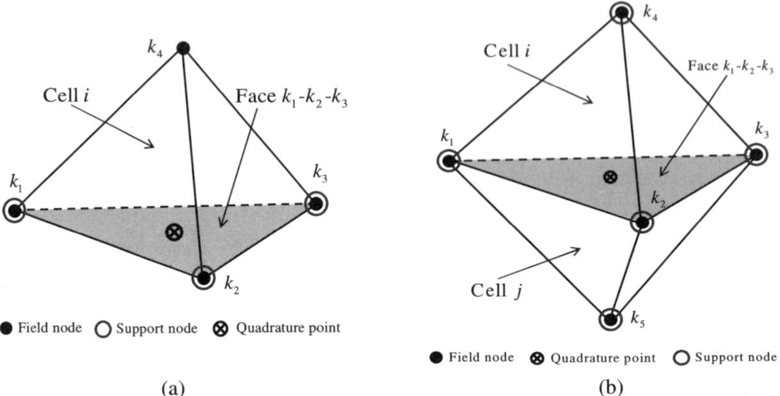

FIGURE 8.18 Face-based T-schemes, where the quadrature point is located on the face k_1-k_2-k_3: (a) face-based T3-scheme selecting the three vertices of the home face (k_1, k_2 and k_3) as the support nodes; (b) face-based T5-scheme selecting the three vertices of the home face (k_1, k_2 and k_3) plus the two remote nodes (k_4 and k_5) of the two neighboring cells (cell i and cell j) as the support nodes.

8.2.2 Evaluation of cell-based smoothed strains

In the scheme of the CS-PIM for 3D problems, the smoothing domains used are exactly the background four-node tetrahedral cells, and hence no additional treatments are needed for smoothing domain construction. The smoothed strains are computed by using the generalized gradient smoothing operation over the cell-based smoothing domains. Therefore, for 3D CS-PIM formulations, the number of quadrature cells equals to the number of smoothing domains, which is also exactly that of the background tetrahedral cells, i.e. $N_q = N_s = N_c$.

Performing the generalized gradient smoothing operation, the smoothed strains can be obtained for each smoothing domain as described in Section 4.5.2. Finally the discretized system equation can be derived with the same form as Equation (8.5), and the sub-stiffness matrix can be obtained using

$$\bar{\mathbf{K}}_{IJ} = \sum_{i=1}^{N_c} V_i^s \bar{\mathbf{B}}_I^{\mathrm{T}} \mathbf{c} \bar{\mathbf{B}}_J \qquad (8.11)$$

where $V_i^s = \int_{\Omega_i^s} d\Omega$ is the volume of the smoothing domain i, and

$$\bar{\mathbf{B}}_I(\mathbf{x}_i) = \begin{bmatrix} \bar{\phi}_{Ix}(\mathbf{x}_i) & 0 & 0 \\ 0 & \bar{\phi}_{Iy}(\mathbf{x}_i) & 0 \\ 0 & 0 & \bar{\phi}_{Iz}(\mathbf{x}_i) \\ \bar{\phi}_{Iy}(\mathbf{x}_i) & \bar{\phi}_{Ix}(\mathbf{x}_i) & 0 \\ 0 & \bar{\phi}_{Iz}(\mathbf{x}_i) & \bar{\phi}_{Iy}(\mathbf{x}_i) \\ \bar{\phi}_{Iz}(\mathbf{x}_i) & 0 & \bar{\phi}_{Ix}(\mathbf{x}_i) \end{bmatrix} \tag{8.12}$$

$$\bar{\phi}_{Il} = \frac{1}{V_i^s} \sum_{m=1}^{n_{seg}} \left[\sum_{n=1}^{n_G} W_n^G \left(\phi_I(\mathbf{x}_{m,n}) n_{l,m} \right) \right] \quad (l = x,\, y,\, z) \tag{8.13}$$

in which n_{seg} refers to the number of triangular surfaces of a smoothing domain, and n_G is the number of Gauss points located within each surface triangle. For 3D CS-PIM using tetrahedral cells, we shall have $N_{seg} = 4$. When linear shape functions are used, $n_G = 1$, an when RPIM shape functions are used, $n_G = 3$.

8.2.3 Numerical implementation

8.2.3.1 Possible 3D CS-PIM models

In the scheme of CS-PIM models, the calculation of the system stiffness matrix is performed using the generalized strain smoothing technique over the cell-based smoothing domains, i.e. the background tetrahedral cells, which converts the volume integration into the area integration over each triangular faces of the tetrahedrons. As discussed in Section 8.2.1, we use RPIM shape functions to approximate the points of interest and three CS-PIM/CS-RPIM models have been developed in this book depending on the use of different face-based T-schemes to select support nodes.

CS-PIM-Te3: same as FEM-Te4

A CS-PIM-Te3 model uses linear PIM shape functions and the three support nodes at the three vertices of the cell face. This is the simplest 3D CS-PIM model and is exactly the same as the FEM-Te4, and hence it is a fully compatible model. Therefore, the CS-PIM-Te3 model will not be discussed any further in this book.

CS-RPIM-Te5

The displacement field in the CS-RPIM-Te5 model is constructed using the RPIM shape functions with the support nodes selected using the face-based T5-scheme. Since the RPIM shape functions can always be created without the singularity problem, no special treatment is needed. The displacement field in the CS-RPIM-Te5 model is incompatible.

Because the support nodes for interior and boundary cells will be different, we need to find out the average number of support nodes for a cell-based smoothing domain, for which we consider the 3D Lame problem presented using 173 nodes (see Example 6.3.2). The number of the support nodes for a smoothing domain is 6-8 and the average number is found 7.5 for this model using the mesh of 173 nodes, which is about 1.9 times that of the FEM-Te4. Therefore, the average number of nonzero entries in a row in the (global) stiffness matrix of a CS-RPIM-Te5 model (n_{ze}) is about 1.9 times that of the FEM-Te4 counterpart using the same mesh, as listed in TABLE 8.13.

CS-RPIM-Te5-Cd

The CS-RPIM-Tr5-Cd uses the 3D condensed RPIM shape functions to construct the displacement field, where the support nodes are selected using the face-based T5-scheme. As described in Section 3.7, 3D condensed RPIM shape functions can be formulated using five field nodes selected using face-based T5-scheme plus another six virtual nodes. Thus compared to the RPIM shape functions using only five field nodes (as used in the CS-RPIM-Te5 model), this condensed RPIM shape functions are designed to improve the performance of the results without increasing the DOFs of the model. The displacement field in the CS-RPIM-Te5-Cd is incompatible.

Because the support nodes of a smoothing domain and the DOFs of the CS-RPIM-Te5-Cd is the same as the CS-RPIM-Te5 model, the average number of nonzero entries in a row in the stiffness matrix in the CS-RPIM-Te5-Cd model is the same as the CS-RPIM-Te5 model.

CS-RPIM-Te2L

In a CS-RPIM-Te2L model, the RPIM shape functions are created using support nodes selected by the face-based T2L-scheme, as described in Section 1.6.3. This type of T-scheme is well controlled and with a lot of freedom, works well for virtually randomly distributed nodes, but less efficient for much more nodes used in the interpolation. The displacement field in the CS-RPIM-Te2L model is incompatible.

The number of the support nodes for a smoothing domain is 14-43 and the average number is 26, that is about 6.5 times that of the FEM-Te4. Therefore, the average number of nonzero entries in a row in the stiffness matrix of a CS-RPIM-Te2L model is about 6.5 times that of the FEM-Te4 counterpart using the same mesh, as listed in TABLE 8.13.

TABLE 8.13 3D CS-PIM models and overall characteristics in relation to the FEM-Te4 model using the same mesh

Numerical method	Support nodes in an integral cell/element	Estimated average number of nonzero entries	Estimated Solver CPU time
FEM-Te4	4	n_{ze}	t_{CPU}
NS-PIM-Te4	6-20 (10.9)	$2.7\,n_{ze}$	$2.7\,\bar{n}_{iter}t_{CPU}$
FS-PIM-Te4	4-5 (4.8)	$1.2\,n_{ze}$	$1.2\,\bar{n}_{iter}t_{CPU}$
CS-RPIM-Te5-Cd	6-8 (7.5)	$1.9\,n_{ze}$	$1.9\,\bar{n}_{iter}t_{CPU}$
CS-RPIM-Te5	6-8 (7.5)	$1.9\,n_{ze}$	$1.9\,\bar{n}_{iter}t_{CPU}$
CS-RPIM-Te2L	14-43 (26.0)	$6.5\,n_{ze}$	$6.5\,\bar{n}_{iter}t_{CPU}$

In TABLE 8.13, we assume that an iterative solver is used and the CPU time of such a solver can be given by Equation 6.3.

8.2.3.2 Condition number of 3D CS-PIM models

TABLE 8.14 lists the condition numbers of the global stiffness matrixes for various 3D CS-PIM models in relation to the FEM-Te4 using the same tetrahedral meshes for the 3D Lame problem. Again we find the ratio of condition number of different methods with respect to the linear FEM-Te4 model is almost independent of the model size. In the following estimation of computation cost, we will use the relative $\bar{n}_{iter}$ given in the last column in TABLE 8.14. Similar to the results for 2D problems shown in TABLE 8.2, we can find that CS-RPIM-Te5 and CS-RPIM-Te5-Cd models have smaller condition numbers, together with the linear NS-PIM-Te4 and FS-PIM-Te4 models, while the CS-RPIM-Te2L has a bigger condition numbers than the FEM-Te4. Therefore, in the following study, the CS-RPIM-Te2L model will not be used, because it is unlikely to be very efficient for large condition number and the number of nonzero entries in the stiffness matrix. We also find that compared to the CS-RPIM-Te5, the CS-RPIM-Te5-Cd uses the same number of support nodes but has smaller condition numbers, owing probably to the constraints used in the condensation process, which plays a kind of "regularization" role. The performance of these two models will be investigated in details in the following study.

TABLE 8.14 Condition number of the global stiffness matrixes for various 3D CS-PIM models for the Lame problem in relation to the FEM-Te4 model using the same mesh

Numerical method	Mesh-1 of 96 nodes		Mesh-2 of 173 nodes	
	$cond(\mathbf{K})$	$\bar{n}_{iter}$	$cond(\mathbf{K})$	$\bar{n}_{iter}$
FEM-Te4	3.284e+02	1.00	5.274e+02	1.00
NS-PIM-Te4	1.074e+02	0.57	1.638e+02	0.56
FS-PIM-Te4	2.476e+02	0.87	3.704e+02	0.84
CS-RPIM-Te5-Cd	1.339e+02	0.64	1.977e+02	0.61
CS-RPIM-Te5	2.472e+02	0.87	4.035e+02	0.87
CS-RPIM-Te2L	1.166e+03	1.88	1.860e+03	1.88

8.2.3.3   Estimation of computational cost for 3D CS-PIM

Combining TABLE 8.13 and TABLE 8.14 we shall have a rough estimation of computational cost for the CS-PIM models in relation to the standard FEM-Te4 using the same mesh, and the results are given in TABLE 8.15. It is seen that 1) all the CS-RPIM models will cost more than the FEM-Te4 using the same mesh; 2) the CS-RPIM-Te5-Cd model costs less CPU time than the CS-RPIM-Te5 model, and is comparable to that of linear FEM with only 20% more. If the CS-RPIM-Te5-Cd solution can be 20% more accurate than the FEM model, it can be more computationally *efficient*. The CS-RPIM-Te5 model needs almost double the solution accuracy in order to be comparable with the standard FEM-Te4 model. The CS-RPIM-Te2L will likely to be less comparable, because it has to produce results that are more accurate by as much as 12 times. Hence it will not be used in the example problems. This simple estimation shows that the two models, CS-RPIM-Te5-Cd and CS-RPIM-Te5, are potentially at least comparable to the FEM-Te4 models.

TABLE 8.15 Estimation of computational cost for 3D CS-PIM models in relation to the FEM-Te4 model using the same mesh

Numerical method	Support nodes in an integral cell/element	Estimated Solver CPU time
FEM-Te4	4	t_{CPU}
NS-PIM-Te4	6-20 (10.9)	$1.5\,t_{CPU}$
FS-PIM-Te4	4-5 (4.8)	$1.0\,t_{CPU}$
CS-RPIM-Te5-Cd	6-8 (7.5)	$1.2\,t_{CPU}$
CS-RPIM-Te5	6-8 (7.5)	$1.7\,t_{CPU}$
CS-RPIM-Te2L	14-43 (26.0)	$12.2\,t_{CPU}$

8.2.4   Numerical examples for 3D solids

Example 8.2.1   A 3D linear patch test

The 3D standard patch test described in Example 6.3.1 is first studied using the CS-RPIM models. TABLE 8.16 lists the displacement norm errors (defined in Equation 5.88) of the CS-RPIM solutions for the patch test using both the regular and irregular meshes shown in FIGURE 6.30. All the CS-RPIM models can pass these tests exactly (to the machine accuracy). Note that the displacement field in all the studied CS-RPIM models is incompatible, but it is linearly conforming and capable to reproduce exactly (to the machine accuracy) linear field. This confirms Remark 8.3 for 3D problems.

TABLE 8.16 Error norm in displacements of numerical results for the 3D standard patch test obtained using various CS-RPIM models

3D CS-RPIM models	Regular mesh	Irregular mesh
CS-RPIM-Te5 (displacement incompatible)	4.241E-16	3.672E-14
CS-RPIM-Te5-Cd (displacement incompatible)	4.045E-12	2.265E-12
CS-RPIM-Te2L (displacement incompatible)	4.317E-14	7.763E-12

Example 8.2.2 A 3D Lame problem

The 3D Lame problem described in Example 6.3.2 is now studied again using the two potential CS-RPIM models: CS-RPIM-Te5 and CS-RPIM-Te5-Cd.

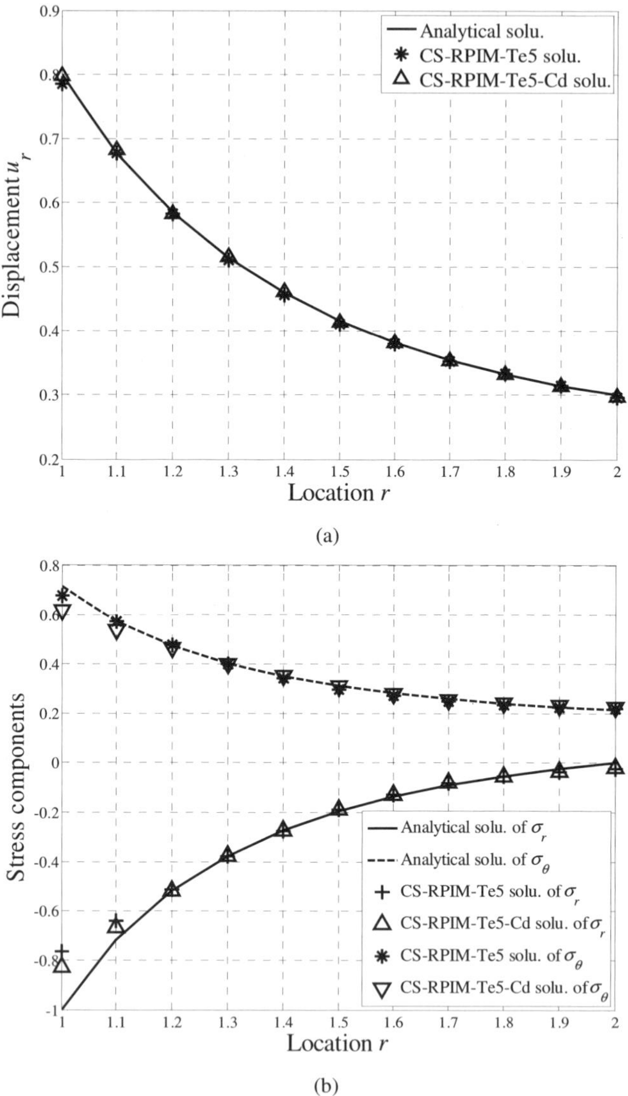

FIGURE 8.19 Displacement and stress distribution along the *x*-axis of the one-eighth model of the 3D Lame problem: (a) radial displacement; (b) radial and tangential stress components.

First, the problem domain (one-eighth model) is presented using linear tetrahedral cells with a total of 1,138 irregularly distributed nodes, as shown in FIGURE 6.31b, and the computed displacement and stress components along the x-axis are plotted in FIGURE 8.19. It is found that the CS-RPIM results agree well with the analytical solutions.

To investigate the convergence property of the CS-RPIM models, four models of 173, 317, 729, and 1,304 irregularly distributed nodes are used. For each of these four models the errors of the numerical results in both displacement and energy norms are computed. For comparison purpose, the linear FEM, the NS-PIM-Te4 and the FS-PIM-Te4 are also employed in this study, using the same sets of node distributions. The convergence of the solution errors in displacement norm is plotted in FIGURE 8.20. The errors in the displacement norm for the numerical results obtained using the same set of triangular mesh (Mesh-4 of 1,304 nodes) are listed in TABLE 8.17. Taking the computational cost given in TABLE 8.15 into consideration for these methods, the estimated computational efficiency measured in displacement errors is calculated and listed in the last column in TABLE 8.17. We can note the following observations.

1) The CS-RPIM-Te5 model performs worse than the CS-RPIM-Te5-Cd model, but better than those of FEM-Te4 and NS-PIM-Te4 in terms of accuracy and convergence rate.

2) In terms of accuracy, the CS-RPIM-Te5-Cd model provides much more accurate results than the FEM-Te4, and it performs even better than the FS-PIM-Te4 for this problem.

3) In terms of convergence rate, the CS-PIM models generally achieve higher convergence rates than the FEM-Te4. The CS-RPIM-Te5-Cd stands out quite clearly in this regard showing superconvergence.

4) When the finest Mesh-4 is used for this problem, the CS-RPIM-Te4-Cd is about 2 times more accurate, and the computational efficiency is about 1.6 times that of the FEM-Te4. Compared to the FS-PIM-Te4, the CS-RPIM-Te4-Cd is about 1.2 more accurate, but they are even in terms of the computational efficiency in displacement norm measure for this problem.

5) Among all these models, the CS-RPIM-Te4-Cd and FS-PIM-Te4 are equally good for this problem, in terms of computational efficiency measured in displacement norm.

TABLE 8.17 Estimated computational efficiency of different methods measured in *displacement error* for the numerical results for the Lame problem with the same set of tetrahedral mesh (Mesh-4 of 1,304 nodes)

Numerical method	Solution error	Error ratio to the FEM-Te4	Efficiency
FEM-Te4	1.8381E-02	1.00	1.0
NS-PIM-Te4	3.5123E-02	1.91	0.35
FS-PIM-Te4	1.1838E-02	0.64	1.6
CS-RPIM-Te5-Cd	9.6249E-03	0.52	1.6
CS-RPIM-Te5	1.2092E-02	0.66	0.89

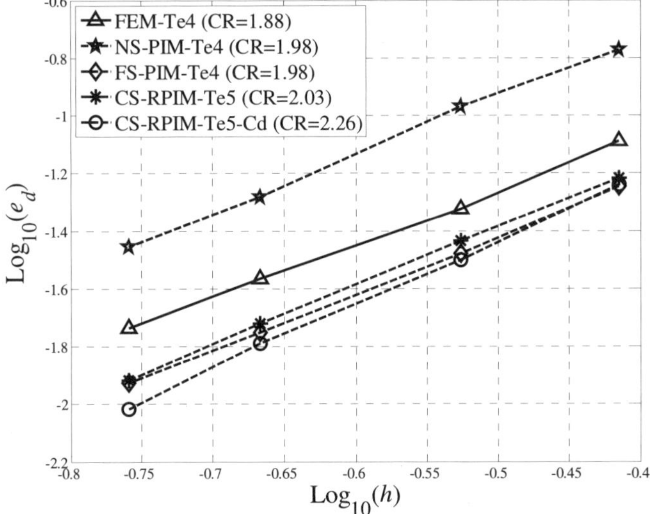

FIGURE 8.20 Comparison of convergence rates and accuracy of the numerical results in displacement norm obtained using CS-PIM models, together with the linear FEM and other S-PIM models for the 3D Lame problem.

FIGURE 8.21 plots the solution error of the numerical results in energy norm for meshes of various node densities. The errors in the energy norm for the numerical results obtained using the same set of triangular mesh (Mesh-4) are listed in TABLE 8.18. Considering the computational cost given in TABLE 8.15 for these methods, the estimated computational efficiency measured in energy norm measure are calculated and given in the last column in TABLE 8.18. The following points may be noted for this example.

1) Both CS-PIM models (CS-RPIM-Te5 and CS-RPIM-Te5-Cd) perform better than the linear FEM in terms of both accuracy and convergence rate.

2) In terms of accuracy, the CS-RPIM-Te5-Cd provides more accurate results than the FEM-Te4, which are also better than other S-PIM models.

3) In terms of convergence rate, the CS-RPIM-Te5-Cd achieves a rate of 1.12 that is higher than that of the linear FEM of 0.92, showing *superconvergence* in energy norm. Among all the S-PIM methods, the NS-PIM-Te4 stands out with the convergence rate of 1.44, which is very close to the ideal rate of 1.5 for W^2 models (see, Remark 3.23).

TABLE 8.18 Estimated computational efficiency of different methods measured in *energy error* for the numerical results for the Lame problem with the same set of tetrahedral mesh (Mesh-4 of 1,304 nodes)

Numerical method	Solution error	Error ratio to the FEM-Te4	Efficiency
FEM-Te4	1.5008E-01	1.00	1.0
NS-PIM-Te4	7.7785E-02	0.52	1.3
FS-PIM-Te4	1.2310E-01	0.82	1.2
CS-RPIM-Te5-Cd	6.8244E-02	0.45	1.9
CS-RPIM-Te5	1.2603E-01	0.84	0.70

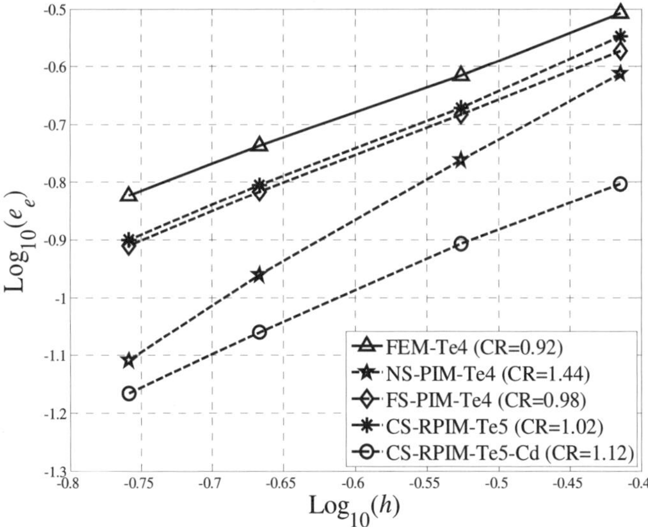

FIGURE 8.21 Comparison of accuracy and convergence rates of the numerical results in energy norm obtained using CS-PIM models, together with the linear FEM and other S-PIM models for the 3D Lame problem.

4)   When the finest Mesh-4 is used, the CS-RPIM-Te5-Cd is about 2.2 times more accurate, and hence about 1.9 times more efficient than the FEM-Te4. Compared to FS-PIM-Te4, the CS-RPIM-Te5-Cd is about 1.8 more accurate, and 1.6 times more efficient in energy norm measure for this problem.

5)   Although the CS-RPIM-Te5 is more accurate than the FEM-Te4, but loses out in efficiency.

6)   Among all these models, the CS-RPIM-Te5-Cd stands out for this problem, in terms of computational efficiency measured in energy norm.

The convergence processes of the strain energy solutions for different CS-PIM models are shown in FIGURE 8.22, together with the FEM-Te4, NS-PIM-Te4 and FS-PIM-Te4 models. The errors in the strain energy solution for the numerical results obtained using the same set of triangular mesh (Mesh-4) are listed in TABLE 8.19. Considering the computational cost given in TABLE 8.15 for these methods, the estimated computational efficiency measured in strain energy solution are calculated and given in the last column in TABLE 8.19. We can find the following points.

1)   CS-PIM models behave softer than the overly-stiff FEM model that gives lower bounds and stiffer than the overly-soft NS-PIM-Te4 that gives upper bounds. All the CS-PIM models are bounded by the FEM-Te4 and NS-PIM-Te4 solutions.

2)   The CS-RPIM-Te5 provides lower bound solutions to this problem, and performs quite similarly as the FS-PIM-Te4.

3)   The CS-RPIM-Te5-Cd model provides very tight upper bound solutions for this problem, compared to the NS-PIM-Te4 model.

4)   In terms of accuracy, the CS-RPIM-Te5-Cd and the CS-RPIM-Te5 are about 1.4 and 2.2 times more accurate than the linear FEM-Te4, respectively.

5)   In terms of computational efficiency, the CS-RPIM-Te5 is 1.3 times, and the CS-RPIM-Te5-Cd is 1.2 times higher than the FEM-Te4.

6)   Among all these models, the FS-PIM-Te4 stands out closely followed by the CS-RPIM-Te5-Cd for this problem, in terms of computational efficiency measured in strain energy solution.

TABLE 8.19 Estimated computational efficiency of different methods measured in the error in *strain energy solution* of the numerical results for the Lame problem with the same set of tetrahedral mesh (Mesh-4 of 1,304 nodes)

Numerical method	Strain energy Solution	Error[*] (%)	Error ratio to FEM-Te4	Efficiency
FEM-Te4	6.1701E-01	-1.86	1.00	1.0
NS-PIM-Te4	6.5268E-01	3.81	2.0	0.33
FS-PIM-Te4	6.2354E-01	-0.82	0.44	2.3
CS-RPIM-Te5-Cd	6.3680E-01	1.29	0.69	1.2
CS-RPIM-Te5	6.2362E-01	-0.81	0.44	1.3

[*] The reference value of the strain energy for the Lame problem is 6.2869698E-01.

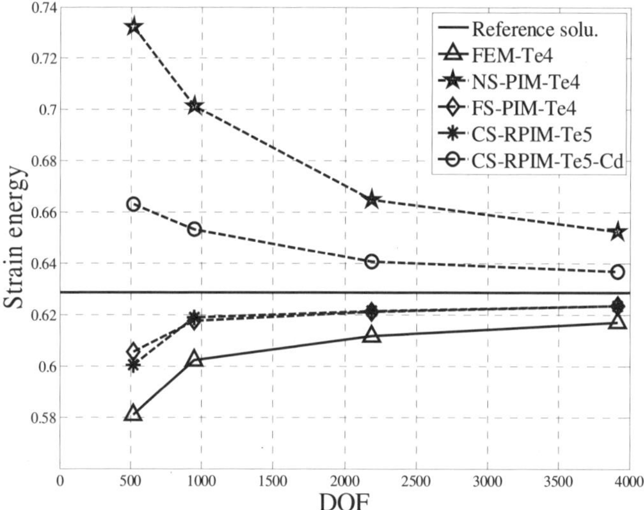

FIGURE 8.22 Bound solutions in strain energy obtained using the CS-RPIM models, together with the linear FEM and other S-PIM models for the 3D Lame problem.

Example 8.2.3 A 3D axletree base

The 3D axletree base problem described in Example 6.4.6 is now studied using the CS-RPIM models. The parameters are taken exactly same as that in Example 6.4.6 and four models with 781, 1828, 2,566 and 3,675 nodes. For this problem, we do not have exact solution, and hence we can only study the performance of these methods in terms of the strain energy solutions.

With the increase of DOF, FIGURE 8.23 plots the convergence process of numerical solutions in strain energy norm. The errors in the strain energy solution for the numerical results obtained using the same set of triangular mesh (Mesh-4) are listed in TABLE 8.20. Considering the computational cost given in TABLE 8.15 for these methods, the estimated computational efficiency measured in strain energy solution are calculated and given in the last column in TABLE 8.20. We found that

1) CS-PIM models behave softer than the overly-stiff FEM-Te4 model that gives lower bounds and stiffer than the overly-soft NS-PIM-Te4 that gives upper bounds. All the CS-PIM models are bounded by the FEM-Te4 and NS-PIM-Te4 solutions.

2) The CS-RPIM-Te5 provides much tighter lower bound solutions to this problem, compared to the FEM-Te4 and FS-PIM-Te4 methods.

3) The computed strain energy of the CS-RPIM-Te5-Cd model is very close to the reference one, showing it has close-to-exact stiffness.

4) In terms of accuracy, the CS-RPIM-Te5 and the CS-RPIM-Te5-Cd are respectively 2.6 and 40 times more accurate than the FEM-Te4 model. The CS-RPIM-Te5-Cd is even 27 times more accurate than the FS-PIM-Te4.

5) In terms of computational efficiency, the CS-RPIM-Te5 is 1.5 times, and the CS-RPIM-Te5-Cd is 33.3 times higher than the FEM-Te4.

6) Among all these models, the CS-RPIM-Te5-Cd stands out clearly for this problem, in terms of computational efficiency measured in strain energy solution.

TABLE 8.20 Estimated computational efficiency of different methods measured in the error in *strain energy solution* of the numerical results for the axletree base problem with the same set of tetrahedral mesh (Mesh-4 of 3,675 nodes)

Numerical method	Strain energy Solution	Error[*] (%)	Error ratio to FEM-Te4	Efficiency
FEM-Te4	1.7479E-03	-16.3	1.00	1.0
NS-PIM-Te4	2.4388E-03	16.8	1.03	0.65
FS-PIM-Te4	1.8528E-03	-11.3	0.69	1.4
CS-RPIM-Te5-Cd	2.0795E-03	-0.40	0.025	33.3
CS-RPIM-Te5	1.9568E-03	-6.28	0.39	1.5

[*] The reference value of the strain energy for the 3D axletree base is 2.0878887E-03, which is obtained using FEM with a very fine mesh (ten-node tetrahedral mesh of 238,041 elements).

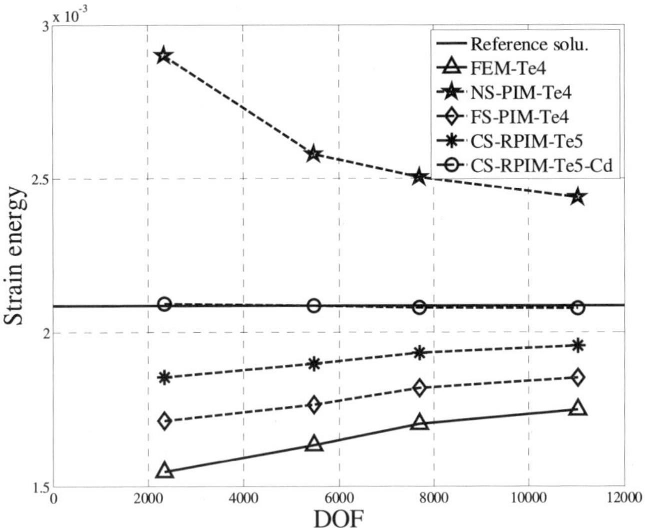

FIGURE 8.23 Bound solutions obtained using the CS-RPIM models for the 3D axletree base problem.

8.3 Concluding remarks

This chapter presents in detail the CS-PIM with various models. It is a W^2 formulation based on the GS-Galerkin weak form. It has a number of important properties in contrary to the standard FEM. Before leaving this chapter we mention the following additional issues.

Remark 8.5 CS-PIM models: bound property

1) The stiffness of the CS-PIM models (except CS-PIM-Tr2) is largely in between those of NS-PIM and ES-PIM.

2) The CS-PIM models (except CS-PIM-Tr2) produces usually (not always) upper bound solutions, and they are often softer models.

3) Because the CS-PIM models have quite close-to-exact stiffness, their solutions are very close to the exact solution. Also because of this, these solutions can sometimes become lower bounds, depending on the settings (loading and BCs, background cells, etc.) of the problem.

Remark 8.6 Extension to general n-sided polygonal cells/elements

In the above discussions on CS-PIM, we focused on the use of triangular background cells. However, the same cell-based smoothing idea can be applied for general n-sided cells, as practiced in the CS-FEM [18-20].

Remark 8.7 CS-PIM vs. CS-FEM

Note that the CS-PIM is in general very much different from the CS-FEM x[18-20]. Although they all used cell-based smoothing domains, the interpolations are different. In CS-PIM, the interpolation can be performed beyond the cells, but in the CS-FEM it is within the elements.

8.4 Computer program

We provide now the source codes CSPIM.F90 for various CS-PIM/CS-RPIM models. In Section 8.4.2, the main program and three subroutines particularly for CS-PIM models are detailed, and all the common subroutines for S-PIM are provided in the Appendix.

8.4.1 About the program

Consider Example 6.1.2 using Mesh-1 with 120 nodes shown in FIGURE 6.9. TABLE 8.21 lists the three subroutines which are coded particularly for CS-PIM models.

TABLE 8.21 Particular subroutines used in CSPIM.F90 and their functions

Subroutines	Function explanation
FormB_CSPIM	Form the strain-displacement matrix for a cell-based smoothing domain
Nodalstress_CSPIM	Obtain nodal stress and strain components for the CS-PIM models
Enernorm_error_CSPIM	Compute the global energy norm error and strain energy for CS-PIM results

By adjusting the values of the two switches shown in TABLE 8.22, five CS-PIM/CS-RPIM models can be established using the same set of source codes. The meanings of these two switches are described in TABLE 6.30.

TABLE 8.22 Choices of switches for different CS-PIM/CS-RPIM models

CS-PIM/CS-RPIM models	NTscheme	NPIMSF
CS-PIM-Tr2 (FEM-Tr3)[*]	1	1
CS-PIM-Tr4-CT	5	3
CS-PIM-Tr4-Iso	5	4
CS-RPIM-Tr4-Cd	5	5
CS-RPIM-Tr4	5	2
CS-RPIM-Tr2L	6	2

[*]Note in the CS-PIM scheme, the cell-based T3-scheme and the edge-based T2-scheme will create the same linear PIM shape functions. So the edge-based T2-scheme has not been coded in the following program and the CS-PIM-Tr2, i.e. FEM-Tr3, model can be established by assigning NTscheme=1 and NPIMSF=1.

8.4.2 Source codes in FORTRAN 90

Program 8.1 Main code of "CSPIM.F90"

```
PROGRAM CSPIM
!================================================================
! The main program for conducting the 2D CS-PIM/CS-RPIM methods
!================================================================
Use Parameters
Use Variables
Implicit real*8 (a-h, o-z)
Character*20 DATAINPUT, DATAOUTPUT
Open (5, FILE=' DATAINPUT')
Open (10, FILE=' DATAOUTPUT', status='unknown')

Call input
Call C_materialM
Call Cell_information
Gk=0.d0; force=0.d0
Loop_on_cells: do ia=1,numcell
    Bmat=0.d0; ndexnow=0; nvlist=0
    Call formB_CSPIM(ia)
    area_inte=area_cell(ia)
    Call StiffM_Intedomain(ia,area_inte,gpk)
```

```
      Call Form_GK
End do Loop_on_cells
Call Natural_BC
Call Essential_BC
Call Solver_LAE(Gk,force,nx*numnode,nx*numn)
Call Dispnorm_error
Call Nodalstress_CSPIM
Call Enernorm_error_CSPIM


End PROGRAM CSPIM
```

Program 8.2 Source code of "FormB_CSPIM"

```
Subroutine FormB_CSPIM_BOOk(ia)
!================================================================
! Form the strain-displacement matrix of the ia_th cell-based smoothing domain
!
! Input:   x, nxcell, area_cell, kcell, NameEdge_cell and nodefour.
! Output: Bmat, nvlist and ndexnow.
!================================================================
use Parameters
use Variables
implicit real*8 (a-h,o-z)
! work array
dimension xg(nx,2),xxg(nx,ngauss),wei(ngauss)
dimension xgg(nx,3),onxy(nx,2),nv(ndom),gpos(nx),phi(ndom)


ss=area_cell(ia); kcb=kcell(ia)
n1=nxcell(1,ia); n2=nxcell(2,ia); n3=nxcell(3,ia)
xn1=x(1,n1); yn1=x(2,n1); xn2=x(1,n2); yn2=x(2,n2); xn3=x(1,n3); yn3=x(2,n3)
! Loop on the three edges of the ia_th cell
Loop_on_edges: do i1=1,nce
   i1cell=NameEdge_cell(i1,ia); n4edge=nodefour(4,i1cell); xg=0.d0
   select case(i1)
```

```
case(1)
   xg(1,1)=xn1; xg(2,1)=yn1; xg(1,2)=xn2; xg(2,2)=yn2
   xL=((xn1-xn2)**2+(yn1-yn2)**2)**0.5; onx=(yn2-yn1)/xL; ony=(xn1-xn2)/xL
case(2)
   xg(1,1)=xn2; xg(2,1)=yn2; xg(1,2)=xn3; xg(2,2)=yn3
   xL=((xn3-xn2)**2+(yn3-yn2)**2)**0.5; onx=(yn3-yn2)/xL; ony=(xn2-xn3)/xL
case(3)
   xg(1,1)=xn1; xg(2,1)=yn1; xg(1,2)=xn3; xg(2,2)=yn3
   xL=((xn1-xn3)**2+(yn1-yn3)**2)**0.5; onx=(yn1-yn3)/xL; ony=(xn3-xn1)/xL
endselect
xxg=0.d0; wei=0.d0
call Line_gauss(xg,xxg,wei,det,Ngauss)
ndex=0; nv=0
select case(NTscheme)
case(1)
   ndex=3
   do ik=1,ndex
   nv(ik)=nxcell(ik,ia)
   enddo
case(5)
   do ik=1,4
      ika=nodefour(ik,i1cell)
      if (ika>0) then
      ndex=ndex+1
      nv(ndex)=ika
      endif
   enddo
case(6)
   if (n4edge>0) then
      call edge_basedT2L(i1cell,ndex,nv)
   else
      ndex=3
      do ik=1,ndex
      nv(ik)=nxcell(ik,ia)
      enddo
   endif
endselect
! Loop on the Gauss points located on the i1_th edge
```

```
Loop_on_GP: do i2=1,ngauss
     gpos(1)=xxg(1,i2); gpos(2)=xxg(2,i2); weight=wei(i2)
     padx=onx*weight*det/ss; pady=ony*weight*det/ss; phi=0.d0
     if (n4edge>0) then
         select case (NPIMSF)
         case(1); call PPIM_SF2D(gpos,nv,ndex,phi)
         case(2); call RPIM_SF2D(gpos,nv,ndex,phi)
         case(3); call PPIM_CT_SF2D(gpos,nv,ndex,xL,phi)
         case(4); call Iso_PPIM_SF2D(gpos,nv,ndex,phi)
         case(5); call Con_RPIM_SF2D(gpos,nv,ndex,phi)
         endselect
     else
         call PPIM_SF2D(gpos,nv,ndex,phi)
     endif
     kp=i1+i2
  if (kp.eq.2) then
     ndexnow=ndex
     Inner1_1: do j2=1,ndex
         nvlist(j2)=nv(j2)
         k1=nx*j2-1; k2=nx*j2-0
         bmat(1,k1)=bmat(1,k1)+padx*phi(j2); bmat(2,k2)=bmat(2,k2)+pady*phi(j2)
         bmat(3,k1)=bmat(3,k1)+pady*phi(j2); bmat(3,k2)=bmat(3,k2)+padx*phi(j2)
     enddo Inner1_1
  else
     ndexlast=ndexnow
     Inner1_2: do j3=1,ndex
         ka=nv(j3); kindex=0
         Inner1_2_1: do j4=1,ndexlast
         kb=nvlist(j4)
         if (ka.eq.kb) then
         kindex=1; kc=nx*j4-1; kd=nx*j4-0
         bmat(1,kc)=bmat(1,kc)+padx*phi(j3); bmat(2,kd)=bmat(2,kd)+pady*phi(j3)
         bmat(3,kc)=bmat(3,kc)+pady*phi(j3); bmat(3,kd)=bmat(3,kd)+padx*phi(j3)
         endif
         enddo Inner1_2_1
     if (kindex.eq.0) then
     ndexnow=ndexnow+1; nvlist(ndexnow)=ka; kc=nx*ndexnow-1; kd=nx*ndexnow
     bmat(1,kc)=bmat(1,kc)+padx*phi(j3); bmat(2,kd)=bmat(2,kd)+pady*phi(j3)
```

```
        bmat(3,kc)=bmat(3,kc)+pady*phi(j3); bmat(3,kd)=bmat(3,kd)+padx*phi(j3)
      endif
      enddo Inner1_2
      endif
  enddo Loop_on_GP
end do Loop_on_edges


Return
End subroutine FormB_CSPIM_BOOK
```

Program 8.3 Source code of "Nodalstress_CSPIM"

```
Subroutine Nodalstress_CSPIM
!==================================================================
! Obtain nodal stress and strain components  for the CS-PIM models
!
! Input:   Cmat, area_cell, numneig, neighbour and
!             force(storing the displacemet solutions).
! Output: stress_cell, strain_cell, stress_node and strain_node.
!==================================================================
use Parameters
use Variables
implicit real*8 (a-h,o-z)
! wrok array
dimension Dinv(3,3),ne(nx*nndom)

Dinv=0.d0; Dinv=Cmat
call inversion(3,3,Dinv)
stress_node=0.d0; strain_node=0.d0; stress_cell=0.d0; strain_cell=0.d0
! Computing stress and strain solutions for each background cell
Loop_on_cells: do ia=1,numcell
  weight=area_cell(ia)
  Bmat=0.d0; ndexnow=0; nvlist=0
  call formB_CSPIM(ia)
  do ih=1,ndexnow
```

```fortran
      n1=nx*ih-1; n2=nx*ih-0
      ne(n1)=nx*nvlist(ih)-1
      ne(n2)=nx*nvlist(ih)-0
    enddo
    do ij=1,3
    do ik=1,3
    do il=1,nx*ndexnow
      mn=ne(il)
      stress_cell(ij,ia)=stress_cell(ij,ia)+Cmat(ij,ik)*Bmat(ik,il)*force(mn)
    enddo
    enddo
    enddo
    do iq=1,3
    do ir=1,3
      strain_cell(iq,ia)=strain_cell(iq,ia)+Dinv(iq,ir)*stress_cell(ir,ia)
    enddo
    enddo
  enddo Loop_on_cells
! Computing nodal stress and strain components
Loop_on_nodes: do i=1,numnode
  ia=numneig(i)
  ss=0.; ss1=0.; ss2=0.; ss3=0.; sn1=0.; sn2=0.; sn3=0.
  Inner: do j=1,ia
    ib=neighbour(j,i); ss=ss+area_edge(ib)
    ss1=ss1+stress_cell(1,ib)*area_cell(ib); ss2=ss2+stress_cell(2,ib)*area_cell(ib)
    ss3=ss3+stress_cell(3,ib)*area_cell(ib); sn1=sn1+strain_cell(1,ib)*area_cell(ib)
    sn2=sn2+strain_cell(2,ib)*area_cell(ib); sn3=sn3+strain_cell(3,ib)*area_cell(ib)
  enddo Inner
  stress_node(1,i)=ss1/ss; stress_node(2,i)=ss2/ss; stress_node(3,i)=ss3/ss
  strain_node(1,i)=sn1/ss; strain_node(2,i)=sn2/ss; strain_node(3,i)=sn3/ss
enddo Loop_on_nodes

!output of the nodal stress results for the first 20 nodes
write(10,11) 'Field No.','Stress Sigma_xx','Stress Sigma_yy','Stress Tau_xy'
do i=1,20
write(10,12) i,(stress_node(j,i),j=1,3)
```

```
enddo
11 format(a10,3a18)
12 format(i10,3e18.8)

Return
End subroutine Nodalstress_CSPIM
```

Program 8.4 Source code of "Enernorm_error_CSPIM"

```
Subroutine Enernorm_error_CSPIM
!==================================================================
! Compute the global energy norm error and strain energy for CS-PIM results
!
! Input:   area_cell, center, stress_cell and strain_cell;
! Output: Enernorm (the global displacement norm error) and
!          StrainE (the computed strain energy).
!==================================================================
use Parameters
use Variables
implicit real*8 (a-h,o-z)
! wrok array
dimension Dinv(3,3),ne(nx*nndom),gpos(nx)
dimension stressex(3),strainex(3),errstress(3),errstrain(3)

ai=ylength**3/12.d0
Dinv=0.d0; Dinv=Cmat
call inversion(3,3,Dinv)
StrainE=0.; Enernorm=0.
Loop_on_cells: do ia=1,numcell
  weight=area_cell(ia)
  gpos=0.d0
  gpos(1)=center(1,ia)
  gpos(2)=center(2,ia)
  stressex=0.d0; strainex=0.d0
```

```
stressex(1)=-pLoad*(xlength-gpos(1))*gpos(2)/ai
stressex(3)=pLoad/2./ai*(ylength**2/4.-gpos(2)**2)
do iq=1,3
do ir=1,3
   strainex(iq)=strainex(iq)+Dinv(iq,ir)*stressex(ir)
enddo
enddo
errstress=0.d0; errstrain=0.d0
do im=1,3
   errstress(im)=stress_cell(im,ia)-stressex(im)
   errstrain(im)=strain_cell(im,ia)-strainex(im)
enddo
do ip=1,3
   StrainE=StrainE+weight*0.5*strain_cell(ip,ia)*stress_cell(ip,ia)
enddo
do ip=1,3
   Enernorm=Enernorm+weight*errstrain(ip)*errstress(ip)
enddo
end do Loop_on_cells

Enernorm=dsqrt(Enernorm)
write(10,11) 'Global error in energy norm:',Enernorm
write(10,12) 'Calculated strain energy:',StrainE
11 format(a29,e18.8)
12 format(a26,e21.8)

Return
End subroutine Enernorm_error_CSPIM
```

8.4.3 Computed results

Running the present program with NTscheme=5 and NPIMSF=3, the CS-PIM-Tr4-CT will be established and the following output file can be obtained.

Output file of "DATAOUTPUT"

Field No.	Disp. U_x	Disp. U_y
1	-0.36800000E-05	-0.30000000E-05
2	0.36800000E-05	-0.30000000E-05
3	-0.73600000E-05	-0.27000000E-04
4	0.73600000E-05	-0.27000000E-04
5	0.20247077E-05	-0.47128710E-04
6	0.63357872E-04	-0.52272508E-04
7	-0.64140284E-04	-0.52786517E-04
8	0.10643191E-03	-0.74608351E-04
9	-0.10687781E-03	-0.75069528E-04
10	0.00000000E+00	-0.75000000E-04
11	0.00000000E+00	-0.75000000E-04
12	-0.79448261E-04	-0.17841858E-03
13	0.92806919E-04	-0.18616224E-03
14	-0.20528354E-03	-0.16585486E-03
15	0.21216594E-03	-0.16900283E-03
16	0.23871591E-03	-0.14394370E-03
17	-0.23794075E-03	-0.14437768E-03
18	-0.29176987E-03	-0.37476981E-03
19	0.31274529E-03	-0.38432332E-03
20	0.46488000E-03	-0.32978568E-03

Global error in displacement norm: 0.46129533E-01

Field No.	Stress Sigma_xx	Stress Sigma_yy	Stress Tau_xy
1	0.52664116E+03	0.87684699E+01	-0.12299380E+03
2	-0.50391605E+03	-0.28945223E+01	-0.12113166E+03
3	0.19378930E+04	-0.17801697E+02	-0.13003590E+03
4	-0.18900251E+04	0.14532535E+02	-0.13080938E+03
5	0.18113285E+02	0.17722182E+01	-0.12615258E+03
6	0.10787051E+04	0.43960526E+01	-0.10701193E+03
7	-0.11143970E+04	-0.94616469E+00	-0.11564513E+03
8	0.30059424E+04	-0.19134399E+02	-0.85357531E+02
9	-0.31589240E+04	0.24434216E+02	-0.10058895E+03
10	0.37288033E+04	-0.74063121E+02	-0.88776940E+02
11	-0.44724794E+04	0.87086468E+02	-0.11774106E+03
12	-0.71236289E+03	0.55750807E+01	-0.15887428E+03

13	0.74695581E+03	-0.24400053E+01	-0.14404417E+03
14	-0.20973843E+04	0.15262039E+01	-0.10174755E+03
15	0.19945880E+04	0.35305678E+01	-0.85601603E+02
16	0.34716817E+04	-0.15239145E+02	-0.48142492E+02
17	-0.31494106E+04	0.14521167E+02	-0.57566538E+02
18	-0.20980807E+04	0.24885755E+01	-0.14138379E+03
19	0.21166431E+04	-0.24358043E+01	-0.12219998E+03
20	0.29769869E+04	-0.22359623E+01	-0.69975524E+02

Global error in energy norm:　　0.33350208E+00

Calculated strain energy:　　0.89941761E+01

8.5 References

1. Liu, G. R., *Meshfree Methods: Moving beyond the Finite Element Method*, 2nd Edition, CRC press, Boca Taton, USA, 2009.

2. Liu, G. R., Zhang, G. Y., Dai, K. Y, Wang, Y. Y., Zhong, Z. H., Li, G. Y. and Han, X., A linearly conforming point interpolation method (LC-PIM) for 2D solid mechanics problems, *International Journal of Computational Methods*, 2(4): 645-665, 2005.

3. Zhang, G. Y., Liu, G. R., Wang, Y. Y., Huang, H. T., Zhong, Z. H., Li, G. Y. and Han, X., A linearly conforming point interpolation method (LC-PIM) for three-dimensional elasticity problems, *International Journal for Numerical Methods in Engineering*, 72: 1524-1543, 2007.

4. Liu, G. R. and Zhang, G. Y., Upper bound solution to elasticity problems: A unique property of the linearly conforming point interpolation method (LC-PIM). *International Journal for Numerical Methods in Engineering*, 74: 1128-1161, 2008.

5. Liu, G. R., Li, Y., Dai, K. Y., Luan M.T. and Xue, W., A Linearly conforming radial point interpolation method for solid mechanics problems, *International Journal of Computational Methods*, 3: 401-428, 2006.

6. Zhang, G. Y., Liu, G. R., Nguyen-Thoi, T., Song, C. X., Han, X., Zhong, Z. H. and Li, G. Y., The upper bound property for solid mechanics of the linearly conforming radial point interpolation method (LC-RPIM). *International Journal of Computational Methods*, 4(3): 521-541, 2007.

7. Liu, G. R. and Zhang, G. Y., Edge-based smoothed point interpolation methods, *International Journal of Computational Methods*, 5(4): 621-646, 2008.

8. Liu, G. R., Nguyen-Thoi, T. and Lam, K. Y. An edge-based smoothed finite element method (ES-FEM) for static, free and forced vibration analyses in solids. *Journal of Sound and Vibration*, 320: 1100-1130, 2009.

9. Liu, G. R., Wang, Z., Zhang, G. Y., Zong, Z. and Wang, S., An edge-based smoothed point interpolation method for material discontinuity. *Mechanics of Advanced Materials and Structures*, 19(1-3): 3-17, 2012.

10. He, Z. C., Cheng, A. G., Zhang, G. Y., Zhong, Z. H. and Liu, G. R., Dispersion error reduction for acoustic problems using the edge-based smoothed finite element method (ES-FEM). *International Journal for Numerical Methods in Engineering*, 86: 1322-1338, 2011.

11. Cui, X. Y., Liu, G. R., Li, G. Y., Zhang, G. Y., Zheng, G., Analysis of plates and shells using an edge-based smoothed finite element method, *Computational Mechanics*, 45: 141-156, 2010.

12. Liu, G. R. and Zhang, G. Y. A normed G space and weakened weak (W^2) formulation of a cell-based smoothed point interpolation method. *International Journal of Computational Methods*, 6(1): 147-179, 2009.

13. Zhang, G. Y. and Liu, G. R., Meshfree cell-based smoothed point interpolation method using isoparametric PIM shape functions and condensed RPIM shape functions. *International Journal of Computational Methods*, 8(4): 705-730, 2011.

14. Liu, G. R., Jiang, Y., Chen, L., Zhang, G. Y. and Zhang, Y. W., A singular cell-based smoothed radial point interpolation method for fracture problems. *Computers & Structures*, 89(13-14): 1378-1396, 2011.

15. Liu, G. R. and Gu, Y. T., *An introduction to meshfree method methods and their programming*, Springer, 2005.

16. Liu, G. R. and Gu, Y. T., A point interpolation method for two-dimensional solids, *Int. J. Numer. Methods Eng.*, 50: 937-951, 2001.

17. Wang, J. G. and Liu, G. R., A point interpolation meshless method based on radial basis functions, *International Journal for Numerical Methods in Engineering*, 54: 1623-1648, 2002.

18. Liu, G. R., Dai K. Y. and Nguyen-Thoi, T., A smoothed finite element method for mechanics problems, *Computational Mechanics*, 39: 859-877, 2007.

19. Dai, K. Y., Liu, G. R. and Nguyen-Thoi, T., An n-sided polygonal smoothed finite element method (nSFEM) for solid mechanics. *Finite elements in analysis and design*; 43: 847-860, 2007.

20. Liu, G. R. and Nguyen-Thoi, T., *Smoothed Finite Element Methods*, CRC press, Boca Taton, USA, 2010.

Chapter 9

The Cell-based Smoothed Alpha Radial Point Interpolation Method (CS-αRPIM)

In the previous chapters, three major models of smoothed point interpolation methods, i.e. NS-PIM/NS-RPIM, ES-PIM/ES-RPIM and CS-PIM/CS-RPIM, have been presented for 2D and 3D solid mechanics problems. It has been found that these three types of S-PIMs have their own attractive properties. The NS-PIMs use the node-based strain smoothing operation, have superconvergence property in energy norm, and can provide upper bound solutions in energy norm, but behave overly-soft and cannot be directly used in dynamic analysis [1-6]. The ES-PIMs use the edge/face-based strain smoothing operation, can provide very accurate and superconvergent results in both displacement and energy norms, produce models with "close-to-exact" stiffness, are both specially and temporally stable, and hence works very well for dynamic analysis [7-12]. It is regarded as the "star" performer. The CS-PIMs use the cell-based strain smoothing operation, provide very accurate numerical results with super-convergence property in both displacement and energy norms, possess models with the stiffness being stiffer than the overly-soft NS-PIMs and generally softer than the ES-PIMs, can be both specially and temporally stable and works well also for dynamic analysis [1, 13-15].

Obtaining the exact solution, at least in a norm, using a discrete numerical method is a very attractive but challenging idea in the field of computational methods. Some effective efforts have been made by Liu's group aiming to obtain the exact solution in a norm using discrete numerical models [16-18]. The so-called alpha finite element method using four-node quadrilateral elements (αFEM-Q4) was developed for finding a nearly exact solution in strain energy norm using coarse meshes [16]. This was achieved by scaling the gradient of strains using a real factor $\alpha \in [0,1]$. It was found that the resultant strain energy function having a simple polynomial form in terms of α. A general procedure has also been presented to obtain nearly exact or best possible solutions using meshes with the same aspect ratio. An exact-α approach is devised for overestimation problems, and a zero-α approach for underestimation problems. However, the αFEM-Q4 cannot provide exact solutions for all elasticity problems. In addition, the quadrilateral elements used cannot be generated in a fully automated manner for complicated domains. Making use of the NS-FEM providing upper bound solutions, and the FEM giving lower bound solutions, another type of αFEM for obtaining the exact solution has also been presented, and has been applied for both 2D problems using three-node triangular elements (αFEM-T3) and 3D problems using four-node tetrahedral elements (αFEM-T4) [17]. The essential idea of this approach is to introduce a factor $\alpha \in [0,1]$ to establish a continuous function of strain energy with contributions from both the standard FEM and the NS-FEM. This novel combined formulation makes the best use of the upper bound property of the NS-FEM and the lower bound property of the standard FEM. The αFEM is equipped with a parameter α that can be tuned for desired properties. Using meshes with the same aspect ratio, approaches have also been proposed to obtain the nearly exact solution in strain energy for any given linear elasticity problem. The αFEM-T3 and the αFEM-T4 are very easy to implement for practical problems of complicated geometry, and the existing FEM codes can largely be made use of.

In the previous studies on S-PIMs, it has been shown an important fact that different types of smoothing operations, node-based, edge/face-based and cell-based, can lead to different softening effects in the numerical model. In addition, for a numerical model using a particular strain smoothing operation, the softening effect also depends on the order of the PIM shape functions used, as we discussed in Section 6.4.2. Generally, when high order PIM shape functions are used, the displacements approximated in a smoothing domain are closer to the exact solution, which reduces the stiffening effect and vice versa. In Chapter 8, the CS-RPIM-Tr4-Cd and the CS-RPIM-Te5-Cd have been introduced for 2D and 3D problems respectively. We found that these two models can generally provide very tight upper bound solutions in strain energy, at least for the problems studied so far. It has also been discussed in Section 8.1.4 that when the linear PIM shape functions are used in the CS-PIM, the constructed numerical model will be exactly the same as the linear FEM with the lower bound property. Making use of both the CS-RPIM-Cd (including CS-RPIM-Tr4-Cd and the CS-RPIM-Te5-Cd) and linear FEM models, this chapter introduces a novel cell-based alpha radial point interpolation method using three-node triangular elements (CS-αRPIM-Tr4) for 2D problems and CS-αRPIM-Te5 using four-node tetrahedral elements for 3D problems, which originated in [19]. In these CS-αRPIM methods, the gradient (strain) field is obtained using the cell-based smoothing operation, which is the same as other CS-PIM/CS-RPIM models, but the displacement field is constructed using the novel αPIM shape functions as described in Section 3.6. In this chapter, the αPIM shape functions are created by combining the condensed RPIM shape functions with the linear PIM shape functions via a scaling factor $\alpha \in [0, 1]$. Through adjusting the value of α, we can further change the softening effect for desired purposes. We can also find out a particular value of α leading to "nearly exact" solution in strain energy. Using meshes with the same aspect ratio, we present an approach to obtain the nearly exact solution in strain energy for given linear elasticity problems.

The difference between the present methods and those two αFEM methods (αFEM-Q4 and αFEM-T3 or αFEM-T4) is as follows [19]. In the αFEM-Q4, the strain field is obtained by scaling the gradient of the bilinear strain field using a factor α [16]. In the αFEM-T3 or αFEM-T4, the strain field is obtained via combining the node-based smoothed strains and the compatible strains using a factor α. Both these two αFEM methods manipulate the strain field. However, in the present CS-αRPIM models, we manipulate the displacement field using the αPIM shape functions that are equipped with an adjustable parameter α. The CS-αRPIM models work very well with triangular/tetrahedral meshes which can be generated automatically for complicated problem domains, and can provide nearly exact solutions in strain energy for given linear elasticity problem.

9.1 CS-αRPIM-Tr4 for 2D solids

9.1.1 Approximation of displacement field

The problem domain is first represented by a set of N_c nonoverlapping and seamless (NOSL) triangular background cells with N_n field nodes, as in any S-PIM model, such that $\overline{\Omega} = \bigcup_{i=1}^{N_c} \overline{\Omega}_i^c$. In the CS-αRPIM, the points of interest (quadrature points) are always located on the edges of triangles, because of the cell-based smoothing. Refer to Equation (3.125), the displacement field is approximated using the αPIM shape functions as follows.

$$u^h(\mathbf{x}) = \sum_{i=1}^{n^{(\alpha)}} \phi_i^{(\alpha)}(\mathbf{x})u_i = \mathbf{\Phi}_S^{(\alpha)}(\mathbf{x})\mathbf{d}_s \tag{9.1}$$

where $\mathbf{\Phi}_S^{(\alpha)}$ is the matrix of αPIM shape functions, which can be formulated following the general way as described in Section 3.6. In the scheme of CS-αRPIM-Tr4, the αPIM shape functions are particularly constructed via combining the condensed RPIM shape functions selecting support nodes using the edge-based T4-scheme and the linear PIM shape functions selecting support nodes using the edge-based T2-scheme, i.e.

$$\phi_i^{(\alpha)}(\mathbf{x}) = \begin{cases} \alpha\phi_{Ii}(\mathbf{x}) + (1-\alpha)\phi_{IIi}^c(\mathbf{x}) & \forall i \in S^{(I)} \bigcap S^{(II)} \\ \alpha\phi_{Ii}(\mathbf{x}) & \forall i \in S^{(I)} \setminus S^{(II)} \\ (1-\alpha)\phi_{IIi}^c(\mathbf{x}) & \forall i \in S^{(II)} \setminus S^{(I)} \end{cases} \tag{9.2}$$

where $\phi_{Ii}(\mathbf{x})$ and $\phi_{IIi}^c(\mathbf{x})$ refer to linear PIM shape function and condensed RPIM shape function respectively, subscripts (*I*) and (*II*) stands for the nodal shape functions for the local nodes in set $S^{(I)}$ and set $S^{(II)}$, and $n^{(\alpha)}$ in Equation (9.1) is the number of the nodes in the set $S^{(\alpha)} = S^{(I)} \bigcup S^{(II)}$.

9.1.2 Properties of αPIM shape functions

As discussed in Section 3.6.3, the αPIM shape functions combined using the condensed RPIM and linear PIM shape functions possess the following properties.

Remark 9.1 The partitions of unity property

The αPIM shape functions preserve the partitions of unity of the original two sets of PIM shape functions. In the present case, these two sets of PIM shape functions are both of the partitions of unity property, and hence the αPIM shape functions are also of the partitions of unity property.

Remark 9.2 The Kronecker Delta function property

The condensed RPIM shape functions and linear PIM shape functions are created with the support nodes selected using, respectively, the edge-based T4-scheme and the edge-based T2-scheme, which ensures that the point of interest is located in the common area supported by both the two sets of $S^{(I)}$ and $S^{(II)}$.

Therefore, the present αPIM shape functions possess the Kronecker Delta function property.

Remark 9.3 Linear consistency property

The order of consistency of the αPIM shape functions is determined by the lower order of consistency of the two sets of original shape functions. The present αPIM shape functions are thus of linear consistency.

9.1.3 Evaluation of cell-based smoothed strains

In the CS-αRPIM models, the gradient smoothing domains are exactly the background cells, and the strain field is obtained using the cell-based strain smoothing operation. The process of evaluating the cell-based smoothed strains is the same as that for other CS-PIM/CS-RPIM models presented in Chapter 8, and is omitted here. The only difference is the shape functions used. Thus for the present CS-αRPIM models we have $N_q = N_s = N_c$.

9.1.4 CS-αRPIM formulations

For static problems, the discretized algebraic system equation follows Equation (5.79) and has the form of

$$\overline{\mathbf{K}}^{\text{CS-αRPIM}}\overline{\mathbf{d}} = \tilde{\mathbf{f}} \tag{9.3}$$

where $\overline{\mathbf{K}}^{\text{CS-αRPIM}}$ is the global *smoothed* stiffness matrix whose entries are given by

$$\overline{\mathbf{K}}_{IJ}^{\text{CS-αRPIM}} = \int_{\Omega} \overline{\mathbf{B}}_I^{\text{T}}\mathbf{c}\overline{\mathbf{B}}_J \mathrm{d}\Omega = \sum_{k=1}^{N_c}\int_{\Omega_k^s} \overline{\mathbf{B}}_I^{\text{T}}\mathbf{c}\underbrace{\overline{\mathbf{B}}_J}_{\text{constant in }\Omega_k^s}\mathrm{d}\Omega = \sum_{k=1}^{N_c}A_k^s\overline{\mathbf{B}}_I^{\text{T}}\mathbf{c}\overline{\mathbf{B}}_J \tag{9.4}$$

in which $A_k^s = \int_{\Omega_k^s} \mathrm{d}\Omega$ is the area of cell-based smoothing domain Ω_k^s, and the smoothed strain-displacement matrix $\overline{\mathbf{B}}_I$ is computed using Equation (5.71).

9.2 CS-αRPIM-Te5 for 3D solids

To extend the CS-αRPIM from solving 2D problem to 3D problem is straightforward. In 3D case, however, we use four-node tetrahedral cells and the method is termed as CS-αRPIM-Te5.

The 3D αRPIM shape functions are created by combing the 3D condensed RPIM shape functions selecting support nodes using the face-based T5-scheme and the 3D linear PIM shape functions selecting support nodes using the face-based T3-scheme. The formulation has the same form as Equation (9.2), and the created 3D shape functions also possess these properties discussed in Section 9.1.2.

The 3D problem domain is discretized using four-node tetrahedral background cells and the smoothed strains are obtained using the cell-based strain smoothing operation, which is as same as these 3D CS-PIM/CS-RPIM models described in Chapter 8.

For 3D static problems, the discretized algebraic system equation has the similar form as Equation (9.3) and the sub-stiffness matrix can be obtained as

$$\overline{\mathbf{K}}_{IJ}^{\text{CS-}\alpha\text{RPIM}} = \sum_{i=1}^{N_c} V_i^s \overline{\mathbf{B}}_I^{\text{T}} \mathbf{c} \overline{\mathbf{B}}_J \tag{9.5}$$

where $V_i^s = \int_{\Omega_i^s} \mathrm{d}\Omega$ is the volume of the smoothing domain i.

In terms of stability the CS-αRPIM should be at least more stable than the CS-RPIM-Cd counterpart. Therefore, it can also be used for dynamic problems, and the formulation is largely the same as those presented in Section 8.1.3.

9.3 Numerical implementation

9.3.1 Macro flowchart of the CS-αRPIM

The macro flowchart of the CS-αRPIM codes is quite similar to other CS-PIM models described in Section 8.1.4.1, except that the αPIM shape functions are used with a scaling factor α. Therefore, an additional procedure for finding out the possible best α_{exact} and computing the nearly exact solution is needed. The procedure can be summarized as follows:

1) Discretize the problem domain into two set of coarse meshes using triangular/tetrahedral cells with the same aspect ratio.

2) Set one array of a number of factors $\alpha \in \overline{0:1}$, for example $\alpha = \begin{bmatrix} 0.0 & 0.2 & 0.4 & 0.6 & 0.8 & 1.0 \end{bmatrix}^{\mathrm{T}}$.

3) Loop over the two sets of mesh created in step 1).

    (a) Loop over the array of α.

    (b) Follow the procedure described in Section 8.1.4.1, form the discretized algebraic system equation with the current α and save the computed solution (in strain energy).

    (c) End the loop over the array of α, and a solution-α curve will be obtained.

4) End the loop over the two sets of coarse meshes.

5) Find out the particular α_{exact} as the intersection point of these two solution-α curves.

6) Use the α_{exact} and a finer mesh to compute the solution that is expected to be nearly exact.

9.3.2 Meshes with the same aspect ratio

In the present study, meshes with the same aspect ratio are used to find out the α_{exact} for computing the nearly exact strain energy. Such a process was proposed in [16, 17], and is briefed as follows.

For regular meshes (that can only be used for regular problem domains), meshes with the same aspect ratio implies that the ratio of the cell numbers along the coordinate directions of these meshes must be identical. For example, for a rectangular-shape 2D domain, three meshes of (20×4), (40×8) and (80×16) are three meshes with the same aspect ratio of 4.

For irregular meshes which can be used to discretize complicated problem domains, the meshes with the same aspect ratio are obtained via dividing, in a nested manner, each triangular cell into 2^2, 2^4 and 2^6 etc. equal cells and each tetrahedral cell into 2^3, 2^6 and 2^9 etc. equal cells.

Routines for creating the meshes with the same aspect ratio are available in many automatic programs and commercial software, and hence can be implemented without any technical difficulty. Note that we do not require all the cells in a particular mesh to have the same aspect ratio, but only the cells in two consequent meshes to have the same aspect ratio.

9.3.3 Some possible CS-αRPIM models

In this chapter, two CS-αRPIM models are presented for 2D and 3D problems. In these two models we will use cell-based strain smoothing technique and the displacement field is approximated using the αPIM shape functions that is the combination of the condensed RPIM shape functions and the linear PIM shape functions. We first note the following.

Remark 9.4 Combination of CS-RPIM-Cd and linear FEM models

The CS-αRPIM model is a combination of the CS-RPIM-Cd model that is "softer" and the linear FEM model that is "stiffer". A CS-αRPIM-Tr4 model becomes a CS-RPIM-Tr4-Cd model when $\alpha=0$ and becomes a FEM-Tr3 model when $\alpha=1$. Similarly, a CS-αRPIM-Te5 becomes a CS-RPIM-Te5-Cd when $\alpha=0$ and become the FEM-Te4 when $\alpha=1$. Therefore, a CS-αRPIM model can have a property between a soft CS-RPIM-Cd and a stiff FEM model.

CS-αRPIM-Tr4 for 2D problems

In the scheme of the CS-αRPIM-Tr4, the displacement field is constructed using the 2D αPIM shape functions which are a combination of the condensed RPIM shape functions and linear PIM shape functions with a factor $\alpha \in [0,1]$. For these two types of basis shape functions, the edge-based T4-scheme and the edge-based T2-scheme are used, respectively, to select the local support nodes. The displacement field in the CS-αRPIM-Tr4 is thus incompatible, except $\alpha=1$.

For the 2D cantilever beam with the same meshes as shown in FIGURE 6.4, we study the overall characteristics of the CS-αRPIM-Tr4 model in relation to the standard base model FEM-Tr3. Because the same edge-based T4-scheme is used to select the support nodes for both the CS-RPIM-Tr4-Cd and the CS-αRPIM-Tr4 models, these two models shall have the same number of local

TABLE 9.1 Overall characteristics of the CS-αRPIM-Tr4 model in relation to the FEM-Tr3 model using the same mesh

Numerical method	Support nodes in an integral cell/element	Estimated average number of nonzero entries	Estimated Solver CPU time
FEM-Tr3	3	n_{ze}	t_{CPU}
CS-RPIM-Tr4-Cd	5-6 (5.8)	$1.9\,n_{ze}$	$1.9\,\overline{n}_{iter}t_{CPU}$
CS-αRPIM-Tr4	5-6 (5.8)	$1.9\,n_{ze}$	$1.9\,\overline{n}_{iter}t_{CPU}$

support nodes in an integral cell. Hence, the estimated average number of nonzero entries (n_{ze}) in the resultant stiffness matrix, and the number of the iterations $\bar{n}_{iter}$ needed in an iteration solver should also be the same, as listed in TABLE 9.1.

CS-αRPIM-Te5 for 3D problems

In the scheme of the CS-αRPIM-Te5, the displacement field is constructed using the 3D αPIM shape functions that are a combination of condensed RPIM shape functions and linear PIM shape functions via a factor $\alpha \in [0, 1]$. For these two types of basis shape functions, the face-based T5-scheme and the face-based T3-scheme are used, respectively, to select the local support nodes. The displacement field in the CS-αRPIM-Te5 is thus incompatible, except $\alpha=1$.

Using the 3D Lame problem with the tetrahedral mesh of 173 nodes, we study the overall characteristics of the CS-αRPIM-Te5 model in relation to the standard base model FEM-Te4. As the same face-based T5-scheme is used to select the support nodes for both the CS-RPIM-Te5-Cd and the CS-αRPIM-Te5 models, these two models shall have the same number of local support nodes in an integral cell, and the same estimated average number of nonzero entries, as shown in TABLE 9.2.

TABLE 9.2 Overall characteristics of the CS-αRPIM-Te5 model in relation to the FEM-Te4 model using the same mesh.

Numerical method	Support nodes in an integral cell/element	Estimated average number of nonzero entries	Estimated Solver CPU time
FEM-Te4	4	n_{ze}	t_{CPU}
CS-RPIM-Te5-Cd	6-8 (7.5)	$1.9\,n_{ze}$	$1.9\,\bar{n}_{iter}t_{CPU}$
CS-αRPIM-Te5	6-8 (7.5)	$1.9\,n_{ze}$	$1.9\,\bar{n}_{iter}t_{CPU}$

9.3.4 Condition number of the CS-αRPIM models

TABLE 9.3 lists the condition numbers of the global stiffness matrixes computed for the CS-αRPIM-Tr4 model in relation to the base model FEM-Tr3 using the same triangular meshes for the cantilever beam problem. Results of the CS-αRPIM-Tr4 are computed with the $\alpha=0.6654$, which is obtained using the general numerical procedure given in Section 9.3.1 with two set of meshes of the same aspect ratio and will be described in Example 9.4.2. For the purpose of comparison, results of the CS-RPIM-Tr4-Cd model are also provided. We find that the condition number of the CS-αRPIM-Tr4 is smaller than the counterpart of the FEM-Tr3 and bigger than that of the CS-RPIM-Tr4-Cd. As discussed previously, being the combination of the CS-RPIM-Tr4-Cd and the FEM-Tr3 models, the condition number of the CS-αRPIM-Tr4 is expected to be in between these two models.

TABLE 9.4 shows the condition number for the 3D numerical models using the same tetrahedral mesh for the Lame problem. Results of the CS-αRPIM-Te5 are computed using $\alpha=0.5646$ which is obtained by following the general procedure given in Section 9.3.1 and will be detailed in Example 9.5.2. From this table, we can find that the condition number of the CS-αRPIM-Te5 is smaller than the FEM-Te4 and bigger than the CS-RPIM-Te5-Cd, as expected.

As the ratio between the condition numbers of different methods with respect to the linear FEM model is found almost independent of the mesh size, we will use the relative numbers given in last column in TABLE 9.3 and TABLE 9.4 in the estimation of the computation cost.

TABLE 9.3 Condition number of the global stiffness matrixes for the CS-αRPIM-Tr4 model for the 2D cantilever in relation to the FEM-Tr3 model using the same mesh

Numerical method	Mesh-1 of 120 nodes		Mesh-2 of 248 nodes	
	$cond(\mathbf{K})$	$\bar{n}_{iter}$	$cond(\mathbf{K})$	$\bar{n}_{iter}$
FEM-Tr3	1.645e+08	1.00	2.399e+08	1.00
CS-RPIM-Tr4-Cd	5.415e+07	0.57	8.362e+07	0.59
CS-αRPIM-Tr4	1.004e+08	0.78	1.461e+08	0.78

TABLE 9.4 Condition number of the global stiffness matrixes for the CS-αRPIM-Te5 model for the 3D Lame problem in relation to the FEM-Te4 model using the same mesh

Numerical method	Mesh-1 of 96 nodes		Mesh-2 of 173 nodes	
	$cond(\mathbf{K})$	$\bar{n}_{iter}$	$cond(\mathbf{K})$	$\bar{n}_{iter}$
FEM-Te4	3.284e+02	1.00	5.274e+02	1.00
CS-RPIM-Te5-Cd	1.339e+02	0.64	1.977e+02	0.61
CS-αRPIM-Te5	1.919e+02	0.76	2.856e+02	0.74

9.3.5 Estimation of computational cost

Combining TABLE 9.1 (TABLE 9.2 for 3D) and TABLE 9.3 (TABLE 9.4 for 3D), we shall have a rough estimation of the computational cost for the CS-αRPIM models in relation to the linear FEM model using the same mesh. The results are shown in TABLE 9.5 (TABLE 9.6 for 3D). Compared to the linear FEM models, the CS-αRPIM models cost more time (50% and 40% for 2D and 3D problems respectively) due to using more support nodes for field approximation. Compared to the CS-RPIM-Cd models, the present CS-αRPIM models consume more CPU time which is caused by the larger condition numbers.

TABLE 9.5 Estimation of computational cost for 2D CS-αRPIM-Tr4 model in relation to the FEM-Tr3 model using the same mesh.

Numerical method	Support nodes in an integral cell/element	Estimated Solver CPU time
FEM-Tr3	3	t_{CPU}
CS-RPIM-Tr4-Cd	5-6 (5.8)	1.1 t_{CPU}
CS-αRPIM-Tr4	5-6 (5.8)	1.5 t_{CPU}

TABLE 9.6 Estimation of computational cost for 3D CS-αRPIM-Te5 model in relation to the FEM-Te4 model using the same mesh.

Numerical method	Support nodes in an integral cell/element	Estimated Solver CPU time
FEM-Te4	4	t_{CPU}
CS-RPIM-Te5-Cd	6-8 (7.5)	1.2 t_{CPU}
CS-αRPIM-Te5	6-8 (7.5)	1.4 t_{CPU}

9.3.6 Evaluation of nodal strain (stress)

In the scheme of CS-αRPIM models, the nodal strain (stress) can be obtained via the area-weighted averaging process with the cell-based smoothed strains, which is as same as that for the CS-PIM models presented in Chapter 8 (see Equation (8.10)).

9.3.7 Rank analysis for CS-αRPIM models

As same as other CS-PIM models, CS-αRPIM models use cell-based smoothed strains and hence satisfy the requirements of independence of smoothing domains and the minimum number of smoothing domains listed in TABLE 2.2. On the other hand, the αPIM shape functions possess the partition of unity property as discussed in Section 9.1.2 and the stiffness of a CS-αRPIM model is SPD for solids of stable materials after constraining all the rigid movements.

Remark 9.5 CS-αRPIM models: spatially stable

The CS-αRPIM models, including CS-αRPIM-Tr4 and CS-αRPIM-Te5, are spatially stable. A CS-αRPIM model possesses only "legal" zero-energy modes that represent the physic rigid motions; there exist no spurious unphysical zero-energy modes. There is no deformed zero-energy mode existing in a CS-αRPIM model, and any finite deformation in the model (except the rigid motions) will result in a finite amount of strain energy in a CS-αRPIM model.

9.3.8 Temporal stability analysis

Based on the temporal stability analysis of CS-PIM models in Section 8.1.4.7 and Example 8.1.5, we find that the CS-RPIM-Tr4-Cd model may have spurious nonzero-energy modes using coarse mesh. We can however always make it stable (see, Remark 8.2). It is known that the linear FEM model is always temporally stable. Hence we can conclude that CS-αRPIM models may be temporally instable with a smaller value of α, and will be temporally stable with a bigger value of α. This implies that we now have an additional means to make the CS-αRPIM stable: by tuning α.

Remark 9.6 CS-αRPIM models: temporally stable

We can always make a CS-αRPIM model temporally stable. This can be achieved by either using not too coarse mesh, and/or using a larger value of the factor α.

9.4 Numerical examples for 2D solids

Several numerical examples are studied in this section to examine the CS-αRPIM models. The materials used are all linear elastic with Young's modulus $E=3.0\times10^7$ Pa and passion's ratio $\nu=0.3$, and the units used are based on the international standard unit system unless specified explicitly. Numerical results are assessed using the error indicators in the displacement and energy norms defined in Equations (5.88) and (5.94).

Example 9.4.1 A 2D linear patch test

The linear patch test described in Example 6.1.1 is first carried using the CS-αRPIM models. TABLE 9.7 lists the displacement norm errors (defined in Equation (5.88)) of the CS-αRPIM-Tr4 solutions for the patch test using both the regular and irregular meshes shown in FIGURE 6.5. Regardless of the values of the factor α, the CS-αRPIM-Tr4 model can always pass these tests exactly (to the machine accuracy), despite the fact that the displacement field in the CS-αRPIM-Tr4 (with $\alpha \neq 1$) is incompatible. We also note that because of the use of the RBFs in the RPIM shape functions, a loss about up to 3 digits accuracy is observed. This is due to the large condition number in the moment matrix in creating the RPIM shape functions [1].

TABLE 9.7 Error norm in displacements of numerical results for the standard patch test obtained using the CS-αRPIM-Tr4 model

	α=0.0	α=0.2	α=0.4	α=0.6	α=0.8	α=1.0
Regular mesh	2.65E-12	9.08E-13	3.57E-13	1.47E-13	5.19E-14	7.37E-16
Irregular mesh	3.67E-11	1.95E-11	8.51E-12	3.82E-12	1.43E-12	1.24E-15

Remark 9.7 CS-αRPIM-Tr4: linearly conforming and 2nd order accuracy

This standard patch test example demonstrates numerically that the CS-αRPIM-Tr4 is linearly conforming: it can reproduce linear fields exactly regardless of the incompatible displacement field used and different α. Together with the stability, the CS-αRPIM-Tr4 solution will converge to the exact solution of any well-posed linear elasticity problem. This implies also that the model is at least 2nd order accuracy: the solution error in displacement is on the terms of 2nd order and above, regardless of the α value. This is consistent with Remark 3.23.

Example 9.4.2 Rectangular cantilever

The benchmark rectangular cantilever problem described in Example 6.1.2 is again studied using the CS-αRPIM-Tr4. For comparison, the FEM-Tr3, NS-PIM-Tr3, ES-PIM-Tr3 and CS-RPIM-Tr4-Cd models are also adopted in this analysis with the same set of triangular meshes shown in FIGURE 6.9.

Following the procedure given in Section 9.3.1, we found α_{exact}=0.6654 at the intersection of two strain energy curves using two irregular meshes with the same aspect ratio (Mesh-1 of 120 nodes and Mesh-2 of 425 nodes), as shown in FIGURE 9.1. This particular value of α will be used in the following study for the rectangular cantilever. We can find that the intersection of the two curves is very close to the exact strain energy, meaning the CS-αRPIM-Tr4 model with this particular value will possess a very close-to-exact stiffness.

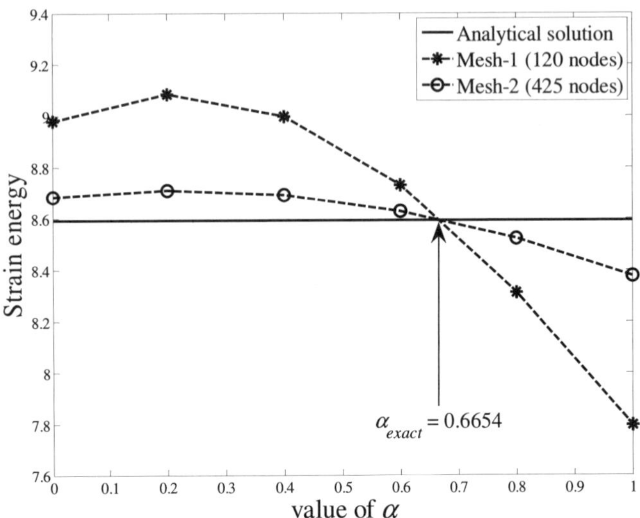

FIGURE 9.1 The strain energy curves of the CS-αRPIM-Tr4 solutions using two meshes with the same aspect ratio interest at α_{exact}=0.6654 for the rectangular cantilever problem.

Using mesh-2 (with 399 nodes) in FIGURE 6.9, this cantilever problem is studied using the CS-αRPIM-Tr4 and the calculated numerical results along two particular lines are plotted in FIGURE 9.2. For the displacement along the neutral line, the CS-αRPIM-Tr4 results with α_{exact}=0.6654 agree very well with the analytical solution and have much better performance than that of the FEM-Tr3 and CS-RPIM-Tr4-Cd. For the shear stress along the middle line, the CS-αRPIM-Tr4 performs better than the FEM-Tr3 and a little worse than the CS-RPIM-Tr4-Cd. The CS-αRPIM-Tr4 lay in between those of CS-RPIM-Tr4-Cd and FEM-Tr3.

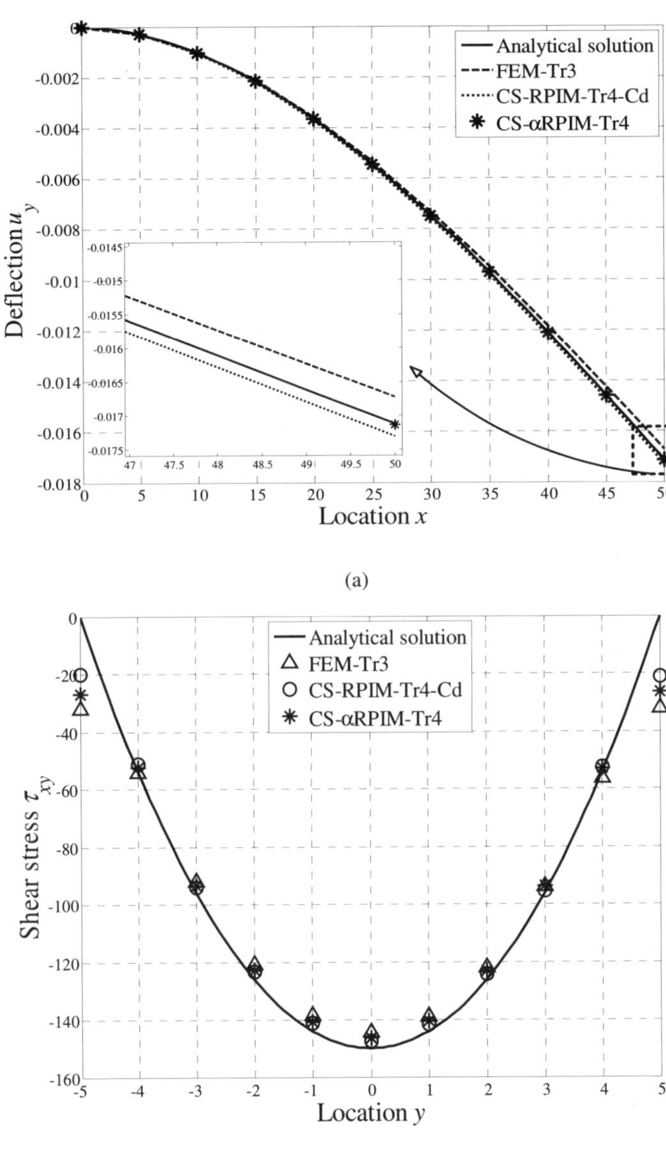

(a)

(b)

FIGURE 9.2 Comparison of numerical results obtained using different numerical models for the rectangular cantilever problem with the same triangular mesh: (a) deflection along the neutral line of the cantilever; (b) shear stress along the line of $x=L/2$ of the cantilever.

The convergence of the solution error in displacement norm for the cantilever problem is plotted in FIGURE 9.3. The errors in the displacement norm for the numerical results obtained using the same set of triangular mesh (Mesh-4) are listed in TABLE 9.8. Taking the computational cost given in TABLE 9.5 into consideration, the estimated computational efficiency measured in displacement error are calculated and listed in the last column in TABLE 9.8. We can now clearly observe the following.

1) The CS-αRPIM-Tr4 is about 15 times more efficient than the standard base model FEM-Tr3, and 7 times than the CS-RPIM-Tr4-Cd model. It still loses to the "star" performer ES-PIM-Tr3 by 6.5 times for this problem in displacement norm measure.

2) In terms of accuracy, the CS-αRPIM-Tr4 is 23 times more accurate than the FEM-Tr3 and 10 times more accurate than the CS-RPIM-Tr4-Cd. Compared to the ES-PIM-Tr3, it is about 5 times less accurate when Mesh-4 is used.

3) The CS-αRPIM-Tr4 performs particularly well when coarse mesh is used. This may be because the α_{exact} is determined using coarse meshes.

4) The rate of convergence of the CS-αRPIM-Tr4 is lower than the other models. This is also partially because of the good accuracy for coarse meshes. When the mesh is refined, the accuracy cannot increase very fast.

5) The ES-PIM-Tr3 stands out clear in this case.

TABLE 9.8 Estimated computational efficiency of different methods measured in *displacement norm* error for the numerical results for the cantilever beam problem with the same set of triangular mesh (Mesh-4 of 1,696 nodes)

Numerical method	Solution error	Error ratio to FEM-Tr3	Efficiency
FEM-Tr3	5.2334E-03	1.00	1.0
NS-PIM-Tr3	5.5340E-03	1.06	0.8
ES-PIM-Tr3	4.6937E-05	0.0090	101.0
CS-RPIM-Tr4-Cd	2.3951E-03	0.46	2.0
CS-αRPIM-Tr4	2.2444E-04	0.043	15.5

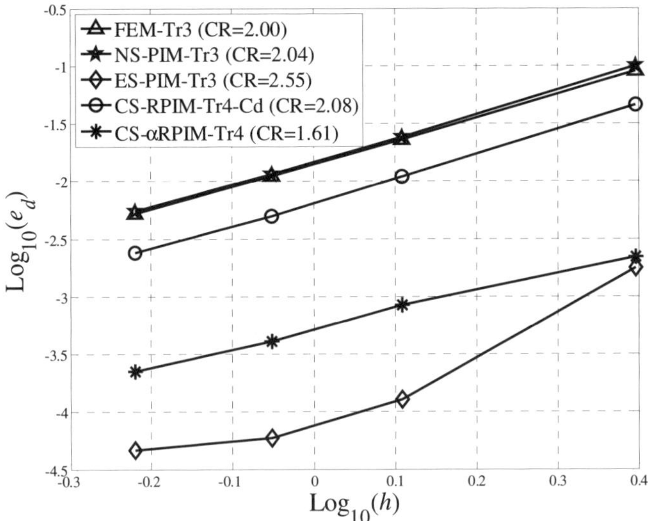

FIGURE 9.3 Comparison of convergence rates and accuracy of the numerical results in displacement norm obtained using different models for the rectangular cantilever problem.

FIGURE 9.4 shows the convergence of the solution error in energy norm (defined in Equation (5.94)) for different methods. The errors in the energy norm for the numerical results obtained using the same set of triangular mesh (Mesh-4) are listed in TABLE 9.9. Considering the computational cost given in TABLE 9.5, the estimated computational efficiency measured in energy norm measure are calculated and given in the last column in TABLE 9.9. The following points can be found.

1) The CS-αRPIM-Tr4 is about 2 times more efficient than the base model FEM-Tr3 and the CS-RPIM-Tr4-Cd model. It loses to the "star" performer ES-PIM-Tr3 by 4 times for this problem in energy norm measure.

2) In terms of accuracy, the CS-αRPIM-Tr4 is about 3 times more accurate than the base models of FEM-Tr3 and CS-RPIM-Tr4-Cd.

3) The rate of convergence of the CS-αRPIM-Tr4 is a little better than the FEM-Tr3, and about the same as the CS-RPIM-Tr4-Cd.

4) The ES-PIM-Tr3 stands out clear in this case, followed by the NS-PIM-Tr3.

TABLE 9.9 Estimated computational efficiency of different methods measured in *energy norm* error for the numerical results for the cantilever beam problem with the same set of triangular mesh (Mesh-4 of 1,696 nodes)

Numerical method	Solution error	Error ratio to FEM-Tr3	Efficiency
FEM-Tr3	2.4406E-01	1.00	1.0
NS-PIM-Tr3	5.0076E-02	0.21	4.0
ES-PIM-Tr3	2.8075E-02	0.12	7.6
CS-RPIM-Tr4-Cd	2.3207E-01	0.95	0.96
CS-αRPIM-Tr4	8.6208E-02	0.35	1.9

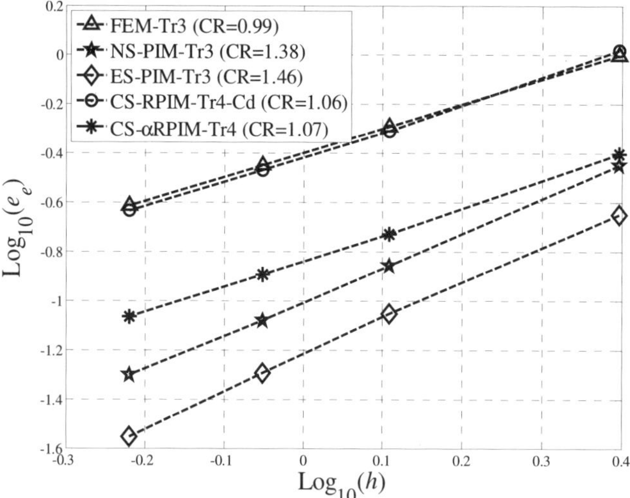

FIGURE 9.4 Comparison of convergence rates and accuracy of the numerical results in energy norm obtained using different models for the rectangular cantilever problem.

The convergence process of the strain energy solution for the CS-αRPIM-Tr4 model is shown in FIGURE 9.5, together with the FEM-Tr3, NS-PIM-Tr3, ES-PIM-Tr3 and CS-RPIM-Tr4-Cd models. The errors in the strain energy solution for the numerical results obtained using the same set of triangular mesh (Mesh-4) are listed in TABLE 9.10. Taking the computational cost given in TABLE 9.5 into consideration, the estimated computational efficiency measured in strain energy solution is given in the last column in TABLE 9.10. We can find for this case that

1) the CS-αRPIM-Tr4 is about 26 times more efficient than the base model FEM-Tr3 and about 12 times than the CS-RPIM-Tr4-Cd model. It loses to the "star" performer ES-PIM-Tr3 by 5 times for this problem in strain solution;

2)   the CS-αRPIM-Tr4 gives an upper bound solution, while the ES-PIM-Tr3 gives a lower bound. Both solutions are very close to the exact solution; and

3)   the ES-PIM-Tr3 stands out clear in this case, followed by CS-αRPIM-Tr4.

In summary, for this cantilever problem with exact solutions in polynomial forms, the CS-αRPIM-Tr4 is the 2nd best in terms of efficiency. It loses to the "star" performer ES-PIM-Tr3. However, the CS-αRPIM-Tr4 can be tuned for upper bound solutions, but ES-PIM-Tr3 gives usually only the lower bound solution. Therefore, this important role of the CS-αRPIM-Tr4 cannot be played by ES-PIM-Tr3.

TABLE 9.10 Estimated computational efficiency of different methods measured in the error in *strain energy solution* of the numerical results for the cantilever beam problem with the same set of triangular mesh (Mesh-4 of 1,696 nodes)

Numerical method	Strain energy Solution	Error[*] (%)	Error ratio to FEM-Tr3	Efficiency
FEM-Tr3	8.5474	-0.5341	1.00	1.0
NS-PIM-Tr3	8.6398	0.5411	1.01	0.8
ES-PIM-Tr3	8.5930	-0.0035	0.0066	137.7
CS-RPIM-Tr4-Cd	8.6127	0.2258	0.42	2.2
CS-αRPIM-Tr4	8.5945	0.0136	0.025	26.7

[*] The analytical value of the strain energy for the cantilever beam problem is 8.59333333. Negative indicates lower bound, and positive indicates upper bound.

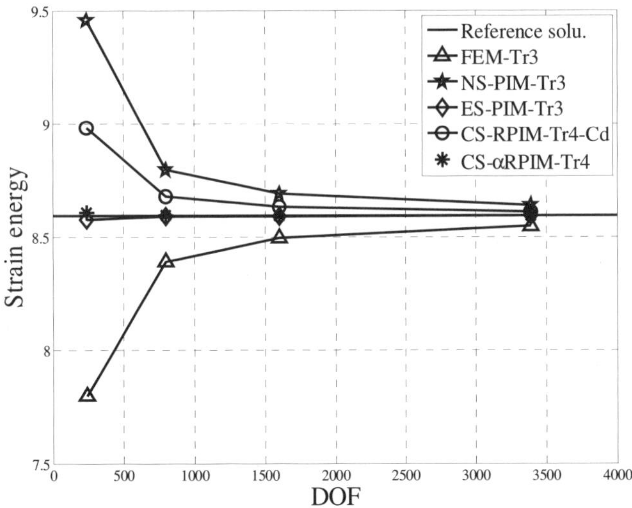

FIGURE 9.5 Converging process of the numerical results (in strain energy norm) for the problem of rectangular cantilever.

Example 9.4.3 Infinite solid with a circular hole

The infinite 2D solid with a circular hole problem described in Example 6.1.3 is again studied using the CS-αRPIM-Tr4.

Following the procedure given in Section 9.3.1, we found α_{exact}=0.4453 at the intersection of two strain energy curves using two irregular meshes with the same aspect ratio (Mesh-1 of 577 nodes and Mesh-2 of 2,218 nodes), as shown in FIGURE 9.6. The α_{exact} will be used in the following study for the problem of infinite solid with a circular hole. For this benchmark problem with exact strain energy, we found again that the intersection of the two curves is very close to the exact strain energy.

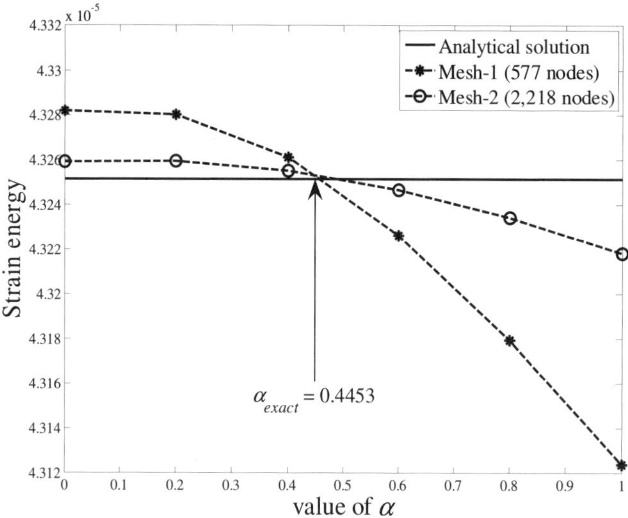

FIGURE 9.6 The strain energy curves of the CS-αRPIM-Tr4 solutions using two meshes with the same aspect ratio interest at α_{exact}=0.4453 for the problem of infinite 2D solid with a circular hole.

Using the meshes of different nodal density shown in FIGURE 6.14, the convergence of the solution errors for the CS-αRPIM-Tr4 model are computed using Equations (5.88) and (5.94). The convergence of the solution error in displacement norm is plotted in FIGURE 9.7. The errors in the displacement norm for the numerical results obtained using the same set of triangular mesh (Mesh-4) are listed in TABLE 9.11. Taking the computational cost given in TABLE 9.5 into consideration, the estimated computational efficiency measured in displacement error are calculated and listed in the last column in TABLE 9.11. We can note the following observations.

1) The CS-αRPIM-Tr4 is about 3.7 times more efficient than the standard base model FEM-Tr3, and 1.5 times than the CS-RPIM-Tr4-Cd model. It is a little more efficient even than the "star" performer ES-PIM-Tr3 for this problem in displacement norm measure.

2) The CS-αRPIM-Tr4 performs relatively better when coarse meshes are used. This may be because the α_{exact} is determined using coarse meshes.

3) In terms of accuracy, the CS-αRPIM-Tr4 performs the best among these five methods measured in displacement norm. It is 5.5 times and 2 times more accurate than the FEM-Tr3 and CS-RPIM-Tr4-Cd respectively. It is even 1.4 times more accurate than the ES-PIM-Tr3.

TABLE 9.11 Estimated computational efficiency of different methods measured in *displacement norm* error for the numerical results for the infinite solid with hole problem with the same set of triangular mesh (Mesh-4 of 3,578 nodes)

Numerical method	Solution error	Error ratio to the FEM-Tr3	Efficiency
FEM-Tr3	1.4435E-03	1.00	1.0
NS-PIM-Tr3	1.1520E-03	0.80	1.0
ES-PIM-Tr3	3.7620E-04	0.26	3.5
CS-RPIM-Tr4-Cd	5.1579E-04	0.36	2.5
CS-αRPIM-Tr4	2.6349E-04	0.18	3.7

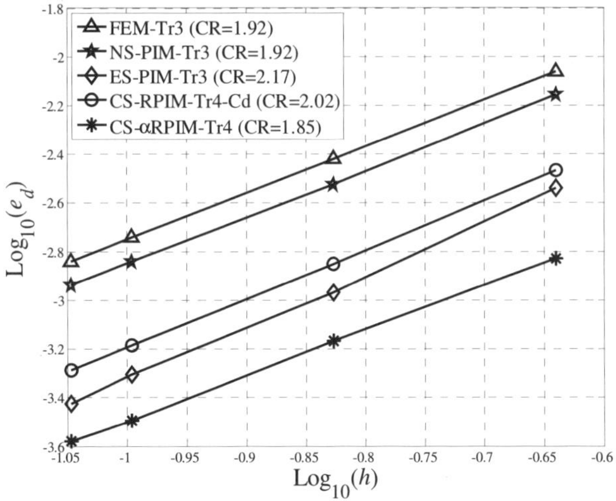

FIGURE 9.7 Comparison of convergence rates of the numerical results in displacement norm obtained using different methods for the problem of infinite solid with circular hole.

4) The convergence rate of the CS-αRPIM-Tr4 is a little lower than the other models. This is also partially because of the good accuracy for coarse meshes. When the mesh is refined, the accuracy cannot increase very fast.

5) The CS-αRPIM-Tr4 stands out clear in this case, closely followed by ES-PIM-Tr3.

FIGURE 9.8 plots the solution errors of the numerical results in energy norm for meshes of various node densities. The errors in the strain energy norm for the numerical results obtained using the same set of triangular mesh (Mesh-4) are listed in TABLE 9.12. Considering the computational cost given in TABLE 9.5, the estimated computational efficiency measured in energy norm measure are calculated and given in the last column in TABLE 9.12. The following points may be noted for this example.

1) The CS-αRPIM-Tr4 is about 4 times more efficient than the standard base model FEM-Tr3 and the CS-RPIM-Tr4-Cd model. It performed even better than the "star" performer ES-PIM-Tr3 by about 30% for this problem in energy norm measure.

2) Results of the CS-αRPIM-Tr4 are 5 times more accurate than that of the FEM-Tr3 and the CS-RPIM-Tr4-Cd. It is 1.8 times more accurate than the counterpart of ES-PIM-Tr3.

3) The rate of convergence of the CS-αRPIM-Tr4 is much higher than the FEM-Tr3 and CS-RPIM-Tr4-Cd, but lower than ES-PIM-Tr3 and NS-PIM-Tr3.

4) The CS-αRPIM-Tr4 stands out for this case.

TABLE 9.12 Estimated computational efficiency of different methods measured in *energy norm* error for the numerical results for the infinite solid with hole problem with the same set of triangular mesh (Mesh-4 of 3,578 nodes)

Numerical method	Solution error	Error ratio to FEM-Tr3	Efficiency
FEM-Tr3	1.3831E-04	1.00	1.00
NS-PIM-Tr3	5.0516E-05	0.37	2.3
ES-PIM-Tr3	4.2694E-05	0.31	2.9
CS-RPIM-Tr4-Cd	1.2637E-04	0.91	1.00
CS-αRPIM-Tr4	2.3601E-05	0.17	3.9

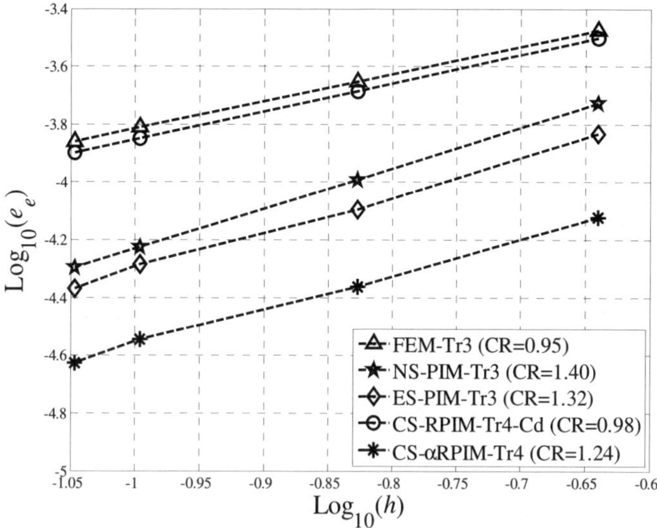

FIGURE 9.8 Comparison of convergence rates of the numerical results in energy norm obtained using different methods for the problem of infinite solid with circular hole.

The convergence process of the strain energy solution for the CS-αRPIM-Tr4 model is shown in FIGURE 9.9, together with the FEM-Tr3, NS-PIM-Tr3, ES-PIM-Tr3 and CS-RPIM-Tr4-Cd models. The errors in the strain energy solution for the numerical results obtained using the same set of triangular mesh (Mesh-4) are listed in TABLE 9.13. Considering the computational cost given in TABLE 9.5, the estimated computational efficiency measured in strain energy solution are calculated and given in the last column in TABLE 9.13. We found that

1) the CS-αRPIM-Tr4 is about 9 times more efficient than the standard base model FEM-Tr3 and about 2.5 times than the CS-RPIM-Tr4-Cd model. It is even 3 times more efficient than the ES-PIM-Tr3 for this problem in strain solution;

2) the CS-αRPIM-Tr4 gives a very tight upper bound solution, which is much closer to the exact one than other methods, and

3) the CS-αRPIM-Tr4 stands out clear in this case, showing much better performance than other methods.

In summary, for this case with exact solutions in trigonometric function forms, the CS-αRPIM-Tr4 is the best in terms of efficiency. It is even better than the "star" performer ES-PIM-Tr3. Compared to the NS-PIM-Tr3, the CS-αRPIM-Tr4 provides a much tighter upper bound solution, which is very close to the exact one.

TABLE 9.13 Estimated computational efficiency of different methods measured in the error in *strain energy solution* of the numerical results for the infinite solid with hole problem with the same set of triangular mesh (Mesh-4 of 3,578 nodes)

Numerical method	Strain energy Solution	Error[*] (%)	Error ratio to FEM-Tr3	Efficiency
FEM-Tr3	4.3230E-05	-0.0495	1.00	1.0
NS-PIM-Tr3	4.3265E-05	0.0314	0.63	1.3
ES-PIM-Tr3	4.3245E-05	-0.0148	0.30	3.0
CS-RPIM-Tr4-Cd	4.3257E-05	0.0130	0.26	3.5
CS-αRPIM-Tr4	4.3253E-05	0.0037	0.075	8.9

[*] The analytical value of the strain energy for the infinite solid with hole problem is 4.32513989E-05.

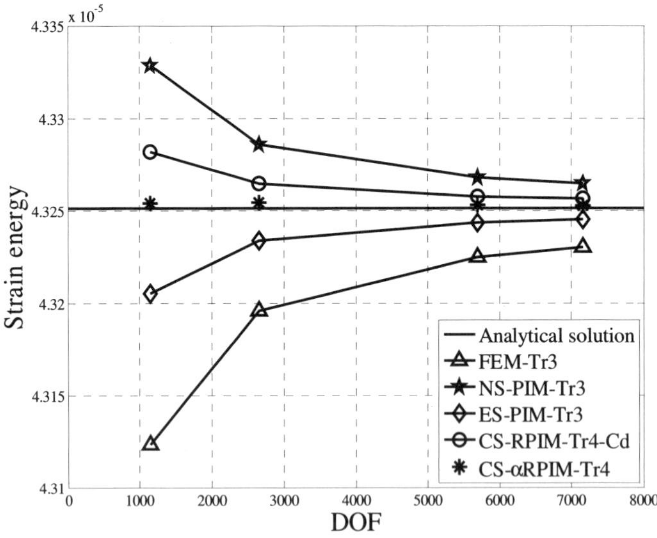

FIGURE 9.9 Converging process of the numerical results (in strain energy norm) for the problem of infinite plate with circular hole.

Example 9.4.4 A mechanical part: 2D connecting rod

The rim problem described in Example 6.4.4 is studied here using the CS-αRPIM-Tr4 model.

First we found α_{exact} =0.2778 at the intersection of two strain energy curves using two irregular meshes with the same aspect ratio (Mesh-1 of 319 nodes and Mesh-2 of 1,115 nodes), as shown in FIGURE 9.10. This particular value of α will be used in the following study for the rim problem.

Compared to the previous two cases with exact strain energy, we may find that the intersection of the two curves is a little far from the reference strain energy for this case. As we mentioned in Example 6.4.4, the reference strain energy is obtained using the FEM with a very fine mesh of six-node triangular elements and hence it is a lower bound of the exact one. Therefore, the exact solution should be higher than the reference one and will be closer to the intersection.

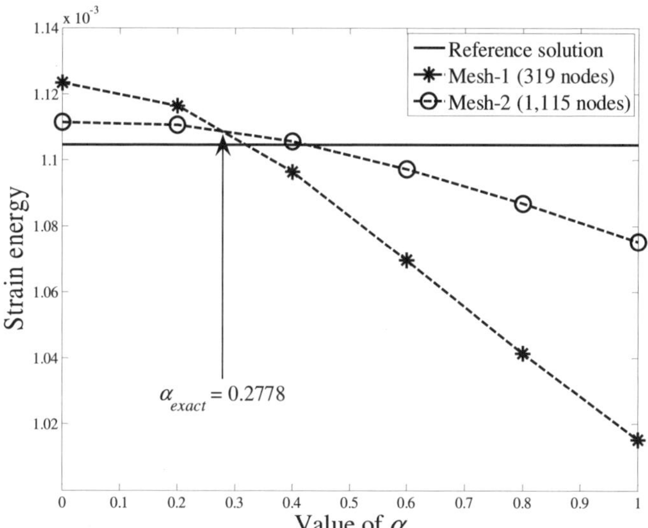

FIGURE 9.10 The strain energy curves of the CS-αRPIM-Tr4 solutions using two meshes with the same aspect ratio interest at α_{exact}=0.2778 for the connecting rod problem.

Using the same sets of triangular meshes shown in FIGURE 6.46, this problem is studied using different methods and the convergence processes of computed strain energy of the numerical results are plotted in FIGURE 9.11. The errors in the strain energy solution for the numerical results obtained using the same set of triangular mesh (Mesh-4) are listed in TABLE 9.14. Considering the computational cost given in TABLE 9.5, the estimated computational efficiency measured in strain energy solution are calculated and given in the last column in TABLE 9.14. For this mechanical model with complicated shape, we found that

1)   the CS-αRPIM-Tr4 is about 7.9 times more efficient than the standard base FEM-Tr3 but a little less efficient than the CS-RPIM-Tr4-Cd, and

2)   the CS-αRPIM-Tr4 gives a very tight upper bound solution for this case.

TABLE 9.14 Estimated computational efficiency of different methods measured in the error in *strain energy solution* of the numerical results for the connecting rod problem with the same set of triangular mesh (Mesh-4 of 2,155 nodes)

Numerical method	Strain energy Solution	Error[*] (%)	Error ratio to FEM-Tr3	Efficiency
FEM-Tr3	1.0814E-03	-2.105	1.00	1.0
NS-PIM-Tr3	1.1165E-03	1.073	0.51	1.6
ES-PIM-Tr3	1.0955E-03	-0.828	0.39	2.3
CS-RPIM-Tr4-Cd	1.1070E-03	0.213	0.10	9.1
CS-αRPIM-Tr4	1.1066E-03	0.177	0.084	7.9

[*] The reference value of the strain energy for the connecting rod problem is 1.1046479E-03, which is obtained using FEM with a very fine mesh (six-node triangular mesh of 192,420 elements and 387,977 nodes).

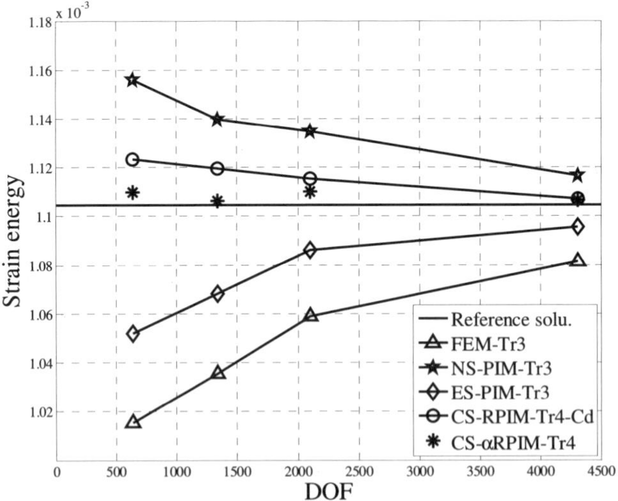

FIGURE 9.11 Converging process of the numerical results (in strain energy norm) for the connecting rod problem.

9.5 Numerical examples for 3D solids

Example 9.5.1 A 3D linear patch test

The 3D linear patch test described in Example 6.3.1 is again studied using the CS-αRPIM-Te5 model. Using different α values, TABLE 9.15 lists the displacement norm errors (defined in Equation (5.85)) of the CS-αRPIM-Te5 solutions for the patch test using both the regular and irregular meshes shown in FIGURE 6.30. The CS-αRPIM-Te5 with all these α values can pass the patch test exactly (to the machine accuracy). Note that the displacement field in the CS-αRPIM-Te5 model (with $\alpha\neq1$) is incompatible, but it is linearly conforming and capable to reproduce exactly (to the machine accuracy) linear field. This confirms Remark 9.7 for 3D problems. We note again that because of the use of the RBFs in the RPIM shape functions, a loss about up to 3 digits accuracy is observed. This is due to the large condition number in the moment matrix in creating the RPIM shape functions [1].

TABLE 9.15 Error norm in displacements of numerical results for the 3D standard patch test obtained using the CS-αRPIM-Te5 model

	$\alpha=0.0$	$\alpha=0.2$	$\alpha=0.4$	$\alpha=0.6$	$\alpha=0.8$	$\alpha=1.0$
Regular mesh	3.89E-12	1.65E-12	6.29E-13	2.44E-13	8.09E-14	3.34E-16
Irregular mesh	1.89E-12	1.45E-12	8.61E-13	4.24E-13	1.58E-13	3.59E-16

Example 9.5.2 A 3D Lame problem

The 3D Lame problem with known exact solution described in Example 6.3.2 is now studied again using the CS-αRPIM-Te5 model.

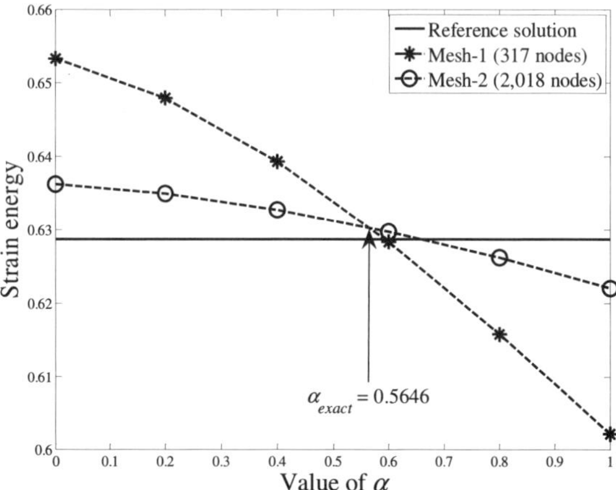

FIGURE 9.12 The strain energy curves of the CS-αRPIM-Te5 solutions using two meshes with the same aspect ratio interest at α_{exact}=0.5646 for the Lame problem.

First, we find $\alpha_{exact}=0.5646$ at the intersection of two strain energy curves using two irregular meshes with the same aspect ratio (Mesh-1 of 317 nodes and Mesh-2 of 2,018 nodes), as shown in FIGURE 9.12. This particular value of α will be used in the following further study for the Lame problem.

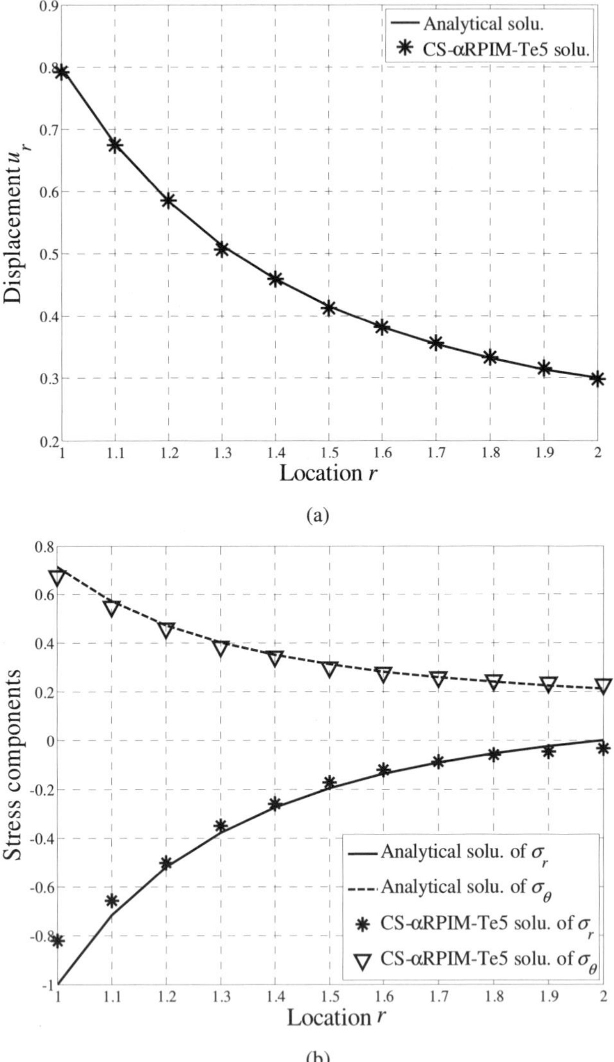

FIGURE 9.13 Displacement and stress distribution along the x-axis of the one-eighth model of the 3D Lame problem: (a) radial displacement; (b) radial and tangential stress components.

The problem domain (one-eighth model) is presented using linear tetrahedral cells with a total of 1,138 irregularly distributed nodes, as shown in FIGURE 6.31b, and the computed displacement and stress components along the x-axis are plotted in FIGURE 9.13. It is found that the CS-αRPIM-Te5 solutions agree well with the analytical solutions in terms of both displacement and stress components.

To investigate the convergence property of the CS-αRPIM-Te5, four models of 173, 317, 729, and 1,304 irregularly distributed nodes are used. For each of these four models the errors of the numerical results in both displacement and energy norms are computed. For comparison purpose, the FEM-Te4, the NS-PIM-Te4, the FS-PIM-Te4 and CS-RPIM-Te5-Cd are also employed in this study, using the same sets of tetrahedral meshes.

The convergence of the solution errors in displacement norm for the Lame problem is plotted in FIGURE 9.14. The errors in the displacement norm for the numerical results obtained using the same set of triangular mesh (Mesh-4) are listed in TABLE 9.16. Taking the computational cost given in TABLE 9.6 into consideration for these methods, the estimated computational efficiency measured in displacement error are calculated and listed in the last column in TABLE 9.16. We can now clearly observe the following.

1) The CS-αRPIM-Te5 is about 1.6 times more efficient than the standard base model FEM-Te4, and as same as the CS-RPIM-Te5-Cd and FS-PIM-Te4 for this problem in displacement norm measure.

2) In terms of accuracy, the CS-αRPIM-Te5 is 2.3 times more accurate than the FEM-Te4, is 1.2 times more accurate than the CS-RPIM-Te5-Cd, and is the best among these five models.

3) The rate of convergence of the CS-αRPIM-Te5 is about the same as CS-RPIM-Te5-Cd, and is among the highest ones.

4) The CS-αRPIM-Te5 performs best in this case.

TABLE 9.16 Estimated computational efficiency of different methods measured in *displacement norm* error for the numerical results for the Lame problem with the same set of tetrahedral mesh (Mesh-4 of 1,304 nodes)

Numerical method	Solution error	Error ratio to the FEM-Te4	Efficiency
FEM-Te4	1.8381E-02	1.00	1.00
NS-PIM-Te4	3.5123E-02	1.91	0.35
FS-PIM-Te4	1.1838E-02	0.64	1.6
CS-RPIM-Te5-Cd	9.6249E-03	0.52	1.6
CS-αRPIM-Te5	8.0086E-03	0.44	1.6

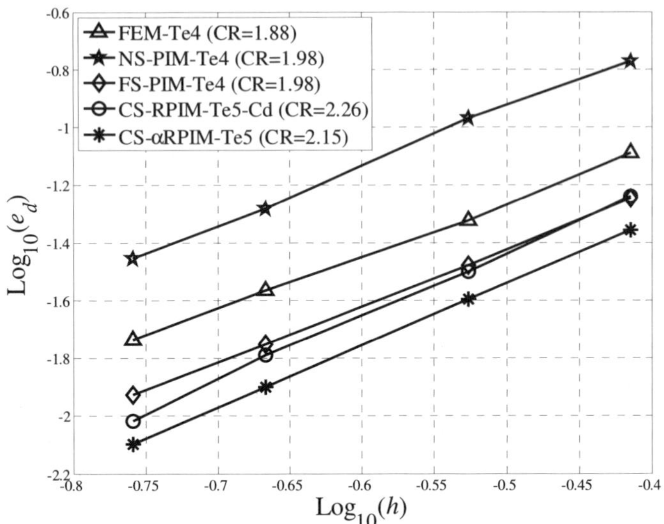

FIGURE 9.14 Comparison of convergence rates of the numerical results in displacement norm obtained using different methods for the 3D Lame problem.

FIGURE 9.15 shows the convergence of the solution errors in energy norm (defined in Equation (5.94)) for different methods. The errors in the energy norm for the numerical results obtained using the same set of triangular mesh (Mesh-4) are listed in TABLE 9.17. Considering the computational cost given in TABLE 9.6 and TABLE 8.15 for these methods, the estimated computational efficiency measured in energy norm measure is calculated and given in the last column in TABLE 9.17. We note the following.

1) The CS-αRPIM-Te5 is about 20% more efficient than the standard base model FEM-Te4. However, it is about 40% less efficient than the CS-RPIM-Te5-Cd.

2) The convergence rate of the CS-αRPIM-Te5 is a little higher than the FEM-Te4 and FS-PIM-Te4, but is lower than the NS-PIM-Te4 and CS-RPIM-Te5-Cd.

3) The CS-RPIM-Te5-Cd performs best in this case, followed by CS-αRPIM-Te5.

TABLE 9.17 Estimated computational efficiency of different methods measured in *energy norm* error for the numerical results for the Lame problem with the same set of tetrahedral mesh (Mesh-4 of 1,304 nodes)

Numerical method	Solution error	Error ratio to the FEM-Te4	Efficiency
FEM-Te4	1.5008E-01	1.00	1.0
NS-PIM-Te4	7.7785E-02	0.52	1.3
FS-PIM-Te4	1.2310E-01	0.82	1.2
CS-RPIM-Te5-Cd	6.8244E-02	0.45	1.9
CS-αRPIM-Te5	9.1704E-02	0.61	1.2

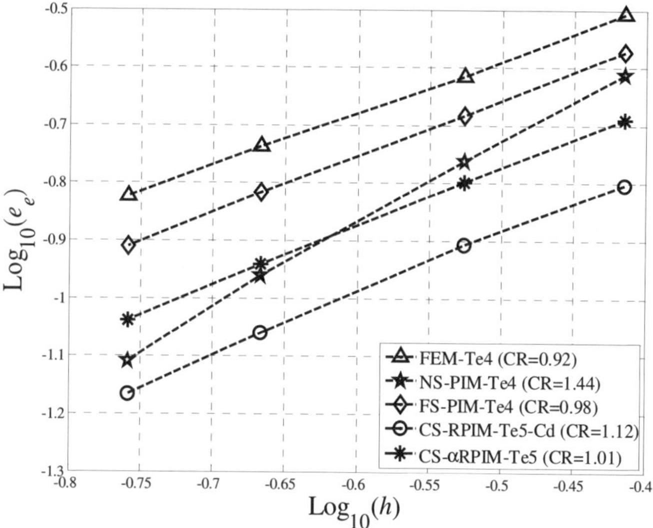

FIGURE 9.15 Comparison of accuracy and convergence rates of the numerical results in energy norm obtained using different numerical models for the 3D Lame problem.

The bound properties of the CS-αRPIM-Te5 solutions are also studied and the results are plotted in FIGURE 9.16, together with the FEM-Te4, NS-PIM-Te4, FS-PIM-Te4 and CS-RPIM-Te5-Cd models. The errors in the strain energy solution for the numerical results obtained using the same set of tetrahedral mesh (Mesh-4) are listed in TABLE 9.18. Considering the computational cost given in TABLE 9.6 for these methods, the estimated computational efficiency measured in strain energy solution is given in the last column in TABLE 9.18. We found for this case the following.

1)　The CS-αRPIM-Te5 is about 47.6 times more efficient than the standard base model FEM-Te4 and about 40 times than the CS-RPIM-Te5-Cd.

2)　The CS-αRPIM-Te5 solution is very close to the exact solution as shown in FIGURE 9.16.

3)　The CS-αRPIM-Te5 stands out distinctively in this case.

In summary, for this 3D Lame problem with exact solutions, the CS-αRPIM-Te5 is the best in overall efficiency. It loses by 40% only to the CS-RPIM-Te5-Cd when measured in energy norm.

TABLE 9.18　Estimated computational efficiency of different methods measured in the *error in strain energy* solution of the numerical results for the Lame problem with the same set of tetrahedral mesh (Mesh-4 of 1,304 nodes)

Numerical method	Strain energy Solution	Error* (%)	Error ratio to FEM-Te4	Efficiency
FEM-Te4	6.1701E-01	-1.86	1.00	1.0
NS-PIM-Te4	6.5268E-01	3.81	2.0	0.33
FS-PIM-Te4	6.2354E-01	-0.82	0.44	2.3
CS-RPIM-Te5-Cd	6.3680E-01	1.29	0.69	1.2
CS-αRPIM-Te5	6.2853E-01	-0.027	0.015	47.6

* The reference value of the strain energy for the Lame problem is 6.2869698E-01.

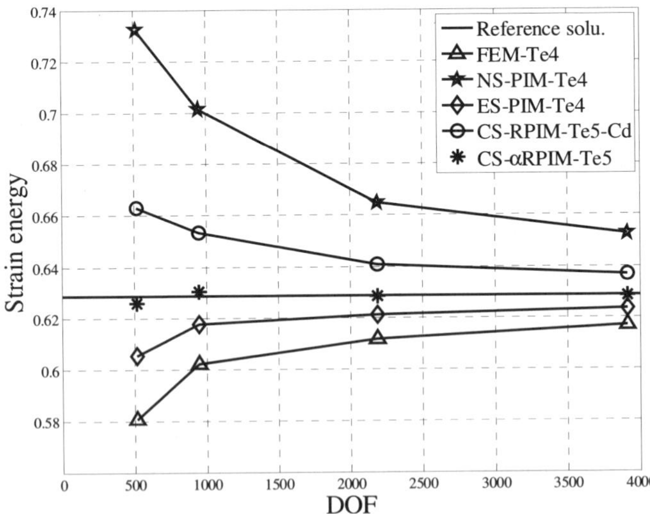

FIGURE 9.16　Bound solutions obtained using different numerical models for the 3D Lame problem.

9.6 Concluding remarks

This chapter presents the novel CS-αRPIM model which is a combination of the "softer" CS-RPIM-Cd and the "stiffer" linear FEM models. We note the following remarks.

Remark 9.8 Both upper and lower bound property

Changing the value of $\alpha \in [0, 1]$, we can get the CS-αRPIM with different softness, hence obtain both upper and lower bound solutions.

Remark 9.9 Superconvergence measured in energy norm

Numerical results obtained using CS-αRPIM models are generally of superconvergence when measured in energy norm.

Remark 9.10 High accuracy

Numerical results obtained using CS-αRPIM models are generally of high accuracy measured in both displacement and energy norms.

Remark 9.11 High efficiency

The CS-αRPIM models have higher efficiency than those two base models: linear FEM and CS-RPIM-Cd, and perform even better than the star performer ES-PIM-Tr3.

Remark 9.12 Close-to-exact strain energy

Computed strain energy of the CS-αRPIM results is very close to the exact one.

Remark 9.13 Search for α_{exact}: an open topic

This chapter presents a very simple and practical procedure to search for the α_{exact}. If a more efficient technique can be devised, the CS-αRPIM model can then be applied in much wider fields, as we can control the softness and hence different properties of the model as we wish. More work is, however, needed in this direction of research and development.

9.7 References

1. Liu, G. R., *Meshfree Methods: Moving beyond the Finite Element Method*, 2nd Edition, CRC press, Boca Taton, USA, 2009.

2. Liu, G. R., Zhang, G. Y., Dai, K. Y, Wang, Y. Y., Zhong, Z. H., Li, G. Y. and Han, X., A linearly conforming point interpolation method (LC-PIM) for 2D solid mechanics problems, *International Journal of Computational Methods*, 2(4): 645-665, 2005.

3. Zhang, G. Y., Liu, G. R., Wang, Y. Y., Huang, H. T., Zhong, Z. H., Li, G. Y. and Han, X., A linearly conforming point interpolation method (LC-PIM) for three-dimensional elasticity problems, *International Journal for Numerical Methods in Engineering*, 72: 1524-1543, 2007.

4. Liu, G. R., Li, Y., Dai, K. Y., Luan M.T. and Xue, W., A Linearly conforming radial point interpolation method for solid mechanics problems, *International Journal of Computational Methods*, 3: 401-428, 2006.

5. Liu, G. R. and Zhang, G. Y., Upper bound solution to elasticity problems: A unique property of the linearly conforming point interpolation method (LC-PIM). *International Journal for Numerical Methods in Engineering*, 74: 1128-1161, 2008.

6. Zhang, G. Y., Liu, G. R., Nguyen-Thoi, T., Song, C. X., Han, X., Zhong, Z. H. and Li, G. Y., The upper bound property for solid mechanics of the linearly conforming radial point interpolation method (LC-RPIM). *International Journal of Computational Methods*, 4(3): 521−541, 2007.

7. Liu, G. R., Zhang, G. Y., Edge-based smoothed point interpolation methods, *International Journal of Computational Methods*, 5(4): 621-646, 2008.

8. Liu, G. R., Nguyen-Thoi, T. and Lam, K. Y. An edge-based smoothed finite element method (ES-FEM) for static, free and forced vibration analyses in solids. *Journal of Sound and Vibration*, 320: 1100-1130, 2009.

9. Nguyen-Thoi, T., Liu, G. R., Lam, K. Y. and Zhang, G. Y., A Face-based Smoothed Finite Element Method (FS-FEM) for 3D linear and nonlinear solid mechanics problems using 4-node tetrahedral elements. *International Journal for Numerical Methods in Engineering*, 78: 324-353, 2009.

10. Liu, G. R., Wang, Z., Zhang, G. Y., Zong, Z. and Wang, S., An edge-based smoothed point interpolation method for material discontinuity. *Mechanics of Advanced Materials and Structures*, 19(1-3): 3-17, 2012.

11. He, Z. C., Cheng, A. G., Zhang, G. Y., Zhong, Z. H. and Liu, G. R., Dispersion error reduction for acoustic problems using the edge-based smoothed finite element method (ES-FEM). *International Journal for Numerical Methods in Engineering*, 86: 1322-1338, 2011.

12. Nguyen-Thoi, T., Liu, G. R., Vu-Do, H. C. and Nguyen-Xuan, H., A face-based smoothed finite element method (FS-FEM) for visco-elastoplastic analyses of 3D solids using tetrahedral mesh. *Computer Methods in Applied Mechanics and Engineering*, 198: 3479-3498, 2009.

13. Liu, G. R., Zhang, G. Y. A normed G space and weakened weak (W^2) formulation of a cell-based smoothed point interpolation method. *International Journal of Computational Methods*, 6(1): 147-179, 2009.

14. Zhang, G. Y. and Liu, G. R., Meshfree cell-based smoothed point interpolation method using isoparametric PIM shape functions and condensed RPIM shape functions. *International Journal of Computational Methods*, 8(4): 705-730, 2011.

15. Liu, G. R., Jiang, Y., Chen, L., Zhang, G. Y. and Zhang, Y. W., A singular cell-based smoothed radial point interpolation method for fracture problems. *Computers & Structures*, 89(13-14): 1378-1396, 2011.

16. Liu, G. R., Nguyen-Thoi, T., Lam, K. Y., A novel FEM by scaling the gradient of strains with factor α (αFEM). *Computational Mechanics*, 43: 369-391, 2009.

17. Liu, G. R., Nguyen-Thoi, T., Lam, K. Y. A novel Alpha Finite Element Method (αFEM) for exact solution to mechanics problems using triangular and tetrahedral elements. *Computer Methods in Applied Mechanics and Engineering*, 197: 3883-3897, 2008.

18. Liu, G. R. and Nguyen-Thoi, T., *Smoothed Finite Element Methods*, CRC press, Boca Taton, USA, 2010.

19. Liu, G. R., Zhang, G. Y., Zong, Z. and Li, M., Meshfree cell-based smoothed alpha radial point interpolation method (CS-αRPIM) for solid mechanics problems. *International Journal of Computational Methods*, 10(4), 31 pages, DOI: 10.1142/S0219876213500205, 2013.

Chapter 10

Strain-constructed Point Interpolation Method (SC-PIM)

In the previous chapters, we have introduced NS-PIMs, ES-PIMs, CS-PIM and CS-αPIMs, which are W^2 models based on the GS-Galerkin formulation [1-8]. It is in fact a special and the simplest strain-constructed model. In this chapter, we present more general SC-models using PIM shape functions known as SC-PIM, based on the general SC-Galerkin formulation [1, 9-11]. We shall take much bolder approaches to construct the strain fields by simple means of point interpolation using smoothed and compatible strains at critical points over the node-based, edge-based and/or cell-based smoothing domains.

We present and examine a total of six novel schemes for constructing strain fields, based on a set of triangular integration cells. All these schemes are performed under conditions of 1) no additional degrees of freedom are added in any way in the SC-model compared to the FEM model; 2) the formulation procedure should remain as simple as those in the standard FEM; and 3) the model is stable and the solutions converge to the exact solution. All these schemes proposed will be studied in great detail, via mainly numerical means, and a number of strain-constructed point interpolation method (SC-PIM) models are found with excellent performances. The theoretical proof on the stability and convergence of the SC-Galerkin models has been given in Chapters 4 and 5. This chapter considers only 2D problems.

10.1 Formulation of SC-PIM

The formulation of SC-PIM is briefly introduced, following the work presented in [9].

10.1.1 Displacement field construction

Same as any S-PIM models, in the SC-PIM the problem domain is first represented by N_c nonoverlapping and seamless (NOSL) triangular background cells with N_n scattered field nodes, such that $\overline{\Omega} = \bigcup_{i=1}^{N_c} \overline{\Omega}_i^c$ and $\Omega_i^c \cap \Omega_j^c = \varnothing$, $\forall i \neq j$. Based on the triangular background cells, a small number of field nodes in a local support domain can then be selected for constructing PIM shape functions [12-15], as presented in Chapter 3. T-schemes detailed in Section 1.6.3 are used to perform the node selection. For various SC-PIM models presented in this chapter, the points of interests (the quadrature points) are always located inside the triangular background cells or on the boundaries of the problem domain. Therefore, cell-based T-schemes are used to select nodes for constructing polynomial PIM shape functions. When the cell-based T3-scheme is used we have linear PIM shape functions; and when the cell-based T6/3-scheme is used we obtain quadratic PIM shape functions. Using a set of small number of local support nodes selected using a proper cell-based T-scheme, PIM shape functions can then be created using the techniques detailed in Section 3.2.

Note that only polynomial PIM shape functions are used in this chapter to approximate the displacement field for our SC-PIM models. The formulation can be easily extended to the models by using RPIM shape functions.

10.1.2 Strain field construction

Strain field construction in SC-PIM models follows largely the steps outlined in Section 4.5.3.

1) Creation of quadrature/integration cells.

Based on the triangular background cells, a set of NOSL triangular integration cells are then constructed, such that $\overline{\Omega} = \bigcup_{i=1}^{N_q} \overline{\Omega}_i^q$ and $\Omega_i^q \cap \Omega_j^q = \varnothing$, $\forall i \neq j$. The integration cells are usually created in nested fashion by subdividing the background cells into smaller cells in various ways. Such a nested division helps to satisfy these conditions of strain norm equivalence and strain convergence (see Section 4.4).

2) Determine the strain $\breve{\varepsilon}$ at the vertices of the integration cells.

These strains are either the compatible strains $\tilde{\varepsilon}$ (from a continuous displacement field) or the smoothed strains $\overline{\varepsilon}$ obtained using a local smoothing domain [2, 6-8]. This also helps to make sure the strain norm equivalence and strain convergence conditions.

3) Construct the strain field $\hat{\varepsilon}$ in the quadrature cell by interpolation.

In this book, we use only the simplest linear interpolation to construct the strain field over each integration cell [9-11, 16], which is

$$\hat{\varepsilon}(\mathbf{x}) = \sum_{i=1}^{3} \mathbf{\Phi}_i(\mathbf{x})\breve{\varepsilon}_i \tag{10.1}$$

where $\mathbf{\Phi}_i$ is a diagonal matrix of linear PIM shape functions and $\breve{\varepsilon}_i$ is the strain at the vertices of the integration cells.

As triangular quadrature/integration cells are used, we can use the *area coordinates* in the process of integration of weak form, which facilitates easy integration and hence no numerical integration is needed, as discussed in Section 4.5.3. Based on the above steps, we design the following schemes for the strain field construction.

10.1.2.1 Scheme A (1 to 1 division)

For Scheme A, each triangular cell is used as a single integration cell, as shown in FIGURE 10.1. Thus the total number of the integration (quadrature) cells N_q equals the number of background cells N_c : a simple 1 to 1 division. In each triangular integration cell, these three strains at the vertices are obtained using node-based smoothing domains which are the same as those in the NS-PIM, except that the strains obtained are only for the node (not for the entire smoothing domain). The strain field in an integration cell is linearly interpolated using

$$\begin{aligned}
\hat{\varepsilon}(\mathbf{x}) &= \sum_{i=1}^{3} \mathbf{\Phi}_i(\mathbf{x})\breve{\varepsilon}_i \\
&= \sum_{i=1}^{3} \mathbf{\Phi}_i(\mathbf{x})\overline{\varepsilon}_{ni}
\end{aligned} \tag{10.2}$$

where $\bar{\varepsilon}$ is the smoothed strain with the subscript *"n"* standing for node-based smoothing strain. The strain energy potential can then be obtained using Equation (4.32). Note that the strain field constructed in Scheme A is continuous in the entire problem domain.

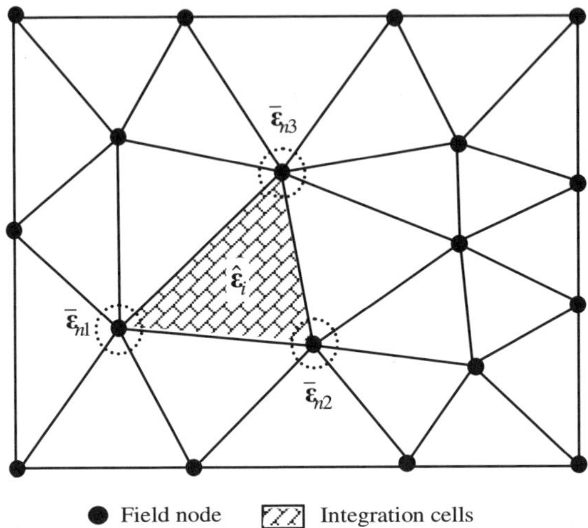

FIGURE 10.1 Integration cell *i* over which the strain field is constructed. Scheme A: each triangular background cell is used as a single integration cell and the strain field over it is constructed by linear interpolation with the node-based smoothed strains at these three vertices.

10.1.2.2 Scheme B (1 to 3 division)

FIGURE 10.2 shows the Scheme B for the construction of integration cells, where a triangular background cell is divided into three integration cells, thus $N_q = 3N_c$: a 1 to 3 division. Each integration cell is formed by connecting two end-points of an edge and the centroid of the neighboring triangle.

The strain fields within each integration cell are calculated using Equation (10.1). There are three editions for Scheme B that are listed in TABLE 10.1, where $\bar{\varepsilon}_e$ refers to smoothed strain obtained for the mid-edge point of the edge using the edge-based smoothing domains as those in the ES-PIM. Note that the strain field constructed in Scheme B2 is continuous in the entire problem domain, while for Scheme B1 and B3, the constructed strain field is not continuous and the discontinuity will occur at the field nodes and the centroid of the background triangles, respectively.

TABLE 10.1 Strains used for forming the constructed strain field: Scheme B

Scheme B	Strain at vertex 1	Strain at vertex 2	Strain at vertex 3
Scheme B1	$\breve{\varepsilon}_1 = \tilde{\varepsilon}$	$\breve{\varepsilon}_2 = \overline{\varepsilon}_e$	$\breve{\varepsilon}_3 = \overline{\varepsilon}_e$
Scheme B2	$\breve{\varepsilon}_1 = \tilde{\varepsilon}$	$\breve{\varepsilon}_2 = \overline{\varepsilon}_{n2}$	$\breve{\varepsilon}_3 = \overline{\varepsilon}_{n3}$
Scheme B3	$\breve{\varepsilon}_1 = \overline{\varepsilon}_e$	$\breve{\varepsilon}_2 = \overline{\varepsilon}_{n2}$	$\breve{\varepsilon}_3 = \overline{\varepsilon}_{n3}$

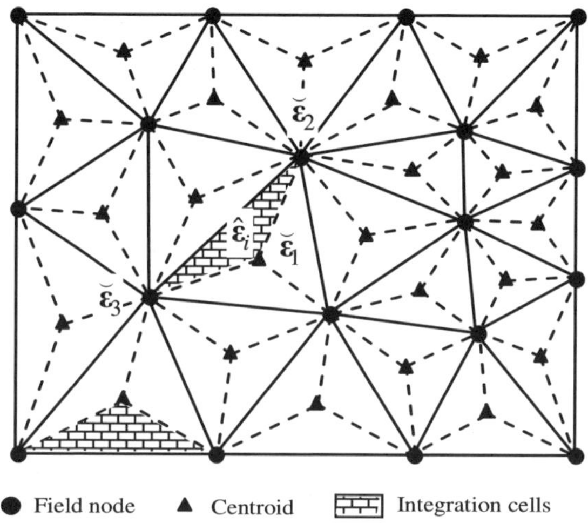

FIGURE 10.2 Integration cell *i* over which the strain field is constructed. Scheme B: a triangular background cell is further divided into three integration cells, and each integration cell is formed by connecting two vertices and the centroid of the background cell. Strain field within an integration cell is linearly interpolated.

10.1.2.3 Scheme C (1 to 4 division)

Scheme C is shown in FIGURE 10.3. Each triangular element is further divided into four integration cells, thus $N_q = 4N_c$: a 1 to 4 division. For each integration cell, the constructed strain fields are formed via linear interpolation using Equation (10.1) with node-based smoothed strains ($\overline{\varepsilon}_n$) and edge-based smoothed strains ($\overline{\varepsilon}_e$). The edge-based smoothed strains are obtained using edge-based smoothing domains as those in the ES-PIM, except that the strains obtained are only for the midpoint of the edge. The assignments of strains at the vertices of each integration cell are listed in TABLE 10.2. Note that the strain field constructed in Scheme C is also continuous in the entire problem domain.

TABLE 10.2 Strains used for forming the constructed strain field: Scheme C

Four integration cells in each triangular background cell	Smoothed strains used for strain field construction in each integration cell
Integration cell ($\hat{\varepsilon}_1$)	$\breve{\varepsilon}_1 = \bar{\varepsilon}_{n1}, \breve{\varepsilon}_2 = \bar{\varepsilon}_{e1}, \breve{\varepsilon}_3 = \bar{\varepsilon}_{e3}$
Integration cell ($\hat{\varepsilon}_2$)	$\breve{\varepsilon}_1 = \bar{\varepsilon}_{n2}, \breve{\varepsilon}_2 = \bar{\varepsilon}_{e2}, \breve{\varepsilon}_3 = \bar{\varepsilon}_{e1}$
Integration cell ($\hat{\varepsilon}_3$)	$\breve{\varepsilon}_1 = \bar{\varepsilon}_{n3}, \breve{\varepsilon}_2 = \bar{\varepsilon}_{e3}, \breve{\varepsilon}_3 = \bar{\varepsilon}_{e2}$
Integration cell ($\hat{\varepsilon}_4$)	$\breve{\varepsilon}_1 = \bar{\varepsilon}_{e1}, \breve{\varepsilon}_2 = \bar{\varepsilon}_{e2}, \breve{\varepsilon}_3 = \bar{\varepsilon}_{e3}$

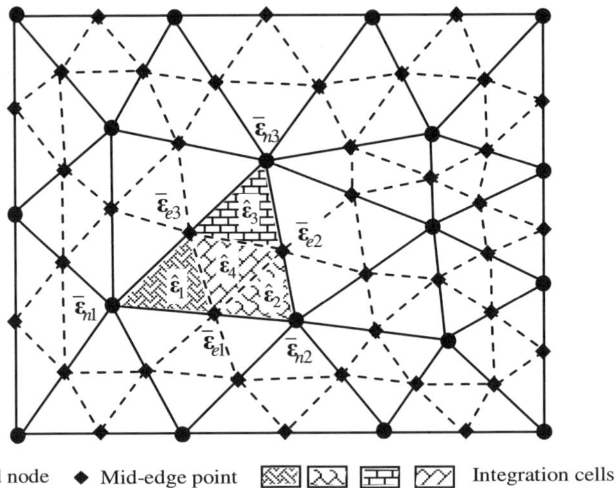

FIGURE 10.3 Four integration cells over which the strain field is constructed. Scheme C: each triangular element is further divided into four integration cells and the strain fields in each integration cell are linearly interpolated using the smoothed strains associated with the nodes and edges.

10.1.2.4 Scheme D (1 to 6 division)

The construction of integration cells in Scheme D is illustrated in FIGURE 10.4, where a triangular cell is further divided into six triangular integration cells, thus $N_q = 6N_c$: a 1 to 6 division. Each of them is formed by connecting the field node, the centroid of the triangle and the corresponding mid-edge point. The strain field within each integration cell is linearly interpolated using Equation (10.1) with the three strains at the three vertices which are the smoothed strain or compatible strains:

$$\breve{\boldsymbol{\varepsilon}}_1 = \overline{\boldsymbol{\varepsilon}}_n, \ \breve{\boldsymbol{\varepsilon}}_2 = \tilde{\boldsymbol{\varepsilon}}, \ \breve{\boldsymbol{\varepsilon}}_3 = \overline{\boldsymbol{\varepsilon}}_e \tag{10.3}$$

Note the constructed strain field in Scheme D is continuous in the entire problem domain.

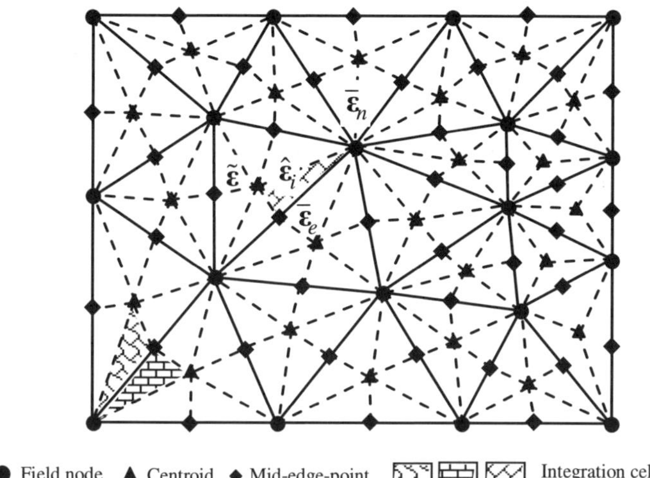

FIGURE 10.4 Integration cell *i* over which the strain field is constructed. Scheme D: a triangular background cell is further divided into six integration cells and each one is formed by connecting the field node, the centroid of the triangular element and the corresponding mid-end-point of the edge. Strain fields within each integration cell are linearly interpolated using the smoothed and the compatible strains associated with the three vertices.

10.1.3 Stiffness matrix for SC-PIM

Substituting the assumed displacement Equation (5.67) and the constructed strain Equation (10.1) into the SC-Galerkin weak form, Equation (5.44), the linear system of equations of the SC-PIM can be finally obtained as

$$\hat{\mathbf{K}}^{\text{SC-PIM}}\hat{\mathbf{d}} = \tilde{\mathbf{f}} \tag{10.4}$$

where $\hat{\mathbf{d}}$ is the nodal displacements for all the nodes in the entire problem domain, the force vector $\tilde{\mathbf{f}}$ is calculated using Equation (5.81) and $\hat{\mathbf{K}}^{\text{SC-PIM}}$ is the stiffness matrix of the SC-PIM. Based on Equation (4.35), the entries of the stiffness matrix can be obtained using

$$
\begin{aligned}
\hat{\mathbf{K}}_{IJ}^{\text{SC-PIM}} &= \int_{\Omega} \hat{\mathbf{B}}_I^{\text{T}} \mathbf{c} \hat{\mathbf{B}}_J \, \mathrm{d}\Omega \\
&= \sum_{k=1}^{N_q} \int_{\Omega_k^q} \hat{\mathbf{B}}_I^{\text{T}} \mathbf{c} \hat{\mathbf{B}}_J \, \mathrm{d}\Omega \\
&= \sum_{k=1}^{N_q} \frac{A_k^q}{6} \left(\breve{\mathbf{B}}_1^{\text{T}} \mathbf{c} \breve{\mathbf{B}}_1 + \breve{\mathbf{B}}_2^{\text{T}} \mathbf{c} \breve{\mathbf{B}}_2 + \breve{\mathbf{B}}_3^{\text{T}} \mathbf{c} \breve{\mathbf{B}}_3 \right. \\
&\qquad\qquad \left. + \breve{\mathbf{B}}_1^{\text{T}} \mathbf{c} \breve{\mathbf{B}}_2 + \breve{\mathbf{B}}_1^{\text{T}} \mathbf{c} \breve{\mathbf{B}}_3 + \breve{\mathbf{B}}_2^{\text{T}} \mathbf{c} \breve{\mathbf{B}}_3 \right)
\end{aligned}
\tag{10.5}
$$

in which $A_k^q = \int_{\Omega_k^q} \mathrm{d}\Omega$ is the area of integration cell Ω_k^q, and the strain-displacement matrix $\breve{\mathbf{B}}$ is either for the compatible strains or for the smoothed strains depending on the scheme used to construct the strain field (see Section 10.1.2).

Note that the sub-matrix of stiffness $\hat{\mathbf{K}}_{IJ}^{\text{SC-PIM}}$ defined in Equation (10.5) needs to be computed only when nodes I and J share a same integration cell. Otherwise, it is zero. Hence, the global stiffness matrix will be sparse for an SC-PIM model, in addition to the obvious symmetry. Based on the discussion given in Section 5.7.6, the global stiffness matrix $\hat{\mathbf{K}}^{\text{SC-PIM}}$ of a SC-PIM model is SPD. The SC-PIM solution will be stable, unique and converge to the exact solution when the cells are refined (together with the integration cells).

Remark 10.1 Features of the global stiffness matrix of SC-PIM models

The global stiffness matrix of a SC-PIM model is SPD and sparse. It is banded if the field nodes are properly numbered as that in the FEM.

10.2 Numerical implementation

10.2.1 Macro flowchart of the SC-PIM

The macro flowchart for constructing the global matrices and vectors of SC-PIM code can be briefly given as follows.

1) Loop over all the integration cells.

2) Obtain the associated compatible strains and smoothed strains at the vertexes of the integration cell.

3) Obtain the constructed strain field by interpolation.

4) Compute the stiffness matrix and force vector of the integration cell.

5) Assemble them to the global matrices and vectors.

6) End the loop on the integration cells.

10.2.2 Possible SC-PIM models

Based on the displacement field constructed using different order of polynomial PIM shape functions and various strain fields constructed, we introduce the following SC-PIM models in this book, as detailed in TABLE 10.3.

TABLE 10.3 Definition of SC-PIM models

SC-PIM models	T-scheme for displacement field construction	Scheme for strain field construction
Scheme A-Tr3	Cell-based T3-scheme	Scheme A
Scheme A-Tr6/3	Cell-based T6/3-scheme	Scheme A
Scheme B1-Tr3	Cell-based T3-scheme	Scheme B1
Scheme B1-Tr6/3	Cell-based T6/3-scheme	Scheme B1
Scheme B2-Tr3	Cell-based T3-scheme	Scheme B2
Scheme B2-Tr6/3	Cell-based T6/3-scheme	Scheme B2
Scheme B3-Tr3	Cell-based T3-scheme	Scheme B3
Scheme B3-Tr6/3	Cell-based T6/3-scheme	Scheme B3
Scheme C-Tr3	Cell-based T3-scheme	Scheme C
Scheme C-Tr6/3	Cell-based T6/3-scheme	Scheme C
Scheme D-Tr3	Cell-based T3-scheme	Scheme D
Scheme D-Tr6/3	Cell-based T6/3-scheme	Scheme D

To find out the rough number of support nodes for each integral cell of SC-PIM model, the cantilever beam problem presented using 248 irregular nodes, as shown in FIGURE 6.4, is studied again. TABLE 10.4 lists the support nodes (the minimum, the maximum and the average number) for the integral cell of each SC-PIM model, in relation to the FEM-Tr3 using the same mesh. We may find that the support nodes in an integral cell for all the SC-PIM models will be more than the counterpart of FEM-Tr3.

TABLE 10.4 SC-PIM schemes and overall characteristics in relation to the FEM-Tr3 model using the same mesh

Numerical method	Support nodes in an integral cell	Estimated average number of nonzero entries	Estimated Solver CPU time
FEM-Tr3	3	n_{ze}	t_{CPU}
Scheme A-Tr3	7-13 (11.1)	$3.7\,n_{ze}$	$3.7\,\bar{n}_{iter}t_{CPU}$
Scheme A-Tr6/3	9-23 (18.1)	$6.0\,n_{ze}$	$6.0\,\bar{n}_{iter}t_{CPU}$
Scheme B1-Tr3	3-4 (3.9)	$1.3\,n_{ze}$	$1.3\,\bar{n}_{iter}t_{CPU}$
Scheme B1-Tr6/3	3-8 (7.3)	$2.4\,n_{ze}$	$2.4\,\bar{n}_{iter}t_{CPU}$
Scheme B2-Tr3	6-11 (9.4)	$3.1\,n_{ze}$	$3.1\,\bar{n}_{iter}t_{CPU}$
Scheme B2-Tr6/3	7-20 (15.7)	$5.2\,n_{ze}$	$5.2\,\bar{n}_{iter}t_{CPU}$
Scheme B3-Tr3	6-11 (9.4)	$3.1\,n_{ze}$	$3.1\,\bar{n}_{iter}t_{CPU}$
Scheme B3-Tr6/3	7-20 (15.7)	$5.2\,n_{ze}$	$5.2\,\bar{n}_{iter}t_{CPU}$
Scheme C-Tr3	4-8 (6.5)	$2.2\,n_{ze}$	$2.2\,\bar{n}_{iter}t_{CPU}$
Scheme C-Tr6/3	4-15 (11.3)	$3.8\,n_{ze}$	$3.8\,\bar{n}_{iter}t_{CPU}$
Scheme D-Tr3	4-8 (6.7)	$2.2\,n_{ze}$	$2.2\,\bar{n}_{iter}t_{CPU}$
Scheme D-Tr6/3	4-15 (11.5)	$3.8\,n_{ze}$	$3.8\,\bar{n}_{iter}t_{CPU}$

10.2.3 Condition number of SC-PIM models

TABLE 10.5 lists the condition numbers of the global stiffness matrixes for various SC-PIM models in relation to the FEM-Tr3 using the same triangular meshes for the cantilever beam problem. Except the condition number of Scheme B1-Tr6/3 is a slightly bigger than the counterpart of the linear FEM model, all other SC-PIM models have smaller (51%-83%) condition numbers than the FEM-Tr3, even though they use much more support nodes.

We can find that the ratio of the condition numbers for different methods in relation to the linear FEM-Tr3 model has little change with the refinement of the mesh. In the following analysis, we will use the relative $\bar{n}_{iter}$ given in last column in TABLE 10.5.

TABLE 10.5 Condition number of the global stiffness matrixes for various SC-PIM schemes for the cantilever beam in relation to the FEM-Tr3 model using the same mesh

	Mesh-1 of 120 nodes		Mesh-2 of 248 nodes	
	$cond(\mathbf{K})$	$\overline{n}_{iter}$	$cond(\mathbf{K})$	$\overline{n}_{iter}$
FEM-Tr3	1.645e+08	1.00	2.399e+08	1.00
Scheme A-Tr3	4.205e+07	0.51	6.241e+07	0.51
Scheme A-Tr6/3	4.458e+07	0.52	8.072e+07	0.58
Scheme B1-Tr3	1.185e+08	0.85	1.733e+08	0.85
Scheme B1-Tr6/3	1.690e+08	1.01	2.695e+08	1.06
Scheme B2-Tr3	5.484e+07	0.58	8.084e+07	0.58
Scheme B2-Tr6/3	6.634e+07	0.64	9.825e+07	0.64
Scheme B3-Tr3	5.253e+07	0.57	7.793e+07	0.57
Scheme B3-Tr6/3	7.139e+07	0.66	1.077e+08	0.67
Scheme C-Tr3	6.748e+07	0.64	1.014e+08	0.65
Scheme C-Tr6/3	1.014e+08	0.79	1.653e+08	0.83
Scheme D-Tr3	7.242e+07	0.66	7.065e+07	0.67
Scheme D-Tr6/3	9.396e+07	0.76	1.387e+08	0.76

10.2.4 Estimation of computational cost for SC-PIM

Combining TABLE 10.4 and TABLE 10.5 we shall have a rough estimation of computational cost for the SC-PIM models in relation to the standard FEM-Tr3 model using the same mesh, as shown in TABLE 10.6. We found that all the SC-PIM models will cost more time than the linear FEM due to more support nodes are used for each integral cell. Compared to the linear SC-PIM models, the quadratic SC-PIM models will cost even more time.

Therefore, in terms of computational cost, the quadratic SC-PIM models are not recommended in the practical analysis.

TABLE 10.6 Estimation of computational cost for SC-PIM schemes in relation to the FEM-Tr3 model using the same mesh

Numerical methods	Support nodes in an integral cell	Estimated Solver CPU time
FEM-Tr3	3	t_{CPU}
Scheme A-Tr3	7-13 (11.1)	$1.9\,t_{CPU}$
Scheme A-Tr6/3	9-23 (18.1)	$3.5\,t_{CPU}$
Scheme B1-Tr3	3-4 (3.9)	$1.1\,t_{CPU}$
Scheme B1-Tr6/3	3-8 (7.3)	$2.5\,t_{CPU}$
Scheme B2-Tr3	6-11 (9.4)	$1.8\,t_{CPU}$
Scheme B2-Tr6/3	7-20 (15.7)	$3.3\,t_{CPU}$
Scheme B3-Tr3	6-11 (9.4)	$1.8\,t_{CPU}$
Scheme B3-Tr6/3	7-20 (15.7)	$3.5\,t_{CPU}$
Scheme C-Tr3	4-8 (6.5)	$1.4\,t_{CPU}$
Scheme C-Tr6/3	4-15 (11.3)	$3.2\,t_{CPU}$
Scheme D-Tr3	4-8 (6.7)	$1.5\,t_{CPU}$
Scheme D-Tr6/3	4-15 (11.5)	$2.9\,t_{CPU}$

10.3 Numerical examples

Several numerical examples are studied in this section to examine these SC-PIM models. The materials used are all linear elastic with Young's modulus E=3.0×10^7 Pa and passion's radio v=0.3, and the units used are based on the international standard unit system. Numerical results are evaluated using the error indicators in displacement and energy norms which are defined in Equation (5.88) and Equation (5.95), respectively.

Example 10.3.1 Linear patch test

The linear patch test described in Example 6.1.1 is again studied using all these SC-PIM schemes listed in TABLE 10.3. Using both the regular and irregular triangular meshes shown in FIGURE 6.5, errors in displacement norm of the numerical results obtained using these SC-PIM schemes are listed in TABLE 10.7. It can be found that all these SC-PIM schemes can pass the linear patch test to machine precision. This example demonstrates numerically that these SC-PIM schemes can reproduce linear fields exactly, and hence with the stability ensured the solution will converge to the exact solution [17, 18]. It also implies that all these models are at least 2nd order accuracy: the error in the displacement will be the terms of 2nd order and higher. We shall then expect a convergence ratio of at least 2.

TABLE 10.7 Error norm in displacements of numerical results for the standard patch test obtained using various schemes of SC-PIM

SC-PIM models	Regular mesh	Irregular mesh
Scheme A-Tr3 (displacement compatible)	1.697E-14	4.497E-15
Scheme A-Tr6/3 (displacement incompatible)	2.279E-14	1.801E-14
Scheme B1-Tr3 (displacement compatible)	9.642E-16	1.707E-15
Scheme B1-Tr6/3 (displacement incompatible)	1.282E-14	1.280E-14
Scheme B2-Tr3 (displacement compatible)	1.870E-15	2.214E-15
Scheme B2-Tr6/3 (displacement incompatible)	4.871E-15	7.664E-15
Scheme B3-Tr3 (displacement compatible)	3.234E-15	2.279E-15
Scheme B3-Tr6/3 (displacement incompatible)	1.305E-14	1.685E-14
Scheme C-Tr3 (displacement compatible)	1.622E-15	2.122E-15
Scheme C-Tr6/3 (displacement incompatible)	1.506E-14	1.595E-14
Scheme D-Tr3 (displacement compatible)	3.976E-15	1.349E-15
Scheme D-Tr6/3 (displacement incompatible)	8.542E-15	1.058E-14

Example 10.3.2 Rectangular cantilever

The benchmark problem of a rectangular cantilever described in Example 6.1.2 is now studied using the SC-PIM schemes with the set of linear triangular meshes shown in FIGURE 6.9. For the purpose of comparison, we study this problem also using other numerical methods, including FEM-Tr3, and three GS-Galerkin models (NS-PIM-Tr3, ES-PIM-Tr3 and CS-PIM-Tr4-CT). The FEM-Tr3 is chosen as the "bottom line": any SC-PIM scheme should perform at least as good as the liner FEM, or has some special properties. The NS-PIM is chosen for comparison, because we know it gives upper bound solutions, and it is the softest GS-Galerkin model fund so far [19-21]. The ES-PIM and CS-PIM are chosen because they have been found as excellent performers that produce very accurate numerical results [7, 8].

FIGURE 10.5 plots the convergence of the solutions in displacement norm for the cantilever solved using various SC-PIM schemes. We can observe the following findings.

1) The solutions of all the SC-PIM schemes converge with reducing nodal spacing at similar convergence rate that is close to the theoretical value of 2.0 for both the linear weak and weakened weak (W^2) formulations (see Remark 3.23).

2) In terms of accuracy, quadratic models (T6/3) of the SC-PIM perform, in general, better than the corresponding linear ones (T3).

3) Among all these six SC-PIM schemes, Scheme B1-Tr6/3 performs the best, with a convergence rate of 3.21 that is even higher than the theoretical rate of 3 of the quadratic FEM models. The accuracy of Scheme B1-Tr6/3 is about 10-20 times better than all the other schemes, when finest mesh is used, as shown in FIGURE 10.5.

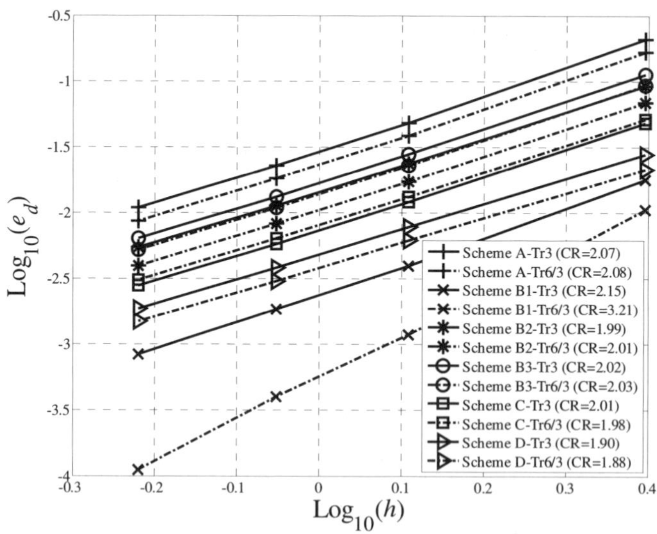

FIGURE 10.5 Comparison of convergence rates and accuracy of the numerical results in *displacement norm* for the cantilever problem obtained using various SC-PIM schemes.

FIGURE 10.6 shows the comparison of the convergence between all the linear SC-PIM schemes and other four numerical methods. TABLE 10.8 compares the computational efficiency of three particular linear SC-PIM schemes (Scheme B1, Scheme C and Scheme D) with the linear FEM and other GS-Galerkin models. In the table, the efficiency is roughly calculated by considering both the displacement error for the numerical results obtained using the same mesh and the computational cost in TABLE 10.6. We mention the following points.

1) In terms of convergence rate, the displacement solutions of all these methods achieve similar convergence rate closing to the theoretical value of 2.0.

2) In terms of accuracy, Scheme A is a little worse than the FEM measured in the displacement norm; Schemes B2, B3 and NS-PIM are about the same as FEM and all other schemes. When Mesh-4 is used, Scheme B1, Scheme C and Scheme D give 6.3 times, 1.9 times and 2.8 times, respectively, more accurate solution compared to the FEM-Tr3.

3) In terms of efficiency, the three SC-PIM schemes are more efficient than both the linear FEM and NS-PIM models. The SC-PIM with scheme B1-Tr3 is even 2.6 times more efficient than the CS-PIM-Tr4-CT.

4) As expected, the "star performer" ES-PIM gives the best accuracy and strong superconvergence in displacement norm with a rate of 2.55. It is much more accurate than the linear FEM, and it is even better than the SC-PIM Scheme B1-Tr3.

TABLE 10.8 Estimated computational efficiency of different methods measured in *displacement norm* error for the numerical results for the cantilever beam problem with the same set of triangular mesh (Mesh-4 of 1,696 nodes)

Numerical method	Solution error	Error ratio to FEM-Tr3	Efficiency
FEM-Tr3	5.2334E-03	1.00	1.00
NS-PIM-Tr3	5.5340E-03	1.06	0.8
ES-PIM-Tr3	4.6937E-05	0.0090	101.0
CS-PIM-Tr4-CT	1.9940E-03	0.38	2.2
Scheme B1-Tr3	8.4197E-04	0.16	5.7
Scheme C-Tr3	2.7918E-03	0.53	1.3
Scheme D-Tr3	1.8778E-03	0.36	1.9

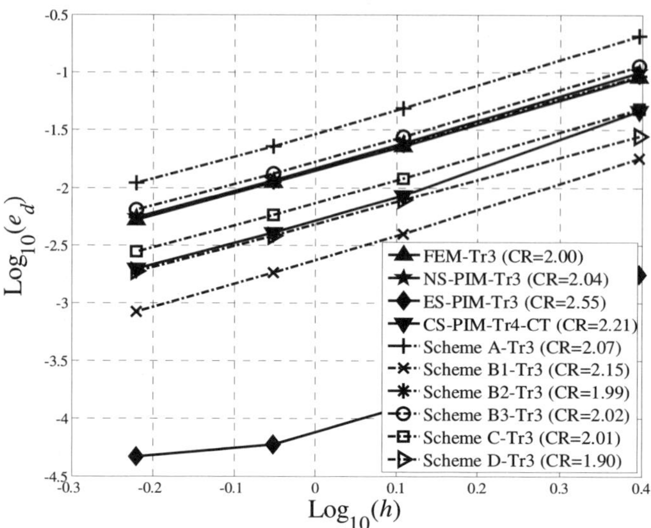

FIGURE 10.6 Comparison of convergence rates and accuracy of the numerical results in *displacement norm* for the cantilever problem obtained using linear FEM, three GS-Galerkin models and linear SC-PIM schemes.

FIGURE 10.7 plots the convergence of the solutions in energy norm for the same cantilever solved using these SC-PIM schemes. We can observe the following.

1) In terms of convergence rate, all these six SC-PIM schemes converge monotonically and have convergence rates higher than 1.0. Theoretically, the standard weak formulation (FEM) has a theoretical rate of 1.0, and a W^2 formulation can has a theoretical rate of 1.0 and an ideal rate of 1.5 (see Remarks 3.22 and 3.23). This example shows the fact that the rates of all these W^2 methods are between 1.0 and 1.7. Schemes A and C give rates higher than the ideal rate of 1.5 of the GS-Galerkin models [1, 4, 5].

2) In terms of accuracy, the quadratic (T6/3) models of these schemes do not show superiority than those of linear (T3) editions, except for Scheme A. The SC-PIM with Scheme B1-Tr6/3 that performs the best in the displacement norm is found the worst in energy norm. This shows the usual complementary phenomenon: an accurate displacement solution model is usually less accurate in stress solution.

3) Among all these six schemes, Scheme C stands out in terms of both accuracy and convergence rate, especially the linear model.

4) In terms of accuracy and convergence rate, the quadratic models (T6/3) of the SC-PIM are not necessarily better than the corresponding linear ones (T3), when measuring in energy norm.

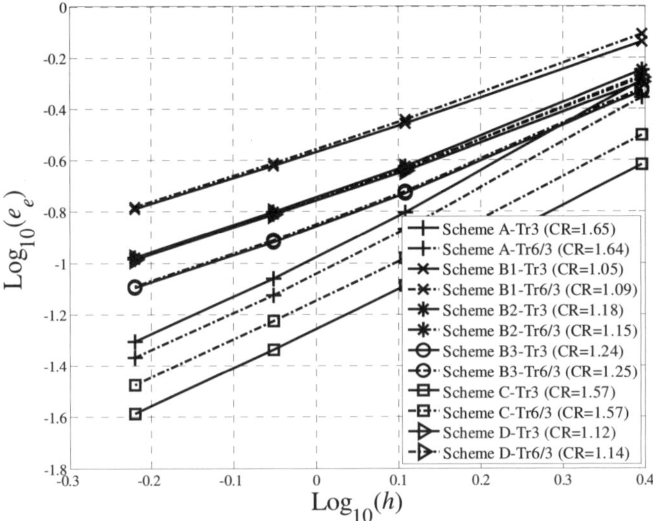

FIGURE 10.7 Comparison of convergence rates and accuracy of the numerical results in energy norm for the cantilever problem obtained using various SC-PIM schemes.

FIGURE 10.8 shows the comparison of the convergence status between these linear SC-PIM schemes and other four numerical methods. TABLE 10.9 compares the computational efficiency of three linear SC-PIM schemes (Scheme B1, Scheme C and Scheme D) with the linear FEM and other GS-Galerkin models. In the table, the efficiency is roughly calculated by considering both the energy norm error for the numerical results obtained using the same Mesh-4 and the computational cost in TABLE 10.6. We note the following points.

1) In terms of the convergence rate, except Scheme B1 that has about the same convergence rate as the FEM, all other five SC-PIM schemes achieve higher convergence rates showing superconvergence, which are similar as other three S-PIM methods.

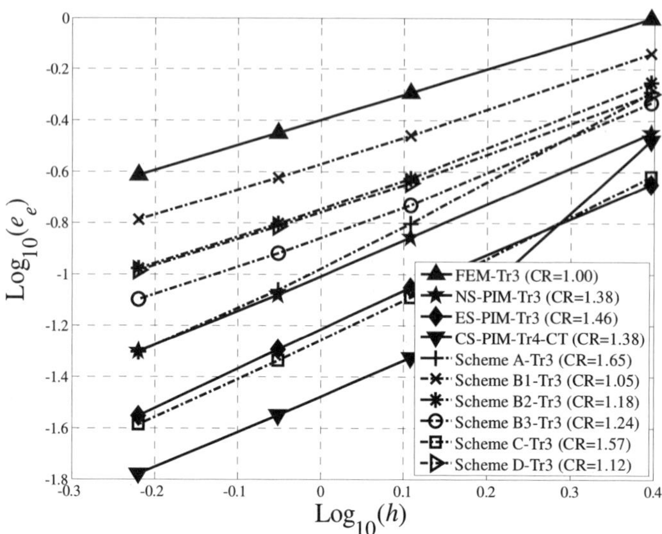

FIGURE 10.8 Comparison of convergence rates and accuracy of the numerical results in energy norm for the cantilever problem obtained using linear FEM, three GS-Galerkin models and linear SC-PIM schemes.

2) In terms of accuracy, all the six schemes of SC-PIM have much higher accuracy compared to the FEM, when the error is measured in energy norm. When the finest Mesh-4 is used, results of the Scheme C-Tr3 will be about 9 times more accurate than the linear FEM.

3) In terms of efficiency, we may find that Scheme B1-Tr3 and Scheme D-Tr3 have similar performance, which are respectively 1.4 and 1.6 times more efficient than the FEM-Tr3. Scheme C-Tr3 is 6.5 times more efficient than the linear FEM, which is comparable to the ES-PIM-Tr3.

4) In terms of both accuracy, convergence and efficiency, the CS-PIM-Tr4-CT stands out clearly followed by the SC-PIM with Scheme C-Tr3, and the ES-PIM-Tr3.

Remark 10.2 Superconvergence in both displacement and energy norms

The SC-PIM models show the weak superconvergence in displacement norm and strong superconvergence in energy norm.

TABLE 10.9 Estimated computational efficiency of different methods measured in *energy norm* error for the numerical results for the cantilever beam problem with the same set of triangular mesh (Mesh-4 of 1,696 nodes)

Numerical method	Solution error	Error ratio to FEM-Tr3	Efficiency
FEM-Tr3	2.4406E-01	1.00	1.00
NS-PIM-Tr3	5.0076E-02	0.21	4.0
ES-PIM-Tr3	2.8075E-02	0.12	7.6
CS-PIM-Tr4-CT	1.6635E-02	0.068	12.0
Scheme B1-Tr3	1.6285E-01	0.67	1.4
Scheme C-Tr3	2.5948E-02	0.11	6.5
Scheme D-Tr3	1.0338E-01	0.42	1.6

FIGURE 10.9a plots the converging process of the strain energy solutions to the exact solution for the cantilever obtained using different SC-PIM schemes. FIGURE 10.9b shows the comparison of the converging process between the linear SC-PIM solutions and those of other numerical methods. Measured in strain energy, TABLE 10.10 compares the computational efficiencies of three linear SC-PIM schemes (Scheme B1, Scheme C and Scheme D) with those of the linear FEM and other GS-Galerkin models. In the table, the efficiency is roughly calculated by considering both the strain energy errors for the numerical results obtained using the same Mesh-4 and the computational cost in TABLE 10.6. Through the table and these two figures, we found the following points.

1) The calculated strain energies from all these methods converge monotonically to the exact one with the increase of DOFs for these meshes.

2) All the W^2 models, including the schemes of SC-PIM, NS-PIM, ES-PIM and CS-PIM, give upper bound solution in energy norm with respect to the FEM solution, meaning that all these W^2 models are "softer" than the FEM counterpart.

3) FEM, SC-PIM with Scheme B1 and linear ES-PIM give lower bound solutions. The SC-PIM with Scheme B1 performs much softer than the overly-stiff FEM model. It is less accurate than the "star performer" ES-PIM-T3, but they are quite comparable. All the other five SC-PIM schemes, NS-PIM and CS-PIM provide upper bound solutions to the exact one.

4) The quadratic SC-PIM schemes (T6/3) generally give tighter bound solutions than those linear ones (T3). As we discussed in Remark 6.15, a higher order approximation in the displacement field will push the model closer to the exact model, and hence leads to tighter bound solutions.

5) For this problem, the SC-PIM with Scheme D gives the tightest upper bound among all these W^2 models studied so far. Hence SC-PIM with scheme D and ES-PIM give the best (tightest) pair of bounds to the exact solution.

6) The SC-PIM with Schemes A and B3 are found softer even than the NS-PIM-Tr3 that was found as the softest (spatially) stable GS-Galerkin model.

7) The "softness-ranking" of all these models for this problem is Scheme A, Scheme B3, NS-PIM-Tr3, Scheme B2, Scheme C, CS-PIM-Tr4-CT, Scheme D, ES-PIM-Tr3, Scheme B1 and FEM-Tr3.

8) In terms of efficiency, the ES-PIM-Tr3 stands out clearly, followed by SC-PIM of Scheme B1-Tr3, which is 5 times more efficient than the linear FEM. Scheme C-Tr3 and Scheme D-Tr3 are comparable to the CS-PIM-Tr4-CT model, and are more efficient than the linear FEM and NS-PIM models.

9) The SC-PIM with scheme A becomes now the softest spatially-stable model ever found so far.

TABLE 10.10 Estimated computational efficiency of different methods measured in the error in *strain energy solution* of the numerical results for the cantilever beam problem with the same set of triangular mesh (Mesh-4 of 1,696 nodes)

Numerical method	Strain energy Solution	Error[*] (%)	Error ratio to FEM-Tr3	Efficiency
FEM-Tr3	8.5474	-0.5341	1.00	1.00
NS-PIM-Tr3	8.6398	0.5411	1.01	0.8
ES-PIM-Tr3	8.5930	-0.0035	0.0066	137.7
CS-PIM-Tr4-CT	8.6097	0.1908	0.36	2.3
Scheme B1-Tr3	8.5853	-0.0935	0.18	5.1
Scheme C-Tr3	8.6164	0.2684	0.50	1.4
Scheme D-Tr3	8.6086	0.1777	0.33	2.0

[*] The analytical value of the strain energy for the cantilever beam problem is 8.59333333.
Negative indicates lower bound, and positive indicates upper bound.

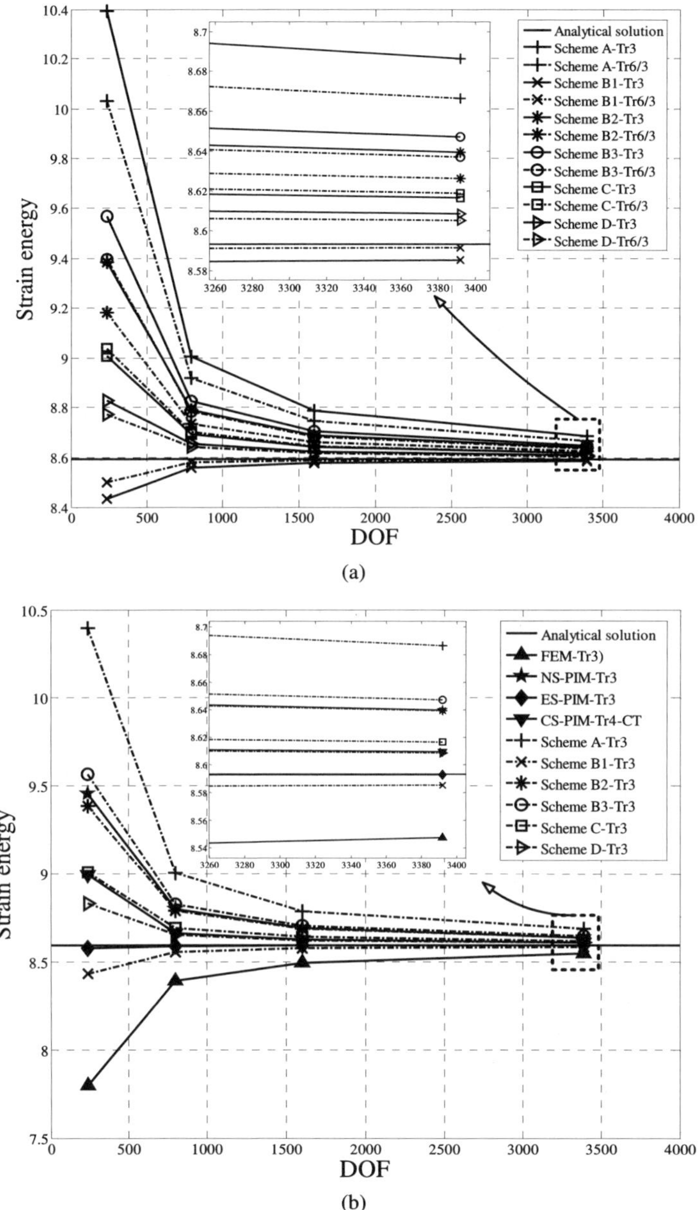

FIGURE 10.9 Converging process of the numerical results in strain energy solution for the cantilever: (a) comparison among these SC-PIM schemes; (b) comparison between linear SC-PIM schemes and other methods.

Example 10.3.3 Infinite solid with a circular hole

The problem of an infinite solid with a central circular hole described in Example 6.1.3 is again studied [22]. Using the same set of triangular meshes shown in FIGURE 10.10, the problem is analyzed using the SC-PIM schemes together with other four numerical methods.

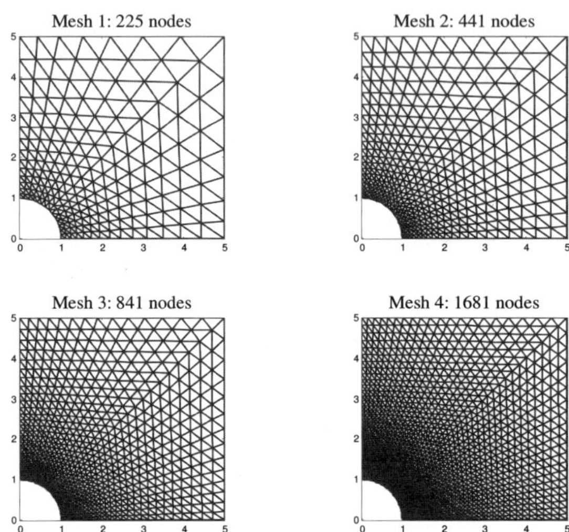

FIGURE 10.10 Domain discretization of the 2D solid with circular hole using a set of triangular meshes.

FIGURE 10.11 plots the convergence of the solutions in displacement norm for this problem solved using different SC-PIM schemes. We summarize the following findings.

1) Similar as the previous example, the solutions of all these SC-PIM schemes converge monotonically with convergence rates from 1.71 to 2.55, in terms of displacement norm. Scheme B1-Tr6/3 registers the highest rate of 2.55 and ranks the 2nd in accuracy. Scheme D-Tr6/3 produces the most accurate solution but with the lowest convergence rate of 1.71. Note that Scheme D-Tr6/3 will be less efficient due to larger bandwidth.

2)   The quadratic models (T6/3) perform generally better than those linear ones (T3) in terms of displacement norm.

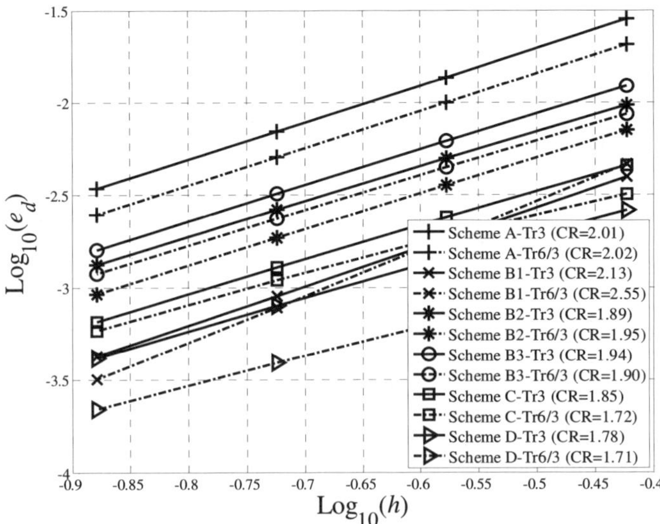

FIGURE 10.11 Comparison of convergence rates and accuracy of the numerical results in displacement norm for the infinite solid with hole obtained using various SC-PIM schemes.

FIGURE 10.12 compares the solution convergence in displacement norm between the linear SC-PIM schemes and other four methods. We note the following findings.

1)   Scheme A is again found less accurate than the FEM-Tr3, and all other linear SC-PIM schemes are at least better than the FEM.

2)   Schemes B2 and B3 are about the same as the linear FEM and NS-PIM, and Scheme B1, Scheme C and Scheme D give much more accurate solution compared to FEM.

3)   ES-PIM and CS-PIM stand out clearly, followed by SC-PIM with scheme D and then Scheme B1.

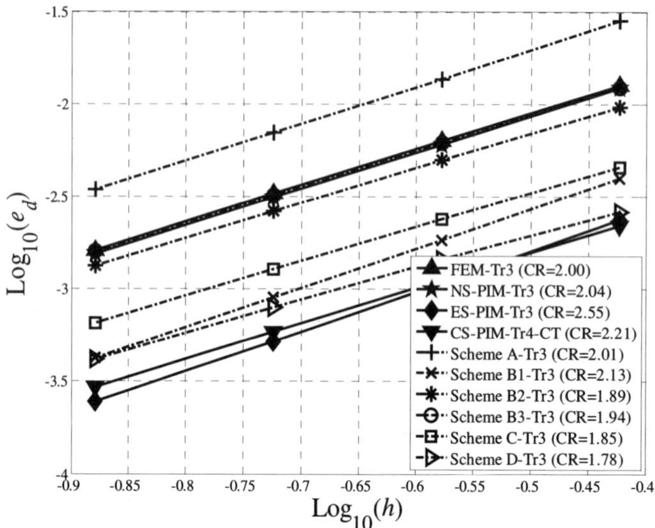

FIGURE 10.12 Comparison of convergence rates and accuracy of the numerical results in displacement norm for the infinite solid with hole obtained using linear FEM, three GS-Galerkin models and linear SC-PIM schemes.

FIGURE 10.13 plots the convergence of the solutions in energy norm for the problem of infinite solid with a hole solved using SC-PIM schemes. We can have the following findings.

1) Scheme C stands out again in terms of both accuracy and convergence rate among these six schemes. Scheme B2-Tr3 is found the least accurate in energy norm (the complimentary phenomenon). Two Scheme D models deliver an average performance.

2) The quadratic models (T6/3) do not necessarily perform better than those linear ones (T3) in terms of energy norm.

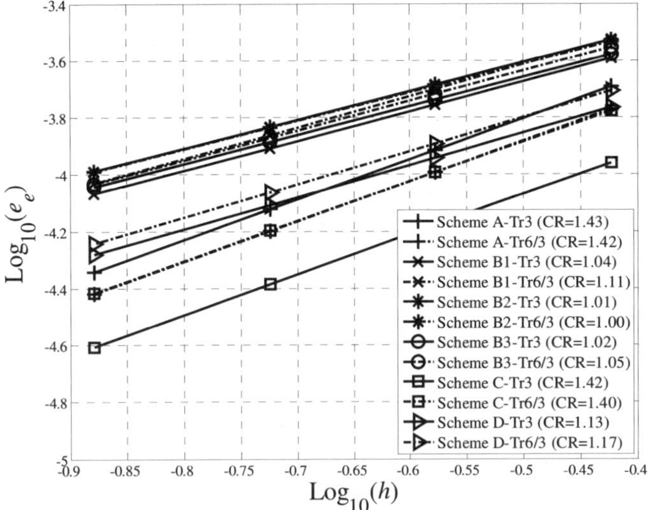

FIGURE 10.13 Comparison of convergence rates and accuracy of the numerical results in energy norm for the infinite solid with hole obtained using various SC-PIM schemes.

FIGURE 10.14 shows the comparison of the convergence of the solutions in energy norm between these linear SC-PIM schemes and other four numerical methods. We have the following points.

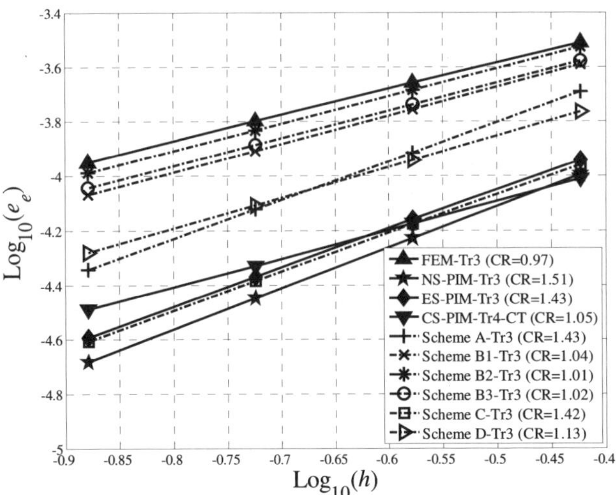

FIGURE 10.14 Comparison of convergence rates and accuracy of the numerical results in energy norm for the infinite solid with hole obtained using linear FEM, three GS-Galerkin models and linear SC-PIM schemes.

1) All these six SC-PIM schemes and these three GS-Galerkin methods, have better accuracy and higher convergence rates than the linear FEM.

2) In terms of convergence rate, NS-PIM-Tr3, ES-PIM-Tr3 and SC-PIM with Scheme C achieve rates around the ideal value of 1.5 (see Remark 3.23).

3) Scheme C outperforms all the other SC-PIM schemes and has similar performance as linear NS-PIM and ES-PIM.

FIGURE 10.15a plots the processes of the strain energy converging to the exact solution for the problem of infinite solid with a hole obtained using these SC-PIM schemes. FIGURE 10.15b compares the solution converging processes of linear SC-PIM schemes with those of other numerical methods. We note the following findings.

1) All the methods converge monotonically to the exact one with the increase of DOFs.

2) All these W^2 models, including the SC-PIM schemes, NS-PIM, ES-PIM and CS-PIM, give upper bound solutions in energy norm with respect to the FEM solution.

3) FEM, Scheme B1 and linear ES-PIM give lower bound solutions, and all the other five SC-PIM schemes together with the NS-PIM and CS-PIM provide upper bound solutions to the exact one.

4) The quadratic SC-PIM schemes generally give tighter bound solutions than those linear ones.

5) Scheme B1 performs much softer than the overly-stiff linear FEM and is comparable to the ES-PIM.

6) Among the upper bound models, Schemes C and D have tighter bounds that are comparable to the CS-PIM.

7) The "softness-ranking" of all these models for this case is Scheme A, Scheme B3, NS-PIM-Tr3, Scheme B2, Scheme C, Scheme D, CS-PIM-Tr4-CT, ES-PIM-Tr3, Scheme B1 and FEM-Tr3.

8) Scheme A is found still the softest (spatially) stable model that is even softer than the NS-PIM-Tr3.

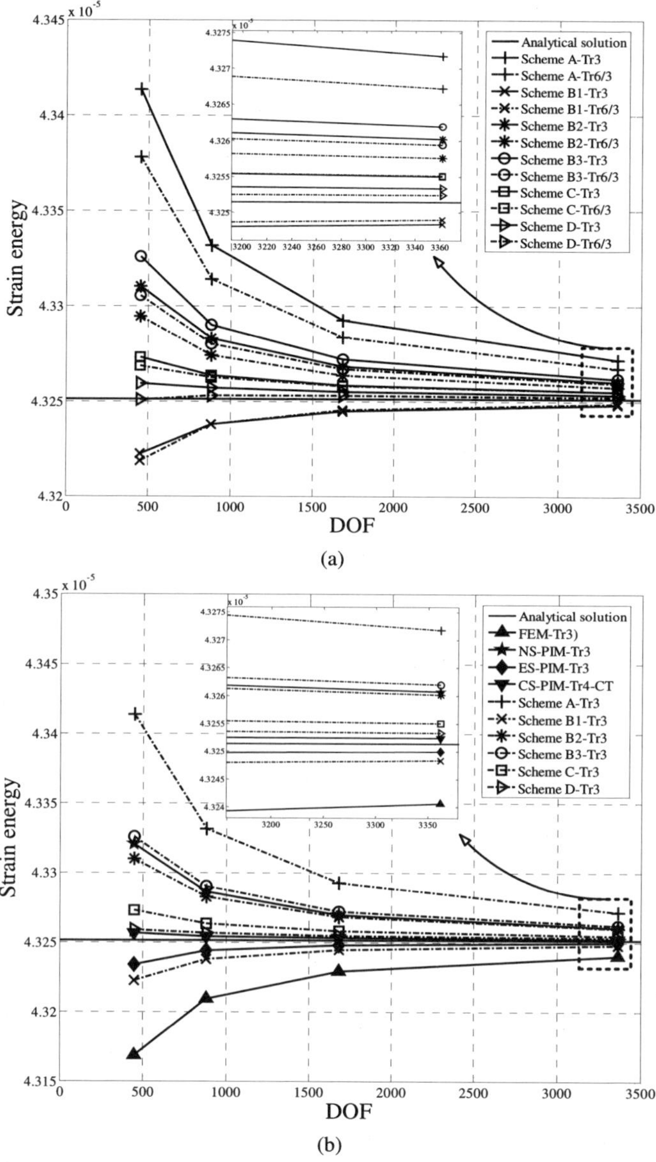

FIGURE 10.15 Converging process of the numerical results (in strain energy norm) for the problem of infinite solid with hole: (a) comparison among all these SC-PIM schemes; (b) comparison between SC-PIM schemes and other methods.

10.4 Concluding remarks

In this Chapter, a strain-constructed PIM or SC-PIM method has been introduced. In the SC-PIM the point interpolation method is used for constructing both the displacement and strain fields. A total of six schemes are devised for the construction of the strain fields over triangular integration cells using strains at the vertices. These strains are assigned as the node-based smoothed strains, edge-based and/or compatible strains. A number of numerical examples have been used to investigate in great detail on these six schemes. Through these investigations, note the following remarks.

Remark 10.3 SC-PIM: linearly conforming and at least 2nd order accuracy

All the six schemes A, B1, B2, B3, C and D are found linearly conforming; they all can pass the standard patch test with both continuous (T3) and discontinuous displacement fields (T6/3). With the ensured stability, the numerical solutions of these schemes are guaranteed to converge to the exact solution. The rate of convergence is at least 2, and the accuracy is of 2nd order.

Remark 10.4 SC-PIM: more accurate than FEM-Tr3

In terms of accuracy in displacement, all the SC-PIM schemes can obtain better accuracy than the FEM-Tr3, except Scheme A that performs a litter worse. In terms of energy error, all these six schemes of linear SC-PIM give much better accuracy and generally higher convergence rates compared to the linear FEM. The convergence rate is from 1.0 to 1.7.

Remark 10.5 SC-PIM with Schemes C, B1 and D: good performers

Among these six schemes, SC-PIM with Scheme C performs, in overall, the best in terms of both displacement and energy norms, followed by Scheme B1 and D. These three schemes are generally comparable to these star performers ES-PIM and CS-PIM models.

Remark 10.6 SC-PIM with Schemes A: the softest model

Scheme A is found even softer than the NS-PIM, and it is the softest model found so far. It can provide upper bound solutions, as long as the mesh is not too coarse.

Remark 10.7 Softness-ranking

The "softness-ranking" of all the models studied in this chapter is roughly: Scheme A, Scheme B3, NS-PIM-Tr3, Scheme B2, Scheme C, Scheme D/CS-PIM-Tr4-CT, ES-PIM-Tr3, Scheme B1 and FEM-Tr3.

Remark 10.8 SC-PIM: T6/3 models are comparable to T3 models

The quadratic SC-PIM models are found not much better than linear models, but are in general more expensive. We recommend the use of linear SC-PIM models. Further explore for bilinear SC-PIM models may be worthwhile.

Remark 10.9 Extension to 3D problem

In this chapter only the formulation for 2D problems is introduced. The same idea can also be extended for 3D problems.

Remark 10.10 Extension to general n-sided polygonal cells/elements

In the above discussions on SC-PIM, we focused on the use of triangular background cells/elements. The same idea can be applied for general n-sided cells/elements, if needed.

10.5 References

1. Liu, G. R., *Meshfree Methods: Moving beyond the Finite Element Method*, 2nd Edition, CRC press, Boca Taton, USA, 2009.

2. Liu, G. R., A generalized Gradient smoothing technique and the smoothed bilinear form for Galerkin formulation of a wide class of computational methods. *International Journal of Computational Methods*, 5(2): 199-236, 2008.

3. Liu, G. R., On the G space theory. *International Journal of Computational Methods*, 6(2): 257-289, 2009.

4. Liu, G. R., A G space theory and a weakened weak (W^2) form for a unified formulation of compatible and incompatible methods, Part I Theory, *International Journal for Numerical Methods in Engineering*, 81: 1093-1126, 2010.

5. Liu, G. R., A G space theory and a weakened weak (W^2) form for a unified formulation of compatible and incompatible methods, Part II Application to solid mechanics problems, *International Journal for Numerical Methods in Engineering*, 81: 1127-1156, 2010.

6. Liu, G. R., Zhang, G. Y., Dai, K. Y, Wang, Y. Y., Zhong, Z. H., Li, G. Y. and Han, X., A linearly conforming point interpolation method (LC-PIM) for 2D solid mechanics problems, *International Journal of Computational Methods*, 2(4): 645-665, 2005.

7.   Liu, G. R. and Zhang, G. Y., Edge-based smoothed point interpolation methods, *International Journal of Computational Methods*, 5(4): 621-646, 2008.

8.   Liu, G. R. and Zhang, G. Y., A normed G space and weakened weak (W^2) formulation of a cell-based smoothed point interpolation method. *International Journal of Computational Methods*, 6(1): 147-179, 2009.

9.   Zhang, G. Y., Liu, G. R. and Xu, X., A strain-constructed point interpolation method (SC-PIM) and strain field construction schemes for solid mechanics problems using triangular mesh. *Applied Mathematics and Computation*, 219: 2067-2086, 2012.

10.   Liu, G. R. and Zhang, G. Y., A novel scheme of strain-constructed point interpolation method for static and dynamic mechanics problems. *International Journal of Applied Mechanics*, 1(1): 233-258, 2009.

11.   Liu, G. R., Xu, X., Zhang G. Y. and Gu, Y. T., An extended Galerkin weak form and a point interpolation method with continuous strain field and superconvergence using triangular mesh, *Computational Mechanics*, 43: 651-673, 2009.

12.   Liu, G. R., *Meshfree Methods: Moving beyond the Finite Element Method*, 1st edition, CRC press, Boca Taton, USA, 2002.

13.   Liu, G. R. and Gu, Y. T., *An introduction to meshfree method methods and their programming*, Springer, 2005.

14.   Liu, G. R. and Gu, Y. T., A point interpolation method for two-dimensional solids, *International Journal for Numerical Methods in Engineering*, 50: 937-951, 2001.

15.   Wang, J. G. and Liu, G. R., A point interpolation meshless method based on radial basis functions, *International Journal for Numerical Methods in Engineering*, 54: 1623-1648, 2002.

16.   Liu G. R., Xu X., Zhang G. Y. and Nguyen-Thoi, T., A superconvergent point interpolation method (SC-PIM) with piecewise linear strain field using triangular mesh, *International Journal for Numerical Method in Engineering*, 77: 1439-1467, 2009.

17.   Zienkiewicz, O. C. and Taylor R. L., *The Finite Element Method*, 5th ed., Butterworth Heimemann, Oxford, 2000.

18.   Liu, G. R. and Quek, S. S. *The finite element method: a practical course.* Butterworth Heinemann: Oxford, 2002.

19.   Liu, G. R. and Zhang, G. Y., Upper bound solution to elasticity problems: A unique property of the linearly conforming point interpolation method (LC-PIM). *International Journal for Numerical Methods in Engineering*, 74: 1128-1161, 2008.

20.   Liu, G. R., Nguyen-Thoi, T., Nguyen-Xuan, H. and Lam, K. Y., A node-based smoothed finite element method (NS-FEM) for upper bound solutions to solid mechanics problems, *Computers and Structures*, 87: 14-26, 2009.

21.   Zhang, G. Y., Liu, G. R., Nguyen-Thoi, T., Song, C. X., Han, X., Zhong, Z. H. and Li, G. Y., The upper bound property for solid mechanics of the linearly conforming radial point interpolation method (LC-RPIM). *International Journal of Computational Methods*, 4(3): 521-541, 2007.

22.   Timoshenko, S. P. and Goodier, J. N., *Theory of Elasticity*, 3rd ed., McGraw-Hill, New York, 1970.

Chapter 11

S-PIM for Heat Transfer and Thermoelasticity Problems

Heat transfer refers to the transition of thermal energy in media that can be solids or fluids. It is a very important physical phenomenon in most of the systems in engineering and science [1, 2]. As one of the typical field problems with temperature as the field variable, the heat transfer problem is described by the second law of thermodynamics leading to a set of PDEs known as Helmholtz equations. It is one of the most fundamental topics in the analysis and design of mechanical, electrical, civil, biological and chemical systems. The Helmholtz equation is a rather a quite general PDEs that governs a number of other physical problems, including torsional deformation of bars, irrotational flow and acoustic problems. Therefore, the formulations and procedures presented in this chapter shall, in principle, applicable also these problems. In fact the S-FEM has been found particularly effective for acoustic problems [3-9].

Compared to solid mechanics problems of vector fields of primary variables (displacements), the temperature field in heat transfer problems is scalar. In this sense, heat transfer problems are easier to deal with numerically. A large variety of numerical methods have been used to analyze the problems of heat transfer, such as the finite element method (FEM) [10-13], the finite difference method (FDM) [14, 15] and types of meshfree methods [16-19].

In this chapter, we provide a detailed formulation for the recent S-PIM models for general heat transfer and thermoelastic problems. First, we will give a brief description about the setting of general heat transfer problems. We will then present the field variable approximation, the gradient smoothing operation and the corresponding weakened weak (W^2) forms for S-PIM models. The discretized system equations for S-PIM are then derived. Formulations about the thermoelastic problems (analyses of displacement and stress fields in solid media resultant from the temperature changes) will also be introduced. Finally, the S-PIM models will be applied for 1D, 2D and 3D problems to demonstrate some of the attractive properties found in the study.

11.1   Heat transfer problems

11.1.1   Problem statements

Consider a "conductive" heat transfer problem in a three-dimensional (3D) anisotropic medium Ω bounded by a surface Γ. The problem is then governed by a special case of the (dynamic) Helmholtz equations that have the following parabolic form [20]

$$\frac{\partial}{\partial x}\left(k_x \frac{\partial T}{\partial x} \right) + \frac{\partial}{\partial y}\left(k_y \frac{\partial T}{\partial y} \right) + \frac{\partial}{\partial z}\left(k_z \frac{\partial T}{\partial z} \right) + Q_v = \rho c_{hc} \frac{\partial T}{\partial t} \qquad (11.1)$$

where T [°C] is the temperature, t [s] is time, k_x, k_y and k_z [W/m°C] are thermal conductivity coefficients in three principle directions, Q_v [W/m^3] is the internal heat generation per unit volume often known as heat source, ρ [kg/m^3] is the density of the material and c_{hc} [J/kg°C] is the specific heat capacity.

The Fourier's law describes the heat conduction within the media in the following way [1]

$$\begin{Bmatrix} q_x \\ q_y \\ q_z \end{Bmatrix} = - \begin{bmatrix} k_{11} & k_{12} & k_{13} \\ k_{21} & k_{22} & k_{23} \\ k_{31} & k_{32} & k_{33} \end{bmatrix} \begin{Bmatrix} \dfrac{\partial T}{\partial x} \\[2mm] \dfrac{\partial T}{\partial y} \\[2mm] \dfrac{\partial T}{\partial z} \end{Bmatrix} \qquad (11.2)$$

where q_x, q_y, q_z [W/m^2] are the rate of heat flow per unit area in x, y, z direction (known as heat flux) and k_{ij} is the symmetric thermal conductivity tensor representing the material property of the medium. The "−" sign reflects the simple fact that the heat flows in the reverse direction of the temperature gradient.

For an isotropic medium with constant thermal properties, the Fourier's law can be written simply as

$$\begin{cases} q_x = -k\dfrac{\partial T}{\partial x} \\[2mm] q_y = -k\dfrac{\partial T}{\partial y} \\[2mm] q_z = -k\dfrac{\partial T}{\partial z} \end{cases} \tag{11.3}$$

and Equation (11.1) can then be expressed as follows

$$k\left(\frac{\partial^2 T}{\partial x^2}+\frac{\partial^2 T}{\partial y^2}+\frac{\partial^2 T}{\partial z^2}\right)+Q_v = \rho c_{hc}\frac{\partial T}{\partial t} \tag{11.4}$$

or in a concise form of

$$\nabla^2 T + \frac{Q_v}{k} = \frac{1}{\alpha_T}\frac{\partial T}{\partial t} \tag{11.5}$$

where $\alpha_T = k/\rho c_{hc}$ is the thermal diffusivity and ∇^2 is the Laplace operator defined as

$$\nabla^2 = \frac{\partial^2}{\partial x^2}+\frac{\partial^2}{\partial y^2}+\frac{\partial^2}{\partial z^2} \tag{11.6}$$

With constant thermal properties and no internal heat generation, Equation (11.5) reduces further to the following Fourier equation

$$\nabla^2 T = \frac{1}{\alpha_T}\frac{\partial T}{\partial t} \tag{11.7}$$

which is a typical diffusion equation.

With constant thermal properties and steady-state heat transfer, Equation (11.5) reduces to the following Poisson's equation.

$$\nabla^2 T + \frac{Q_v}{k} = 0 \qquad (11.8)$$

For constant thermal properties, no internal heat generation, and steady-state heat transfer, the heat conduction equation (11.5) reduces to the following Laplace's equation

$$\nabla^2 T = 0 \qquad (11.9)$$

For well-posedness, the heat conduction equation must be subjected to proper initial conditions and boundary conditions. The initial condition specifies the temperature at an initial time,

$$T(x, y, z, 0) = T_0(x, y, z) \qquad (11.10)$$

Boundary conditions of heat conduction take several forms and some frequently encountered conditions are as follows.

Temperature condition

$$T = T_{\Gamma_1}(x, y, z, t) \quad \text{on } \Gamma_1 \qquad (11.11)$$

The surface temperature on the boundary Γ_1 is prescribed to be $T_{\Gamma 1}$, which may vary with position and time. Prescribed temperature is an example of a Dirichlet boundary condition.

Heat flux condition

$$-k_x \frac{\partial T}{\partial x} n_x - k_y \frac{\partial T}{\partial y} n_y - k_z \frac{\partial T}{\partial z} n_z = q_{\Gamma_2} \quad \text{on } \Gamma_2 \qquad (11.12)$$

where n_x, n_y and n_z are the direction cosines of the outward normal to the surface. This condition describes that the rate of heat flow across the boundary Γ_2 is specified to be $q_{\Gamma 2}$ (positive into the surface). This condition can also be expressed as

$$q_x n_x + q_y n_y + q_z n_z = q_{\Gamma_2} \quad \text{on } \Gamma_2 \tag{11.13}$$

or for isotropic media:

$$-k \frac{\partial T}{\partial n} = q_{\Gamma_2} \quad \text{on } \Gamma_2 \tag{11.14}$$

Heat exchange conditions

There are two types of heat exchange conditions on the boundary of the medium, and the convective heat exchange condition is as follows

$$-k_x \frac{\partial T}{\partial x} n_x - k_y \frac{\partial T}{\partial y} n_y - k_z \frac{\partial T}{\partial z} n_z = h_T \left(T - T_e \right) \quad \text{on } \Gamma_3 \tag{11.15}$$

where h_T [W/(m^2°C)] is a convection heat transfer coefficient and T_e is the temperature of an adjacent fluid. This condition describes that the rate of heat flow across a boundary Γ_3 is proportional to the difference between the surface temperature T and a convective exchange temperature T_e of an adjacent medium. The convective exchange temperature T_e may be a function of a boundary coordinate and/or time. The convective heat exchange condition also takes the form

$$q_x n_x + q_y n_y + q_z n_z = h_T \left(T - T_e \right) \quad \text{on } \Gamma_3 \tag{11.16}$$

or for isotropic media:

$$-k \frac{\partial T}{\partial n} = h_T \left(T - T_e \right) \quad \text{on } \Gamma_3 \tag{11.17}$$

The second type of heat exchange condition is the radiation heat exchange condition that takes the following form for an isotropic medium

$$-k \frac{\partial T}{\partial n} = \sigma_T \varepsilon_T T_s - \alpha_A q_e \quad \text{on } \Gamma_3 \tag{11.18}$$

where σ_T is the Stefan-Boltzmann constant, ε_T is the surface emissivity, T_s is the surface temperature, α_A is the surface absorptivity coefficient, and q_e is the incident radiant thermal energy. The radiation heat exchange involves the nonlinear problem and will not be discussed in this work.

11.1.2 Steady-state heat transfer

For an isotropic medium, the steady-state heat transfer problem can be described using the following conduction equation and boundary conditions

$$k\nabla^2 T + Q_v = 0 \tag{11.19}$$

$$T = T_{\Gamma_1}(x, y, z, t) \quad \text{on } \Gamma_1 \tag{11.20}$$

$$-k\frac{\partial T}{\partial n} = q_{\Gamma_2} \quad \text{on } \Gamma_2 \tag{11.21}$$

$$-k\frac{\partial T}{\partial n} = h_T (T - T_e) \quad \text{on } \Gamma_3 \tag{11.22}$$

11.1.3 Temperature field approximation

The approximation of the temperature field function is largely the same as what we do for a component of the displacement field function for solid mechanics problems. Consider a media with domain Ω that is discretized into N_c linear cells, i.e. four-node tetrahedral for 3D and three-node triangular cells for 2D, which contains N_n nodes and N_f faces or N_{eg} edges. Then the assumed temperature can be approximated as follows

$$T^h(\mathbf{x}) = \sum_{I \in S_n} \phi_I(\mathbf{x})\, T_I = \mathbf{\Phi}_s(\mathbf{x})\mathbf{T}^e \tag{11.23}$$

where ϕ_I is the shape function for node I, T_I is the temperature of node I and S_n is the set of local support nodes involved in the approximation which contains n nodes. In the above equation, $\mathbf{\Phi}_s$ and $\mathbf{T}^e$ are shape functions matrix and vector of nodal temperatures respectively.

11.1.4 Generalized gradient smoothing operation

When the assumed temperature function is continuous, the derivatives of the temperature can be obtained in a piecewise fashion as

$$\begin{cases} \dfrac{\partial T^h}{\partial x} = \displaystyle\sum_{I \in S_n} \dfrac{\partial \phi_I}{\partial x} \, T_I \\[2ex] \dfrac{\partial T^h}{\partial y} = \displaystyle\sum_{I \in S_n} \dfrac{\partial \phi_I}{\partial y} \, T_I \\[2ex] \dfrac{\partial T^h}{\partial z} = \displaystyle\sum_{I \in S_n} \dfrac{\partial \phi_I}{\partial z} \, T_I \end{cases} \qquad (11.24)$$

or in matrix form, we have the expression for the temperature gradient

$$\mathbf{g} = \nabla T^h = \mathbf{B}^{temp} \mathbf{T}^e \qquad (11.25)$$

where $\mathbf{g}^{\mathrm{T}} = \{ g_x \quad g_y \quad g_z \} = \left\{ \dfrac{\partial T^h}{\partial x} \quad \dfrac{\partial T^h}{\partial y} \quad \dfrac{\partial T^h}{\partial z} \right\}$ is the vector collects the

derivatives of temperature and $\mathbf{B}^{temp}$ is the temperature-gradient matrix

$$\mathbf{B}^{temp} = \begin{bmatrix} \dfrac{\partial \phi_1}{\partial x} & \dfrac{\partial \phi_2}{\partial x} & \cdots & \dfrac{\partial \phi_n}{\partial x} \\[2ex] \dfrac{\partial \phi_1}{\partial y} & \dfrac{\partial \phi_2}{\partial y} & \cdots & \dfrac{\partial \phi_n}{\partial y} \\[2ex] \dfrac{\partial \phi_1}{\partial z} & \dfrac{\partial \phi_2}{\partial z} & \cdots & \dfrac{\partial \phi_n}{\partial z} \end{bmatrix} \qquad (11.26)$$

Performing the generalized gradient smoothing operation by using a proper set of the volume (area) of the smoothing domain [21-24], the "smoothed" temperature gradients within the smoothing domain k are obtained as follows [25-29]

$$\overline{\mathbf{g}}_k = \frac{1}{V_k^s} \int_{\Gamma_k^s} \mathbf{L}_q T^h d\Gamma \qquad (11.27)$$

where V_k^s is the volume (area) of smoothing domain k, Γ_k^s is the corresponding boundary surface (edges) and $\mathbf{L}_q$ is the vector of unit outward normal which has the form of

$$\mathbf{L}_q = \mathbf{n} = \left\{ n_x \quad n_y \quad n_z \right\}^{\mathrm{T}}$$ (11.28)

By substituting Equation (11.23) into Equation (11.27), we have

$$\overline{\mathbf{g}}_k = \frac{1}{V_k^s} \int_{\Gamma_k^s} \mathbf{L}_q \mathbf{\Phi}_s \mathbf{T}^e \, \mathrm{d}\Gamma = \sum_{I \in S_s} \overline{\mathbf{B}}_I^{temp} \left(\mathbf{x}_k \right) T_I$$ (11.29)

where S_s is the set of field nodes involved in constructing the smoothed gradient field within the smoothing domain V_k^s and $\overline{\mathbf{B}}^{temp}$ is the smoothed temperature-gradient matrix

$$\overline{\mathbf{B}}_I^{temp} \left(\mathbf{x}_k \right) = \begin{bmatrix} \overline{\phi}_{Ix}^{temp} \left(\mathbf{x}_k \right) \\ \overline{\phi}_{Iy}^{temp} \left(\mathbf{x}_k \right) \\ \overline{\phi}_{Iz}^{temp} \left(\mathbf{x}_k \right) \end{bmatrix}$$ (11.30)

in which the elements are defined as

$$\overline{\phi}_{Il}^{temp} \left(\mathbf{x}_k \right) = \frac{1}{V_k^s} \int_{\Gamma_k^s} \phi_I \left(\mathbf{x}_k \right) n_l \left(\mathbf{x}_k \right) \mathrm{d}\Gamma, \quad \left(l = x, \, y, \, z \right)$$ (11.31)

The integration in Equation (11.31) can be obtained using a numerical integration scheme, such as the Gauss integration procedure as follows

$$\overline{\phi}_{Il}^{temp} = \frac{1}{V_k^s} \sum_{m=1}^{n_{seg}} \left[\sum_{n=1}^{n_G} W_n^G \phi_I \left(\mathbf{x}_{m,n} \right) n_{l,m} \right], \quad \left(l = x, \, y, \, z \right)$$ (11.32)

where n_{seg} is the number of the surfaces (edges) of a smoothing domain, n_G is the number of Gauss points located on each surface (edge) and W_n^G is the corresponding weight number of Gauss integration scheme. Note Equation (11.29) is used regardless of the continuity of the assumed temperature function, as discussed in Chapters 3-5.

11.1.5 Discrete system equations

The differential equation, for example Equation (11.19), needs to be satisfied at any point in the problem domain. However, for an assumed temperature field using Equation (11.23), it is generally not possible. We need to use weak or W^2 formulation for stable and convergent solutions. In the weak formulation, the corresponding energy functional is defined as

$$\Pi(T) = \int_\Omega \left\{ \frac{1}{2}k\left[\left(\frac{\partial T}{\partial x}\right)^2 + \left(\frac{\partial T}{\partial y}\right)^2 + \left(\frac{\partial T}{\partial z}\right)^2 \right] - Q_v T \right\} d\Omega$$
$$+ \int_{\Gamma_2} q_{\Gamma_2} T d\Gamma + \int_{\Gamma_3} h_T \left(\frac{T^2}{2} - TT_e \right) d\Gamma \tag{11.33}$$

Replacing the gradient of the temperature ($\mathbf{g}$) with the generalized smoothed gradient ($\bar{\mathbf{g}}$) obtained using a proper set of N_s smoothing domains, we shall have the "smoothed" energy functional for the W^2 formulation:

$$\bar{\Pi}(T) = \sum_{i=1}^{N_s} \frac{1}{2} V_i^s \bar{\mathbf{g}}^T k \bar{\mathbf{g}} - \int_\Omega Q_v T d\Omega$$
$$+ \int_{\Gamma_2} q_{\Gamma_2} T d\Gamma + \int_{\Gamma_3} h_T \left(\frac{T^2}{2} - TT_e \right) d\Gamma \tag{11.34}$$

where V_i^s is the area of the ith smoothing domain. The stationary conditions of the energy functional defined in the above equation give the following GS-Galerkin weak form [21-27]

$$\delta\Pi(T) = \sum_{i=1}^{N_s} V_i^s \left(\delta\bar{\mathbf{g}}\right)^T k \bar{\mathbf{g}} - \int_\Omega Q_v \delta T d\Omega + \int_{\Gamma_2} q_{\Gamma_2} \delta T d\Gamma$$
$$+ \int_{\Gamma_3} h_T T \delta T d\Gamma - \int_{\Gamma_3} h_T T_e \delta T d\Gamma = 0 \tag{11.35}$$

By substituting Equations (11.23) and (11.29) into the above equation and invoking the arbitrariness of virtual nodal temperature, we have the following discretized algebraic system equation

$$\left[\overline{\mathbf{K}}^{temp} + \mathbf{K}^3\right]\mathbf{T} = \tilde{\mathbf{f}}^{temp} \tag{11.36}$$

where $\overline{\mathbf{K}}^{temp}$ is the global conductance matrix related to conduction or stiff matrix with entries of

$$\overline{\mathbf{K}}_{IJ}^{temp} = \sum_{i=1}^{N_s} V_i^s \left(\overline{\mathbf{B}}_I^{temp}(\mathbf{x}_k)\right)^{\mathrm{T}} k\left(\overline{\mathbf{B}}_J^{temp}(\mathbf{x}_k)\right) \tag{11.37}$$

$\mathbf{K}^3$ is the global conductance matrix related to convection or the stiff matrix generated by the convective heat exchange condition, which is assembled using

$$\mathbf{K}_{IJ}^3 = \int_{\Gamma_3} \mathbf{\Phi}_I^{\mathrm{T}} h_T \mathbf{\Phi}_J \, \mathrm{d}\Gamma \tag{11.38}$$

$\mathbf{T}$ is the vector which collects all the nodal temperature over the problem domain, and $\tilde{\mathbf{f}}^{temp}$ is the heat loading vector with the following form

$$\tilde{\mathbf{f}}_I^{temp} = \int_{\Omega} Q_v \mathbf{\Phi}_I^{\mathrm{T}} \, \mathrm{d}\Omega + \int_{\Gamma_2} T_{\Gamma_2} \mathbf{\Phi}_I^{\mathrm{T}} \, \mathrm{d}\Gamma + \int_{\Gamma_3} h_T T_e \mathbf{\Phi}_I^{\mathrm{T}} \, \mathrm{d}\Gamma \tag{11.39}$$

11.1.6 Transit-state heat transfer

The smoothed energy functional for the transit-state heat transfer problem can be given as

$$\begin{aligned}
\overline{\Pi}(T) = \sum_{i=1}^{N_s} \frac{1}{2} V_k^s \overline{\mathbf{g}}^{\mathrm{T}} k \overline{\mathbf{g}} - \int_{\Omega} Q_v T \, \mathrm{d}\Omega + \int_{\Omega} \rho c_{hc} \frac{\partial T}{\partial t} T \, \mathrm{d}\Omega \\
+ \int_{\Gamma_2} q_{\Gamma_2} T \, \mathrm{d}\Gamma + \int_{\Gamma_3} h_T \left(\frac{T^2}{2} - T T_e\right) \mathrm{d}\Gamma
\end{aligned} \tag{11.40}$$

Following the similar way presented in the previous section, the discretized system equations for the transit-state heat transfer problem are as follows

$$\mathbf{M}^{temp}\dot{\mathbf{T}} + \left[\overline{\mathbf{K}}^{temp} + \mathbf{K}^3\right]\mathbf{T} = \tilde{\mathbf{f}}^{temp} \tag{11.41}$$

where $\mathbf{M}^{temp}$ is the global capacitance matrix with the entries of

$$\mathbf{M}_{IJ}^{temp} = \int_{\Omega} \mathbf{\Phi}_I^{\mathrm{T}} \rho c_{hc} \mathbf{\Phi}_J \mathrm{d}\Omega \tag{11.42}$$

Typical weighted residual formulation can be obtained using a generalized two-level time difference approximations:

$$\left\{ \mathbf{M}^{temp} / \Delta t + \left[\bar{\mathbf{K}}^{temp} + \mathbf{K}^3 \right] \theta^{time} \right\} \mathbf{T}_{n+1}$$
$$+ \left\{ -\mathbf{M}^{temp} / \Delta t + \left[\bar{\mathbf{K}}^{temp} + \mathbf{K}^3 \right] (1 - \theta^{time}) \right\} \mathbf{T}_n = \bar{\mathbf{f}}^{temp} \tag{11.43}$$

where Δt is the time interval and θ^{time} is the time marching coefficient. Different values of θ^{time} lead to different time difference approximations: $\theta^{time} = 0$ for forward-difference approximation (or Euler difference approximation); $\theta^{time} = 1/2$ for central-difference approximation (or Grank-Nicholson difference approximation); $\theta^{time} = 2/3$ for Galerkin difference approximation and $\theta^{time} = 1$ for backward-difference approximation.

Then the solution of Equation (11.43) yields

$$\left\{ \mathbf{M}^{temp} / \Delta t + \left[\bar{\mathbf{K}}^{temp} + \mathbf{K}^3 \right] \theta^{time} \right\} \mathbf{T}_{n+1} =$$
$$\bar{\mathbf{f}}^{temp} + \left\{ \mathbf{M}^{temp} / \Delta t + \left[\bar{\mathbf{K}}^{temp} + \mathbf{K}^3 \right] (\theta^{time} - 1) \right\} \mathbf{T}_n \tag{11.44}$$

Setting the following two working matrices

$$\mathbf{K}_{t1} = \mathbf{M}^{temp} / \Delta t + \left[\bar{\mathbf{K}}^{temp} + \mathbf{K}^3 \right] \theta^{time}$$
$$\mathbf{K}_{t2} = \mathbf{M}^{temp} / \Delta t + \left[\bar{\mathbf{K}}^{temp} + \mathbf{K}^3 \right] (\theta^{time} - 1) \tag{11.45}$$

Then Equation (11.44) can be written as

$$\mathbf{K}_{t1} \mathbf{T}_{n+1} = \bar{\mathbf{f}}^{temp} + \mathbf{K}_{t2} \mathbf{T}_n \tag{11.46}$$

where the load vector $\bar{\mathbf{f}}^{temp}$ can be obtained as

$$\bar{\mathbf{f}}^{temp} = \tilde{\mathbf{f}}_n^{temp} \left(1 - \theta^{time} \right) + \tilde{\mathbf{f}}_{n+1}^{temp} \theta^{time} \tag{11.47}$$

11.2 Thermoelastic problems

11.2.1 Modeling of the thermal strain and stress

Linear thermoelasticity deals with displacement and stress field in elastic solids caused by the thermal expansion due to the temperature change in the solids. Suppose the change of temperature is $\Delta T = T - T_0$, the resultant strain of (linear) thermal expansion for an isotropic medium becomes

$$\boldsymbol{\varepsilon}_0 = \left\{ \alpha_{exp}\Delta T \quad \alpha_{exp}\Delta T \quad \alpha_{exp}\Delta T \quad 0 \quad 0 \quad 0 \right\}^{\mathrm{T}} \tag{11.48}$$

where α_{exp} [/°C] is the thermal expansion coefficient and $\boldsymbol{\varepsilon}_0$ can be treated as the initial strain. Using the generalized Hook's law, the stresses related to the strains are obtained as

$$\boldsymbol{\sigma}^{ther} = \mathbf{c}\left(\tilde{\boldsymbol{\varepsilon}} - \boldsymbol{\varepsilon}_0 \right) \tag{11.49}$$

where $\boldsymbol{\sigma}^{ther}$ is called thermal stress, $\mathbf{c}$ is the matrix of material constants, and $\tilde{\boldsymbol{\varepsilon}}$ is the mechanical strain. By replacing the compatible strain $\tilde{\boldsymbol{\varepsilon}}$ with the smoothed strain $\overline{\boldsymbol{\varepsilon}}$ constructed in Chapter 4, the smoothed thermal stress can be expressed as

$$\overline{\boldsymbol{\sigma}}^{ther} = \mathbf{c}\left(\overline{\boldsymbol{\varepsilon}} - \boldsymbol{\varepsilon}_0 \right) \tag{11.50}$$

11.2.2 Discrete system equations

For an isotropic solid medium, the linear steady-state thermoelasticity problem without considering the thermal-mechanical coupling effects and the inertia can be formulated by following two steps.

First, we obtain the present temperature field by solving the heat transfer equations (with proper thermal boundary conditions) presented in Section 11.1.2. After obtaining the temperature field, we can then get the distribution of the initial strains over the problem domain, as defined in Equation (11.48).

Second, by taking the initial strain into account, together with the usual mechanical forces $\mathbf{b}$ and mechanical boundary conditions, we can now consider the corresponding mechanical problem governed by the following equations.

$$\mathbf{L}_d^{\mathrm{T}}\boldsymbol{\sigma}^{ther} + \mathbf{b} = \mathbf{0} \tag{11.51}$$

$$\mathbf{u} = \mathbf{u}_\Gamma \qquad \text{on } \Gamma_u \tag{11.52}$$

$$\mathbf{L}_n^{\mathrm{T}}\boldsymbol{\sigma}^{ther} = \mathbf{t}_\Gamma \qquad \text{on } \Gamma_t \tag{11.53}$$

The energy functional corresponding to the above equations is as

$$\Pi_P(\mathbf{u}) = \int_\Omega \left(\frac{1}{2}\tilde{\boldsymbol{\varepsilon}}^{\mathrm{T}}\mathbf{c}\tilde{\boldsymbol{\varepsilon}} - \tilde{\boldsymbol{\varepsilon}}^{\mathrm{T}}\mathbf{c}\boldsymbol{\varepsilon}_0 - \mathbf{u}^{\mathrm{T}}\mathbf{b} \right) d\Omega - \int_{\Gamma_t} \mathbf{u}^{\mathrm{T}}\mathbf{t}_\Gamma d\Gamma \tag{11.54}$$

Using the generalized smoothed strains obtained using a proper set of N_s smoothing domains, we shall have the "smoothed" energy functional for the W^2 formulation

$$\overline{\Pi}_P(\mathbf{u}) = \sum_{i=1}^{N_s} \frac{1}{2}V_i^s\overline{\boldsymbol{\varepsilon}}^{\mathrm{T}}\mathbf{c}\overline{\boldsymbol{\varepsilon}} - \sum_{i=1}^{N_s} V_i^s\overline{\boldsymbol{\varepsilon}}^{\mathrm{T}}\mathbf{c}\boldsymbol{\varepsilon}_0 - \int_\Omega \mathbf{u}^{\mathrm{T}}\mathbf{b}d\Omega - \int_{\Gamma_t} \mathbf{u}^{\mathrm{T}}\mathbf{t}_\Gamma d\Gamma \tag{11.55}$$

The stationary condition of the energy functional defined in the above equation leads to:

$$\begin{aligned} \delta\overline{\Pi}_P(\mathbf{u}) &= \sum_{i=1}^{N_s} V_i^s\left(\delta\overline{\boldsymbol{\varepsilon}}\right)^{\mathrm{T}}\mathbf{c}\overline{\boldsymbol{\varepsilon}} - \sum_{i=1}^{N_s} V_i^s\left(\delta\overline{\boldsymbol{\varepsilon}}\right)^{\mathrm{T}}\mathbf{c}\boldsymbol{\varepsilon}_0 \\ &\quad - \int_\Omega \left(\delta\mathbf{u}\right)^{\mathrm{T}}\mathbf{b}d\Omega - \int_{\Gamma_t} \left(\delta\mathbf{u}\right)^{\mathrm{T}}\mathbf{t}_\Gamma d\Gamma = 0 \end{aligned} \tag{11.56}$$

By substituting the approximated displacement and smoothed strains into the above equation and invoking the arbitrariness of virtual nodal temperature, we have the following discretized algebraic system equation:

$$\overline{\mathbf{K}}\overline{\mathbf{d}} = \overline{\mathbf{f}}^{ther} \tag{11.57}$$

where $\overline{\mathbf{K}}$ is the global stiffness matrix as same as that defined for mechanical problems with the form of

$$\overline{\mathbf{K}}_{IJ} = \sum_{k=1}^{N_s} V_k^s \overline{\mathbf{B}}_I^{\mathrm{T}}\mathbf{c}\overline{\mathbf{B}}_J \tag{11.58}$$

$\mathbf{d}$ is the vector of displacements and $\overline{\mathbf{f}}^{\,ther}$ is the force vector has the following form

$$\overline{\mathbf{f}}_I^{\,ther} = \sum_{k=1}^{N_s} \overline{\mathbf{B}}_I^{\,temp} \mathbf{c}\boldsymbol{\varepsilon}_0 \mathrm{d}\Omega + \int_\Omega \boldsymbol{\Phi}_I^{\mathrm{T}} \mathbf{b}\mathrm{d}\Omega + \int_{\Gamma_t} \boldsymbol{\Phi}_I^{\mathrm{T}} \mathbf{t}_\Gamma \mathrm{d}\Gamma \tag{11.59}$$

The first part in the right hand of the above equation is induced by the thermal strains and is called thermal load.

11.3 Numerical examples

A number of benchmark problems have been studied to examine the performance of present S-PIMs. In the study, the following *equivalent energy norm* for heat transfer problem is used as error indicators to measure the error in the numerical results quantitatively [10].

$$U_e = \int_\Omega \mathbf{g}^{\mathrm{T}} \mathbf{k} \mathbf{g} \mathrm{d}\Omega = \begin{cases} \displaystyle\sum_{k=1}^{N_s} V_k^s \left(\overline{\mathbf{g}}^{\mathrm{T}} \mathbf{k} \overline{\mathbf{g}}\right) & \text{(for S-PIM)} \\[4mm] \displaystyle\sum_{i=1}^{N_e} V_i^e \left(\mathbf{g}^{\mathrm{T}} \mathbf{k} \mathbf{g}\right) & \text{(for FEM)} \end{cases} \tag{11.60}$$

where $\mathbf{k}$ is the vector of (anisotropic) thermal conductivity coefficients, V_k^s is the volume of the kth smoothing domain, and V_k^e is the volume of the kth element. Note that the integration in Equation (11.60) is conducted by the summation over smoothing domains for smoothed point interpolation methods and over elements for FEM.

Example 11.3.1 1D thermal fin

First, a 1D thermal fin problem with length L and uniform cross-section is considered. The fin is subjected to a uniform inner heating and two types of boundary conditions, as illustrated in FIGURE 11.1. The exact solution of temperature for this simple problem is given by

$$T(x) = -\frac{Q_v}{2k} x^2 + \frac{q_{\Gamma_2} + Q_v}{k} x + T_{\Gamma_1} \tag{11.61}$$

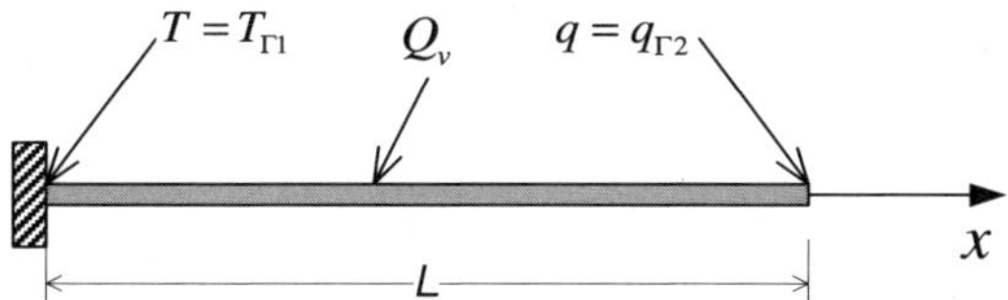

FIGURE 11.1 A 1D thermal fin problem with length L and uniform cross-section. The fin is subjected to a uniform inner heating and two types of boundary conditions.

In the computation, the parameters are taken as L=1.0m, k=5.0W/m°C, Q_v=100.0 W/m^3, $T_{\Gamma 1}$=0.0°C and $q_{\Gamma 2}$=200W/m^2. With these parameters, the exact value of the equivalent energy can be obtained using Equation (11.60) as 1.2667E+04.

The computed equivalent energy values of both linear NS-PIM and FEM solutions are plotted against DOF (also the number of field nodes) in FIGURE 11.2. It can be found NS-PIM produces exactly the same results as FEM when the problem domain is discretized with only two nodes. In this case, two smoothing domains are generated with respect to these two nodes which share the same gradient field. Hence the gradient smoothing has no effect at all, and

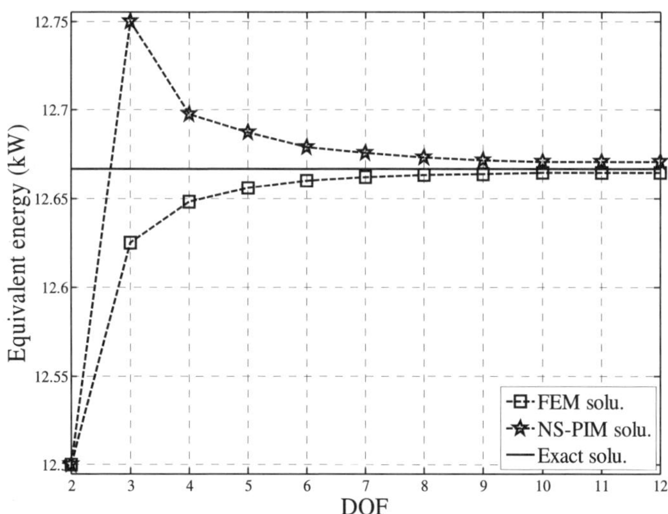

FIGURE 11.2 Solutions (in equivalent energy norm) converging to the exact solution for the problem of 1D thermal fin problem obtained using linear NS-PIM and FEM with the same set of meshes.

the discretized system equations of NS-PIM are exactly the same as that of FEM. When the number of field nodes is bigger than two, meaning there are more than two different elements, the gradient smoothing operation will take effect and NS-PIM gives an upper bound solution in the equivalent energy norm with respect to both FEM and exact solutions owning to the softening effect. On the contact, FEM gives lower bound solutions due to the overly-stiff behavior. With the increase of DOF, NS-PIM and FEM solutions approach to the exact one monotonically from above and lower, respectively.

The upper bound property of NS-PIM solutions in the equivalent energy norm for heat transfer problem is consistent with the counterpart for elasticity problem discussed in Chapter 6. This simple 1D numerical problem also indicates that for heat transfer problem, NS-PIM can also provide upper bound solutions in the norm of equivalent energy. This property will be further studied in the following problems.

Example 11.3.2 2D heat conduction with convection

The second problem describes a 2D domain as shown in FIGURE 11.3. This is a typical problem used for benchmarking by NAFEMS [30] and has been studied by other researchers, such as Huang and Usmani by using adaptive FEM technique [31] and Pepper using one meshless method [32]. The setting of the problem is as follows. The left edge of the domain is insulated; a fixed temperature of 100°C is applied along the lower side; a surface convection of 0°C is set along the right and upper sides with h_T=750W/(m^2°C) and k=52W/m°C. Due to the severe discontinuity in boundary conditions at point A,

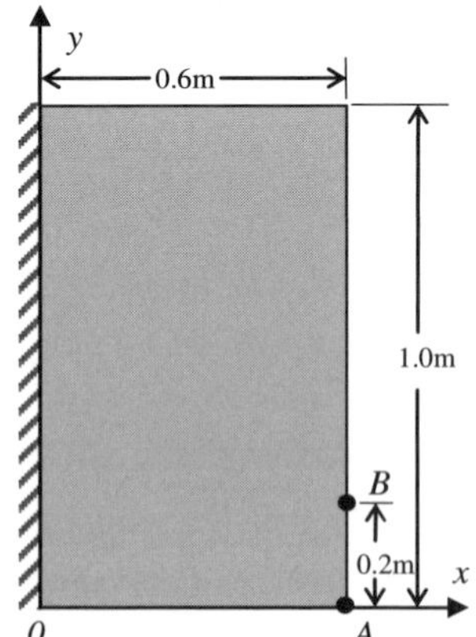

FIGURE 11.3 A 2D heat transfer problem with convection on part of the boundary.

there exists quasi-singular type behavior around this location. The analytical solution of the temperature at point B is 18.2535°C and this point is also used for comparison.

TABLE 11.1 lists the computed temperature results at point B obtained using linear FEM and two NS-PIM models with the same set of three-node triangular mesh. It can be clearly seen that compared to the linear FEM, the NS-PIM models provide more accurate results. Between the two models of NS-PIM, the quadratic one with T6/3-scheme shows even better performance.

TABLE 11.1 Comparison of computed temperature at point B for the 2D heat transfer problem with convection

Method	Temperature at point B (°C)	Relative error (%)[*]	Number of elements	Number of nodes
FEM-Tr3	18.4262	0.946	534	300
NS-PIM-Tr3	18.4223	0.925	534	300
NS-PIM-Tr6/3	18.3926	0.762	534	300

[*] The analytical temperature at point B is 18.2535°C.

Example 11.3.3 2D thermoelastic problem

Following, a 2D thermoelastic problem is studied. A 2D rectangular plate with length L and width W is considered, as shown in FIGURE 11.4. It is fixed at the left end and subjected to different types of thermal boundary conditions: adiabatic condition along the upper edge, temperature condition along the left edge, heat flow condition along the right edge, and exchange condition along the lower edge.

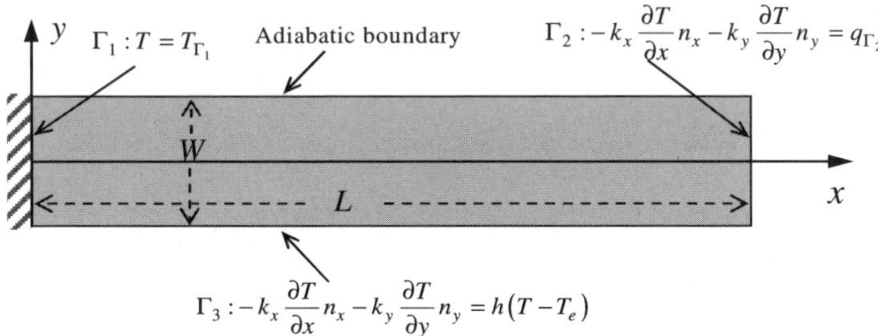

FIGURE 11.4 A 2D cantilever solid with anisotropic material and subjected to four types of thermal boundary conditions.

The material of the problem is anisotropic and the parameters are taken as: L=0.05m, W=0.01m, E=2.225×10^{11}Pa, v=0.3, k_x=15 W/(m°C), k_y=10 W/(m°C),

$T_{\Gamma 1}$=0°C, $q_{\Gamma 2}$=-4000W/m^2, h_T=1500W/(m^2°C), T_e=200°C and α_{exp} =1.02×10^{-5}/°C. As no analytical solution is available for this problem, a reference solution is obtained using FEM with a very fine mesh (total 329,217 nodes) for comparison purpose.

First, the problem is calculated using both NS-PIM models and FEM with the same triangular mesh of 369 nodes. The contours of the computed temperature gradient in y direction are plotted in FIGURE 11.5. We can find that the gradient field of the NS-PIM models is smoother than that of FEM and agrees better with the reference ones. Compared to the linear NS-PIM-Tr3, the quadratic NS-PIM-Tr6/3 performs a little better, mainly showing on the computed gradient fields around the lower left corner. FIGURE 11.6 shows the distribution of thermal stress in x direction obtained using different methods. We may find again that NS-PIM models perform better than the linear FEM. For the two models of the NS-PIM, the difference between the results is very small.

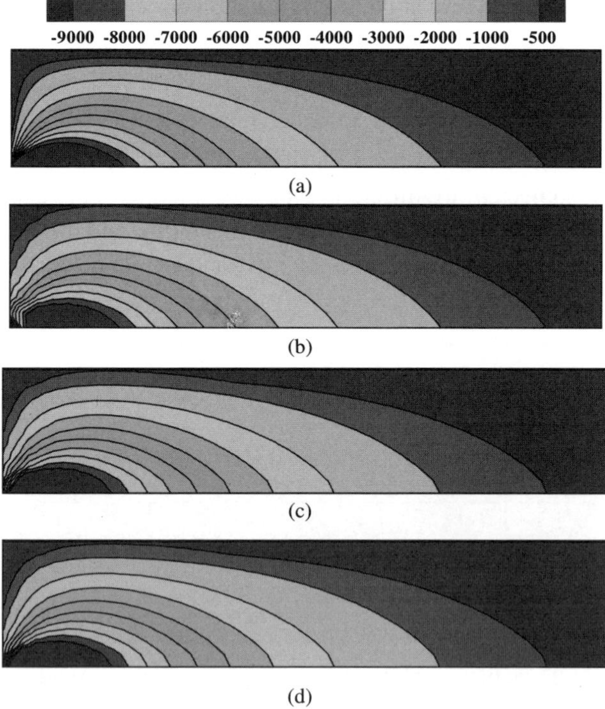

FIGURE 11.5 Contours of temperature gradient in y direction: (a) reference solution; (b) FEM-Tr3 solution; (c) NS-PIM-Tr3 solution; (d) NS-PIM-Tr6/3 solution.

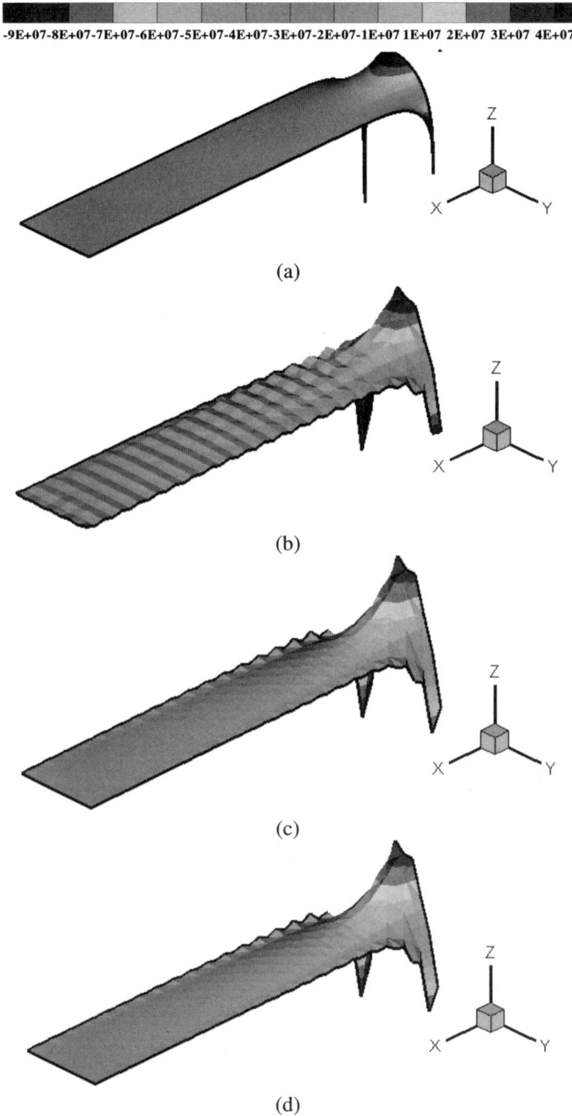

FIGURE 11.6 Contours of thermal stress in x direction: (a) reference solution; (b) FEM-Tr3 solution; (c) NS-PIM-Tr3 solution; (d) NS-PIM-Tr6/3 solution.

Second, we study this problem with a set of triangular meshes and calculate the equivalent energy of the numerical results using Equation (11.60). Against the increase of DOF, FIGURE 11.7 plots converging process of the calculated equivalent energies. We summarize the following points.

1) With the increase of DOF, all the numerical solutions converge to the reference one.

2) All the S-PIM models give upper bound corresponding to the FEM, showing their softer stiffness owing to the gradient smoothing operation introduced.

3) For this case with homogeneous temperature condition ($T_{\Gamma 1}$=0°C), the NS-PIM models give upper bound solutions and other models give lower bound solutions. This phenomenon is similar to that for elasticity problems, i.e. NS-PIM models can provide upper bound solutions in strain energy norm for those problem with homogeneous essential boundary conditions contrasting to FEM and ES-PIM which obtain lower bound solutions.

4) The ES-PIM models give much tighter lower bound solutions than the FEM-Tr3 showing better accuracy of the results.

5) Compared with the linear models with T3-scheme, the quadratic NS-PIM and ES-PIM models with T6/3-scheme can obtain more accurate results and give tighter bounds to the reference one. This outcome is consistent with that for solid mechanics problems and the reason has been discussed in Section 6.4.2.

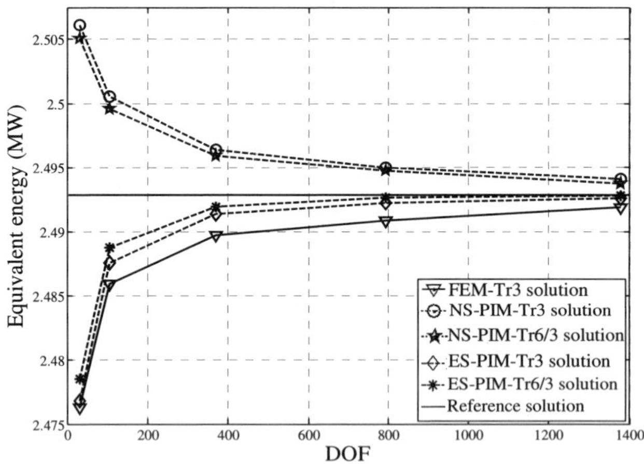

FIGURE 11.7 Solutions (in equivalent energy norm) converging to the reference solution for the 2D thermoelastic problem obtained using different methods and same set of triangular meshes (The reference solution is obtained using FEM with a mesh of total 329,217 nodes).

To further investigate the bound property, we plot the convergence process of computed thermal strain energies, as shown in FIGURE 11.8, obtained using the following equation for the numerical solutions of S-PIM and FEM.

$$
U_{ts} = \begin{cases} \left[\displaystyle\sum_{i=1}^{N_s} V_i^s \left(\overline{\boldsymbol{\varepsilon}}^{num}(\mathbf{x}_c) - \boldsymbol{\varepsilon}_0(\mathbf{x}_c) \right)^{\mathrm{T}} \mathbf{c} \left(\overline{\boldsymbol{\varepsilon}}^{num}(\mathbf{x}_c) - \boldsymbol{\varepsilon}_0(\mathbf{x}_c) \right) \right]^{\frac{1}{2}}, \text{ for S-PIM} \\[2em] \left[\displaystyle\sum_{i=1}^{N_e} V_i^e \left(\tilde{\boldsymbol{\varepsilon}}^{num}(\mathbf{x}_c) - \boldsymbol{\varepsilon}_0(\mathbf{x}_c) \right)^{\mathrm{T}} \mathbf{c} \left(\tilde{\boldsymbol{\varepsilon}}^{num}(\mathbf{x}_c) - \boldsymbol{\varepsilon}_0(\mathbf{x}_c) \right) \right]^{\frac{1}{2}}, \text{ for FEM} \end{cases}
\tag{11.62}
$$

where $\mathbf{x}_c$ is the "center" of the smoothing domain or that of the element.

This time we find that NS-PIM models give lower bound, and FEM gives upper bound solutions to the reference one. The reason is that for the thermoelastic study, the thermal strain ($\boldsymbol{\varepsilon}_0$) acts as a kind of constraints, which makes the problem behave like a "displacement-driving" problems, where inhomogeneous displacement boundary conditions are imposed. We know that for "displacement-driving" problems, the stiff FEM model will produce upper bound, and the soft NS-PIM will produce lower bound. This results in the role-exchange of providing bound solutions by the NS-PIM and FEM. Note that for this type of problems, NS-PIM and FEM solutions converge to the reference one from opposite sides and hence we can still obtain a bound of the exact one.

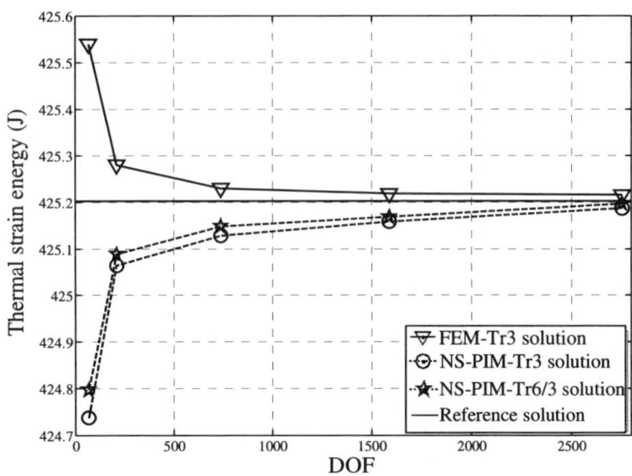

FIGURE 11.8 Solutions (in thermal strain energy) converging to the reference solution for the 2D thermoelastic problem obtained using different methods and same set of triangular meshes (The reference solution is obtained using FEM with a very mesh with a total of 329,217 nodes).

Finally, a transit-state analysis is conducted for this problem with following parameters: L=0.5m, W=0.1m, E=2.225×10^{11}Pa, v=0.3, ρ=3000kg/m^3, k_x=50.0 W/(m°C), k_y=50.0W/(m°C), $T_{\Gamma1}$=0°C, $q_{\Gamma2}$=-4000W/m^2, h_T=1500W/(m^2°C), T_e=200°C, Q_v=0W/m^3, c_{hc}=50J/(kg°C) and T_0=25°C. In the computing, the time increment is set to be Δt=0.1seconds. FIGURE 11.9 shows the time history of equivalent energy of numerical results obtained using ES-PIM and FEM models. Compared to the results of FEM, the ES-PIM solutions are much closer to the reference one and the results of quadratic ES-PIM are in very good agreement with the reference especially. All the transient responses will arrive at the steady state in about 160 seconds.

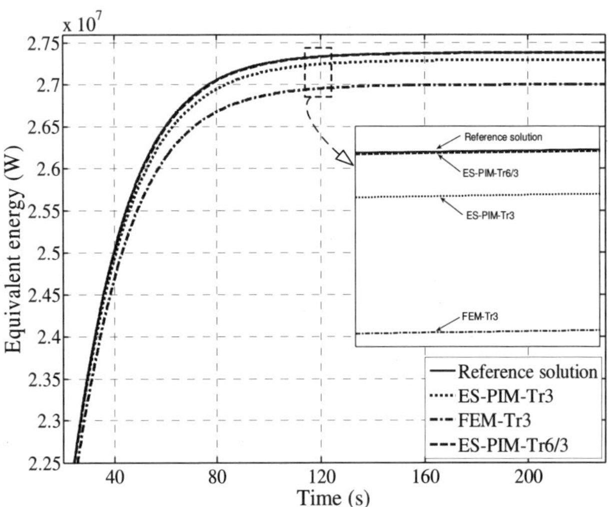

FIGURE 11.9 Time history of equivalent energy for the 2D transit-state analysis obtained using ES-PIM and FEM models.

Example 11.3.4 3D engine pedestal

A practical engine pedestal with complex geometry and manufactured by the plasma deposition-layered technique [33] is studied here. As shown in FIGURE 11.10, this pedestal part is made of superalloy material and subjected to three types of boundary conditions, including temperature condition on the upper surface of the part, heat flow condition on the inner surface of the cylindrical hole and heat exchange condition on the outer surface of the circular plate. This problem has also been studied by other researchers [28, 34].

In the computation, the parameters are taken as: ρ=3000kg/m^3, k_x=30W/(m°C), k_y=40W/(m°C), k_z=50W/(m°C), Q_v=0W/m^3, $T_{\Gamma1}$=0°C, $q_{\Gamma2}$=-

6000W/m^2, h_T=1000W/(m^2°C), T_e=500°C, c_{hc}=50J/(kg°C), T_0=25°C and Δt=0.0002 seconds. A reference solution is obtained using FEM with a very fine mesh (total 12,344 nodes) for comparison purpose.

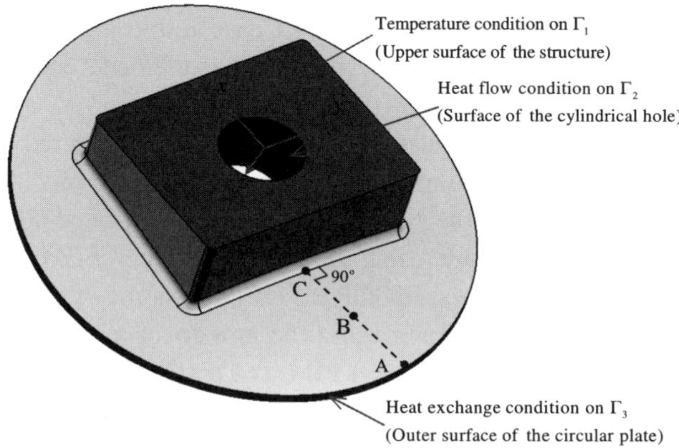

FIGURE 11.10 3D engine pedestal with complex geometry and subjected to three types of thermal boundary conditions.

First, we consider the steady-state of this heat transfer problem and the convergence property of the numerical solutions is studied with a set of four-node tetrahedral meshes. FIGURE 11.11 shows the converging process of the

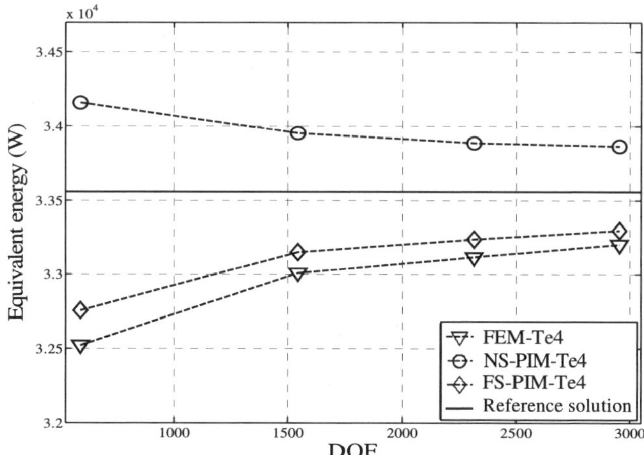

FIGURE 11.11 Solutions (in equivalent energy norm) converging to the reference solution for the 3D engine pedestal problem obtained using NS-PIM, FS-PIM and FEM models with the same set of tetrahedral meshes (The reference solution is obtained using FEM with a very fine mesh with a total of 12,344 nodes).

numerical solutions in the equivalent energy norm, with the refinement of the model. For this 3D problem with complicated geometry and boundary conditions, we can find again that NS-PIM provides upper bound, while FS-PIM and FEM provide lower bound solutions. All the numerical results converge to the reference one with the increase of DOF. Compared to the solution of FEM, the counterpart of FS-PIM is much tighter to the reference one and provides more accurate results.

Then the transit-state of this heat transfer problem is considered and the time history of the temperature at three sample points is monitored. As shown in FIGURE 11.10, the three sample points (A, B and C) are located along a straight line. We studied this case by using both FS-PIM and FEM with a same tetrahedral mesh of 1,117 nodes. FIGURE 11.12 shows the temperature history of the three sample points obtained using the numerical models, together with the reference solution. It is clear that FS-PIM-Te4 performs better than the linear FEM, although the improvement for this 3D case is not as much as the 2D cases. The pictures also show that the system reaches the steady state after about 8 seconds.

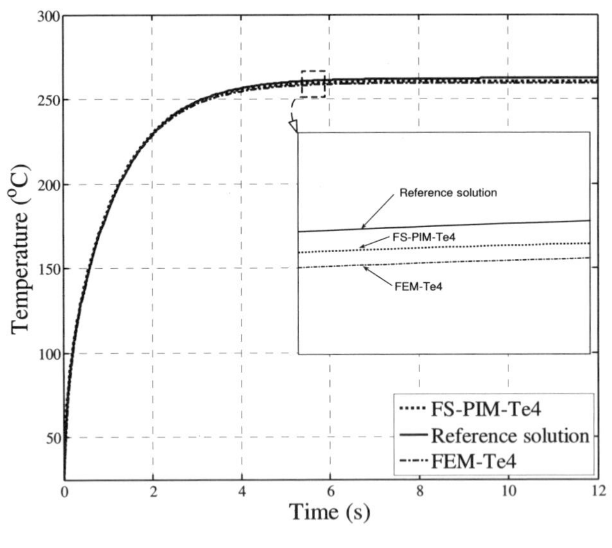

(a)

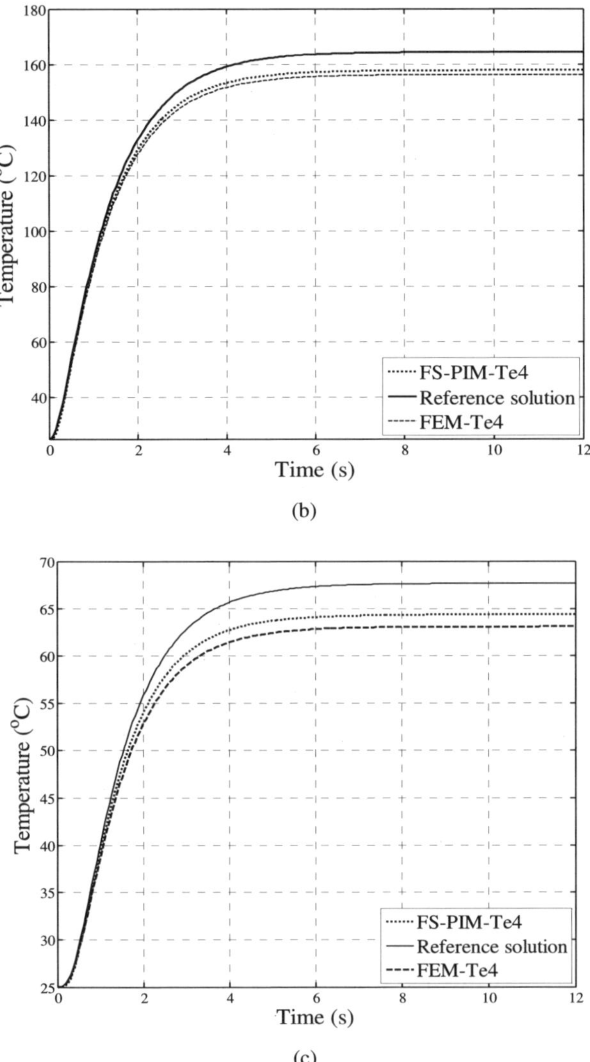

FIGURE 11.12 Time history of the temperature at three sample points on the 3D engine pedestal problem: (a) temperature history of point A; (b) temperature history of point B; (c) temperature history of point C.

Example 11.3.5 3D manufacturing system

Finally, a practical temperature-controlled system of the plasma deposition dieless manufacturing (PDM) is studied [33, 28]. The manufacturing system conducts a complicated process of rapid, transient, high-temperature, multi-parameters and combined with different physical and chemical responses. FIGURE 11.13 shows the model of the system and different types of boundary conditions: temperature condition on the bottom surface of the substrate, heat flow condition on the four side faces of the substrate and heat exchange condition on the inner surface of the seven tubes, where cooling waters will go through the substrate. The substrate and the turbine are made of same material.

In the computation, the parameters are taken as: ρ=6000kg/m^3, k_x=k_y=k_z=40W/(m°C), Q_v=0W/m^3, $T_{\Gamma 1}$=25°C, $q_{\Gamma 2}$=-20W/m^2, h_T=1500W/(m^2°C), T_e=100°C, c_{hc}=50J/(kg°C) and Δt=0.02 seconds. The initial temperature of the substrate is T_0=25°C and one layer of turbine is deposited with an initial temperature of 2000°C. A reference solution is obtained using FEM with a very fine mesh (total 12,859 nodes) for comparison purpose.

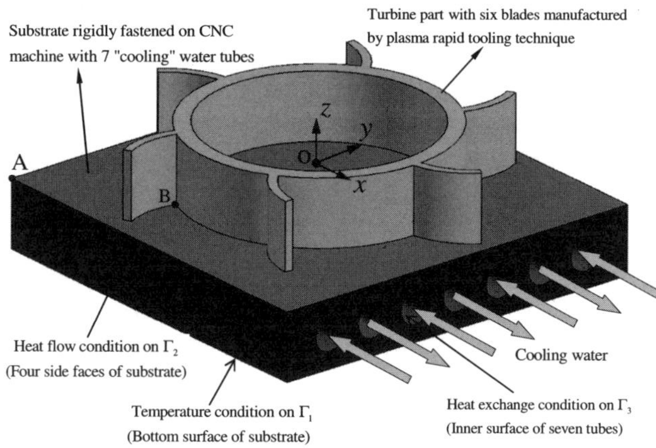

FIGURE 11.13 Simplified model of the 3D manufacturing system with complex geometry and subjected to three types of thermal boundary conditions.

For simplicity, we divide the whole model into two parts along the y-z plane and only one part is computed using both FS-PIM and FEM with same set of four-node tetrahedral meshes. FIGURE 11.14 shows the temperature distributions when the system reaches the steady state after 150 seconds. It can be found that the numerical results of FS-PIM agree well with the reference one and are more accurate than that of linear FEM. FIGURE 11.15 shows the temperature evolutions at three sample points A, B and O, which are marked in FIGURE 11.13. Compared to the reference results obtained by the fine mesh, the FS-PIM-Te4 provides much more accurate solutions than the linear FEM does with the same tetrahedral mesh.

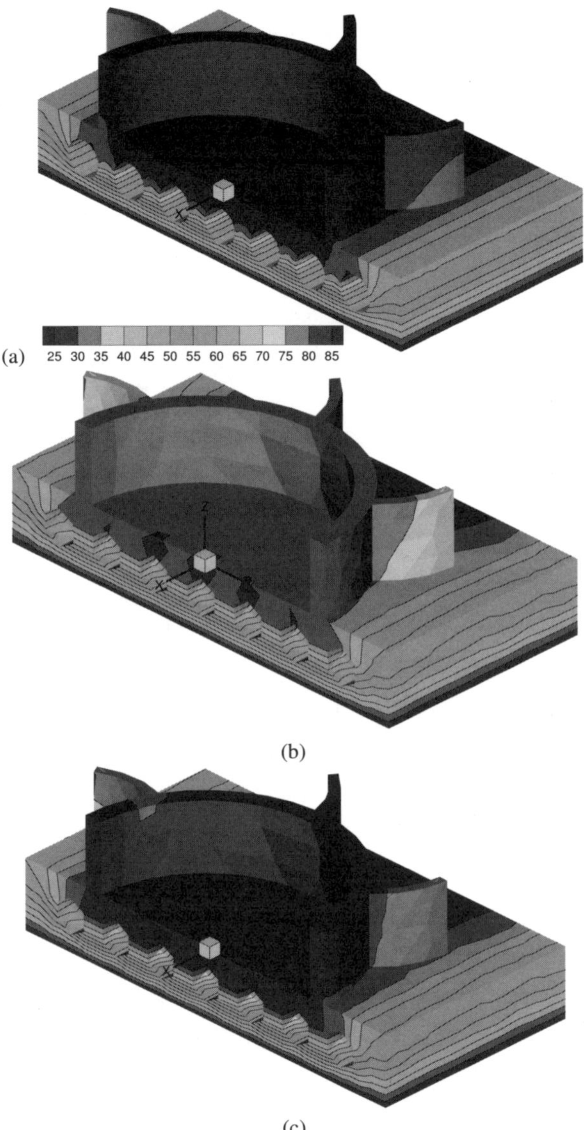

FIGURE 11.14 Contour of temperature distribution for the 3D manufacturing system, where the reference solution is obtained using FEM with a very fine mesh with a total of 12,859 nodes and the numerical solutions are obtained using FS-PIM and FEM models with the same four-node tetrahedral mesh of 1,167 nodes: (a) reference solution; (b) FEM-Te4 solution and (c) FS-PIM-Te4 solution.

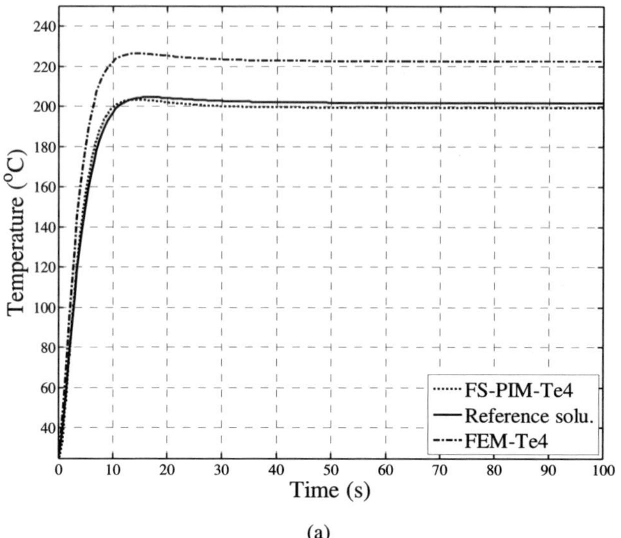

(a)

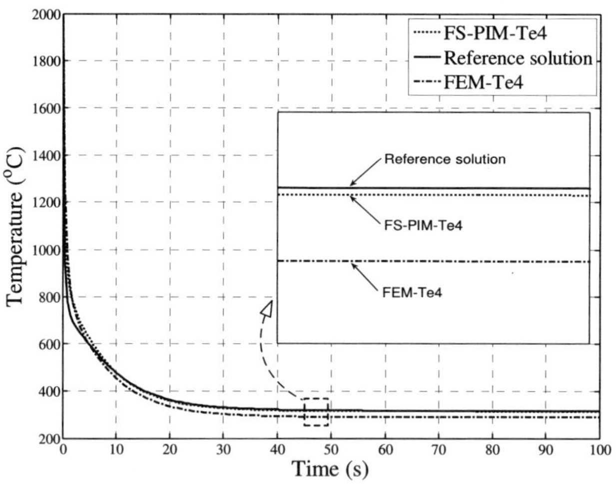

(b)

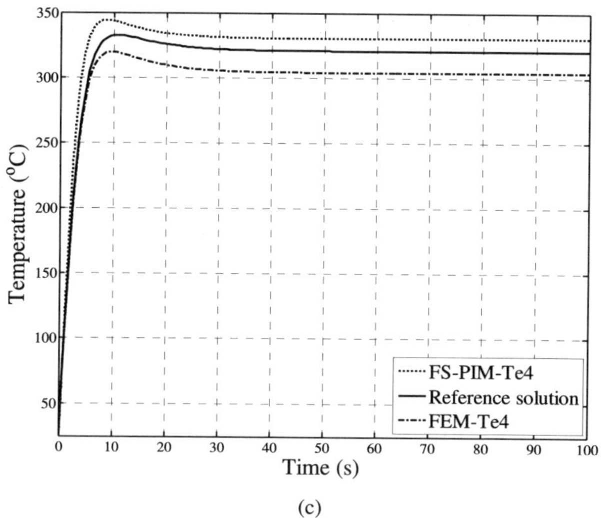

FIGURE 11.15 Temperature history of the three sample points for the 3D manufacturing system: (a) temperature history of point A; (b) temperature history of point B; (c) temperature history of point O.

11.4   Concluding remarks

S-PIM has been formulated for heat transfer and thermoelastic problems in this chapter. Some S-PIM models including NS-PIM, ES-PIM and FS-PIM are used to study a number of example problems. The implementation is straightforward and the S-PIM models have been found working well for various 1D, 2D and 3D heat transfer problems.

Remark 11.1 Easy implementation

Compared to the standard FEM, the S-PIM formulations for heat transfer problems change only the stiffness (conductivity) matrix of the discrete model.

Remark 11.2 Bound property

For the numerical results in the equivalent energy norm, the NS-PIM can provide upper bound to the exact solution for a flux driving heat transfer problem, as long as the mesh is sufficiently fine to provide sufficient softening effects. For the thermoelastic problems, the NS-PIM models will provide lower bound solutions, while the FEM gives upper bound solutions and they can bound the exact solution from both sides.

Remark 11.3 High accuracy

The S-PIM models can generally provide numerical solutions of better accuracy than the linear FEM models by using the same set of triangular/tetrahedral meshes.

Remark 11.4 ES-PIM/FS-PIM is temporally stable

The ES-PIM/FS-PIM model is temporally stable and hence works well for the transit problem, which is consistent with the conclusions for solid mechanics problems.

11.5   References

1.   Thomas, L. C., *Heat Transfer*, Prentice-Hall, International edition, 1992.

2.   Incropera, F.P. and DeWitt, D. P., *Fundamentals of Heat and Mass Transfer*, 5th ed., Wiley, New York, 2002.

3.   He, Z. C., Liu, G. R., Zhong, Z. H., Wu, S. C., Zhang, G. Y., Cheng, A. G., An edge-based smoothed finite element method (ES-FEM) for analyzing three-dimensional acoustic problems. *Computer Methods in Applied Mechanics and Engineering*. 199: 20-33, 2009.

4.   He, Z. C., Liu, G. R., Zhong, Z. H., Zhang, G. Y. and Cheng, A. G., Dispersion free analysis of acoustic problems using the alpha finite element method. *Computational Mechanics*, 46(6): 867-881, 2010.

5.   He, Z. C., Liu, G. R., Zhong, Z. H., Zhang, G. Y. and Cheng, A. G., Coupled analysis of 3D structural-acoustic problems using the edge-based smoothed finite element method/finite element method. *Finite Elements in Analysis and Design*, 46(12): 1114-1121, 2010.

6.   He, Z. C., Liu, G. R., Zhong, Z. H., Cui, X. Y., Zhang, G .Y. and Cheng, A. G., A coupled edge-/face-based smoothed finite element method for structural-acoustic problems. *Applied Acoustics*, 71(10): 955-964, 2010.

7. He, Z. C., Liu, G. R., Zhong, Z. H., Zhang, G. Y. and Cheng, A. G., A coupled ES-FEM/BEM method for fluid-structure interaction problems. *Engineering Analysis with Boundary Elements*, 35(1): 140-147, 2011.

8. He, Z. C., Cheng, A. G., Zhang, G. Y., Zhong, Z. H. and Liu, G. R., Dispersion error reduction for acoustic problems using the edge-based smoothed finite element method (ES-FEM). *International Journal for Numerical Methods in Engineering*, 86: 1322-1338, 2011.

9. Liu, G. R. and Nguyen-Thoi, T., Smoothed Finite Element Methods, CRC press, Boca Taton, USA, 2010.

10. Zienkiewicz, O. C. and Taylor, R. L., *The Finite Element Method*, 5th ed., Butterworth Heinemann, Oxford, UK, 2000.

11. Huebner, K. H., Dewhirst D. L., Smith, D. E., and Byrom, T. G., *Finite Element Method for Engineers*, 4th ed., John Wiley, New York, 2001.

12. Reddy, J. N., *The Finite Element Method in Heat Transfer and Fluid Dynamics*, CRC press, Boca Raton, 2000.

13. Huang Hou-Cheng and Usmani, A. S., Finite Element Analysis for Heat Transfer: Theory and Software, Springer-Verlag, New York, 1994.

14. Dusinberre, G. M., *Heat-transfer Calculations by Finite Differences*, International Textbook Co., Scranton, USA, 1961.

15. Croft, D. R. and Lilley D. G., *Heat Transfer Calculations using Finite Difference Equations*, Applied Science Publishers, London, UK, 1977.

16. Sladek, J., Sladek, V. and Zhang, Ch., Transient heat conduction analysis n functionally gradient aterials by the meshless local boundary integral equation method. *Computational Materials Sciences*, 28: 494-504, 2003.

17. Qian, L. F. and Batra, R. C., Three-Dimensional transient heat conduction in a functionally graded thick plate with a higher-order plate theory and a Meshless local Petrov-Galerkin method. *Computational Mechanics*, 35: 214-226, 2005.

18. Wang, H., Qin, Qing-Hua and Kang, Y. L., A meshless method for transient heat conduction in functionally graded materials. *Computational Mechanics*, 38(1): 51-60, 2006.

19. Chen, C. S., Rashed, Y. F. and Golberg, M. A., A mesh-free method for linear dissusion equations. *Numerical Heat Transfer Part B-Fundamentals*, 33(4): 469-486, 1998.

20. Rao, S. S., *The finite element method in engineering*. Pergamon Press, 1992.

21. Liu, G. R., *Meshfree Methods: Moving beyond the Finite Element Method*, 2nd Edition, CRC press, Boca Taton, USA, 2009.

22. Liu, G. R., A generalized Gradient smoothing technique and the smoothed bilinear form for Galerkin formulation of a wide class of computational methods. *International Journal of Computational Methods*, 5(2): 199-236, 2008.

23. Liu, G. R., A G space theory and a weakened weak (W^2) form for a unified formulation of compatible and incompatible methods, Part I Theory, *International Journal for Numerical Methods in Engineering*, 81: 1093-1126, 2009.

24. Liu, G. R., A G space theory and a weakened weak (W^2) form for a unified formulation of compatible and incompatible methods, Part II Application to solid mechanics problems, *International Journal for Numerical Methods in Engineering*, 81: 1127-1156, 2009.

25. Wu, S. C., Liu, G. R., Zhang, H. O. and Zhang, G. Y., A node-based smoothed point interpolation method (NS-PIM) for three-dimensional thermoelastic problems. *Numerical Heat Transfer, Part A: Applications*, 54: 1121-1147, 2008

26. Wu, S. C., Liu, G. R., Zhang, H. O. and Zhang, G. Y., A node-based smoothed point interpolation method (NS-PIM) for thermoelastic problems with solution bounds. *International Journal of Heat and Mass Transfer*, 52: 1464-1471, 2009.

27. Wu, S. C., Liu, G. R., Zhang, H. O., Xu, X. and Li, Z. R., A node-based smoothed point interpolation method (NS-PIM) for three-dimensional heat transfer problems. *International Journal of Thermal Sciences*, 48(7): 1367-1376, 2009.

28. Wu, S. C., Liu, G. R., Zhang, H. O., Nguyen-Thoi, T. and Zhang, G. Y., An edge-based smoothed point interpolation method (ES-PIM) for heat transfer analyses of rapid manufacturing system. *International Journal of Heat and Mass Transfer*, 53(9-10): 1938-1950, 2010.

29. Cheng, J., Liu, G. R., Li, T. C., Wu, S. C. and Zhang, G. Y., An ES-PIM method with cell death and birth technique for simulating heat transfer in the concrete dam construction process. *Journal of Engineering Mechanics-ASCE*, 138(1): 133-142, 2012.

30. Barlow, J. and Davies, G. A. O., Selected FE benchmarks in structural and thermal analysis. *Technical Report FEBSTA REV 1, NAFEMS*, 1986.

31. Huang, H. C. and Usmani, A. S., *Finite Element Analysis for Heat Transfer*, Springer-Verlag, London, UK, 1994.

32. Pepper, D. W., *Meshless methods*, in *Handbook of Numerical Heat Transfer*, 2nd ed., edited by Minkowycz, W. J., Sparrow, E. M. and Murthy, J. Y., Johnm Wiley & Sons, Inc, New Jersey, 2006.

33. Wu, S. C., Zhang H. O. and Wang, G. L., Numerical and experimental evaluation of the temperature and stress fields during plasma deposition dieless manufacturing. Proc. Of 15th ISEM, Pittsburgh, edited by Rajurkar, K. P., U.S.A., April 2007.

34. Wang, G. L., Wu, S. C. and Zhang H. O., Numerical simulation of temperature field on complicated parts during plasma deposition dieless manufacturing, *Transactions of the China Welding Institution*, 28: 49-52, 2007.

Chapter 12

Singular CS-RPIM for Fracture Mechanics Problems

Fracture mechanics is an important topic in ensuring the mechanical performance of the materials and structural components. The fracture process in material is intimately connected with some of the very important parameters in fracture mechanics, such as the stress intensity factors (SIFs) and the energy release rate [1].

For line cracks in isotropic and elastic bulk materials, it is well known that the stresses and strains near the crack tip are of square-root singularity: $\sigma_{ij} \sim 1/\sqrt{r}$ or $\varepsilon_{ij} \sim 1/\sqrt{r}$ (where r is the radial distance from the crack tip) [1]. To capture this singularity, numerical simulations have been carried out using special schemes in the finite element method (FEM) [2-5] and newly developed meshfree methods [6-12] settings. When the FEM is used, a layer of eight-node quarter-point element or the six-node quarter-point element (collapsed quadrilateral) is often used at the crack tip to model the singularity [2-4]. To make the singular elements compatible with other standard elements, the quadratic elements need to be used for the whole domain. Otherwise, transition elements or additional treatments are needed to ridge between the crack tip elements and the standard elements [13, 14]. Recently, a so-called extended finite element method (XFEM) has been developed to model arbitrary discontinuities in meshes using local enrichment functions [15-18]. Because the enrichment is performed only partially in the elements at the crack tip, the partitions-of-unity in the XFEM can be lost in the transition zones [19].

In our previous work, a simple and effective construction of singular stress field in the schemes of ES-FEM (i.e. ES-PIM-Tr3) and NS-FEM (i.e. NS-PIM-Tr3) has been successfully developed using triangular cells [20-27]. Because triangular cells can be used, the automation removing the worries related to difficulty in re-meshing. Also because triangular cells can be used with good accuracy, the usual accuracy problems are resolved. Considering the stress singularity in the vicinity of the crack tip, a five-node singular crack tip element has been developed. As no derivatives of shape functions are needed in the S-PIM and S-FEM models, this type of singular element can be easily formulated and implemented without the need of mapping that involves the Jacobian matrix. Therefore, the mesh sensitivity issue is well resolved, meaning that the quality of the shape of the triangular cells is no longer very important. Using such a singular S-PIM and S-FEM, the stress intensity factors can also be easily evaluated by an appropriate treatment in the domain of the interaction integral. Intensive numerical studies have shown that the singular ES-FEM can significantly improve the accuracy of the computed stress intensity factors and energy release rate in comparison with the standard FEM and singular FEM. In terms of the *J*-integral or the energy release rate, the singular NS-FEM can produce upper bound solutions, corresponding to the lower bound solutions produced by the displacement based compatible FEM.

Following the work on singular NS-FEM and singular ES-FEM reported in [20-27], this chapter presents another powerful singular CS-RPIM (sCS-RPIM) to study linear fracture problems. The formulation follows largely the work of [28]. In the sCS-RPIM, a five-node singular element is used to simulate the singular stress field near the crack tip, which is similar to those singular elements used in singular NS-FEM and singular ES-FEM. The interaction integral method [29-31] is used to compute the stress intensity factors. Compared to the singular NS-FEM and singular ES-FEM, the present method uses the background cells as the smoothing domains directly, which makes the implementation more straightforward. The RPIM shape functions used are naturally immune from the interpolation singularity problem and provide more freedom for the selection of local support nodes [32]. The singular CS-RPIM works well with triangular elements which can be generated automatically for complicated domain. This property is also very important for crack propagation problems [25].

12.1 Formulation of the singular CS-RPIM

We consider only 2D problems with line crack and the formulation is introduced by mainly emphasizing the difference between the sCS-RPIM and the general CS-RPIM presented in Chapter 8.

12.1.1 Approximation of displacement field

The problem domain is first represented by a set of N_c nonoverlapping and seamless (NOSL) triangular background cells with N_n field nodes, as in any S-PIM model, such that $\boxed{\Omega} = \bigcup_{i=1}^{N_c} \boxed{\Omega}_i^c$. Same as the CS-PIM discussed in Chapter 8, these background cells serve also as smoothing domains [33, 34]. In the sCS-RPIM, a layer of singular cells are created, as shown in FIGURE 12.1, to simulate the singularity of stress field near the crack tip. The singular cell is formed by adding in one node on each edge directly connecting to the crack tip. The location of the added node is usually (do not have to) at the one quarter length of the edge from the crack tip, as shown in FIGURE 12.1. The arbitrariness of the location of the edge node is a distinct feature of the singular S-PIM models comparing to the singular FEM models. A detailed examination and discussion on this is reported in [21]. We note also that the number of the edge nodes needed is minimum. As shown in FIGURE 12.1, only seven more nodes are added. In addition, there is no need for the so-called "transition

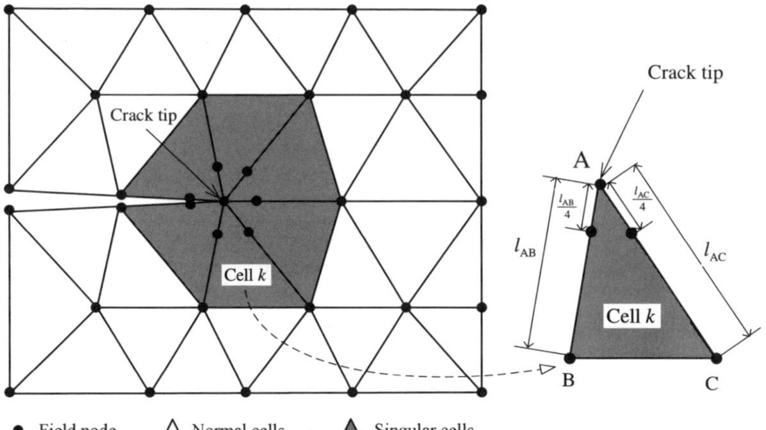

FIGURE 12.1 2D problem domain is discretized with a set of three-node triangular cells with a layer of five-node singular cells. The singular cell is formed by adding one node on the edge of the three-node cell directly connecting to the crack tip.

elements", even though we use mixed order of interpolation here, because of the use of W^2 formulation in the present sCS-RPIM.

12.1.1.1 Displacement approximation in normal cells

For all the normal cells, the displacement field is approximated using the RPIM shape functions with the support nodes selected using the edge-based T4-scheme, as shown in FIGURE 8.1b. This procedure is exactly same as that in the CS-RPIM-Tr4 for 2D problems, as described in Chapter 8.

12.1.1.2 Displacement approximation in singular cells

Considering the stress singularity at the crack tip, shape functions are constructed with a term of $\sqrt{r}$ for the displacement approximation. This shall lead to a desired $1/\sqrt{r}$ singular stress field [20-27].

For singular cells in the sCS-PIM, the points of interest (quadrature points) may locate on the edges connecting to the crack tip (singular edge), on the remote edges of the crack tip node (base edge), or on the interior lines parallel to the base edge (parallel lines), as shown in FIGURE 12.2. At these three types of quadrature points, we use different basis function for interpolation to construct the displacement field.

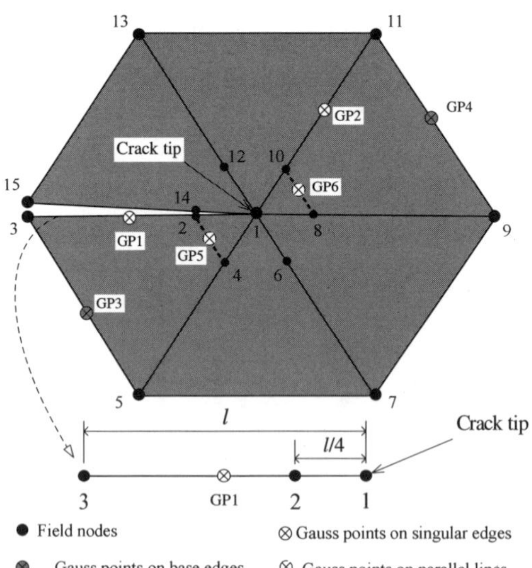

FIGURE 12.2 Five-node singular triangular cells and the possible locations of the Gauss points.

For the quadrature points located on the singular edges, for example, GP1 and GP2 in FIGURE 12.2, the displacement is constructed using

$$u^h(\mathbf{x}) = a_0 + a_1 r + a_2 \sqrt{r} \tag{12.1}$$

where r is the radial coordinate originated at the crack tip, and a_0, a_1 and a_2 are interpolation coefficients corresponding to the given point $\mathbf{x}$ on the singular edge. Values of these coefficients can be determined by enforcing Equation (12.1) to pass through the nodal values of the three nodes (nodes 1, 2 and 3) on the edge, as shown in FIGURE 12.2.

$$\begin{cases} u_1 = a_0 + a_1 r_1 + a_2 \sqrt{r_1} \\ u_2 = a_0 + a_1 r_2 + a_2 \sqrt{r_2} \\ u_3 = a_0 + a_1 r_3 + a_2 \sqrt{r_3} \end{cases} \tag{12.2}$$

where u_i (i=1, 2, 3) are the three nodal displacements. Solving Equation (12.2) for the coefficients and substituting them into Equation (12.1), we can obtain

$$u(\mathbf{x}) = \left[\underbrace{1 + 2\frac{r}{l} - 3\sqrt{\frac{r}{l}}}_{\phi_1} \quad \underbrace{-4\frac{r}{l} + 4\sqrt{\frac{r}{l}}}_{\phi_2} \quad \underbrace{2\frac{r}{l} - \sqrt{\frac{r}{l}}}_{\phi_3} \right] \begin{Bmatrix} u_1 \\ u_2 \\ u_3 \end{Bmatrix} \tag{12.3}$$

where l is the length of the project edge, and the singular shape functions $(\phi_i, i = 1, 2, 3)$ corresponding to the three nodes can be written as

$$\begin{cases} \phi_1 = 1 + 2\frac{r}{l} - 3\sqrt{\frac{r}{l}} \\ \phi_2 = -4\frac{r}{l} + 4\sqrt{\frac{r}{l}} \\ \phi_3 = 2\frac{r}{l} - \sqrt{\frac{r}{l}} \end{cases} \tag{12.4}$$

For the quadrature points on the base edges, such as GP3 and GP4 in FIGURE 12.2, the displacement will be constructed using the usual RPIM shape functions with support nodes selected using the edge-based T4-scheme, which is exactly same as that in normal cells.

For the quadrature points located on the interior parallel lines, such as GP5 and GP6 in FIGURE 12.2, the displacement will be linearly interpolated using the functions values at the two end-points of the edge. For example, the displacement field at GP5 will be constructed using linear interpolation with the displacement values at node 2 and node 4.

It is clear that the constructed displacement field in the sCS-RPIM is not continuous. The present method, which is a W^2 formulation based on the normed G space theory, will however be stable and convergent to the exact solution [32, 33, 35-37], as proven in Chapter 2. This property allows us to treat crack-tip area differently from the rest with ease.

12.1.2 Evaluation of cell-based smoothed strains

Because the background cells are also used as smoothing domains, we have $\boxed{\Omega} = \bigcup_{k=1}^{N_s} \boxed{\Omega}_k^s$ where $\Omega_i^s \cap \Omega_j^s = \varnothing$, $\forall i \neq j$, $\Omega_k^s = \Omega_k^c$, and $N_q = N_s = N_c$ in the CS-RPIM. In the sCS-RPIM, however, the singular cells needs to be further divided into more sub-smoothing domains (SSD). Suppose the problem domain is discretized using N_C^N normal cells and N_C^S singular cells, then we have $N_q = N_s = N_c^N + k^s N_c^S$, where k^s is the number of sub-smoothing domains per singular cell.

12.1.2.1 Smoothed strains in normal cells

For the normal cells shown in FIGURE 12.1, the cell-based smoothed strains can be obtained, following the operation described in Section 4.5.2. Note that this procedure is exactly same as that in other CS-PIM/CS-RPIM models as presented in Chapter 8.

12.1.2.2 Smoothed strains in singular cells

For the singular cells around the crack tip, we divide each singular cell into more smoothing domains and conduct the cell-based strain smoothing operation in each sub-smoothing domain to capture the stress singularity. In the present study, four schemes have been presented as shown in FIGURE 12.3.

Scheme A

In Scheme A, each singular cell is further divided into two sub-smoothing domains. As shown in FIGURE 12.3, singular cell 1-5-7 is divided into sub-smoothing domains 1-4-6 and 4-5-7-6. Thus for this scheme, we have $N_q = N_s = N_c^N + 2N_c^S$.

Scheme B

In Scheme B, four virtual nodes are added at the mid-edge points and each singular cell will be further divided into four sub-smoothing domains. As shown in FIGURE 12.3, singular cell 1-5-7 consists of four sub-smoothing domains, i.e. 1-M3-M4, M3-4-6-M4, 4-M1-M2-6 and M1-5-7-M2, where M1~M4 are the four virtual nodes. Thus for this scheme, we have $N_q = N_s = N_c^N + 4N_c^S$.

Scheme C

In Scheme C, two virtual nodes will be added at the mid-edge points and each singular cell will be further divided into three sub-smoothing domains. As shown in FIGURE 12.3, singular cell 1-5-7 is consists of three sub-smoothing domains, i.e. 1-4-6, 4-M1-M2-6 and M1-5-7-M2, where M1 and M2 are virtual node. Thus for this scheme, we have $N_q = N_s = N_c^N + 3N_c^S$.

Scheme D

In Scheme D, two virtual nodes will be added at the mid-edge points and each singular cell will be further divided into three sub-smoothing domains. As shown in FIGURE 12.3, singular cell 1-5-7 is consists of three sub-smoothing

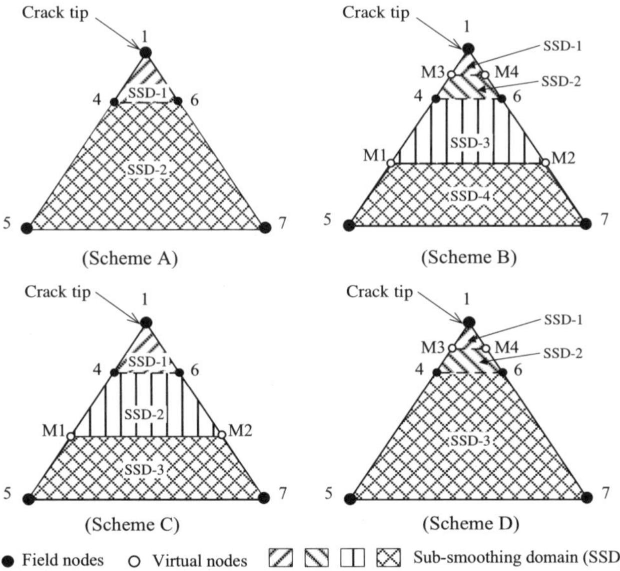

FIGURE 12.3 Four schemes of smoothing domain construction in a singular cell. The five-node singular cell is further divided into sub-smoothing domains. The virtual nodes are located at mid-edge points: M1 is the mid-point of edge 4-5, M2 is the mid-point of edge 6-7, M3 is the mid-point of edge 1-4 and M4 is the mid-point of edge 1-6.

domains, i.e. 1-M3-M4, M3-4-6-M4 and 4-5-7-6. Thus for this scheme, we have $N_q = N_s = N_c^N + 3N_c^S$.

After the smoothing domains in each singular cell have been constructed, the cell-based smoothed strain can be obtained following exactly the procedure described in Section 4.5.2.

Note that the treatment on the singular cells around the crack tip is crucial in the fracture mechanics. By dividing the singular cell further into smaller smoothing domains in different ways, different softening effect can be introduced and we can thus obtain numerical models with different stiffness. Numerical results in the following study will give us a clearer picture on this. Note that in Schemes B, C and D, function values at the virtual nodes will be interpolated using singular shape functions of the actual field nodes given in Equation (12.4), and there is no additional DOFs introduced.

12.1.3 Discretized system equations

For static problems, the discretized algebraic system equation follows Equation (8.5) and has the form as

$$\overline{\mathbf{K}}^{\text{sCS-RPIM}}\overline{\mathbf{d}} = \tilde{\mathbf{f}} \tag{12.5}$$

where $\overline{\mathbf{K}}^{\text{sCS-RPIM}}$ is the global "smoothed" stiffness matrix whose entries are given by

$$\overline{\mathbf{K}}_{IJ}^{\text{sCS-RPIM}} = \sum_{k=1}^{N_s} \int_{\Omega_k^s} \overline{\mathbf{B}}_I^{\text{T}}\mathbf{c}\underbrace{\overline{\mathbf{B}}_J}_{\text{constant in } \Omega_k^s} \, d\Omega = \sum_{k=1}^{N_s} A_k^s \overline{\mathbf{B}}_I^{\text{T}}\mathbf{c}\overline{\mathbf{B}}_J \tag{12.6}$$

in which $A_k^s = \int_{\Omega_k^s} d\Omega$ is the area of the cell-based smoothing domain Ω_k^s , and the smoothed strain-displacement matrix $\overline{\mathbf{B}}_I$ is computed using Equation (5.71). Note that there is no smoothing operation for $\tilde{\mathbf{f}}$, and hence it is essentially the same as that in the standard FEM.

12.1.4 Possible singular CS-RPIM models

Based on different schemes for constructing the smoothed strains in the singular cells and the displacement field construction, we can have different versions of sCS-RPIM models. The following are some of the sCS-RPIM models developed so far.

sCS-RPIM-Tr4-A

In this model, Scheme A is used for the singular cells: the displacement field in normal cells will be constructed using RPIM shape functions with the support nodes selected using the edge-based T4-scheme, and the displacement field in singular cells is constructed by following the way described in Section 12.1.1. The smoothed strains in normal cells are obtained using cell-based strain smoothing operation and the strain fields in singular cells are constructed using Scheme A presented in Section 12.1.2.

sCS-RPIM-Tr4-B

In this model, we use Scheme B to construct the strain field in singular cells.

sCS-RPIM-Tr4-C

The Scheme C is used to construct strain fields in singular cells for this model.

sCS-RPIM-Tr4-D

We use the Scheme D to construct strain fields in singular cells.

sCS-FEM-Tr3

In this model, the displacement along the edges connecting to the crack tip will be constructed using singular shape functions as in the sCS-RPIM-Tr4, and all the other treatments will be the same as the linear FEM. In singular cells, the strain fields will be constructed using Scheme A as presented in Section 12.1.2 by performing the cell-based strain smoothing operation. In normal cells, compatible strains are used which are as same as in the FEM model.

Note that sCS-FEM-Tr3 and sCS-RPIM-Tr4-A have same displacement and strain fields in singular cells, however, they have different displacement and strain fields in normal cells.

12.2 Stress intensity factor evaluation

Stress intensity factor is frequently used in the fracture mechanics to predict the stress intensity near the crack tip caused by a remote load [38, 39]. The integral method is one of the effective methods used to compute stress intensity factor in fracture mechanics.

12.2.1 *J*-integral

The *J*-integral provides a way to calculate the strain energy release rate (or energy per unit fracture surface area) [40, 41]. The *J*-integral uses the fact that an energetic contour path integral (*J*) is independent of the path around a crack. The theoretical concept was presented in 1967 by Cherepanov [41] and in 1968 by Jim Rice [42] independently. Under quasistatic conditions and for linear elastic materials, the *J*-integral is equal to the strain energy release rate for a crack in a body subjected to monotonic loading [43].

For linear elastic materials under the assumption of small displacement gradient, the general form of *J*-integral for a 2D crack problem can be written as [44]

$$J = -\int_{\Gamma_J} \left(\sigma_{ij} \frac{\partial u_i}{\partial x_1} - W_S \delta_{1j} \right) n_j \mathrm{d}\Gamma, \quad i=1,2; \; j=1,2 \tag{12.7}$$

where W_S is the strain energy density, Γ_J is an arbitrary line-path enclosing the crack tip located at the origin of the coordinate system showing in FIGURE 12.4a, n_j is the outward unit normal on Γ_J, σ_{ij} and u_i are stress and displacement components. We know, in theory mentioned earlier, that the *J* value in the above equation is path independent; numerically however, we will observe some path dependence.

To achieve a better numerically path independency, we often use a so-called area-path in lieu of the line-path for the integration [29-31]. Computing such area-path integration, the integrand in Equation (12.7) is multiplied with a smoothing weighting function (q_w) as [29]

$$J = -\int_{\Gamma_J} \left(\sigma_{ij} \frac{\partial u_i}{\partial x_1} - W_S \delta_{1j} \right) n_j q_w \mathrm{d}\Gamma \tag{12.8}$$

As shown in FIGURE 12.4b, where a typical 2D sharp-cracked body is surrounded by an assumed closed area Ω_J with the contour of $\Gamma_J = \Gamma_{J1} \cup \Gamma_{J-} \cup \Gamma_{J2} \cup \Gamma_{J+}$. The segments Γ_{J+} and Γ_{J-} are the boundaries of the upper and lower crack face, respectively. We define a vector **m** as the outward normal to Γ_J, such that **n=m** on Γ_{J2} and **n=-m** on Γ_{J1}. In such a closed contour, the *J*-integral can be defined in area-integration form as [29]

$$J = \int_{\Omega_J} \left(\sigma_{ij} \frac{\partial u_i}{\partial x_1} - W_S \delta_{1j} \right) \frac{\partial q_w}{\partial x_j} \, d\Omega \tag{12.9}$$

where q_w is now a sufficiently smoothing function defined in Ω_J. In the following Section 12.2.3, we will introduce on how q_w is defined in the scheme of present sCS-RPIM.

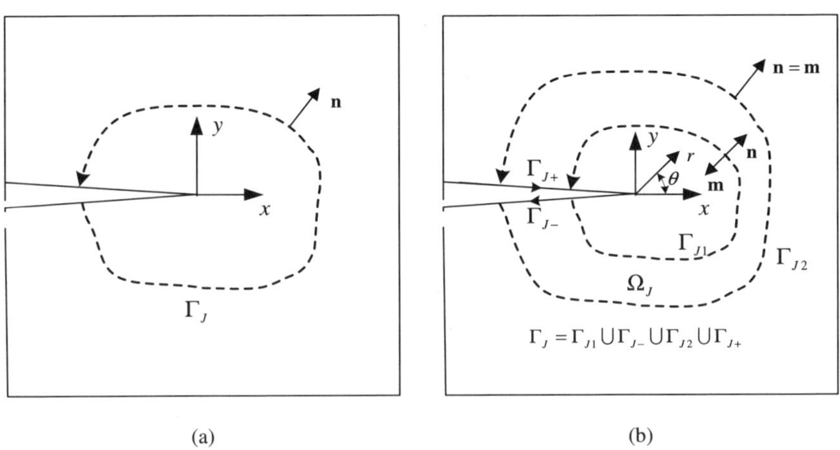

FIGURE 12.4 Closed paths around the crack tip used for computation of mixed mode stress intensity factors in 2D spaces: (a) line-path; (b) area-path.

12.2.2 Domain interaction integral

To evaluate the stress intensity factors for mixed modes effectively, we use the area-path interaction integral method [30, 31, 45] in the present study.

In the linear elastic fracture mechanics concepts, for a general mixed mode problem, we will have the following relationship between the J-integral and these two stress intensity factors.

$$J = \frac{K_I^2}{E^*} + \frac{K_{II}^2}{E^*} \tag{12.10}$$

where K_I and K_{II} are the stress intensity factors for Mode I and Mode II respectively, and E^* is defined in terms of material parameters (Young's modulus E and Poisson's ration v) as

$$E^* = \begin{cases} E & \text{(Plane stress)} \\ \dfrac{E}{1-v} & \text{(Plane strain)} \end{cases} \tag{12.11}$$

For pure Mode I, Equation (12.10) becomes

$$J = \frac{K_I^2}{E^*} \tag{12.12}$$

and for pure Mode II, it becomes

$$J = \frac{K_{II}^2}{E^*} \tag{12.13}$$

For a general fracture analysis of mixed mode, two states of the cracked body are needed to extract the values of K_I and K_{II} from the J-integral. Let state 1 $\left(\sigma_{ij}^{(1)}, \varepsilon_{ij}^{(1)}, u_i^{(1)}\right)$ correspond to the present state and state 2 $\left(\sigma_{ij}^{(2)}, \varepsilon_{ij}^{(2)}, u_i^{(2)}\right)$ represent an auxiliary state which may be a state of pure Mode I or pure Mode II. There are two conditions for the stress intensity factors regarding to the auxiliary state.

$$\begin{cases} K_I^{(2)} = 1 \ \& \ K_{II}^{(2)} = 0, & \text{(for pure Mode I)} \\ K_I^{(2)} = 0 \ \& \ K_{II}^{(2)} = 1, & \text{(for pure Mode II)} \end{cases} \tag{12.14}$$

In detail, the parameters $\left(\sigma_{ij}^{(2)}, \varepsilon_{ij}^{(2)}, u_i^{(2)}\right)$ corresponding to the auxiliary state can be evaluated as follows [38]. For pure Mode I, we have

$$\begin{cases} u_1^{(2)} = \dfrac{K_I^{(2)}\left(1+v\right)\sqrt{r}}{E\sqrt{2\pi}}\cos\dfrac{\theta}{2}\left(\kappa - \cos\theta\right) \\[4mm] u_2^{(2)} = \dfrac{K_I^{(2)}\left(1+v\right)\sqrt{r}}{E\sqrt{2\pi}}\sin\dfrac{\theta}{2}\left(\kappa - \cos\theta\right) \end{cases} \tag{12.15}$$

$$\begin{cases} \sigma_{11}^{(2)} = \dfrac{K_I^{(2)}}{\sqrt{2\pi r}} \cos\dfrac{\theta}{2}\left(1-\sin\dfrac{\theta}{2}\sin\dfrac{3\theta}{2}\right) \\[2ex] \sigma_{22}^{(2)} = \dfrac{K_I^{(2)}}{\sqrt{2\pi r}} \cos\dfrac{\theta}{2}\left(1+\sin\dfrac{\theta}{2}\sin\dfrac{3\theta}{2}\right) \\[2ex] \sigma_{12}^{(2)} = \dfrac{K_I^{(2)}}{\sqrt{2\pi r}} \cos\dfrac{\theta}{2}\sin\dfrac{\theta}{2}\cos\dfrac{3\theta}{2} \\[2ex] \sigma_{12}^{(2)} = \sigma_{21}^{(2)} \end{cases}$$

and for pure Mode II, we have

$$\begin{cases} u_1^{(2)} = \dfrac{K_{II}^{(2)}(1+v)\sqrt{r}}{E\sqrt{2\pi}}\sin\dfrac{\theta}{2}(\kappa+2+\cos\theta) \\[2ex] u_2^{(2)} = \dfrac{-K_{II}^{(2)}(1+v)\sqrt{r}}{E\sqrt{2\pi}}\cos\dfrac{\theta}{2}(\kappa-2+\cos\theta) \\[2ex] \sigma_{11}^{(2)} = \dfrac{-K_{II}^{(2)}}{\sqrt{2\pi r}}\sin\dfrac{\theta}{2}\left(2+\cos\dfrac{\theta}{2}\cos\dfrac{3\theta}{2}\right) \\[2ex] \sigma_{22}^{(2)} = \dfrac{K_{II}^{(2)}}{\sqrt{2\pi r}}\cos\dfrac{\theta}{2}\sin\dfrac{\theta}{2}\cos\dfrac{3\theta}{2} \\[2ex] \sigma_{12}^{(2)} = \dfrac{K_{II}^{(2)}}{\sqrt{2\pi r}}\cos\dfrac{\theta}{2}\left(1-\sin\dfrac{\theta}{2}\sin\dfrac{3\theta}{2}\right) \\[2ex] \sigma_{12}^{(2)} = \sigma_{21}^{(2)} \end{cases} \tag{12.16}$$

where (r,θ) are the polar coordinates measured from the origin of the coordinate system located at the crack tip, and κ is defined as

$$\kappa = \begin{cases} \dfrac{3-v}{1+v} & \text{(Plane stress)} \\[2ex] 3-4v & \text{(Plane strain)} \end{cases} \tag{12.17}$$

In linear elastic fracture mechanics, the superimposition of fields can be performed. Thus we have

$$\begin{cases} \sigma_{ij}^{(1+2)} = \sigma_{ij}^{(1)} + \sigma_{ij}^{(2)} \\[4pt] \varepsilon_{ij}^{(1+2)} = \varepsilon_{ij}^{(1)} + \varepsilon_{ij}^{(2)} \\[4pt] u_i^{(1+2)} = u_i^{(1)} + u_i^{(2)} \end{cases} \tag{12.18}$$

and similarly for stress intensity factors we have

$$\begin{cases} K_I^{(1+2)} = K_I^{(1)} + K_I^{(2)} \\[4pt] K_{II}^{(1+2)} = K_{II}^{(1)} + K_{II}^{(2)} \end{cases} \tag{12.19}$$

Then based on Equations (12.8) and (12.18), the *J*-integral for mixed states of 1 and 2 can be derived as

$$J^{(1+2)} = \int_{\Gamma_J} \left[\frac{1}{2} \left(\sigma_{ij}^{(1)} + \sigma_{ij}^{(2)} \right) \left(\varepsilon_{ij}^{(1)} + \varepsilon_{ij}^{(2)} \right) \delta_{1j} \right. \\ \left. - \left(\sigma_{ij}^{(1)} + \sigma_{ij}^{(2)} \right) \frac{\partial \left(u_i^{(1)} + u_i^{(2)} \right)}{\partial x_1} \right] n_j \, d\Gamma \tag{12.20}$$

Expanding and rearranging Equation (12.20) yields

$$J^{(1+2)} = J^{(1)} + J^{(2)} + I^{(1,2)} \tag{12.21}$$

where $I^{(1,2)}$ is termed the interaction integral for states 1 and 2, and has the form of

$$I^{(1,2)} = \int_{\Gamma_J} \left(W_s^{(1,2)} \delta_{1j} - \sigma_{ij}^{(1)} \frac{\partial u_i^{(2)}}{\partial x_1} - \sigma_{ij}^{(2)} \frac{\partial u_i^{(1)}}{\partial x_1} \right) n_j \, d\Gamma \tag{12.22}$$

in which $W_s^{(1,2)}$ is the interaction strain energy density defined by

$$W_s^{(1,2)} = \sigma_{ij}^{(1)} \varepsilon_{ij}^{(2)} = \sigma_{ij}^{(2)} \varepsilon_{ij}^{(1)} \tag{12.23}$$

Then based on Equation (12.10), the *J*-integral containing states 1 and 2 has the form of

$$J^{(1+2)} = \frac{(K_I^{(1+2)})^2}{E^*} + \frac{(K_{II}^{(1+2)})^2}{E^*} \tag{12.24}$$

Substituting Equation (12.19) into Equation (12.24), we have

$$J^{(1+2)} = J^{(1)} + J^{(2)} + \frac{2}{E^*}(K_I^{(1)}K_I^{(2)} + K_{II}^{(1)}K_{II}^{(2)}) \tag{12.25}$$

Comparing Equations (12.21) and (12.25), we arrive at

$$I^{(1,2)} = \frac{2}{E^*}(K_I^{(1)}K_I^{(2)} + K_{II}^{(1)}K_{II}^{(2)}) \tag{12.26}$$

Choosing the asymptotic field (state 2) as pure Mode I or pure Mode II, the stress intensity factors for each mode can be finally computed by substituting Equation (12.14) into Equation (12.26) as

$$\begin{cases} K_I^{(1)} = \dfrac{E^*}{2} I^{(1,\,\text{Mode I})} \\[2mm] K_{II}^{(1)} = \dfrac{E^*}{2} I^{(1,\,\text{Mode II})} \end{cases} \tag{12.27}$$

Thus after the interaction integral in Equation (12.22) has computed, the stress intensity factors can be easily obtained using Equation (12.27).

In this study, we will use area-path integral in place of line-path integral. To this end, we multiply a smoothing weighting function q_w on the integrand in Equation (12.22) and have

$$I^{(1,2)} = \lim_{\Gamma_{J1} \to 0} \left\{ \int_{\Gamma_J} \left(W_s^{(1,2)} \delta_{1j} - \sigma_{ij}^{(1)} \frac{\partial u_i^{(2)}}{\partial x_1} - \sigma_{ij}^{(2)} \frac{\partial u_i^{(1)}}{\partial x_1} \right) n_j q_w \mathrm{d}\Gamma \right\} \tag{12.28}$$

where $\Gamma_J = \Gamma_{J1} \cup \Gamma_{J-} \cup \Gamma_{J2} \cup \Gamma_{J+}$ is the boundary contour of the integration area Ω_J as shown in FIGURE 12.4b. The weighting function q_w in Equation (12.28) is defined inside the domain Ω_J and takes the value 1 on Γ_{J1}, vanishes on Γ_{J2} and is sufficiently smooth in Ω_J.

Using the divergence theorem and passing to the limit as the contour Γ_J is shrunk to the crack tip, we can obtain the following expression for the area-path interaction integral in domain form:

$$I^{(1,2)} = \int_{\Omega_J} \left(-W_s^{(1,2)} \delta_{1j} + \sigma_{ij}^{(1)} \frac{\partial u_i^{(2)}}{\partial x_1} + \sigma_{ij}^{(2)} \frac{\partial u_i^{(1)}}{\partial x_1} \right) \frac{\partial q_w}{\partial x_j} \, d\Omega \qquad (12.29)$$

Using Equation (12.29), $I^{(1, \text{Mode I})}$ and $I^{(1, \text{Mode II})}$ can be obtained, and finally the stress intensity factors can be calculated using Equation (12.27).

12.2.3 Determination of area-path

To perform the domain integration in Equation (12.29) numerically in different numerical models, we need to determine a particular area-path of Ω_J. Generally, the domain Ω_J is set to be the collection of all background cells which at least have a node within the circle centered at the crack tip with radius of $r_J = \alpha_J \times h_J$, where α_J is a dimensionless parameter and h_J is the characteristic length (average cell length) for the cells around the crack tip [46]. Here we denote the cells forming domain Ω_J as N_J, as shown in FIGURE 12.5. Then we divide the set of cells N_J into two groups: cell groups N_J^{in} and N_J^{out}, which are illustrated in FIGURE 12.5. The cell group N_J^{in} includes the cells whose three vertices are all located inside the circle, and the cell group N_J^{out} includes those belong to N_J but have at least one node located outside the circle.

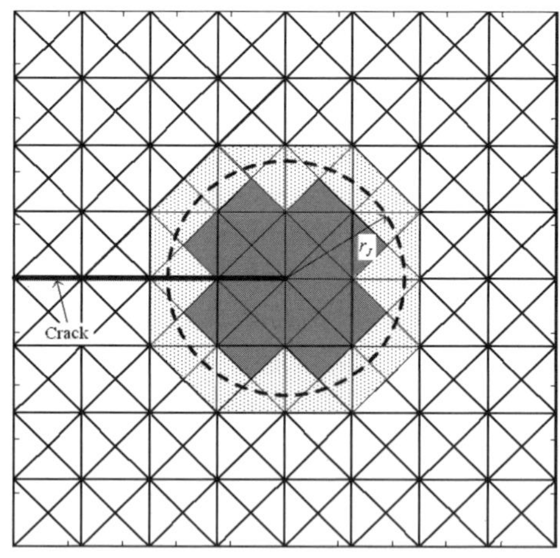

FIGURE 12.5 Closed paths around the crack tip used for computation of mixed mode stress intensity factors in 2D spaces.

The weighting function q_w in Equation (12.29) is chosen as a piecewisely linear function passing through all the nodes belonging to the cells of N_J. In detail, for a node affiliated to a cell of N_J but located outside the circle of radius r_J, the weighting function at this node will be set to zero; for a node located inside the circle, the nodal weighting function value is unit. As the gradient of the weighting function is used in Equation (12.29), we found that the cell set N_J^{in} contributes nothing to the domain interaction integral. Therefore, Equation (12.29) can be finally expressed as the summation of domain integral over the cells belong to set N_J^{out} as

$$I^{(1,2)} = \sum_{k \in N_j^{\text{out}}} \int_{\Omega_{J,k}^{\text{out}}} \left(-W_S^{(1,2)} \delta_{1j} + \sigma_{ij}^{(1)} \frac{\partial u_i^{(2)}}{\partial x_1} + \sigma_{ij}^{(2)} \frac{\partial u_i^{(1)}}{\partial x_1} \right) \frac{\partial q_w}{\partial x_j} \, d\Omega \qquad (12.30)$$

where $\Omega_{J,k}^{\text{out}}$ is the kth cell in the cell set N_J^{out}.

12.3 Numerical examples

Example 12.3.1 Rectangular plate with an edge crack under tension

A benchmark problem of an edge crack in a rectangular plate loaded by tension is first studied. The plate is of width W=1mm and height H=2mm, and the crack

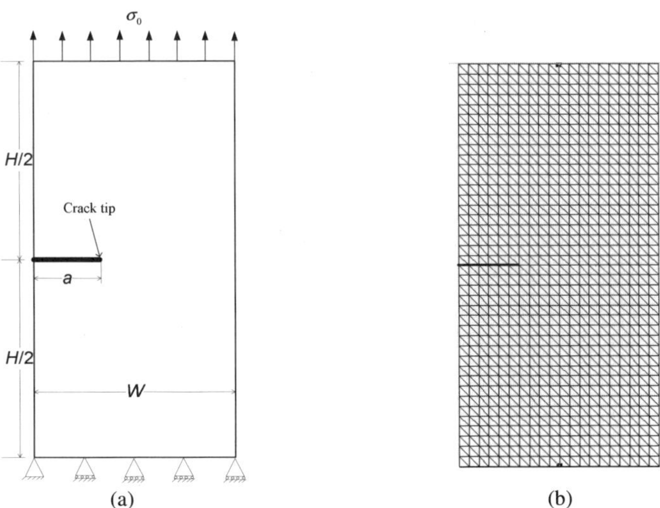

FIGURE 12.6 Plate with an edge crack under remote tension: (a) dimensions of the model; (b) domain discretization (pattern of 21×41) using three-node triangular cells.

length is $a=0.3$mm. The tension load is of $\sigma_0 = 1.0$Pa and the material parameters are Young's modulus $E = 1 \times 10^3$MPa and Poisson's ratio $v = 0.3$. The displacements in the y-axis direction are fixed along the bottom edge and the plate is clamped at the bottom left corner, as shown in FIGURE 12.6a. In this study, plane strain condition is considered.

First, we study the influence of the number of Gauss points used in the process of numerical integration along each edge of smoothing domains for computing the smoothed strains (see Section 5.7.3). The sCS-RPIM-Tr4-B model is constructed using the mesh of 21×41 shown in FIGURE 12.6b to discretize the domain. TABLE 12.1 shows strain energy and K_I computed using different number of Gauss points (n_G) along each edge of smoothing domains. It can be clearly found out that when more than four Gauss points are used, both the strain energy and K_I are almost constant. Thus in this work, $n_G = 4$ is used for the following studies.

TABLE 12.1 Computed strain energy and SIFs (K_I) obtained using sCS-RPIM-Tr4-B and the mesh of 21×41 with different number of Gauss points n_G

n_G	Strain energy ($\times 10^{-3}$)	K_I
1	1.14196283	1.593119
2	1.15993944	1.646992
3	1.15983803	1.646683
4	1.15983861	1.646683
5	1.15983867	1.646683
6	1.15983867	1.646683
7	1.15983867	1.646683

In linear fracture mechanics, the computed SIFs should be, theoretically, of domain independence, as mentioned earlier. TABLE 12.2 shows the computed (normalized) SIFs using sCS-RPIM-Tr4-D with different meshes and the size of the area-path controlled by the dimensionless parameter (α_J). We found that when α_J is no less than 3.0, the computed SIFs are almost the same. Thus $\alpha_J = 3.0$ is used in the present study to determine the dimension of the interaction integral domain.

TABLE 12.2 Computed SIFs (K_I) obtained using sCS-RPIM-Tr4-D with different meshes and values of the dimensionless parameter (α_I)

α_I	Three-node triangle meshes					
	31×61	41×81	51×101	61×121	71×141	81×161
2.0	1.5982	1.6051	1.6079	1.6093	1.6102	1.6107
2.5	1.5984	1.6050	1.6077	1.6090	1.6098	1.6103
3.0	1.5985	1.6051	1.6079	1.6093	1.6100	1.6105
3.5	1.5985	1.6051	1.6079	1.6093	1.6101	1.6106
4.0	1.5986	1.6052	1.6080	1.6093	1.6101	1.6106
4.5	1.5986	1.6051	1.6078	1.6092	1.6100	1.6105

Using a series of meshes with different densities including 21×41, 31×61, 41×81, 51×101, 61×121, 71×141 and 81×161, we further study the convergence property of the numerical solutions. For the purpose of comparison, numerical solutions of the FEM-Tr3 and the singular FEM using Tr6 elements are also provided. Here, we use the strain energy and normalized SIFs as the index parameters. As the analytical value of the stain energy is not available, a reference one (1.16367×10^{-3}) is obtained using the singular FEM-Tr6 with a very fine mesh (with 340,630 nodes). The exact value of SIF $\left(K_I^0\right)$ for this case can be obtained by [1]

$$K_I^0 = C_g\sigma_0\sqrt{\pi a} \tag{12.31}$$

where C_g is a finite geometry correction factor given by

$$C_g = 1.12 - 0.231(\frac{a}{W_S}) + 10.55(\frac{a}{W_S})^2 - 21.72(\frac{a}{W_S})^3 + 30.39(\frac{a}{W_S})^4 \tag{12.32}$$

The computed strain energy using different methods and meshes are listed in TABLE 12.3 and the convergence process is plotted in FIGURE 12.7. TABLE 12.4 lists in detail the comparison between sCS-RPIM-Tr4-A and the singular FEM using quadratic Tr6 elements. We can find the following points.

1) With the increase of DOFs, the computed strain energy for all the methods will converge to the reference one.

2) Except sCS-RPIM-Tr4-A that gives upper bound solution, all the other numerical methods provide lower bound solutions. Compared to the linear FEM model, all the CS-RPIM models give tighter bound solutions showing softer stiffness. Compared to the normal models without special measure for

the singularity, i.e. FEM-Tr3 and CS-RPIM-Tr4, the singular models provide much tighter solution bound.

3)	From the zoom in picture in FIGURE 12.7 and TABLE 12.4, we can find that sCS-RPIM-Tr4-A and the singular FEM using Tr6 elements provide much tighter upper and lower bound solutions for this case. In detail, the former provides even tighter bound solutions and has smaller relative error of the results than the latter does.

4)	In terms of accuracy, sCS-RPIM-Tr4-A stands out clearly, which provides a very tight upper bound solution in strain energy norm.

TABLE 12.3 Computed strain energy ($\times 10^{-3}$) using different methods for the rectangular plate with an edge crack under tension

Method	Triangular mesh						
	21×41	31×61	41×81	51×101	61×121	71×141	81×161
CS-RPIM-Tr4	1.1367	1.1453	1.1497	1.1524	1.1543	1.1556	1.1566
sCS-RPIM-Tr4-A	1.1647	1.1642	1.1640	1.1639	1.1639	1.1638	1.1638
sCS-RPIM-Tr4-B	1.1570	1.1591	1.1602	1.1608	1.1613	1.1616	1.1619
sCS-RPIM-Tr4-C	1.1579	1.1597	1.1606	1.1612	1.1616	1.1619	1.1621
sCS-RPIM-Tr4-D	1.1520	1.1591	1.1613	1.1622	1.1626	1.1629	1.1630
sCS-FEM-Tr3	1.1569	1.1589	1.1601	1.1607	1.1612	1.1615	1.1618
FEM-Tr3	1.1207	1.1341	1.1411	1.1455	1.1484	1.1505	1.1521
Singular FEM-Tr6	1.1620	N/A	1.1629	N/A	1.1631	N/A	1.1633

TABLE 12.4 Comparison on the strain energy of the numerical results obtained using sCS-RPIM-Tr4-A and singular FEM-Tr6 for the rectangular plate with an edge crack under tension

Mesh	sCS-RPIM-Tr4-A		Singular FEM-Tr6	
	Strain energy ($\times 10^{-3}$)	Relative error (%)	Strain energy ($\times 10^{-3}$)	Relative error (%)
21×41	1.164677	0.0865	1.162022	-0.142
41×81	1.164039	0.0317	1.162867	-0.069
61×121	1.163881	0.0182	1.163142	-0.0455
81×161	1.163814	0.0124	1.163274	-0.0340

[*] The reference strain energy for this problem is 1.16367×10^{-3} which is obtained using the singular FEM with Tr6 elements and a very fine mesh of 340,630 nodes.

[*] For the singular FEM-Tr6, the comparison is conducted using the meshes with the same DOFs.

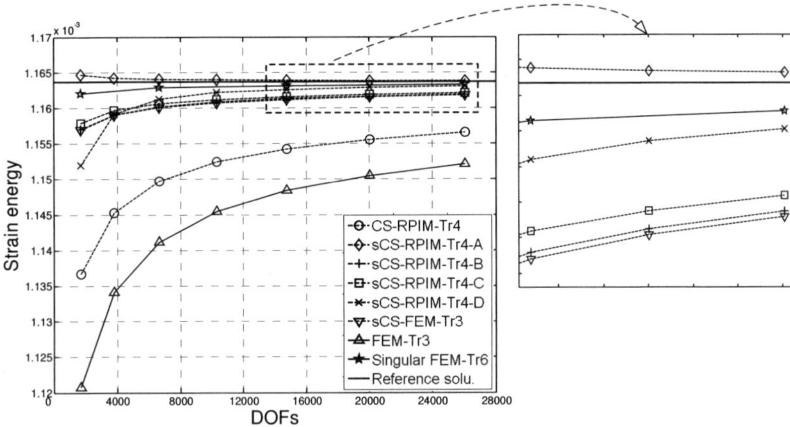

FIGURE 12.7 Converging process of the numerical results in strain energy norm for the rectangular plate with an edge crack under tension.

TABLE 12.5 lists the normalized SIFs $\left(K_I / K_I^0\right)$ obtained using different methods and FIGURE 12.8 shows the converging process of the computed SIFs against the increase of DOFs. For this case we have the following points.

1) With the increase of DOFs, the computed SIFs for all the numerical methods will converge to the exact one, i.e., the normalized values of SIFs will approach to 1.0.

2) The sCS-RPIM-Tr4-A is again the only one providing upper bound solutions in terms of SIFs. Compared to the linear FEM-Tr3 model, all the CS-RPIM models can give tighter bound solutions showing softer stiffness. Compared to the normal models without special measure for the singularity, the singular models provide much tighter solution bound.

3) Among all the numerical methods studied, the singular FEM with Tr6 elements gives tightest solution (lower) bound, which is even a little bigger than he exact one when using a relatively fine mesh (81×161).

4) Considering the SIF, the singular FEM-Tr6 stands out followed by sCS-RPIM-Tr4-A, which performs the best among all the present sCS-RPIM models.

TABLE 12.5 Normalized SIF $\left(K_I / K_I^0\right)$ obtained using different methods for the rectangular plate with an edge crack under tension

Method	Triangular mesh						
	21×41	31×61	41×81	51×101	61×121	71×141	81×161
CS-RPIM-Tr4	0.9591	0.9722	0.9790	0.9832	0.9861	0.9881	0.9897
sCS-RPIM-Tr4-A	1.0023	1.0013	1.0011	1.0009	1.0008	1.0008	1.0007
sCS-RPIM-Tr4-B	0.9901	0.9931	0.9949	0.9960	0.9967	0.9972	0.9977
sCS-RPIM-Tr4-C	0.9913	0.9941	0.9956	0.9965	0.9972	0.9977	0.9980
sCS-RPIM-Tr4-D	0.9782	0.9920	0.9960	0.9978	0.9987	0.9991	0.9995
sCS-FEM-Tr3	0.9871	0.9918	0.9941	0.9954	0.9962	0.9968	0.9973
FEM-Tr3	0.9316	0.9536	0.9649	0.9718	0.9764	0.9798	0.9823
Singular FEM-Tr6	0.9981	N/A	0.9994	N/A	0.9998	N/A	1.0001

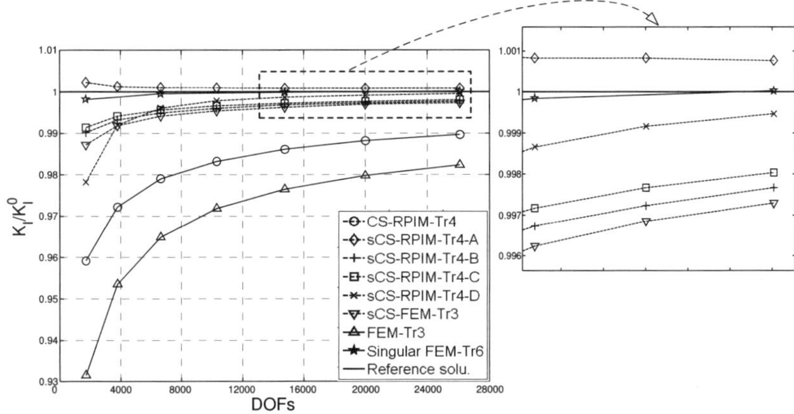

FIGURE 12.8 Converging process of the normalized SIFs $\left(K_I / K_I^0\right)$ for the rectangular plate with an edge crack under tension.

Example 12.3.2 Rectangular plate with an edge crack under shear

Example 12.3.1 is a case of simple pure Mode I. We now consider a slightly more complicated case of mixed-mode of Model I and Model II. As shown in FIGURE 12.9, a rectangular plate with an edge crack is subjected to remote shear, the displacement in the y-axis direction are fixed along the bottom edge and the plate is fully clamped at the bottom left corner.

The plate is of width $W=7$mm and height $H=16$mm, and the crack length is $a=3.5$mm. The shear load is of $\tau_0 = 1.0$ Pa and the material parameters are Young's modulus $E = 3 \times 10^7$ Pa and Poisson's ratio $v = 0.25$. In this study, plane strain condition is considered.

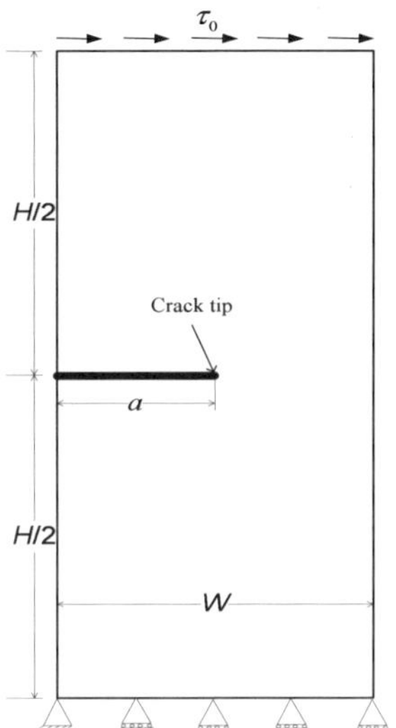

FIGURE 12.9 Plate with edge crack under remote shear.

Using a series of meshes having the same pattern as used in Example 12.3.1, we study this problem using various CS-RPIM models together with the FEM-Tr3 and the singular FEM using quadratic Tr6 elements. Reference values for both the strain energy and the SIFs (K_I^0 and K_{II}^0) are obtained using the singular FEM-Tr6 and a very fine mesh (total 401,677 nodes).

The computed strain energy using different methods and meshes are listed in TABLE 12.6 and the convergence process is plotted in FIGURE 12.10. TABLE 12.7 lists in detail the comparison between sCS-RPIM-Tr4-A and the singular FEM-Tr6. For this case we summarize the following points.

1) The computed strain energy for all the methods will converge to the reference one from below, with the increase of DOFs.

2) All the sCS-RPIM models provide tighter bound solution than the linear FEM-Tr3 does. The four Schemes (A, B, C and D) of the singular CS-RPIM can even give tighter lower bound solutions than the singular FEM-Tr6.

3) Among all the methods, sCS-RPIM-Tr4-A stands out clearly, which provides more accurate solutions than the singular FEM-Tr6.

TABLE 12.6 Computed strain energy ($\times 10^{-5}$) using different methods for the rectangular plate with an edge crack under shear

Method	Triangular mesh						
	21×41	31×61	41×81	51×101	61×121	71×141	81×161
CS-RPIM-Tr4	8.3748	8.5214	8.6042	8.6593	8.6995	8.7306	8.7557
sCS-RPIM-Tr4-A	8.7028	8.7468	8.7760	8.7981	8.8159	8.8308	8.8437
sCS-RPIM-Tr4-B	8.5929	8.6747	8.7225	8.7555	8.7805	8.8006	8.8173
sCS-RPIM-Tr4-C	8.6039	8.6821	8.7280	8.7599	8.7842	8.8038	8.8201
sCS-RPIM-Tr4-D	8.2165	8.5548	8.6758	8.7369	8.7745	8.8009	8.8210
sCS-FEM-Tr3	8.4018	8.5348	8.6008	8.6422	8.6717	8.6945	8.7129
FEM-Tr3	7.9678	8.2278	8.3638	8.4494	8.5093	8.5542	8.5894
Singular FEM-Tr6	8.6391	N/A	8.7153	N/A	8.7571	N/A	8.7864

TABLE 12.7 Comparison on the strain energy of the numerical results obtained using sCS-RPIM-Tr4-A and singular FEM-Tr6 for the rectangular plate with an edge crack under shear

Mesh	sCS-RPIM-Tr4-A		Singular FEM using Tr6 elements	
	Strain energy ($\times 10^{-5}$)	Relative error (%)	Strain energy ($\times 10^{-5}$)	Relative error (%)
21×41	8.702759	-2.08	8.639063	-2.80
41×81	8.775995	-1.26	8.715257	-1.94
61×121	8.815871	-0.811	8.757142	-1.47
81×161	8.843731	-0.497	8.786392	-1.14

[*] The reference strain energy for this problem is 8.88794×10^{-5} which is obtained using the singular FEM with Tr6 elements and a very fine mesh (total 401,677 nodes).

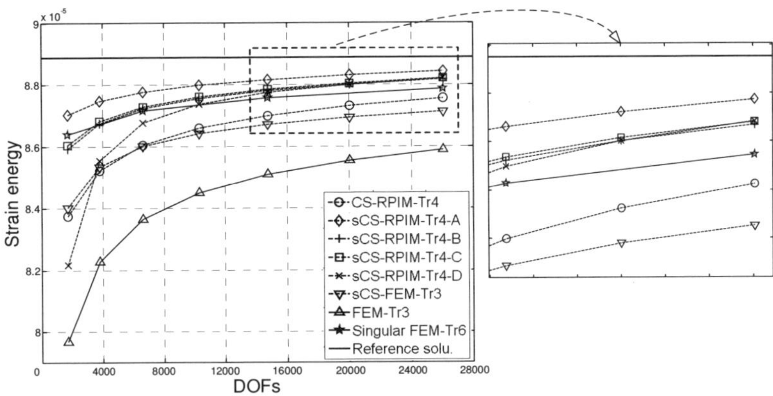

FIGURE 12.10 Converging process of the numerical results in strain energy norm for the rectangular plate with an edge crack under shear.

TABLE 12.8 and TABLE 12.9 list the normalized SIFs K_I / K_I^0 and K_{II} / K_{II}^0 respectively. The converging processes about the normalized SIFs are plotted in FIGURE 12.11 and FIGURE 12.12. The following points can be found for this case.

1) All the CS-RPIM models provide much tighter bound solutions than the linear FEM-Tr3. The singular CS-RPIM models give tighter bound results than the normal CS-RPIM-Tr4.

2) For the computed K_I / K_I^0, sCS-RPIM-Tr4-A and the singular FEM-Tr6 provide very tight bounds from above and below respectively, when fine meshes are used.

3) For the computed K_{II} / K_{II}^0, results of sCS-RPIM-Tr4-A and sCS-FEM-T3 are much closer to the reference one than other methods. When fine meshes are used, results of these two methods can cross the reference one.

4) For this case of mixed modes, sCS-RPIM-Tr4-A stands out again, followed by singular FEM-Tr6 and the sCS-FEM-Tr3.

TABLE 12.8 Normalized SIF $\left(K_I / K_I^0\right)$ obtained using different methods for the rectangular plate with an edge crack under shear

Method	Triangular mesh						
	21×41	31×61	41×81	51×101	61×121	71×141	81×161
CS-RPIM-Tr4	0.9360	0.9567	0.9673	0.9737	0.9781	0.9812	0.9835
sCS-RPIM-Tr4-A	0.9997	1.0005	1.0007	1.0007	1.0007	1.0007	1.0006
sCS-RPIM-Tr4-B	0.9793	0.9871	0.9907	0.9927	0.9941	0.9950	0.9957
sCS-RPIM-Tr4-C	0.9814	0.9885	0.9918	0.9936	0.9948	0.9956	0.9962
sCS-RPIM-Tr4-D	0.9355	0.9752	0.9874	0.9925	0.9951	0.9966	0.9975
sCS-FEM-Tr3	0.9942	0.9960	0.9969	0.9975	0.9979	0.9982	0.9984
FEM-Tr3	0.8874	0.9251	0.9440	0.9553	0.9629	0.9682	0.9723
Singular FEM-Tr6	0.9972	N/A	0.9987	N/A	0.9991	N/A	0.9994

[*] The reference value of SIF $\left(K_I^0\right)$ for this problem is 34.02, which is obtained using the singular FEM-Tr6 and a very fine mesh (total 401,677 nodes).

TABLE 12.9 Normalized SIF $\left(K_{II} / K_{II}^0\right)$ obtained using different methods for the rectangular plate with an edge crack under shear

Method	Triangular mesh						
	21×41	31×61	41×81	51×101	61×121	71×141	81×161
CS-RPIM-Tr4	0.9815	0.9862	0.9891	0.9909	0.9921	0.9930	0.9937
sCS-RPIM-Tr4-A	1.0064	1.0030	1.0016	1.0009	1.0004	1.0000	1.0000
sCS-RPIM-Tr4-B	0.9886	0.9913	0.9930	0.9942	0.9947	0.9951	0.9955
sCS-RPIM-Tr4-C	0.9883	0.9911	0.9928	0.9938	0.9945	0.9950	0.9953
sCS-RPIM-Tr4-D	0.9590	0.9849	0.9928	0.9959	0.9974	0.9981	0.9985
sCS-FEM-Tr3	0.9993	0.9997	0.9998	1.0001	1.0000	1.0002	1.0002
FEM-Tr3	0.9363	0.9556	0.9661	0.9727	0.9772	0.9804	0.9828
Singular FEM-Tr6	0.9964	N/A	0.9974	N/A	0.9979	N/A	0.9982

[*] The reference value of SIF $\left(K_{II}^0\right)$ for this problem is 4.5, which is obtained using the singular FEM-Tr6 and a very fine mesh (total 401,677 nodes).

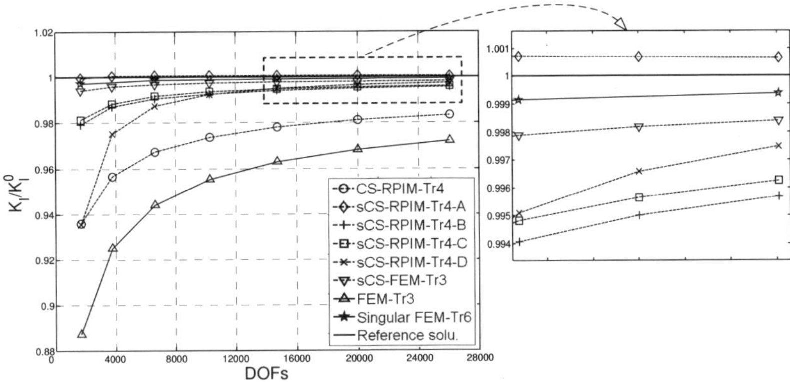

FIGURE 12.11 Converging process of the normalized SIFs $\left(K_I / K_I^0\right)$ for the rectangular plate with an edge crack under shear.

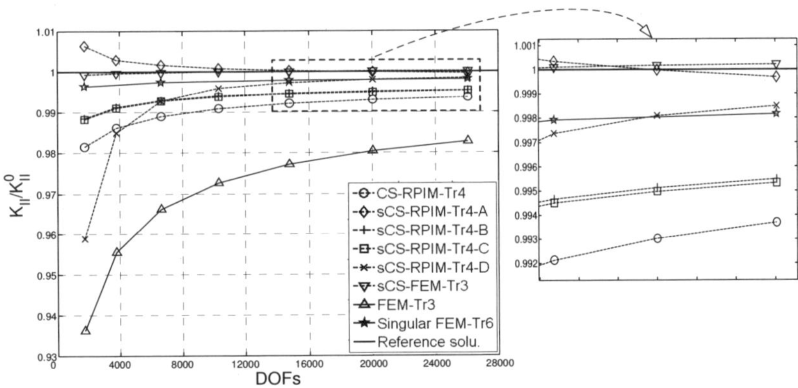

FIGURE 12.12 Converging process of the normalized SIFs $\left(K_{II} / K_{II}^0\right)$ for the rectangular plate with an edge crack under shear.

Example 12.3.3 Rectangular plate with an inclined crack under tension

Finally a rectangular plate with an inclined crack of two tips is considered, as shown in FIGURE 12.13. The plate is of 10mm×10mm. A crack is located at the center of the plate with the length of $2a = \sqrt{2}$ mm and an angle θ from the horizontal line. The tension load is of $\sigma_0 = 1.0$Pa and the material parameters are Young's modulus $E = 3 \times 10^7$ Pa and Poisson's ratio $v = 0.3$. In this study, plane strain condition is considered and two values of the θ (45° and 60°) are used.

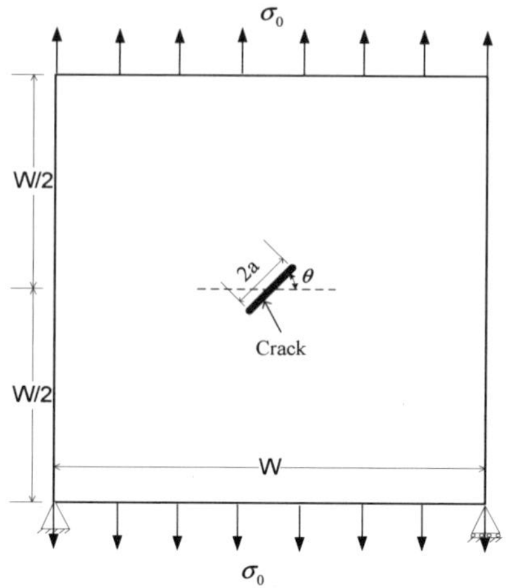

FIGURE 12.13 Plate with an inclined crack under remote tension.

In this study, five set of triangular meshes (37×37, 61×61, 85×85, 109×109 and 121×121) are used to discretize the problem domain, as illustrated in FIGURE 12.14. Reference values of the strain energy for this problem are obtained using the singular FEM-Tr6 with very fine meshes (766,305 and 578,582 nodes for $\theta = 45^0$ and $\theta = 60^0$ respectively). The exact SIFs for this case are provided as [1]

$$K_I^0 = \sigma_0 \sqrt{\pi a}\,\cos^2\theta$$
$$K_{II}^0 = \sigma_0 \sqrt{\pi a}\,\sin\theta\cos\theta$$

(12.33)

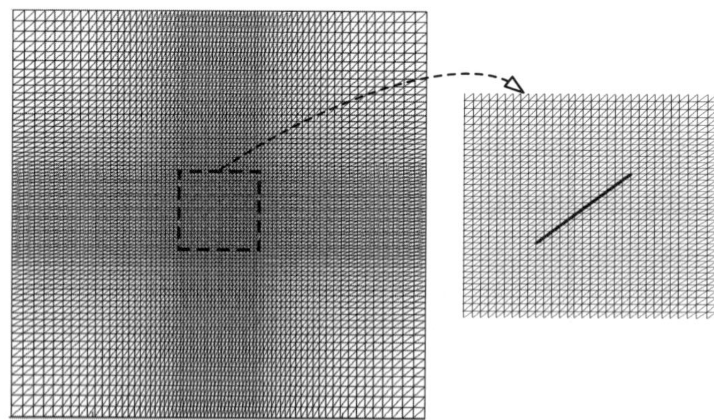

FIGURE 12.14 Illustration of domain discretization using three-node triangular mesh for the plate with an inclined crack under tension.

For the case of $\theta = 45^0$, the computed strain energy using different methods and meshes are listed in TABLE 12.10 and the convergence process is plotted in FIGURE 12.15. TABLE 12.11 lists in detail the comparison between sCS-RPIM-A, sCS-RPIM-D and the singular FEM-Tr6. The following points can be found.

1) With the increase of DOFs, the computed strain energy for all the methods will converge to the reference one.

2) All the singular CS-RPIM models give much tighter bound solution than the FEM-Tr3 and the normal CS-RPIM-Tr4, which shows the effectiveness of the proposed singularity approximation and the stiffness softening by the cell-based strain smoothing operation.

3) The sCS-RPIM-Tr4-A provides a very tight upper bound solution and all the other methods give lower bound solutions in strain energy norm.

4) As detailed in TABLE 12.11, sCS-RPIM-Tr4-A, sCS-RPIM-Tr4-D and the singular FEM-Tr6 provide much tighter bound solutions than other methods.

5) The sCS-RPIM-Tr4-D stands out for this case, followed closely by the singular FEM-Tr6 and sCS-RPIM-Tr4-A.

TABLE 12.10 Computed strain energy ($\times 10^{-6}$) using different methods for the rectangular plate with a 45° inclined crack under tension

Method	Triangular mesh				
	37×37	61×61	85×85	109×109	121×121
CS-RPIM-Tr4	1.5832	1.5850	1.5857	1.5861	1.5862
sCS-RPIM-Tr4-A	1.5876	1.5876	1.5875	1.5875	1.5875
sCS-RPIM-Tr4-B	1.5868	1.5870	1.5871	1.5872	1.5872
sCS-RPIM-Tr4-C	1.5869	1.5871	1.5872	1.5873	1.5873
sCS-RPIM-Tr4-D	1.5871	1.5874	1.5874	1.5874	1.5874
sCS-FEM-Tr3	1.5869	1.5871	1.5872	1.5873	1.5873
FEM-Tr3	1.5821	1.5842	1.5851	1.5856	1.5858
Singular FEM-Tr6	1.5870	1.5873	1.5874	1.5874	1.5874

[*] The reference strain energy for this problem is 1.58745×10^{-6} which is obtained using the singular FEM with Tr6 elements and a very fine mesh (total 766,305 nodes).

TABLE 12.11 Comparison on the strain energy of the numerical results obtained using sCS-RPIM-Tr4-A, sCS-RPIM-Tr4-D and singular FEM-Tr6 for the rectangular plate with a 45° inclined crack under tension

Mesh	sCS-RPIM-Tr4-A		sCS-RPIM-Tr4-D		Singular FEM-Tr6	
	Strain energy $(\times 10^{-6})$	Relative error (%)	Strain energy $(\times 10^{-6})$	Relative error (%)	Strain energy $(\times 10^{-6})$	Relative error (%)
37×37	1.58764	0.0120	1.58709	-0.0227	1.5869076	-0.0342
61×61	1.58756	0.00693	1.58735	-0.00630	1.5873046	-0.00916
85×85	1.58753	0.00504	1.58740	-0.00315	1.5873678	-0.00518
109×109	1.58751	0.00378	1.58742	-0.00189	1.5873928	-0.00360
121×121	1.58750	0.00315	1.58743	-0.00126	1.5874062	-0.00276

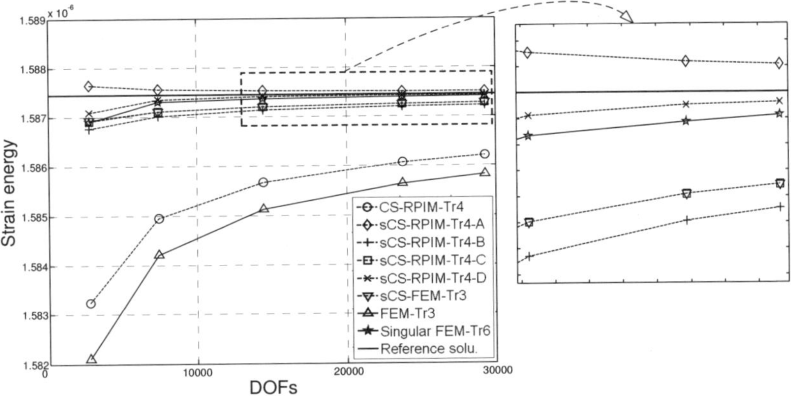

FIGURE 12.15 Converging process of the numerical results in strain energy norm for the rectangular plate with a 45° inclined crack under tension.

TABLE 12.12 and TABLE 12.13 list the normalized SIFs for the case of $\theta = 45^0$. The converging processes about the normalized SIFs are plotted in FIGURE 12.16 and FIGURE 12.17. We found the following points for this case.

1) All the singular CS-RPIM models provide much tighter bound solutions than both the linear FEM-Tr3 and the normal CS-RPIM-Tr4.

2) For the computed K_I / K_I^0, all the methods give lower bound solutions. Among these methods, the singular FEM-Tr6 provides the tightest solution, followed by sCS-RPIM-Tr4-C and sCS-RPIM-Tr4-A.

3) For the computed K_{II} / K_{II}^0, except sCS-RPIM-Tr4-A gives a tight upper bound solution, all the other methods have lower bound.

4) For the computed SIFs, the singular FEM-Tr6 and sCS-RPIM-Tr4-A stand out.

TABLE 12.12 Normalized SIF $\left(K_I / K_I^0\right)$ obtained using different methods for the rectangular plate with a 45° inclined crack under tension

Method	Triangular mesh				
	37×37	61×61	85×85	109×109	121×121
CS-RPIM-Tr4	1.0095	1.0123	1.0145	1.0159	1.0163
sCS-RPIM-Tr4-A	1.0187	1.0192	1.0196	1.0197	1.0198
sCS-RPIM-Tr4-B	1.0163	1.0173	1.0182	1.0187	1.0189
sCS-RPIM-Tr4-C	1.0190	1.0195	1.0198	1.0201	1.0202
sCS-RPIM-Tr4-D	1.0172	1.0181	1.0188	1.0192	1.0193
sCS-FEM-Tr3	1.0122	1.0158	1.0173	1.0181	1.0183
FEM-Tr3	0.9968	1.0076	1.0135	1.0140	1.0142
Singular FEM-Tr6	1.0181	1.0193	1.0201	1.0203	1.0203

[*] The exact value of the normalized SIF is 1.0216.

TABLE 12.13 Normalized SIF $\left(K_{II} / K_{II}^0\right)$ obtained using different methods for the rectangular plate with a 45° inclined crack under tension

Method	Triangular mesh				
	37×37	61×61	85×85	109×109	121×121
CS-RPIM-Tr4	0.9976	1.0009	1.0035	1.0050	1.0055
sCS-RPIM-Tr4-A	1.0143	1.0130	1.0120	1.0118	1.0117
sCS-RPIM-Tr4-B	1.0071	1.0080	1.0087	1.0091	1.0092
sCS-RPIM-Tr4-C	1.0059	1.0073	1.0080	1.0084	1.0085
sCS-RPIM-Tr4-D	1.0064	1.0091	1.0093	1.0099	1.0095
sCS-FEM-Tr3	1.0088	1.0095	1.0099	1.0101	1.0102
FEM-Tr3	0.9966	1.0033	1.0043	1.0050	1.0056
Singular FEM	1.0100	1.0103	1.0105	1.0106	1.0106

[*] The exact value of the normalized SIF is 1.0112.

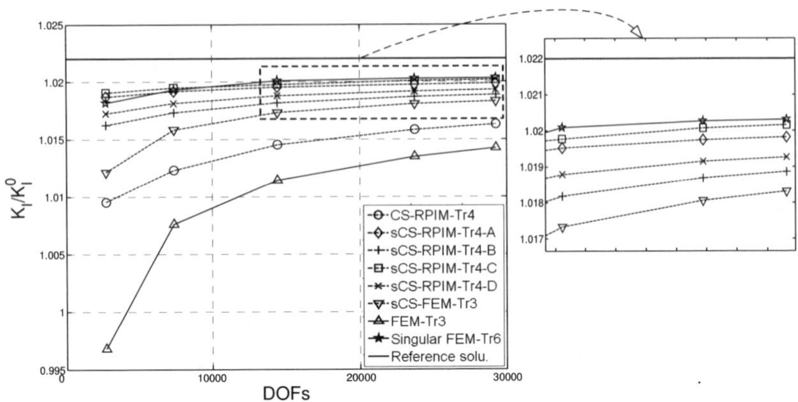

FIGURE 12.16 Converging process of the normalized SIFs $\left(K_I / K_I^0\right)$ for the rectangular plate with a 45° inclined crack under tension.

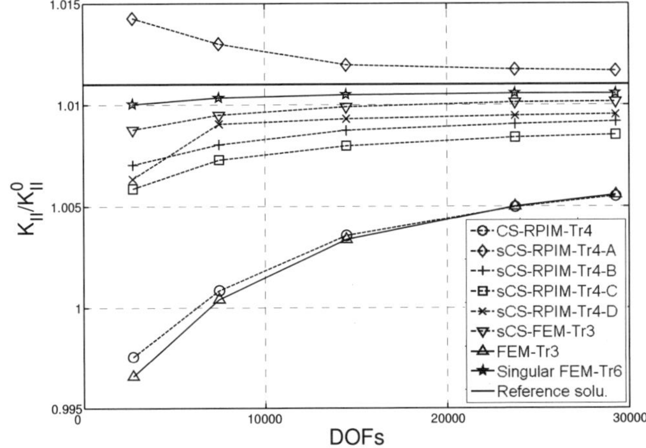

FIGURE 12.17 Converging process of the normalized SIFs $\left(K_{II} / K_{II}^0\right)$ for the rectangular plate with a 45° inclined crack under tension.

For the case of $\theta = 60^0$, the computed strain energy using different methods and meshes are listed in TABLE 12.14 and the convergence process is plotted in FIGURE 12.18. TABLE 12.15 lists in detail the comparison between sCS-RPIM-Tr4-A, sCS-RPIM-Tr4-D and the singular FEM-Tr6. Some points may be drawn from the numerical results.

1) With the increase of DOFs, the computed strain energy for all the methods will converge to the reference one from above (for sCS-RPIM-Tr4-A and sCS-FEM-Tr3) and below (for other methods).

2) All the singular CS-RPIM models give much tighter bound solution than the FEM-Tr3 and the normal CS-RPIM-Tr4.

3) The sCS-FEM-Tr3 provides a very tight upper bound solution in strain energy norm, which is even closer to the reference one than the counterpart of sCS-RPIM-Tr4-A.

4) As detailed in TABLE 12.15, the singular FEM-Tr6 provides the tightest lower bound solution, followed by sCS-RPIM-Tr4-D.

5) The sCS-FEM-Tr3, singular FEM-Tr6 and sCS-RPIM-Tr4-D stand out for this case.

TABLE 12.14 Computed strain energy ($\times 10^{-6}$) using different methods for the rectangular plate with a 60° inclined crack under tension

Method	Triangular mesh				
	37×37	61×61	85×85	109×109	121×121
CS-RPIM-Tr4	1.57272	1.57362	1.57400	1.57422	1.57429
sCS-RPIM-Tr4-A	1.57523	1.57511	1.57506	1.57504	1.57503
sCS-RPIM-Tr4-B	1.57466	1.57476	1.57481	1.57484	1.57486
sCS-RPIM-Tr4-C	1.57478	1.57483	1.57486	1.57488	1.57489
sCS-RPIM-Tr4-D	1.57446	1.57480	1.57488	1.57492	1.57493
sCS-FEM-Tr3	1.57498	1.57496	1.57495	1.57495	1.57495
FEM-Tr3	1.57228	1.57334	1.57380	1.57405	1.57414
Singular FEM-Tr6	1.57462	1.57487	1.57491	1.57493	1.57494

TABLE 12.15 Comparison on the strain energy of the numerical results obtained using sCS-RPIM-Tr4-A, sCS-RPIM-Tr4-D and singular FEM-Tr6 for the rectangular plate with a 60° inclined crack under tension

Mesh	sCS-RPIM-Tr4-A		sCS-RPIM-Tr4-D		Singular FEM-Tr6	
	Strain energy ($\times 10^{-6}$)	Relative error (%)	Strain energy ($\times 10^{-6}$)	Relative error (%)	Strain energy ($\times 10^{-6}$)	Relative error (%)
37×37	1.5752303	0.0183	1.5744618	-0.0305	1.5746244	-0.0202
61×61	1.5751126	0.0108	1.5748073	-0.00857	1.5748745	-0.00430
85×85	1.5750665	0.00789	1.5748883	-0.00342	1.5749146	-0.00175
109×109	1.5750420	0.00634	1.5749186	-0.00150	1.5749297	-0.00079
121×121	1.5750335	0.00560	1.5749270	-0.00096	1.5749375	-0.00030

[*] The reference strain energy for this problem is 1.5749422×10^{-6} which is obtained using the singular FEM-Tr6 and a very fine mesh (total 578,582 nodes).

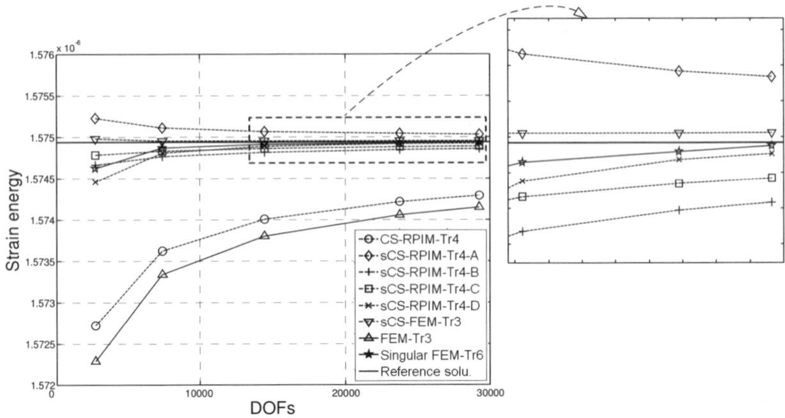

FIGURE 12.18 Converging process of the numerical results in strain energy norm for the rectangular plate with a 60° inclined crack under tension.

TABLE 12.16 and TABLE 12.17 list the normalized SIFs for the case of $\theta = 60^0$. The converging processes about the normalized SIFs are plotted in FIGURE 12.19 and FIGURE 12.20. We have the following points.

1) All the singular CS-RPIM models provide much tighter bound solutions than the linear FEM-Tr3 and the normal CS-RPIM-Tr4.

2) For the computed K_I / K_I^0, the sCS-RPIM-Tr4-A gives upper bound solutions, and all the other methods give lower bound solutions. The singular FEM-Tr6 and sCS-RPIM-Tr4-D provide much tighter lower bound solutions than others.

3) For the computed K_{II} / K_{II}^0, except sCS-RPIM-Tr4-A gives a tight upper bound solution, all the other methods have lower bound. The singular FEM-Tr6 gives the tightest lower bound solution, followed by sCS-FEM-Tr3 and sCS-RPIM-Tr4-D.

4) For the computed SIFs, the singular FEM-Tr6 stands out, followed by sCS-RPIM-Tr4-A, sCS-RPIM-Tr4-D and sCS-FEM-Tr3.

TABLE 12.16 Normalized SIF $\left(K_I / K_I^0\right)$ obtained using different methods for the rectangular plate with a 60° inclined crack under tension

Method	Triangular mesh				
	37×37	61×61	85×85	109×109	121×121
CS-RPIM-Tr4	1.0128	1.0153	1.0176	1.0185	1.0189
sCS-RPIM-Tr4-A	1.0246	1.0235	1.0228	1.0225	1.0225
sCS-RPIM-Tr4-B	1.0176	1.0194	1.0203	1.0207	1.0210
sCS-RPIM-Tr4-C	1.0176	1.0192	1.0204	1.0207	1.0207
sCS-RPIM-Tr4-D	1.0174	1.0212	1.0217	1.0218	1.0219
sCS-FEM-Tr3	1.0121	1.0192	1.0202	1.0209	1.0212
FEM-Tr3	0.9982	1.0085	1.0128	1.0148	1.0156
Singular FEM-Tr6	1.0119	1.0208	1.0217	1.0218	1.0218

[*] The exact value of the normalized SIF is 1.0219.

TABLE 12.17 Normalized SIF $\left(K_{II} / K_{II}^0\right)$ obtained using different methods for the rectangular plate with a 60° inclined crack under tension

Method	Triangular mesh				
	37×37	61×61	85×85	109×109	121×121
CS-RPIM-Tr4	1.0010	1.0039	1.0059	1.0072	1.0077
sCS-RPIM-Tr4-A	1.0144	1.0134	1.0128	1.0126	1.0126
sCS-RPIM-Tr4-B	1.0093	1.0100	1.0104	1.0106	1.0107
sCS-RPIM-Tr4-C	1.0088	1.0101	1.0105	1.0107	1.0109
sCS-RPIM-Tr4-D	1.0086	1.0109	1.0111	1.0113	1.0113
sCS-FEM-Tr3	1.0096	1.0112	1.0115	1.0117	1.0118
FEM-Tr3	0.9938	1.0009	1.0040	1.0056	1.0064
Singular FEM-Tr6	1.0026	1.0102	1.0118	1.0119	1.0119

[*] The exact value of the normalized SIF is 1.0120.

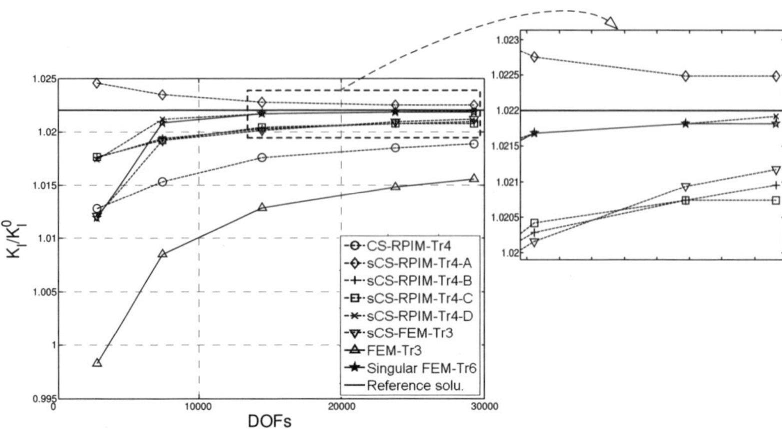

FIGURE 12.19 Converging process of the normalized SIFs $\left(K_I / K_I^0\right)$ for the rectangular plate with a 60° inclined crack under tension.

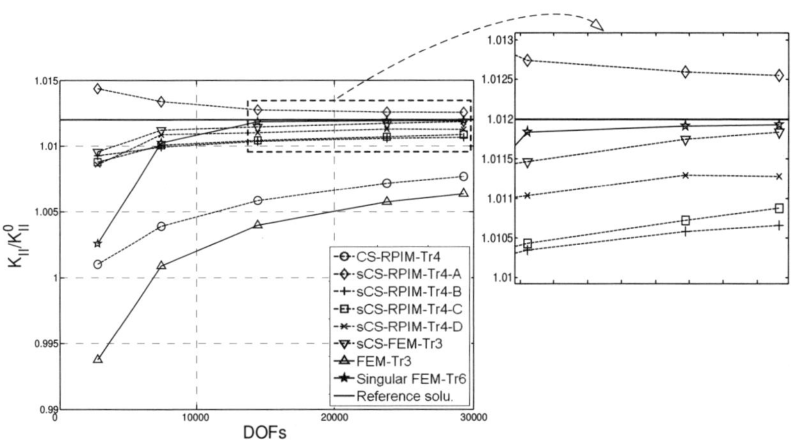

FIGURE 12.20 Converging process of the normalized SIFs $\left(K_{II} / K_{II}^0\right)$ for the rectangular plate with a 60° inclined crack under tension.

12.4 Concluding remarks

We have presented the singular CS-RPIM (or sCS-RPIM) models for the fracture analysis in linear elastic materials, by introducing a layer of five-node singular cells near the crack tip. For the singular cells, the displacement fields along the edges radiated from the crack tip are approximated using the enriched shape functions, and the strain fields are constructed using the cell-based strain smoothing operation with some different schemes. Three benchmark numerical examples have been studied and the results are compared with the linear FEM and the singular FEM using Tr6 elements. We may have the following conclusions.

Remark 12.1 Upper and lower bound property

By dividing a singular cell into more sub-smoothing domains, we have different schemes of the sCS-RPIM with different softness, and hence can obtain both upper and lower bound solutions in terms of strain energy and SIFs. The sCS-RPIM-Tr4-A and sCS-RPIM-Tr4-D can provide very tighter bound solutions than other models.

Remark 12.2 High accuracy

All the sCS-RPIM models have higher accuracy than both the linear FEM-Tr3 and the normal CS-RPIM-Tr4.

Remark 12.3 sCS-RPIM-Tr4-A: outstanding performer

Among the singular CS-RPIM models proposed, sCS-RPIM-Tr4-A stands out clearly. The numerical solutions of sCS-RPIM-Tr4-A have much better accuracy than the normal CS-RPIM-Tr4 and linear FEM, and are comparable to that of the singular FEM-Tr6. It can generally provide, not always, very tight upper bound solutions for both strain energy and SIFs.

12.5 References

1.	Anderson, T. L., *Fracture Mechanics: Fundamentals and Applications*. CRC press, 1995.

2.	Henshell, R. D. and Shaw, K. G., Crack tip finite elements are unnecessary. *International Journal for Numerical Methods in Engineering*, 9: 495-507, 1975.

3.	Barsoum, R. S., On the use of isoparametric finite elements in linear fracture mechanics. *International Journal for Numerical Methods in Engineering*, 10: 551-564, 1976.

4.	Barsoum, R. S., Triangular quarter-point elements as elastic and perfectly-plastic crack tip elements. *International Journal for Numerical Methods in Engineering*, 11: 85-98, 1977.

5.   Kim, J. H. and Paulino, G. H., Finite element evaluation of mixed mode stress intensity factors in functionally graded materials. *International Journal for Numerical Methods in Engineering*, 53(8): 1903-1935, 2002.

6.   Rabczuk, T. and Belytschko, T., Cracking particles: a simplified meshfree method for arbitrary evolving cracks. *International Journal for Numerical Methods in Engineering*, 61(13): 2316-2343, 2004.

7.   Rabczuk, T. and Belytschko, T., A three dimensional large deformation meshfree method for arbitrary evolving cracks. *Computer Methods in Applied Mechanics and Engineering*, 196(29-30): 2777-2799, 2007.

8.   Rabczuk, T., Zi, G., A meshfree method based on the local partition of unity for cohesive cracks. *Computational Mechanics*, 39(6): 743-760, 2007.

9.   Klein, P. A., Foulk, J. W., Chen, E. P., Simmer, S. A. and Gao, H. J., Physics-based modeling of brittle fracture: cohesive formulations and the application of meshfree methods. *Theoretical and Applied Fracture Mechanics*, 37(1-3): 99-166, 2001.

10. Hao, S., Liu, W. K. and Chang, C. T., Computer implementation of damage models by finite element and meshfree methods. *Computer Methods in Applied Mechanics and Engineering*, 187(3-4), 401-440, 2000.

11. Liu, W. K., Hao, S., Belytschko, T., Li, S. F. and Chang, C. T., Multiple scale meshfree methods for damage fracture and localization. *Computational Materials Science*, 16(1-4): 197-205, 1999.

12. Rabczuk, T., Areias, P. M. A. and Belytschko, T., A meshfree thin shell method for non-linear dynamic fracture. *International Journal for Numerical Methods in Engineering*, 72(5): 524-548, 2007.

13. Lynn, P.P. and Ingraffea, A.R., Transition elements to be used with quarter-point crack-tip elements. *International Journal for Numerical Methods in Engineering*, 15: 1427-1445, 1980.

14. Liu, G. R. and Quek, S. S., *The Finite Element Method: a practical course.* Betterworth Heinemann, Oxford, 2002.

15. Belytschko, T. and Black, T., Elastic crack growth in finite elements with minimal remeshing. *International Journal for Numerical Methods in Engineering*, 45(5): 601-620, 1999.

16. Belytschko, T., Moes, N., Usui, S., Parimi, C., Arbitrary discontinuities in finite elements. *International Journal for Numerical Methods in Engineering*, 50(4): 993-1013, 2001.

17. Song, J. H., Areias, P. M. A. and Belytschko, T., A method for dynamic crack and shear band propagation with phantom nodes. *International Journal for Numerical Methods in Engineering*, 67(6): 868-893, 2006.

18. Moes, N. and Dolbow, J., Belytschko, T., A finite element method for crack growth without remeshing. *International Journal for Numerical Methods in Engineering*, 46(1): 131-150, 1999.

19. Chessa, J., Wang, H. and Belytschko, T., On the construction of blending elements for local partition of unity enriched finite elements. *International Journal for Numerical Methods in Engineering*, 57: 1015-1038, 2003.

20. Chen, L., Liu, G. R., Nourbakhshnia, N. and Zeng, K., A singular edge-based smoothed finite element method (ES-FEM) for bimaterial interface cracks. *Computational Mechanics*, 45(2-3): 109-125, 2010.

21. Liu, G. R., Nourbakhshnia, N., Chen, L. and Zhang, Y. W., A novel general formulation for singular stress field using the ES-FEM method for the analysis of mixed-mode cracks. *International Journal of Computational Methods*, 7(1): 191-214, 2010.

22. Liu, G. R., Nourbakhshnia, N. and Zhang, Y. W., A novel singular ES-FEM method for simulating singular stress fields near the crack tips for linear fracture problems. *Engineering Fracture Mechanics*, 78(6): 863-876, 2011.

23. Chen, L., Liu, G. R., Jiang, Y., Zeng, K. Y. and Zhang, J., A singular edge-based smoothed finite element method (ES-FEM) for crack analysis in anisotropic media. *Engineering Fracture Mechanics*, 78(1): 85-109, 2011.

24. Liu, G. R., Chen, L., Nguyen-Thoi, T., Zeng, K. and Zhang, G. Y., A novel singular node-based smoothed finite element method (NS-FEM) for upper bound solutions of fracture problems. *International Journal for Numerical Methods in Engineering*, 83(11): 1466-1497, 2010.

25. Nourbakhshnia, N. and Liu, G. R., Aquasi-static crack growth simulation based on the singular ES-FEM. *International Journal for Numerical Methods in Engineering*,88(5): 473-492, 2011.

26. Nguyen-Xuan, H., Liu, G. R., Nourbakhshnia, N. and Chen, L., A novel singular ES-FEM for crack growth simulation. *Engineering Fracture Mechanics*, 84: 41-66, 2012.

27. Chen, L., Rabczuk, T., Bordas, S. P. A., Liu, G. R. Zeng, K. Y. and Kerfriden, P., Extended finite element method with edge-based strain smoothing (ESm-XFEM) for linear elastic crack growth. *Computer Methods in Applied Mechanics and Engineering*, 209: 250-265, 2012.

28. Liu, G. R., Jiang, Y., Chen, L., Zhang, G. Y. and Zhang, Y. W., A singular cell-based smoothed radial point interpolation method for fracture problems. *Computers & Structures*, 89(13-14), 2011.

29. Li, F. Z., Shih, C. F. and Needleman, A., A comparison of methods for calculating energy release rates. *Engineering Fracture Mechanics*, 21(2): 405-421, 1985.

30. Shih, C. and Asaro, R., Elastic-plastic analysis of cracks on biomaterial interfaces: Part I-small scale yielding, *Journal of Applied Mechanics*, 55: 299-316, 1985.

31. Yau, J., Wang, S. and Corten, H., A mixed mode crack analysis of isotropic solids using conservation laws of elasticity. *Journal of Applied Mechanics*, 47: 335-341, 1980.

32. Liu, G. R., *Meshfree Methods: Moving beyond the Finite Element Method*, 2nd ed., CRC press, Boca Taton, USA, 2009.

33. Liu, G. R. and Zhang, G. Y. A normed G space and weakened weak (W^2) formulation of a cell-based smoothed point interpolation method. *International Journal of Computational Methods*, 6(1): 147-179, 2009.

34. Zhang, G. Y. and Liu, G. R., Meshfree cell-based smoothed point interpolation method using isoparametric PIM shape functions and condensed RPIM shape functions. *International Journal of Computational Methods*, 8(4): 705-730, 2011.

35. Liu, G. R., On a G space theory. *International Journal of Computational Methods*, 6(2): 257-289, 2009.

36. Liu, G. R., A G space theory and a weakened weak (W^2) form for a unified formulation of compatible and incompatible methods: Part I theory. *International Journal for Numerical Methods in Engineering*, 81: 1093-1126, 2010.

37. Liu, G. R., A G space theory and a weakened weak (W^2) form for a unified formulation of compatible and incompatible methods: Part II applications to solid mechanics problems. *International Journal for Numerical Methods in Engineering*, 81: 1127-1156, 2010.

38. Tada, B. H., Paris, P. C. and Irwin, G. R., *The stress Analysis of Cracks Handbook*, 3rd ed., American Society of Mechanical Engineers, 2000.

39. Wikipedia contributors. "Stress intensity factor." *Wikipedia, The Free Encyclopedia*. Wikipedia, The Free Encyclopedia, 17 Feb. 2010. Web. 12 Apr. 2010.

40. Wikipedia contributors. "J integral." *Wikipedia, The Free Encyclopedia*. Wikipedia, The Free Encyclopedia, 17 Mar. 2010. Web. 12 Apr. 2010.

41. Cherepanov, G. P., The propagation of cracks in a continuous medium, *Journal of Applied Mathematics and Mechanics*, 31(3): 503-512, 1967.

42. Rice, J. R., A path independent integral and the approximate analysis of strain concentration by notches and cracks. *Journal of Applied Mechanics*, 35: 379-386, 1968.

43. Yoda, M., The J-integral fracture toughness for Mode II, *International Journal of Fracture*, 16(4): 175-178, 1980.

44. Moran, B. and Shih, C. F., Crack tip and associated domain integrals from momentum and energy balance. *Engineering Fracture Mechanics*, 27(6): 615-641, 1987.

45. Belytschko, T. and Black, T., Elastic crack growth in finite elements with minimal remeshing. *International Journal for Numerical Methods in Engineering*, 45(5): 601-620, 1999.

46. Sukumar, N. and Prevost, J. H., Modeling quasi-elastic crack growth with the extended finite element method Part I: Computer implementation. *International Journal of Solids and Structures*, 40(26): 7513-7537, 2003.

Chapter 13

Adaptive Analysis Using S-PIMs

13.1 Introduction

In the practices of stress analysis of engineering systems using either FEM or meshfree methods, the analyst needs to be constantly concerned about the density and the distribution of the density of the element mesh or background cells, because this can directly affect the numerical solution obtained. Therefore, the experience of the analyst becomes crucially important, in order to obtained accurate solution at least cost. Even for an experienced analyst, it is often very difficult to make an optimal decision. To alleviate this difficulty, the use of adaptive analysis techniques is one of the ideal means, which empowers the computer to make the decision automatically for the analyst. It is the opinion of the authors that a robust adaptive analysis technique is crucial for future automated modeling and simulation of engineering systems. As demonstrated in the previous chapters, S-PIM models can work very well with triangular/ tetrahedral background cells and we do not even demand too much on the mesh quality, which is essential for an adaptive analysis, because the refinement can be performed automatically without human intervention. In this chapter, we discuss some simple techniques for the adaptive analysis using S-PIMs. Since S-PIMs work well for triangular/tetrahedral cells, many of the adaptive techniques developed in the past for linear FEM are adoptable with some simple modifications using S-PIM solution properties.

Adaptive techniques were introduced for FEM by Bahuška and Rheinbolt in the late 1970s [1, 2]. In the years that follow, a lot of work has been done to develop various adaptive procedures for the FEM [3-29] as well as different meshfree methods [30-39]. Generally, there are two important issues in an adaptive analysis: 1) posteriori error estimation and 2) domain refinement scheme. The former requires a cheap and reasonably good error estimator or indicator to measure the local and global errors in the current refinement stage. The adaptive procedure determines whether a refinement is required, which part of the domain should be refined, and how the refinement is conducted. To conduct a posteriori error estimation, two solution values, a computed value and a reference value, are usually required. The first is the raw data from the computations for the problem, and the second is derived via postprocessing (e.g., smoothing or projection). In general, there are two distinct types of error indicators which have been widely used in the adaptive analysis, i.e. the recovery based error indicator and the residual based error indicator [3].

In the recovery based error indicator suggested by Zienkiewicz and Zhu [4], simple averaging and the L^2 projection recovery were used to assess errors. A recovery process based on the super-convergence concept can be found in [15, 16]. Other recovery techniques can be found in [17-20]. Most of the residual based error estimators or indicators make use of the residual of the numerical approximation, either explicitly or implicitly. This type of error estimators was introduced by Bahuška and Rheinboldt [1, 2] and has been further developed by many researchers [21-26]. There are also some recent surveys and textbooks available on error estimation and adaptivity. The volumes edited by Brebbia and Aliabadi [27] and by Bahuška et al. [28] have reviewed adaptive techniques for the FEM and the Boundary Element Method (BEM). Books by Szabo and Bahuška [29] and by Zienkiewicz and Taylor [3] contain chapters on error estimation and adaptive analyses.

In many meshfree methods, the field nodes can be moved, inserted and deleted in easier manners [30], and they are particularly attractive for adaptive analysis. With the rapid development of meshfree methods in recent years, many effective adaptive procedures have been developed. Duarte and Oden [31] derived an error estimate that involves only the computation of interior residuals and the residuals for Neumann boundary conditions for the *h-p* cloud method. For the EFG method, Chung and Belytschko have proposed an error estimate based on the difference between the values of projected stresses and stresses of the EFG solution [32]. More in-depth study about refinement procedures in EFG has been conducted by Lee and Zhou [33, 34]. In the context of the reproducing kernel particle method (RKPM), Liu et al. [35] have proposed approaches based on the residual errors. You et al. [36] have developed an approach by utilizing the reproducing kernel as a low-pass filter and the corresponding high-pass filter is used to identify the locations of high gradient. Liu and Tu [30, 37] have also developed an error indicator for the strain energy in individual triangular background cells. For each cell, computing cell energy and reference cell energy are generated based on a same stress field by using two different integration schemes. The difference between the two energy values is then used as the basic measure of error level. The residual based error estimations recently were also applied to the meshfree point interpolation methods (PIM) by Kee et al. [38] and Zhang et al. [39].

Based on the strategy, the refinement techniques can be broadly classified into three categories: *h*-type, *p*-type, and *r*-type refinements. In an *h*-type refinement, the orders of the shape functions or the interpolations are kept the same, but sizes of the element or the background cells are reduced. In a *p*-type refinement, the element size is kept the same but the order of the shape functions or interpolations are increased. In an *r*-type refinement, the locations of the nodes are adjusted in an "optimal" fashion to achieve the adaptive effects. All these refinement strategies are possible for S-PIM models. In this chapter, we use essentially the *h*-type refinement with only triangular/tetrahedral background cells for its simplicity and applicability for generally complicated geometries.

13.2 Adaptive analysis using S-PIMs

In this section, a simple but efficient adaptive procedure including a novel error indicator and a simple *h*-type refinement strategy is introduced and implemented in the framework of NS-PIM and ES-PIM models for both 2D and 3D solid problems.

13.2.1 Error indicator based on cell energy error

In linear FEM using triangular elements, the computed stresses/strains do not possess inter-element continuity and have a Heaviside-type discrepancy along element boundaries. Thus the improved (or recovered) value of stress solution can be obtained via smoothing the inter-element discontinuity on strains, and hence it is always computable. The difference between the computed "raw" and the improved solution form a basis for the recovery based error estimation in the FEM practice. The similar ideal is adoptable to S-PIM models.

In the NS-PIM/ES-PIM models, there are two types of domains, i.e. triangular background cells Ω_i^c and smoothing domains Ω_i^s, and they are overlaid as shown in FIGURE 6.2 and FIGURE 7.2. When the Heaviside type of smoothing functions and the generalized gradient smoothing operation are used (see Section 4.5.2), there are discontinuities of stresses or strains along smoothing domain boundaries. Making use of this discrepancy, an efficient adaptive procedure for NS-FEM, i.e. NS-PIM-Tr3, has been developed using the ideal of recovery based error estimation [40].

In an adaptive analysis, an (computable) error indicator is usually constructed based on the following definition of "error": a proper measure of the difference between the exact/reference solution and the approximate one. Therefore, the calculation of "exact" or reference/improved solution is generally needed in such an adaptive procedure. For example, recovery solutions are often used as improved one. In this section, we define an error indicator which uses the difference between the computed values at different identities (field nodes, background cells or smoothing cells) in a particular norm. Using this error indicator, adaptive analysis is performed using NS-PIM and ES-PIM, and very good results have been obtained for a number of 2D and 3D benchmark problems [41, 43].

Refinement is usually required at locations with steep gradient of stresses or singularity point, where the gradient of stresses (or strain energy and Von Mises stresses etc.) is much higher than other places of the problem domain. Making use of this fact, our error indicator uses the difference of the computed strain energy at the nodes (vertices) of the background cell. For 2D problems using three-node triangular cells, the error indicator e^{In} for cell k is defined as

$$e_k^{In} = \max\left(U_{PE}^{(12)}, U_{PE}^{(23)}, U_{PE}^{(13)}\right) \tag{13.1}$$

where U_{PE} is the strain energy on the difference of the nodal strains at two nodes:

$$\begin{cases} U_{PE}^{(12)} = \dfrac{1}{2}\left(\Delta\varepsilon^{(12)}\right)^{\mathrm{T}}\mathbf{c}\Delta\varepsilon^{(12)} \\[2mm] U_{PE}^{(23)} = \dfrac{1}{2}\left(\Delta\varepsilon^{(23)}\right)^{\mathrm{T}}\mathbf{c}\Delta\varepsilon^{(23)} \\[2mm] U_{PE}^{(13)} = \dfrac{1}{2}\left(\Delta\varepsilon^{(13)}\right)^{\mathrm{T}}\mathbf{c}\Delta\varepsilon^{(13)} \end{cases} \tag{13.2}$$

In Equation (13.2), $\Delta\varepsilon$ is defined as

$$\begin{cases} \Delta\boldsymbol{\varepsilon}^{(12)} = \boldsymbol{\varepsilon}^{(1)} - \boldsymbol{\varepsilon}^{(2)} \\ \Delta\boldsymbol{\varepsilon}^{(23)} = \boldsymbol{\varepsilon}^{(2)} - \boldsymbol{\varepsilon}^{(3)} \\ \Delta\boldsymbol{\varepsilon}^{(13)} = \boldsymbol{\varepsilon}^{(1)} - \boldsymbol{\varepsilon}^{(3)} \end{cases} \tag{13.3}$$

where $\boldsymbol{\varepsilon}^{(1)}$, $\boldsymbol{\varepsilon}^{(2)}$ and $\boldsymbol{\varepsilon}^{(3)}$ are the computed strain vectors at the three nodes of the triangular cell k. It is clear that we are trying to get the information related to the 2nd derivatives of the displacement field for the cell.

For NS-PIM, the nodal strain values can be computed using the displacement solutions directly. For ES-PIM models, nodal strain values will be obtained via averaging the raw strains of the smoothing domains around this node, as presented in Section 7.4.4. Alternatively, e_k^{In} can be evaluated using the raw strain values on these three edges (instead of at the nodes) of the kth cell [44].

It is clear that the procedure to calculate the present error indicator is very simple: we first obtain the difference of the nodal strain values, then calculate the corresponding strain energy, and finally we take the maximum value of these strain energy values as the error indicator of this cell. Extension of this calculation to 3D problems using tetrahedral cells is straightforward. We note that our error estimator is very simple, cheap to compute and easy to use.

13.2.2 Local refinement criteria

In usual situations, the refinement in each adaptive step is always performed for a small part of the background cells. In our adaptive scheme, we simply use a threshold value $\eta \in [0, 1]$ to select the cells for refinement. First, we obtain the error indicate values using Equation (13.1) for each background cell, and then sort out in descending order according to these values. This process selects $N_{re} = N_c \eta$ cells for refinement.

It is clear that by changing the value of η, the speed of the adaptive refinement can be effected: a big value of η allows more cells to be refined at each stage. This can reduced the number of refinement stages, but may not give the optimal quality of the refined cells. The choice is the analyst's.

13.2.3 Refinement strategy

In this study, we use the simple h-type refinement strategy by adding "new" field nodes in the cells that meet the local refinement criteria. Thus the information stored in the "old" nodes can be kept.

For 2D problems discretized using triangular cells, we have initially divided all background cells into two groups: internal cells and boundary cells. An internal cell has no edges on the boundaries of the problem domain, and a boundary cell has at least one edge on the boundaries. For an internal cell to be refined, only one new node is added at the centroid and the cell will be divided into three triangles, as shown in FIGURE 13.1a. For a boundary cell subjected to refinement, three new nodes will be added at the mid-edge locations and this cell will be further divided into four triangular cells, as shown in FIGURE 13.1b.

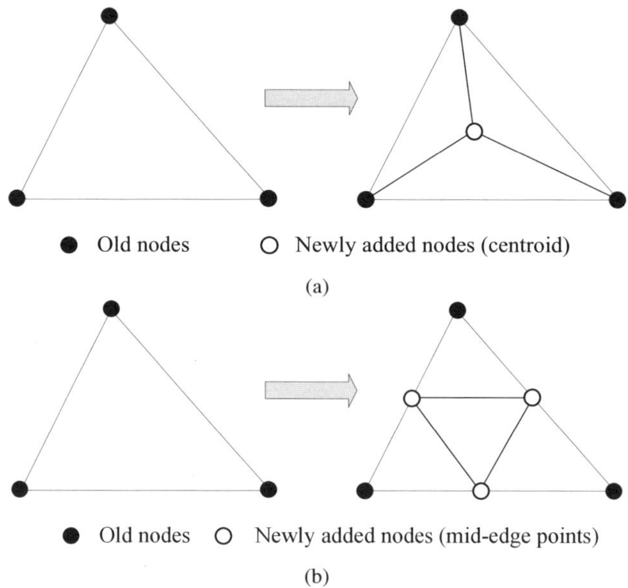

FIGURE 13.1 Illustration of the *h*-type refinement strategy for 2D triangular cells: (a) refinement scheme for internal cells; (b) refinement scheme for boundary cells.

For 3D problem domain discretized using tetrahedral cells, several refinement techniques have been developed in recent years by means of bisection. In our study, six new nodes will be added at the locations of mid-edge points and the target cell will be further divided into eight new cells, as show in FIGURE 13.2. For the target cell N_1-N_2-N_3-N_4, six new nodes (N_5~N_{10}) at the mid-edge positions are added and we now have eight tetrahedral cells, i.e. N_1-N_5-N_7-N_8, N_2-N_6-N_5-N_9, N_3-N_7-N_6-N_{10}, N_4-N_9-N_8-N_{10}, N_5-N_7-N_8-N_6, N_5-N_9-N_6-N_8, N_7-N_{10}-N_8-N_6 and N_6-N_9-N_{10}-N_8. It seems quite complicated for 3D cases, but all the operations can be done automatically without human intervention, due to the use of tetrahedrons.

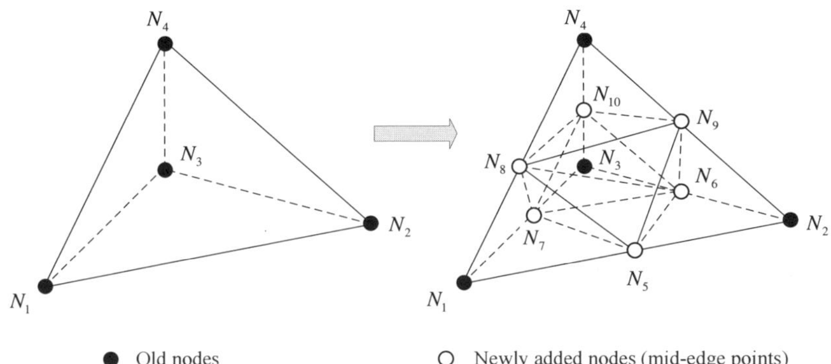

FIGURE 13.2 Illustration of the *h*-type refinement strategy for 3D tetrahedral cells, where six new nodes are added and the target cell is further divided into eight tetrahedral cells.

13.2.4 Cell regeneration

After the *h*-type refinement, we will get a new set of cells which is generally not in very good quality. To improve the quality, a process of local cell generator is incorporated as "cosmetic" touching-up.

Two automatic cell generators, Triangle [45] and Tegen [46], are used to generate triangular and tetrahedral cells for 2D and 3D problem domains. These two generators are based on the well-known Delaunay technology [47-49]. We coded these two cell generators in our S-PIM, and hence complicated 2D and 3D domains can be easily represented using a set of vertices, segments and facets, with the so-called piecewise linear complex [50].

For example, for triangular cells, we can use the incremental insertion algorithm to perform cheap "cosmetic" operations to improve the cell quality. FIGURE 13.3 illustrates the operation consisting of inserting node and flipping edges. In our adaptive analysis process, the refinement and cosmetic operations are all performed automatically (again) without manual operation.

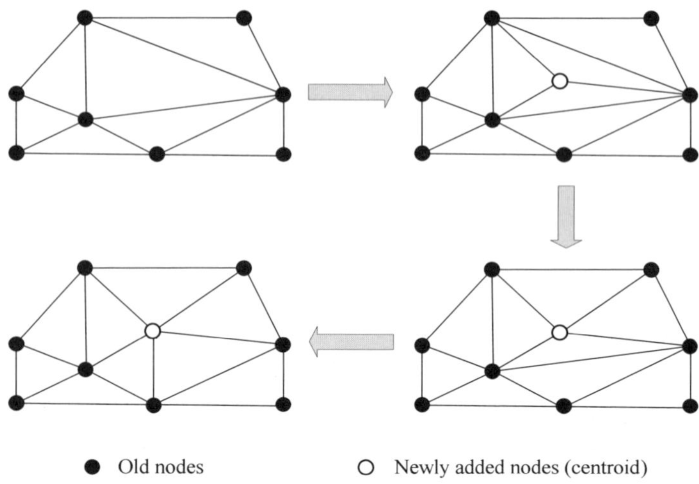

FIGURE 13.3 Refinement with cosmetic operations for 2D triangular cells.

Tetgen uses the radius-edge ratio to measure the quality of a tetrahedron, where the radius refers to the circumradius of the tetrahedron and the edge refers to the shortest edge of the tetrahedron. If a tetrahedron has a large radius-edge ratio, it is then regarded badly-shaped. To generate the best possible shaped cells for each adaptive step, Tetgen uses an incremental point insertion approach extended from a classical Delaunay refinement scheme proposed by Shewchuk [45, 51].

13.2.5 General adaptive procedure

FIGURE 13.4 shows the general procedure for adaptive analyses using S-PIM models with triangular/tetrahedral background cells. In the analysis, a maximum number of adaptive steps is predefined by the user to simply control the adaptive procedure.

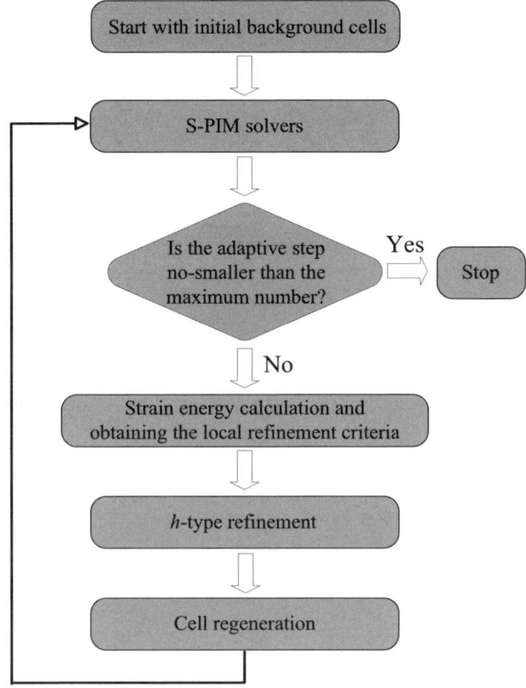

FIGURE 13.4 Flow chart of the general adaptive procedure.

13.3 Numerical examples

Example 13.3.1 A 2D L-shaped solid

An L-shaped solid subjected to uniform tensile in the horizontal direction is first studied. Dimensions and boundary conditions of the problem are shown in FIGURE 13.5. The solid is constrained in x and y direction along the right and upper boundary edges, respectively. It is clear that there will be stress concentration at the reentrant corner. In the present study, the plane stress condition is considered, the units used are based on the international standard unit system and the parameters are taken as $E=3.0\times10^7$, $v=0.3$, $a=5$ and $P=10$.

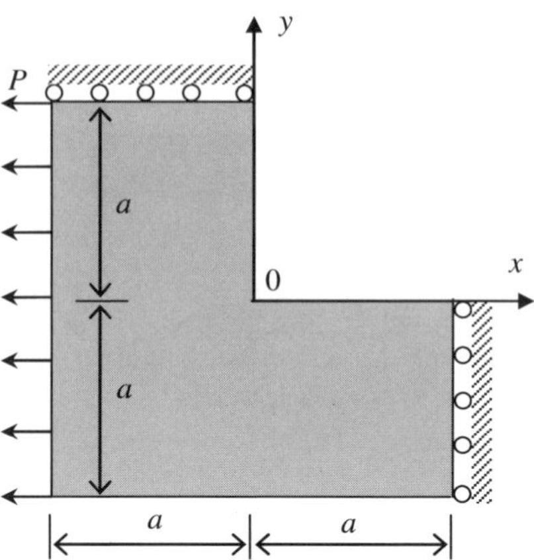

FIGURE 13.5 L-shaped solid subjected to uniform tensile.

The adaptive analysis is conducted using different S-PIM models, as well as the linear FEM. Started from the same uniform cells of 121 nodes, the adaptive analysis is conducted for four steps for each numerical method. Values of the local refinement criteria η are taken as 0.2, 0.3 and 0.15 for the FEM-Tr3, NS-PIM-Tr3, and ES-PIM models, respectively. For the comparison purpose, numerical results obtained using the uniform meshes are also provided.

The patterns of the background cells at the initial and each adaptive stage are plotted in FIGURE 13.6 and FIGURE 13.7 for NS-PIM-Tr3 and ES-PIM-Tr3, respectively. It can be clearly found that the cell refinement concentrates around the origin, where exactly the singularity exists. The figures also show that more cells are involved in the refinement for the NS-PIM-Tr3 using bigger value of the local refinement criteria η, which is consistent with the discussions given in Section 13.2.2.

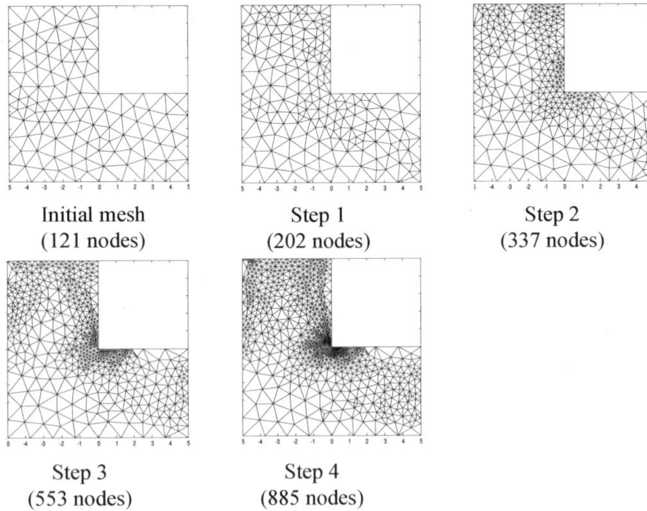

FIGURE 13.6 Triangular background cells generated during the adaptive analysis using the NS-PIM-Tr3 for the L-shaped solid problem (η=0.3).

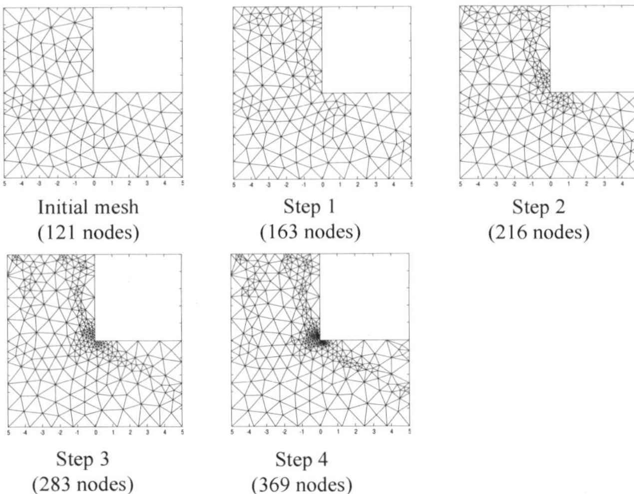

FIGURE 13.7 Triangular background cells generated during the adaptive analysis using the ES-PIM-Tr3 for the L-shaped solid problem (η=0.15).

FIGURE 13.8 shows the convergence process of the numerical results to the reference strain energy of the problem (5.18963×10^{-4}), which is obtained using FEM with a fine mesh of 13,654 nodes. The following points can be found.

1) With the increase of DOF, all the solutions converge to the reference one.

2) The FEM-Tr3 and ES-PIM models give lower bound solutions, and the NS-PIM-Tr3 gives upper bound solutions for this case, which is consistent with our previous studies.

3) For all the methods used in this case, results of the adaptive scheme converge much faster than those of the uniform meshes and provide much tighter bound to the reference solution, which demonstrates the effectiveness of the present adaptive procedure.

4) Among the methods used, two ES-PIM models provide much more accurate results. The ES-RPIM-Tr2L performs the best, but requires more computational resources, which will be four times expensive compared to the ES-PIM-Tr3 as we discussed in Chapter 7. Considering the time consuming problem, only the linear models will be conducted in the following studies.

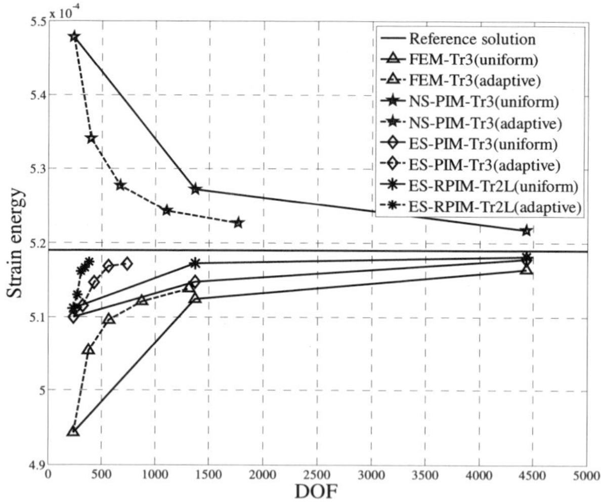

FIGURE 13.8 Converging process of the numerical results (in strain energy) obtained using uniform and adaptive meshes for the L-shaped solid problem.

Example 13.3.2 A 2D Infinite solid with a circular hole

The infinite 2D solid with a circular hole problem described in Example 6.1.3, is again studied. The problem has a stress concentration around the circular hole, and adaptive analysis is now performed. For comparison, results obtained using uniform meshes are also provided. Stared from the same uniform mesh of 63 nodes, the adaptive analysis is conducted for three steps for each numerical method with the same value of the local refinement criteria $\eta = 0.2$.

The background cell patterns at each adaptive step are plotted in FIGURE 13.9 and FIGURE 13.10 for the NS-PIM-Tr3 and ES-PIM-Tr3, respectively. The figures show that the adaptive procedure performs the refinement near the regions of stress concentration, which demonstrates that the present error indicator can automatically identify the stress concentration region properly.

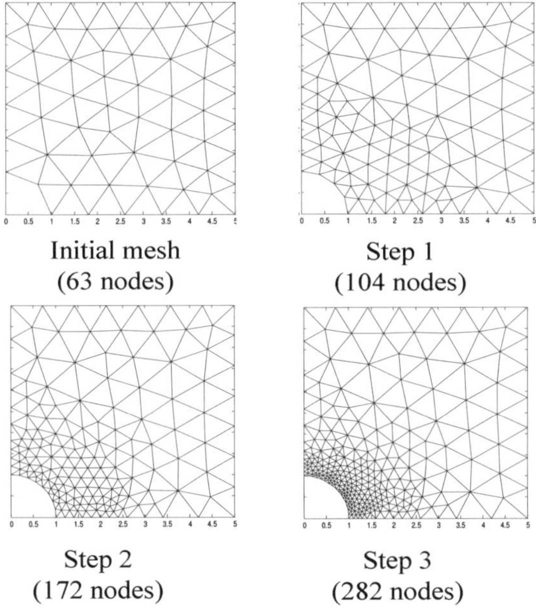

FIGURE 13.9 Triangular background cells generated during the adaptive analysis using the NS-PIM-Tr3 for the infinite solid with a circular hole problem (η=0.2).

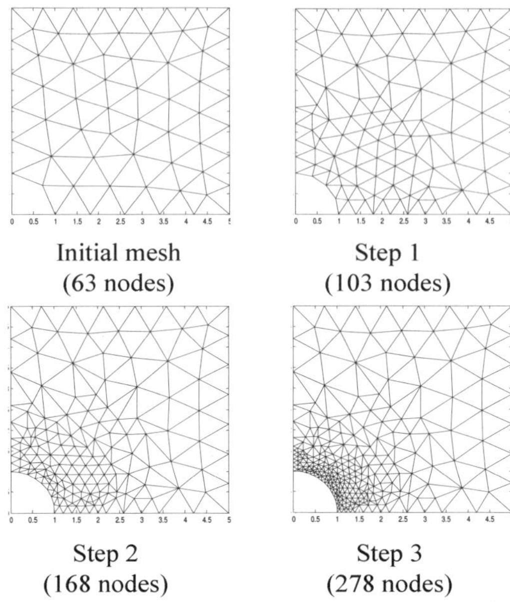

Initial mesh
(63 nodes)

Step 1
(103 nodes)

Step 2
(168 nodes)

Step 3
(278 nodes)

FIGURE 13.10 Triangular background cells generated during the adaptive analysis using the ES-PIM-Tr3 for the infinite solid with a circular hole problem (η=0.2).

FIGURE 13.11 shows the convergence process of the numerical results in strain energy. For this case we can find the following points.

1) With the increase of DOF, results of the NS-PIM-Tr3 converge from the upper side, and the results of the FEM-Tr3 and the ES-PIM-Tr3 converge from the lower side.

2) For all the methods, results obtained using the adaptive procedure converge much faster than those of the uniform meshes.

3) The ES-PIM-Tr3 provides more accurate results with both uniform and adaptive meshes, which is consistent with our previous conclusions in Chapter 7.

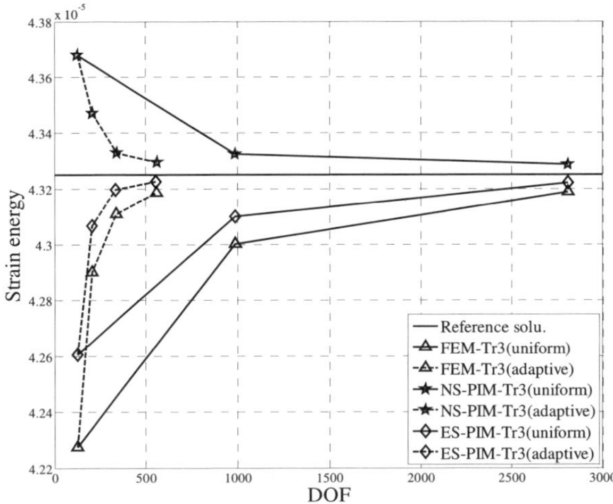

FIGURE 13.11 Converging process of the numerical results (in strain energy) obtained using uniform and adaptive meshes for the infinite solid with a circular hole problem.

Example 13.3.3 A 3D column footing under biaxial bending

A 3D problem involving the biaxial bending of a column footing is now considered. As shown in FIGURE 13.12, the column footing is fixed on the bottom face of $z=0$. The uniform pressure T_x is applied on the face ABED and

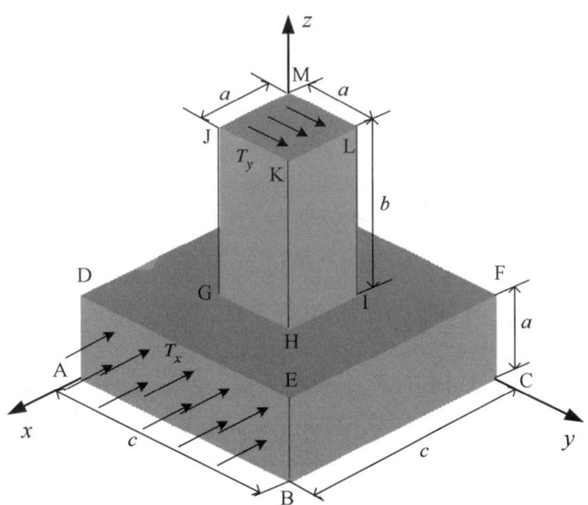

FIGURE 13.12 3D column footing under biaxial bending.

the uniform shear T_y is applied on the top surface JKLM. In the computation, the units used are based on the international standard unit system and the parameters are taken as $E=1$, $v=0.3$, $a=1$, $b=2$, $c=3$, $T_x = -1$ and $T_y = 1$.

For the purpose of comparison, results obtained using uniform meshes (shown in FIGURE 13.13) are also provided. The adaptive analysis is conducted for two steps with the same local refinement criteria $\eta = 0.1$ for all the methods. FIGURE 13.14 shows the mesh patterns at each adaptive step using the FS-PIM-Te4 method. More refinements have been generated around the root edges of the column, where stress concentration exists.

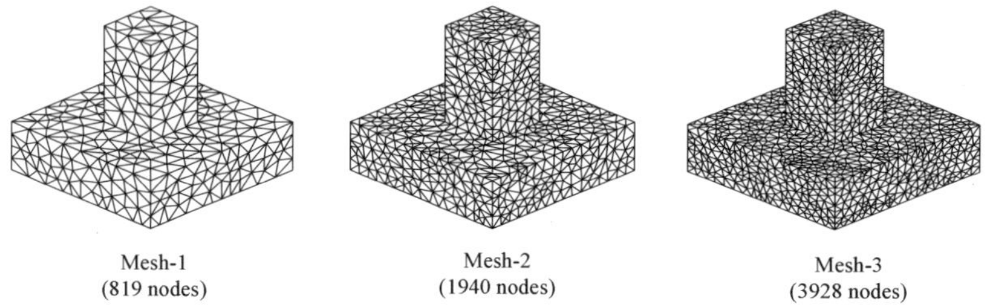

FIGURE 13.13 Uniform tetrahedral background cells used for the 3D column footing problem.

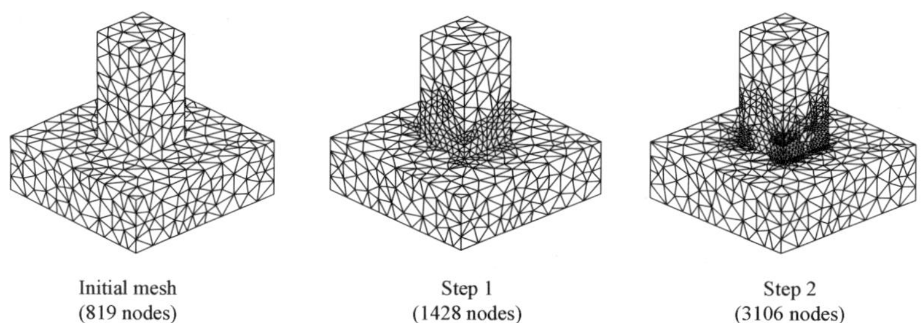

FIGURE 13.14 Tetrahedral background cells generated during the adaptive analysis using the FS-PIM-Te4 for the 3D column footing problem (η=0.1).

FIGURE 13.15 shows the convergence process of the numerical results to the reference strain energy of the problem (27.051145), which is obtained using the FEM with a very fine mesh. For this 3D case we find that with the increase of DOF, results of the NS-PIM-Te4 converge from the upper side, and the results of the FEM-Te4 and the FS-PIM-Te4 converge from the lower side. For all the methods, results of the adaptive procedure converge much faster than those of the uniform meshes and hence provide much tighter bound to the reference solution. Compared to the other two methods, the FS-PIM-Te4 provides more accurate results.

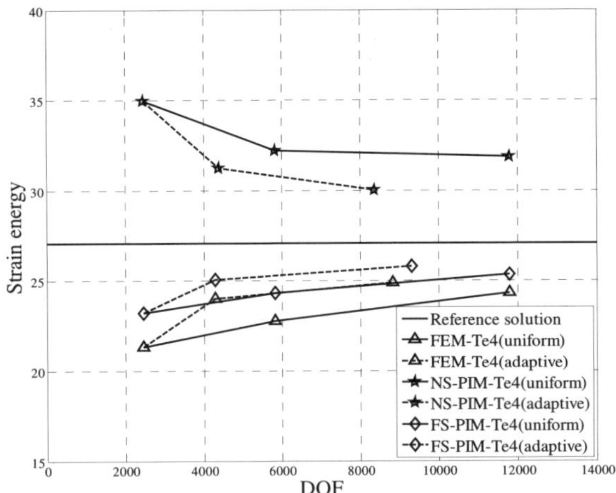

FIGURE 13.15 Converging process of the numerical results (in strain energy) obtained using uniform and adaptive meshes for the 3D column footing problem.

Example 13.3.4 A 3D cube without one-eighth part under tensile load

The second 3D case studied is a rigid cube with one-eighth part removed and subjected to uniform tensile load in three plane surfaces, as shown in FIGURE 13.16. Displacement boundary conditions are applied as follows: $u=0$ on face JGFK, $v=0$ on face IDGJ and $w=0$ on face DEFG. Based on the international standard unit system, the parameters are taken as $E=3.0\times10^{7}$, $\nu=0.3$, $a=2$, $b=1$ and $T_x = T_y = T_z = 100$.

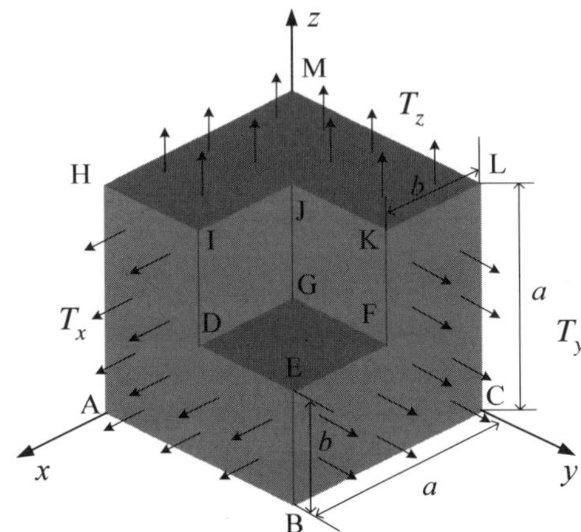

FIGURE 13.16 3D cube without one-eighth part subjecting to uniform tensile load.

The adaptive analysis is conducted for three steps with the same value of the local refinement criteria $\eta = 0.05$ for all the methods. FIGURE 13.17 shows the background cell patterns at each adaptive step using the FS-PIM-Te4 method. It can be clearly found that more refinements have been generated reentrant corner edges (lines GD, GF and GJ) where stress singularity exists.

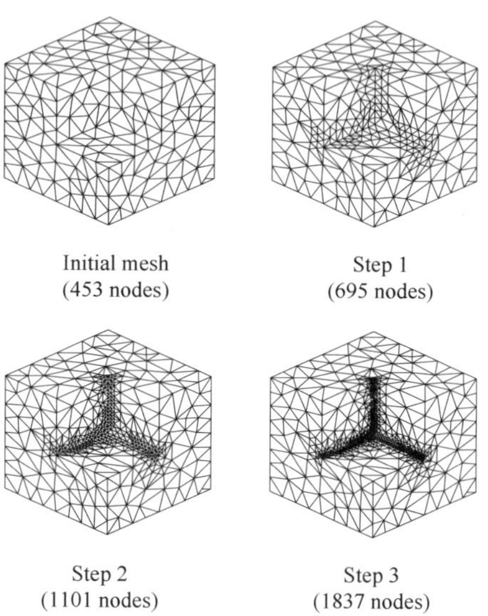

Initial mesh
(453 nodes)

Step 1
(695 nodes)

Step 2
(1101 nodes)

Step 3
(1837 nodes)

FIGURE 13.17 Tetrahedral background cells generated during the adaptive analysis using the FS-PIM-Te4 for the 3D cube problem (η=0.05).

A reference of strain energy (3.60464×10^{-3}) has been obtained using the FEM with a very fine mesh, and FIGURE 13.18 shows the convergence process of the numerical results to the reference value. Again we can find the NS-PIM-Te4 provides upper bound solutions and the other two methods give lower bound solutions. For all these three methods, results obtained using the adaptive process converge much faster than those of uniform meshes. FS-PIM-Te4 stands out clearly for providing more accurate results.

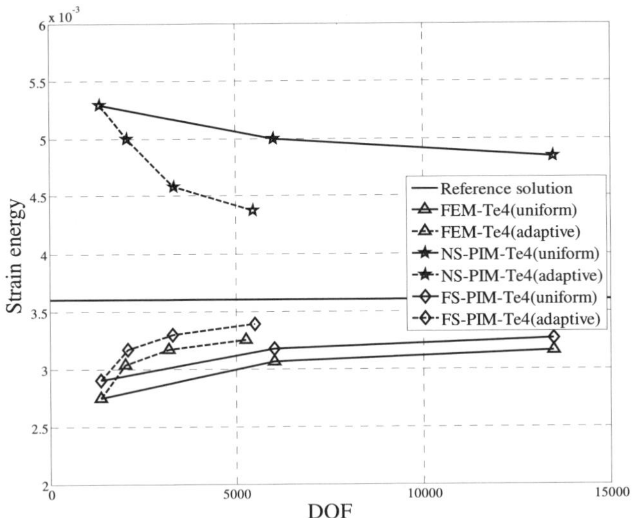

FIGURE 13.18 Converging process of the numerical results (in strain energy) obtained using uniform and adaptive meshes for the 3D cube problem.

13.4 Concluding remarks

In this section, an error indicator for S-PIM models has been introduced. Using the simple h-type refinement strategy, an adaptive procedure for NS-PIM/ES-PIM methods is developed for both 2D and 3D problems using triangular and tetrahedral meshes. We now end this chapter with the following remarks.

Remark 13.1 The error estimator is simple and effective

Making use of the smoothing cell structure, the present error estimator is obtained based on only the computed solutions at the "center" of the smoothing cells. It is computable, cheap to compute and easy to use. The numerical results of some benchmark problems have shown that the present error indicator can accurately catch the stress concentration/singularity regions.

Remark 13.2 Effective adaptive procedure

A simple *h*-type refinement strategy has been adopted by adding new nodes in the cells to be refined. The numerical tests have shown the present adaptive procedure can significantly improve the convergence property of the results.

Remark 13.3 Robust and practical adaptive analysis

The S-PIM models work well with the simplest triangular/tetrahedral cells, which can be generated for very complicated geometries automatically at each refinement stage. It clearly has the potential to be further developed into a robust automatic solution tool for practical engineering problems with generally complicated geometries. The work presented in this chapter is just to demonstrate this enormous potential of the S-PIMs in automatic computation.

13.5 References

1. Bahuŝka, I. and Rheinboldt, C., A-Posteriori error estimates for the finite element method. *International Journal for Numerical Methods in Engineering*, 12: 1597-1615, 1978.

2. Bahuŝka, I. and Rheinboldt, C., Adaptive approaches and reliability estimates in finite element analysis. *Computer Methods in Applied Mechanics and Engineering*, 17/18: 519-540, 1979.

3. Zienkiewicz, O. C. and Taylor R. L., *The Finite Element Method*, 5th ed., Butterworth Heimemann, Oxford, 2000.

4. Zienkiewicz, O. C. and Zhu, J. Z., A simple error estimator and adaptive procedure for practical engineering analysis. *International Journal for Numerical Methods in Engineering*, 24: 337-357, 1987.

5. Zhu, J. Z. and Zienkiewicz, O. C., Adaptive techniques in the finite element method, *Commun. Appl. Num. Meth.*, 4: 197-204, 1988.

6. Zienkiewicz, O. C., Zhu, J. Z. and Gong, N. G., Effective and practical h-p version adaptive analysis procedures for the finite element method, *International Journal for Numerical Methods in Engineering*, 28: 879-891, 1989.

7. Zienkiewicz, O. C. and Zhu, J. Z., The three R's of engineering analysis and error estimation and adaptivity, *Comp. Meth. Appl. Mech. Eng.*, 82: 95-113, 1990.

8. Atamaz-Sibai, W. and Hinton, E., Adaptive mesh refinement with the Morley plate element, *Proc. of NUMETA 90 Conf. Swansea*, 2: 1044-1057, Elsevier Appl Sci, London, 1990.

9. Selman, A., Hinton, E. and Atamaz-Sibai, W., Edge effects in Mindlin-Reissner plates using adaptive mesh refinement, *Engineering Computations*, 7:217-227, 1990.

10. Hinton, E., .zak.a, M. and Rao, N. V. R., Adaptive analysis of thin shells using facet elements, *Int. Report CR/950/90,* Univ. College of Swansea, 1990.

11. Onate, E., Castro, J. and Kreiner, R., Error estimations and mesh adaptivity techniques for plate and shell problems, *3rd Int. Conf. Quality Assurance and Standards in Finite Element Methods, Stratford-upon-Avon,* NAFEMS, 1991.

12. Wu, J., Zhu, J. Z., Smelter, J. and Zienkiewicz, O. C., Error estimation and adaptivity in Navier-Stokes incompressible flow, *Computational Mechanics,* 6: 259-271, 1990.

13. Ladeveze, P. and Leguillon, D., Error estimate procedure in the finite element method and applications. *SIAM Journal on Numerical Analysis,* 20: 483-509, 1983.

14. Ladeveze, P. and Rougeot, P. H., New advances on a posteriori error on constitutive relation in f.e. analysis. *Computer Methods in Applied Mechanics and Engineering* 150: 239-249, 1997.

15. Zienkiweicz, O. C. and Zhu, J. Z., The superconvergent patch recovery and a posteriori error estimates. Part 1: The recovery technique. *International Journal for Numerical Methods in Engineering,* 33: 1331-1364, 1992.

16. Zienkiweicz, O. C., Zhu, J. Z. The superconvergent patch recovery and a posteriori error estimates. Part 2: Error estimates and adaptivity. *International Journal for Numerical Methods in Engineering,* 33: 1365-1382, 1992.

17. Wiberg, N. E., Zeng, L. F. and Li, X. D., Error estimation and adaptivity in elastodynamics. *Computer Methods in Applied Mechanics and Engineering,* 101: 369-395, 1992.

18. Wiberg, N. E. and Abdulwahab, F., Patch recovery based on the superconvergent and equilibrium. *International Journal for Numerical Methods in Engineering,* 36: 2703-2724, 1993.

19. Wiberg, N. E., Abdulwahab, F. and Li, X. D., Error estimation and adaptive procedures based on superconvergent patch recovery (SPR) techniques, *Archives of Computational Methods in Engineering,* 4: 203-242, 1997.

20. Blacker, T. and Belytschko, T., Superconvergent patch recovery with equilibrium and conjoint interpolation enhancements. *International Journal for Numerical Methods in Engineering,* 37: 517-536, 1994.

21. Bank, R. E. and Weiser, A., Some a posteriori error estimators for elliptic partial differential equations. *Mathematics of Computation,* 44: 283-301, 1985.

22. Oden, J. T., Demkowicz, L., Rachowicz, W. and Westermann, T. A., Toward a universal h-p adaptive finite element strategy. Part 2: A posteriori error estimation. *Computer Methods in Applied Mechanics and Engineering,* 77: 113-180, 1989.

23. Johnson, C. and Hansbo, P., Adaptive finite element methods in computational mechanics. *Computer Methods in Applied Mechanics and Engineering,* 101: 143-181, 1992.

24. Ainsworth, M. and Oden, J. T., A unified approach to a posteriors error estimation using element residual methods. *Numerische Mathematik,* 65: 23-50, 1993.

25. Bahuška, I., Srouboulis, T. and Upadhyay, C. S., A model study of the quality of a posteriori error estimators for linear elliptic problems. Error estimation in the

interior of patchwise uniform grids of triangles. *Computer Methods in Applied Mechanics and Engineering*, 114: 307-378, 1994.

26. Paulino, G. H., Gray, L. J. and Zarikian, V., Hypersingular residuals-A new approach for error estimation in the boundary element method. *International Journal for Numerical Methods in Engineering*, 39: 2005-2029, 1996.

27. Brebbia, C. A. and Aliabadi, M. H. (eds). *Adaptive Finite and Boundary Element Methods.* Computational Mechanics Publications, Southampton and Elsevier Spplied Science, London, 1993.

28. Bahuška, I., Zienkiewicz, O. C., Gago, J. and Oliveira, A. (eds). *Accuracy Estimates and Adaptive Refinements in Finite Element Computations*, Wiley, Chichester, 1986.

29. Szabo, B. and Bahuška, I., *Finite Element Analysis.* Wiley, New York, USA, 1991.

30. Liu, G. R., *Meshfree Methods: Moving beyond the Finite Element Method*, 2nd Edition, CRC press, Boca Taton, USA, 2009.

31. Durate, C. A. and Oden, J. T., An H-p adaptive method using clouds. *Computer Methods in Applied Mechanics and Engineering,* 139: 237-262, 1996.

32. Chung, H. J., Belytschko T., An error estimate in the EFG method. *Computational Mechanics*, 21: 91-100, 1998.

33. Lee, C. K. and Zhou, C. E., On error estimation and adaptive refinement for element free Galerkin method: part I: stress recovery and a posteriori error estimation. *Computers & Structures,* 82: 413-428, 2003.

34. Lee, C. K. and Zhou, C. E. On error estimation and adaptive refinement for element free Galerkin method: part II: adaptive refinement. *Computers & Structures,* 82: 429-443, 2003.

35. Liu, W. K., Uras, R. A. and Chen, Y., Enrichment of the finite element method with the reproducing kernel particle method. *Journal of Applied Mechanics-Transactions of the ASME,* 64: 861-870, 1997.

36. You, Y., Chen, J. S. and Lu, H., Filters, reproducing kernel, and adaptive meshfree method. *Computational Mechanics,* 31: 316-326, 2003.

37. Liu, G. R. and Tu, Z. H., An adaptive procedure based on background cells for meshless methods. *Computer Methods in Applied Mechanics and Engineering,* 191: 1923-1943, 2002.

38. Kee, B. B. T., Liu, G. R., Zhang, G. Y. and Lu, C. A residual based error estimator using radial basis functions. *Finite Elements in Analysis and Design*, 44(9-10): 631-645, 2008.

39. Zhang, G. Y., Liu, G. R. and Li, Y. An efficient adaptive analysis procedure for certified solutions with exact bounds of strain energy for elasticity problems. *Finite Elements in Analysis and Design*, 44(14): 831-841, 2008.

40. Nguyen-Thoi, T., Liu, G. R., Nguyen-Xuan, H. and Nguyen-Tran, C., Adaptive analysis using the node-based smoothed finite element method (NS-FEM). *Communications in Numerical Methods in Engineering*, 27(2): 198-218, 2011.

41. Tang, Q., Liu, G. R., Zhong, Z. H. and Zhang, G. Y., A three-dimensional adaptive analysis using the meshfree node-based smoothed point interpolation method (NS-PIM). *Engineering Analysis with Boundary Elements*, 35(10): 1123-1135, 2011.

42. Tang, Q., Zhong, Z. H., Zhang, G. Y. and Xu, X., An efficient adaptive analysis procedure for node-based smoothed point interpolation method (NS-PIM). *Applied Mathematics and Computation*, 217(21): 8387-8402, 2011.

43. Tang, Q., Zhang, G. Y., Liu, G. R., Zhong, Z. H. and He, Z. C., An efficient adaptive analysis procedure using the edge-based smoothed point interpolation method (ES-PIM) for 2D and 3D problems. *Engineering Analysis with Boundary Elements,* 36: 1424-1443, 2012.

44. Chen, L., Zhang, G. Y., Zhang, J., Nguyen-Thoi, T. and Tang, Q., An adaptive edge-based smoothed point interpolation method for mechanics problems. *International Journal of Computer Mathematics*, 88(11): 2379-2402, 2011.

45. Shewchuk, J. R., Triangle: Engineering a 2D quality mesh generator and delaunay triangulator, *Applied Computational Geometry: Towards Geometric Engineering.* 1148: 203-222, 1996.

46. Hang, S., Adaptive tetrahedral mesh generation by constrained Delaunay refinement, *International Journal of Numerical Methods in Engineering*, 75: 856-880, 2008.

47. Chew, P. L., Guaranteed-quality Delaunay meshing in 3D, In: *Proc. of the 13th annual ACM symposium on Computational Geometry*, Nice, France, 391-393, 1997.

48. Frey, P. J., George, P. L., *Mesh Generation. Application to Finite Elements*, Hermes Science Publishing, Oxford, 2000.

49. Weatherill, N. P., Hassan, O., Efficient three-dimentional Delaunay triangulation with automatic point creation and imposed boundary constraints, *International Journal of Numerical Methods in Engineering,* 37: 2005-2039, 1994.

50. Miller, G. L., Talmor, D., Teng, S. H., Walkington, N., Wang, H., Control Volume meshes using Sphere packing: Generation, Refinement and Coarsening, In: *Proc. of 5th international Meshing Roundtable*, Pittsburgh, 47-61, 1996.

51. Shewchuk, J. R., *Delaunay refinement mesh generation*, Ph.D. Thesis, Department of Computer Science, Carnegie Mellon University, Pittsburgh, PA. Available as Technical report CMU-CS-97-137, 1997.

Appendix

Program Codes Library

In the appendix, all the common subroutines, which are used in the programs of NSPIM.F90, ESPIM.F90 and CSPIM.F90, are provided in detail. In Appendix 1, the function of the subroutines and the meaning of the major variables and arrays are explained. In Appendix 2, an example input file is presented, which is for the demonstration case of Example 6.1.2 with triangular background mesh of 120 nodes (see Mesh-1 of FIGURE 6.9). Two modules of "Parameters" and "Variables" are presented in Appendix 3 and all the common subroutines in FORTRAN 90 are listed in Appendix 4.

Appendix 1: Description of the subroutines

APPENDIX 1.1 Common subroutines and their functions

No.	Subroutines	Function explanation
1	Input	Input data from an external file and find out the nodes with essential and natural boundary conditions
2	C_materialM	Compute the material matrix
3	Cell_information	Obtain necessary information of the background cell
4	StiffM_Intedomain	Compute the stiffness matrix for an integration domain
5	Form_GK	Form the global stiffness matrix by assembling the local stiffness matrixes
6	Natural_BC	Enforce the nature boundary conditions
7	Essential_BC	Apply the essential boundary conditions
8	Solver_LAE	Solve the linear algebraic equations
9	Dispnorm_error	Compute the global displacement norm error
10	PPIM_SF2D	Create normal polynomial PIM shape functions
11	PPIM_CT_SF2D	Create bilinear polynomial PIM shape functions with Coordinate Transformation (CT) technique

12	Iso_PPIM_SF2D	Create isoparametric polynomial PIM shape functions
13	RPIM_SF2D	Create normal Radial PIM (RPIM) shape functions
14	Con_RPIM_SF2D	Create condensed RPIM shape functions
15	Poly_Basis2D	Form polynomial basis functions
16	Radial_Basis2D	Form radial basis functions (RBF) augmented with polynomial terms
17	Cell_basedT2L	Find the support nodes using the cell-based T2L-scheme
18	Edge_basedT2L	Find the support nodes using the edge-based T2L-scheme
19	GaussEli_Solver	Solve linear algebraic equations using the Gauss elimination method
20	Band_solver	Solve banded linear algebraic equations
21	Gausspointcoe_line	Get the positions and weighting coefficients for Gauss quadrature in 1D domain
22	Line_gauss	Obtain locations and weighting coefficients for the Gauss points located on a section of line
23	Determinant	Compute the determinant of a square matrix
24	Brinv	Obtain the inversion of a matrix using the Gauss-Jodon method
25	Inversion	Get the inversion of a positive definite matrix
26	Indexx	Reorder an array in the ascending order according to the array values

APPENDIX 1.2 Global variables and parameters defined in the "Parameters" module

Variable	Type	Usage	Meaning
e	Long real	Input	Young's modulus of the material
v	Long real	Input	Possion's ratio of the material
Pi	Long real	Input	Circumference ratio
ep	Long real	Input	Tolerance
xlength	Long real	Input	Length of the beam in x direction
ylength	Long real	Input	Length of the beam in y direction
PLoad	Long real	Input	Total value of the load applied
Rc	Long real	Input	Shape parameter of MQ-RBF
q	Long real	Input	Shape parameter of MQ-RBF
Npoly	Integer	Input	Number of monomials used in the augmented RBF
Numnode	Integer	Input	Total number of field nodes
Numcell	Integer	Input	Total number of background cells
NTscheme	Integer	Input	T-schemes used for support nodes selection
NPIMSF	Integer	Input	Types of PIM shape functions
nx	Integer	Parameter	Dimension of the problem
nce	Integer	Parameter	Number of nodes per background cell
Ngauss	Integer	Parameter	Number of Gauss points on each segment of the smoothing domain boundaries Ngauss=1 is used for linear S-PIM models Ngauss=2 can be used for higher order S-PIM models
Mgauss	Integer	Parameter	Number of Gauss points on one edge of triangular cells with natural boundary conditions Mgauss=2 is used for the reference program
numn	Integer	Parameter	The maximum number of field nodes
nume	Integer	Parameter	The maximum number of background cells
numED	Integer	Parameter	The maximum number of edges of the triangular cells
numBC	Integer	Parameter	The maximum number of field nodes with boundary conditions
nre	Integer	Parameter	The maximum number of neighboring cells for a field node
ndom	Integer	Parameter	The maximum number of support nodes for an approximation

nndom	Integer	Parameter	The maximum number of support nodes used for constructing a local stiffness matrix corresponding to an integration domain
Dc	Long real	Compute	The characteristic length of triangular cells
Numedge	Integer	Compute	Total number of edges of the background cells
NumEBC	Integer	Compute	Number of nodes with essential boundary conditions
NumNBC	Integer	Compute	Number of nodes with natural boundary conditions
Ndexnow	Integer	Compute	Number of support nodes for constructing a local stiffness matrix corresponding to an integration domain

APPENDIX 1.3 Global array variables defined in the "Variables" module

Array	Type	Usage	Meaning
x (nx,numnode)	Long real	Input	coordinates of the field nodes $x(1,i)=xi$; $x(2,i)=yi$
nxcell (nce,numcell)	Integer	Input	Node ID of background cells nxcell(1,i): the first node (n1) of cell_i nxcell(2,i): the second node (n2) of cell_i nxcell(3,i): the third node (n3) of cell_i
npEBC (3,numEBC)	Integer	Compute	Index for nodes with essential boundary conditions npEBC(1,i): ID of the field nodes npEBC(2,i): index in x direction: "1" means prescribed and "0" means non-prescribed npEBC(3,i): index in y direction: "1" means prescribed and "0" menas non-prescribed
pEBC (2,numEBC)	Long real	Compute	Prescribed displacement values for the nodes with essential boundary conditions pEBC(1,i): prescribed displacement in x direction pEBC(2,i): prescribed displacement in y direction
npNBC (numNBC)	Integer	Compute	The list of the nodes with natural boundary conditions
Cmat (3,3)	Long real	Compute	The matrix of elastic constants
numneig (numnode)	Integer	Compute	Number of neighboring cells for each node
neighbour (nre,numnode)	Integer	Compute	ID of the neighboring cells for each node
nodein (numnode)	Integer	Compute	Index for each field node nodein(i)=0 means node_i is located inside nodein(i)=1means node_i is located on the boundaries
kcell (numcell)	Integer	Compute	Index for each background cell kcell(i)=0 means cell_i is an interior cell and has three neighboring cells kcell(i)=1 means cell_i is a boundary cell and has two neighboring cells kcell(i)=2 means cell_i is a boundary cell and has one neighboring cell
kcell2 (nce,numcell)	Integer	Compute	Index for three edges of each background cell kcell2(1,i): index for the edge of n1-n2; value "0" means inside and "1" means on the boundaries. kcell2(2,i): index for the edge of n2-n3; value "0" means inside and "1" means on the boundaries. kcell2(3,i): index for the edge of n3-n1; value "0" means inside and "1" means on the boundaries.

center (nx,numcell)	Long real	Compute	Coordinates of the centroid of each background cell
area_cell (numcell)	Long real	Compute	Area of each background cell which is also the area of cell-based smoothing domains
area_node (numnode)	Long real	Compute	Area of each node-based smoothing domain
area_edge (numedge)	Long real	Compute	Area of each edge-based smoothing domain
nodesix (6,numcell)	Integer	Compute	Support nodes selected using cell-based T6-scheme
nodefour (4,numedge)	Integer	Compute	Support nodes selected using edge-based T4-scheme
Node_edge (2,numedge)	Integer	Compute	ID of two end nodes of each edge
Neig_edge (2,numedge)	Integer	Compute	ID of neighboring cells of each edge
NameEdge_cell (nce,numcell)	Integer	Compute	ID of the three edges of each cell NameEdge_cell(1,i): ID of the first edge (n1-n2) of cell_i NameEdge_cell(2,i): ID of the second edge (n2-n3) of cell_i NameEdge_cell(3,i): ID of the first edge (n3-n1) of cell_i
NoSmoo_node (numnode)	Integer	Compute	Number of the neighboring edge-based smoothing domains for each node
NaSmoo_node (nre,numnode)	Integer	Compute	ID of the neighboring edge-based smoothing domains for each node
nvlist (ndexnow)	Integer	Compute	ID of the support nodes used to construct the strain-displacement matrix for an integration domain
nvglobal (nx*ndexnow)	Integer	Compute	Index array for assembling the global stiffness matrix
Bmat (3,nx*ndexnow)	Long real	Compute	Strain-displacement matrix for each integration domain
Gpk(nx*ndexnow* nx*ndexnow)	Long real	Compute	Local stiffness matrix of an integration domain
Gk(nx*numnode,n x*numnode)	Long real	Compute	Global stiffness matrix
Force (nx*numnode)	Long real	Compute	Global force vector which is used as the displacement vector after solving the linear algebraic equations
stress_node (3,numnode)	Long real	Compute	The array of nodal stress components
strain_node (3,numnode)	Long real	Compute	The array of nodal strain components
stress_cell (3,numcell)	Long real	Compute	The array of stress components for each cell
strain_cell (3,numcell)	Long real	Compute	The array of strain components for each cell
stress_edge (3,numedge)	Long real	Compute	The array of stress components for each edge-based smoothing domain
strain_edge(3,num edge)	Long real	Compute	The array of strain components for each edge-based smoothing domain

Appendix 2: A demonstration input file

The following gives the input file "DATAINPUT", which contains the prescribed values of various parameters and variables and the background mesh information for the demonstration example. By assigning different values to the switches of "NTscheme" and "NPIMSF" in the input file, various S-PIM models can be established, as described in TABLEs 6.31, 7.13 and 8.22. For the current contents of the DATAINPUT file, the NS-PIM-Tr3 model will be established as NTscheme=1 and NPIMSF=1. Note that to decrease the length of the file, the list of nodes coordinates and triangular cell has been folded and presented in two columns.

Input file of "DATAINPUT"

```
* e and v: Young's modulus and Possion's ratio
3.0d07,0.3d0
* NTscheme: T-schemes uesed for support nodes selection
1
* NPIMSF: Types of PIM shape functions
1
* npoly: number of monomials used in the augmented RBF
3
* Rc and q: shape parameters for MQ_RBF
4.0d0,1.03d0
* xlength,ylength and p: parameters for the problem
50.d0,10.d0,-1000.d0
* Number of field nodes
120
* Number of background cells
186
* Coordinates of field nodes
```

1	0.000000000000000	1.00000000000000	61	24.9999641896000	2.61550401549000
2	0.000000000000000	-1.00000000000000	62	26.1904665447000	0.200777285891000
3	0.000000000000000	3.00000000000000	63	26.1904761905000	5.00000000000000
4	0.000000000000000	-3.00000000000000	64	26.1904761905000	-5.00000000000000
5	1.90907754846000	4.88358110929E-2	65	27.3809692827000	-2.32580017829000
6	1.73368595938000	2.00000000000000	66	27.3809511946000	2.61543644198000
7	1.73360070432000	-2.00000000000000	67	28.5714550010000	0.200726870966000
8	1.53721338235000	3.58766149513000	68	28.5714285714000	5.00000000000000
9	1.53719633134000	-3.58790661820000	69	28.5714285714000	-5.00000000000000
10	0.000000000000000	5.00000000000000	70	29.7620022434000	-2.32594527803000
11	0.000000000000000	-5.00000000000000	71	29.7619495825000	2.61535486124000
12	3.84113279984000	-1.07934953171000	72	30.9525486321000	0.200472646435000
13	3.92190592934000	1.26092676417000	73	30.9523809524000	5.00000000000000
14	3.32055394285000	-3.23425563713000	74	30.9523809524000	-5.00000000000000
15	3.35621225219000	3.30508253287000	75	32.1433894087000	-2.32673524911000
16	2.38095238095000	5.00000000000000	76	32.1431675222000	2.61460255459000
17	2.38095238095000	-5.00000000000000	77	33.3344123200000	0.197945518319000
18	5.62566158641000	-2.75226267255000	78	33.3333333333000	5.00000000000000
19	5.67733750211000	2.93139946895000	79	33.3333333333000	-5.00000000000000

                    Smoothed Point Interpolation Methods

20	4.76190476190000	5.00000000000000	80	34.5266815785000	-2.33097877751000
21	4.76190476190000	-5.00000000000000	81	34.5270429830000	2.60914039006000
22	6.41352239706000	0.145232905516000	82	35.7237991328000	0.183092177686000
23	7.14285714286000	5.00000000000000	83	35.7142857143000	5.00000000000000
24	7.14285714286000	-5.00000000000000	84	35.7142857143000	-5.00000000000000
25	8.17565959235000	-2.40448316569000	85	36.9193808600000	-2.34778000613000
26	8.12798029186000	2.66554111031000	86	36.9204661317000	2.61039937764000
27	9.31123186882000	0.162513514641000	87	38.1482717495000	0.139529895455000
28	9.52380952381000	5.00000000000000	88	38.0952380952000	5.00000000000000
29	9.52380952381000	-5.00000000000000	89	38.0952380952000	-5.00000000000000
30	10.6462592304000	-2.37292687925000	90	39.3411536244000	-2.39033184471000
31	10.6466333397000	2.66358924264000	91	39.3377510153000	2.62384175798000
32	11.8387722091000	0.179909453733000	92	40.6591461363000	0.110878725455000
33	11.9047619048000	5.00000000000000	93	40.4761904762000	5.00000000000000
34	11.9047619048000	-5.00000000000000	94	40.4761904762000	-5.00000000000000
35	13.0791423598000	-2.34059214214000	95	41.7700580620000	2.67956860775000
36	13.0739333728000	2.66816919954000	96	41.7914086903000	-2.46854456117000
37	14.2665403109000	0.173320606559000	97	42.8571428571000	5.00000000000000
38	14.2857142857000	5.00000000000000	98	42.8571428571000	-5.00000000000000
39	14.2857142857000	-5.00000000000000	99	43.5662336762000	8.120849742040E-2
40	15.4740285610000	-2.32691452760000	100	44.3335987929000	-2.82812462707000
41	15.4701193778000	2.64303057636000	101	44.3691625201000	2.86626001107000
42	16.6632459848000	0.216321079460000	102	45.2380952381000	5.00000000000000
43	16.6666666667000	5.00000000000000	103	45.2380952381000	-5.00000000000000
44	16.6666666667000	-5.00000000000000	104	46.0785630297000	-1.09919330188000
45	17.8566765479000	-2.32601279511000	105	46.1126693112000	1.16556261076000
46	17.8553426903000	2.62300078828000	106	46.6610967917000	-3.25870407587000
47	19.0466454881000	0.203178478410000	107	46.6707759760000	3.27122035051000
48	19.0476190476000	5.00000000000000	108	47.6190476191000	5.00000000000000
49	19.0476190476000	-5.00000000000000	109	47.6190476191000	-5.00000000000000
50	20.2379961422000	-2.32581981974000	110	48.1206576129000	1.106155147940E-2
51	20.2375778213000	2.61715183824000	111	48.2663140406000	-2.00000000000000
52	21.4283171970000	0.201135083758000	112	48.2663992957000	2.00000000000000
53	21.4285714286000	5.00000000000000	113	48.4628036687000	3.58790661820000
54	21.4285714286000	-5.00000000000000	114	48.4627866177000	-3.58765413947000
55	22.6190274476000	-2.32577937840000	115	50.0000000000000	1.00000000000000
56	22.6189059129000	2.61579004851000	116	50.0000000000000	-1.00000000000000
57	23.8094618735000	0.200828159840000	117	50.0000000000000	3.00000000000000
58	23.8095238095000	5.00000000000000	118	50.0000000000000	-3.00000000000000
59	23.8095238095000	-5.00000000000000	119	50.0000000000000	5.00000000000000
60	24.9999989429000	-2.32577545200000	120	50.0000000000000	-5.00000000000000

* Triangular background cell information

1	22	12	18	94	46	41	42
2	22	13	12	95	41	36	37
3	114	106	109	96	36	31	32
4	114	111	106	97	31	26	27
5	107	113	108	98	26	19	22
6	107	112	113	99	19	15	13
7	8	15	16	100	15	6	13
8	8	6	15	101	6	5	13
9	9	14	7	102	5	7	12
10	9	17	14	103	21	18	14
11	6	8	3	104	24	25	18
12	8	10	3	105	29	30	25
13	16	10	8	106	34	35	30
14	117	113	112	107	39	40	35
15	113	119	108	108	44	45	40
16	117	119	113	109	49	50	45

17	111	114	118	110	54	55	50
18	114	120	118	111	59	60	55
19	109	120	114	112	64	65	60
20	4	9	7	113	69	70	65
21	9	11	17	114	74	75	70
22	4	11	9	115	79	80	75
23	105	99	104	116	84	85	80
24	25	27	22	117	89	90	85
25	30	32	27	118	94	96	90
26	35	37	32	119	98	100	96
27	40	42	37	120	103	106	100
28	45	47	42	121	116	110	111
29	50	52	47	122	115	112	110
30	55	57	52	123	102	101	107
31	60	62	57	124	97	95	101
32	65	67	62	125	93	91	95
33	70	72	67	126	88	86	91
34	75	77	72	127	83	81	86
35	80	82	77	128	78	76	81
36	85	87	82	129	73	71	76
37	90	92	87	130	68	66	71
38	96	99	92	131	63	61	66
39	100	104	99	132	58	56	61
40	110	105	104	133	53	51	56
41	101	99	105	134	48	46	51
42	95	92	99	135	43	41	46
43	91	87	92	136	38	36	41
44	86	82	87	137	33	31	36
45	81	77	82	138	28	26	31
46	76	72	77	139	23	19	26
47	71	67	72	140	20	15	19
48	66	62	67	141	1	5	6
49	61	57	62	142	2	7	5
50	56	52	57	143	17	21	14
51	51	47	52	144	21	24	18
52	46	42	47	145	24	29	25
53	41	37	42	146	29	34	30
54	36	32	37	147	34	39	35
55	31	27	32	148	39	44	40
56	26	22	27	149	44	49	45
57	19	13	22	150	49	54	50
58	5	12	13	151	54	59	55
59	7	14	12	152	59	64	60
60	14	18	12	153	64	69	65
61	18	25	22	154	69	74	70
62	25	30	27	155	74	79	75
63	30	35	32	156	79	84	80
64	35	40	37	157	84	89	85
65	40	45	42	158	89	94	90
66	45	50	47	159	94	98	96
67	50	55	52	160	98	103	100
68	55	60	57	161	103	109	106
69	60	65	62	162	118	116	111
70	65	70	67	163	116	115	110
71	70	75	72	164	115	117	112
72	75	80	77	165	108	102	107
73	80	85	82	166	102	97	101
74	85	90	87	167	97	93	95

75	90	96	92	168	93	88	91
76	96	100	99	169	88	83	86
77	100	106	104	170	83	78	81
78	106	111	104	171	78	73	76
79	111	110	104	172	73	68	71
80	110	112	105	173	68	63	66
81	112	107	105	174	63	58	61
82	107	101	105	175	58	53	56
83	101	95	99	176	53	48	51
84	95	91	92	177	48	43	46
85	91	86	87	178	43	38	41
86	86	81	82	179	38	33	36
87	81	76	77	180	33	28	31
88	76	71	72	181	28	23	26
89	71	66	67	182	23	20	19
90	66	61	62	183	20	16	15
91	61	56	57	184	3	1	6
92	56	51	52	185	1	2	5
93	51	46	47	186	2	4	7

Appendix 3: Source codes of two modules

Module 1: MODULE Parameters

```
MODULE Parameters
!==================================================================
!    The Module for storing the parameters and variables used in the program
!==================================================================
implicit real*8 (a-h,o-z)
Common/material/e,v,Pi,ep
Common/problem/xlength,ylength,pLoad,Dc
Common/PIMSF/Rc,q,Npoly,NTscheme,NPIMSF
Common/mesh/numnode,numcell,numedge,numEBC,numNBC,Ndexnow
Parameter (nx=2,nce=3)
Parameter (ngauss=2,mgauss=2)
Parameter (nre=26,ndom=50,nndom=10*ndom)
Parameter (numn=1000,nume=2*numn,numED=3*numn,numbc=0.2*numn)
END MODULE Parameters
```

Module 2: MODULE Variables

```
MODULE Variables
!==================================================================
!    The Module for the definition of the array variables used in the program
!==================================================================
Use Parameters
Implicit real*8 (a-h,o-z)
```

```
Dimension x(nx,numn),nxcell(nce,nume),Cmat(3,3)
Dimension npEBC(3,numBC),pEBC(2,numBC),npNBC(numBC)
Dimension numneig(numn),neighbour(nre,numn),nodein(numn)
Dimension kcell(nume),kcell2(nce,nume),center(nx,nume)
Dimension area_cell(nume),area_node(numn),area_edge(numED)
Dimension nodesix(6,nume),nodefour(4,numED)
Dimension Node_edge(2,numED),Neig_edge(2,numED),NameEdge_cell(nce,nume)
Dimension NoSmoo_node(numn),NaSmoo_node(nre,numn)
Dimension Bmat(3,nx*nndom),nvlist(nndom),nvglobal(nx*nndom)
dimension Gpk(nx*nndom*nx*nndom),GK(nx*numn,nx*numn),force(nx*numn)
Dimension stress_node(3,numn),strain_node(3,numn)
Dimension stress_cell(3,nume),strain_cell(3,nume)
Dimension stress_edge(3,numED),strain_edge(3,numED)
END MODULE Variables
```

Appendix 4: Source codes of the common subroutines

Program 1: source code of "Input"

```
Subroutine input
!=============================================================
! Input data from an external file and find out the nodes with boundary conditions
!
! Input:   the file of DATAINPUT including parameter values and mesh information;
! Output: Pi, ep, e, v, NTscheme, NPIMSF, npoly, Rc, q, xlength, ylength, pLoad,
!          numnode, numcell, x, nxcell, numEBC, npEBC, pEBC, numNBC and npNBC.
!=============================================================
use Parameters
use Variables
implicit real*8 (a-h,o-z)
character description*70

Pi=3.141592653589793d0; ep=1.0D-15
! Input parameters from file 'DATAINPUT'
read(5,11) description; read(5,*)  e,v
read(5,11) description; read(5,*)  NTscheme
read(5,11) description; read(5,*)  NPIMSF
read(5,11) description; read(5,*)  npoly
read(5,11) description; read(5,*)  Rc,q
read(5,11) description; read(5,*)  xlength,ylength,pLoad
read(5,11) description; read(5,*)  numnode
read(5,11) description; read(5,*)  numcell
! Input mesh information from file ' DATAINPUT '
read(5,11) description
x=0.d0
do i=1,numnode
```

```fortran
   read(5,*) ia,(x(j,ia),j=1,nx)
enddo
read(5,11) description
do i=1,numcell
   read(5,*) ia,(nxcell(j,ia),j=1,nce)
enddo
! Find out the nodes with essential boundary conditons
numEBC=0; npEBC=0; pEBC=0.d0
do i=1,numnode
  xi=abs(x(1,i))
  if (xi<1.0e-5) then
  numEBC=numEBC+1; npEBC(1,numEBC)=i
  endif
enddo
ai=ylength**3/12.
do i=1,numEBC
  j=npEBC(1,i)
  npEBC(2,i)=1; npEBC(3,i)=1
  pEBC(1,i)=-pLoad*x(2,j)/6./e/ai*((6*xlength-x(1,j)*3)*x(1,j)+&
          &(2+v)*(x(2,j)**2-ylength**2/4.))
  pEBC(2,i)=pLoad/6./e/ai*(3*v*x(2,j)**2*(xlength-x(1,j))+&
          &(4+5*v)*ylength**2*x(1,j)/4.+(3*xlength-x(1,j))*x(1,j)**2)
enddo
! Find out the nodes with natural boundary conditions
numNBC=0; npNBC=0
do i=1,numnode
  xi=abs(x(1,i)-xlength)
  if (xi<1.0e-5) then
  numNBC=numNBC+1; npNBC(numNBC)=i
  endif
enddo
! Compute the characteristic length of triangular cells
Dc=(xlength*ylength/numcell*4./(3.**0.5))**0.5
11   format(a50)

Close (5)
Return
End  subroutine input
```

Program 2: source code of "C_materialM"

```fortran
Subroutine C_materialM
!=============================================================
! Compute the material matrix for the plane stress condition
!
! Input:   e and v;
! Output: Cmat
```

```
!====================================================================
use Parameters
use Variables
implicit real*8 (a-h,o-z)

Cmat=0.d0; work=e/(1.-v**2)
Cmat(1,1)=1.*work; Cmat(1,2)=v*work; Cmat(2,1)=v*work; Cmat(2,2)=1.*work
Cmat(3,1)=Cmat(1,3); Cmat(3,2)=Cmat(2,3); Cmat(3,3)=(1.-v)/2.*work

Return
End subroutine C_materialM
```

Program 3: source code of "Cell_information"

```
Subroutine Cell_information
!====================================================================
! Obtainning necessary informations based on the triangular background cells
!
! Input:    x, nxcell, numnode, numcell;
! Output: numneig, neighbour, nodein, kcell, kcell2, center, area_cell,
!         area_node, area_edge, nodesix, nodefour, Node_edge, Neig_edge,
!         NameEdge_cell, NoSmoo_node, NaSmoo_node and numedge.
!====================================================================
use Parameters
use Variables
implicit real*8 (a-h,o-z)
! work array
dimension neighbor(nce,numcell),index_neighbor(nce,numcell)
dimension cellang(nce,numcell),mbb(10*nce),rri(numnode),nrr(numnode)

! Compute the data of numneig, neighbour, area_cell and center
numneig=0; neighbour=0; area_cell=0.d0; center=0.d0; cellang=0.d0
Loop_on_cells01: do i=1,numcell
  n1=nxcell(1,i); n2=nxcell(2,i); n3=nxcell(3,i)
  x1=x(1,n1) ; x2=x(1,n2) ; x3=x(1,n3); y1=x(2,n1) ; y2=x(2,n2) ; y3=x(2,n3)
  numneig(n1)=numneig(n1)+1;                     numneig(n2)=numneig(n2)+1;
numneig(n3)=numneig(n3)+1
  k1=numneig(n1); k2=numneig(n2); k3=numneig(n3)
  neighbour(k1,n1)=i; neighbour(k2,n2)=i; neighbour(k3,n3)=i
  area_cell(i)=(x1*y2+x2*y3+x3*y1)*0.5-(x2*y1+x3*y2+x1*y3)*0.5
  center(1,i)=(x1+x2+x3)/3.d0; center(2,i)=(y1+y2+y3)/3.d0
  aa=sqrt((x1-x2)**2+(y1-y2)**2)
  bb=sqrt((x2-x3)**2+(y2-y3)**2)
  cc=sqrt((x1-x3)**2+(y1-y3)**2)
  rf12=(x2-x1)/aa; bt12=(y2-y1)/aa; rf13=(x3-x1)/cc;  bt13=(y3-y1)/cc
  rf21=(x1-x2)/aa;  bt21=(y1-y2)/aa; rf23=(x3-x2)/bb; bt23=(y3-y2)/bb
  rf31=(x1-x3)/cc;  bt31=(y1-y3)/cc; rf32=(x2-x3)/bb; bt32=(y2-y3)/bb
```

```fortran
cosn1=rf12*rf13+bt12*bt13; cosn2=rf21*rf23+bt21*bt23; cosn3=rf31*rf32+bt31*bt32
cellang(1,i)=acos(cosn1)*180.d0/pi
cellang(2,i)=acos(cosn2)*180.d0/pi
cellang(3,i)=acos(cosn3)*180.d0/pi
end do    Loop_on_cells01
! Compute the data of area_node and nodein
area_node=0.d0; nodein=0
Loop_on_nodes01: do i=1,numnode
  kk=numneig(i); area=0.; sumang=0.
  Inner: do j=1,kk
      j1=neighbour(j,i); area=area+area_cell(j1)/3.
      n1=nxcell(1,j1); n2=nxcell(2,j1); n3=nxcell(3,j1)
      if (i.eq.n1) then; sumang=sumang+cellang(1,j1)
      else if (i.eq.n2) then; sumang=sumang+cellang(2,j1)
      else; sumang=sumang+cellang(3,j1)
      endif
  enddo Inner
  differ=abs(sumang-360.d0)
  if (differ>1.0E-10) then; nodein(i)=1
  else; nodein(i)=0
  endif
  area_node(i)=area
end do Loop_on_nodes01
! Compute the data of kcell, kcell2 and nodesix
kcell=0; kcell2=0; nodesix=0; neighbor=0; index_neighbor=0
Loop_on_cells02: do i=1,numcell
  n1=nxcell(1,i); n2=nxcell(2,i); n3=nxcell(3,i)
  x1=x(1,n1); x2=x(1,n2); x3=x(1,n3); y1=x(2,n1); y2=x(2,n2); y3=x(2,n3)
  mcc=0; mbb=0
  Inner1: do ik=1,nce
    ian=nxcell(ik,i); numian=numneig(ian)
    Inner1_1: do ib=1,numian
        ibc=neighbour(ib,ian)
        if (ibc/=i) then
           if (ik.eq.1) then; mcc=mcc+1; mbb(mcc)=ibc
           else ; inde=0
                Inner1_1_1: do ic=1,mcc
                ic1=mbb(ic)
                if (ibc.eq.ic1) then; inde=1
                endif
                enddo Inner1_1_1
                if (inde.eq.0) then
                mcc=mcc+1; mbb(mcc)=ibc
                endif
           endif
        endif
    enddo Inner1_1
  enddo Inner1
```

```fortran
      kp=0
      Inner2: do ia=1,mcc
        ibc=mbb(ia); kk=0; nnn=0
        Inner2_1: do id=1,nce
            ka=nxcell(id,ibc)
            if (ka.eq.n1) then; kk=kk+1; nnn=nnn+1
            else if (ka.eq.n2) then; kk=kk+1; nnn=nnn+2
            else if (ka.eq.n3) then; kk=kk+1; nnn=nnn+3
            endif
        enddo Inner2_1
        if (kk.eq.2) then
            kp=kp+1; neighbor(kp,i)=ibc
            select case(nnn)
            case(3); index_neighbor(kp,i)=1; kcell2(1,i)=1
            case(5); index_neighbor(kp,i)=2; kcell2(2,i)=1
            case(4); index_neighbor(kp,i)=3; kcell2(3,i)=1
            endselect
        endif
        if (kp>=nce) then; goto 01
        endif
      enddo Inner2
01 select case(kp)
   case(1); kcell(i)=2
   case(2); kcell(i)=1
   case(3); kcell(i)=0
   endselect
   do i1=1,nce
     nodesix(i1,i)=nxcell(i1,i)
   enddo
   num=nce
   select case(kcell(i))
   case(0)
   Inner3_1: do i2=1,nce
       ic=neighbor(i2,i)
       Inner3_1_1: do i3=1,nce
           id=nxcell(i3,ic)
           if (id/=n1 .and. id/=n2 .and. id/=n3) then
           num=num+1; nodesix(num,i)=id
           endif
       enddo Inner3_1_1
   enddo Inner3_1
   case(1)
   Inner3_2: do i2=1,2
     ic=neighbor(i2,i)
     Inner3_2_1: do i3=1,nce
         id=nxcell(i3,ic)
         if (id/=n1 .and. id/=n2 .and. id/=n3) then
         num=num+1; nodesix(num,i)=id
```

```fortran
       endif
     enddo Inner3_2_1
   enddo Inner3_2
   ip1=nodesix(1,i); ip2=nodesix(2,i); ip3=nodesix(3,i)
   ip4=nodesix(4,i); ip5=nodesix(5,i)
   rri=0.d0; nrr=0; xxi=center(1,i); yyi=center(2,i)
   do k=1,numnode
       rri(k)=((xxi-x(1,k))**2+(yyi-x(2,k))**2)**0.5
   enddo
   call indexx(numnode,rri,nrr)
   Inner3_3: do kk=1,6
     nkk=nrr(kk)
     nkk2=(nkk-nodesix(1,i))*(nkk-nodesix(2,i))*(nkk-nodesix(3,i))&
         &*(nkk-nodesix(4,i))*(nkk-nodesix(5,i))*(nkk-nodesix(6,i))
     if (nkk2/=0) then
     num=num+1; nodesix(num,i)=nkk
     endif
   enddo Inner3_3
   endselect
end do Loop_on_cells02
Loop_on_cells03: do i=1,numcell
  iee=kcell(i)
  if (iee.eq.2) then
  Inner4: do j=1,nce
    jee=neighbor(j,i); ijee=kcell(jee)
    if (jee>0 .and. jee<=numcell .and. ijee/=2) then
    do k=1,6
    nodesix(k,i)=nodesix(k,jee)
    enddo
    endif
  enddo Inner4
  endif
enddo Loop_on_cells03
! Compute the data of Node_edge, Neig_edge and numedge
Node_edge=0; Neig_edge=0; numedge=0
Loop_on_cells04: do i=1,numcell
  n1=nxcell(1,i); n2=nxcell(2,i); n3=nxcell(3,i)
  select case(kcell(i))
  case(0); num_neighbor=3
  case(1); num_neighbor=2
  case(2); num_neighbor=1
  endselect
  if (i.eq.1) then
    Node_edge(1,1)=n1; Node_edge(2,1)=n2; Node_edge(1,2)=n2
    Node_edge(2,2)=n3; Node_edge(1,3)=n3; Node_edge(2,3)=n1
    Inner5_1: do ij=1,nce
        ij_cell=neighbor(ij,i); ij_index=index_neighbor(ij,i)
        if (ij_cell>0) then
```

```fortran
        numEdge=numEdge+1
        select case(ij_index)
            case (1)
            Node_edge(1,numEdge)=n1; Node_edge(2,numEdge)=n2
            Neig_edge(1,numEdge)=i; Neig_edge(2,numEdge)=ij_cell
            case (2)
            Node_edge(1,numEdge)=n2; Node_edge(2,numEdge)=n3
            Neig_edge(1,numEdge)=i; Neig_edge(2,numEdge)=ij_cell
            case (3)
            Node_edge(1,numEdge)=n3; Node_edge(2,numEdge)=n1
            Neig_edge(1,numEdge)=i; Neig_edge(2,numEdge)=ij_cell
        endselect
        endif
    enddo Inner5_1
    Inner5_2: do ik=1,nce
        index_edge=kcell2(ik,i)
        if (index_edge.eq.0) then
        numEdge=numEdge+1
        select case(ik)
            case(1)
            Node_edge(1,numEdge)=n1; Node_edge(2,numEdge)=n2
            Neig_edge(1,numEdge)=i; Neig_edge(2,numEdge)=0
            case(2)
            Node_edge(1,numEdge)=n2; Node_edge(2,numEdge)=n3
            Neig_edge(1,numEdge)=i; Neig_edge(2,numEdge)=0
            case(3)
            Node_edge(1,numEdge)=n3; Node_edge(2,numEdge)=n1
            Neig_edge(1,numEdge)=i; Neig_edge(2,numEdge)=0
        endselect
        endif
    enddo Inner5_2
else
    Inner5_3: do j=1,num_neighbor
        ja=neighbor(j,i); jb=index_neighbor(j,i)
        if (ja>i) then
        numEdge=numEdge+1
        select case(jb)
            case (1)
            Node_edge(1,numEdge)=n1; Node_edge(2,numEdge)=n2
            Neig_edge(1,numEdge)=i; Neig_edge(2,numEdge)=ja
            case (2)
            Node_edge(1,numEdge)=n2; Node_edge(2,numEdge)=n3
            Neig_edge(1,numEdge)=i; Neig_edge(2,numEdge)=ja
            case (3)
            Node_edge(1,numEdge)=n3; Node_edge(2,numEdge)=n1
            Neig_edge(1,numEdge)=i; Npeig_edge(2,numEdge)=ja
        endselect
        endif
```

```fortran
          enddo Inner5_3
     Inner5_4: do jk=1,nce
          index_edge=kcell2(jk,i)
          if (index_edge.eq.0) then
          numEdge=numEdge+1
          select case(jk)
              case(1)
              Node_edge(1,numEdge)=n1; Node_edge(2,numEdge)=n2
              Neig_edge(1,numEdge)=i;  Neig_edge(2,numEdge)=0
              case(2)
              Node_edge(1,numEdge)=n2; Node_edge(2,numEdge)=n3
              Neig_edge(1,numEdge)=i;  Neig_edge(2,numEdge)=0
              case(3)
              Node_edge(1,numEdge)=n3; Node_edge(2,numEdge)=n1
              Neig_edge(1,numEdge)=i;  Neig_edge(2,numEdge)=0
          endselect
          endif
     enddo Inner5_4
  endif
end do Loop_on_cells04
! Compute the data of area_edge, NoSmoo_node, NaSmoo_node, nodefour and
!   NameEdge_cell
area_edge=0.d0; NoSmoo_node=0; NaSmoo_node=0; nodefour=0; NameEdge_cell=0
Loop_on_edges: do i=1,numEdge
  ie1=Neig_edge(1,i); ie2=Neig_edge(2,i)
  in1=node_edge(1,i); in2=node_edge(2,i); in12=in1+in2
  NoSmoo_node(in1)=NoSmoo_node(in1)+1;
  NoSmoo_node(in2)=NoSmoo_node(in2)+1
  k1=NoSmoo_node(in1); k2=NoSmoo_node(in2)
  NaSmoo_node(k1,in1)=i; NaSmoo_node(k2,in2)=i
  n1e1=nxcell(1,ie1); n2e1=nxcell(2,ie1); n3e1=nxcell(3,ie1)
  n12e1=n1e1+n2e1;   n23e1=n2e1+n3e1;   n13e1=n1e1+n3e1
  if (in12.eq.n12e1) then; NameEdge_cell(1,ie1)=i
  elseif (in12.eq.n23e1) then; NameEdge_cell(2,ie1)=i
  elseif (in12.eq.n13e1) then; NameEdge_cell(3,ie1)=i
  endif
  if (ie2>0) then
    area_edge(i)=area_edge(i)+area_cell(ie1)/3.+area_cell(ie2)/3.
    n1e2=nxcell(1,ie2); n2e2=nxcell(2,ie2); n3e2=nxcell(3,ie2)
    n12e2=n1e2+n2e2;   n23e2=n2e2+n3e2;   n13e2=n1e2+n3e2
    if (in12.eq.n12e2) then
        NameEdge_cell(1,ie2)=i
        nodefour(1,i)=n3e2; nodefour(2,i)=n1e2; nodefour(4,i)=n2e2
        elseif (in12.eq.n23e2) then
        NameEdge_cell(2,ie2)=i
        nodefour(1,i)=n1e2; nodefour(2,i)=n2e2; nodefour(4,i)=n3e2
        elseif (in12.eq.n13e2) then
        NameEdge_cell(3,ie2)=i
```

```
      nodefour(1,i)=n2e2; nodefour(2,i)=n3e2; nodefour(4,i)=n1e2
    endif
    nodefour(3,i)=nxcell(1,ie1)+nxcell(2,ie1)+nxcell(3,ie1)-in12
  else
    area_edge(i)=area_edge(i)+area_cell(ie1)/3.
    nodefour(1,i)=nxcell(1,ie1); nodefour(2,i)=nxcell(2,ie1); nodefour(3,i)=nxcell(3,ie1)
  endif
enddo Loop_on_edges


Return
End subroutine Cell_information
```

Program 4: source code of "StiffM_Intedomain"

```fortran
Subroutine StiffM_Intedomain(ia,area_inte,stiffM_inteD)
!=====================================================================
! Compute the stiffness matrix for the ia_th integration domain
!
! Input:   Bmat, area_inte(the area of the ia_th integration domain);
! Output: stiffM_inteD(local stiffness matrix for the ia_th integration domain)
!=====================================================================
use Parameters
use Variables
implicit real*8 (a-h,o-z)
dimension stiffM_inteD(nx*ndexnow,nx*ndexnow)
! work array
dimension Bnoo(3,nx*ndexnow)

stiffM_inteD=0.d0
do i=1,3
do j=1,nx*ndexnow
  Bnoo(i,j)=Bmat(i,j)
enddo
enddo
do 10 i=1,nx*ndexnow
do 10 j=1,nx*ndexnow
  do 20 m=1,3
  do 20 n=1,3
  stiffM_inteD(i,j)=stiffM_inteD(i,j)+Bnoo(m,i)*Cmat(m,n)*Bnoo(n,j)*area_inte
  20 continue
10 continue
nvglobal=0
do i=1,ndexnow
  n1=nx*i-1; n2=nx*i-0
  nvglobal(n1)=nx*nvlist(i)-1; nvglobal(n2)=nx*nvlist(i)
end do
```

Return
End subroutine StiffM_Intedomain

Program 5: source code of "Form_GK"

Subroutine Form_GK
```
!===============================================================
! Form the global stiffness matrix by assembling the local stiffness matrixes
!
! Input:   nvglobal, gpk and gk;
! Output: gk
!===============================================================
use Parameters
use Variables
implicit real*8 (a-h,o-z)

do 10 ik=1,nx*ndexnow
do 10 jk=1,nx*ndexnow
   m=nvglobal(ik); n=nvglobal(jk)
   ngpk=(jk-1)*nx*ndexnow+ik
   gk(m,n)=gk(m,n)+gpk(ngpk)
10 continue

Return
End subroutine Form_GK
```

Program 6: source code of "Natural_BC"

Subroutine Natural_BC
```
!===============================================================
! Enforce the nature boundary conditions using the Gauss integration scheme
!
! Input:   x, npNBC, numNBC,
! Output: force
!===============================================================
use Parameters
use Variables
implicit real*8 (a-h,o-z)
! work array
dimension gpos(nx),xg(nx,2),xxg(nx,mgauss),wei(mgauss)
dimension cy(numBC),nnbc1(numBC),nnbc2(numBC),GPcoe_line(2,Mgauss)

cy=0.d0; nnbc1=0; nnbc2=0
do i=1,numNBC
   cy(i)=x(2,npNBC(i))
enddo
```

```
call indexx(numNBC,cy,nnbc1)
do i=1,numNBC
   j=nnbc1(i); k=npNBC(j); nnbc2(i)=k
enddo
npNBC=0; npNBC=nnbc2
call Gausspointcoe_line(Mgauss,GPcoe_line)
ai=(1./12.)*ylength**3
Loop1: do i=1,numNBC-1
  in=npNBC(i); jn=npNBC(i+1); xg=0.d0
  xg(1,1)=x(1,in); xg(2,1)=x(2,in)
  xg(1,2)=x(1,jn); xg(2,2)=x(2,jn)
  call Line_gauss(xg,xxg,wei,det,Mgauss)
  Loop2: do ick=1,mgauss
    gpos(1)=xxg(1,ick); gpos(2)=xxg(2,ick); weight=wei(ick); ajac=det
    Ty=(pLoad/(2.*ai))*(ylength**2/4.-gpos(2)**2)
    dis1=((gpos(1)-xg(1,1))**2+(gpos(2)-xg(2,1))**2)**0.5
    dis2=((gpos(1)-xg(1,2))**2+(gpos(2)-xg(2,2))**2)**0.5
    f1=dis2/(dis1+dis2); f2=dis1/(dis1+dis2)
    Force(nx*in)=Force(2*in)+weight*ajac*f1*Ty
    Force(2*jn)=Force(2*jn)+weight*ajac*f2*Ty
  end do Loop2
end do Loop1


Return
End subroutine Natural_BC
```

Program 7: source code of "Essential_BC"

```
Subroutine Essential_BC
!================================================================
! Apply the essential boundary conditions
!
! Input:   numEBC, npEBC, pEBC, GK and force;
! Output: GK and froce.
!================================================================

use Parameters
use Variables
implicit real*8 (a-h,o-z)

Loop1: do i=1,numEBC
   ip=npEBC(1,i); ia=npEBC(2,i); ib=npEBC(3,i)
   ux=pEBC(1,i); uy=pEBC(2,i)
   if (ia.eq.1) then
     ja=nx*ip-1
     Loop1_1: do j=1,nx*numnode
        if (j.ne.ja) then
```

```
            force(j)=force(j)-Gk(j,ja)*ux; Gk(j,ja)=0.; Gk(ja,j)=0.
            else
            force(ja)=ux; Gk(ja,ja)=1.
            endif
        enddo Loop1_1
      endif
      if (ib.eq.1) then
        jb=nx*ip-0
        Loop1_2: do j=1,nx*numnode
            if (j.ne.jb) then
            force(j)=force(j)-Gk(j,jb)*uy; Gk(j,jb)=0.; Gk(jb,j)=0.
            else
            force(jb)=uy; Gk(jb,jb)=1.
            endif
        enddo Loop1_2
      endif
    enddo Loop1

    return
    End subroutine Essential_BC
```

Program 8: source code of "Solver_LAE"

```
Subroutine Solver_LAE(ak,fp,neq,nmat)
!==================================================================
! Solving linear algebraic equations; it calls subroutine 'Band_solver' for banded linear
!     equations and 'GaussEli_Solver' for general ones.
!
! Input:   ak: coefficient matrix of the equations;
!          neq: number of equations;
!          nmat: the amx number of rows of ak.
! Input and output:
!          fp: input the right hand side of the equations and output the solutions.
!==================================================================
use Parameters
implicit real*8 (a-h,o-z)
dimension ak(nmat,nmat),fp(nmat)
real(8), allocatable :: tp(:,:)
real(8), allocatable :: stfp(:,:)
allocate (tp(1:neq,1:nmat))
allocate (stfp(1:neq,1:neq))

stfp=0.d0; tp=0.d0; stfp=ak
ni=nEq; Lp=0
do 10 i=1,ni
  do j=ni,i,-1
      if(stfp(i,j).ne.0.) then
```

```fortran
         if(abs(j-i).gt.Lp) Lp=abs(j-i)
         go to 20
         endif
      enddo
   20 continue
      do j=1,i
         if(stfp(i,j).ne.0.) then
         if(abs(j-i).gt.Lp) Lp=abs(j-i)
         go to 10
         endif
      enddo
10 continue
ilp=2*lp+1; nm=nEq
if(ilp.lt.nEq) then
    call Band_solver(stfp,fp,tp,nm,lp,ilp,nmat)
else
    call GaussEli_Solver(nEq,nmat,ak,fp,kwji)
endif

deallocate (tp)
deallocate (stfp)
Return
End subroutine Solver_LAE
```

Program 9: source code of "Dispnorm_error"

```fortran
Subroutine Dispnorm_error
!=============================================================
! Compute the global displacement norm error
!
! Input:   force(storing the displacemet solutions)
! Output: Dispnorm(the global displacement norm error)
!=============================================================
use Parameters
use Variables
implicit real*8 (a-h,o-z)
! work array
dimension Exact_Dis(nx*numnode)

Exact_Dis=0.d0; ai=ylength**3/12.d0
do i=1,numnode
  j=2*i-1; k=2*i-0
  Exact_Dis(j)=-pLoad*x(2,i)/6./e/ai*((6*xlength-x(1,i)*3)*x(1,i)+&
          &(2+v)*(x(2,i)**2-ylength**2/4.))
  Exact_Dis(k)=pLoad/6./e/ai*(3*v*x(2,i)**2*(xlength-x(1,i))+&
          &(4+5*v)*ylength**2*x(1,i)/4.+(3*xlength-x(1,i))*x(1,i)**2)
enddo
```

```
sumexact_u=0.; sumexact_v=0.; sumerror_u=0.; sumerror_v=0.
do i=1,numnode
  j=nx*i-1; k=nx*i-0
  weight=area_node(i)
  sumexact_ux=sumexact_ux+Exact_Dis(j)**2
  sumexact_uy=sumexact_uy+Exact_Dis(k)**2
  sumerror_ux=sumerror_ux+(Exact_Dis(j)-force(j))**2
  sumerror_uy=sumerror_uy+(Exact_Dis(k)-force(k))**2
enddo
Dispnorm=((sumerror_ux+sumerror_uy)/(sumexact_ux+sumexact_uy))**0.5
write(10,10) 'Global error in displacement norm:',Dispnorm
10 format(a40,e18.8)

Return
End subroutine Dispnorm_error
```

Program 10: source code of "PPIM_SF2D"

```
Subroutine PPIM_SF2D(gpos,nv,ndex,phi)
!================================================================
! Creating Polynomial PIM shape functions
!
! Input:    nv(ID of support nodes);
!           ndex(number of support nodes);
!           gpos(coordinates of the point of interest).
! Output:  phi(values of PIM shape functions)
!================================================================
use Parameters
use Variables
implicit real*8 (a-h,o-z)
dimension gpos(nx),nv(ndex),phi(ndex)
! work array
dimension xinter(ndex),yinter(ndex),polym(ndex,ndex),qrk(ndex)

do i=1,ndex
  xinter(i)=x(1,nv(i))
  yinter(i)=x(2,nv(i))
enddo
polym=0.d0
select case(ndex)
case(3)
do i=1,ndex
  polym(i,1)=1.; polym(i,2)=xinter(i); polym(i,3)=yinter(i)
enddo
case(4)
do i=1,ndex
  polym(i,1)=1.; polym(i,2)=xinter(i); polym(i,3)=yinter(i)
```

```fortran
enddo
case(6)
do i=1,ndex
   polym(i,1)=1.; polym(i,2)=xinter(i); polym(i,3)=yinter(i)
   polym(i,4)=xinter(i)**2; polym(i,5)=xinter(i)*yinter(i); polym(i,6)=yinter(i)**2
enddo
endselect
qrk=0.d0
call Poly_Basis2D(gpos(1),gpos(2),qrk,ndex)
call brinv(polym,ndex,Lp)
qrk=matmul(qrk,polym)
do i=1,ndex
   phi(i)=qrk(i)
enddo


Return
End subroutine PPIM_SF2D
```

Program 11: source code of "PPIM_CT_SF2D"

```fortran
Subroutine PPIM_CT_SF2D(gpos,nv,ndex,edge_length,phi)
!===============================================================
! Creating bilinear Polynomial PIM shape functions with Coordinate Transformation
(CT)
!
! Input:    nv(ID of support nodes);
!           ndex(number of support nodes);
!           gpos(coordinates of the point of interest);
!           edge_length: the length of the edge with the point of interest.
! Output: phi(values of PIM shape functions)
!===============================================================
use Parameters
use Variables
implicit real*8 (a-h,o-z)
dimension gpos(nx),nv(ndex),phi(ndex)
! work array
dimension xinterA(ndex),yinterA(ndex),xinterB(ndex),yinterB(ndex)
dimension xinter(ndex),yinter(ndex),qrk(ndex),polym(ndex,ndex),polym2(ndex,ndex)
dimension polymA(ndex,ndex),polymB(ndex,ndex)
dimension polymAA(ndex,ndex),polymBB(ndex,ndex)

xori=gpos(1); yori=gpos(2)
do i=1,ndex
   xinter(i)=x(1,nv(i)); yinter(i)=x(2,nv(i))
enddo
polym=0.d0
select case(ndex)
```

```fortran
case(3)
do i=1,ndex
  polym(i,1)=1.; polym(i,2)=xinter(i); polym(i,3)=yinter(i)
enddo
case(4)
do i=1,ndex
  polym(i,1)=1.; polym(i,2)=xinter(i)
  polym(i,3)=yinter(i); polym(i,4)=xinter(i)*yinter(i)
enddo
case(6)
do i=1,ndex
  polym(i,1)=1.; polym(i,2)=xinter(i); polym(i,3)=yinter(i)
  polym(i,4)=xinter(i)**2; polym(i,5)=xinter(i)*yinter(i); polym(i,6)=yinter(i)**2
enddo
endselect
polym2=polym
Deter=16*(edge_length/2.)**4
call Determinant(polym2,4,deter4)
deter4=abs(deter4)

if (deter4 < Deter) then
  angle=0.5*acos(deter4/deter)
  onxx=sin(angle); onyy=cos(angle)
  xinterA=0.d0; yinterA=0.d0; polymA=0.d0
  do ja=1,ndex
      xinterA(ja)=+(xinter(ja)-xori)*onyy+(yinter(ja)-yori)*onxx
      yinterA(ja)=-(xinter(ja)-xori)*onxx+(yinter(ja)-yori)*onyy
  enddo
  do i=1,ndex
      polymA(i,1)=1.; polymA(i,2)=xinterA(i)
      polymA(i,3)=yinterA(i); polymA(i,4)=xinterA(i)*yinterA(i)
  enddo
  xinterB=0.d0; yinterB=0.d0; polymB=0.d0
  do ja=1,ndex
      xinterB(ja)=(xinter(ja)-xori)*onyy-(yinter(ja)-yori)*onxx
      yinterB(ja)=(xinter(ja)-xori)*onxx+(yinter(ja)-yori)*onyy
  enddo
  do i=1,ndex
      polymB(i,1)=1.; polymB(i,2)=xinterB(i)
      polymB(i,3)=yinterB(i); polymB(i,4)=xinterB(i)*yinterB(i)
  enddo
  polymAA=polymA; polymBB=polymB
  deterA=0.d0; deterB=0.d0
  call Determinant(polymAA,4,deterA)
  call Determinant(polymBB,4,deterB)
  deterA=abs(deterA); deterB=abs(deterB)
  polym=0.d0
  if (deterA>=deterB) then
```

```fortran
    polym=polymA
    gpos1=+(gpos(1)-xori)*onyy+(gpos(2)-yori)*onxx
    gpos2=-(gpos(1)-xori)*onxx+(gpos(2)-yori)*onyy
  else
    polym=polymB
    gpos1=(gpos(1)-xori)*onyy-(gpos(2)-yori)*onxx
    gpos2=(gpos(1)-xori)*onxx+(gpos(2)-yori)*onyy
  endif
  gpos(1)=gpos1; gpos(2)=gpos2; qrk=0.d0
  call Poly_Basis2D(gpos(1),gpos(2),qrk,ndex)
  call brinv(polym,ndex,Lp)
  qrk=matmul(qrk,polym)
  do i=1,ndex
     phi(i)=qrk(i)
  enddo
else
  qrk=0.d0
  call Poly_Basis2D(gpos(1),gpos(2),qrk,ndex)
  call brinv(polym,ndex,Lp)
  qrk=matmul(qrk,polym)
  do i=1,ndex
     phi(i)=qrk(i)
  enddo
endif

Return
End subroutine PPIM_CT_SF2D
```

Program 12: source code of "Iso_PPIM_SF2D"

```fortran
Subroutine Iso_PPIM_SF2D(gpos,nv,ndex,phi)
!================================================================
! Creating isoparametric Polynomial PIM shape functions
!
! Input:    nv(ID of support nodes);
!           ndex(number of support nodes);
!           gpos(coordinates of the point of interest);
! Output: phi(values of PIM shape functions)
!================================================================
use Parameters
use Variables
implicit real*8 (a-h,o-z)
dimension gpos(nx),nv(ndex),phi(ndex)

epss=1.0E-6
nn1=nv(1); nn2=nv(2); nn3=nv(3); nn4=nv(4)
x1=x(1,nn1); y1=x(2,nn1); x2=x(1,nn2); y2=x(2,nn2)
```

```fortran
x3=x(1,nn3); y3=x(2,nn3); x4=x(1,nn4); y4=x(2,nn4)
x21=x2-x1; y21=y2-y1; x32=x3-x2; y32=y3-y2; x34=x3-x4; y34=y3-y4
x41=x4-x1; y41=y4-y1; x14=x1-x4; y14=y1-y4
x2134=x21-x34; y2134=y21-y34; x3241=x32-x41; y3241=y32-y41
ax=x2134*y34-x34*y2134
bx=gpos(1)*y2134+x14*y34+x2134*y4-x34*y14-x4*y2134-gpos(2)*x2134
cx=gpos(1)*y14+x14*y4-x4*y14-x14*gpos(2)
ay=x3241*y41-x41*y3241
by=gpos(1)*y3241+x21*y41+x3241*y1-x41*y21-x1*y3241-gpos(2)*x3241
cy=gpos(1)*y21+x21*y1-x1*y21-x21*gpos(2)
xx=0.d0
if (abs(ax)<epss .and. abs(bx)<epss) then
  write(*,*) 'Wrong: a=b=0'; stop
else if (abs(ax)<epss) then
  xx=-cx/bx
else
  xx11=(-bx+(bx**2-4.d0*ax*cx)**0.5)/(2.d0*ax)
  xx12=(-bx-(bx**2-4.d0*ax*cx)**0.5)/(2.d0*ax)
  if (xx11>=0. .and. xx11<=1.) then; xx=xx11
  else if (xx12>=0. .and. xx12<=1.) then; xx=xx12
  endif
endif
if (abs(xx)<epss) then
  write(*,*) 'xx=0'; stop
endif

yy=0.d0
if (abs(ay)<epss .and. abs(by)<epss) then
  write(*,*) 'Wrong: a=b=0'; stop
else if (abs(ay)<epss) then
  yy=-cy/by
else
  yy11=(-by+(by**2-4.d0*ay*cy)**0.5)/(2.d0*ay)
  yy12=(-by-(by**2-4.d0*ay*cy)**0.5)/(2.d0*ay)
  if (yy11>=0. .and. yy11<=1.) then; yy=yy11
  else if (yy12>=0. .and. yy12<=1.) then; yy=yy12
  endif
endif
if (abs(yy)<epss) then
  write(*,*) 'yy=0'; stop
endif

phi(1)=(1.d0-xx)*(1.d0-yy); phi(2)=xx*(1.d0-yy)
phi(3)=xx*yy; phi(4)=(1.d0-xx)*yy

Return
End subroutine Iso_PPIM_SF2D
```

Program 13: source code of "RPIM_SF2D"

```fortran
Subroutine RPIM_SF2D(gpos,nv,ndex,phi)
!==============================================================
! Creating Radial PIM (RPIM) shape functions
!
! Input:    nv(ID of support nodes);
!           ndex(number of support nodes);
!           gpos(coordinates of the point of interest).
! Output: phi(values of PIM shape functions)
!==============================================================
use Parameters
use Variables
implicit real*8 (a-h,o-z)
integer kwji
dimension gpos(nx),nv(ndex),phi(ndex)
! work array
dimension xv(nx,ndex),rk(ndex+npoly),G0(ndex+npoly,ndex+npoly)

mg=ndex+npoly; phi=0.d0; G0=0.d0; xv=0.d0
do i=1,ndex
  xv(1,i)=x(1,nv(i)); xv(2,i)=x(2,nv(i))
enddo
Loop1: do i=1,ndex
  nn=nv(i)
  call Radial_Basis2D(x(1,nn),x(2,nn),xv,rk,ndex)
  do j=1,ndex
  G0(i,j)=rk(j)
  enddo
  if (npoly.gt.0) then
    select case(npoly)
    case(3)
      G0(i,ndex+1)=1.; G0(i,ndex+2)=x(1,nn); G0(i,ndex+3)=x(2,nn)
      G0(ndex+1,i)=1.; G0(ndex+2,i)=x(1,nn); G0(ndex+3,i)=x(2,nn)
    case(4)
      G0(i,ndex+1)=1.; G0(i,ndex+2)=x(1,nn); G0(i,ndex+3)=x(2,nn)
      G0(i,ndex+4)=x(1,nn)*x(2,nn)
      G0(ndex+1,i)=1.; G0(ndex+2,i)=x(1,nn); G0(ndex+3,i)=x(2,nn)
      G0(ndex+4,i)=x(1,nn)*x(2,nn)
    case(6)
      G0(i,ndex+1)=1.; G0(i,ndex+2)=x(1,nn); G0(i,ndex+3)=x(2,nn)
      G0(i,ndex+4)=x(1,nn)**2; G0(i,ndex+5)=x(1,nn)*x(2,nn);G0(i,ndex+6)=x(2,nn)**2
      G0(ndex+1,i)=1.; G0(ndex+2,i)=x(1,nn); G0(ndex+3,i)=x(2,nn)
      G0(ndex+4,i)=x(1,nn)**2;                          G0(ndex+5,i)=x(1,nn)*x(2,nn);
G0(ndex+6,i)=x(2,nn)**2
    endselect
  endif
end do Loop1
```

```
call Radial_Basis2D(gpos(1),gpos(2),xv,rk,ndex)
call GaussEli_Solver(mg,mg,G0,rk,kwji)
if (kwji.eq.1) then
  write(*,*) 'Fail...'; pause
endif
do i=1,ndex
   phi(i)=rk(i)
enddo


Return
End subroutine RPIM_SF2D
```

Program 14: source code of "Con_RPIM_SF2D"

```
Subroutine Con_RPIM_SF2D(gpos,nv,ndex,phi)
!================================================================
! Creating condensed RPIM shape functions
!
! Input:   nv(ID of support nodes);
!          ndex(number of support nodes);
!          gpos(coordinates of the point of interest).
! Output: phi(values of PIM shape functions).
!================================================================
use Parameters
use Variables
implicit real*8 (a-h,o-z)
integer kwji
dimension gpos(nx),nv(ndex),phi(ndex)
! work array
dimension xv(nx,ndex+4),rk(ndex+npoly+4),G0(ndex+npoly+4,ndex+npoly+4)

nndex=ndex+4; mg=nndex+npoly; phi=0.d0; G0=0.d0; xv=0.d0
do i=1,ndex
   xv(1,i)=x(1,nv(i)); xv(2,i)=x(2,nv(i))
enddo
xv(1,5)=(xv(1,1)+xv(1,2))/2.d0; xv(2,5)=(xv(2,1)+xv(2,2))/2.d0
xv(1,6)=(xv(1,2)+xv(1,3))/2.d0; xv(2,6)=(xv(2,2)+xv(2,3))/2.d0
xv(1,7)=(xv(1,3)+xv(1,4))/2.d0; xv(2,7)=(xv(2,3)+xv(2,4))/2.d0
xv(1,8)=(xv(1,1)+xv(1,4))/2.d0; xv(2,8)=(xv(2,1)+xv(2,4))/2.d0
Loop1: do i=1,nndex
   call Radial_Basis2D(xv(1,i),xv(2,i),xv,rk,nndex)
   do j=1,nndex
   G0(i,j)=rk(j)
   enddo
   if (npoly.gt.0) then
     select case(npoly)
     case(3)
```

```
    G0(i,nndex+1)=1.; G0(i,nndex+2)=xv(1,i); G0(i,nndex+3)=xv(2,i)
    G0(nndex+1,i)=1.; G0(nndex+2,i)=xv(1,i); G0(nndex+3,i)=xv(2,i)
  case(4)
    G0(i,nndex+1)=1.; G0(i,nndex+2)=xv(1,i); G0(i,nndex+3)=xv(2,i)
    G0(i,nndex+4)=xv(1,i)*xv(2,i)
    G0(nndex+1,i)=1.; G0(nndex+2,i)=xv(1,i); G0(nndex+3,i)=xv(2,i)
      G0(nndex+4,i)=xv(1,i)*xv(2,i)
  case(6)
    G0(i,nndex+1)=1.; G0(i,nndex+2)=xv(1,i); G0(i,nndex+3)=xv(2,i)
    G0(i,nndex+4)=xv(1,i)**2; G0(i,nndex+5)=xv(1,i)*xv(2,i)
    G0(i,nndex+6)=xv(2,i)**2
    G0(nndex+1,i)=1.; G0(nndex+2,i)=xv(1,i); G0(nndex+3,i)=xv(2,i)
    G0(nndex+4,i)=xv(1,i)**2; G0(nndex+5,i)=xv(1,i)*xv(2,i)
    G0(nndex+6,i)=xv(2,i)**2
  endselect
  endif
end do Loop1
call Radial_Basis2D(gpos(1),gpos(2),xv,rk,nndex)
call GaussEli_Solver(mg,mg,G0,rk,kwji)
if (kwji.eq.1) then
  write(*,*) 'Fail...'; pause
endif
phi(1)=rk(1)+0.5*rk(5)+0.5*rk(8); phi(2)=rk(2)+0.5*rk(5)+0.5*rk(6)
phi(3)=rk(3)+0.5*rk(6)+0.5*rk(7); phi(4)=rk(4)+0.5*rk(7)+0.5*rk(8)

Return
End subroutine Con_RPIM_SF2D
```

Program 15: source code of "Poly_Basis2D"

```
Subroutine Poly_Basis2D(xi,yi,rk,ndex)
!================================================================
! Form the polynomial basis functions.
!
! Input:   xi and yi(coordinates of the point of interest);
!          ndex: number of monomials terms.
! Output: rk(polynomial basis functions).
!================================================================
implicit real*8 (a-h,o-z)
dimension rk(ndex)

select case(ndex)
case(3)
  rk(1)=1.; rk(2)=xi; rk(3)=yi
case(4)
  rk(1)=1.; rk(2)=xi; rk(3)=yi; rk(4)=xi*yi
case(6)
```

```
  rk(1)=1.; rk(2)=xi; rk(3)=yi; rk(4)=xi**2; rk(5)=xi*yi; rk(6)=yi**2
endselect

Return
End  subroutine Poly_Basis2D
```

Program 16: source code of "Radial_Basis2D"

```
Subroutine Radial_Basis2D(xi,yi,xv,rk,ndex)
!=============================================================
! Creating radial basis functions (RBF) augmented with polynomial terms.
!
! Input:   xi and yi: coordinates of the point of interest);
!              ndex: number of support nodes;
!              xv: coordinates of the support nodes.
! Output: rk(radial basis functions).
!=============================================================
use Parameters
implicit real*8 (a-h,o-z)
dimension xv(2,ndex),rk(ndex+npoly)

do i=1,ndex
   rr=(xi-xv(1,i))**2+(yi-xv(2,i))**2; rk(i)=(rr+(rc*dc)**2)**q
enddo
if (npoly.gt.0) then
select case(npoly)
   case(3)
   rk(ndex+1)=1.; rk(ndex+2)=xi; rk(ndex+3)=yi
   case(4)
   rk(ndex+1)=1.; rk(ndex+2)=xi; rk(ndex+3)=yi; rk(ndex+4)=xi*yi
   case(6)
   rk(ndex+1)=1.; rk(ndex+2)=xi; rk(ndex+3)=yi
   rk(ndex+4)=xi**2; rk(ndex+5)=xi*yi; rk(ndex+6)=yi**2
endselect
endif

Return
End  subroutine Radial_Basis2D
```

Program 17: source code of "Cell_basedT2L"

```
Subroutine cell_basedT2L(ne1,ndex,nv)
!=============================================================
! Find the support nodes using the cell-based T2L-scheme
!
! Input:   nxcell, numneig neighbour and ne1(ID of the targeted cell).
```

```fortran
! Output: nv (ID of the support nodes) and ndex (number of the support nodes)
!==============================================================
use Parameters
use Variables
implicit real*8 (a-h,o-z)
dimension nv(ndom)
! work array
dimension nvall(ndom*3)

ndex=nce
do i=1,nce
  nv(i)=nxcell(i,ne1)
enddo
ndexall=0
Loop1: do i=1,nce
  numin=numneig(nxcell(i,ne1))
  Loop1_1: do j=1,numin
    jn=neighbour(j,in)
    do k=1,nce
    kn=nxcell(k,jn); ndexall=ndexall+1; nvall(ndexall)=kn
    enddo
  enddo Loop1_1
enddo Loop1
Loop2: do i=1,ndexall
  in=nvall(i); indexx=0
  Loop2_1: do j=1,ndex
    jn=nv(j)
    if (in.eq.jn) then
    indexx=indexx+1
    endif
  enddo Loop2_1
  if (index.eq.0) then
    ndex=ndex+1; nv(ndex)=in
  endif
enddo Loop2

Return
End subroutine cell_basedT2L
```

Program 18: source code of "Edge_basedT2L"

```fortran
Subroutine edge_basedT2L(ne1,ndex,nv)
!==============================================================
! Find the support nodes using the edge-based T2L-scheme
!
! Input:   nxcell, nodefour, numneig neighbour and ne1(ID of the targeted edge)
! Output: nv (ID of the support nodes) and ndex (number of the support nodes).
```

```fortran
!=================================================================
use Parameters
use Variables
implicit real*8 (a-h,o-z)
dimension nv(ndom)
! work array
dimension nvall(ndom*3)

ndex=0; nv=0; ndexall=0; nvall=0
Loop1: do i=1,4
  ndex=ndex+1; ndexall=ndexall+1
  nv(ndex)=nodefour(i,ne1)
  nvall(ndexall)=nodefour(i,ne1)
enddo Loop1
Loop2: do i=1,ndex
  iaN=numneig(nv(i))
  Loop2_1: do j=1,iaN
    iaNE=neighbour(j,nv(i))
    do k=1,nce
       ka=nxcell(k,iaNE); ndexall=ndexall+1; nvall(ndexall)=ka
    enddo
  enddo Loop2_1
enddo Loop2
do 10 i=1,ndexall
  in=nvall(i)
  indexx=0
  do j=1,ndex
     jn=nv(j)
     if (in.eq.jn) then; indexx=indexx+1; goto 10
     endif
  enddo
  if (indexx.eq.0) then
      ndex=ndex+1; nv(ndex)=in
  endif
10 continue

Return
End subroutine edge_basedT2L
```

Program 19: source code of "FormB_NSPIM"

```fortran
Subroutine GaussEli_Solver(n,mk,a,b,kwji)
!=================================================================
! Solving linear algebraic equations ax=b using Gauss elimination method.
!     kwji=1 means no solution and kwji=0 menas having solution.
!=================================================================
use Parameters
```

```fortran
implicit real*8 (a-h,o-z)
dimension a(mk,n),b(mk),m(2*mk)

kwji=0
do 10 i=1,n
10 m(i)=i
do 20 k=1,n
  p=0.0
  do 30 i=k,n
  do 30 j=k,n
     if(abs(a(i,j)).le.abs(p)) goto 30
     p=a(i,j); io=I; jo=j
 30 continue
  if(abs(p)-ep) 100,100,110
 100 kwji=1
     return
 110 if(jo.eq.k) goto 120
   do 40 i=1,n
   t=a(i,jo); a(i,jo)=a(i,k)
   40 a(i,k)=t
   j=m(k); m(k)=m(jo); m(jo)=j
 120 if(io.eq.k) goto 130
   do 50 j=k,n
   t=a(io,j); a(io,j)=a(k,j)
   50 a(k,j)=t
   t=b(io); b(io)=b(k); b(k)=t
 130 p=1.0/p
   in=n-1
  if(k.eq.n) goto 140
   do 60 j=k,in
   60 a(k,j+1)=a(k,j+1)*p
 140 b(k)=b(k)*p
   if(k.eq.n) goto 20
   do 70 i=k,in
   do 80 j=k,in
   80 a(i+1,j+1)=a(i+1,j+1)-a(i+1,k)*a(k,j+1)
   70 b(i+1)=b(I+1)-a(i+1,k)*b(k)
 20  continue
do 90 i1=2,n
i=n+1-i1
do 90 j=i,in
90 b(i)=b(i)-a(i,j+1)*b(j+1)
do 1 k=1,n
i=m(k)
 1 a(1,i)=b(k)
do 2 k=1,n
 2 b(k)=a(1,k)
```

```fortran
Return
End subroutine GaussEli_Solver
```

Program 20: source code of "Band_solver"

```fortran
SUBROUTINE Band_solver(A,F,B,N,L,IL,nmat)
!===========================================================
! Solving banded linear algebraic equations
!===========================================================
implicit real*8 (a-h,o-z)
DIMENSION A(N,N),F(N),B(N,nmat)
real(8), allocatable :: d(:,:)
allocate (d(1:n,1:1))

M=1; LP1=L+1
LOOP1: DO I=1,N
  LOOP1_1: DO K=1,IL
  B(I,K)=0.
  IF(I.LE.LP1) B(I,K)=A(I,K)
  IF(I.GT.LP1.AND.I.LE.(N-L)) B(I,K)=A(I,I+K-LP1)
  IF(I.GT.(N-L).AND.(I+K-LP1).LE.N) B(I,K)=A(I,I+K-LP1)
  ENDDO LOOP1_1
ENDDO LOOP1
DO I=1,N
  D(I,1)=F(I)
ENDDO
IT=1
IF (IL.NE.2*L+1) THEN
  IT=-1; WRITE(*,11); RETURN
ENDIF
11 FORMAT(1X,'***FAIL***')
LS=L+1
LOOP2: DO K=1,N-1
  P=0.0
  DO 10 I=K,LS
     IF (ABS(B(I,1)).GT.P) THEN
     P=ABS(B(I,1)); IS=I
     END IF
  10 CONTINUE
  IF (P+1.0.EQ.1.0) THEN
     IT=0; WRITE(*,11); RETURN
  END IF
  DO 20 J=1,M
     T=D(K,J); D(K,J)=D(IS,J); D(IS,J)=T
  20 CONTINUE
  DO 30 J=1,IL
     T=B(K,J); B(K,J)=B(IS,J); B(IS,J)=T
```

```fortran
 30 CONTINUE
    DO J=1,M
       D(K,J)=D(K,J)/B(K,1)
    ENDDO
    DO J=2,IL
       B(K,J)=B(K,J)/B(K,1)
    ENDDO
    DO 40 I=K+1,LS
       T=B(I,1)
       DO J=1,M
       D(I,J)=D(I,J)-T*D(K,J)
       ENDDO
       DO J=2,IL
       B(I,J-1)=B(I,J)-T*B(K,J)
       ENDDO
       B(I,IL)=0.0
 40 CONTINUE
    IF (LS.NE.N) LS=LS+1
 enddo LOOP2
 IF (ABS(B(N,1))+1.0.EQ.1.0) THEN
    IT=0; WRITE(*,11); RETURN
 END IF
 DO J=1,M
    D(N,J)=D(N,J)/B(N,1)
 ENDDO
 JS=2
 DO 50 I=N-1,1,-1
   DO 60 K=1,M
   DO 60 J=2,JS
 60 D(I,K)=D(I,K)-B(I,J)*D(I+J-1,K)
   IF (JS.NE.IL) JS=JS+1
 50 CONTINUE
 if(it.le.0) write(*,*) "Band_solver failed"
 DO I=1,N
   F(I)=D(I,1)
 ENDDO
 deallocate (d)

 RETURN
 END SUBROUTINE Band_solver
```

Program 21: source code of "Gausspointcoe_line"

```fortran
Subroutine Gausspointcoe_line(NumG,GPcoe_line)
!================================================================
! This subroutine returns the positions and weighting coefficients for Gauss
!      integration scheme in 1D domain
```

```fortran
!
! Input:    NumG(number of Gauss points)
! Output: GPcoe_line(positions and weighting coefficients for Gauss quadrature)
!==================================================================
use Parameters
use Variables
implicit real*8 (a-h,o-z)
dimension GPcoe_line(2,NumG)

GPcoe_line=0.d0
select case(NumG)
case (1)
   GPcoe_line(1,1)=0.d0; GPcoe_line(2,1)=2.d0
case(2)
   GPcoe_line(1,1)=-.577350269189626d0; GPcoe_line(1,2)=-GPcoe_line(1,1)
   GPcoe_line(2,1)=1.0d0; GPcoe_line(2,2)=GPcoe_line(2,1)
case(3)
   GPcoe_line(1,1)=-.774596669241483d0; GPcoe_line(1,2)=0.0d0
   GPcoe_line(1,3)=-GPcoe_line(1,1); GPcoe_line(2,1)=.555555555555556d0
   GPcoe_line(2,2)=.888888888888889d0; GPcoe_line(2,3)=GPcoe_line(2,1)
case(4)
   GPcoe_line(1,1)=-.861136311594053d0; GPcoe_line(1,2)=-.339981043584856d0
   GPcoe_line(1,3)=-GPcoe_line(1,2); GPcoe_line(1,4)=-GPcoe_line(1,1)
   GPcoe_line(2,1)=.347854845137454d0; GPcoe_line(2,2)=.652145154862546d0
   GPcoe_line(2,3)=GPcoe_line(2,2); GPcoe_line(2,4)=GPcoe_line(2,1)
case(5)
   GPcoe_line(1,1)=-.906179845938664d0; GPcoe_line(1,2)=-.538469310105683d0
   GPcoe_line(1,3)=-.0d0; GPcoe_line(1,4)=-GPcoe_line(1,2)
   GPcoe_line(1,5)=-GPcoe_line(1,1); GPcoe_line(2,1)=.236926885056189d0
   GPcoe_line(2,2)=.478628670499366d0; GPcoe_line(2,3)=.568888888888889d0
   GPcoe_line(2,4)=GPcoe_line(2,2); GPcoe_line(2,5)=GPcoe_line(2,1)
case(6)
   GPcoe_line(1,1)=-.932469514203152d0; GPcoe_line(1,2)=-.661209386466265d0
   GPcoe_line(1,3)=-.238619186083197d0; GPcoe_line(1,4)=-GPcoe_line(1,3)
   GPcoe_line(1,5)=-GPcoe_line(1,2); GPcoe_line(1,6)=-GPcoe_line(1,1)
   GPcoe_line(2,1)=.171324492379170d0; GPcoe_line(2,2)=.360761573048139d0
   GPcoe_line(2,3)=.467913934572691d0; GPcoe_line(2,4)=GPcoe_line(2,3)
   GPcoe_line(2,5)=GPcoe_line(2,2); GPcoe_line(2,6)=GPcoe_line(2,1)
end select

Return
End subroutine Gausspointcoe_line
```

Program 22: source code of "Line_gauss"

```fortran
Subroutine Line_gauss(xg,xxg,wei,det,NumG)
!==================================================================
```

```
! Obtain locations and weighting coefficients for the Gauss points located in a section of
!     line.
!
! Input:   xg(coordinates of the two end points of the segment).
! Output: xxg(coordinates of the Gauss points on the segment);
!            wei(weighting coefficients)
!================================================================
use Parameters
use Variables
implicit real*8 (a-h,o-z)
dimension xg(nx,2),xxg(nx,NumG),wei(NumG)
! work array
dimension GPcoe_line(2,NumG)

call Gausspointcoe_line(NumG,GPcoe_line)
x1=xg(1,1); y1=xg(2,1); x2=xg(1,2); y2=xg(2,2)
xL=((x1-x2)**2+(y1-y2)**2)**0.5
x0=(x1+x2)/2.; y0=(y1+y2)/2.
do i=1,NumG
   xx=x0+(x0-x1)*GPcoe_line(1,i); yy=y0+(y0-y1)*GPcoe_line(1,i)
   xxg(1,i)=xx; xxg(2,i)=yy; wei(i)=GPcoe_line(2,i)
enddo
det=xL/2.

Return
End subroutine Line_gauss
```

Program 23: source code of "Determinant"

```
Subroutine Determinant(a,n,det)
!================================================================
! To get the determinant of a square matrix
!
! Input:   a(the square matrix) and n(the dimension of the matrix)
! Output: det(determinant of the matrix a)
! Note:    the matrix a will be destroyed after the calculation.
!================================================================
implicit real*8 (a-h,o-z)
dimension a(n,n)
double precision a,det,f,d,q

f=1.0; det=1.0
Loop1: do k=1,n-1
   q=0.0
   do 100 i=k,n
   do 100 j=k,n
     if (abs(a(i,j))>q) then
```

```fortran
            q=abs(a(i,j)); is=i; js=j
        endif
100 continue
    if (q+1.0.eq.1.0) then
        det=0.0; return
    endif
    if (is/=k) then
        f=-f
        do 200 j=k,n
            d=a(k,j); a(k,j)=a(is,j); a(is,j)=d
200     continue
    endif
    if (js/=k) then
        f=-f
        do 300 i=k,n
            d=a(i,js); a(i,js)=a(i,k); a(i,k)=d
300     continue
    endif
    det=det*a(k,k)
    do 400 i=k+1,n
        d=a(i,k)/a(k,k)
        do 500 j=k+1,n
        500 a(i,j)=a(i,j)-d*a(k,j)
    400 continue
end do Loop1
det=f*det*a(n,n)

Return
End subroutine Determinant
```

Program 24: source code of "Brinv"

```fortran
Subroutine brinv(a,n,L)
!===============================================================
! To get the inversion of a matrix using the Gauss-Jodon method
!
! Input:   n(dimension of the matrix) and L(L=0 notes failure).
! Input and output:  a(the objective matrix)
!===============================================================
implicit real*8 (a-h,o-z)
dimension a(n,n),is(n),js(n)
double precision a,t,d

L=1; is=0; js=0
Loop1: do k=1,n
    d=0.
    do 10 i=k,n
```

```
   do 10 j=k,n
     pa=a(i,j)
     if (abs(pa).gt.d) then
        d=abs(a(i,j)); is(k)=I; js(k)=j
     endif
10 continue
   if (d+1.0 .eq. 1.0) then
     L=0; write(*,20); return
   endif
20 format(1x,'Error**not inv')
   do 30 j=1,n
        t=a(k,j); a(k,j)=a(is(k),j); a(is(k),j)=t
30 continue
   do 40 i=1,n
        t=a(i,k); a(i,k)=a(i,js(k)); a(i,js(k))=t
40 continue
   a(k,k)=1./a(k,k)
   do 50 j=1,n
        if (j.ne.k) then; a(k,j)=a(k,j)*a(k,k)
        endif
50 continue
   do 70 i=1,n
     if (i.ne.k) then
        do 60 j=1,n
             if (j.ne.k) then; a(i,j)=a(i,j)-a(i,k)*a(k,j)
             endif
        60 continue
     endif
70 continue
   do 80 i=1,n
        if(i.ne.k) then; a(i,k)=-a(i,k)*a(k,k)
        endif
80 continue
end do Loop1

Loop2: do k=n,1,-1
   do 90 j=1,n
        t=a(k,j); a(k,j)=a(js(k),j); a(js(k),j)=t
90 continue
   do 100 i=1,n
        t=a(i,k); a(i,k)=a(i,is(k)); a(i,is(k))=t
100 continue
end do Loop2

Return
End subroutine brinv
```

Program 25: source code of "Inversion"

```fortran
Subroutine Inversion(n,ma,a)
!===============================================================
! To get the inversion of a positive definite matrix using the Gauss-Jodon method
!
! Input:   n(dimension of the matrix) and ma(the max number of rows of the matrix)
! Input and output:  a(the objective matrix, must be definite but may be asymmetric)
!
!===============================================================
use Parameters
implicit real*8 (a-h,o-z)
dimension A(ma,n)

do 100 k=1,n
  c=a(k,k)
  if (dabs(c).le.ep) pause
  c=1.0/c; a(k,k)=1.0
  do j=1,n
     a(k,j)=a(k,j)*c
  enddo
do 100 i=1,n
  if (i.eq.k) goto 100
  c=a(i,k); a(i,k)=0.0
  do j=1,n
     a(i,j)=a(i,j)-a(k,j)*c
  enddo
100 continue

Return
End subroutine Inversion
```

Program 26: source code of "Indexx"

```fortran
Subroutine indexx(n,arr,indx)
!===============================================================
! To reorder an array (arr) in the ascending order according to the values
! Input:   n(length of the array), arr(the objective array)
! Output: indx(the ascending order)
!===============================================================
implicit none
integer n,indx(n)
integer i,indxt,ir,itemp,j,jstack,k,l,istack(n)
double precision arr(n)
double precision a

do j=1,n
  indx(j)=j
```

```
enddo
jstack=0; l=1; ir=n
do while (1)
if(ir-l<7)then
  do j=l+1,ir
  indxt=indx(j); a=arr(indxt)
  do i=j-1,l,-1
  if(arr(indx(i))<=a)exit
  indx(i+1)=indx(i)
  enddo
  indx(i+1)=indxt
  enddo
  if(jstack.eq.0)return
  ir=istack(jstack); l=istack(jstack-1); jstack=jstack-2
else
  k=(l+ir)/2.; itemp=indx(k); indx(k)=indx(l+1); indx(l+1)=itemp
  if(arr(indx(l))>arr(indx(ir)))then
  itemp=indx(l); indx(l)=indx(ir); indx(ir)=itemp
  endif
  if(arr(indx(l+1))>arr(indx(ir)))then
  itemp=indx(l+1); indx(l+1)=indx(ir); indx(ir)=itemp
  endif
  if(arr(indx(l))>arr(indx(l+1)))then
  itemp=indx(l); indx(l)=indx(l+1); indx(l+1)=itemp
  endif
  i=l+1; j=ir; indxt=indx(l+1); a=arr(indxt)
  do while (1)
  i=i+1
  do while(arr(indx(i))<a)
  i=i+1
  enddo
  j=j-1
  do while(arr(indx(j))>a)
  j=j-1
  enddo
  if(j<i)exit
  itemp=indx(i); indx(i)=indx(j); indx(j)=itemp
  enddo
  indx(l+1)=indx(j); indx(j)=indxt; jstack=jstack+2
  if(ir-i+1>=j-l)then
  istack(jstack)=ir; istack(jstack-1)=i; ir=j-1
  else
  istack(jstack)=j-1; istack(jstack-1)=l; l=i
  endif
endif
enddo

Return
End subroutine indexx
```

Index

Cell-based smoothed radial point interpolation method (CS-RPIM), 402, 405

Cell-based smoothing domain, 113, 133, 180, 227, 398, 436, 497, 566

Central-difference approximation, 537

Certified solution, 23, 325

Characteristic length, 29, 126, 263, 574

Close-to-exact, 184, 233, 341, 375, 380, 395, 407, 415, 461, 474

Coercivity, 211, 225

Collocation method, 13, 21, 130

Common subroutines, 42, 327, 381, 448, 623, 631

Compact support, 96, 112, 130

Compatibility, 19, 23, 71, 96, 103, 137, 150, 217, 246

Compatibility equations, 5, 8

Compatible strain, 5, 22, 25, 166, 174

Completeness, 83

Condensed RPIM shape functions, 146, 326, 402, 411, 433, 464, 650

Condition number, 18, 26, 232, 251, 281, 295, 354, 403, 469, 506

Conductance matrix, 536

Connecting rod problem, 317, 425, 485

Consistency
    kth-order (C^k), 102
    Linear (C^1), 96, 102, 111, 123, 183, 465
    zero-order (C^0), 102

Constitutive equation, 4, 6

Constrained solid, 12, 49, 171, 253, 407

Constructed strain, 22, 25, 57, 76, 166, 169, 172, 206, 500

Convection heat transfer coefficient, 531

Convergence, 16, 30, 87, 156, 173, 183, 210, 213, 227, 263

Convergence rate, 151, 153, 156, 227, 163, 360, 411, 477, 510

Coordinate mapping, 19, 97, 137, 141

Coordinate transformation (CT), 32, 121, 133, 143, 217, 326, 401, 623

Crack, 42, 559, 561, 568, 574, 578

CS-FEM, 448

CS-PIM, *see also* Cell-based smoothed point interpolation method
    CS-PIM-Te3, 435
    CS-PIM-Tr2, 396, 401, 407, 447, 449
    CS-PIM-Tr4-CT, 401, 403, 409, 411
    CS-PIM-Tr4-Iso, 402, 409, 411, 449

CSPIM.F90, 448, 623

CS-RPIM, *see also* Cell-based smoothed radial point interpolation method
    CS-RPIM-Te2L, 436
    CS-RPIM-Te5, 436, 440
    CS-RPIM-Te5-Cd, 436,
    CS-RPIM-Tr2L, 403, 409, 449
    CS-RPIM-Tr4, 402, 405, 409, 411, 449
    CS-RPIM-Tr4-Cd, 402. 409, 449

CS-αRPIM, *see also* Cell-based smoothed alpha radial point interpolation method
    CS-αRPIM-Te5, 463, 466, 469, 487
    CS-αRPIM-Tr4, 463, 468, 470, 472

D

Damping, 99, 345, 379, 399, 431

Damping matrix, 99, 346, 399

Deflection, 259, 284, 359, 407, 472

Deformation, 1, 11, 16, 254, 356, 407, 472

Degree of freedom (DOF), 17, 23, 151, 167, 169, 207, 232, 250

Delaunay triangulation algorithm, 27

Delta function property, 96, 102, 111, 122, 132, 145, 150, 204, 217, 226, 464

Determinant, 135, 624, 659

Differential operator, 5, 182

Diffusion equation, 529

Discontinuous function, 56, 63, 116, 118, 208

Discrete equation, 23, 26

Discrete numerical methods, 13, 76, 95, 461

Discretized system equation, 21, 41, 176, 191, 241, 293, 434, 528, 536, 566